Lobsters: Biology, Fisheries and Aquaculture

E. V. Radhakrishnan • Bruce F. Phillips
Gopalakrishnan Achamveetil
Editors

Lobsters: Biology, Fisheries and Aquaculture

Editors
E. V. Radhakrishnan
ICAR-Central Marine Fisheries Research Institute
Cochin, Kerala, India

Bruce F. Phillips
School of Molecular and Life Sciences
Curtin University
Perth, WA, Australia

Gopalakrishnan Achamveetil
ICAR-Central Marine Fisheries Research Institute
Cochin, Kerala, India

ISBN 978-981-32-9096-9   ISBN 978-981-32-9094-5 (eBook)
https://doi.org/10.1007/978-981-32-9094-5

© Springer Nature Singapore Pte Ltd. 2019
This work is subject to copyright. All rights are reserved by the Publisher, whether the whole or part of the material is concerned, specifically the rights of translation, reprinting, reuse of illustrations, recitation, broadcasting, reproduction on microfilms or in any other physical way, and transmission or information storage and retrieval, electronic adaptation, computer software, or by similar or dissimilar methodology now known or hereafter developed.
The use of general descriptive names, registered names, trademarks, service marks, etc. in this publication does not imply, even in the absence of a specific statement, that such names are exempt from the relevant protective laws and regulations and therefore free for general use.
The publisher, the authors, and the editors are safe to assume that the advice and information in this book are believed to be true and accurate at the date of publication. Neither the publisher nor the authors or the editors give a warranty, expressed or implied, with respect to the material contained herein or for any errors or omissions that may have been made. The publisher remains neutral with regard to jurisdictional claims in published maps and institutional affiliations.

This Springer imprint is published by the registered company Springer Nature Singapore Pte Ltd.
The registered company address is: 152 Beach Road, #21-01/04 Gateway East, Singapore 189721, Singapore

Foreword

This book is a welcome addition to the knowledge of lobster research. The book complements other books published on lobster research and management as it focuses on Indian lobster fisheries and aquaculture developments where there have been nearly 350 research papers and reports and 19 PhD awards. The book has 15 chapters covering many aspects of the biology of a number of spiny and slipper lobster species occurring in India and the status of lobster fisheries and aquaculture in India. The lobster fisheries in the Indian Ocean Rim countries and lobster aquaculture developments in Vietnam and Indonesia are also examined. The final chapter provides some perspectives and future directions of research.

The book is timely as the Second International Indian Ocean Expedition (IIOE) is currently underway (2015–2020), 50 years after the original IIOE (1959–1965), with some of the original lobster research on the biology and distribution of phyllosoma larvae being undertaken on the plankton samples collected during the first IIOE.

Many of the chapters are contributed by the authors from the Central Marine Fisheries Research Institute (ICAR-CMFRI) which has been collecting fishery and biological data on lobsters since 1950 when lobster fishing began on a subsistence scale, followed by some industrial fishing for lobsters in different parts of India. Unfortunately the development of some of these lobster fisheries was followed by overfishing due to lack of enforcement of regulations. Therefore it is important that these cases are documented and lessons learnt from them. CMFRI also initiated aquaculture research on lobsters in 1975 with the establishment of a field laboratory in Chennai. The book provides a valuable addition to our knowledge of the biology, fisheries and aquaculture of spiny and slipper lobsters.

Supervising Scientist (Invertebrate), Nick Caputi
Science and Resource Assessment,
Sustainability and Biosecurity,
Department of Primary Industries and Regional Development
Western Australian Fisheries and Marine Research Laboratories,
North Beach, WA, Australia

Preface

Lobster is one of the most important commercially harvested marine resources in the world because of its higher economic importance. The high-value crustaceans support some of the most profitable fisheries in many countries of the world. World capture fisheries production of lobsters touched an all-time high of 0.3 million tonnes in 2016 with an additional 2000 tonnes from aquaculture. The major lobster-producing countries are Canada, USA, UK, Australia, Indonesia, Cuba, Brazil and Mexico. With over 260 species of extant lobsters under 54 genera identified till date, they constitute one of the prominent groups under the suborder Macrura Reptantia, owing to their large size and reasonably dense population forming commercially important fisheries in many parts of the world. Apart from their economic importance, they play a key role in maintaining and balancing the marine ecosystem, acting both as a benthic predator and as a prey. The world trade in lobsters grew substantially from 110,000 tonnes in 2001 to 170,000 tonnes in 2014 valued at US$3.3 billion. The scientific investigation on clawed and spiny lobsters gained importance as their fisheries became more profitable and the need for management of the resource more inevitable. Biological research on scyllarid lobsters was on a low key as they are not as commercially important as the nephropid and palinurid lobsters. However, failure of spiny lobster fisheries in some parts of the world due to overexploitation and poor management triggered commercial interest shifting towards the scyllarids. They form an important by-product in trawl fisheries, and wherever they are directly targeted, their volume has declined sharply and the fishery collapsed with no sign of recovery even after many years.

Lobsters were known to humans since the seventeenth century, with the first recorded lobster catch in 1605. The first scientific publication (monograph) on clawed lobster biology was compiled and published by Francis Hobart Herrick in 1895. Bruce F. Phillips and J. Stanley Cobb organized the First International Workshop on Lobster and Rock Lobster Ecology and Physiology (ICWL) in 1977 in Perth, Australia, which was followed by ten international workshops in different countries. The conference proceedings were published in several journals. The first comprehensive book (*The Biology and Management of Lobsters*) devoted to clawed and spiny lobsters (two volumes) edited by J.S. Cobb and B.F. Phillips was published by Academic Press in 1980. However, there was a big gap after the publication of this book, and in 1994 B.F. Phillips, J.S. Cobb and J. Kittaka edited and published a book focusing mainly on *Spiny Lobster Biology, Aquaculture and Management*. This volume was followed by a book on *Biology of the Lobster: Homarus americanus* (edited by Jan Robert Factor), and in 2000, B.F. Phillips and J. Kittaka updated an earlier volume published on spiny lobster fisheries and culture. Then in 2006, B.F. Phillips edited and published a book entitled *Lobsters: Biology, Management, Aquaculture and Fisheries*. The second edition of the book by B.F. Phillips, *Lobsters: Biology, Management, Aquaculture and Fisheries*, was published in 2013. In general, clawed and spiny lobsters received the maximum attention in these books. In 2007, Kari Lavalli and Ehud Spanier edited and published the first multi-authored book, *The Biology and Fisheries of the Slipper Lobster*, focusing on scyllarid lobster biology and fisheries. Two chapters, 'Global Review of Spiny Lobster Aquaculture' and 'Slipper Lobsters', published in the book *Recent Advances and New Species in Aquaculture* in 2011 and edited by Ravi Fotedar and B.F. Phillips, provide additional information on aquaculture of lobsters.

Lobster research in India probably dates back to the late nineteenth century when Alcock and Anderson (1894) documented the first report on deep sea crustaceans from the Bay of Bengal and Laccadive Sea (Lakshadweep), based on the collections by the exploratory cruise, H.M. Indian marine survey steamer 'Investigator'. A series of papers on the biology and distribution of phyllosoma larvae were published based on the plankton samples collected from the Indian Ocean by the DANA Expedition and the International Indian Ocean Expedition (IIOE). The Central Marine Fisheries Research Institute (CMFRI) under the Indian Council of Agricultural Research has been collecting fishery and biological data on lobsters since 1950 and is the only statistical database on lobsters in India. Lobster fishing, which began on a subsistence scale in the 1950s, flourished with the establishment of seafood processing plants along the southwest coast of India and export. In Kanyakumari district of Tamil Nadu, on the southwest coast of India, once the largest producer of lobsters in India, the catch has declined dramatically due to uncontrolled exploitation. With the expansion of trawl fishing, the northwestern states of Maharashtra and Gujarat became the leading lobster producers. However, landing in this part of the coast has also declined due to overharvesting, in the absence of any specific fishing regulations. Research investigation on the biology, fishery and population dynamics of lobsters from the southern (1960s) and the Bombay (Mumbai) coast (late 1980s and early 1990s) provide a volume of information on the stock position of lobsters. On the recommendation of CMFRI, Minimum Legal Size (MLS) for export of four species of lobsters was notified by the Ministry of Commerce and Industry, Government of India. However, in the absence of strict enforcement of fishing regulations by the state governments, the resource is facing extreme fishing pressure.

Lobster aquaculture research, specifically in Australia, is on the threshold of a phenomenal breakthrough with success in commercial-level hatchery production of the spiny lobster *Panulirus ornatus* and slipper lobsters of the genus *Thenus*. In India, aquaculture research on lobsters was initiated by ICAR-Central Marine

Fisheries Research Institute (CMFRI), Cochin in 1975 with the establishment of a Field Laboratory at Kovalam, Chennai. The centre has been conducting research on physiology, breeding, puerulus settlement, reproductive biology, feeding and growth of tropical spiny and slipper lobsters since then. Captive maturation and breeding of the spiny lobster *P. homarus homarus* and the slipper lobster *T. unimaculatus* have been achieved with successful seed production of the latter. The technical and economic feasibility of Lobster culture in indoor system were tested. Sea cage culture for value addition to lobsters, incidentally caught during fishing operations, on the southern and northwestern coasts involving local fishermen had a positive impact on the livelihood of coastal impoverished fishermen. Apart from CMFRI, lobster research has also been carried out by the Tamil Nadu Veterinary and Animal Sciences University and National Institute of Ocean Technology.

Although several books on biology, fisheries, management and aquaculture have been published on lobsters with scientific contributions from international experts, only little attention has been given to Indian lobster fisheries and aquaculture developments. The Indian authors have published nearly 350 research papers and reports and 19 researchers have been awarded PhD by various Universities. One of us (EVR) is most fortunate to have been mentored by the pre-eminent researchers who pioneered research on lobsters. We believe that it is time to compile and review available biological information and bring out a volume with focus on research carried out in India. Earlier books have brought out a large volume of information on biology, ecology, fisheries, management and aquaculture of lobsters. The book opens with a brief introduction, general biology and life history of nephropid, palinurid and scyllarid lobsters followed by an updated taxonomy of world's marine lobsters, molecular phylogeny, a checklist of Indian lobster fauna and global distribution of palinurid lobsters. The book also reviews the status of lobster fisheries in India with a brief description of fisheries and management in Indian Ocean Rim countries. The global status and future challenges for managing major lobster fisheries has also been discussed. The chapters on Reproductive biology, Lobster mariculture, Health management in aquaculture, Larval culture of scyllarid lobsters and Aquaculture developments in Vietnam and Indonesia provide insights into aquaculture prospects of lobsters. Post harvest handling, processing, marketing and export of lobsters is covered in the following chapter. The last chapter discusses on prospects and challenges and future directions on research of this valuable resource. A bibliography of publications on lobsters has been included at the end. Since this volume is the most recent addition to a series of books published, the new developments in taxonomy, phylogeny, aquaculture and future challenges to major lobster fisheries across the globe have also been included.

We the Indian editors (EVR and AG) especially would like to thank Bruce Phillips of Curtin University, the doyen of lobster research in the world and editor of several books on lobsters, for joining as one of the editors of this book and providing guidance and support. The first editor (EVR) would like to thank Bruce Phillips, Tin Yam Chan, Clive Jones and Kaori Wakabayashi for contributing their expertise and vast knowledge in their respective fields. Thanks are also due to all the authors who immensely supported and contributed their vast experience in chapters authored by them. We also thank the Central Marine Fisheries Research Institute for its support in the publication of this book.

It is our earnest hope that this volume will spur greater research interest amongst scientists and graduate research students in better understanding of lobster life history, physiology, genetics and pathology.

Cochin, Kerala, India	E. V. Radhakrishnan
Perth, WA, Australia	Bruce F. Phillips
Cochin, Kerala, India	Gopalakrishnan Achamveetil

Contents

1	**Introduction to Lobsters: Biology, Fisheries and Aquaculture**.......... E. V. Radhakrishnan, Joe K. Kizhakudan, and Bruce F. Phillips	1
2	**Updated Checklist of the World's Marine Lobsters** .... Tin-Yam Chan	35
3	**Lobster Fauna of India** .... E. V. Radhakrishnan, Joe K. Kizhakudan, Lakshmi Pillai S, and Jeena N. S	65
4	**Applications of Molecular Tools in Systematics and Population Genetics of Lobsters**.. Jeena N. S, Gopalakrishnan A, E. V. Radhakrishnan, and Jena J. K	125
5	**Ecology and Global Distribution Pattern of Lobsters** .... E. V. Radhakrishnan, Bruce F. Phillips, Lakshmi Pillai S, and Shelton Padua	151
6	**Food, Feeding Behaviour, Growth and Neuroendocrine Control of Moulting and Reproduction** .... E. V. Radhakrishnan and Joe K. Kizhakudan	177
7	**Lobster Fisheries and Management in India and Indian Ocean Rim Countries** .... E. V. Radhakrishnan, Joe K. Kizhakudan, Saleela A, Dineshbabu A. P, and Lakshmi Pillai S	219
8	**A Review of the Current Global Status and Future Challenges for Management of Lobster Fisheries** .... Bruce F. Phillips and Mónica Pérez-Ramírez	351
9	**Reproductive Biology of Spiny and Slipper Lobster** .... Joe K. Kizhakudan, E. V. Radhakrishnan, and Lakshmi Pillai S	363
10	**Breeding, Hatchery Production and Mariculture**............................... E. V. Radhakrishnan, Joe K. Kizhakudan, Vijayakumaran M, Vijayagopal P, Koya M, and Jeena N. S	409

11 **Culture of Slipper Lobster Larvae (Decapoda: Achelata: Scyllaridae) Fed Jellyfish as Food**.. 519
    Kaori Wakabayashi, Yuji Tanaka, and Bruce F. Phillips

12 **Lobster Aquaculture Development in Vietnam and Indonesia**........... 541
    Clive M. Jones, Tuan Le Anh, and Bayu Priyambodo

13 **Health Management in Lobster Aquaculture**.. 571
    E. V. Radhakrishnan and Joe K. Kizhakudan

14 **Post-harvest Processing, Value Addition and Marketing of Lobsters** 603
    Vijayakumaran M, E. V. Radhakrishnan, G. Maheswarudu, T. K. Srinivasa Gopal, and Lakshmi Pillai S

15 **Perspectives and Future Directions for Research** 635
    Gopalakrishnan A, E. V. Radhakrishnan, and Bruce F. Phillips

Bibliography of Lobster Fauna of India 647

Index.. 667

About the Editors

E. V. Radhakrishnan joined the Central Marine Fisheries Research Institute, Kochi, under the Indian Council of Agricultural Research as scientist in 1976, after completing his post-graduate studies in marine biology from the University of Kerala. Throughout his scientific career spanning for more than 35 years, he has been studying lobster biology, physiology, ecology and fisheries and exploring the possibilities of aquaculture of lobsters. He has been instrumental in developing aquaculture laboratories at regional and research centres of CMFRI. He held the position of Head of the Crustacean Fisheries Division for 10 years and as Emeritus Scientist of ICAR for 2 years. He was the principal investigator of two World Bank-funded projects, 'Mud Crab Breeding and Seed Production' and 'A Value Chain on Oceanic Tuna Fisheries in Lakshadweep Sea'.

Dr. Radhakrishnan organized a series of stakeholder awareness workshops on lobster conservation in the maritime states of Tamil Nadu, Maharashtra and Gujarat under a participatory management project. The Ministry of Commerce issued a notification on minimum legal size for export of lobsters based on his studies and recommendation by CMFRI.

Dr. Radhakrishnan has been the co-editor of the *Handbook of Marine Prawns of India* and editor of research manuals and handbooks including contribution of chapters to several national and international books on lobsters. He has authored more than 200 peer-reviewed publications and technical reports.

Bruce F. Phillips was a scientist studying spiny lobsters with CSIRO for 28 years, before spending 3 years as the chief scientist for the Australian Fisheries Management Authority in Canberra.

Since 1996 he has been a research fellow and later an adjunct professor at Curtin University, in Western Australia. He is currently in the School of Molecular and Life Sciences and associated with several research projects developing spiny lobster aquaculture and enhancement.

Dr. Phillips has been the editor of 12 books on rock (spiny) lobster biology, management and aquaculture, including the contribution of many of the chapters in these volumes. He has also edited two books on eco-labelling in fisheries. In 2017 he published a two-volume book on *Climate Change Impacts on Fisheries and Aquaculture: A Global Analysis*.

Gopalakrishnan Achamveetil is a conservation geneticist with 29 years of research experience in the field of genetic characterization and gene banking of marine fauna. He started his career as scientist at the National Bureau of Fish Genetic Resources (NBFGR), Lucknow, in 1989 and later served as head of NBFGR Kochi Unit. Since 2013, he has been holding the position of director of CMFRI, Kochi. Dr. Gopalakrishnan has also served as the vice chairman of the Indian branch of the Asian Fisheries Society, the chairman of Genetics and Biodiversity – of the E – Consultation of Task Force (TF) Members for Development of the Network of Aquaculture Centres in Asia-Pacific (NACA) and a member of the Indo-Norwegian Working Group for Fisheries and Animal Sciences.

His areas of specialization include genetic stock identification of fishes using DNA markers, DNA barcoding of fishes using mtDNA markers, development of protocol for cryopreservation of milt of indigenous fishes for conservation, captive breeding of indigenous fishes, fish reproduction and fish genetic stock identification. He developed milt cryopreservation and captive breeding protocols of six threatened fish species and was instrumental in the development of six fish cell lines that will be of use in viral studies.

He has won several accolades during his service including a Fellowship of the National Academy of Agricultural Sciences, the DBT-CREST Award and the Asian Fisheries Society Gold Medal Award. He has authored more than 150 peer-reviewed articles and technical reports and 9 books.

Introduction to Lobsters: Biology, Fisheries and Aquaculture

E. V. Radhakrishnan, Joe K. Kizhakudan, and Bruce F. Phillips

Abstract

Marine lobsters are high-valued crustaceans occupying a variety of habitats in tropical, subtropical and temperate oceans, from continental shelves and slopes to deep sea ridges, remote seamounts, lagoons and even estuaries. The highly diverse form and size are suited to their wide distribution in a range of shallow coastal and oceanic habitats. They are economically important, forming sustained profitable fisheries in many countries. In 2016, world capture production was 3,08,926 t, with an additional 2000 t from aquaculture. Apart from their economic importance, they are an important component of the marine ecosystems, playing pivotal roles as a prey and a predator. Lobster has been one of the most extensively studied group, and literature search has yielded more than 15,000 entries including research papers, technical reports and popular articles on taxonomy, biology, physiology, ecology, fisheries and aquaculture of clawed, spiny and slipper lobsters. The knowledge base created on individual species is critical in our understanding of the life history strategies and ecology of larvae, juveniles and adults. Advances in molecular techniques have been instrumental in resolving the conflicting hypotheses of evolutionary relationships among different taxa. Recent success in breeding and seed production has prompted researchers to develop commercially viable technologies for aquaculture of a few species, which is expected to relieve the mounting pressure on the natural resource. Lobster has been subjected to intensive exploitation due to high price in the international markets and except in a few countries where impeccable management strategies have ensured sustained production, the resource has been under heavy

E. V. Radhakrishnan (✉) · J. K. Kizhakudan
ICAR-Central Marine Fisheries Research Institute, Cochin, Kerala, India
e-mail: evrkrishnan@gmail.com

B. F. Phillips
School of Molecular and Life Sciences, Curtin University, Perth, WA, Australia
e-mail: B.Phillips@curtin.edu.au

© Springer Nature Singapore Pte Ltd. 2019
E. V. Radhakrishnan et al. (eds.), *Lobsters: Biology, Fisheries and Aquaculture*,
https://doi.org/10.1007/978-981-32-9094-5_1

pressure due to poor enforcement of fishing and marketing regulations. The need to develop alternate management strategies, including co-management, is emphasised so that the resource could be conserved and their sustainability ensured. This volume attempts to present available information on biology, fisheries and aquaculture of lobsters with special focus on lobster research in India.

Keywords
External morphology · Life cycle · Biology · Spawning · Nephropid · Palinurid · Scyllarid lobsters

1.1 Introduction

Lobsters are an ecologically and commercially valuable marine crustacean resource harvested by many countries in the world. It contributes significantly to the economy of at least a few countries. Few invertebrates have attracted as much attention as lobsters due to their commercial importance and the growing research interest in the fundamental and applied biology of this unique group of crustaceans (Radhakrishnan 1989; Poore 2004). For the past seven decades, lobsters have been the subject of physiological, biochemical and molecular research and more than 15,000 research papers have been published on lobster biology, ecology, physiology, fisheries and aquaculture during the previous and current century. In fact, the clawed American lobster, *Homarus americanus*, was the most intensively studied species. The beginning of the twenty-first century witnessed an upsurge in molecular research, and these studies have redrawn the phylogenetic relationships among the different groups of lobsters (Tsang et al. 2008; Karasawa et al. 2013; Lavery et al. 2014).

Although many research findings on lobsters have been published in various journals, the first comprehensive book on lobsters was published in 1980 (*The Biology and Management of Lobsters*, Vol. 1 & 2, Edited by J.S. Cobb and B.F. Phillips, Academic Press). Austin Williams published *Lobsters of the World – An Illustrated Guide* in 1988, which was followed by L.B. Holthuis' invaluable *Marine Lobsters of the World: An Annotated and Illustrated Catalogue of the Species of Interest to Fisheries Known to Date*, FAO Species Catalogue No. 125, Vol. 13 in 1991. A series of books with almost a 14-year gap were published (*Spiny Lobster Management*, B.F. Phillips, J.S. Cobb and Jiro Kittaka, Fishing News Books, 1994; *The Biology of the Lobster Homarus americanus* by Jan Robert Factor, Academic Press, 1995; the second edition of spiny mobster management entitled *Spiny Lobsters: Fisheries and Culture* by B.F. Phillips & Jiro Kittaka, 2000; *Lobsters: Biology, Management, Fisheries and Aquaculture* by B.F. Phillips, Oxford, Blackwell, 2006; the first book on slipper lobsters, *The Biology and Fisheries of Slipper Lobsters*, Kari Lavalli & Ehud Spanier, Taylor & Francis, 2007 and the second edition of the earlier book, *Lobsters: Biology, Management, Aquaculture and Fisheries* by B.F. Phillips in 2013, Wiley-Blackwell).

Apart from publication of these books, lobster scientists from across the globe decided to meet every 3 or 4 years in different locations around the world to review the research carried out during the previous 3 years and develop strategies to meet the new challenges confronting lobster fisheries across the world. So far, 11 International Conferences and Workshops on Lobster Biology and Management (ICWL) were held from 1977 onwards though the second workshop was held only after 8 years. The first Workshop held at Perth, Australia, in 1977 was attended by 34 scientists from 6 countries who discussed issues on lobster ecology and physiology. The aims of the ICWL are to review recent advances in biology, ecology, fisheries and aquaculture of clawed, spiny and slipper lobsters to identify gaps in current knowledge and future research priorities and to encourage collaborative studies for future research (Briones-Fourzan and Lozano-Alvarez 2015). After Perth, subsequent ICWLs were held in Saint Andrews, Canada (1985), La Habana, Cuba (1990), Sanriku, Japan (1993), Queenstown, New Zealand (1997), Key West, USA (2000), Hobart, Australia (2004), Charlottetown, Canada (2007), Bergen, Norway (2011), Cancun, Mexico (2014) and Maine, USA (2017). The 12th Conference will be held at Perth, Australia, in 2020, the birth place of the first lobster conference. An interim workshop was held at Chennai, India, in 2010. The proceedings of some of these conferences were published as special volumes in international scientific journals (e.g., *Canadian Journal of Aquatic and Fisheries Sciences*, Volume 43, *Crustaceana*, Volume 66, *Marine and Freshwater Research*, Volume 48 (8) and 52 (8) and the *New Zealand Journal of Marine and Freshwater Research*, Volume, 39). The papers presented at the Chennai workshop were published in *Journal of Marine Biological Association of India*, Volume 52. Three reviews and 23 selected papers presented in the 10th ICWL at Mexico were published in the *ICES Journal of Marine Science, Journal du Conseil*, Volume 72 (Supplement 1), 2015. Significant advances in lobster research have happened since the publication of the last book in 2013, especially in taxonomy with additions of new species and changes in phylogenetic relationships based on molecular research. Further, new information on larval diets, feeding and pathogens has been generated, which helped in improvements in breeding and hatchery production technology of tropical lobsters. Seed production and farming of slipper lobsters have gained momentum with advancement of technologies and prospects for commercial farming of a few tropical species have emerged.

The Indian Ocean (FAO area 51 and 57) is one of the most productive ecosystems inhabited by lobsters representing almost all the groups: nephropid, palinurid and slipper lobsters. The fishers in countries bordering the Indian Ocean have been fishing for lobsters for more than 100 years, and research on ecology and biology of lobsters probably helped them to understand the behaviour and distribution pattern in relation to the ocean environment and their habitat. Lobster research in India dates back to the turn of the nineteenth century when R.I.M.S.S *Investigator* carried out coastal and deep sea surveys in the Bay of Bengal, Laccadive (Lakshadweep) Sea and the Andaman Sea. These resource surveys brought to light many new and interesting lobster fauna to science (Wood-Mason 1872; Alcock and Anderson 1894; Alcock 1901; Nobili 1903; Borradaile 1906; Balss 1925).

The Central Marine Fisheries Research Institute under the Indian Council of Agricultural Research has carried out pioneering work on biology, fisheries and aquaculture of lobsters. The institute has a strong database of gear and species-wise lobster landings across the country from 1950 onwards. The research information delivered in this volume, especially on the fisheries in India, is based on the institute's database. Nearly 350 research papers have been published on Indian lobster fauna in national and international journals and 19 theses were awarded PhDs by various universities. Kathirvel (2004) brought out a bibliography of Indian lobster fauna, compiling research papers, technical reports, popular articles and reports published by various agencies. An updated bibliography, mainly the research papers published by researchers on Indian lobster fauna and the titles of doctoral theses on lobsters, is listed in this book. However, a comprehensive book on lobsters reviewing the voluminous work carried out by lobster researchers in India has not been published. Therefore, this volume, apart from a special focus on lobster fisheries of India, also attempts to bring out the current taxonomic position of world fauna of lobsters and available information on molecular taxonomy, general biology, world distribution of palinurids, reproductive biology of spiny and slipper lobsters, diseases and recent developments in aquaculture from a global perspective.

The introductory chapter provides an outline of what is expected in this book and also from each chapter. It also deals with general biology, the external morphology of three major groups, homarid, palinurid and scyllarid lobsters, and the life cycle of homarid, spiny and slipper lobsters. Recent advances in molecular genetics have redefined phylogenetic relationships among different groups of marine lobsters. Several new species have been added after publication of an annotated checklist of marine lobsters with 248 valid species by Chan (2010). In Chap. 2, Tin Yam Chan presents the current taxonomic status and an updated checklist of global fauna of lobsters. The taxonomy and the distribution map of Indian lobster fauna are presented in Chap. 3. A review of molecular taxonomy and phylogenetics of lobsters with recent work carried out on Indian species is presented in Chap. 4. Chapter 5 deals with the ecology and distribution of world fauna of spiny lobsters, and Chap. 6 discusses the interrelationships of feeding, growth, reproduction and hormonal control of moulting and reproduction. Lobster fisheries and management in India and Indian Ocean Rim countries are presented in Chap. 7. Bruce Phillips and Monica Perez Ramirez deal with the current global status and challenges for management of lobster fisheries in Chap. 8. The large volume of work carried out on the reproductive biology of spiny and slipper lobsters with a special focus on Indian species is reviewed in Chap. 9. Lobster breeding and hatchery production is an exciting and at the same time a challenging field, spanning more than 100 years of research. The historical and recent developments in spiny and slipper lobster aquaculture with future challenges in making lobster farming a reality is discussed in Chap. 10. Identifying and developing a suitable feed for lobster larvae have been the greatest challenge for lobster aquaculturists. The art of culturing scyllarid lobster larvae by exclusively feeding on jellyfish is presented in Chap. 11. There have been two primary sectors of research and development on lobster aquaculture: (i) developing controlled breeding and hatchery production technology and (ii) utilising naturally settling lobster seed with a view on commercialising the sector. While the recent research outcome has revolutionised the hatchery-based aquaculture of spiny and slipper lobsters, Vietnam and Indonesia have embarked upon commercial aquaculture of lobsters using natural seed supply. The recent developments in lobster aquaculture in Vietnam and Indonesia and the future challenges are discussed in Chap. 12. The range of diseases found in aquaculture is one among the major problems faced by aquaculturists all over the world. Pathogens and disease management are part of aquaculture production systems and, therefore, health management in lobster aquaculture is given special focus in Chap. 13. The postharvest processing, value addition and marketing of lobsters are considered in Chap. 14 and the perspectives and future directions in lobster research in Chap. 15.

1.2 What Are Lobsters?

'Lobsters' generally refers to the clawed lobsters of the Family Nephropidae, though there are other related groups, the spiny/rock lobsters, slipper lobsters, reef lobsters and the blind lobsters, which all come under the Suborder Macrura Reptantia of Order Decapoda under Class Malacostraca of Subphylum Crustacea and Phylum Arthropoda. Currently, 260 valid species (including 4 valid subspecies) of extant marine lobsters in 6 families and 54 genera are recognised (Chan, Chap. 2 in this book). The clawed lobsters are more related to the reef lobsters and have a sister relationship with the three families of freshwater crayfish (Crandall et al. 2000; Porter et al. 2005; Tsang et al. 2008; Toon et al. 2009). Spiny lobsters play an important role, both as prey and predator, serving as prey for sharks, finfish and other marine species and as a keystone predator of a diverse assemblage of benthic and infaunal species (Lipcius and Cobb 1994; Toller 2003). Indiscriminate and unregulated fishing due to high demand for this special seafood has led to stock collapse in some countries. There are some well-managed fisheries in the world due to enforcement of strict regulatory measures and those fisheries are sustainable.

Lobsters are marine animals and are generally poikilosmotic (Philipps et al. 1980) over their tolerated salinity range and very few enter brackish waters. For example, the mudspiny lobster *P. polyphagus* can tolerate wide ranges of salinity 17–50 ppt (Kasim 1986). Lobsters can stay outside water for a prolonged period if their gills are sufficiently moistened, and this advantage is exploited by the live lobster export trade to transport lobsters to long distances. While some species are solitary in nature, others are gregarious. *H. gammarus* is typically found on rocky substrates but may also burrow into cohesive mud or form depressions in sand. *N. norvegicus* spends much of its time inside burrows constructed in muddy substrates. Some of the species attain large sizes. The American lobster, *H. americanus*, attains a total body length of 64 cm and weighs more than 20 kg and the Southern rock lobster, *Sagmariasus* (*Jasus*) *verreauxi*, also grows to a total length of 60 cm. The smallest lobsters are in the Family Scyllaridae. An adult specimen of *Eduarctus* (*Scyllarus*) *martensii* measures a total body length of 2.5 cm (Holthuis 1991).

The world annual production of lobsters from capture fisheries is an average 2,73,051 t, excluding a small volume of 'mud shrimps' and 'ghost shrimps', which are not currently considered as lobsters (2005–2015) (FAO 2016). The global market for lobsters is over US$4 billion a year with the Asia-Pacific region contributing about 75% of the total revenue (Musa Aman 2012). Among the 260 extant species (including four subspecies) of marine lobsters distributed worldwide, only three families are commercially important: the Nephropidae (clawed lobsters), Palinuridae (spiny or rock lobsters) and Scyllaridae (slipper lobsters). The clawed lobster fishery is supported by four genera: *Homarus*, *Nephrops*, *Metanephrops* and *Nephrospsis*. In 2015, while the American lobster, *H. americanus*, constituted 49% of the total world production, the Norwegian lobster, *N. norvegicus*, contributed 15%, the Caribbean spiny lobster, *P. argus*, 12% and others, 24% of the total production. Generally known as 'rock', 'spiny' or 'crayfishes', depending upon local or regional trade requirements, palinurid lobsters support valuable seafood trade and exports in many countries. Among the 63 species (59 species + 4 subspecies) of palinurids under the 12 genera, 34 species under 6 genera (*Jasus*, *Projasus*, *Sagmariasus*, *Panulirus*, *Palinurus* and *Puerulus*), contribute to lobster fisheries. Lobsters, of the genus *Panulirus*, support the largest fisheries. The scyllarid lobsters also have several different names, 'slipper', 'shovel-nosed', 'sand' and 'Moreton Bay bug' as in Australia, and they are found throughout the tropical and temperate world oceans. Generally, they form incidental catch in commercial trawl fisheries and are rarely targeted, except for some species valuable in the aquarium trade. Occasionally, they constitute a subsistence gillnet fishery in some areas. The commercially valuable scyllarid genera are *Thenus*, *Scyllarides* and *Ibacus*.

1.3 The Life Cycle of Homarid, Spiny and Slipper Lobsters

Nephropid lobsters exhibit an abbreviated larval development. The prelarva is followed by three larval stages and the fourth stage termed as the postlarva is planktonic and soon settles at the bottom and changes to a benthic lifestyle. The larval phase is completed within 3 weeks. On the other hand, spiny and slipper lobsters have a prolonged larval phase lasting several months. Lobsters undergo five different phases in their life history (Philipps et al. 1980): adult, egg, larva (phyllosoma), postlarva (puerulus or nisto) and juvenile. The juvenile phase has recently been divided into an early benthic phase and an older juvenile phase (subadults), occupying almost similar adult habitats (Herrnkind and Butler 1986). The sexes are mostly separate and some are hermaphrodites, whereas abnormal development (psuedohermaphroditism) has also been also reported (Radhakrishnan et al. 1990). Larval behaviour and the ecology of many species are still not clearly understood though fairly good information is available on recruitment processes and the environmental influences on these processes of some shallow water species (Booth and Phillips 1994). Information on larval distribution patterns is essential to understand the recruitment processes (Booth et al. 2005).

1.3.1 Spawning–Hatching Cycle of Nephropid Lobsters

In nephropids, the male deposits the spermatophore in the thelycum or the seminal receptacle on the ventral side of the newly moulted female (Aiken and Waddy 1980). Most observations support that fertilisation is external in nephropid lobsters (Talbot and Helluy 1995; see review by Aiken and Waddy 1980; Phillips et al. 1980; Aiken et al. 2004) except that of Farmer (1974) who thought it may be internal in

H. gammarus. In general, the nephropids have a 2-year reproductive cycle, spawning in early autumn, hatching the eggs the next year in summer, and then moulting (and probably mating) in late summer or early autumn (Ennis 1980, 1984; Campbell 1983; Comeau and Savoie 2002; Agnalt et al. 2007). However, deviation from the normal cycle has also been reported (Aiken and Waddy 1982; Comeau and Savoie 2002). In *H. americanus* from southern St. Lawrence, Canada, 20% of multiparous females ranging between 65 and 109 mm CL have been observed to spawn in successive years instead of the generally accepted 2-year cycle, and some could even moult and spawn during the same summer. Similarly, up to 20% of primiparous females could also moult and spawn (for the first time) in the same year instead of spawning the following year. The European lobster, *H. gammarus*, also has a pluri-annual reproductive cycle similar to *H. americanus*, spawning every 2 or 3 years (Aiken and Waddy 1986). The Norwegian lobster, *N. norvegicus*, reproduces biannually, spawning in summer and the eggs hatching out either in late winter or early summer. The incubation period varies from 6 months in the Mediterranean population to 10 months in Icelandic lobsters (Sarda 1995).

In *H. americanus*, hatching and release of eggs occur after 9–12 months incubation by the ovigerous female. Ovigerous females migrate to deeper water (>200 m) during the winter months (Campbell and Stasko 1986). Egg-bearing females return to shallow water during the following summer to hatch their eggs. Hatching takes place from late May through September. Observations made on spawning in *H. americanus* in the Massachusetts State Lobster Hatchery for 10 years show that the lowest temperature at which hatching occurs was at 12.2 °C, but usually hatching takes place at temperatures between 15 and 20 °C (Hughes and Matthiessen 1962).

The eggs hatch at night and the larvae swim to the surface of the ocean and drift away along with the currents, feeding on zooplankton. *H. gammarus* hatches out as prelarva and attaches to the maternal pleopods after hatching (Mehrtens 2011). The prelarva on release undergoes three moults before settling as postlarva on the benthic substratum (Rotzer and Haug 2015). The larval phase lasts for 15–35 days. The larva of *H. americanus* hatches out at the prelarval stage and undergoes a series of four moults before metamorphosing into the postlarval stage (Ennis 1975). Stage 1 larva is approximately 8 mm long and Stage II is 9 mm and closely resembles Stage 1. The main difference between Stage 1 and Stage II larvae is the presence of four pairs of pleopods on the second to fifth abdominal segments. Stage III larvae is 11 mm long and the prominent feature is the complete tail fan with uropods. The final postlarva (Stage IV) resembles a miniature adult and is marked by the transition from a pelagic to a benthic life style.

In *H. americanus*, the duration of postlarval stage is about 11 days at 22 °C. The settling behaviour starts 2–6 days after moulting to the postlarval stage (Cobb et al. 1989). The postlarvae prefer substrates with crevices and macroalgal cover and if suitable substratum is found, they may even settle one day after moulting to the postlarva and burrows within 34 h (Botero and Atema 1982). The shelter-restricted juvenile measures 4–14 mm CL. Usually one lobster is seen in a shelter. Sometimes, more numbers of juveniles with in a contiguous shelter space may also be found (Lawton and Lavalli 1995). The emergent juvenile stage (15-mm CL) lobsters are mostly with limited movements outside the shelter. The vagile juvenile stage

(25–40 mm CL) is the size at which they attain physiological maturity and are still found in shallower benthic environment. More extensive movements outside the shelter and foraging for food were observed. The adolescent stage is reached when the lobster measures an average 50 mm CL but is still to reach functional maturity. The lobster attains adult stage at a carapace length more than 50 mm CL and is mostly nocturnal. Seasonal reproductively mediated movements from shallow to deeper waters were observed in sexually mature lobsters (Lawton and Lavalli 1995).

Postlarval *H. gammarus* spends nearly 2 years in burrows. Their preferred habitat is gravels ranging from coarse sand to fine shingle, though juveniles have also been known to form burrows in cohesive mud. They feed on marine worms, small crabs, urchins and gastropods as well as retaining the ability to filter-feed on plankton. They leave their burrows for crevices in rocky substrate at about 15 mm CL to begin life as an adult.

1.3.2 Spawning and Hatching in Palinurid Lobsters

Palinurids reproduce every year after attaining sexual maturity and may even spawn several times in a year. The Caribbean spiny lobster, *P. argus*, breeds year round with peak spawning activity during the spring (March–June) followed by a smaller peak during the autumn months (September–October) (Briones-Fourzan et al. 2008). Berry (1970) observed four broods/year in *P. homarus rubellus* from eastern Africa. *P. homarus* in Oman waters follows the general pattern of breeding by the tropical palinurids that they either breed throughout the year or had an extended spawning period with multiple spawning (Al-Marzouqi et al. 2008). Laboratory-held adult females of *P. cygnus* breed continuously, averaging six spawnings in a year, when held at a constant temperature of 25 °C and given abundant food (Chittleborough 1976). *P. penicillatus* in Solomon Islands is estimated to produce at least four broods per year and even as many as 11 broods per year in the most active size class (95–105 mm CL) (Prescott 1988). Vijayakumaran et al. (2005) report an average four broods per year with a maximum of seven spawnings in a year by a single *P. homarus homarus* lobster under captive conditions. Although egg-bearing lobsters are found throughout the year, peak spawning activity has been reported in certain months. While some species such as *P. homarus homarus* spawn in shallow coastal waters, species such as *P. ornatus* in Torres Strait was reported to migrate to long distances across the Gulf of Papua and breed in certain specific breeding grounds at the eastern limit of the Gulf (Moore and MacFarlane 1984). Mass movement of lobsters to specific breeding grounds has also been recorded for *P. argus* (Herrnkind 1980), *P. cygnus* (Phillips 1983) and (*Jasus*) *Sagmariasus verreauxi* (Booth 1984, 1997). On attaining maturity, adults of *P. argus* in Cuban waters are believed to migrate to deeper offshore reefs to breed (Cruz et al. 1986; Davis and Dodrill 1980). In palinurid lobsters, fertilization is external. In Stridentes subgroup of palinuridae, males deposit the spermatophore as a sticky mass on the thoracic sternal plate of the females, which hardens on contact with seawater, whereas in the Silentes subgroup, the gelatinous spermatophore sticks to the sternum of the female temporarily and falls off rapidly (Berry 1970; Berry and Heydorn 1970; MacDiarmid 1989; George 2005). Sexually mature females release the ova, which get fertilised on contact with the sperms released from the spermatophore. Females carry the fertilised eggs on the pleonal pleopods during which the embryonic development takes place. The newly deposited eggs in palinurids are bright orange in colour, which turn into coral red and then finally to brown just before hatching. The eyespot of the larva is visible through the transparent egg shell at this stage.

There is a close relationship between egg size and fecundity, with higher fecundity in females producing smaller eggs. Adult females of the same size have higher fecundity when the eggs are smaller and vice versa, presumably an adaptation to offset high larval loss during their prolonged developmental phase in oceanic waters (Sekiguchi et al. 2007). In tropical palinurids, specifically in the genus *Panulirus*, fecundity is high and the eggs are smaller whereas in temperate species, eggs are few and larger. In many species under the genus *Panulirus*, the larvae hatch out directly as the phyllosoma larvae, which are free swimming. Sometimes these larvae of spiny lobster have also been described as 'naupliosoma', 'prenaupliosoma' and 'prephyllosoma' (Phillips and Sastry 1980). In some species of palinurids such as those under the genus *Jasus*, the larva hatches out as naupliosoma. In *Jasus edwardsii*, the larva hatches out as a nonfeeding naupliosoma stage lasting 0.5–1 h, before metamorphosing into Stage 1 phyllosoma (Tong et al. 2000). The naupliosoma is round in shape and the first three pairs of thoracic limbs are folded under ventrally (Batham, 1967). Capture of naupliosoma stage of *J. edwardsii* in the plankton suggests that this is a normal stage in this species (MacDiarmid 1985). In *P. homarus*, the larva in most instances hatched out as the free-swimming phyllosoma stage (Radhakrishnan and Vijayakumaran 1995). Though there are instances of larvae hatching out as naupliosoma as in *P. homarus* under laboratory conditions, it was believed to be due to handling/transportation stress (Vijayakumaran et al. 2014). The naupliosoma had their pereipods wrinkled so that they were unable to swim freely and settled to the bottom of the tank and died (Radhakrishnan, personal observation). In *J. lalandii*, Silberbaur (1971) observed the naupliosoma after 8–12 h of hatching settling at the bottom of the tank and moulting into the first phyllosoma stage after 5–20 min.

Unlike the short larval phase of nephropid lobsters, the life cycle of *Panulirus* species is complex and includes a prolonged oceanic larval phase, lasting several months before metamorphosing into the postlarva (puerulus). The life history of a generalised palinurid is depicted in Fig. 1.1. The egg-bearing lobsters move to deeper breeding grounds and the eggs hatch out releasing the phyllosoma larvae. The larvae carried away by the currents to oceanic waters undergo final metamorphosis to the postlarva (puerulus) and return to the coastal areas for settlement. In the western rock lobster *Panulirus cygnus*, the larval phase lasts for 9–11 months (Phillips et al. 1979) and in *Panulirus ornatus*, the larval phase is estimated to be only 4–7 months (Dennis et al. 2001) though it is still shorter under laboratory rearing conditions (refer Chap. 10). The phyllosoma larva of *Panulirus* sp. is transparent with a dorsoventrally flattened bilobed body and several plumose appendages, morphologically adapted for a prolonged planktonic existence (Fig. 1.2). The dispersive ability of phyllosoma larvae is

Fig. 1.1 Life cycle of a palinurid lobster. (Sketch by: Joe K Kizhakudan, CMFRI)

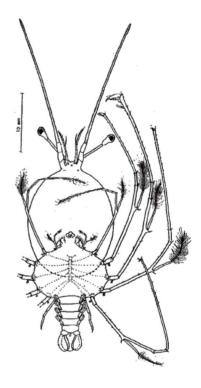

Fig. 1.2 Diagrammatic sketch of late-stage phyllosoma larva of *Panulirus gracilis* (from Johnson 1971). (Reproduced from FAO, 1991 Fig.13). FAO Species catalogue, Vol.13. Marine Lobsters of the world. FAO Fisheries Synopsis No. 125, Vol.13

evident from the estimated duration of nearly 24 months in *J. edwardsii*, probably the longest for a crustacean larva (Booth 1994). In the laboratory, the phyllosoma larvae of *J. edwardsii* have been reared to settlement in 212–274 days (Kittaka et al. 2005). The shorter larval duration in culture was possible due to sustained elevated water temperature of the rearing water and the availability of unrestricted amount of live feed (Ritar 2001). The larval phase of palinurids varies between 65 and 359 days depending upon species, and this information was available only after completion of the larval phase in the laboratory (Kittaka, 1994). Earlier information on phyllosoma larva morphology and their distribution was mainly based on descriptions of larvae from plankton collections (Gurney 1936; Prasad and Tampi 1959, 1965, 1966; Johnson 1971a, b; Berry 1974; Prasad et al. 1975; Kathirvel 1989). Assignment of larvae to a particular species and also to a specific stage in development was based on the presence or absence of certain appendages or spines and sometimes resulted in incorrect identifications. For example, the Stage I larvae described as that of *P. polyphagus* by Deshmukh (1968) belong to the *P. homarus–P. ornatus–P. versicolor* group, as the author's figure had subexopodal spines on leg 4 (Berry 1974), which was proved to be absent in later investigations (Prasad et al. 1975). The early stage larvae of almost all tropical *Panulirus* species of lobsters look morphologically similar whereas in *Palinurus* species, such as *P. elephas* and *P. delagoae*, the larvae hatch in an advanced stage of development. Identification of planktonic phyllosoma specimens from planktonic collections is very difficult due to the morphological similarity between species and difference within species (Konishi et al. 2006). Several lobster species share similar distribution ranges (Holthuis 1991), and their long-lived teleplanic larvae (Chittleborough and Thomas 1969) may be transported to oceanic waters far away from the adult distributional range. However, recent advances in molecular phylogeny have significantly improved species identification of lobster phyllosoma larvae. Silberman and Walsh (1992), using restriction fragment length polymorphism (RFLP) analysis based on polymerase chain reaction (PCR) amplification of 28S rDNA, successfully discriminated between phyllosoma larvae of two northwestern Atlantic *Panulirus* species, *P. argus* and *P. guttatus*. Ptacek et al. (2001) analysed partial nucleotide sequences of two mitochondrial DNA segments (cytochrome oxidase subunit I and 16S ribosomal DNA) of almost all lobster species in the genus *Panulirus*. Yamauchi et al. (2002) reported the nucleotide sequence of the entire mitochondrial DNA of the Japanese spiny lobster *P. japonicus*. Chow et al. (2006) applied nucleotide sequence analysis for *Panulirus* phyllosoma samples collected in the Japanese waters and successfully showed the samples to comprise eight species. Konishi et al. (2006) identified a late stage larvae of the spiny lobster *P. echinatus* collected from Atlantic Ocean using nucleotide sequence analysis of mitochondrial 16S rDNA.

The larval size at hatching varies between different species. The first instar of *Panulirus* species of lobsters measures a mean 1.7 mm body length and the final instar, an average 29.4 mm (Matsuda et al. 2006; Goldstein et al. 2008). Among the palinurids, the first instars of *Palinurus elephas* and *P. delagoae* are larger in size, measuring 2.8–3.9 mm in total length (TL) (Berry 1974; Williamson 1983; Kittaka and Ikegami 1988). The larger size at hatching and shorter time to metamorphose into puerulus is characteristic of these species (Kittaka and Ikegami 1988). The first-stage larva of *P. penicillatus* has a pair of unsegmented eyes, a long cephalothorax, a very short abdomen and three pairs of pereiopods and measures 1.75–1.80 mm in TL, whereas final stage larvae measure 27.0–33.6 mm TL (Matsuda et al. 2006). The phyllosoma larvae of all palinurid lobsters, except *Jasus, Sagmariasus* and possibly *Projasus*, have a biramous (both exopod and endopod) third maxilliped whereas *Jasus, Sagmariasus* and *Projasus* have a uniramous third maxilliped (exopod absent) (Baisre 1994). Immediately after hatching, phyllosoma larvae are highly positively phototactic whereas late-stage larvae exhibit negative phototaxis (Goldstein et al. 2008).

After hatching, the free-swimming phyllosoma larvae of palinurids pass through several metamorphic moults and finally metamorphose into the pelagic postlarva, 'puerulus', in offshore waters. The transparent puerulus is the postlarval transitional stage between the planktonic phyllosoma and the benthic juvenile and resembles the adult in morphology but lacks pigments, sculpturing, pubescence and spination of the carapace of the adult (Booth and Phillips 1994; McWilliam 1995). The first instar of *Panulirus* species of lobsters measure a mean 1.7 mm body length and the final instar, an average 29.4 mm (Matsuda et al. 2006; Goldstein et al. 2008). The puerulus swims towards the shore and settles on to the seaweeds or rocks. The non-feeding puerulus metamorphoses into juveniles within 8–12.8 days depending upon the temperature (Lemmens 1994). The puerulus survive on the energy reserves stored in the hepatopancreas during the final phyllosoma stage, which is crucial for their moulting to the postpuerulus stage and survival.

Information on the pueulus and early juvenile morphology of several species of the genera, *Panulirus, Palinurus, Jasus* and *Projasus*, is known (Bouvier 1913, 1914; Santucci 1926; Orton and Ford 1933; Caroli 1946; Gordon 1953; Berry 1974; Deshmukh 1966; Nishida et al. 1990; Briones-Fourzán and McWilliam 1997; Kittaka et al. 1997; Hunter 1999; Báez and Ruiz 2000). Wild-caught pueruli of spiny lobster measure 7.0–9.9 mm in carapace length. Pueruli obtained from laboratory-reared phyllosoma larvae of *P. penicillatus* and *P. japonicus* are slightly smaller in size compared to wild specimens whereas laboratory-reared pueruli of *P. argus* are comparatively larger than the wild-caught ones (Table 1.1). Briones-Fourzan and McWilliam (1997) compared the morphological features of the pueruli of *P. guttatus* and *P. argus*, the sympatric species distributed in the Caribbean Sea, and found the former larger in size with a more specialised form and, hence, is presumably of a recent evolutionary origin. McWilliam (1995) proposed an evolutionary sequence of four species groups (P1–P4) in the larval and puerulus phases of *Panulirus*. Groups P1 and P2 pueruli have a tapered antennal flagellum and no posterolateral sternal spines and those in Groups P3 and P4 have a spatulate or semispatulate antennal flagellum and posteriorly directed posterolateral sternal spine. The puerulus of *P. guttatus* falls in Groups P3 and P4 but has a relatively short exopod on the third maxilliped (characteristic of Group P2) and a long exopod, with a nonannulate flagellum on the second maxilliped (as in Groups P1 and P2) and, therefore, puerulus of *P. guttatus* was given a special status 'Group P2B'. *P. argus* falls in Group P1.

1 Introduction to Lobsters: Biology, Fisheries and Aquaculture

Table 1.1 Body length measurements of wild-caught and laboratory-reared pueruli of palinurid lobsters

Species	Wild/laboratory reared	Carapace length (mm)	Total length (mm)	Location	Authors
Panulirus homarus	Wild	7.78 ± 1.16 (mean)	20.83 ± 1.77	Tuticorin, east coast of India	Dharani et al. (2009)
Panulirus polyphagus	Wild	9.0	23.2	Chennai, east coast of India	Girijavallabhan and Devarajan (1978)
Panulirus ornatus	Wild	7.0–8.0	–	Vietnam	
Panulirus penicillatus	Wild	–	26.0–27.9	Japan	Tanaka et al. (1984)
			28.5–29.3	Pacific equatorial	Michel (1971)
	Laboratory	9.08	23.08	Japan	Matsuda et al. (2006)
Panulirus japonicus	Wild	7.0–9.9	20.2–26.8	Japan	Nakamura (1994), Nonaka et al. (1958)
	Laboratory	6.0–8.0	17.3	Japan	Kittaka and Kimura (1989), Yamakawa et al. (1989), Sekine et al. (2000)
Panulirus mauritanicus	Wild	8.8–9.0	23.3–24.0	Western Mediterranean	Guerao et al. (2006)
Palinurus elephas	Wild	–	21.0	Mediterranean	Hunter (1999)
	Wild	7.5–8.0	–	Western Mediterranean	Goni and Latrouite (2005)
	Wild	7.0–8.0	–		Diaz et al. (2001)
	Laboratory	10.0	–	Japan	Kittaka (2000), Ceccaldi and Latrouite (2000)
Panulirus cygnus	Wild	7.0–8.0		Western Australia	Booth and Kittaka (2000)
Panulirus guttatus	Wild	10.0–10.5	27–28	Western Central Atlantic coast	Sharp et al. (1997), Briones-Fourzan and McWilliam (1997), Robertson and Butler (2003)
Panulirus argus	Wild	5.4–6.5	16–19	Western Central Atlantic coast	Briones-Fourzan and McWilliam (1997)
Panulirus argus	Laboratory	6.2–6.75	15.7–17.9		Goldstein et al. (2008)

1.3.3 Spawning and Hatching in Scyllarid Lobsters

The majority of temperate species of scyllarids exhibits a seasonal reproductive cycle with a clear peak spawning activity during spring and summer (Oliveira et al. 2008). Most species of the genus *Scyllarides* spawn multiple broods during the reproductive season (Spanier and Lavalli 2006). Unlike most of the tropical palinurids, which breed throughout the year, *T. unimaculatus* from Mumbai waters exhibits a well-defined breeding season from October to January (Radhakrishnan et al. 2005). Successive spawning within a season is believed to occur in *T. orientalis* and *T. indicus* (Jones 1988), *S. latus* (Spanier and Lavalli 1998), *S. nodifer* and *S. depressus* (Lyons 1970).

In a few scyllarids, absence of spermatophore on the sternum of egg-bearing females has led to speculation that fertilization may be internal (Lavalli et al. 2007). In other species, although the spermatophore is deposited on the sternal plate, it rapidly breaks down in seawater, and, therefore, is seen only occasionally (Kizhakudan 2014).

Generally, eggs of scyllarid lobsters hatch out into phyllosoma, though there are exceptions. In the laboratory, *Ibacus ciliatus* and *I. novemdentatus* hatch as phyllosoma larvae (Wakabayashi et al. 2012, 2016). Larvae of some species such as *Scyllarides latus* hatch out as the nonfeeding naupliosoma (Aktas et al. 2011). The naupliosoma, as in palinurid lobsters, remains only for a few hours and moults into first-stage phyllosoma. Earlier information on phyllosoma larvae of scyllarid lobsters was also from descriptions of larvae from plankton samples (Prasad and Tampi 1957, 1960, 1965, 1968; Johnson 1971a, b; Berry 1974; Kathirvel 1990; Inoue and Sekiguchi 2005). The duration of the larval phase of some species of scyllarids has been estimated from laboratory culture (see Chap. 10) and is, in general, shorter (17–192 days) than the palinurid lobsters. The number of instars varies among species from 4 to 13. 'Giant phyllosomas' (75–80 mm TL) have been recorded from Atlantic, Pacific and Indian Oceans (Richters 1873; Johnson 1971b; Robertson 1968; Prasad et al. 1975; Yoneyama and Takeda 1998), which were assigned to the scyllarid *Parribacus antarcticus*. Palero et al. (2014) confirmed by morphological identification and DNA-barcoding techniques that the 'giant phyllosoma' larvae, collected from Coral Sea near Osprey Reef and earlier from various oceanic regions, belong to *P. antarcticus*. The first larval instar of laboratory-reared phyllosoma of *T. orientalis* measures an average 3.89 mm TL and the final instar (Stage IV instar), 18.22 mm TL (Mikami and Greenwood 1997). The first instar of phyllosoma larva of *T. australiensis* is larger (TL 4.03 mm) (Wakabayashi and Phillips 2016). The scyllarid phyllosoma larvae have a uniramous third maxilliped unlike the majority of palinurid lobster phyllosoma larvae. The cephalic shield of phyllosoma larvae of scyllarid lobsters is wide rather than long (CW: CL ratio >1) whereas in palinurid larvae, it is longer than wide (<1). While no difference in the cephalic shield (CW:CL) ratio (1.2) between Stage I and Stage IV phyllosoma was observed in *T. orientalis*, for phyllosoma reared in the laboratory (Mikami and Greenwood 1997), the CW:CL ratio of Stage X larva (0. 0.57) was lower than Stage I (0.78) of laboratory-reared *P. argus* phyllosoma larvae (Goldstein et al. 2008). Gradual

Fig. 1.3 Life cycle of a generalised scyllarid lobster. (Sketch by: Joe K Kizhakudan, CMFRI)

reduction in CL: CW ratio was observed until Stage III (Stage I, 0.78 and Stage III, 0.60) beyond which it was stable (0.51–0.55) throughout until the final Stage X.

The life cycle of a generalised scyllarid is depicted in Fig. 1.3. The phyllosoma larvae on hatching out are carried away by oceanic currents and after final metamorphosis settles as the nectonic 'nisto' in coastal habitats. The 'nisto' is an intermediate stage between the phyllosoma larva and the adult stage, characteristic of Achelata. The duration of nisto varies between 5 and 24 days, depending upon the species. It is a non-feeding post-larval stage similar to the puerulus of spiny lobsters. The carapace length of nistos of species from different subfamilies vary: Arctidinae (CL: 9–15 mm), Ibacinae (CL: 11–21 mm), Scyllarinae (CL: 3.5–8 mm) and Theninae (CL: 7–9 mm) (Sekiguchi et al. 2007).

For detailed information on the early life history of spiny lobsters, refer Booth and Phillips (1994), and for the slipper lobsters, Sekiguchi et al. (2007).

1.4 External Morphology of Nephropid, Palinurid and Scyllarid Lobsters

Lobsters, in general, have a typical crustacean anatomy with the body bilaterally symmetrical with paired appendages on each segment. The body consists of 19 somites or segments, which is covered by the hard exoskeleton (Fig. 1.4). The body is divided into two parts: an anterior sub-cylindrical cephalothorax (1–13 all fused) and the posterior pleon (14–19 all unfused) (Wahle et al. 2012). Each of the 19 somites bears a pair of appendages, except somite 14 in palinurids. Schram and

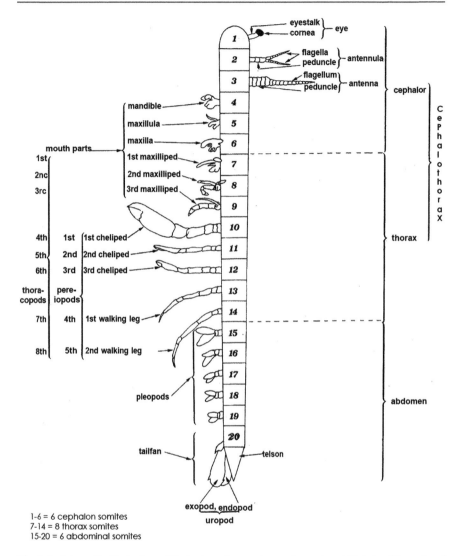

Fig. 1.4 Schematic illustration of the body and appendages of a nephropid lobster. (Reproduced from FAO, 1991 Fig.2). FAO Species catalogue, Vol. 13. Marine Lobsters of the world. FAO Fisheries Synopsis No. 125, Vol.13

Koenemann (2004) suggested that 'pleon' is the most appropriate term for the posterior tagmata of malacostracans and is not just an equal or interchangeable alternative for the term abdomen, as the pleon of malacostracans and the abdomen of other crustaceans exhibit fundamentally different developmental pathways. The cephalon bears the sensory and feeding appendages: the eyes, antennules, antennae, mandibles, maxillules (first maxillae) and maxillae (second maxillae). The thorax bears the feeding appendages (first, second and third maxillipeds), the first pereiopods and walking legs (the remaining pereiopods). The pleon bears the pleopods (swimmerets) and, posteriorly, the tail fan (Wahle et al. 2012). The appendages may be either uniramous (exopod only) or biramous with both exopod and endopod.

1.4.1 Cephalothorax

On the dorsal side, the cephalothorax is covered by the carapace, which is formed by the fusion of five somites of cephalon with eight somites of the thorax in homarid, palinurid and scyllarid lobsters (Wahle et al. 2012). The carapace is either sub-cylindrical as in nephropids, palinurids and some scyllarids or dorsoventrally flattened as in the genus *Thenus* (Holthuis 1991). Laterally, the carapace extends up to the legs and covers the branchial chamber. In nephropids, the carapace bears a rostrum, the median extension of the anterior part of the carapace, in between the eyes (Figs. 1.5a, 1.5b and 1.5c). The carapace is divided into distinct regions with carinae, grooves and spines, which have taxonomic implications. In some groups, such as the palinurids, a part of the antennular somites forms the so-called antennular plate which bears spines; the number and arrangement are of taxonomic importance (Holthuis 1991). In some genera of palinuridae, the lateral margins of the antennular plate is ridge-like, forming a stridulating organ which produces a rasping sound when the inner margin of the antennal peduncle rubs over the ridge. Those without a stridulating organ fall in the Silentes group. Palinurids possess elongated and spiny antennae, a pair of frontal horns over the eyes, whip-like antennules which are sensory and a sub-chela on the fifth pereiopod of females (Fig. 1.6). The Scyllarids have highly modified, broad, flattened antennae, blunted nodules on carapace and

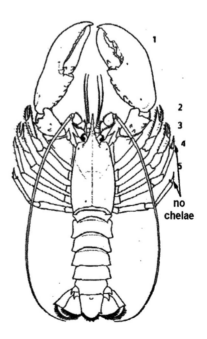

Fig. 1.5a Schematic sketch of *Homarus americanus*. (Reproduced from FAO, 1991 Fig.18). FAO Species catalogue, Vol.13. Marine Lobsters of the world. FAO Fisheries Synopsis No. 125, Vol.13

Fig. 1.5b Schematic sketch of *Homarus gammarus*. (Reproduced from FAO, 1991 Fig.110). FAO Species catalogue, Vol.13. Marine Lobsters of the world. FAO Fisheries Synopsis No. 125, Vol.13

Fig. 1.5c Schematic sketch of *Nephrops norvegicus*. (Reproduced from FAO, 1991 Fig.162). FAO Species catalogue, Vol.13. Marine Lobsters of the world. FAO Fisheries Synopsis No. 125, Vol.13

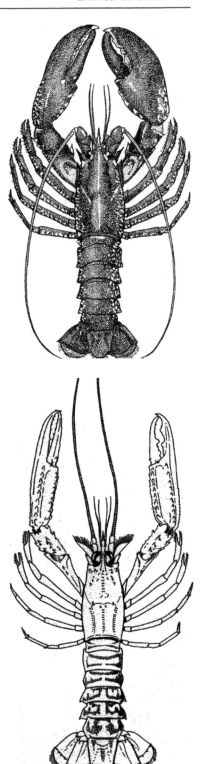

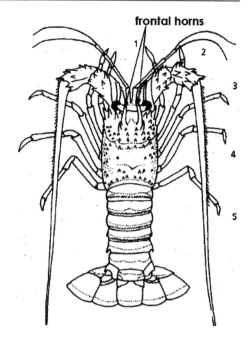

Fig. 1.6 Schematic sketch of a palinurid lobster. (Reproduced from FAO, 1991 Fig.22). FAO Species catalogue, Vol.13. Marine Lobsters of the world. FAO Fisheries Synopsis No. 125, Vol.13

raised elevations above the eye socket. The diagrammatic illustration of dorsal view of the right half of a generalised scyllarid carapace is shown in Fig. 1.7. Spanier et al. (2010) classified the body shape of scyllarids into two groups: one with a cylindrical and vaulted body occupying more of complex substrates and the second group with more of a dorsoventrally flattened body and triangular form, inhabiting softer muddy substrates. The scyllarids belonging to the first group include the genera, *Acantharctus, Scyllarides* (Fig. 1.8) and *Scyllarus,* and those in the second group are *Thenus* (Fig. 1.9), *Evibacus* and *Ibacus* with *Parribacus* falling in an intermediate group. The carapace of Polychelids is either dorsoventrally flattened like scyllarids or has a sub-cylindrical shape with chelae on first four (*Polycheles*) or all five (*Pentacheles*) pereiopods with the first pair larger than the rest, long slender antennae, a small rostrum and a pointed and calcified telson (Fig. 1.10). The sculpturing of the carapace of polychelid species varies considerably with swollen carinae and tubercles in some species and robust spinules and setose in some other species. The cervical grooves are deep and are with branchial, post-orbital, median post-rostral and post-cervical carinae (Lavalli and Spanier 2010).

The ventral cephalothorax has a central sternal plate with grooves. The sternal plate in palinurids is triangular or arrow shaped. Lateral projections may be present on both right and left margins and tubercles on the surface of the sternal plate. In the females the external opening of the oviduct is visible on the coxa of the third pair of pereiopods. In the males, the gonoduct (vas deferens) opens on the coxa of the fifth pereiopods.

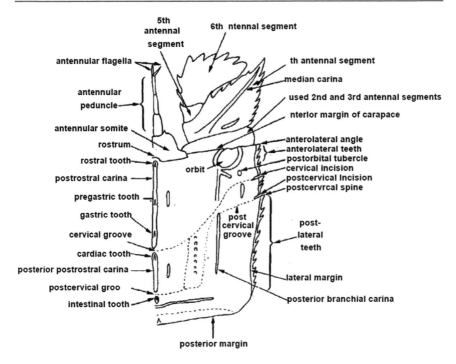

Fig. 1.7 Schematic dorsal view of right half of a scyllarid lobster. (Reproduced from FAO, 1991 Fig.6). FAO Species catalogue, Vol.13. Marine Lobsters of the world. FAO Fisheries Synopsis No. 125, Vol.13

Fig. 1.8 Schematic sketch of *Scyllarides*. (Reproduced from FAO, 1991 Fig.20). FAO Species catalogue, Vol.13. Marine Lobsters of the world. FAO Fisheries Synopsis No. 125, Vol.13

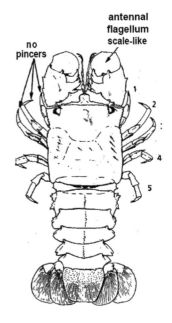

1 Introduction to Lobsters: Biology, Fisheries and Aquaculture   21

Fig. 1.9 Schematic sketch of *Thenus* sp. (Reproduced from FAO, 1991 Fig.323). FAO Species catalogue, Vol.13. Marine Lobsters of the world. FAO Fisheries Synopsis No. 125, Vol.13

Fig. 1.10 Schematic sketch of a polychelid lobster. (Reproduced from FAO, 1991 Fig.19). FAO Species catalogue, Vol.13. Marine Lobsters of the world. FAO Fisheries Synopsis No. 125, Vol.13

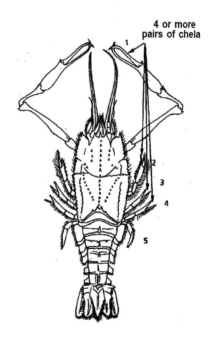

1.4.2 Pleon

The pleon has six separate somites (14–19), which are not fused, but flexible, smooth or grooved and connected to the carapace by the arthrodial membrane (Fig. 1.4). Each somite is covered on the dorsal side by thick chitinous armour, the tergum and the lateral pleuron, which is continuous with the tergum (Fig. 1.11). The shape and colouration of the pleuron are of taxonomic value. The anterior portion of the tergum, which is visible only during flexion of the pleon, is smooth. The ventral surface of each somite has a thin and flexible calcified sternum. Each of the somites is connected to the other by intertergal and intersternal arthrodial membranes, which permit flexibility to the abdomen.

The somites vary greatly in colouration and sculpture and bear teeth with or without denticles on the pleura (Lavalli and Spanier 2010). In Palinurids, it may vary from small squamae arranged in two to three transverse rows (*Jasus* spp.), deep transverse grooves in each somite (*Palinurus* spp.) to uninterrupted/interrupted, crenulated transverse grooves, bands or lines (*Panulirus* spp.). Median ridges or carinae on the abdomen are characteristics of some groups (*Nephropsis* spp., *Linuparus* & *Projasus* spp) (Holthuis 1991). Scyllarid lobsters have their squamae arranged in transverse rows that are obscured by tubercles and fine 'hairs', bearing transverse grooves on elevated median ridges with a humped appearance with median carinae (Lavalli and Spanier 2010). Polychelids possess a triangular-shaped pleon that tapers posteriorly (Galil 2000). Each somite bears a median tergal carina and a sub-median tergal groove. Pleurae are with smooth edges and are rounded. In a majority of species, the second somite is proportionally wider in females than in males relative to its length.

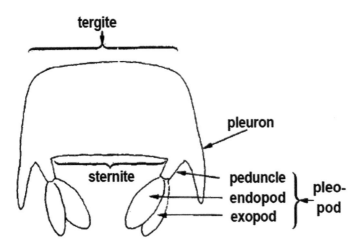

Fig. 1.11 Schematic cross-section through an abdominal somite. (Reproduced from FAO, 1991 Fig.15). FAO Species catalogue, Vol.13. Marine Lobsters of the world. FAO Fisheries Synopsis No. 125, Vol.13

The first pleon segment (14) has no pleopods in palinurid, scyllarid and polychelid lobsters. The appendages on the second to fourth pleonal somites (15–18) are the pleopods or swimmerets. In the male palinurids, the pleopods are uniramous in some species and biramous in some other species and in females they are biramous with an exopod and an endopod. The first pair of pleopods (both exopod and endopod) is leaf-like in females. The other three pairs of pleopods have a leaf-like exopod and a single-segmented peduncular endopod, which bear setae to carry eggs (Fig. 1.12). In the nephropids, the first pair of pleopods is greatly reduced in females, whereas in males the first and second pairs are modified as 'copulatory stylets.' The first pair of the stiffened pleopod of the male is used to transfer sperm from the male to the seminal receptacle of the female. The second pleopod of male has three to five appendages except that it has the appendix masculina, which aids in copulation.

The sixth pleonal somite (somite 19) bears the tail fan, which consists of a central telson and a pair of uropods, which is used for propulsion (Fig. 1.13).

Fig. 1.12 Abdomen of *Panulirus homarus homarus* female showing the leaf-like exopod and endopod. (Photo by: E. V. Radhakrishnan CMFRI)

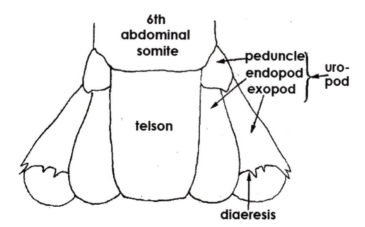

Fig. 1.13 Schematic diagram of tail fan. (Reproduced from FAO, 1991 Fig.17). FAO Species catalogue, Vol.13. Marine Lobsters of the world. FAO Fisheries Synopsis No. 125, Vol.13

1.4.3 Mouthpart Appendages and Pereiopods

The mouthpart appendages are biramous with an exopod and an endopod. The stalked eyes are located on somite 1; the eyes are sometimes reduced with the cornea lacking pigment in some species. Somite 2 carries the antennules (first antenna), each of these consisting of a three-segmented peduncle carrying two flagella. Somite 3 carries the antennae (second antenna), which consists of a peduncle of five segments and a single flagellum. The flagellum is whip-like as in palinurids, which may be with spinules, stiff and strong (Fig. 1.6). The lobster uses the antennae as a defensive shield in the event of an attack by predators. In the Family Scyllaridae, the flagellum is transformed to a single plate-like segment, which makes the antennae six segmented (Fig. 1.7). Somite 4–9 carry the mouth parts, which have a food-handling and ingestion function. Somite 4 carries the mandible, strongly calcified, resembling a molar teeth of vertebrates. Somites 5 and 6 carry the maxillulae (or first maxillae) and maxillae (second maxillae), respectively, and are leaf-like organs. Somites 7–9 carry the first to third maxillipeds; the first is leaf-like, the maxilla, the second and third are leg-like, especially the third.

Somites 10–14 carry the five pairs of pereiopods or true walking legs. In nephropids, the first pereiopod, and sometimes the second and third, often (but not always) ends in a chela or a pincer. The first pair, in some species unequal, is the largest and the robust; the larger left chela in *Homarus* is the crushing claw with molar-like teeth by which it crushes molluscs and other hard objects, and the smaller right one is the cutting claw (Figs. 1.5a, 1.5b and 1.5c). The last two pairs are always nonchelate. In palinurids, the legs are not chelate and the first pereiopod is usually the largest and is also used for feeding. The third walking leg is normally the longest and is never with a true chela. The legs without pincers are called walking legs as they are mainly engaged in locomotion (Holthuis 1991). Pereiopods consist of seven segments and from proximal to distal they are coxa, basis, ischium, merus, carpus, propodus and dactylus (Fig. 1.14).

1.5 Summary

The very objectives of this book are to provide the reader with a comprehensive updated knowledge on the biology, taxonomy, species and distribution, current status of molecular phylogeny, fisheries and aquaculture of lobsters with a global perspective. The fisheries section has mainly dealt with the Indian fisheries and management with a brief description of lobster fisheries in Indian Ocean Rim countries. The chapters also reflect recent developments in controlled breeding and aquaculture and the challenges ahead for management of fisheries to make them sustainable.

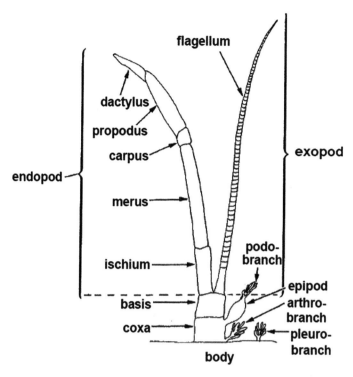

Fig. 1.14 Schematic diagram of a thoracopod (Holthuis 1991). (Reproduced from FAO, 1991 Fig.12). FAO Species catalogue, Vol.13. Marine Lobsters of the world. FAO Fisheries Synopsis No. 125, Vol.13

References

Agnalt, A., Kristiansen, T. S., & Jørstad, K. E. (2007). Growth, reproductive cycle, and movement of berried European lobsters (*Homarus gammarus*) in a local stock off southwestern Norway. *ICES Journal of Marine Science, 64*(2), 288–297. https://doi.org/10.1093/icesjms/fsl020.

Aiken, D. E., & Waddy, S. L. (1980). Reproductive biology. In J. S. Cobb & B. F. Phillips (Eds.), *The biology and management of lobsters, Vol. I. Physiology and behavior* (pp. 215–276). New York: Academic.

Aiken, D. E., & Waddy, S. L. (1982). Cement gland development, ovary maturation and reproductive cycles in the American lobster, *Homarus americanus*. *Journal of Crustacean Biology, 2*, 315–327.

Aiken, D. E., & Waddy, S. L. (1986). Environmental influences on recruitment of the American lobster (*H. americanus*): A perspective. *Canadian Journal of Fisheries and Aquatic Sciences, 43*, 2258–2270.

Aiken, D. E., Waddy, S. L., & Mercer, S. M. (2004). Confirmation of external fertilization in the American lobster, *Homarus americanus*. *Journal of Crustacean Biology, 24*(3), 474–480.

Aktaş, M., Genç, E., & Ayçe Genç, M. (2011). Maturation, spawning and production of phyllosoma larvae of Mediterranean slipper lobster, *Scyllarides latus* (Latreille 1803) in captivity. *Journal of Black Sea/Mediterranean Environment, 17*(3), 275–281.

Alcock, A. (1901). *A descriptive catalogue of the Indian deep-sea Crustacea Decapoda Macrura and Anomala, in the Indian Museum, being a revised account of the deep-sea species collected by the Royal Indian Marine Survey Ship Investigator.* Calcutta. iv+286 pp., pls. 1–3.

Alcock, A., & Anderson, A. R. S. (1894). An account of a recent collection of deep sea Crustacea from the Bay of Bengal and Laccadive Sea. Natural history notes from H. M. Indian Marine Survey Steamer "Investigator", commander C. F. Oldham, R. N., commanding. Series II, No. 14 J. *Asiatic Society of Bengal, 63*(2), 141–185, pl. 9.

Al-Marzouqi, A., Groeneveld, J. C., Al-Nahdi, A., & Al-Hosni, A. (2008). Reproductive season of the Scalloped spiny lobster *Panulirus homarus* along the Coast of Oman: Management implications. *Agricultural Marine Sciences, 13*, 33–42.

Báez, P., & Ruiz, R. (2000). Puerulus y postpuerulus de *Projasus bahamondei* George, 1976 (Crustacea, Decapoda, Palinuridae). *Investigaciones Marinas, 28*, 15–25.

Baisre, J. (1994). Phyllosoma larvae and the phylogeny of Palinuroidea (Crustacea: Decapoda): A review. *Australian Journal of Marine & Freshwater Research, 45*, 925–944.

Balss, H. (1925). *Palinura, Astacura and Thalassinidea. Wissenschaftliche Ergebnisse der Deutschen Tiefsee-Expedition auf dem Dampfer "Validivia" 1898-1899, 20*, 185–216, pls 18–19.

Batham, E. J. (1967). The first three larval stages and feeding behavior of phyllosoma of the New Zealand palinurid crayfish *Jasus edwardsii* (Hutton, 1875). *Transactions of the Royal Society of New Zealand, 9*(6), 53–64.

Berry, P. F. (1970). Mating behaviour, oviposition and fertilization in the spiny lobster *Panulirus homarus* (Linnaeus). *Investment Report of Oceanographic Research Institute South Africa, 24*, 1–16.

Berry, P. F. (1974). Palinurid and scyllarid lobster larvae of the Natal Coast, South Africa. *South African Association of Marine Biological Research Oceanographic Research Investment Report, 34*, 1–44.

Berry, P. F., & Heydorn, A. E. F. (1970). A comparison of the spermatophoric mass and mechanisms of fertilization in South African spiny lobsters (Palinuridae). *Investment Report of Oceanographic Research Institute South Africa, 25*, 1–18.

Booth, J. D. (1984). Movement of packhorse rock lobsters (*Jasus verreauxi*) tagged along the eastern coast of the North Island, New Zealand. *New Zealand Journal of Marine and Freshwater Research, 18*, 275–281.

Booth, J. D. (1994). *Jasus edwardsii* larval recruitment off the east coast of New Zealand. *Crustaceana, 66*, 295–317.

Booth, J. D. (1997). Long-distance movements in *Jasus* spp. and their role in larval recruitment. *Bulletin of Marine Science, 61*, 111–128.

Booth, J. D., & Kittaka, J. (2000). Spiny lobster growout. In B. F. Phillips & J. Kittaka (Eds.), *Spiny lobsters: Fisheries and culture* (2nd ed., pp. 556–585). Oxford: Fishing News Books.

Booth, J. D., & Phillips, B. F. (1994). Early life history of spiny lobster. *Crustaceana, 66*(1), 271–294.

Booth, J. D., Webber, W. R., Sekiguchi, H., & Coutures, E. (2005). Diverse larval recruitment strategies within the Scyllaridae. *New Zealand Journal of Marine and Freshwater Research, 39*, 581–592.

Borradaile, L. A. (1906). Marine crustaceans XIII. The Hippidae, Thalassinidea and Scyllaridea. In J. S. Gardiner (Ed.), *The fauna and geography of the Maldives and Laccadive Archipelago* (pp. 750–754). Cambridge: Cambridge University Press.

Botero, L., & Atema, J. (1982). Behavior and substrate selection during larval settling in the lobster *Homarus americanus*. *Journal of Crustacean Biology, 2*(1), 59–69.

Bouvier, M. E. L. (1913). The post-embryonic development of the spiny lobster. *Nature, 9*, 633–634.

Bouvier, M. E. L. (1914). Recherches sur le développement post-embryonnaire de la Langouste commune (*Palinurus vulgaris*). *Journal of Marine Biological Association United Kingdom, 10*, 179–193.

Briones-Fourzan, P., & Lozano-Alvarez, E. (2015). Lobsters: ocean icons in changing times. *ICES Journal of Marine Science, 72*, i1–i6.

Briones-Fourzán, P., & McWilliam, P. S. (1997). Puerulus of the spiny lobster *Panulirus guttatus* (Latreille, 1804) (Palinuridae). *Marine and Freshwater Research, 48*, 699–706. https://doi.org/10.1071/MF97130.

Briones-Fourzan, P., Candela, J., & Lozano-Alvarez, E. (2008). Postlarval settlement of the spiny lobster *Panulirus argus* along the Caribbean coast of Mexico: Patterns, influence of physical factors, and possible sources of origin. *Limnology and Oceanography, 53*(3), 970–985.

Campbell, A. (1983). Growth of tagged American lobster, *Homarus americanus*, in the Bay of Fundy. *Canadian Journal of Fisheries and Aquatic Sciences, 40*, 1667–1675.

Campbell, A., & Stasko, A. B. (1986). Movements of lobsters (*Homarus americanus*) tagged in the Bay of Fundy, Canada. *Marine Biology, 92*(3), 393–404.

Caroli, E. (1946). Di un puerulus de *Palinurus vulgaris*, pescato nel Golfo di Napoli. *Pubblicazioni della Stazione Zoologica di Napoli, 20*, 152–157.

Ceccaldi, H. J., & Latrouite, D. (2000). The French fisheries for the European spiny lobster *Palinurus elephas*. In B. F. Phillips & J. Kittaka (Eds.), *Spiny lobsters. Fisheries and culture* (Second ed., pp. 200–209). Oxford: Blackwell Scientific.

Chan, T. Y. (2010). Annotated checklist of the world's marine lobsters (Crustacea: Decapoda: Astacidea, Glypheidea, Achelata, Polychelida). *The Raffles Bulletin of Zoology, 23*(Suppl), 153–181.

Chittleborough, R. G. (1976). Breeding of Panulirus longipes cygnus under natural and controlled conditions. *Australian Journal of Marine & Freshwater Research, 27*, 499–516.

Chittleborough, R. G., & Thomas, L. R. (1969). Larval ecology of Western Australian crayfish, with notes upon other *Panulirus* larvae from the eastern Indian Ocean. *Australian Journal of Marine & Freshwater Research, 20*, 199–223.

Chow, S., Suzuki, N., Imai, N., & Yoshimura, T. (2006). Molecular species identification of spiny lobster phyllosoma larvae of the genus *Panulirus* from the northwestern Pacific. *Marine Biotechnology, 8*, 260–267.

Cobb, J. S., Wang, D., & Campbell, D. B. (1989). Timing of settlement by postlarval lobsters (Homarus americanus): Field and laboratory evidence. *Journal of Crustacean Biology, 9*(1), 60–66.

Comeau, M., & Savoie, F. (2002). Maturity and reproductive cycle of the female American lobster, *Homarus americanus*, in the southern Gulf of St Lawrence, Canada. *Journal of Crustacean Biology, 22*, 762–774.

Crandall, K. A., Harris, D. J., & Fetzner, J. W., Jr. (2000). The monophyletic origin of freshwater crayfish estimated from nuclear and mitochondrial DNA sequences. *Proceedings of the Royal Society of London Biological Sciences, 267*(1453), 1679–1686.

Cruz, R., Brito, R., Díaz, E., & Lalana, Y. R. (1986). Ecología de la langosta (*Panulirus argus*) al SE de la Isla de la Juventud. II. Patrones de movimiento. *Revista de Investigaciones Marinas, 7*(3), 19–35.

Davis, G. E., & Dodrill, J. W. (1980). Marine parks and sanctuaries for spiny lobster fishery management. *Proceedings of the Gulf and Caribbean Fisheries Institute, 32*, 194–207.

Dennis, D. M., Pitcher, C. R., & Skewes, T. D. (2001). Distribution and transport pathways of *Panulirus ornatus* (Fabricius, 1776) and *Panulirus* spp., larvae in the Coral Sea, Australia. *Marine and Freshwater Research, 52*, 175–185.

Deshmukh, S. (1966). The puerulus of the spiny lobster *Panulirus polyphagus* (Herbst) and its metamorphosis into the post-puerulus. *Crustaceana, 10*, 137–150.

Deshmukh, S. (1968). On the first phyllosomae of the Bombay spiny lobsters (*Panulirus*) with a note on the unidentified first Panulirus phyllosoma from India (Palinuridae). *Crustaceana (Supplement), 2*, 47–58.

Dharani, G., Maitrayee, G. A., Karthikayulu, S., Kumar, T. S., Anbarasu, M., & Vijayakumaran, M. (2009). Identification of *Panulirus homarus* puerulus larvae by restriction fragment length polymorphism of mitochondrial cytochrome oxidase gene. *Pakistan Journal of Biological Sciences, 12*(3), 281–285.

Diaz, D., Mari, M., Abello, P., & Demestre, M. (2001). Settlement and juvenile habitat of the European spiny lobster *Palinurus elephas* (Crustacea: Decapoda: Palinuridae) in the western Mediterranean Sea. *Scientia Marina, 65*(4), 347–356.

Ennis, G. P. (1975). Observations on hatching and larval release in the lobster Homarus americanus. *Journal of the Fisheries Research Board of Canada, 32*(11), 2210–2213.

Ennis, G. P. (1980). Size-maturity relationships and related observations in Newfoundland populations of the lobster (*Homarus americanus*). *Canadian Journal of Fisheries and Aquatic Sciences, 37*, 945–956.

Ennis, G. P. (1984). Incidence of molting and spawning in the same season in female lobsters. *Homarus americanus. Fisheries Bulletin US, 82*, 529–530.

FAO-Fishstat Database Plus. (2016). Fisheries and Aquaculture Department, Food and Agriculture Organization of the United Nations. Rome. www.fao.org/fishery/statistics/software/fishstat

Farmer, A. S. (1974). Reproduction in *Nephrops norvegicus* (Decapoda: Nephropidae). *Journal of Zoology, 174*, 161–183.

Galil, B. S. (2000). Crustacea Decapoda: review of the genera and species of the family Polychelidae Wood-Mason, 1874. In A. Crosnier (Ed.), *Résultats des campagnes MUSORSTOM, 21* (Mémoires du Muséum national d'histoire naturelle) (Vol. 184, pp. 285–387). Paris: Publications scientifiques du Muséum.

George, R. W. (2005). Comparative morphology and evolution of the reproductive structures in spiny lobsters, *Panulirus. New Zealand Journal of Marine and Freshwater Research, 39*(3), 493–501.

Girijavallabhan, K. G., & Devarajan, K. (1978). On the occurrence of puerulus of spiny lobster *Panulirus polyphagus* (Herbst) along the Madras coast. *Indian Journal of Fisheries, 25*(1&2), 253–254.

Goldstein, J. S., Matsuda, H., Takenouchi, T., & Butler, M. J., IV. (2008). the complete development of larval Caribbean spiny lobster *Panulirus argus* in culture. *Journal of Crustacean Biology, 28*(2), 306–327.

Goni, R., & Latrouite, D. (2005). Review of the biology, ecology and fisheries of *Palinurus* spp. of European waters: *Palinurus elephas* (Fabricius, 1787) and *Palinurus mauritanicus* (Gruvel, 1911). *Cahiers de Biologie Marine, 46*, 137–142.

Gordon, I. (1953). On the puerulus stage of some spiny lobsters (Palinuridae). *Bulletin of the British Museum (Natural Nistory), Zoology, 2*, 17–42.

Guerao, G., Diaz, D., & Abello, P. (2006). Morphology of puerulus and early juvenile stages of the spiny lobster *Palinurus mauritanicus. Journal of Crustacean Biology, 26*(4), 480–494.

Gurney, R. (1936). Larvae of Decapod Crustacea. *Discovery Reports, 12*, 400–440.

Herrnkind, W. F. (1980). Spiny lobsters: Patterns of movement. In J. S. Cobb & B. F. Phillips (Eds.), *The biology and management of lobsters* (Vol. II, pp. 349–401). New York: Academic.

Herrnkind, W. F., & Butler, M. J., IV. (1986). Factors regulating settlement and microhabitat use by juvenile spiny lobsters *Panulirus argus. Marine Ecology Progress Series, 34*, 23–30.

Holthuis, L. B. (1991). Marine lobsters of the World. FAO species catalogue, Vol. 13. *FAO Fisheries Synopsis, Food and Agriculture Organization, Rome, 125*(13), 1–292.

Hughes, J. T., & Matthiessen, G. C. (1962). Observations on the biology of the American lobster, *Homarus americanus. Limnology and Oceanography, 7*, 414–421.

Hunter, E. (1999). Biology of the European spiny lobster, *Palinurus elephas* (Fabricius, 1787) (Decapoda, Palinuridea). *Crustaceana, 72*, 545–565.

Inoue, N., & Sekiguchi, H. (2005). Distribution of Scyllarid Phyllosoma Larvae (Crustacea: Decapoda: Scyllaridae) in the Kuroshio Subgyre. *Journal of Oceanography, 61*(3), 389–398.

Johnson, M. W. (1971a). On palinurid and scyllarid lobster larvae and their distribution in the South China Sea (Decapoda, Palinuridea). *Crustaceana, 21*(3), 247–282.

Johnson, M. W. (1971b). The phyllosoma larvae of slipper lobsters from the Hawaiian Islands and adjacent areas (Decapoda, Scyllaridae). *Crustaceana, 20*, 77–103.

Jones, C. M. (1988). *The biology and behaviour of bay lobsters, Thenus spp. (Decapoda: Scyllaridae), in Northern Queensland, Australia.* PhD thesis, School of Biological Sciences, University of Queensland, p. 311.

Karasawa, H., Schweitzer, C. E., & Feldmann, R. M. (2013). Phylogeny and Systematics of extant and extinct lobsters. *Journal of Crustacean Biology, 33*(1), 78–123.

Kasim, H. M. (1986). Effect of salinity, temperature and oxygen partial pressure on the respiratory metabolism of *Panulirus polyphagus* (Herbst). *Indian Journal of Fisheries, 33*, 66–75.

Kathirvel, M. (1990). On the collection of phyllosoma larvae by *ISAACS-KIDD midwater trawl from the west coast of India*. Proceedings of the First Workshop Scientific Result. FORV Sagar Sampada, 5–7 June, 1989, pp. 141–146.

Kathirvel, M. (2004). A Bibliography of Indian lobsters. *FTF Bibliography Series No.3* (p. 117). Chennai: Fisheries Technocrats Forum.

Kathirvel, M., Suseelan, C., & Rao, P. V. (1989). Biology, population and exploitation of the Indian deep sea lobster, *Puerlus sewelli*. *Fishing Chimes, February, 1989*, 16–25.

Kittaka, J. (1994). Culture of phyllosomas of spiny lobster and its application to studies of larval recruitment and aquaculture. *Crustaceana, 66*(3), 258–270.

Kittaka, J. (2000). Culture of larval spiny lobsters. In B. F. Phillips & J. Kittaka (Eds.), *Spiny lobster management* (2nd ed., pp. 508–532). Oxford: Fishing News Books.

Kittaka, J., & Ikegami, E. (1988). Culture of the palinurid *Palinurus elephas* from egg stage to puerulus. *Nippon Suisan Gakkaishi, 54*(7), 1149–1154.

Kittaka, J., & Kimura, K. (1989). Culture of the Japanese spiny lobster *Panulirus japonicus* from egg to juvenile stage. *Nippon Suisan Gakkaishi, 55*, 963–970.

Kittaka, J., Ono, K., & Booth, J. (1997). Complete development of the green rock lobster, *Jasus verreauxi* from egg to juvenile. *Bulletin of Marine Science, 6*, 57–71.

Kittaka, J., Ono, K., Booth, J. D., & Webber, W. R. (2005). Development of the red rock lobster, *Jasus edwardsii*, from egg to juvenile. *New Zealand Journal of Marine and Freshwater Research, 39*(2), 263–277.

Kizhakudan, J. K. (2014). Reproductive biology of the female shovel-nosed lobster *Thenus unimaculatus* (Burton & Davie, 2007) from north-west coast of India. *Indian Journal of Geo-Marine Sciences, 43*(6), 927–935.

Konishi, K., Suzuki, N., & Chow, S. (2006). A late-stage phyllosoma larva of the spiny lobster *Panulirus echinatus* Smith, 1869 (Crustacea: Palinuridae) identified by DNA analysis. *Journal of Plankton Research, 28*(9), 841–845.

Lavalli, K. L., & Spanier, E. (2010). Infraorder Palinura Latreille, 1802. In F. R. Schram & J. C. von Vaupel Klein (Eds.), *The Crustacea, Traite de Zoologie 9A- Decapoda, Chapter 68*. Leiden: Koninklijke Brill.

Lavalli, K. L., Spanier, E., & Grasso, F. (2007). Behavior and sensory biology of slipper lobsters. In K. L. Lavalli & E. Spanier (Eds.), *The biology and fisheries of slipper lobsters* (Crustacean Issues 17) (pp. 133–181). New York: CRC Press (Taylor & Francis Group).

Lavery, S. D., Farhadi, A., Farahmand, H., Chan, T. Y., Azhdehakosphpour, A., Thakur, V., & Jeffs, A. G. (2014). Evolutionary divergence of geographic subspecies within the scalloped spiny lobster *Panulirus homarus* (Linnaeus, 1758). *PLoS One, 9*(6), 1–13.

Lawton, P., & Lavalli, K. L. (1995). Postlarval, juvenile, adolescent and adult ecology. In J. R. Factor (Ed.), *Biology of the lobster Homarus americanus* (pp. 47–88). San Diego: Academic.

Lemmens, L. W. T. J. (1994). Biochemical evidence for absence of feeding in puerulus larvae of the Western rock lobster *Panulirus cygnus* (Decapoda: Palinuridae). *Marine Biology, 118*(3), 383–391.

Lipcius, R., & Cobb, J. (1994). Introduction. In B. F. Phillips, J. Cobb, & J. Kittaka (Eds.), *Spiny lobster management* (pp. 158–168). Oxford: Fishing New Books.

Lyons, W. G. (1970). Scyllarid lobsters (Crustacea, Decapoda). *Memoirs of the Hourglass Cruises, 1*, 1–74.

MacDiarmid, A. B. (1985). Sunrise release of larvae from the palinurid rock lobster *Jasus edwardsii*. *Marine Ecology Progress Series, 21*, 313–315.

MacDiarmid, A. B. (1989). Moulting and reproduction of the spiny lobster *Jasus edwardsii* (Decapoda: Palinuridae) in northern New Zealand. *Marine Biology, 103*, 303–310.

Matsuda, H., Takenouchi, T., & Goldstein, J. S. (2006). The complete larval development of the pronghorn spiny lobster *Panulirus penicillatus* (Decapoda: Palinuridae) in culture. *Journal of Crustacean Biology, 26*(4), 579–600.

McWilliam, P. S. (1995). Evolution in the phyllosoma and puerulus phases of the spiny lobster genus *Panulirus* White. *Journal of Crustacean Biology, 1*, 542–557.

Mehrtens, F. (2011). *Untersuchungen zu den Entwicklungsbedingungen des Europäischen Hummers Homarus gammarus bei Helgoland in Freiland und Labor*. PhD thesis, University Hamburg.

Michel, A. (1971). Note sur les puerulus de Palinuridae et les larves phyllosomes de *Panulirus homarus* (L). Clef de determination des larves phyllosomes recoltees dans le Pacifique equatorial et sud-tropical (decapodes). *Cahiers O.R.S.T.O.M., Serie Oceanographie, 9*, 459–473.

Mikami, S., & Greenwood, J. G. (1997). Complete development and comparative morphology of larval *Thenus orientalis* and *Thenus* sp. (Decapoda: Scyllaridae) reared in the laboratory. *Journal of Crustacean Biology, 17*(2), 289–308.

Moore, R., & MacFarlane, W. (1984). Migration of the ornate rock lobster, *Panulirus ornatus* (Fabricius) in Papua New Guinea. *Australian Journal of Marine & Freshwater Research, 35*, 197–212.

Musa Aman, D. S. (2012). *Deal signed for world's biggest lobster aquaculture park*. World News, FIS, USA.

Nakamura, K. (1994). Maturation. In B. F. Phillips, J. S. Cobb, & J. Kittaka (Eds.), *Spiny lobster fisheries, management and culture* (pp. 374–383). Cambridge, United States: Blackwell Scientific Publications.

Nishida, S., Quigley, B. D., Booth, J., Nemoto, T., & Kittaka, J. (1990). Comparative morphology of the mouthparts and foregut of the final-stage phyllosoma, puerulus, and postpuerulus of the rock lobster *Jasus edwardsii* (Decapoda: Palinuridae). *Journal of Crustacean Biology, 10*, 293–305.

Nobili, G. (1903). Crestacei di Pondichery, Mahe, Bombay, etc. *Bull Mus Zool Anot Comp Torino, 18*(352), 1–24.

Nonaka, M., Oshima, Y., & Hirano, R. (1958). Rearing of phyllosoma of Ise lobster and moulting. *Suisan Zoshoku (Aquiculture), 5*, 13–15. (In Japanese).

Oliveira, G., Freire, A., & Bertuol, P. R. K. (2008). Reproductive biology of the slipper lobster *Scyllarides deceptor* (Decapoda: Scyllaridae) along the southern Brazilian coast. *Journal of the Marine Biological Association of the United Kingdom, 88*(7), 1433–1440.

Orton, J. H., & Ford, E. (1933). The post-puerulus of *Palinurus vulgaris* Latr. *Proceedings of the Zoological Society of London, 1933*, 181–188.

Palero, F., Guerao, G., Hall, M., Chan, T. Y., & Clark, P. F. (2014). The 'giant phyllosoma' are larval stages of *Parribacus antarcticus* (Decapoda: Scyllaridae). *Invertebrate Systematics, 28*, 258–276.

Phillips, B. F. (1983). Migrations of pre-adult western rock lobsters, *Panulirus cygnus*, in Western Australia. *Marine Biology, 76*, 311–318.

Phillips, B. F., & Sastry, A. N. (1980). Larval ecology. In J. S. Cobb & B. F. Phillips (Eds.), *The biology and management of lobsters* (Vol. II, pp. 11–57). New York: Academic.

Phillips, B. F., Brown, P. A., Rimmer, D. W., & Reid, D. D. (1979). Distribution and dispersal of the phyllosoma larvae of the western rock lobster *Panulirus cygnus*, in the south-eastern Indian Ocean. *Australian Journal of Marine & Freshwater Research, 30*, 773–783.

Phillips, B. F., Cobb, J. S., & George, R. W. (1980). General biology. In J. S. Cobb & B. F. Phillips (Eds.), *The biology and management of Lobsters* (Vol. 1, pp. 1–72). New York: Academic.

Poore, G. C. B. (2004). In G. C. B. Poore (Ed.), *Marine Decapod Crustacea of Southern Australia. A guide to identification*. Clayton: CSIRO Publishing.

Porter, M. L., Perez-Losada, M., & Crandall, K. A. (2005). Model based multi-locus estimater of decapod phylogeny and divergence times. *Molecular Phylogenetics and Evolution, 37*, 355–369.

Prasad, R. R., & Tampi, P. R. S. (1957). Phyllosoma of Mandapam. *Proceedings of the National Institute of Sciences of India. Part B, 23*, 48–67.

Prasad, R. R., & Tampi, P. R. S. (1959). On a collection of palinurid phyllosomas from the Laccadive Seas. *Journal of the Marine Biological Association of India, 2*(2), 143–164.

Prasad, R. R., & Tampi, P. R. S. (1960). Phyllosoma of scyllarid lobsters from the Arabian sea. *Journal of the Marine Biological Association of India, 2*(2), 241–249.

Prasad, R. R., & Tampi, P. R. S. (1965). A preliminary report on the phyllosomas of the Indian Ocean collected by the DANA Expedition 1928-1930. *Journal of the Marine Biological Association of India, 7*(2), 277–283.

Prasad, R. R., & Tampi, P. R. S. (1966). Note on the phyllosoma of *Puerulus sewelli* Ramadan. *Journal of the Marine Biological Association of India, 8*(2), 339–341.

Prasad, R. R., & Tampi, P. R. S. (1968). On the distribution of palinurid and scyllarid lobsters in the Indian Ocean. *Journal of the Marine Biological Association of India, 10*(1), 78–87.

Prasad, R. R., Tampi, P. R. S., & George, M. J. (1975). Phyllosoma larvae from the Indian Ocean collected by the DANA Expedition 1928-1930. *Journal of the Marine Biological Association of India, 17*(2), 56–107.

Prescott, J. (1988). *Tropical spiny lobster: An overview of their biology, the fisheries and the economics with particular reference to the double spined rock lobster Panulirus penicillatus*. SPC Workshop. Pacific Inshore Fisheries Research, New Caledonia. WP 18, 36p.

Ptacek, M. B., Sarver, S. K., Childress, M. J., & Herrnkind, W. F. (2001). Molecular phylogeny of the spiny lobster genus *Panulirus* (Decapoda: Palinuridae). *Marine and Freshwater Research, 52*, 1037–1047.

Radhakrishnan, E. V. (1989). Physiological and biochemical studies in the spiny lobster *Panulirus homarus*. PhD thesis, University of Madras, p 141.

Radhakrishnan, E. V., & Vijayakumaran, M. (1995). Early larval development of spiny lobster, *Panulirus homarus* (Linnaeus, 1758) reared in the laboratory. *Crustaceana, 68*(2), 151–159.

Radhakrishnan, E. V., Vijayakumaran, M., & Shahul Hameed, K. (1990). Incidence of pseudohermaphroditism in the spiny lobster *Panulirus homarus* (Linnaeus). *Indian Journal of Fisheries, 37*(2), 169–170.

Radhakrishnan, E. V., Deshmukh, V. D., Manisseri, M. K., Rajamani, M., Kizhakudan, J. K., & Thangaraja, R. (2005). Status of the major lobster fisheries in India. *New Zealand Journal of Marine and Freshwater Research, 39*, 723–732.

Richters. (1873). Die Phyllosomen. Ein Beitrag zur Entwicklungsgeschichtec der Loricaten. *Zeitschrift fur Wissenschartliche Zoologie, 23*, 623–646.

Ritar, A. J. (2001). The experimental culture of phyllosoma larvae of southern rock lobster (*Jasus edwardsii*) in a flowthrough system. *Aquacultural Engineering, 24*, 149–156.

Robertson, P. B. (1968). A giant scyllarid phyllosoma larva from the Caribbean Sea, with notes on smaller specimens (Palinuridea). *Crustaceana, 2*, 83–97.

Robertson, D. N., & Butler, M. J. I. V. (2003). Growth and size at maturity in the spotted spiny lobster, *Panulirus guttatus*. *Journal of Crustacean Biology, 23*, 265–272.

Rotzer, M. A. I. N., & Haug, J. T. (2015). Larval development of the European Lobster and how small heterochronic shifts lead to a more pronounced metamorphosis. International Journal of Zoology, Article ID 345172, 17. https://doi.org/10.1155/2015/345172

Santucci, R. (1926). Lo stadio natante e la prima forma post-natante dell'Aragosta (*Palinurus vulgaris* Latr.) del Mediterraneo. *Regio Comitato Talassografico Italiano, Memoria, CXXVII*, 1–11.

Sardà, F. (1995). A review (1967–1990) of some aspects of the life history of Nephrops norvegicus. *ICES Marine Science Symposia, 199*, 78–88.

Schram, F. R., & Koenemann, S. (2004). Developmental genetics and arthropod evolution: on body regions of crustacea. In G. Scholtz (Ed.), *Evolutionary developmental biology of Crustacea* (Crustacean Issues) (Vol. 15, pp. 75–92). Lisse: Balkema.

Sekiguchi, H., Booth, J. D., & Webber, W. R. (2007). Early life histories of slipper lobsters. In K. L. Lavalli & E. Spanier (Eds.), *The biology and fisheries of slipper lobsters* (Crustacean Issues 17) (pp. 69–90). New York: CRC Press, Taylor & Francis Group.

Sekine, S., Shima, Y., Fushimi, H., & Nonaka, M. (2000). Larval period and molting in the Japanese spiny lobster *Panulirus japonicus* under laboratory conditions. *Fisheries Science, 66*, 19–24.

Sharp, W. C., Hunt, J. H., & Lyons, W. G. (1997). Life history of the spotted spiny lobster, *Panulirus guttatus*, an obligate reef-dweller. *Marine and Freshwater Research, 48*, 687–698.
Silberbauer, B. I. (1971). The biology of the South African rock lobster *Jasus lalandii* (H. Milne-Edwards) 1. Development. *Investment Report South African Division of Sea Fisheries, 92*, 1–90.
Silberman, J. D., & Walsh, P. J. (1992). Species identification of spiny lobster phyllosome larvae via ribosomal DNA analysis. *Molecular Marine Biology and Biotechnology, 1*, 195–205.
Spanier, E., & Lavalli, K. L. (1998). Natural history of *Scyllarides latus* (Crustacea:Decapoda): A review of the contemporary biological knowledge of the Mediterranean slipper lobster. *Journal of Natural History, 32*, 1769–1986.
Spanier, E., & Lavalli, K. L. (2006). Scyllarides species. In B. F. Phillips (Ed.), *Lobsters: Biology, management, aquaculture and fisheries* (1st ed., pp. 462–489). London: Blackwell Publishing Ltd..
Spanier, E., Lavalli, K. L., & Weihs, D. (2010). Comparative morphology in slipper lobsters: possible adaptations to habitat and swimming, with emphasis on lobsters from the Mediterranean and adjacent seas. *Monografie del Museo Regionale di Scienzie Naturali di Torino, XXXX*, 111–130.
Talbot, R., & Helluy, S. (1995). Reproduction and embryonic development. In J. R. Factor (Ed.), *Biology of the Lobster Homarus americanus* (pp. 177–212). New York: Academic.
Tanaka, T., O. Ishida., and. Kaneko, S. (1984). Puerulus larvae of some spiny lobster (*Panulirus*) collected on the seashore at Chikura, Chiba Prefecture. Suisanzoushoku, 32: 92–101. [In Japanese].
Toller, W. (2003). *Spiny lobster (Panulirus argus) fact sheet*. Department of Planning and Natural Resources. Division of Fish and Wildlife.U.S.V.I. Animal Fact Sheet #14. 2pp. Available from http://bcrc.bio.umass.edu/vifishandwildlife/Education/FactSheet/PDF_Docs/14Lobster.pdf
Tong, L. J., Moss, G. A., Pickering, T. D., & Paewai, M. P. (2000). Temperature effects on embryo and early larval development of the spiny lobster *Jasus edwardsii*, and description of a method to predict larval hatch times. *Marine and Freshwater Research, 51*, 243–248.
Toon, A., Finley, M., Staples, J., & Crandall, K. A. (2009). Decapod phylogenetics and molecular evolution. In J. W. Martin, K. A. Crandall, & D. L. Felder (Eds.), *Crustacean issues 18: Decapod crustacean phylogenetics* (pp. 15–29). Boca Raton: Taylor & Francis/CRC Press.
Tsang, L. M., Ma, K. Y., Ahyong, S. T., Chan, T. Y., & Chu, K. H. (2008). Phylogeny of Decapoda using two nuclear protein-coding genes: Origin and evolution of the Reptantia. *Molecular Phylogenetics and Evolution, 48*, 359–368.
Vijayakumaran, M., Murugan, T. S., Remany, M. C., Mary Leema, T., Jha, D. K., Santhanakumar, J., Venkatesan, R., & Ravindran, R. (2005). Captive breeding of the spiny lobster *Panulirus homarus* (Linnaeus, 1758). *New Zealand Journal of Marine and Freshwater Research, 39*, 325–334.
Vijayakumaran, M., Maharajan, A., Rajalakshmi, S., Jayagopal, P., & Remani, M. C. (2014). Early larval stages of the spiny lobsters, *Panulirus homarus, Panulirus versicolor* and *Panulirus ornatus* cultured under laboratory conditions. *International Journal of Development Research, 4*(2), 377–383.
Wahle, R. A., Tshudy, D., Cobb, J. S., Factor, J., & Jaini, M. (2012). Infraorder Astacidea Latreille, 1802: The marine clawed lobsters. In F. R. Schram & J. C. von Vaupel Klein (Eds.), *Treatise on zoology–anatomy, taxonomy, biology, the crustacea* (Vol. 9, pp. 3–108). Leiden: Brill Academic Publishers.
Wakabayashi, K., & Phillips, B. F. (2016). Morphological descriptions of laboratory reared larvae and post-larvae of the Australian shovel-nosed lobster *Thenus australiensis* Burton & Davie, 2007 (Decapoda, Scyllaridae). *Crustaceana, 89*(1), 97–117.
Wakabayashi, K., Nagai, S., & Tanaka, Y. (2016). The complete larval development of *Ibacus ciliatus* from hatching to the nisto and juvenile stages using jellyfish as the sole diet. *Aquaculture, 450*, 102–107.
Wakabayashi, K., Sato, R., Ishii, H., Akiba, T., Nogata, Y., & Tanaka, Y. (2012). Culture of phyllosomas of *Ibacus novemdentatus* (Decapoda: Scyllaridae) in a closed recirculating system using jellyfish as food. *Aquaculture, 330*, 162–166.

Williamson, D. I. (1983). Crustacea Decapoda: Larvae, VIII. Nephropidea, Palinuridea, and Eryonidea. *Fich Ident Zooplankton, 167/168*, 8 pp.

Wood-Mason, J. (1872). On *Nephropsis stewarti*, a new genus and species of macrourus crustaceans, dredged in deep water off the eastern coast of the Andaman Islands. *Proceedings of the Asiatic Society of Bengal, 1872*, 151.

Yamauchi, M., Miya, M., & Nishida, M. (2002). Complete mitochondrial DNA sequence of the Japanese spiny lobster, *Panulirus japonicus* (Crustacea: Decapoda). *Gene, 295*, 89–96.

Yamakawa, T., Nishimura, M., Matsuda, H., Tsujigadou, A., & Kamiya, N. (1989). Complete larval rearing of the Japanese spiny lobster *Panulirus japonicus*. *Nippon Suisan Gakkaishi, 55*, 745.

Yoneyama, S., & Takeda, M. (1998). Phyllosoma and nisto stage larvae of slipper lobster, *Parribacus*, from the Izu-Kazan Islands, southern Japan. *Bulletin of the National Science Museum. Series A, 24*, 161–175.

Updated Checklist of the World's Marine Lobsters

2

Tin-Yam Chan

Abstract

The checklist of marine lobsters is updated. The current count consists of 6 families, 54 genera and 260 species (with 4 subspecies). Although the present list differs from the latest checklist of 2010, in the addition of nine species and one less genus, six synonymies (one genus, four species, one subspecies) have been discovered while one subspecies has been revived.

Keywords

Crustacea · Decapoda · Lobsters · Marine · Checklist · Taxonomy

2.1 Introduction

Although marine lobsters are generally large sized, with high economic value, new species have continuously been discovered in the last few years after the recent publication of an annotated checklist on world's marine lobsters (Chan 2010). This chapter provides an updated list of all the valid taxa in marine lobsters with a format largely following Chan (2010). After the publication of the previous checklist (Chan 2010), 15 species of marine lobsters have been described (Ahyong et al. 2012; Artüz et al. 2014; Chan et al. 2013; Chang et al. 2014; Giraldes and Smyth 2016; Tsoi et al. 2011; Yang and Chan 2010, 2012; Yang et al. 2011; Yang et al. 2017). Among them, *Stereomastis artuzi* Artüz et al., 2014 is highly likely not a good species. On the other hand, one genus (Chan et al. 2013; Yang et al. 2012), two species (Chan et al. 2013; Groeneveld et al. 2012; Yang et al. 2012) and one subspecies (Lavery et al. 2014) were synonymized with other taxa, while another subspecies was revived (Yang and Chan

T.-Y. Chan (✉)
Institute of Marine Biology, National Taiwan Ocean University, Keelung, Taiwan
e-mail: tychan@mail.ntou.edu.tw

© Springer Nature Singapore Pte Ltd. 2019
E. V. Radhakrishnan et al. (eds.), *Lobsters: Biology, Fisheries and Aquaculture*,
https://doi.org/10.1007/978-981-32-9094-5_2

2012). Altogether 260 valid species (with 4 valid subspecies) of extant marine lobsters in 6 families and 54 genera are now recognized. Marine lobsters are defined as marine members of the suborder Macrura Reptantia Bouvier, 1917 (see Chan 2010). Whether Macrura Reptantia is a natural group, is still unsettled. For example, a recent combined morphology and molecular analysis (Bracken et al. 2014) argued that lobsters are not a monophyletic group, but latest molecular-only phylogeny consistently shows that Macrura Reptantia is monophyletic (Tsang et al. 2008; Toon et al. 2009; Boisselier-Dubayle et al. 2010). The relationships among the four infraorders of lobsters are also far from settled (Tsang et al. 2008; Toon et al. 2009; Boisselier-Dubayle et al. 2010; Karasawa et al. 2013; Bracken et al. 2014). Thus, the higher classification of marine lobsters still follows the scheme of Chan (2010) under the suborder Macrura Reptantia containing four infraorders. The spelling of author's names generally follows Ng et al. (2008) for the world brachyuran crabs list, while the citing of Charles Spence Bate is following Froglia and Clark (2011) and Clark (2018) instead of De Grave and Fransen (2011). An '*' refers to the type species of the genus. Following the style of Chan (2010), only synonyms still used in taxonomic literature after 1960 are given but not including those with changes in generic assignments as well as misidentifications and spelling errors. Taxonomic decisions for these synonymies already discussed in Chan (2010) are not repeated and only those established after Chan (2010) are explained in the remarks of the corresponding higher taxon. If the original name given for a taxon is different from its current generic allocation, the original name is provided at the end of the name in square brackets. Type locality is given only for valid specific and subspecific names. As noted in the previous checklist (Chan 2010), some species still have unsettled taxonomic and nomenclatural issues. The present work makes no attempt to settle them and the most widely used names are adopted (Figs. 2.1, 2.2, 2.3, 2.4, 2.5, and 2.6).

2.2 Checklist of Marine Lobsters

2.2.1 Suborder Macrura Reptantia Bouvier, 1917

2.2.1.1 Infraorder Astacidea Latreille, 1802

The reef lobsters of the superfamily Enoplometopoidea de Saint Laurent, 1988 have consistently shown to be sister to Nephropoidae Dana, 1852 in the latest phylogenetic studies (Karasawa et al. 2013; Bracken et al. 2014).

2.2.1.1.1 Superfamily Enoplometopoidea de Saint Laurent, 1988

2.2.1.1.1.1 Family Enoplometopidae de Saint Laurent, 1988
Following Chan (2010), only one genus and no subgenus is recognized for this family.

Enoplometopus A. Milne-Edwards, 1862
E. antillensis Lütken, 1865—type locality: West Indies

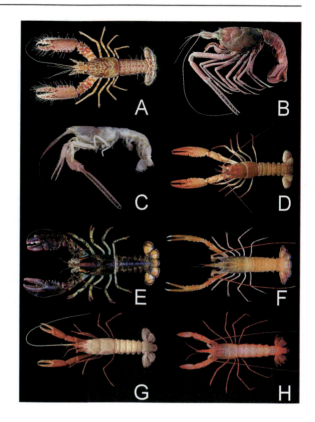

Fig. 2.1 (**a**), *Enoplometopus crosnieri* (Enoplometopidae), Taiwan; (**b**), *Acanthacaris tenuimana* (Nephropidae), Taiwan; (**c**), *Dinochelus ausubeli* (Nephropidae), Philippines; (**d**), *Eunephrops cadenasi* (Nephropidae), French Antilles; (**e**), *Homarus americanus* (Nephropidae), probably Canada (market in Taiwan); (**f**), *Metanephrops formosanus* (Nephropidae), Taiwan; (**g**), *Nephropides caribaeus* (Nephropidae), French Antilles; (**h**), *Nephropsis ensirostris* (Nephropidae), Philippines. (Photo credits: (**d, g**): J. Poupin)

Enoplometopus dentatus Miers, 1880

E. callistus Intès and Le Loeuff, 1970—type locality: Ghana

Enoplometopus biafri Burukovsky, 1972

E. chacei Kensley and Child, 1986—type locality: Philippines

E. crosnieri Chan and Yu, 1998—type locality: Taiwan

  E. daumi Holthuis, 1983 [*Enoplometopus (Enoplometopus) daumi*]—type locality: Moluccas, Indonesia

  E. debelius Holthuis, 1983 [*Enoplometopus (Enoplometopus) debelius*]—type locality: Hawaii

  E. gracilipes (de Saint Laurent, 1988) [*Hoplometopus gracilipes*]—type locality: Tuamotu, French Polynesia

  E. holthuisi Gordon, 1968—type locality: Moluccas, Indonesia

  E. macrodontus Chan and Ng, 2008—type locality: Philippines

  E. occidentalis (Randall, 1840) [*Nephrops occidentalis*]—type locality: West coast of North America, probably an error, should be Hawaii (see Holthuis 1983).

  E. pictus A. Milne-Edwards, 1862—type locality: Reunion

  E. voigtmanni Türkay, 1989 [*Enoplometopus (Hoplometopus) voigtmanni*]—type locality: Maldives

Fig. 2.2 (**a**), *Thaumastocheles dochmiodon* (Nephropidae), Taiwan; (**b**), *Thymopides laurentae* (Nephropidae), Mid-Atlantic Ridge; (**c**), *Laurentaeglyphea neocaledonica* (Glyphaediae), New Caledonia; (**d**), *Neoglyphea inopinata* (Glyphaediae), Philippines; (**e**), *Jasus edwardsii* (Palinuridae), Australia; (**f**), *Justitia longimanus* (Palinuridae), Papua New Guinea; (**g**), *Linuparus trigonus* (Palinuridae), Philippines; (**h**), *Nupalirus japonicus* (Palinuridae), Taiwan. (Photo credits: (**b**): M. Segonzac; (**c**): J.C.Y. Lai; (**d**): J. Forest)

2.2.1.1.2 Superfamily Nephropoidea Dana 1852

Although previous morphological phylogenetic analyses (Tshudy and Sorhannus 2000a, 2000b; Dixon et al. 2003; Schram and Dixon 2004; Ahyong and O'Meally 2004; Ahyong 2006) supported the family status of Thaumastochelidae Bate, 1888, latest morphology (Karasawa et al. 2013) and morphology + molecular (Bracken et al. 2014) analyses agreed with the placement of thaumastocheliforms (definition see Chang et al. 2017) within the family Nephropidae Dana, 1852 by Chan (2010), as revealed in molecular studies (see Tsang et al. 2008; Tshudy et al. 2009).

2.2.1.1.2.1 Family Nephropidae Dana, 1852 [Nephropinae]

Following Chan (2010), the three subfamilies in this family are not recognized. The latest phylogenetic analyses on lobsters by both morphology and molecular characters (Karasawa et al. 2013; Bracken et al. 2014) also do not support these subfamilies. Three new species of nephropids have been described (Ahyong et al. 2012; Chang et al. 2014) after the marine lobster list of Chan (2010). Poore and Dworschak (2017) confirmed that the species *Thaumastochelopsis plantei* Burukovsky, 2005, indeed belongs to the Axiidea genus *Ctenocheles* Kishinouye, 1926. Latest molecular genetic data suggested that *Thaumastocheles dochmiodon* Chan and de Saint Laurent, 1999 may be a polymorphic male form of *Thaumastocheles japonicus* Calman, 1913 (Chang et al. 2014). On the other hand, recent morphological

Fig. 2.3 (**a**), *Palibythus magnificus* (Palinuridae), French Polynesia; (**b**), *Palinurellus wieneckii* (Palinuridae), Christmas Island; (**c**), *Palinurus delagoae* (Palinuridae), Mozambique; (**d**), *Palinustus waguensis* (Palinuridae), Taiwan; (**e**), *Panulirus longipes longipes* (Palinuridae), Taiwan; (**f**), *Projasus bahamondei* (Palinuridae), probably Chile (market in Taiwan); (**g**), *Puerulus gibbosus* (Palinuridae); Mozambique; (**h**), *Arctides regalis* (Scyllaridae), Réunion. (Photo credits: (**a, h**): J. Poupin; (**b**): S. H. Tan)

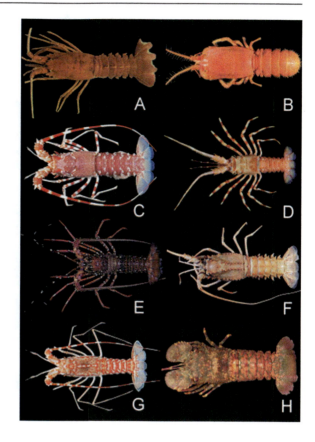

comparisons indicated that the two genera *Thymops* Holthuis, 1974 and *Thymopsis* Holthuis, 1974 may need to be synonymized (Ahyong et al. 2012). However, further studies are needed to determine if they are indeed synonymies (Ahyong et al. 2012; Chang et al. 2014).

Acanthacaris Bate, 1888
  A. caeca A. Milne-Edwards, 1881 [*Phoberus caecus*]—type locality: Grenada, West Indies
  ∗*A. tenuimana* Bate, 1888—type locality: south of New Guinea
   Acanthacaris opipara Burukovsky and Musij, 1976
   Phoberus brevirostris Tung et al., 1985
Dinochelus Ahyong et al., 2010
  ∗*D. ausubeli* Ahyong et al., 2010—type locality: Philippines
Eunephrops Smith, 1885
  ∗*E. bairdii* Smith, 1885—type locality: Gulf of Darien, Colombia
  E. cadenasi Chace, 1939—type locality: south of Cay Sal Bank, Caribbean Sea
  E. manningi Holthuis, 1974—type locality: Florida, USA
  E. luckhursti Manning, 1997—type locality: Bermuda
Homarinus Kornfield et al., 1995

Fig. 2.4 (**a**), *Scyllarides haanii* (Scyllaridae), Taiwan. (**b**), *Ibacus novemdentatus* (Scyllaridae), Taiwan; (**c**), *Parribacus japonicus* (Scyllaridae), Taiwan; (**d**), *Bathyarctus chani* (Scyllaridae), Taiwan; (**e**), *Biarctus vitiensis* (Scyllaridae), Taiwan; (**f**), *Chelarctus aureus* (Scyllaridae), Philippines; (**g**), *Crenarctus bicuspidatus* (Scyllaridae), Madagascar; (**h**), *Eduarctus modestus* (Scyllaridae), Okinawa

H. capensis (Herbst, 1792) [*Cancer (Astacus) capensis*]—type locality: Cape of Good Hope, S. Africa

Homarus Weber, 1795

H. americanus H. Milne Edwards, 1837— type locality: New Jersey, USA

H. gammarus (Linnaeus, 1758) [*Cancer gammarus*]— type locality: Marstrand, Sweden

Homarus vulgaris H. Milne Edwards, 1837

Metanephrops Jenkins, 1972

M. andamanicus (Wood-Mason, 1891) [*Nephrops andamanicus*]—type locality: Andaman Sea

M. arafurensis (de Man, 1905) [*Nephrops arafurensis*]—type locality: Arafura Sea, Indonesia

M. armatus Chan and Yu, 1991— type locality: Taiwan

M. australiensis (Bruce, 1966b) [*Nephrops australiensis*]—type locality: northwest Australia

M. binghami (Boone, 1927) [*Nephrops binghami*]—type locality: north of Glover Reef, west Caribbean Sea

M. boschmai (Holthuis, 1964) [*Nephrops boschmai*]—type locality: Great Australian Bight

Fig. 2.5 (**a**), *Galearctus aurora* (Scyllaridae), Papua New Guinea; (**b**), *Petrarctus rugosus* (Scyllaridae), Taiwan; (**c**), *Remiarctus bertholdii* (Scyllaridae), Philippines; (**d**), *Scammarctus batei arabicus* (Scyllarides), Mozambique; (**e**), *Scyllarus chacei*, Guadeloupe; (**f**), *Thenus orientalis* (Scyllaridae), Taiwan; (**g**), *Cardus crucifer*, Guadeloupe; (**h**), *Pentacheles laevis* (Polychelidae), Taiwan. (Photo credits: (**e, g**): J. Poupin)

Fig. 2.6 (**a**), *Polycheles typhlops* (Polychelidae), Mozambique; (**b**), *Stereomastis panglao* (Polychelidae), Philippines; (**c**), *Willemoesia leptodactyla* (Polychelidae), Taiwan

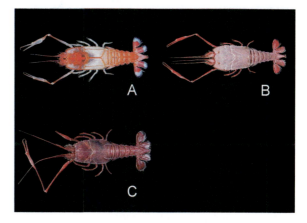

M. challengeri (Balss, 1914) [*Nephrops challengeri*]—type locality: New Zealand
M. formosanus Chan and Yu, 1987— type locality: Taiwan
∗*M. japonicus* (Tapparone-Canefri, 1873) [*Nephrops japonicus*]— type locality: Japan
M. mozambicus Macpherson, 1990—type locality: Madagascar

M. neptunus (Bruce, 1965a) [*Nephrops neptunus*]—type locality: south of Hong Kong, South China Sea

M. rubellus (Moreira, 1903) [*Nephrops rubellus*]—type locality: Brazil

M. sagamiensis (Parisi, 1917) [*Nephrops sagamiensis*]—type locality: Sagami Bay, Japan

Nephrops intermedius Balss, 1921

M. sibogae (de Man, 1916) [*Nephrops sibogae*]—type locality: Kai Isands, Indonesia

M. sinensis (Bruce, 1966a) [*Nephrops sinensis*]—type locality: south of Hainan, South China Sea

M. taiwanicus (Hu, 1983) [*Wongastacia taiwanica*]—type locality: Taiwan. (See Tshudy et al. 2007; Chan et al. 2009)

M. thomsoni (Bate, 1888) [*Nephrops thomsoni*]—type locality: Philippines

M. velutinus Chan and Yu, 1991—type locality: Philippines

Nephropides Manning, 1969

∗*N. caribaeus* Manning, 1969—type locality: Nicaragua, Caribbean Sea

Nephrops Leach, 1814

∗*N. norvegicus* (Linnaeus, 1758) [*Cancer norvegicus*]—type locality: Kullen Peninsula, Sweden

Nephropsis Wood-Mason, 1872

N. acanthura Macpherson, 1990—type locality: Philippines

N. aculeata Smith, 1881 [*Nephropsis aculeatus*]—type locality: Massachusetts, USA

N. agassizii A. Milne-Edwards, 1880a—type locality: Florida, USA but likely wrong and should be north of Yucatan Bank, Mexico (see Holthuis 1974)

N. atlantica Norman, 1882—type locality: Faeroe Channel, Scotland (see also Holthuis 1991)

N. carpenteri Wood-Mason, 1885—type locality: Bay of Bengal

N. ensirostris Alcock, 1901—type locality: north of the Laccadives, Arabian Sea

N. holthuisi Macpherson, 1993—type locality: Ashmore Reef, northwest Australia

Nephropsis macphersoni Watabe and Iizuka, 1999

N. malhaensis Borradaile, 1910—type locality: off Saya de Malha, Western Indian Ocean

N. neglecta Holthuis, 1974—type locality: Florida, USA

N. occidentalis Faxon, 1893—type locality: west of Mexico

N. rosea Bate, 1888—type locality: Bermuda

N. serrata Macpherson, 1993—type locality: northwest Australia

Nephropsis hamadai Watabe and Ikeda, 1994

Nephropsis lyra Zarenkov, 2006

Nephropsis pseudoserrata Zarenkov, 2006

∗*N. stewarti* Wood-Mason, 1872—type locality: Ross Island, Andaman Sea

Nephropsis grandis Zarenkov, 2006

N. suhmi Bate, 1888—type locality: Aru Islands, Indonesia

Nephropsis meteor Zarenkov, 2006

N. sulcata Macpherson, 1990—type locality: Philippines

Thaumastocheles Wood-Mason, 1874

T. bipristis Chang et al., 2014—type locality: Philippines

T. dochmiodon Chan and de Saint Laurent, 1999—type locality: Taiwan

T. japonicus Calman, 1913— type locality: Sagami Bay, Japan

T. massonktenos Chang et al., 2014—type locality: Solomon Islands

∗*T. zaleucus* (Thomson, 1873) [*Astacus Zaleucus*]—type locality: off Sombrero Island, West Indies

Thaumastochelopsis Bruce, 1988

T. brucei Ahyong et al., 2007—type locality: Queensland, Australia

∗*T. wardi* Bruce, 1988—type locality: Queensland, Australia

Thymopides Burukovsky and Averin, 1977

∗*T. grobovi* (Burukovsky and Averin, 1976) [*Bellator grobovi*]—type locality: between Heard Island and Kerguelen Islands, south Indian Ocean

T. laurentae Segonzac and Macpherson, 2003—type locality: Snake Pit, Mid-Atlantic Ridge

Thymops Holthuis, 1974

∗*T. birsteini* (Zarenkov and Semenov, 1972) [*Nephropides birsteini*]—type locality: north of Falkland Islands

T. takedai Ahyong et al., 2012—type locality: Burdwood Bank, Scotia Sea

Thymopsis Holthuis, 1974

∗*T. nilenta* Holthuis, 1974—type locality: south of South Georgia

2.2.1.2 Infraorder Glypheidea Van Straelen, 1925

The relationship of this rare living fossil group to other lobsters is still unsettled in latest studies. The molecular analysis of Boisselier-Dubayle et al. (2010) indicated that it is closer to Achelata Scholtz and Richter, 1995 (and Polychelida Scholtz and Richter, 1995). However, the morphology (Karasawa et al. 2013) and morphology + molecular analyses (Bracken et al. 2014) suggested that Glypheidae is sister to Astacidea (also see Wahle et al. 2012). The original author and date of many taxa in this infraorder also have many confusions (see Chan 2010; Schram 2014) and the opinion of Poore (2016) is followed here.

2.2.1.2.1 Superfamily Glypheoidea Winkler, 1881 [Glyphaeidae]

2.2.1.2.1.1 Family Glypheidae Winkler, 1881 [Glyphaeidae]

Laurentaeglyphea Forest, 2006

∗*L. neocaledonica* (Richer de Forges, 2006) [*Neoglyphea neocaledonica*]—type locality: New Caledonia

Neoglyphea Forest and de Saint Laurent, 1975

∗*N. inopinata* Forest and de Saint Laurent, 1975—type locality: Philippines

2.2.1.3 Infraorder Achelata Scholtz and Richter, 1995

Latest phylogenetic analyses (Boisselier-Dubayle et al. 2010; Karasawa et al. 2013; Bracken et al. 2014) are still inconsistent in the relationship between Palinuroidea

Latreille, 1802 and Polychelidae Wood-Mason, 1875 (see remarks under Polychelida Scholtz and Richter 1995). Thus, following Chan (2010) the infraorder Achelata Scholtz and Richter, 1995 is continuously used here.

2.2.1.3.1 Family Palinuridae Latreille, 1802 [Palinurini]

Other than following the synonymy treatments of Chan (2010), recently Groeneveld et al. (2012) synonymized *Jasus tristani* Holthuis, 1963 with *J. paulensis* (Heller, 1862a) based on genetic evidence. Also molecular genetic analysis (Lavery et al. 2014) revealed that one of the subspecies in *Panulirus homarus* (Linnaeus, 1758), that is, *P. homarus megasculpta* Pesta, 1915, is invalid. The subspecies of *Panulirus argus* (Latreille, 1804), or the so-called *Panulirus argus westonii* Sarver et al., 1998 (unavailable name, see Chan 2010) was formally described as a separated species by Giraldes and Smyth (2016) though without the consideration of the existence of quite a few old synonyms in *P. argus* (see Holthuis 1991; Chan 2010). Other than this, six more species in Palinuridae Latreille, 1802 were described (Tsoi et al. 2011; Chan et al. 2013) after the list of Chan (2010). Although recent molecular analyses (Chow et al. 2011; Iacchei et al. 2016) implied that the eastern Pacific form of *Panulirus penicillatus* (Olivier, 1791) may be a distinct species or subspecies as suggested by George (2005), formal nomenclature action has not yet been taken.

Jasus Parker, 1883

J. caveorum Webber and Booth, 1995—type locality: southeast of Pitcairn Island

J. edwardsii (Hutton, 1875) [*Palinurus edwardsii*]—type locality: New Zealand

Jasus novaehollandiae Holthuis, 1963

J. frontalis (H. Milne Edwards, 1837) [*Palinurus frontalis*]—type locality: Chile and restricted to Juan Fernandez Archipelago (see Holthuis 1991)

∗*J. lalandii* (H. Milne Edwards, 1837) [*Palinurus lalandii*]—type locality: Cape of Good Hope, South Africa

J. paulensis (Heller, 1862a) [*Palinurus paulensis*]—type locality: St. Paul Island, south Indian Ocean

Jasus tristani Holthuis, 1963

Justitia Holthuis, 1946

∗*J. longimanus* (H. Milne Edwards, 1837) [*Palinurus longimanus*]—type locality: Antilles

Palinurus longimanus mauritianus Miers, 1882

Linuparus White, 1847

L. meridionalis Tsoi et al. 2011—type locality: New Caledonia

L. somniosus Berry and George, 1972—type locality: Natal, South Africa

Linuparus andamanensis Mustafa, 1990

L. sordidus Bruce, 1965b—type locality: south of Hong Kong, South China Sea

∗*L. trigonus* (von Siebold, 1824) [*Palinurus trigonus*]—type locality: Japan

Nupalirus Kubo, 1955

N. chani Poupin, 1994 [*Justitia chani*]—type locality: Loyalty Islands

∗*N. japonicus* Kubo, 1955—type locality: Kochi, Japan

N. vericeli Poupin, 1994 [*Justitia vericeli*]—type locality: Tuamotu, French Polynesia

Palibythus Davie, 1990
*P. magnificus Davie, 1990—type locality: Western Samoa
Palinurellus von Martens, 1878
*P. gundlachi von Martens, 1878—type locality: Cuba
P. wieneckii (de Man, 1881) [*Araeosternus wieneckii*]—type locality: Sumatra, Indonesia
Palinurus Weber, 1795
P. barbarae Groeneveld et al., 2006—type locality: Walters Shoals, south of Madagascar
P. charlestoni Forest and Postel, 1964—type locality: Cape Verde Islands
*P. delagoae Barnard, 1926 [*Palinurus gilchristi delagoae*]—type locality: Natal, South Africa
P. elephas (Fabricius, 1787) [*Astacus elephas*]—type locality: 'Americae meridionalis Insulis', likely wrong and should be Italy (see Holthuis 1991)
P. gilchristi Stebbing, 1900—type locality: Cape Province, South Africa
P. mauritanicus Gruvel, 1911 [*Palinurus vulgaris mauritanicus*]—type locality: Mauritania, Cabo Barbas of Western Sahara and St. Louis of Senegal
Palinustus A. Milne-Edwards, 1880b
P. mossambicus Barnard, 1926—type locality: Mozambique
*P. truncatus A. Milne-Edwards, 1880b—type locality: Grenadines
P. unicornutus Berry, 1979—type locality: Natal, South Africa
P. waguensis Kubo, 1963—type locality: Mie, Japan
P. holthuisi Chan and Yu, 1995—type locality: Taiwan
Panulirus White, 1847
P. argus (Latreille, 1804) [*Palinurus argus*]—type locality: "Je la soupçonne des Grandes-Indes", possibly Antilles (see Holthuis 1991)
P. brunneiflagellum Sekiguchi and George, 2005—type locality: Ogasawara (Bonin Islands), Japan
P. cygnus George, 1962 [*Panulirus longipes cygnus*]—type locality: Rottnest Isand, Western Australia
P. echinatus Smith, 1869a—type locality: Brazil
P. femoristriga (von Martens, 1872) [*Palinurus femoristriga*]—type locality: Ambonia, Indonesia
Panulirus albiflagellum Chan and Chu, 1996
P. gracilis Streets, 1871—type locality: Gulf of Tehuantepec, Mexico
P. guttatus (Latreille, 1804) [*Palinurus guttatus*]—type locality: Suriname
P. homarus homarus (Linnaeus, 1758) [*Cancer homarus*]—type locality: Amboina, Indonesia
Palinurus dasypus H. Milne Edwards, 1837
Palinurus burgeri De Haan, 1841
Panulirus burgeri megasculpta Pesta, 1915
P. homarus rubellus Berry, 1974—type locality: southeast coast of Africa (Natal, South Africa; S. Mozambique, S.E. Madagascar)
P. inflatus (Bouvier, 1895) [*Palinurus inflatus*]—type locality: Baja California, Mexico

P. interruptus (Randall, 1840) [*Palinurus interruptus*]—type locality: California, USA

∗*P. japonicus* (von Siebold, 1824) [*Palinurus japonicus*]—type locality: Japan

P. laevicauda (Latreille, 1817) [*Palinurus laevicauda*]—type locality: Brazil

P. longipes longipes (A. Milne-Edwards, 1868) [*Palinurus longipes*]—type locality: Zanzibar

P. longipes bispinosus Borradaile, 1899 [*Panulirus bispinosus*]—type locality: Loyalty Islands

P. marginatus (Quoy and Gaimard, 1825) [*Palinurus marginatus*]—type locality: Hawaii

P. meripurpuratus Giraldes and Smyth 2016—type locality: Brazil

Panulirus argus westonii Sarver et al., 1998 (unavailable name, see Chan 2010)

P. ornatus (Fabricius, 1798) [*Palinurus ornatus*]—type locality: Indian Ocean, possibly Tranquebar, India (see Holthuis 1991)

P. pascuensis Reed, 1954—type locality: Easter Islands

P. penicillatus (Olivier, 1791) [*Astacus penicillatus*]—type locality: unknown (see Holthuis 1991)

P. polyphagus (Herbst, 1793) [*Cancer (Astacus) polyphagus*]—type locality: East Indies

Palinurus fasciatus Fabricius, 1798

Panulirus orientalis Doflein, 1900

P. regius de Brito Capello, 1864—type locality: Cape Verde Islands

Palinurus rissonii Desmarest, 1825

P. stimpsoni Holthuis, 1963—type locality: Hong Kong

P. versicolor (Latreille, 1804) [*Palinurus versicolor*]—type locality: "Cette jolie espèce nous est arrivèe par la frègate le Naturaliste", probably Mauritius and/or Timor (see Holthuis 1991)

Projasus George and Grindley, 1964

P. bahamondei George, 1976—type locality: San Ambrosio Island, southeast Pacific

∗*P. parkeri* (Stebbing, 1902) [*Jasus parkeri*]—type locality: Natal, South Africa

Puerulus Ortmann, 1897a

∗*P. angulatus* (Bate, 1888) [*Panulirus angulatus*]—type locality: north of New Guinea

P. carinatus Borradaile, 1910—type locality: Mozambique

P. gibbosus Chan et al., 2013—type locality: Mozambique

P. mesodontus Chan et al., 2013—type locality: Taiwan

P. quadridentis Chan et al., 2013—type locality: New Caledonia

P. richeri Chan et al., 2013—type locality: Marquesas Islands

P. sericus Chan et al., 2013—type locality: New Caledonia

P. sewelli Ramadan, 1938—type locality: Gulf of Aden

P. velutinus Holthuis, 1963—type locality: Lesser Sunda Islands, Indonesia

Sagmariasus Holthuis, 1991

∗*S. verreauxi* (H. Milne Edwards, 1851) [*Palinurus verreauxi*]—type locality: New South Wales, Australia

2.2.1.3.2 Family Scyllaridae Latreille 1825

Recent molecular analysis (Yang et al. 2012, also see Bracken et al. 2014) suggested that the subfamily Ibacinae Holthuis, 1985 may not be monophyletic, the same for many genera erected in Scyllarinae Latreille, 1825 by Holthuis (2002). Nevertheless, only *Antipodarctus* Holthuis, 2002 is formally synonymized with *Crenarctus* Holthuis, 2002 (Yang et al. 2012; Chan et al. 2013) and with the single species in *Antipodarctus* synonymized with *Crenarctus crenatus* (Whitelegge, 1900). Therefore, for the time being, the scheme of Chan (2010) is continuously followed except for the treatments related to *Antipodarctus* and *Crenarctus*. Four new species in this family, all belong to the subfamily Scyllarinae Latreille, 1825, were described (Yang and Chan 2010, 2012; Yang et al. 2011, 2017) after the list of Chan (2010). Moreover, the subspecies *Chelarctus cultrifer meridionalis* (Holthuis, 1960) is revived according to the recent study by Yang and Chan (2012). It may be needed to point out that even though Genis-Armero et al. (2017) suggested that *Scyllarus subarctus* Crosnier, 1970, and *Scyllarus depressus* (Smith, 1881) may belong to a single species, they have not formally synonymized these two species.

Subfamily Arctidinae Holthuis, 1985
Arctides Holthuis, 1960
A. antipodarum Holthuis, 1960—type locality: New South Wales, Australia
A. guineensis (Spengler, 1799) [*Scyllarus Guineensis*]—type locality: Guinea but very likely wrong (see Holthuis 1991, 2006)
A. regalis Holthuis, 1963—type locality: Hawaii
Scyllarides Gill, 1898
S. aequinoctialis (Lund, 1793) [*Scyllarus aequinoctialis*]—type locality: Jamaica
S. astori Holthuis, 1960—type locality: Galápagos Islands
S. brasiliensis Rathbun, 1906a—type locality: Brazil
S. deceptor Holthuis, 1963—type locality: Brazil
S. delfosi Holthuis, 1960—type locality: Suriname
S. elisabethae (Ortmann, 1894) [*Scyllarus elisabethae*]—type locality: Port Elizabeth, South Africa
S. haanii (De Haan, 1841) [*Scyllarus haanii*]—type locality: Japan
S. herklotsii (Herklots, 1851) [*Scyllarus herklotsii*]—type locality: Ghana
S. latus (Latreille, 1803) [*Scyllarus latus*]—type locality: near Rome, Italy
S. nodifer (Stimpson, 1866) [*Scyllarus nodifer*]—type locality: Florida, USA
S. obtusus Holthuis, 1993b—type locality: Saint Helena
S. roggeveeni Holthuis, 1967—type locality: Easter Islands
S. squammosus (H. Milne Edwards, 1837) [*Scyllarus squammosus*]—type locality: Mauritius
Scyllarus sieboldi De Haan, 1841
S. tridacnophaga Holthuis, 1967—type locality: Gulf of Aqaba, Israel
Subfamily Ibacinae Holthuis, 1985
Evibacus Smith, 1869b
E. princeps Smith, 1869b—type locality: Baja California, Mexico
Ibacus Leach, 1815

I. alticrenatus Bate, 1888—type locality: New Zealand
I. brevipes Bate, 1888—type locality: Kai Islands, Indonesia
Ibacus verdi Bate, 1888
I. brucei Holthuis, 1977—type locality: Queensland, Australia
I. chacei Brown and Holthuis, 1998—type locality: New South Wales, Australia
I. ciliatus (von Siebold, 1824) [*Scyllarus ciliatus*]—type locality: Japan
I. novemdentatus Gibbes, 1850—type locality: unknown (see Holthuis 1985, 1991)
∗*I. peronii* Leach, 1815—type locality: Tasmania, Australia
I. pubescens Holthuis, 1960 [*Ibacus ciliatus pubescens*]— type locality: Philippines.
Parribacus Dana, 1852
∗*P. antarcticus* (Lund, 1793) [*Scyllarus antarcticus*]—type locality: Amboina, Indonesia
P. caledonicus Holthuis, 1960—type locality: New Caledonia
P. holthuisi Forest, 1954—type locality: Tuamotu Archipelago, French Polynesia
P. japonicus Holthuis, 1960—type locality: Tokyo Bay, Japan
P. perlatus Holthuis, 1967—type locality: Easter Island
P. scarlatinus Holthuis, 1960—type locality: Phoenix Archipelago
Subfamily Scyllarinae Latreille, 1825
Acantharctus Holthuis, 2002
A. delfini (Bouvier, 1909) [*Arctus Delfini*]—type locality: Juan Fernandez Island, Chile
∗*A. ornatus* (Holthuis, 1960) [*Scyllarus ornatus*]—type locality: Arabian Peninsula, Oman
A. posteli (Forest, 1963) [*Scyllarus posteli*]—type locality: Pointe Noire, Congo
Antarctus Holthuis, 2002
∗*A. mawsoni* (Bage, 1938) [*Arctus mawsoni*]—type locality: Tasmania
Bathyarctus Holthuis, 2002
B. chani Holthuis, 2002—type locality: New Caledonia
B. faxoni (Bouvier, 1917) [*Scyllarus faxoni*]—type locality: Guadeloupe, West Indies
B. formosanus (Chan and Yu, 1992) [*Scyllarus formosanus*]—type locality: Taiwan
B. ramosae (Tavares, 1997) [*Scyllarus ramosae*]—type locality: Brazil
∗ *B. rubens* (Alcock and Anderson, 1894) [*Arctus rubens*]—type locality: Sri Lanka
B. steatopygus Holthuis, 2002—type locality: Kenya
Biarctus Holthuis, 2002
B. dubius (Holthuis, 1963) [*Scyllarus dubius*]—type locality: Japan, but likely wrong (see Holthuis 2002)
B. pumilus (Nobili, 1906) [*Scyllarus pumilus*]—type locality: Dahlak Archipelago, Red Sea.
Scyllarus Thiriouxi Bouvier, 1914
∗*B. sordidus* (Stimpson, 1860) [*Arctus sordidus*]—type locality: Hong Kong
Scyllarus tutiensis Srikrishnadhas et al. 1991

B. vitiensis (Dana, 1852) [*Arctus vitiensis*]—type locality: Fiji
Scyllarus longidactylus Harada, 1962
Scyllarus amabilis Holthuis, 1963
Chelarctus Holthuis, 2002
C. aureus (Holthuis, 1963) [*Scyllarus aureus*]— type locality: Philippines
C. crosnieri Holthuis, 2002—type locality: Tonga
∗*C. cultrifer cultrifer* (Ortmann, 1897b) [*Arctus cultrifer*]—type locality: Kai Islands, Indonesia
C. cultrifer meridionalis (Holthuis, 1960) [*Scyllarus cultrifer meridionalis*]—type locality: Philippines
C. virgosus Yang and Chan, 2012—type locality: Taiwan
Crenarctus Holthuis, 2002
∗*C. bicuspidatus* (de Man, 1905) [*Arctus bicuspidatus*]—type locality: Flores Sea, Indonesia
C. crenatus (Whitelegge, 1900) [*Arctus crenatus*]—type locality: New South Wales, Australia
Scyllarus aoteanus Powell, 1949
Eduarctus Holthuis, 2002
E. aesopius (Holthuis, 1960) [*Scyllarus aesopius*]—type locality: Philippines
E. lewinsohni (Holthuis, 1967) [*Scyllarus lewinsohni*]—type locality: Gulf of Aqaba, Red Sea
E. marginatus Holthuis, 2002—type locality: Fiji
∗*E. martensii* (Pfeffer, 1881) [*Scyllarus Martensii*]— type locality: Amur, Heilongjiang, China, but highly likely wrong (see Holthuis 1991, 2002)
E. modestus (Holthuis, 1960) [*Scyllarus modestus*]—type locality: Hawaii
E. perspicillatus Holthuis, 2002—type locality: Mozambique
E. pyrrhonotus Holthuis, 2002—type locality: Seychelles
E. reticulatus Holthuis, 2002—type locality: Macclesfield Bank, South China Sea
Galearctus Holthuis, 2002
G. aurora (Holthuis, 1982) [*Scyllarus aurora*]—type locality: Hawaii
G. avulus Yang et al., 2012—type locality: New Caledonia
G. kitanoviriosus (Harada, 1962) [*Scyllarus kitanoviriosus*]—type locality: Osaka Bay, Japan
G. lipkei Yang and Chan, 2010—type locality: Taiwan
G. rapanus (Holthuis Holthuis, 1993a) [*Scyllarus rapanus*]—type locality: Tubuai Archipelago, French Polynesia
∗*G. timidus* (Holthuis, 1960) [*Scyllarus timidus*]—type locality: Philippines
G. umbilicatus (Holthuis, 1977) [*Scyllarus umbilicatus*]—type locality: New South Wales, Australia
Gibbularctus Holthuis, 2002
∗*G. gibberosus* (de Man, 1905) [*Arctus gibberosus*]— type locality: Philippines and Indonesia
Arctus nobilii de Man, 1905
Scyllarus Paulsoni Nobili, 1906
Petrarctus Holthuis, 2002

P. brevicornis (Holthuis, 1946) [*Scyllarus brevicornis*]—type locality: Southern Bungo Strait, Japan
P. demani (Holthuis, 1946) [*Scyllarus demani*]—type locality: Sumatra, Indonesia
P. holthuisi Yang et al., 2008—type locality: Philippines
P. jeppiaari Yang et al., 2017—type locality: Muttom, India
∗*P. rugosus* (H. Milne Edwards, 1837) [*Scyllarus rugosus*]—type locality: Pazhayar, India (see Yang et al. 2017)
Arctus tuberculatus Bate, 1888
P. veliger Holthuis, 2002—type locality: Andaman Sea, south Burma
Remiarctus Holthuis, 2002
∗*R. bertholdii* (Paulson, 1875) [*Scyllarus Bertholdii*]—type locality: China
Scammarctus Holthuis, 2002
∗*S. batei batei* (Holthuis, 1946) [*Scyllarus batei*]—type locality: Philippines
S. batei arabicus Holthuis, 1960 [*Scyllarus batei arabicus*]—type locality: Gulf of Aden
Scyllarus Fabricius, 1775
S. americanus (Smith, 1869b) [*Arctus americanus*]—type locality: Florida, USA
∗*S. arctus* (Linnaeus, 1758) [*Cancer arctus*]—type locality: highly likely near Rome, Italy (see Holthuis 1991)
S. caparti Holthuis, 1952— type locality: Angola
S. chacei Holthuis, 1960—type locality: Suriname
S. depressus (Smith, 1881) [*Arctus depressus*]—type locality: Massachusetts, USA
Scyllarus nearctus Holthuis, 1960
S. paradoxus Miers, 1881 [*Scyllarus (Arctus) arctus*, var. *paradoxus*]—type locality: Senegal
S. planorbis Holthuis, 1969—type locality: Caribbean Sea, off Colombia
S. pygmaeus (Bate, 1888) [*Arctus pygmaeus*]—type locality: Canary Islands
S. subarctus Crosnier, 1970—type locality: Angola
Subfamily Theninae Holthuis, 1985
Thenus Leach, 1816
T. australiensis Burton and Davie, 2007—type locality: Torres Strait, Australia
∗*T. indicus* Leach, 1816—type locality: Indian Ocean (see Burton and Davie 2007)
T. orientalis (Lund, 1793) [*Scyllarus orientalis*]—type locality: Sumatera, Indonesia (see Burton and Davie 2007)
T. parindicus Burton and Davie, 2007—type locality: Moreton Bay, Queensland, Australia
T. unimaculatus Burton and Davie, 2007—type locality: Phuket, Thailand

2.2.1.4 Infraorder Polychelida Scholtz and Richter, 1995

The relationship of the family Polychelidae Wood-Mason, 1875 with Palinuroidea Latreille, 1802 is still unsettled. Recent molecular (Boisselier-Dubayle et al. 2010) and molecular + morphology (Bracken et al. 2014) analyses showed that they form a sister clade. However, morphology-only analysis (Karasawa et al. 2013) suggested that Polychelidae is the most basal extant group in lobsters. For the time being, the scheme of Chan (2010) is continuously followed.

2.2.1.4.1 Family Polychelidae Wood-Mason, 1875

After the list of Chan (2010), only one species in this family has been described and it is *Stereomastis artuzi* Artüz et al., 2014. However, *S. artuzi* is highly likely to belong to the common Mediterranean species *Polycheles typhlops* Heller, 1862b based on the colouration and characteristics shown in the colour photographs of the holotype (Artüz et al. 2014: figs. 2–4, vs. Chan and Yu 1989: pl. 1 A, B; Chan and Yu 1993: 105, unnumbered photo.; Ahyong and Chan 2004: fig. 4H; Ahyong and Chan 2008: fig. 1C). The characteristic interlocking spines at the dorsal orbital sinus (see Ahyong and Chan 2004: fig. 1D) in *P. typhlops* are probably omitted in the poorly prepared line-drawings of Artüz et al. (2014: fig. 5b, c, 7). Therefore, *S. artiuzi* is considered as a synonym of *P. typhlops* at least for the time being.

Cardus Galil, 2000
*C. crucifer (Thomson, 1873) [*Deidamia crucifer*]—type locality: West Indies
Homeryon Galil, 2000
*H. armarium Galil, 2000—type locality: Japan
H. asper (Rathbun, 1906b) [*Polycheles asper*]—type locality: Hawaii
Pentacheles Bate, 1878
P. gibbus Alcock, 1894 [*Pentacheles gibba*]—type locality: Andaman Sea
*P. laevis Bate, 1878—type locality: Moluccas, Indonesia
Pentacheles gracilis Bate, 1878
Polycheles granulatus Faxon, 1893
Pentacheles beaumontii Alcock, 1894
P. obscurus Bate, 1878 [*Pentacheles obscura*]—type locality: New Guinea
Pentacheles carpenteri Alcock, 1894
P. snyderi (Rathbun, 1906b) [*Polycheles snyderi*]—type locality: Hawaii
P. validus A. Milne-Edwards, 1880b—type locality: Antilles
Polycheles demani Stebbing, 1917
Polycheles chilensis Sund, 1920
Polycheles Heller, 1862b
P. amemiyai Yokoya, 1933—type locality: Bungo Strait, Japan
P. baccatus Bate, 1878—type locality: Fiji
P. coccifer Galil, 2000—type locality: Philippines
P. enthrix (Bate, 1878) [*Pentacheles enthrix*]—type locality: Fiji
P. kermadecensis (Sund, 1920) [*Stereomastis kermadecensis*]—type locality: Kermadec Islands
P. martini Ahyong and Brown, 2002—type locality: New South Wales, Australia
P. perarmatus Holthuis, 1952 [*Polycheles typhlops perarmatus*]—type locality: Angola
P. tanneri Faxon, 1893—type locality: Galápagos Islands
*P. typhlops Heller, 1862b—type locality: Sicily
Pentacheles hextii Alcock, 1894
Stereomastis artuzi Artüz et al. 2014
Stereomastis Bate, 1888
S. aculeata (Galil, 2000) [*Polycheles aculeatus*]—type locality: New Caledonia
S. alis (Ahyong and Galil, 2006) [*Polycheles alis*]—type locality: Austral Islands
S. auriculata (Bate, 1878) [*Pentacheles auriculatus*]—type locality: Fiji

S. cerata (Alcock, 1894) [*Pentacheles cerata*]—type locality: Andaman Sea

S. evexa (Galil, 2000) [*Polycheles evexus*]—type locality: Chile

S. galil (Ahyong and Brown, 2002) [*Polycheles galil*]—type locality: northwest Australia

S. helleri (Bate, 1878) [*Polycheles helleri*]—type locality: New Guinea

S. nana (Smith, 1884) [*Pentacheles nanus*]—type locality: southeast of New York, USA

Pentacheles andamanensis Alcock, 1894

Polycheles grimaldii Bouvier, 1905

S. pacifica (Faxon, 1893) [*Polycheles sculptus pacificus*]—type locality: Gulf of Panama

S. panglao Ahyong and Chan, 2008—type locality: Philippines

S. phosphorus (Alcock, 1894) [*Pentacheles phosphorus*]—type locality: Bay of Bengal

S. polita (Galil, 2000) [*Polycheles politus*]—type locality: Philippines

S. sculpta (Smith, 1880) [*Polycheles sculptus*]—type locality: Nova Scotia, Canada

∗*S. suhmi* (Bate, 1878) [*Pentacheles suhmi*]—type locality: Gulf of Penas, Chile

S. surda (Galil, 2000) [*Polycheles surdus*]—type locality: Mozambique

S. talismani (Bouvier, 1917) [*Polycheles sculptus* var. *talismani*]—type locality: Western Sahara

S. trispinosa (de Man, 1905) [*Pentacheles trispinosus*]—type locality: Bali Sea, Indonesia

Willemoesia Grote, 1873

W. forceps A. Milne-Edwards, 1880b—type locality: Cuba

W. inornata Faxon, 1893— type locality: south of Panama

Willemoesia challengeri Sund, 1920

∗*W. leptodactyla* (Thomson, 1873) [*Deidamia leptodactyla*]—type locality: mid-Atlantic, 21°38′N, 44°39′W

Willemoesia indica Alcock, 1901

Willemoesia secunda Sund, 1920

W. pacifica Sund, 1920—type locality: Juan Fernandez Island, Chile

Willemoesia bonaspei Kensley, 1968

2.3 Summary

Following the previous marine lobster checklist, marine lobsters are defined as members of the suborder Macrura Reptantia Bouvier 1917. To date, 260 species (with 4 subspecies) of extant marine lobsters in 4 infraorders, 6 families and 54 genera are recognized. Remarks on their synonyms in the recent literature and information on the type locality of the valid taxa are provided.

Acknowledgements Sincere thanks are extended to L. Corbari of the Muséum national d'Histoire naturelle, Paris, for providing the photograph of *Neoglyphea inopinata*. This is a contribution from grants supported by the Ministry of Science and Technology, Taiwan, R.O.C.

References

Ahyong, S. T. (2006). Phylogeny of the clawed lobsters (Crustacea: Decapoda: Homarida). *Zootaxa, 1109*(1), 14.
Ahyong, S. T., & Brown, D. E. (2002). New species and new records of Polychelidae from Australia (Decapoda: Crustacea). *Raffles Bulletin of Zoology, 50*(1), 53–79.
Ahyong, S. T., & Chan, T. Y. (2004). Polychelid lobsters of Taiwan (Decapoda: Polychelidae). *Raffles Bulletin of Zoology, 51*(1), 171–182.
Ahyong, S. T., & Chan, T. Y. (2008). Polychelidae from the Bohol and Sulu Seas collected by Panglao 2005 (Crustacea: Decapoda: Polychelidae). *Raffles Bulletin of Zoology, 19*, 63–70.
Ahyong, S. T., & Galil, B. S. (2006). Polychelidae from the southern and western Pacific (Decapoda, Polychelida). *Zoosystema, 28*(3), 757–767.
Ahyong, S. T., & O'Meally, D. (2004). Phylogeny of the Decapoda Reptantia: resolution using three molecular loci and morphology. *Raffles Bulletin of Zoology, 52*(2), 673–693.
Ahyong, S. T., Chu, K. H., & Chan, T. Y. (2007). Description of a new species of *Thaumastochelopsis* from the Coral Sea (Crustacea: Decapoda: Nephropoidea). *Bulletin of Marine Science, 80*(1), 201–208.
Ahyong, S. T., Chan, T. Y., & Bouchet, P. (2010). Mighty claws: a new genus and species of lobster from the Philippines deep-sea (Crustacea, Decapoda, Nephropidae). *Zoosystema, 32*(3), 525–535.
Ahyong, S. T., Webber, W. R., & Chan, T. Y. (2012). *Thymops takedai*, a new species of deepwater lobster from the southwest Atlantic Ocean with additional records of "Thymopine" lobsters (Decapoda: Nephropidae). *Crustaceana Monographs, 17*, 49–61.
Alcock, A. (1894). Natural history notes from H.M. Indian marine survey steamer "Investigator", Commander R.F. Hoskyn, R.N., commanding. Ser. II., No. 1. On the results of deep-sea dredging during the season 1890–91. *Annals and Magazine of Natural History, 13*(6), 225–245.
Alcock, A. (1901). *A descriptive catalogue of the Indian deep-sea Crustacea Decapoda Macrura and Anomala, in the Indian Museum, being a revised account of the deep-sea species collected by the Royal Indian Marine Survey Ship Investigator*. Calcutta. iv+286 pp. pls. 1–3.
Alcock, A., & Anderson, A. R. S. (1894). An account of a recent collection of deep sea Crustacea from the Bay of Bengal and Laccadive Sea. Natural history notes from H. M. Indian Marine Survey Steamer "Investigator", Commander C. F. Oldham, R. N., commanding. Series II, No. 14. *Journal of the Asiatic Society of Bengal, 63*(2), 141–185, pl. 9.
Artüz, M. L., Kubanç, C., & Kubanç, S. N. (2014). *Stereomastis artuzi* sp. nov., a new species of Polychelidae (Decapoda, Polychelida) described from the Sea of Marmara, Turkey. *Crustaceana, 87*(10), 1243–1257.
Bage, F. (1938). Crustacea Decapoda (Natantia and Reptantia in part). Australasian Antarctic Expedition 1911–14. *Scientific Reports (C), 2*(6), 5–13, pl. 4.
Balss, H. (1914). Ostasiatische Decapoden. II. Die Natantia und Reptantia. In: Doflein, F. (Ed.), Beiträge zur Naturgeschichte Ostasiens. *Abh. Bay. Akad. Wiss. II*, Suppl., 10: 1–101, figs. 1–51, pl. 1.
Balss, H. (1921). Diagnosen neuer Decapoden aus den Sammlungen der Deutschen Tiefsee-Expedition und der japanischen Ausbeute Dofleins und Haberers. *Zoologischer Anzeiger, 52*, 175–178.
Barnard, K. H. (1926). Report on a collection of Crustacea from Portuguese West Africa. *Transactions of the Royal Society of South Africa, 13*(2): 119–129, pls. 10–11.
Bate, C. S. (1878). XXXII. on the *Willemoesia* group of Crustacea. *Annals and Magazine of Natural History, 2*(5), 273–283, pl. 13.

Bate, C. S. (1888). Report on the Crustacea Macrura collected by H.M.S. Challenger during the years 1873–76. *Report on the Scientific Results of the Voyage of H.M.S. Challenger, 24*, i–xc, 1–942, figs. 1–76, pls. 1–150.

Berry, P. F. (1974). A revision of the *Panulirus homarus* group of spiny lobsters (Decapoda, Palinuridae). *Crustaceana, 27*(1), 31–42.

Berry, P. F. (1979). A new species of deep-water palinurid lobster (Crustacea, Decapoda, Palinuridae) from the East coast of southern Africa. *Annals of the South African Museum, 78*, 93–100.

Berry, P. F., & George, R. W. (1972). A new species of the genus *Linuparus* (Crustacea, Palinuridae) from south-east Africa. *Zoologische Mededelingen, 46*, 17–23.

Boisselier-Dubayle, M.-C., Bonillo, C., Cruaud, C., Couloux, A., Richer de Forges, B., & Vidal, N. (2010). The phylogenetic position of the 'living fossils' *Neoglyphea* and *Laurentaeglyphea* (Decapoda: Glypheidea). *Comptes Rendus Biologies, 333*, 755–759.

Boone, L. (1927). Crustacea from tropical East American seas. Scientific results of the first oceanographic expedition of the "Pawnee", 1925. *Bulletin of the Bingham Oceanographic Collection, 1*(2), 1–147.

Borradaile, L. A. (1899). On the Stomatopoda and Macrura brought by Dr. Willey from the South Seas. In: A. Willey (Ed.), *Zoological results based on material from New Britain, New Guinea, Loyalty Islands and elsewhere, collected during the years 1895, 1896 and 1897, 4*, 395–428, pls. 36–39.

Borradaile, L. A. (1910). Penaeidea, Stenopodidea, and Reptantia from the Western Indian Ocean. The Percy Sladen Trust Expedition to the Indian Ocean in 1905, under the leadership of Mr. J. Stanley Gardiner. *Transactions of the Linnean Society of London, 13*(2), 257–264, pl. 16.

Bouvier, E. L. (1895). Sur une collection de Crustacés décapodes recueillis en Basse-Californie par M. Diguet. *Bulletin du Muséum National d'Histoire Naturelle. Paris, 1*, 6–9.

Bouvier, E. L. (1905). Sur les Palinurides et les Eryonides recueillis dans l'Atlantique orientale par les expéditions françaises et monégasques. *Comptes rendus de l'Académie des Sciences Paris, 140*, 479–482.

Bouvier, E. L. (1909). *Arctus Delfini* sp. nov. *Revista Chilena de Historia Natural, 13*, 213–215.

Bouvier, E. L. (1914). Sur la faune carcinologique de l'île Maurice. *Comptes rendus de l'Académie des Sciences Paris, 159*, 698–704.

Bouvier, E. L. (1917). Crustacés décapodes (Macroures marcheurs) provenant des campagnes des yachts HIRONDELLE et PRINCESSE-ALICE (1885–1915). *Résult. Camp. Sci. Monaco, 50*, 1–140, pls. 1–11.

Bracken, H., Ahyong, S. T., Wilkinson, R., Felmann, R., Schweitzer, C., Breinholt, J., Palero, F., Chan, T. Y., Tsang, L. M., Chu, K. H., Bendall, M., Kim, D., Felder, D., Martin, J., Robles, R., & Crandall, K. (2014). The emergence of the lobsters: Phylogenetic relationships, morphological evolution and divergence time comparisons of an ancient group (Decapoda: Achelata, Astacidea, Glypheidea, Polychelida). *Systematic Biology, 63*(4), 457–479.

Brown, D. E., & Holthuis, L. B. (1998). The Australian species of the genus *Ibacus* (Crustacea: Decapoda: Scyllaridae), with the description of a new species and addition of new records. *Zoologische Mededelingen Leiden, 72*, 113–141.

Bruce, A. J. (1965a). On a new species of *Nephrops* (Decapoda, Reptantia) from the South China Sea. *Crustaceana, 9*(3), 274–284.

Bruce, A. J. (1965b). A new species of the genus *Linuparus* White, from the South China Sea (Crustacea Decapoda). *Zoologische Mededelingen Leiden, 41*, 1–13.

Bruce, A. J. (1966a). *Nephrops sinensis* sp. nov., a new species of lobster from the South China sea. *Crustaceana, 10*(2), 155–166.

Bruce, A. J. (1966b). *Nephrops australiensis* sp. nov., a new species of lobster from northern Australia (Decapoda, Reptantia). *Crustaceana, 10*(3), 245–258.

Bruce, A. J. (1988). *Thaumastochelopsis wardi* gen. et sp. nov., a new blind deep-sea lobster from the Coral Sea (Crustacea: Decapoda: Nephropidea). *Invertebrate Taxonomy, 2*, 903–914.

Burton, T. E., & Davie, P. J. F. (2007). A revision of the shovel-nosed lobsters of the genus *Thenus* (Crustacea: Decapoda: Scyllaridae), with descriptions of three new species. *Zootaxa, 1429*, 1–38.

Burukovsky, R. N. (1972). *Enoplometopus biafri*, new lobster species of the family Nephropidae (Decapoda, Crustacea). *Trudy Atlantniro, 42*, 180–189.

Burukovsky, R. N. (2005). On finding a juvenile lobster of the genus *Thaumastochelopsis* (Decapoda, Thaumastochelidae) from Madagascar shelf. *Zoologicheskii Zhurnal, 84*(4), 510–513.

Burukovsky, R. N., & Averin, B. S. (1976). *Bellator grobovi* gen. et sp. n., a new representative of the family Nephropidae (Decapoda, Crustacea) from the Herd Island region in the Subantarctic. *Zoologicheskii Zhurnal, 55*, 269–299.

Burukovsky, R. N., & Averin, B. S. (1977). A replacement name, *Thymopides*, proposed for the preoccupied generic name *Bellator* (Decapoda, Nephropidae). *Crustaceana, 32*, 216.

Burukovsky, R. N., & Musij, Y. I. (1976). *Acanthacaris opipara* Burukovsky et Musij, sp. n., a new abyssal lobster (Crustacea, Decapoda, Neophoberinae). *Zoologicheskii Zhurnal, 55*(12), 1811–1815.

Calman, W. T. (1913). A new species of the Crustacean genus *Thaumastocheles*. *Annals and Magazine of Natural History, 12*(8), 229–233.

Chace, F. A., Jr. (1939). Preliminary descriptions of one new genus and seventeen new species of decapod and stomatopod Crustacea. Reports on the scientific results of the first Atlantis Expedition to the West Indies, under the auspices of the University of Havana and Harvard University. *Memorias de la Sociedad Cubana de Historia Natural, 13*(1), 31–54.

Chan, T. Y. (2010). Annotated checklist of the world's marine lobsters (Crustacea: Decapoda: Astacidea, Glypheidea, Achelata, and Polychelida). *Raffles Bulletin of Zoology, 23*, 153–181.

Chan, T. Y., & Chu, K. H. (1996). On the different forms of *Panulirus longipes femoristriga* (von Martens, 1872) (Crustacea: Decapoda: Palinuridae), with description of a new species. *Journal of Natural History, 30*, 367–387.

Chan, T. Y., & de Saint Laurent, M. (1999). The rare lobster genus *Thaumastocheles* (Decapoda: Thaumastochelidae) from the Indo-West Pacific, with description of a new species. *Journal of Crustacean Biology, 19*(4), 891–901.

Chan, T. Y., & Ng, P. K. L. (2008). *Enoplometopus* A. Milne-Edwards, 1862 (Crustacea: Decapoda: Nephropidea) from the Philippines, with description of one new species and a revised key to the genus. *Bulletin of Marine Science, 83*(2), 347–365.

Chan, T. Y., & Yu, H. P. (1987). *Metanephrops formosanus* sp. nov., a new species of lobster (Decapoda, Nephropidae) from Taiwan. *Crustaceana, 52*(2), 172–186.

Chan, T. Y., & Yu, H. P. (1989). Two blind lobster of the genus *Polycheles* (Crustacea: Decapoda: Eryonoidea) from Taiwan. *Bulletin of the Institute of Zoology, Academia Sinica, 28*, 165–170.

Chan, T. Y., & Yu, H. P. (1991). Studies of the *Metanephrops japonicus* group (Decapoda, Nephropidae), with description of two new species. *Crustaceana, 60*(1), 18–51.

Chan, T. Y., & Yu, H. P. (1992). *Scyllarus formosanus*, a new slipper lobster (Decapoda, Scyllaridae) from Taiwan. *Crustaceana, 62*(2), 121–127.

Chan, T. Y., & Yu, H. P. (1993). *The illustrated lobsters of Taiwan*. Taipei: Southern Materials Center, Inc. 248pp.

Chan, T. Y., & Yu, H. P. (1995). The rare lobster genus *Palinustus* A. Milne Edwards, 1880 (Decapoda: Palinuridae), with description of a new species. *Journal of Crustacean Biology, 15*(2), 376–394.

Chan, T. Y., & Yu, H. P. (1998). A new reef lobster of the genus *Enoplometopus* A. Milne-Edwards, 1862 (Decapoda, Nephropidae) from the western and southern Pacific. *Zoosystema, 20*(2), 183–192.

Chan, T. Y., Ho, K. H., Li, C. P., & Chu, K. H. (2009). Origin and diversification of the clawed lobster genus *Metanephrops* (Crustacea: Decapoda: Nephropidae). *Molecular Phylogenetics and Evolution, 50*, 411–422.

Chan, T. Y., Ahyong, S. T., & Yang, C. H. (2013). Priority of the slipper lobster genus *Crenarctus* Holthuis, 2002, over *Antipodarctus* Holthuis, 2002 (Crustacea, Decapoda, Scyllaridae). *Zootaxa, 3701*(4), 471–472.

Chan, T. Y., Ma, K. Y., & Chu, K. H. (2013). The deep-sea spiny lobster genus *Puerulus* Ortmann, 1897 (Crustacea, Decapoda, Palinuridae), with descriptions of five new species. In: Ahyong,

S.T., Chan, T.Y., Corbari. L & Ng, P.K.L. (eds). Tropical Deep-sea Benthos, vol. 27. *Mémoires du Muséum national d'Histoire naturelle, 204*, 191–230.

Chang, S. C., Chan, T. Y., & Ahyong, S. T. (2014). Two new species of the rare lobster genus *Thaumastocheles* Wood-Mason, 1874 (Reptantia: Nephropidae) discovered from recent deep-sea expeditions in the Indo-West Pacific. *Journal of Crustacean Biology, 34*(1), 107–122.

Chang, S. C., Tshudy, D., Sorhannus, U., Ahyong, S. T., & Chan, T. Y. (2017). Evolution of the thaumastocheliform lobsters (Crustacea, Decapoda, Nephropidae). *Zoologica Scripta, 46*, 372–387.

Chow, S., Jeffs, A., Miyake, Y., Konishi, K., Okazaki, M., Suzuki, N., Abdullah, M. F., Imai, H., Wakabayasi, T., & Sakai, M. (2011). Genetic Isolation between the Western and Eastern Pacific populations of pronghorn spiny lobster *Panulirus penicillatus*. *PLoS One, 6*(12), 1–9.

Clark, P. F. (2018). Spence bate: what's in a name. *Zootaxa, 4497*(3), 429–438.

Crosnier, A. (1970). Crustacés décapodes brachyoures et macroures recueillis par l'«Undaunted» au sud de l'Angola. Description de *Scyllarus subarctus* sp. nov. *Bulletin du Muséum National d'Histoire Naturelle. Paris, 41*(5), 1214–1227.

Dana, J. D. (1852). Conspectus crustaceorum quae in orbis terrerum circumnavigatione, Carolo Wilkes e classe reipublicae foederatae duce, lexit et descripsit. *Proceedings of the Academy of Natural Sciences of Philadelphia, 6*, 6–28.

Davie, P. J. F. (1990). A new genus and species of marine crayfish, *Palibythus magnificus*, and new records of *Palinurellus* (Decapoda: Palinuridae) from the Pacific Ocean. *Invertebrate Taxonomy, 4*, 685–695.

de Brito Capello, F. (1864). Descripção de tres Especies novas de Crustaceos da Africa occidental e observações ácerca do Penoeus Bocagei. Johnson. Especie nova dos Mares de Portugal. *Memorias da Academia das sciencias de Lisboa, 3*(2), 1–11, pl. 1.

De Grave, S., & Fransen, C. H. J. M. (2011). Carideorum Catalogus: The Recent Species of the Dendrobranchiate, Stenopodidean Procarididean and Caridean Shrimps (Crustacea: Decapoda). *Zoologische Mededelingen. Leiden, 85*(9), 195–589.

De Haan, H. M. (1833–1849). Crustacea. In: P. F. von Siebold (Ed.), *Fauna Japonica, sive Descriptio animalium, quae in itinere per Japoniam, jussu et auspiciis superiorum, qui summum in India Batavia imperium tenent, suscepto, annis 1823–1830 collegit, notis, observationibus a adumbrationibus illustravit*. fasc. 1–8: pp. i–xxi+vii–xvii+ix–xvi+1–243, pls. 1–55, A–Q, circ., pl. 2. Lugduni Batavorum. (For publication dates see Sherborn & Jentink, 1895; Holthuis, 1953; Holthuis & Sakai, 1970).

de Man, J. G. (1881). Carcinological studies in the Leyden Museum. No. 1. *Notes from the Leyden Museum, 3*, 121–144.

de Man, J. G. (1905). Diagnoses of new species of macrurous decapod Crustacea from the "Siboga-Expedition". I. *Tijdschr. Nederlandsch Tijdschrift voor de Dierkunde (2), 9*, 587–614.

de Man, J. G. (1916). Families Eryonidae, Palinuridae, Scyllaridae and Nephropidae the Decapoda of Siboga Expedition. Part III. *Siboga Expedition, Mon. 39*(a2), 1–222, pls. 1–4.

de Saint Laurent, M. (1988). Enoplometopoidea, nouvelle superfamille de crustacés décapods Astacidea. *Comptes rendus de l'Académie des Sciences Paris (3), 307*, 59–62.

Desmarest, A. G. (1825). *Considérations générales sur la classe des Crustacés et description des espèces de ces animaux, qui vivent dans la mer, sur les côtes, ou dans les eaux douces de la France*. F. G. Levrault, Paris et Strasbourg. xix+446 pp., pls. 1–56, 5 tables.

Dixon, C. J., Ahyong, S. T., & Schram, F. R. (2003). A new hypothesis of decapod phylogeny. *Crustaceana, 76*(8), 935–975.

Doflein, F. (1900). Weitere Mitteilungen über dekapode Crustaceen der k. bayerischen Staatssammulungen. *S. B. Bayer. Akad. Wiss., 30*, 125–145, figs. 1–3.

Fabricius, J. C. (1775). *Systema Entomologiae, sistens insectorum classes, ordines, genera, species, adiectis synonymis, locis, descriptionibus, observationibus*. Flensburg, Leipzig: Kortius. 832 pp.

Fabricius, J. C. (1787). *Mantissa Insectorum sistens eorum species nuper detectas adiectis Characteribus genericis, Differentiis specificis, Emendationibus, Observationibus*, I. Hafniae. xx+348 pp.

Fabricius, J. C. (1798). *Supplementum Entomologiae systematicae*. Hafniae: Proft et Storch. 573 pp.

Faxon, W. (1893). Reports on the Dredging Operations off the West Coast of Central America to the Galapagos, to the West Coast of Mexico, and in the Gulf of California, in Charge of Alexander Agassiz, carried on by the U. S. Fish Commission Steamer "Albatross" during 1891, Lieut.-Commander Z. L. Tanner, U. S. N., Commanding. VI. Preliminary Descriptions of New Species of Crustacea. *Bulletin of the Museum of Comparative Zoology, 24*(7), 149–200.

Forest, J. (1954). Scyllaridae. Crustacés Décapodes Marcheurs des îles de Tahiti et des Tuamotu II. *Bulletin du Muséum national d'histoire naturelle Paris (2), 26*, 345–352.

Forest, J. (1963). Sur deux *Scyllarus* de l'Atlantique tropical africain: *S. paradoxus* Miers et *S. posteli* sp. nov. Remarques sur les *Scyllarus* de l'Atlantique oriental. *Bulletin de l'Institut océanographique de Monaco, 60*(1259), 1–20.

Forest, J. (2006). *Laurentaeglyphea*, un nouveau genre pour la seconde es pèce actuelle de Glyphéides récemment découverte (Crustacea Décapoda Glyheidae). *Comptes rendus de l'Académie des Sciences Paris, 329*(10), 841–846.

Forest, J., & Postel, E. (1964). Su une espèce nouvelle de langouste des îles du Cap Vert, *Palinurus charlestoni* sp. nov. *Bulletin du Muséum national d'histoire naturelle Paris (2), 36*(1), 100–121.

Forest, J., & de Saint Laurent, M. (1975). Prèsence dans la fauna actuelle d'un reprèsentant du groupe mèsozoïque des Glyphèides: *Neoglyphea inopinata* gen. nov., sp. nov. (Crustacea Decapoda Gylpheidae). *Comptes rendus de l'Académie des Sciences Paris (D), 281*, 155–158.

Froglia, C., & Clark, P. F. (2011). The forgotten narrative of H.M.S. *Challenger* and the implications for decapod nomenclature. *Zootaxa, 2788*, 45–56.

Galil, B. S. (2000). Crustacea Decapoda: Review of the genera and species of the family Polycheles Wood Mason, 1874. In: Crosnier, A. (ed.), Résultats des Campagnes MUSORSTOM, vol. 21. *Mémoires du Museum National d'Histoire Naturelle Paris, 184*, 285–387.

Genis-Armero, R., Guerao, G., Abelló, P., González-Gordillo, J. I., Cuesta, J. A., Corbari, L., Clark, P. F., Capaccioni-Azzati, R., & Palero, F. (2017). Possible amphi-Atlantic dispersal of *Scyllarus* lobsters (Crustacea: Scyllaridae): molecular and larval evidence. *Zootaxa, 4306*(3), 325–338.

George, R. W. (1962). Description of *Panulirus cygnus* sp. nov., the commerical crayfish (or spiny lobster) of Western Australia. *Journal of the Royal Society of Western Australia, 45*, 100–110, pls. 1–2.

George, R. W. (1976). A new species of spiny lobster, *Projasus bahamondei* (Palinuridae "Silentes"), from the South East Pacific region. *Crustaceana, 30*(1), 27–32, pl. 1.

George, R. W. (2005). Tethys sea fragmentation and speciation of *Panulirus* spiny lobsters. *Crustaceana, 78*, 1281–1309.

George, R. W., & Grindley, J. R. (1964). *Projasus*—a new generic name for Parker's crayfish, *Jasus parkeri* Stebbing (Palinuiridae: "Silentes"). *Journal of the Royal Society of Western Australia, 47*, 87–90.

Gibbes, L. R. (1850). On the carcinological collections of the cabinets of natural history in the United States. With an enumeration of the species contained therein, and descriptions of new species. *Proceedings of the American Association for the Advancement of Science, 3*, 165–201.

Gill, T. (1898). The crustacean genus *Scyllarides*. *Science (n. ser.), 7*(160), 98–99.

Giraldes, B. W., & Smyth, D. M. (2016). Recognizing *Panulirus meripurpuratus* sp. nov. (Decapoda: Palinuridae) in Brazil—Systematic and biogeographic overview of *Panulirus* species in the Atlantic Ocean. *Zootaxa, 4107*(3), 353–366.

Gordon, I. (1968). Description of the holotype of *Enoplometopus dentatus* Miers, with notes on other species of the genus (Decapoda). *Crustaceana, 15*(6), 79–97.

Groeneveld, J. C., Griffiths, C. L., & van Dalsen, A. P. (2006). A new species of spiny lobster, *Palinurus barbarae* (Decapoda, Palinuridae) from Walters Shoals on the Madagascar Ridge. *Crustaceana, 79*(7), 821–833.

Groeneveld, J. C., von der Heyden, S., & Matthee, C. A. (2012). High connectivity and lack of mtDNA differentiation among two previously recognized spiny lobster species in the southern Atlantic and Indian Oceans. *Marine Biology Research, 8*, 764–770.

Grote, A. R. (1873). *Deidamia. Nature, 8*, 485.
Gruvel, A. (1911). Contribution á l'étude générale systématique et économique des Palinuridae. Mission Gruvel sur la côte occidentale d'Afrique (1909–1910). Résultats scientifiques et économiques. *Annales de l'Institut Océanographique, 3*(4), 5–56, pls. 1–6.
Harada, E. (1962). On the genus *Scyllarus* (Crustacea Decapoda: Reptantia) from Japan. *Publications of the Seto Marine Biological Laboratory, 10*, 109–132.
Heller, C. (1862a). Neue Crustaceen, gesammelt während der Weltumseglung der k.k. Fregatte Novara. Zweiter vorläufiger Bericht. *Verhandlungen der Kaiserlich-Königlichen Zoologisch-Botanischen Gesellschaft in Wien, 12*, 519–528.
Heller, C. (1862b). Beiträge zur näheren Kenntnis der Macrouren. *Sitzungsberichte der Kaiserlichen Akademie der Wissenschaften in Wien – mathematisch-naturwissenschaftliche Classe, 45*(1), 389–426, pls. 1–2.
Herbst, J. F. W. (1782–1804). *Versuch einer Naturgeschichte der Krabben und Krebse nebst einer Systematischen Beschreibung ihrer Verschiedenen Arten, Vol. 1–3*. Gottlieb August Lange, Berlin & Stralsund. 515 pp., pls. 1–62.
Herklots, J. A. (1851). *Additamenta ad Faunam Carcinologicam Africae occidentalis*. 31 pp., pls. 1–2.
Holthuis, L. B. (1946). Biological results of the Snellius Expedition XIV. The Decapoda Macrura of the Snellius Expedition I. The Stenopodidae, Nephropidae, Scyllaridae and Palinuridae. *Temminckia, 7*, 1–178, pls. 1–11.
Holthuis, L. B. (1952). Crustacés Décapodes Macrures. *Rés. Sci. Expéd. Ocean. Belge Eaux Côt. Afr. Atl. Sud (1948–1949), 3*(2), 1–88.
Holthuis, L. B. (1960). Preliminary descriptions of one new genus, twelve new species and three new subspecies of Scyllarid lobsters. (Crustacea Decapoda Macrura). *Proceedings of the Biological Society of Washington, 73*, 147–154.
Holthuis, L. B. (1963). Preliminary descriptions of some new species of Palinuridea (Crustacea Decapoda, Macrura, Reptantia). *Proceedings of the Koninklijke Nederlandse Akademie van Wetenschappen/C, 66*, 54–60.
Holthuis, L. B. (1964). On some species of the genus *Nephrops* (Crustacea: Decapoda). *Zoologische Mededelingen. Leiden, 39*, 71–78.
Holthuis, L. B. (1967). Some new species of Scyllaridae. *Proceedings of the Koninklijke Nederlandse Akademie van Wetenschappen (c), 70*, 305–308.
Holthuis, L. B. (1969). A new species of shovel-nose lobster, *Scyllarus planorbis*, from the southwestern Caribbean and northern South America. *Bulletin of Marine Science, 19*(1), 149–158.
Holthuis, L. B. (1974). The lobsters of the superfamily Nephropidea of the Atlantic Ocean (Crustacea: Decapoda). *Bulletin of Marine Science, 24*(4), 723–884.
Holthuis, L. B. (1977). Two new species of scyllarid lobsters (Crustacea Decapoda, Palinuridea) from Australia and the Kermadec Islands, New Zealand. *Zoologische Mededelingen Leiden, 52*, 191–200.
Holthuis, L. B. (1982). A new species of *Scyllarus* (Crustacea Decapoda Palinuridea) from the Pacific Ocean. *Bulletin du Muséum national d'histoire naturelle Paris (4), 3*(A3), 847–853.
Holthuis, L. B. (1983). Notes on the genus *Enoplometopus* with description of a new subgenus and two new species (Crustacea Decapoda Axiidae). *Zoologische Mededelingen Leiden, 56*(22), 281–298.
Holthuis, L. B. (1985). A revision of the family Scyllaridae (Crustacea: Decapoda: Macrura). I. Subfamily Ibacinae. *Zoologische Verhandelingen. Leiden, 218*, 1–130.
Holthuis, L. B. (1991). Marine lobsters of the world. *FAO Fisheries Synopsis, 125*(13), 1–292.
Holthuis, L. B. (1993a). *Scyllarus rapanus*, a new species of locust lobster from the South Pacific (Crustacea, Decapoda, and Scyllaridae). *Bulletin du Muséum national d'histoire naturelle Paris (4), 15*(A1–4), 179–186.
Holthuis, L. B. (1993b). *Scyllarides obtusus* spec. nov., the scyllarid lobster of Saint Helena, Central South Atlantic (Crustacea: Decapoda Reptantia: Scyllaridae). *Zoologische Mededelingen. Leiden, 67*, 505–515.

Holthuis, L. B. (2002). The Indo-Pacific scyllarine lobsters (Crustacea, Decapoda, Scyllaridae). *Zoosystema, 24*(3), 499–683.

Holthuis, L. B. (2006). Revision of the genus *Arctides* Holth uis, 1960 (Crustacea, Decapoda, Scyllaridae). *Zoosystema, 28*(2), 417–433.

Hu, C. H. (1983). Discovery fossil lobster from the Kuechulin Formation (Miocene), Southern Taiwan. *Ann. Taiwan Mus., 26*, 129–136.

Hutton, F. (1875). Descriptions of two new species of crustaceans from New Zealand: *Sesarma pentagona* and *Palinurus edwardsii*. *Transactions and Proceedings of the Royal Society of New Zealand, 7*, 279–280.

Iacchei, M., Gaither, M. R., Bowen, B. W., & Toonen, R. J. (2016). Testing dispersal limits in the sea: range-wide phylogeography of the pronghorn spiny lobster *Panulirus penicillatus*. *Journal of Biogeography, 43*, 1032–1044.

Intès, A., & Le Loeuff, P. (1970). Sur une nouvelle espèce du genre *Enoplometopus* A. Milne Edwards du Golfe de Guniée: *Enoplometopus callistus* nov. sp. (Crustacea, Decapoda, Homaridea). *Bulletin du Muséum national d'histoire naturelle Paris (2), 41*(6), 1442–1447.

Jenkins, R. J. F. (1972). *Metanephrops*, a new genus of late Pliocene to recent lobsters (Decapoda, Nephropidae). *Crustaceana, 22*(2), 161–177.

Karasawa, H., Schweitzer, C. E., & Feldmann, R. M. (2013). Phylogeny and Systematics of extant and extinct lobsters. *Journal of Crustacean Biology, 33*(1), 78–123.

Kensley, B. (1968). Deep sea decapods Crustacea from west of Cape Point, South Africa. *Annals of the South Africa Museum, 50*(12), 283–323.

Kensley, B., & Child, C. A. (1986). A new species of *Enoplometopus* (Thalassinidea: Axiidae) from the northern Philippines. *Journal of Crustacean Biology, 6*(3), 520–524.

Kishinouye, K. (1926). Two rare and remarkable forms of macrurous Crustacea from Japan. *Japanese Journal of Zoology, 11*, 63–70.

Kornfield, I., Williams, A. B., & Steneck, R. S. (1995). Assignment of *Homarus capensis* (Herbst, 1792), the Cape lobster of South Africa, to the new genus *Homarinus* (Decapoda: Nephropidae). *Fish Bulletin, 93*(1), 97–102.

Kubo, I. (1955). Systematic studies on the Japanese macrurous decapod Crustacea. 5. A new palinurid, *Nupalirus japonicus*, gen. and sp. nov. *Journal of the Tokyo University of Fisheries, 41*(2), 185–188.

Kubo, I. (1963). Systematic studies on the Japanese macrurous decapods Crustacea, 6. A new and an imperfectly known species of palinurid lobster. *Journal of the Tokyo University of Fisheries, 49*, 63–71.

Latreille, P. A. (1802). *Histoire naturelle, générale et particulière, des Crustacés et des Insectes* (Vol. 3). Paris: F. DuFart. 467 pp.

Latreille, P. A. (1803). *Histoire naturelle, générale et particulière, des Crustacés et des Insectes* (Vol. 6). Paris: F. DuFart. 392 pp, pls. 44–57.

Latreille, P. A. (1804). Des langoustes du Muséum national d'Histoire naturelle. *Annales du Muséum d'histoire naturelle. Paris, 3*, 388–395.

Latreille, P. A. (1817). Langouste, *Palinurus*, Fab. *Nouveau Dictionnaire d'Histoire naturelle, 17*, 291–295.

Latreille, P. A. (1825). *Familles naturelles du règne animal, exposées succinctement et dans un Ordre analytique, avec l'Indication de leurs genres*. Baillière: J.-B. 570 pp.

Lavery, S. D., Farhadi, A., Farahmand, H., Chan, T. Y., Azhdehakosphpour, A., Thakur, V., & Jeffs, A. G. (2014). Evolutionary divergence of geographic subspecies within the scalloped spiny lobster *Panulirus homarus* (Linnaeus, 1758). *PLoS One, 9*(6), 1–13.

Leach, W. E. (1813–1814). Crustaceology. In: Brewster, D. (ed.). *The Edinburgh Encyclopaedia, 7*(2), 385–437.

Leach, W. E. (1815). *The Zoological Miscellany; Being descriptions of new or interesting animals. Illustrated with coloured figures drawn from nature by R. P. Nodder &c., vol. 2(12)*. London: E. Nodder & Son. pp. 145–154, pls. 116–120.

Leach, W. E. (1816). XXXI. A tabular view of the external characters of four classes of animals, which Linné arranged under Insecta, with the distribution of the genera composing three of these classes into orders &c. and descriptions of several new genera and species. *Transactions of the Linnean Society of London, 11*[for 1815] (2), 306–400 + "Errata".

Linnaeus, C. (1758). *Systema Naturae per Regna Tria Naturae, Secundum Classes, Ordines, Genera, Species, cum Characteribus, Differentiis, Synonymis, Locis, ed. 10, vol. 1*. iii + 824 pp.

Lund, N. T. (1793). Slaegten *Scyllarus*. Iagttagelser til insekternes Historie. I. *Skr. naturh. Selsk. Kbh., 2*(2), 17–22, pl. 1.

Lütken, C. (1865). *Enoplometopus antillensis* Ltk., en ny vestindisk Hummer-Art. *Vidensk Medd naturhist Foren København, 6*, 265–268.

Macpherson, E. (1990). Crustacea Decapoda: On a collection of Nephropidae from the Indian Ocean and Western Pacific. In: Crosnier A. (ed.). Résultats des Campagnes MUSORSTOM, vol. 6. *Mémoires du Muséum National d'Histoire Naturelle Paris (A), 145*, 289–329.

Macpherson, E. (1993). New record for the genus *Nephropsis* Wood-Mason (Crustacea, Decapoda, Nephropidae) from northern Australia, with description of two new species. *The Beagle: Northern Territory Museum of Arts and Sciences, 10*(1), 55–66.

Manning, R. B. (1969). A new genus and species of lobster (Decapoda, Nephropidae) from the Caribbean Sea. *Crustaceana, 17*, 303–309.

Manning, R. B. (1997). *Eunephrops luckhursti*, a new deep-sea lobster from Bermuda (Crustacea: Decapoda: Nephropidae). *Proceedings of the Biological Society of Washington, 110*(2), 256–262.

Miers, E. J. (1880). On a collection of crustacea from the Malaysian region. Part. III. Crustacea Anomura and Macrura (except Penaeidae). *Annals and Magazine of Natural History, 5*(5), 370–384.

Miers, E. J. (1881). On a collection of Crustacea made by Baron Hermann Maltzam [sic] at Goree Island, Senegambia. *Annals and Magazine of Natural History.* (5), 8 (45, Sept. 1881): 204–220; (46, Oct. 1881): 259–281, pls. 13–14; (47, Nov. 1881): 364–377, pls. 15–16.

Miers, E. J. (1882). On some crustaceans collected at Mauritius. *Proceedings of the Zoological Society, London, 1882*, 339–342, 538–543, pls. 20, 36.

Milne Edwards, H. (1834–1837). *Histoire naturelle des Crustacés comprenant l'anatomie, la physiologie et la classification de ces animaux*. Librairie Encyclopédique de Roret, Paris. Vol. 1: xxxv + 468 pp. Vol. 2: 531 pp. Atlas, 1837: 32 pp., pls. 1–42. Vol. 3, 1840: 638 pp.

Milne Edwards, H. (1851). Observations sur le squelette tégumentaire des Crustacés décapodes, et sur la morphologie de ces animaux. *Annales des Sciences Naturelles Zoologie Paris, 16*(3), 221–291, pls. 8–11.

Milne-Edwards, A. (1862). Faune carcinologique de l'île de la Réunion: annexe F. In L. Maillard (Ed.), *Notes sur l'île de la Réunion* (pp. 1–16). Paris: Dentu.

Milne-Edwards, A. (1868). Description de quelques Crustacés nouveaux provenant des voyages de M. Alfred Grandidier à Zanzibar et à Madagascar. *Nouvelles annales du Muséum d'histoire naturelle Paris, 4*, 69–92, pls. 19–21.

Milne-Edwards, A. (1880a). Note sur une nouvelle espèce de Crustacé aveugle provenant des grandes profondeurs de la mer. *Annales des Sciences Naturelles Zoologie Paris (6), 9*(2), 1.

Milne-Edwards, A. (1880b). Études préliminaires sur les Crustacés. Ière Partie. Reports on the results of dredging, under the supervision of Alexander Agassiz, in the Gulf of Mexico, and in the Caribbean Sea, 1877, 78, 79, by the United States Coast Survey steamer "Blake", Lieut.-Commander C. D. Sigsbee, U. S. N., and Commander J. R. Bartlett, U. S. N., commanding. VIII. *Bulletin of the Museum of Comparative Zoology, 8* (1), 1–68, pls. 1–2.

Milne-Edwards, A. (1881). Description de quelques Crustacés Macroures provenant des grandes profondeurs de la mer des Antilles. *Annales des Sciences Naturelles Zoologie Paris (6), 11*(4), 1–16.

Moreira, C. (1903). Campanhas de pesca do hiate "Annie", dos Srs. Bandeira & Bravo. Estudos preliminares. Crustaceos. *Bol. Soc. Nac. Agric. Brasil, 7*(1–3), 60–67.

Mustafa, A. F. (1990). *Linuparus andamanensis*, a new spear-lobster from Andamans. *Journal of the Andaman Science Association, 6*(2), 177–180.

Ng, P. K. L., Guinot, D., & Davie, P. J. F. (2008). Systema Brachyurorum: Part I. An annotated checklist of extant brachyuran crabs of the world. *Raffles Bulletin of Zoology, 17*, 1–286.

Nobili, G. (1906). Diagnoses préliminaires de 34 espéces et variétés nouvelles, et de 2 genres nouveaux de Décapodes de la Mer Rouge. *Bulletin du Museum national d'histoire naturelle. Paris, 11*, 393–411.

Norman, A. M. (1882). Report on the Crustacea exploration of the Faroe Channel, during the summer of 1880, in H. M.'s hired ship "Knight Errant". *Proceedings of the Royal Society of Edinburgh, 11*, 683–689.

Olivier, A. G. (1791). Écrevisse, Astacus. In: Olivier, A. G. (ed.). *Insectes. Encycl. méth. Hist. nat., 6*, 327–349.

Ortmann, A. (1894). Crustacean. In: Semon R. (Ed.), *Zoologische Forschungsreisen in Australien und dem Malayischen Archipel, vol. 5. Denkschriften der Medicinisch-Naturwissenschaftlichen Gesellschaft zu Jena, 8*, 3–80, pls. 1–3.

Ortmann, A. (1897a). On a new species of the Palinurid-genus *Linuparus* found in the upper Cretaceous of Dakota. *American Journal of Science, 4*, 290–297.

Ortmann, A. E. (1897b). Carcinologische Studien. *Zoologische Jahrbücher System, 10*, 258–372, pl. 17

Parisi, B. (1917). I Decapodi Giapponesi del Museo di Milano. V. Galatheidea e Reptantia. *Atti della Società italiana di scienze naturali e del Museo civico di storia naturale di Milano, 56*, 1–24, figs. 1–7.

Parker, T. J. (1883). On the structure of the head in *Palinurus* with special reference to the classification of the genus. *Nature, 29*, 189–190.

Paulson, O. (1875). *Investigations on the Crustacea of the Red Sea, with Notes on Crustacea of the Adjacent Seas. Part. I. Podophthalmata and Edriophthalmata (Cumacea)*. Typografia S. V. Kulzhenko, Kiev. xiv + 144 pp., pls. 1–21.

Pesta, O. (1915). Bemerkungen zu einigen Langusten (Palinuridae) und ihrer geographischen Verbreitung. *S. B. Akad Wiss Wien, 124*(1–2), 3–12, pl. 1.

Pfeffer, G. (1881). Die Panzerkrebse des Hamburger Museums. *Verh naturw Ver Hamburg (2), 5*, 22–55.

Poore, G. C. B. (2016). The names of the higher taxa of Crustacea Decapoda. *Journal of Crustacean Biology, 36*(2), 248–255.

Poore, G. C. B., & Dworschak, P. C. (2017). Family, generic and species synonymies of recently published taxa of ghost shrimps (Decapoda, Axiidea, Eucalliacidae and Ctenochelidae): cautionary tales. *Zootaxa, 4294*(1), 119–125.

Poupin, J. (1994). The genus *Justitia* Holthuis, 1946, with description of *J. chani* and *J. vericeli* spp. nov. (Crustacea: Decapoda: Palinuridae). *Journal of the Taiwan Museum, 47*(1), 37–56.

Powell, A. W. B. (1949). New species of Crustacea from New Zealand of the genera *Scyllarus* and *Ctenocheles* with notes on *Lyreidus tridentatus*. *Records of the Aukland Museum, 3*(6), 368–371, pl. 68.

Quoy, J. R. C., & Gaimard, P. (1825). Des Crustacés. In L. de Freycinet (Ed.), *Voyage autour du monde entrepris par ordre du Roi, sous le ministère et conformément auz instructions de S. Exc. M. le Vocomte du Bouchage, secrétaire d'état au département de la Marine, exécuté sur les corvettes de S. M. L'Uranie et la Physicienne, pendant les années 1817, 1818, 1819 et 1820* (pp. 517–541), pls. 76–82. Zool.

Ramadan, M. M. (1938). The Astacura and Palinura. *Scientific reports/John Murray Expedition, 5*(5), 123–145.

Randall, J. W. (1840). Catalogue of the crustacea brought by Thomas Nuttall and J.K. Townsend, from the west coast of North America and the Sandwich Islands, with description of such species as are apparently new, among which are included several species of different localities, previously existing in the collection of the Academy. *Journal of the Academy of Natural Sciences of Philadelphia, 8*(1), 106–147, pls. 3–7.

Rathbun, M. J. (1906a). A new *Scyllarides* from Brazil. *Proceedings of the Biological Society Washington, 19*, 113.

Rathbun, M. J. (1906b). The Brachyura and Macrura of the Hawaiian Islands. *Bulletin of the United States Fish Commission, 23*(3), 827–930, pls. 3–24.

Reed, E. P. (1954). Palinuridae. *Scientia Valparaiso, 21*, 131–139.

Richer de Forges, B. (2006). Découverte en mer du Corail d'une deuxième espéce de Glyphéide (Crustacea, Decapoda, Glypheoidea). *Zoosystema, 28*(1), 17–29.

Sarver, S., Silberman, J., & Walsh, P. (1998). Mitochondrial DNA sequence evidence supporting the recognition of two subspecies or species of the Florida spiny lobster *Panulirus argus*. *Journal of Crustacean Biology, 18*, 177–186.

Scholtz, G., & Richter, S. (1995). Phylogenetic systematics of the reptantian Decapoda (Crustacea, Malacostraca). *Zoological Journal of the Linnean Society, 113*, 289–328.

Schram, F. R. (2014). [Book review] Charbonnier, S., A. Garassino, G. Schweigert, and M. Simpson. 2013. A worldwide review of fossil and extant glypheid and litogastrid lobsters (Crustacea, Decapoda, Glypheoidea). Mémoires du Muséum national d'Histoire naturelle 28. Publications Scientifiques du Muséum, Paris. *Journal of Crustacean Biology, 34*, 399–401.

Schram, F. R., & Dixon, C. J. (2004). Decapod phylogeny: addition of fossil evidence to a robust morphological cladistic data set. *Bulletin of the Mizunami Fossil Museum, 31*, 1–19.

Segonzac, M., & Macpherson, E. (2003). A new deep-sea lobster of the genus *Thymopides* (Crustacea: Decapoda: Nephropidae) collected near the hydrothermal vent Snake Pit, Mid-Atlantic Ridge. *Cahiers de Biologie Marine, 44*, 361–367.

Sekiguchi, H., & George, R. W. (2005). Description of *Panulirus brunneiflagellum* new species with notes on its biology, evolution, and fisheries. *New Zealand Journal of Marine and Freshwater Research, 39*, 563–570.

Siebold, G. T., de [err. Pro P. F. von]. (1824). *De Historia naturalis in Japonia statu, nec non de augment emolumentisque in decursu perscrutationum expectandis dissertation, cui accedunt Spicilegia Faunae Japonicae*. Bataviae. 16 pp.

Smith, S. I. (1869a). Notice of the Crustacea collected by Prof. C. F. Hartt on the coast of Brazil in 1867, list of the described species of Brazilian Podophthalmia. *Transactions Connecticut Academy of Arts and Sciences. 2*(1), 1–41, pl. 1.

Smith, S. I. (1869b). Descriptions of a genus and two new species of Scyllaridae, and of a new species of *Aethra* from North America. *Amer. Journal of Science (2), 48*(142), 118–121.

Smith, S. I. (1880). Notice of a new species of the "*Willemoesia* Group of Crustacea", recent Eryontidae. *Proceedings of the United States National Museum, 2*, 345–353, pl. 7.

Smith, S. I. (1881). Preliminary notice of the Crustacea dredged, in 64 to 325 fathoms, off the south coast of New England, by the United States Fish Commission in 1880. *Proceedings of the United States National Museum, 3*, 413–452.

Smith, S. I. (1884). XV. Report on the Decapod Crustacea of the *"Albatross"* Dredgings of the East-coast of the United States in 1883. *Reports on U. S. Fish Commission, 10*(1882), 345–426, pls. 1–10.

Smith, S. I. (1885). Description of a new crustacean allied to *Homarus* and *Nephrops*. *Proceedings of the United States National Museum, 8*, 167–170.

Spengler, L. (1799). Beskrivelse af en nye Art Kraebs, *Scyllarus Guineensis*. *Kongelige Danske videnskabernes selskabs skrifter, 5*(2), 333–340, pl. 1.

Srikrishnadhas, B., Rahman, M. K., & Anandasekaran, A. S. M. (1991). A new species of scyllarid lobster *Scyllarus tutiensis* (Scyllaridae: Decapoda) from the Tuticorin Bay in the Gulf of Mannar. *Journal of the Marine Biological Association of India, 33*, 418–421.

Stebbing, T. R. R. (1900). South African Crustacea. *Marine Investigations in South Africa, 1*, 14–66, pls. 1–4.

Stebbing, T. R. R. (1902). South African Crustacea, Part II. *Marine Investigations in South Africa, 12*, 1–92, pls. 5–16.

Stebbing, T. R. R. (1917). South African Crustacea, Part IX. *Annals of the South African Museum, 17*(1), 23–46, pls. 1–8.

Stimpson, W. (1860). Prodromus descriptionis animalium evertebratorum, quae in expeditione ad Oceanum Pacificum Septentrionalem, a Republica Federata missa, C. Ringgold et J. Rogers ducibus, observavit et descripsit. *Proceedings of the Academy of Natural Sciences of Philadelphia, 1860*, 22–47.

Stimpson, W. (1866). Description of new genera and species of macrurous Crustacea from the coasts of North America. *Proceedings of Chicago Academy of Sciences, 1*, 46–48.

Streets, T. H. (1871). Descriptions of five new species of Crustacea from Mexico. *Proceedings of the Academy of Natural Sciences of Philadelphia, 1871*, 225–227, pl. 2.

Sund, O. (1920). The "*Challenger*" Eryonidea (Crustacea). *Annals and Magazine of Natural History (9), 6*, 220–226.

Tapparone-Canefri, C. (1873). Intorno ad una nuova specie di *Nephrops*, genera di Crostacei Decapodi Macruri. *Memorie della Reale accademia delle scienze di Torino (2), 28*, 325–329.

Tavares, M. S. (1997). *Scyllarus ramosae*, new species from the Brazilian continental slope with notes on congeners occurring in the area (Decapoda: Scyllaridae). *Journal of Crustacean Biology, 17*(4), 716–724.

Thomson, C. W. (1873). Notes from the "Challenger". *Nature, 8*, 28–30, 51–53, 109–110, 246–249, 266–267, 347–349, 400–403.

Toon, A., Finley, M., Staples, J., & Crandall, K. A. (2009). Decapod phylogenetics and molecular evolution. In J. W. Martin, K. A. Crandall, & D. L. Felder (Eds.), *Crustacean issues 18: Decapod crustacean phylogenetics* (pp. 15–29). Boca Raton: Taylor & Francis/CRC Press.

Tsang, L. M., Ma, K. Y., Ahyong, S. T., Chan, T. Y., & Chu, K. H. (2008). Phylogeny of Decapoda using two nuclear protein-coding genes: Origin and evolution of the Reptantia. *Molecular Phylogenetics and Evolution, 48*, 359–368.

Tshudy, D., & Sorhannus, U. (2000a). *Jagtia kunradensis*, a new genus and species of clawed lobster (Decapoda: Nephropidae) from the Upper Cretaceous (Upper Maastrichtian) Maastricht Formation, The Netherlands. *Journal of Paleontology, 74*(2), 224–229.

Tshudy, D., & Sorhannus, U. (2000b). Pectinate claws in decapod crustaceans: Convergence in four lineages. *Journal of Paleontology, 74*(3), 474–486.

Tshudy, D., Chan, T. Y., & Sorhannus, U. (2007). Morphology based cladistic analysis of *Metanephrops*, the most diverse extant genus of clawed lobster (Nephropoidae). *Journal of Crustacean Biology, 27*(3), 463–476.

Tshudy, D., Robles, R., Chan, T. Y., Ho, K. C., Chu, K. H., Ahyong, S. T., & Felder, D. L. (2009). Phylogeny of marine clawed lobster families Nephropidae Dana 1852 and Thaumastochelidae Bate 1888 based on mitochondrial genes. In J. W. Martin, K. A. Crandall, & D. L. Felder (Eds.), *Crustacean issues 18: Decapod crustacean phylogenetics* (pp. 357–368). Boca Raton: Taylor & Francis/CRC Press.

Tsoi, K. H., Chan, T. Y., & Chu, K. H. (2011). Phylogenetic and biogeographic analysis of the spear lobsters *Linuparus* (Decapoda, Palinuridae), with the description of a new species. *Zoologischer Anzeiger, 250*, 302–315.

Tung, Y. M., Wang, B. Y., & Li, Z. C. (1985). A new species of Nephropsidea from the deep water of East China Sea. *Acta Zootaxonomica Sinica, 10*(4), 379–380.

Türkay, M. (1989). *Enoplometopus* (*Hoplometopus*) *voigtmanni* n. sp., ein neuter Riffhummer von den Malediven. *Senckenbergiana maritima, 20*(5/6), 225–235.

Van Straelen, V. (1925). Contribution à l'étude des Crustacés décapodes de la période Jurassique. *Mémoires couronnés et mémoires des savants étrangers Belgique (2), 7*(1), 1–462, pls. 461–410.

von Martens, E. (1872). Ueber cubanische Crustaceen nach den Sammlungen Dr. J. Gundlach's. *Archiv für Naturgeschichte, 38*(1), 77–147, pls. 4–5.

von Martens, E. (1878). Einige Crustaceen und Mollusken. *Sitzungsberichte der Gesellschaft Naturforschender Freunde zu Berlin, 1878*, 131–135.

Wahle, R. A., Tshudy, D., Stanley Cobb, J., Factor, J., & Jaini, M. (2012). Infraorder Astacidea Latreille, 1802 p.p.: The marine clawed lobsters. In F. R. Schram & J. C. von Vaupel Klein (Eds.), *The Crustaea, complementary to the volumes translated from the French of the Traité de Zoologie, vol. 9 (B). Eucarida: Decapoda: Astacidea p.p. (Enoplometopodidea, Nephropoidea), Glyphelidea, Axiidea, Gebiidea, and Anomura* (pp. 3–108). Leiden: Koninklijke Bill NV.

Watabe, H., & Iizuka, E. (1999). A new species of the bathyal lobster *Nephropsis* (Crustacea: Decapoda: Nephropidae) from Australian waters, with redescription of *N. holthuisi*. *Species Diversity, 4*, 371–380.

Watabe, H., & Ikeda, H. (1994). *Nephropsis hamadai*, a new nephropid lobster (Decapoda: Nephropidae) from bathyal depth in Sagami Nada (Central Japan). *Crustacean Research, 23*, 102–107.

Webber, W. R., & Booth, J. D. (1995). A new species of *Jasus* (Crustacea: Decapoda: Palinuridae) from the eastern South Pacific Ocean. *New Zealand Journal of Marine and Freshwater Research, 29*, 613–622.

Weber, F. (1795). *Nomenclator entomologicus secundum Entomologiam systematicum ill. Fabricii, adjectis speciebus recens detectis et varietatibus.* Chilonii & Hamburgi. viii + 171 pp.

White, A. (1847). *List of species in the collections of the British Museum.* viii+1–143 pp. London: British Museum.

Whitelegge, T. (1900). Crustacea. Part I. Scientific results of the trawling expedition of H. M. C. S. "Thetis", off the coast of New South Wales, in February and March, 1898. *Australian Museum Memoir, 4*(2), 135–199, pls. 33–35.

Winkler, T. C. (1881). Études carcinologique sur les genres "*Pemphix*", "*Glyphea*" et "*Araeosternus*". *Archives du Musée Teyler (2), 2*, 73–124.

Wood-Mason, J. (1872). On *Nephropsis stewarti*, a new genus and species of macrourus crustaceans, dredged in deep water off the eastern coast of the Andaman Islands. *Proceedings of the Asiatic Society of Bengal, 1872*, 151.

Wood-Mason, J. (1874). Blind Crustacea. *Proceedings of the Asiatic Society of Bengal, 1874*, 180–181.

Wood-Mason, J. (1875). On the genus *Deidamia* Willemoes-Suhm. *Annals and Magazine of Natural History (4), 15*, 131–135.

Wood-Mason, J. (1885). Natural history zoological notes from H. M. S. Indian Marine Survey Steamer "Investigator", Commander A. Carpenter, R. N. commanding. *Proceedings of the Asiatic Society of Bengal, 1885*, 69–72.

Wood-Mason, J. (1891). Note on the Results of the last season's deep-sea dredging. In: Wood-Mason, J. & Alcock, A. (eds.). Natural History Notes from H. M. Indian Marine Survey Steamer "Investigator", Commander R. F. Hoskyn, R. N., Commanding. N°21. *Annals and Magazine of Natural History, 6*(7), 186–202.

Yang, C. H., & Chan, T. Y. (2010). A new slipper lobster of the genus *Galearctus* Holthuis, 2002 (Decapoda: Scyllaridae) from Taiwan and Japan. In: Fransen C.H.J.M, De Grave, S. & Ng, P.K.L. (eds.). Studies on Malacostraca: Lipke Bijdeley Holthuis Memorial Volume. *Crustaceana Mon, 14*, 735–745.

Yang, C. H., & Chan, T. Y. (2012). On the taxonomy of the slipper lobster *Chelarctus cultrifer* (Ortmann, 1897) (Crustacea: Decapoda: Scyllaridae), with description of a new species. *Raffles Bulletin of Zoology, 60*(2), 449–460.

Yang, C. H., Chen, I. S., & Chan, T. Y. (2008). A new slipper lobster of the genus *Petrarctus* (Crustacea, Decapoda, Scyllaridae) from the west Pacific. *Raffles Bulletin of Zoology, 19*, 71–81.

Yang, C. H., Chen, I. S., & Chan, T. Y. (2011). A new slipper lobster of the genus *Galearctus* Holthuis, 2002 (Crustacea: Decapoda: Scyllaridae) from the New Caledonia. *Zoosystema, 32*(2), 207–217.

Yang, C. H., Bracken-Grissom, H., Kim, D., Crandall, K. A., & Chan, T. Y. (2012). Phylogenetic relationships, character evolution, and taxonomic implications within the slipper lobsters (Crustacea: Decapoda: Scyllaridae). *Molecular Phylogenetics and Evolution, 62*, 237–250.

Yang, C. H., Kumar, A. B., & Chan, T. Y. (2017). A new slipper lobster of the genus *Petrarctus* Holthuis, 2002 (Crustacea, Decapoda, Scyllaridae) from Southwest coast of India. *Zootaxa, 4329*(5), 477–486.

Yokoya, Y. (1933). On the distribution of Decapod Crustaceans inhabiting the continental shelf around Japan, chiefly based upon the materials collected by the S.S. Sôyô-Maru, during the year 1923–1930. *Journal of the College of Agriculture Tokyo, 12*, 1–226.

Zarenkov, N. A. (2006). Nephropid lobsters from the Indian Ocean with descriptions of four new species (Crustacea: Decapoda: Nephropidae). *Senkenbergiana maritima, 36*(1), 83–98.

Zarenkov, N. A., & Semenov, V. N. (1972). A new species of the genus *Nephropides* from the South-West Atlantic. *Zoologicheskii Zhurnal, 51*, 599–601.

Lobster Fauna of India

3

E. V. Radhakrishnan, Joe K. Kizhakudan, Lakshmi Pillai S, and Jeena N. S

Abstract

Decapods are a highly diverse group of crustaceans under Phylum Arthropoda, which are distributed throughout a range of depths and latitudes of the world's aquatic ecosystems and terrestrial habitats. Conventional classic taxonomy was mainly dependent upon the morphological features of an organism for species delineation. The interrelationships of the controversial major clades of reptant decapods are resolved through an integrative taxonomic approach combining molecular techniques such as gene sequencing and DNA barcoding with classical morphological features. Molecular analytical tools are increasingly being applied in the current century to delineate the phylogenetic interrelationships of groups within the raptantian decapods.

Marine lobsters are members of the suborder Macrura Reptantia with 4 infraorders, 6 families, 54 genera and 260 extant species (including 4 subspecies) (Chan, Chap. 2). The closely related families of Palinuridae and Scyllaridae share several morphological characters, more specifically their phyllosoma larvae. The lobster fauna of India is so diverse that 38 species belonging to three infraorders (Astacidea, Achelata and Polychelida) and five families are distributed in the seas surrounding the Indian subcontinent. The southern coast (both the west and east coasts) of India is species rich as the diverse habitats and the physical environment favour the settlement of a wide variety of species. Though the palinurid lobsters and the scyllarid *Thenus unimaculatus* are distributed almost throughout the Indian coast, some distinctive species has preference to certain regions. A new deepwater species of scyllarid *Petrarctus jeppiari* was recently added to the scyllarid fauna of India.

E. V. Radhakrishnan (✉) · J. K. Kizhakudan · Lakshmi Pillai S · Jeena N. S
ICAR-Central Marine Fisheries Research Institute, Cochin, Kerala, India
e-mail: evrkrishnan@gmail.com

© Springer Nature Singapore Pte Ltd. 2019
E. V. Radhakrishnan et al. (eds.), *Lobsters: Biology, Fisheries and Aquaculture*,
https://doi.org/10.1007/978-981-32-9094-5_3

Keywords

Lobster fauna · Distribution · Indian subcontinent · Palinurid · Scyllarid · *Petrarctus jeppiari*

3.1 Introduction

Lobsters are the most highly priced crustaceans, with a good market and great demand around the world fetching foreign exchange for the country. Palinurid lobsters are known by different names in different countries: 'ebi' in Japan, 'langouste' for spiny lobsters in France and crayfish/crawfish in UK and South Africa (Patek et al. 2006). Scyllarid lobsters, on the other hand, have a variety of interesting names: slipper lobsters, shovel-nosed lobsters, bugs, locust lobsters, butterfly lobsters and more in various languages (Webber and Booth 2007). The traditional classification mostly depends upon the morphological comparisons of organisms whereas molecular systematics provides a powerful tool to study phylogenetic interrelationships among organisms (Palero et al. 2009). Taxonomic descriptions of almost all the living marine lobster species up to 1991 were published in a monumental work published by FAO: FAO Species Catalogue Vol. 13, *Marine Lobsters of the World* (Holthuis 1991). The suborder Macrura Reptantia includes the superfamilies Nephropoidea (clawed lobsters), Palinuroidea (spiny and slipper lobsters), Eryonoidea (blind lobsters) and the living fossil Glypheoidea under the traditional classification. Phillips et al. (1980) did not consider reef lobsters of the genus *Enoplometopus* as lobsters. However, Thalassinidea was included among the lobsters in the marine lobster catalogue by Holthuis (1991). Application of molecular markers in phylogenetic analysis brought out dramatic changes in the taxonomy of lobsters (Ptacek et al. 2001; Tsang et al. 2008; Toon et al. 2009; Palero et al. 2009; Lavery et al. 2014; Singh et al. 2017).

The family Scyllaridae Latreille, 1825, most closely related to the family Palinuridae belonging to the infraorder Achelata (formerly Palinura), is quite unique in morphology, having a dorso-short and flattened antennae in comparison with a cylindrical antenna, bearing a long multi-articulated and whip-like flagellum) (Holthuis 1985). Four subfamilies (Arctidinae, Ibacinae, Scyllarinae and Theninae) were created under the family Scyllaridae, accounting for almost all the species recognized at that time. Much of the present classification of the family Scyllaridae was formalized by Dr. Lipke Holthuis of the Netherlands (Lavalli and Spanier 2007). Holthuis (2002) subdivided the single genus *Scyllarus* of the subfamily Scyllarinae into 13 new genera each with a suffix 'arctus' to indicate the close relationship between these genera. A key to identify the Indo-Pacific genera of Scyllarinae and the species coming under each genus with photographs was also provided. The family Scyllaridae now has 19 genera and 92 species under 4 subfamilies.

The lobsters belonging to the family Polychelidae are often termed as 'deepsea blind lobsters' and the world fauna is represented by 38 extant species. The systematics of this family, especially at the genus level, is still under consideration (Ahyong and Brown 2002; Ahyong and Chan 2004; Ahyong and Galil 2006).

Recent advances in morphological and molecular phylogenetic investigations have helped to resolve taxonomic ambiguity and knowledge in evolutionary interrelationships in lobsters (Bracken-Grissom et al. 2014). The infraorder Palinuridea is no more in existence under the new classification system, and the infraorder Achelata is used for the species of the superfamily Palinuroidea Latreille, 1802, which is now becoming non-existent (Chan 2010). The family Synaxidae is a junior synonym of Palinuridae, based on the molecular phylogenetic analysis of Achelata and Palinuridae (Palero et al. 2009; Tsang et al. 2009). The suborder Macrura Reptantia has now been classified into four infraorders: Astacidea, Glypheidea, Achelata and Polychelida. Yang et al. (2012) examined the phylogenetic relationships among 54 species from the family Scyllaridae with a special focus on the species-rich subfamily of Scyllarinae and concluded that the subfamilies are monophyletic, except Ibacinae which has paraphyletic relationships among the genera.

Chan (2010) provided a checklist of 6 families, 55 genera and 248 species (with 4 subspecies) of living marine lobsters. The list has been further updated with 260 species (including 4 subspecies) (Chap. 2 of this book). Although the catalogue of marine lobsters of the world by Holthuis (1991) has included almost all the species recognized at that time, many new species have been added in the last two decades. Few new records have also been added to the Indian list of marine lobsters with a total number of 38 species.

Kathirvel et al. (2007) listed the species of lobsters recorded from Indian seas following the taxonomic scheme of Holthuis (1991). The species under the families Thalassinidae and Callianassidae coming under the infraorder Thalassinidea have been included in their list, whereas these two families are excluded under the suborder Macrura Reptantia according to the revised scheme (Chan 2010). Premkumar and Daniel (1975) described the pattern of distribution of economically important palinurid lobsters in Indian Ocean. Prasad and Tampi (1957, 1959, 1960, and 1961) and Prasad (1983, 1986) analysed the plankton samples collected by various expeditions (IIOE and DANA) and research cruises from the east and west coasts of India and made attempts to assign various stages of phyllosoma larvae to the spiny and scyllarid lobsters distributed in the Indian Ocean. Kathirvel and James (1990) reported the phyllosoma larvae of a few species of lobsters from the western coast off the Andaman Islands. However, the adults of these species from this region are yet to be reported. The presence of these larvae in this region probably indicates the availability of these species in these oceanic islands (Kathirvel and James 1990). Since adults of *Scyllarus batei batei* have been reported from the Arabian Sea (George 1967) and Bay of Bengal (Vaitheeswaran 2015), they have been listed in the lobster fauna. The phyllosoma larva of *P. wieneckii* was recorded from southwest coast of India but not the adults. George and George (1965) reported the occurrence of *P. mossambicus* from the southwest coast of India. However, the species was later reasigned as *P. waguensis* (Holthuis 1991).

The chapter presents the taxonomy of the lobster fauna of Indian waters following the recent classification by Chan (2010). The identifying characters, habitat and biology of each species and the distribution in world oceans and along the Indian coast are dealt with. For detailed information on the taxonomy and evolution of scyllarids, refer Lavalli and Spanier (2007). The taxonomic scheme proposed by Holthuis (1991) is as follows.

3.2 Suborder: Macrura Reptantia Bouvier, 1917

3.2.1 Infraorder: Astacidea Latreille, 1802

Superfamily: Nephropoidea Dana, 1852
Family: Nephropidae Dana, 1852
There are three subfamilies: Neophoberinae, Thymopinae and Nephropinae
Subfamily: Neophoberinae Glaessner, 1969
There is only one genus and two species in this family: *Acanthacaris* Bate, 1888 (2 species)
Subfamily: Thymopinae Holthuis, 1974
There are 4 genera and 16 species under this family: *Nephropides* Manning, 1969 (1 species), *Nephropsis* Wood-Mason, 1873 (13 species), *Thymops* Holthuis, 1974 (1 species), *Thymopsis* Holthuis, 1974 (1 species)
Subfamily: Nephropinae Dana, 1852
There are 5 genera and 25 species under this family: *Eunephrops* S.I. Smith, 1885 (3 species), *Homarus* Weber, 1795 (3 species), *Metanephrops* Jenkins, 1972 (17 species), *Nephrops* Leach, 1814 (1 species), *Thymopides* Burukovsky and Averin, 1977 (1 species)
Family: Thaumastochelidae Bate, 1888
There are 2 genera under this family: *Thaumastocheles* Wood-mason, 1874 (2 species), *Thaumastochelopsis* Bruce, 1988 (1 species)

3.2.2 Infraorder: Palinuridea Latreille, 1802

Superfamily: Eryonoidea De Haan, 1841
Family: Polychelidae Wood-Mason, 1875
The genera and species under this family are not covered due to taxonomic confusion
Superfamily: Glypheoidea Zittel, 1885
Family: Glypheidae Zittel, 1885
There is only one genus under this family: *Neoglyphea* Forest and De Saint Laurent, 1975 (1 species)
Superfamily: Palinuroidea Latreille, 1802
Family: Palinuridae Latreille, 1802
There are 9 genera under this family: *Jasus* Parker, 1883 (subgenus *Jasus* Parker, 1883) (6 species), subgenus *Sagmariasus* nov. (1 Species), *Justitia* Holthuis, 1946 (3 species), *Linuparus* White, 1847 (3 species), *Palinurus* Weber, 1795 (5 species), *Palinustus* A. Milne-Edwards, 1880 (4 species), *Panulirus* White, 1847

(19 species, 4 subspecies), *Nupalirus* Kubo, 1955 (2 species), *Puerulus* Ortmann, 1897 (3 species)
Family: Synaxidae Bate, 1881
There are 2 genera under this family: *Palibythus* Davie, 1990 (1 species), *Palinurellus* von Martens, 1878 (2 species)
Family: Scyllaridae Latreille, 1825
Subfamily: Arctidinae Holthuis, 1985
There are 2 genera under this subfamily: *Arctides* Holthuis, 1960 (3 species), *Scyllarides* Gill, 1898 (13 species)
Subfamily: Ibacinae Holthuis, 1967
There are 2 genera under this subfamily: *Ibacus* Leach, 1815 (6 species, 2 subspecies), *Parribacus* Dana, 1852 (6 species)
Subfamily: Scyllarinae Latreille, 1825
There is only 1 genus under this family: *Scyllarus* Fabricius, 1775 (40 species)
Subfamily: Theninae Leach, 1815
There is only 1 genus under this family: *Thenus* Leach, 1815 (1 species)

3.2.3 Infraorder: Thalassinidea Latreille, 1831

Family: Thalassinidae Latreille, 1831
There is only 1 genus in this family: *Thalassina* Latreille, 1806 (1 species)
Family: Upogebiidae Borradaile, 1903
There is only 1 genus in this family: *Upogebia* Leach, 1814 (5 species)
Family: Callianassidae Dana, 1802
There is only 1 genus in this family: *Callianassa* Leach, 1814 (9 species)

3.3 The Taxonomic Scheme of Chan

The new taxonomic scheme, *The Annotated Checklist of World's Marine Lobsters* (Crustacea: Decapoda: Astacidea: Glypeidea: Achelata: Polychelida), of Chan (2010) incorporated the results of the phylogenetic analyses of lobsters (Chan and Ng 2008; Tsang et al. 2008, 2009; Palero et al. 2009; Bracken-Grissom et al. 2009; Ahyong 2009). Following this publication, several new species have been described and the taxonomic status of some species was revised based on molecular sequencing of DNA. The list after suggested changes recognizes 260 extant valid species (including the four valid subspecies under the genus *Panulirus*) of marine lobsters in 6 families, 4 subfamilies and 54 genera (for more information, refer Chap. 2). The taxonomic scheme of Chan (2010) is followed here.

The taxonomy and distribution of species of lobsters from Indian waters (Arabian Sea and Bay of Bengal) and the two island systems, the Lakshadweep group of Islands (formerly Laccadives) and Andaman and Nicobar group of Islands (A&N Islands), based on taxonomic records (Heller (1865), Wood-Mason (1872), Thurston (1887 1894), Alcock and Anderson (1894), Alcock (1901,1906), Borradaile (1906),

Lloyd (1907), Powell (1908), Sewell (1913), De Man (1916), Balss (1925), Gravely (1927, 1929), Patil (1953), John and Kurian (1959), Satyanarayana (1961), Chhapgar and Deshmukh (1961, 1964), Kurian (1963,1965,1967), George and Rao (1965a,b), George and George (1965), Jones (1965), Deshmukh (1966), Radhakrishna and Ganapati (1969), Chekunova (1971), Nair et al. (1973), Hossain (1975), Meiyappan and Kathirvel (1978), Thomas (1979), Lalithadevi (1981), Parulekar (1981), Tikader and Das (1985), George et al. (1986), Tikader et al. (1986), Mustafa (1990), Radhakrishnan et al. (1990, 1995, 2011), Holthuis (1991), Srikrishnadhas et al. (1991), Goswamy (1992), Ali et al. (1994), Kathirvel (1998), Thiyagarajan et al. (1998), Kizhakudan (2002), Kathirvel and Nair (2002), Kizhakudan and Thirumilu (2006), Lakshmi Pillai and Thirumilu (2006), Dineshbabu (2008), Kizhakudan et al. (2012a, b), Jeena (2013), Chakraborty et al. (2014), Purushottama and Saravanan (2014), Radhakrishnan and Jayasankar (2014), Vaitheeswaran (2015) and Yang et al. (2017) are described here. The lobster fauna of India is represented by 5 families, 4 subfamilies, 20 genera and 38 species.

Taxonomic Status
Phylum: Arthropoda
Subphylum: Crustacea
Class: Malacostraca
Subclass: Eumalacostraca
Superorder: Eucarida
Order: Decapoda
Suborder: Macrura Reptantia

3.4 Infraorder: Achelata Scholtz and Richter, 1995

3.4.1 Family: Palinuridae Latreille, 1802 [Palinurini]

The lobsters under the family Palinuridae are currently divided into two groups, namely the 'Stridentes' and the 'Silentes', on the basis of the presence or absence of a stridulating organ. The palinurid list consisting of the following 12 genera, is recognized by George and Main (1967):

Silentes Group: *Jasus* and *Projasus* (not available in India).
Stridentes Group: ***Panulirus**, Palinurus, **Linuparus**, Nupalirus, Palibythus, Palinurellus, **Puerulus**, Justitia, Sagmariasus* and ***Palinustus*** (those genera in bold letters are represented in Indian waters).
Diagnosis: Rostrum either absent or visible as a small spine on mid-dorsal margin of carapace. A pair of frontal horns on carapace above the eyes with spines on the dorsal surface; few and scattered hairs on carapace (Holthuis 1991).
Key to families (after Holthuis 1991).
1a. Antennal flagellum modified into a single, flat plate: Scyllaridae
1b. Antennal flagellum long and whip-like or spear-like with numerous small articles

2a. Rostrum either absent or visible as a small spine on mid-dorsal margin of carapace. A pair of frontal horns on carapace above the eyes with spines on the dorsal surface; few and scattered hairs on carapace: Palinuridae.

Key to Genera Occurring in the Area
Two distinct tooth-like frontal horns on anterior margin of carapace, widely separated; antennal flagella flexible; long, whip-like antennules; antennular plate and stridulating organ present: *Panulirus*.

A single tooth-like frontal horn on anterior margin; the pleura of second to fifth abdominal segments ending in two teeth; antennular plate and stridulating organ present; strongly ridged carapace: *Puerulus*.

Frontal horns fused to form a quadrangular median process; antennal flagella inflexible and straight: *Linuparus*.

Narrow, unarmed frontal plate; major supraorbital processes ending in a blunt crenulated margin; anterior margin of carapace with two spines between the supraorbital processes; first peduncular joint of antennae extending beyond end of peduncle of antennules: *Palinustus*.

3.4.1.1 Genus *Panulirus* White, 1847
Twenty-four species (including four subspecies) within this genus in tropical and subtropical waters of the Indian, Pacific and Atlantic Oceans are currently recognized (Chan, refer Chap. 2). *P. homarus megasculpta* lost the subspecies status following molecular genetic evaluation by Lavery et al. (2014). Morphological and molecular studies confirmed that *Panulirus argus* distributed off the Brazilian coast is a new species and was recognized as *Panulirus meripurpuratus* (Giraldes and Smyth 2016). Six species under this genus occur along the Indian coast.

Diagnosis: Two distinct, widely separated tooth-like frontal horns visible between the anterior margin of the carapace; antennal flagella quite flexible; long, whip-like antennule flagella; longer than the peduncle of antennule.

Key to species of *Panulirus* (George and Main 1967) recorded off the Indian coast and the Andaman Nicobar Islands and the Lakshadweep Islands

1. Transverse grooves on abdominal segments 2–5 2
   No transverse grooves on abdominal segments 2–5 or
   with indistinct grooves in juveniles only
   4
2. Squamae varying from minute to well developed on margin of transverse abdominal grooves. Colour brownish-red in specimens with large squamae to olive green in specimens with minute squamae: *P. homarus*
   No squamae on margin of transverse abdominal grooves 3
3. Antennular plate with two pairs of subequal principal spines fused at their bases. Overall colour olive-black: *P. penicillatus*
   Antennular plate with one pair of equal spines; overall colour purplish-red with abdomen covered with conspicuous white spots: *P. longipes*

Antennular plate with one pair of equal spines, each abdominal segment with white bands. Legs with white spots. Colour olive green: *P. polyphagus*
4. Posterior margin of each abdominal segment with conspicuous transverse white band. Longitudinal white stripes on each leg. Juveniles have white antennae. Overall colour black and green: *P. versicolor*

Abdominal segments with no transverse white band but a conspicuous white spot above each pleural spur. Irregular transverse mottling on each leg with no longitudinal stripes. Overall colour bluish green: *P. ornatus*

3.4.1.1.1 *Panulirus homarus* (Linnaeus, 1758) [*Cancer homarus*]

Panulirus dasypus (H. Milne Edwards, 1837) [*Palinurus dasypus*]
Panulirus burgeri De Haan, 1841) [*Palinurus burgeri*]
English name: Scalloped spiny lobster
Diagnosis: Rostrum absent; antennae cylindrical and longer than body; anterior margin of carapace with two frontal horns; antennular plate with four equal, large well-separated spines arranged in a square; pleura of second to fifth abdominal segments ending in a strong tooth with denticles on posterior margin. Transverse groove on each abdominal segment, interrupted or complete in the middle with anterior margin formed into shallow or deep and large scallops. Legs 1–4 without pincers (Holthuis 1991).

Four subspecies of *P. homarus* have been considered based on morphological differentiation. Lavery et al. (2014) using mtDNA and nDNA sequencing showed two of the subspecies *P. homarus rubellus* Berry, 1974 and *P. homarus* 'Brown' to belong to genetically distinct lineages. However, the taxonomic status of *P. homarus* 'Brown' is still not clear and this 'subspecies' may be an allopatric population (Lavery et al. 2014). Farhadi et al. (2017) based on molecular data, proposed that this species may be considered as an isolated and distinct population rather than a subspecies. The study also revealed that the subspecies *P. homarus megasculpta* (Pesta 1915) [*Panulirus burgeri megasculpta*] and *P. homarus homarus* (Linnaeus 1758) are genetically indistinguishable and since there is no evidence of reproductive isolation between these two, the latter has been synonymized with the more widely distributed and genetically distinct *P. homarus homarus*. Further, *P. homarus homarus* has not diverged at all from *P. homarus megasculpta*. However, different morphotypes of the 'megasculpta' form exist in the western Indian Ocean, the spotted form (Fig. 3.1d) from the Gulf region and the non-spotted with deep abdominal sculpta from eastern South Africa (Fig. 3.1a). Therefore, *P. homarus homarus* (Fig. 3.2a) and *P. homarus rubellus* (Fig. 3.2c) are the only two subspecies currently recognized.

The subspecies in India is *P. homarus homarus*, which is the 'microsculpta' form having a medially interrupted abdominal transverse groove with small scallops (Jeena 2013). Based on molecular analytical data, Singh et al. (2017) suggested that *P. homarus rubellus* (Fig. 3.2c) is a separately evolving lineage and, therefore, might even be considered as a separate species.

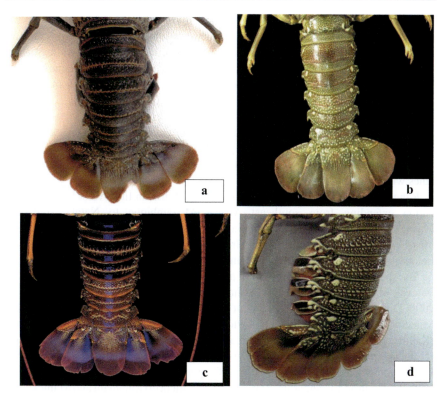

Fig. 3.1 (**a**) The abdominal region: *Panulirus homarus homarus* (megasculpta form), South Africa (**b**) *Panulirus homarus homarus* (microsculpta form) subspecies, India (**c**) *Panulirus homarus rubellus* (megasculpta form) subspecies, South Africa (**d**) *Panulirus homarus homarus* (megasculpta form) Sultanate of Oman

Fig. 3.2 (**a**) *Panulirus homarus homarus* (microsculpta form) (Photo by Abhilash, CMFRI). (**b**) *Panulirus homarus homarus* (megasculpta form) (Photo by Joan Groeneveld, South Africa). (**c**) *Panulirus homarus rubellus* (Photo by Tin Yam Chan, National Taiwan University)

3.4.1.1.1.1 *Panulirus homarus homarus* (Linnaeus, 1758) (Fig. 3.2a)

Diagnosis: Transverse groove on the abdomen with poorly developed squamae along the anterior margin in at least one or sometimes in up to four segments. These specimens may be referred to as the 'microsculpta' form. Antennular plate with four equal, well-separated and large spines. Exopod of third maxilliped absent (Holthuis 1991).

Colour of the first three abdominal segments dark greenish to blackish with numerous small white spots (especially distinct on posterior half of abdomen and larger than the megasculpta form); indistinct white spots and stripes on legs; antennules with white bands.

Range distribution: *Panulirus homarus homarus* has an Indo-West Pacific distribution extending from southeast Africa (Mozambique to Natal), southeast Madagascar, Tanzania, Kenya, Somalia) to Japan including, the Middle East countries (Yemen, Saudi Arabia, Oman, Iran), Pakistan, India, Sri Lanka, Malaysia, Indonesia, Australia, New Caledonia and Japan (Holthuis 1991).

The species has a wide distribution along the entire Indian coast extending from Veraval, Mumbai (Chhapgar and Deshmukh 1961), Netrani Island (Zacharia et al. 2008; Dineshbabu et al. 2011), Calicut (George 1973; Radhakrishnan et al. 2000), Minicoy, Lakshadweep (Meiyappan and Kathirvel 1978; Pillai et al. 1985), Cochin (Kathirvel 1975), Quilon (Balasubramanyan et al. 1961), Vizhinjam (Kuthalingam et al. 1980) to Kanyakumari on the southwest (Miyamoto and Shariff 1961; Balasubramanyan et al. 1960, 1961; George 1965; Vijayanand et al. 2007; Radhakrishnan and Thangaraja 2008) and from Tuticorin (Rajamani and Manickaraja 1997), Mandapam (Gravely 1927; Nair et al. 1973) to Chennai (Radhakrishnan et al. 1990) on the southeast coast and to Kakinada (Lalithadevi 1981) and Digha on the northeast coast (Ramakrishna et al. 2003), and extending up to Andaman and Nicobar Islands (Shanmugham and Kathirvel 1983; Jha et al. 2007; Kumar et al. 2010) (Figs. 3.3a, 3.3b and 3.3c).

Habitat and ecology: Generally a shallow water-dwelling species (1–15 m); can also be found to depths of 90 m; rocky reefs preferred for shelter (Holthuis 1991).

Biology: Measures a maximum of 31 cm in total length and 12 cm in carapace length. Average total length 20–25 cm. *Panulirus homarus homarus* breeds throughout the year. However, peak spawning period is October–December in the population along the Kanyakumari coast (George 1965; Thangaraja and Radhakrishnan 2012) and December–March in Tuticorin (Jawahar et al. 2014).

Status: The species is commercially exploited along the southern and northwestern Indian coast and is exported in live and frozen form.

Panulirus homarus megasculpta (Pesta 1915) (Fig. 3.2b)

Diagnosis: This is a morphotype of *P. homarus homarus* with uninterrupted transverse grooves on the abdomen and large sculpta on the anterior margin of the groove. The species has been synonymized with *P. homarus homarus* (Lavery et al. 2014).

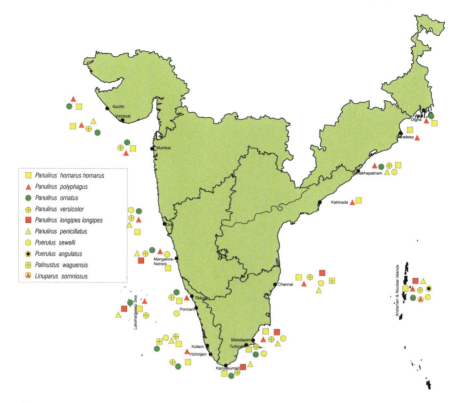

Fig. 3.3a Distribution of lobsters along the Indian coast (*Panulirus* spp., *Puerulus* spp., *Palinustus* sp., *Linuparus* sp.)

Range distribution: Distributed in the northern Arabian Sea (Yemen, Saudi Arabia, Oman).

Panulirus homarus rubellus Berry, 1974 (Fig. 3.2c)

Diagnosis: The abdominal transverse grooves are continuous with larger rounded squamae all along their margins. These specimens may be referred to as 'megasculpta form'.

Colour: Dorsal surface of first three abdominal segments brick-red Minute, pale yellow spots along the posterior margin of all segments.

Range distribution: *P. homarus rubellus* is distributed along the southeast coast of Africa, Mozambique to Natal and southeast Madagascar.

3.4.1.1.2 *Panulirus polyphagus* (Herbst, 1793) [*Cancer (Astacus) polyphagus*] (Fig. 3.4)

Panulirus fasciatus (Fabricius 1798) [*Palinurus fasciatus*]
Panulirus orientalis Doflein, 1900

Diagnosis: Abdominal somites naked, smooth and without transverse grooves; no exopod on third maxilliped; the antennular plate with two strong spines; colour

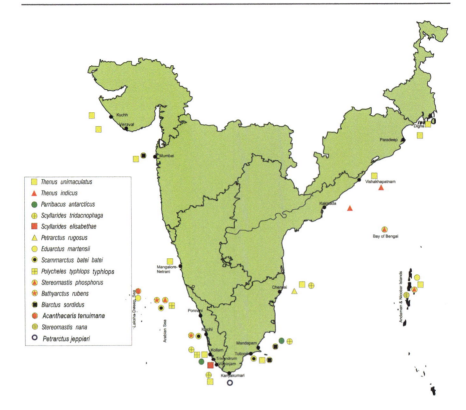

Fig. 3.3b Distribution of lobsters along the Indian coast (*Thenus* spp., *Parribacus* sp., *Scyllarides* sp., *Eduarctus* sp., *Scammarctus* sp., *Stereomastis* sp., *Polycheles* sp., *Bathyarctus* sp., *Biarctus* sp., *Petrarctus* spp., *Acanthacaris* sp.)

greyish-green with transverse white bands on posterior margin of the abdominal somites 2–5. Legs irregularly spotted (Holthuis 1991).

Range distribution: Wide distribution from Pakistan to Vietnam including India, Sri Lanka, the Philippines, Malaysia, Indonesia, northwest coast of Australia and the Gulf of Papua (Holthuis 1991).

In India, the species has been recorded from almost all along the coast from Saurashtra and Bhavnagar coast of Gujarat (Kagwade et al. 1991; Radhakrishnan et al. 2005), Mumbai (Chhapgar and Deshmukh 1961; Deshmukh 1966; Kagwade et al. 1991), Netrani Island in Karnataka (Zacharia et al. 2008; Dineshbabu et al. 2011), Calicut (George 1973; Radhakrishnan et al. 2000), Cochin (Balasubramanyan 1967; Rao and Kathirvel 1971) and Vizhinjam (Balasubramanyan et al. 1961; Kuthalingam et al. 1980) in Kerala, Mandapam and Chennai in Tamil Nadu (Nair et al. 1973; Kagwade et al. 1991) to Kakinada, Andhra Pradesh (Lalithadevi 1981), to Odisha and Digha, West Bengal (Ramakrishna et al. 2003) on the northeast coast (Chopra 1939). They exhibit a discontinuous distribution along the Tamil Nadu coast with single specimen records on

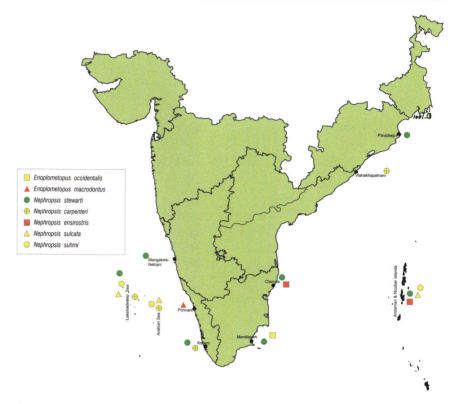

Fig. 3.3c Distribution of lobsters along the Indian coast (*Enoplometopus* spp., *Nephropsis* spp.)

the southwest coast and the Gulf of Mannar, but form a small component of the fishery along the Chennai coast (Fig. 3.3a).

Habitat and ecology: Commonly found in coastal waters on muddy and rocky substrates at a depth of 40 m; occasionally seen at 90 m and is often found near river mouths (Holthuis 1991).

Biology: This species contributes to nearly three-fourth of the total lobster catch of the country. Commercially exploited on the northwest and northeast coasts of India. The size in the fishery ranges from 75 mm to 385 mm total length (TL). In Maharashtra, those between 160 mm and 230 mm TL form the main component of the fisheries (Kagwade 1987). Initial growth rate identical in juveniles of both sexes; measures 85 mm TL in the first year, 145 mm TL in the second year and 205 mm TL in the third year; faster growth rate observed in males. Females attain 50% maturity at 175 mm TL (Kagwade 1988). Peak breeding is in September. High exploitation ratio of 0.85 and 0.82 in males and females, respectively, has resulted in recruitment overfishing in Mumbai waters (Radhakrishnan et al. 2005).

Status: Commercially exploited along the northwest and northeast coasts of India; exported in whole-cooked frozen form.

Fig. 3.4 *Panulirus polyphagus* (Photo by Joe K Kizhakudan, CMFRI)

3.4.1.1.3 *Panulirus ornatus* (Fabricius, 1798) [*Palinurus ornatus*] (Fig. 3.5)

English name: Ornate spiny lobster

Diagnosis: Smooth and naked abdominal somites with colour varying from brownish or greenish-grey with utmost minute indistinct speckles. The large eyespot in the anterior half near the base of the pleura is accompanied by an oblique pale streak placed somewhat median of the eyespot; legs with very sharply defined irregular dark spots of bluish or brownish colour. Antennal flagella distinctly ringed (Holthuis 1991).

Range distribution: Indo-west Pacific ranging from South Africa, coast of East Africa and the Red Sea to southern Japan including India, Sri Lanka, Malaysia, Vietnam, Indonesia, the Solomon Islands, Papua New Guinea, Australia, New Caledonia and Fiji (Holthuis 1991).

The species is widely distributed along the Indian coast from Porbander, Mumbai, Goa, Netrani Island (Zacharia et al. 2008; Dineshbabu 2008), Calicut (Radhakrishnan et al. 2000), Vizhinjam (Kuthalingam et al. 1980), Quilon, Kanyakumari district coast (Vijayanand et al. 2007; Radhakrishnan and Thangaraja 2008), Tuticorin (Rajamani and Manickaraja 1997), Mandapam (Nair et al. 1973), Chennai (Radhakrishnan and Thangaraja 2008; Kagwade et al. 1991), the West Bengal coast (Ramakrishna et al. 2003) and the Andaman and Nicobar Islands (Shanmugham and Kathirvel 1983; Kumar et al. 2010) (Fig. 3.3a).

Fig. 3.5 *Panulirus ornatus* (Photo by Abhilash, CMFRI)

Habitat and ecology: Juveniles found in shallow, sometimes slightly turbid coastal waters; from 1 to 8 m depth, with adult records from depths as great as 50 m; on sandy and muddy substrates and sometimes on rocky bottom often near the mouth of rivers, but also on coral reefs. The species has been reported as solitary but has also been found in larger concentrations.

Biology: Largest and fast growing among the *Panulirus* species; maximum total body length of about 50 cm, but smaller sizes in the fishery (25–30 cm). Fishing both by trawlnets and gillnets and forms a major component of the trawler catch along the east coast of India. Although the species appears throughout the year, highest catch is in May at Tuticorin. The length range of lobsters in the fishery ranges from 113 to 233 mm TL in males and 128 to 452 mm TL in females with 41% falling in the size range of 181–190 mm TL (Rajamani and Manickaraja 1997). At Tuticorin, the inshore fishery for juvenile *P. ornatus* is reported to be detrimental to the stock. Occasionally found along the west coast of Kanyakumari district and forms a small fishery at Tikkoti, Calicut. Breeding population is mostly in deeper waters (40–60 m). Females attain maturity at 90 mm CL. Fecundity of Chennai population (104.4 mm to 145.1 mm CL) ranges from 5,18,181 to 19, 79,522 eggs (Vijayakumaran et al. 2012).

Status: Forms small-scale fishery along the southeast coast of India and exported in live and frozen form.

Fig. 3.6 *Panulirus versicolor* (Photo by Abhilash, CMFRI)

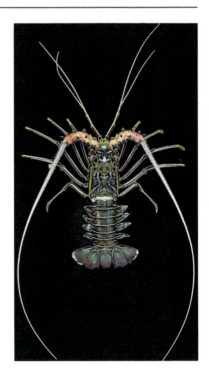

3.4.1.1.4 *Panulirus versicolor* (Latreille, 1804)
[*Palinurus versicolor*] (Fig. 3.6)

English name: Painted spiny lobster

Diagnosis: Four strong spines on the antennular plate are arranged in a quadrangle. Carapace bluish-black in colour; antennal peduncles pink; antennular flagella white; white transverse bands on abdominal somites 2–5; streaks of white lines on legs (Holthuis,1991).

Range distribution: Throughout the Indo-west Pacific region (India, Sri Lanka, Micronesia, Melanesia, Polynesia and northern Australia) and east coast of Japan (Holthuis 1991).

Along the Indian coast, the species has been reported from Veraval, Gujarat (Kizhakudan 2002), Mumbai, Maharashtra (Chhapgar and Deshmukh 1971), Netrani Island, Karnataka (Zacharia et al. 2008; Dineshbabu et al. 2011), Lakshadweep Islands (Rao et al. 1989; Pillai et al. 1985) Calicut (George 1973; Radhakrishnan et al. 2000); Vizhinjam (Kuthalingam et al. 1980); Kanyakumari (Vijayanand et al. 2007), Tuticorin (Rajamani and Manickaraja 1995), Mandapam (Nair et al. 1973) and Chennai (Tamil Nadu) (Radhakrishnan 2013). The species has been recorded from both island systems, Lakshadweep (Rao et al. 1989; Pillai et al. 1985) and Andaman and Nicobar Islands (Shanmugham and Kathirvel 1983; Kumar et al. 2010) (Fig. 3.3aa).

Fig. 3.7 *Panulirus penicillatus* (Photo by R. Thangaraja)

Habitat and ecology: Found mostly on the seaward edge of the coral reef plateau where it utilizes the reef and rocks for shelter; occupies shallow waters to a maximum depth of 15 m (Holthuis 1991). Furthermore, they are nocturnal and only aggregate in very small numbers (Frisch 2007).

Biology: Small-scale fishery reported along the Chennai, Mandapam, Tuticorin and Trivandrum coasts. In Andaman and Nicobar Islands, the species formed 26% of total landings (0.12 t) in 1999–2000 (Kumar et al. 2010). The fecundity of Chennai population (66.0–95.0 mm CL) was estimated to range from 1,70,212 to 7,33,752 eggs (Vijayakumaran et al. 2012).

Status: Forms a small percentage of the total lobster catch and mainly caught along the southeast coast of India.

3.4.1.1.5 *Panulirus penicillatus* (Olivier, 1791)
[*Astacus penicillatus*] (Fig. 3.7)

English name: Pronghorn spiny lobster

Diagnosis: Body colour ranging from yellowish-green to reddish through brown-green to blue-black depending on the habitat. Four strong spines on the antennular plate fused at the base, the anterior pair shorter than the posterior; uninterrupted transverse grooves on the abdomen. Exopod of third maxilliped present with flagellum (Holthuis 1991).

Range distribution: Widely distributed both in the Indo-West Pacific and East Pacific regions from the Red Sea to East Africa, Madagascar and surrounding islands, through the Indian Ocean and South China Sea to Japan, the Philippines, India, Indonesia, Hawaii, Samoa, northern and eastern Australia and as far east as the islands of northwest coast of the USA and Mexico (Holthuis 1991).

Along the Indian coast, the species has been recorded from Gujarat, Daman & Diu, Goa, Netrani Island, (Zacharia et al. 2008; Dineshbabu et al. 2011), Quilon (Satyanarayana 1961), Vizhinjam (Kuthalingam et al. 1980), Mandapam (Nair et al. 1973) and Chennai (Radhakrishnan et al. 1990). The species also has been recorded from Lakshadweep (Pillai et al. 1985; Rao et al. 1989; Meiyappan and Kathirvel 1978) and Andaman & Nicobar Islands (Shanmugham and Kathirvel 1983; Kumar et al. 2010) (Fig. 3.3a).

Habitat and ecology: Commonly inhabits rocky substrates at depths of 1–4 m (maximum 16 m) (Chan 1998). Occupies the outer reef slopes, subtidal zone or surge channels (Holthuis 1991); breeding all year round in the Western Pacific (Chan 1998). Form fishery throughout the year on the southeast coast of Taiwan. While the larger mature females reproduce at least three times a year, smaller females spawn a minimum once a year (Chang et al. 2007). An artisanal fishery for this species exists in the Solomon Islands (Richards et al. 1994). In Saudi Arabia, the species supports a semi-commercial fishery north of Jeddah to the Gulf of Aqaba (Tortell 2004).

Fig. 3.8 (a) *Panulirus longipes longipes* (Photo by Abhilash, CMFRI). (b) *Panulirus longipes bispinosus* (Photo by Tin Yam Chan, National Taiwan University)

Biology: Scanty information is available on the biology of the species as there is only occasional capture from the Indian coast. There is little demand for the species in the live lobster export market.

Status: Few specimens are found in lobster catches along the southeast coast of India.

3.4.1.1.6 *Panulirus longipes* (A. Milne-Edwards, 1868) [*Palinurus longipes*]

English name: Longlegged spiny lobster

There are two valid subspecies, *Panulirus longipes longipes* (A.Milne-Edwards, 1868), the Indian Ocean spotted-leg form (Fig. 3.8a) and *P. longipes bispinosus* Borradaile, 1899, the Pacific form with striped legs (Fig. 3.8b) (Chan and Ng 2001). Antennular flagella are banded in both the species. The name for the species with the white inner antennular flagellum should be *P. femoristriga* (von Martens, 1872), with *P. albiflagellum* Chan and Chu, 1996 as its junior subjective synonym (Chan and Ng 2001). The species found along the Indian coast is *P. longipes longipes*.

3.4.1.1.6.1 *P. longipes longipes* (A. Milne-Edwards, 1868) (Fig. 3.8a)

Diagnosis: The dark-purple abdomen is covered with numerous distinct round spots, white spots on legs, abdominal somites behind the transverse groove naked, exopod of third maxilliped present with flagellum.

Range distribution: Indo-West Pacific; East Africa to Thailand, Taiwan, the Philippines, Indonesia, India and Sri Lanka

Along the Indian coast, the species is distributed on the southwest (Dineshbabu et al. 2011; George and Rao 1965a, Vijayanand et al. 2007), southeast coast of India (Nair et al. 1973; Lakshmi Pillai and Thirumilu 2007) and Andaman and Nicobar Islands (Kumar et al. 2010) (Fig. 3.3a).

Habitat and ecology: The species occupies clear or slightly turbid water at depths of 1–18 m (also reported from 122 m) in rocky areas and coral reefs. The animals are nocturnal and not gregarious (Holthuis 1991).

Biology: Occasionally landed as single specimens and therefore not much information is available on the biology of the species from Indian waters. Attain a maximum total body length of 30 cm (average length 20 to 25 cm). The smallest ovigerous female measures 14 cm TL.

Status: Rare and occasionally found among the lobster catch

Panulirus longipes bispinosus Borradaile, 1899 (Fig. 3.8b)

The pereiopods of this subspecies have longitudinal stripes on legs instead of the spots on the Indian Ocean form, *P. longipes longipes*.

The species is known from south Japan through Micronesia to Papua New Guinea, Vanuatu, Fiji, Tonga, Cook Islands, New Caledonia and east coast of Australia; not present in Indian waters.

3.4.1.2 Genus *Puerulus* Ortmann, 1897

Four species were recognized in this genus, all deepwater forms. Five new species were added to the list (Chan et al. 2014). Two species are known from the Indian coast.

Diagnosis: Antennular plate distinct, a stridulating organ present; carapace with a median ridge behind the cervical groove, often with spines or tubercles, but without submedian rows. Antennal flagella much longer than body length, whip-like. Pereopods long, slender; I to IV not chelate. Abdomen with longitudinal and transverse grooves, pleura terminating ventrally in two strong teeth (Holthuis 1991).

Puerulus sewelli forms a commercially important fishery along the southwest and southeast coast of India and *P. angulatus* is reported to occur off the Nicobar Islands.

3.4.1.2.1 *Puerulus sewelli* Ramadan, 1938 (Fig. 3.9)

Diagnosis: Post-orbital spine not elongated; five post-cervical and two or three intestinal teeth along the median keel of carapace. Carapace as long as abdomen excluding the telson. Fifth pereopod of male not chelate. Body is uniformly orange-brown, with distal parts of teeth on carapace and median keel on abdomen whitish. Basal half of antennal flagella orange-brown and distal half whitish (Holthuis,1991).
Range distribution: Western Indian Ocean including the coasts of Somalia, Gulf of Aden, off Pakistan, India and Myanmar

P. sewelli is distributed both on the southwest, southeast coasts and the Andaman and Nicobar Islands in the depth range of 73–450 m (John and Kurian 1959; Kurian 1963, 1965, 1967; Silas 1969; Rao and George 1973; Oommen and Philip 1974; Kathirvel et al. 1989; Anrose et al. 2010) (Fig. 3.3a).

Habitat and ecology: Known from depths between 73 and 450 m on a substrate of coarse sand, hard mud and shells.
Biology: Total body length a maximum of 20 cm and carapace length about 8 cm. Mean total length about 15 cm. The species is commercially exploited along the southwest and southeast coast of India. The fishery is seasonal, commencing from September–October and extending until February–March. The peak period of abundance is in December to March. During 1999–2000, an average 574 t landed at Sakthikulangara, Kerala. The sizes of *P. sewelli* ranged from 76–80 mm to 186–190 mm TL in males and from 71–75 mm to 201–205 mm in females (Radhakrishnan et al. 2005). Overexploitation resulted in resource decline and the current landing is an average 15 tonnes/annum in Kerala (2007–2012).
Status: Commercially exploited from both southwest and eastcoast of India.

Fig. 3.9 *Puerulus sewelli* (Photo by: Abhilash, CMFRI)

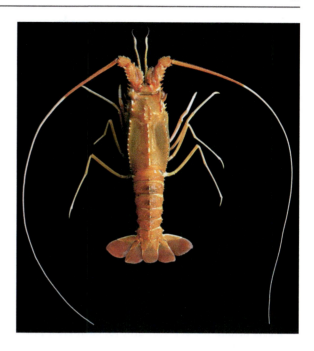

3.4.1.2.2 *Puerulus angulatus* (Bate, 1888)

Diagnosis: Body moderately pubescent; carapace more or less as long as length between abdominal somites 1 and 5; surfaces mostly covered with spinules and sharp granules; supraorbital horn slightly exceeding eye and rarely about middle of antennular plate; dorsal margin generally slightly concave and somewhat crenulate; three or more teeth between the frontal horns and the cervical groove. Eyes smaller and longer than broad; median keel of the carapace with three post-cervical and two intestinal teeth. Pereiopods V not chelate in males (Holthuis 1991).

Colour: Body generally orange-pink with posterior margins of abdominal tergites more or less banded with red. Ventrolateral carapace and sometimes ventral surface of antennal peduncles whitish. Antennal flagellum distinctly banded with orange-pink colour.

Range distribution: Western Indian Ocean, Mozambique, Zanzibar, Somalia, northern Indian Ocean, western Pacific Ocean, Japan, the Philippines, Taiwan, New Guinea, northwestern and eastern Australia, western Tasman Sea.

The species in India is known from the Nicobar Islands, Bay of Bengal (Fig. 3.3a).

Habitat: Deep water, mostly between 200 and 500 m (Chan et al. 2014).
Status: Not a commercial species. Rare.

Though Balss (1925) reported the occurrence of *P. angulatus* off the Nicobar Islands, Chan et al. (2014) have expressed doubts about the occurrence of the species from this region as most reports from this region refer to *P. sewelli* (Kathirvel et al. 1989; Anrose et al. 2010). Further, the species reported from northeast Australia does not belong to the *P. angulatus* group (Chan et al. 2014). However, recently deepsea specimens brought from Port Blair, Andaman and Nicobar Islands were found to be morphologically different from *P. sewelli*. The taxonomic identity of the specimens needs to be confirmed (Radhakrishnan, personal communication).

3.4.1.3 Genus *Linuparus* White, 1847

Linuparus White, 1847, is one of the 12 genera under family Palinuridae (Chan 2010). The genus *Linuparus* comprises a total of four extant species with the addition of a new species, *L. meridionalis* from Australia and Indonesia (Tsoi et al. 2011).

Diagnosis: Frontal horns fused to form a broad two- or four-spined median projection on the anterior margin of the carapace between the eyes; antennal flagella straight, inflexible (Holthuis 1991).

3.4.1.3.1 *Linuparus somniosus* Berry and George, 1972

English Name: African Spear lobster

Diagnosis: Carapace angular dorsally, with one median and two lateral longitudinal crests behind the cervical groove, each crest provided with tubercles; submarginal posterior groove of carapace much wider medially than laterally; antennules slightly extends beyond the antennular peduncle, flagella slightly longer than last segment of antennular peduncle; antennular plate very small, covered by a stridulating organ; each abdominal segment with at most one transverse groove and on each side a longitudinal, tuberculate crest over the bases of the pleura; first five segments with a median crest bearing one or two larger and sometimes some smaller tubercles, crest of sixth segment double. Legs 1–4, without pincers. Vestigial pleopods present on first abdominal segment of female (Holthuis 1991).

Habitat and ecology: Commonly found to inhabit flat soft bottom in depths ranging from 100 to 400 m (Tsoi et al. 2011).

Biology: Maximum total body length about 35 cm, carapace length 14 cm; average carapace length about 10 cm. For more information on the species from the Andaman and Nicobar Islands, refer Ali et al. (1994).

Range distribution: Off the east coast of Africa from Kenya to Natal, South Africa, Strait of Malacca as well as in very shallow waters (20–25 m) off southern Java, Indonesia.

In India, the species is distributed in the Andaman and Nicobar Islands (Ali et al. 1994) (Fig. 3.3a).

Status: Economically insignificant due to their relatively low abundance (Holthuis 1991); does not support commercial fishery in Andaman waters.

Fig. 3.10 *Palinustus waguensis* (Photo by Joe K Kizhakudan, CMFRI)

3.4.1.4 Genus *Palinustus* A. Milne-Edwards, 1880

Four species have been recognized. A new species from Taiwan and Japan, *P. holthuisi*, was added (Chan and Yu 1995). One species is known from the Indian seas.

Diagnosis: Frontal horns end in a broad, bluntly truncated and even crenulated plate; a strong spine is present on the outer margin of each horn. First segment of antennular peduncle reaches beyond the antennal peduncle (Holthuis 1991).

3.4.1.4.1 *Palinustus waguensis* Kubo, 1963 (Fig. 3.10)

English name: Japanese Blunthorn lobster.

Diagnosis: Anterior margin of carapace without a median spine. Anteromedian margin of epistome with tubercles or spinules; anterolateral corner with a small spine or unarmed (Holthuis 1991).

Range distribution: Indo-West Pacific region extending from south Japan to Taiwan, the Philippines, Indonesia, Thailand, Madagascar and India.

Palinustus waguensis is distributed on the southwest and southeast coasts of India, which rarely appear in deep sea and inshore catches (George and George 1965; Kizhakudan and Thiumilu 2006; Lakshmi Pillai and Thirumilu 2006; Chakraborti et al. 2014) (Fig. 3.3a).

Habitat and ecology: Reported from shallow waters in Japan and India. The species has also been caught from depths of 72 and 84 m in India and the Philippines (Holthuis 1991).

Biology: Total body length 5–10 cm. In Chennai, specimens from bottom-set gillnets measured 48–70 mm CL.

Holthuis (1991) opined that the species *P. mossambicus* described by George and George (1965) and George (1973) from southwest coast of India is *P. waguensis*. Deep-sea trawling along the southeast coast of India during 2004 landed about 1.7 t of the species in Chennai (Kizhakudan and Thirumilu 2006). Few specimens were also landed from bottom-set gillnets at Cuddalore (Kizhakudan et al. 2012a) and the Madras Fishing Harbour by indigenous gear (Lakshmi Pillai and Thirumilu 2006). Three specimens were collected from trawl nets at Kollam on the southwest coast of India (Chakraborty et al. 2014).

Status: Generally considered rare. Only few specimens landed along the Indian coast, except for the landing of about 1.7 t of the species at the Fisheries Harbour, Chennai.

3.4.1.5 Family: Scyllaridae Latreille, 1825

The family Scyllaridae originally includes 20 genera distributed in 4 subfamilies, Arctidinae (Holthuis 1985), Ibacinae Holthuis 1985, Scyllarinae Latreille, 1825 and Theninae Holthuis 1985 (Lavalli and Spanier 2007), which underwent considerable revision and now has 19 genera distributed within the 4 subfamilies (Chan, Chap. 2). The subfamily Arctidinae contains two genera *Arctides* Holthuis, 1960 and *Scyllarides* Gill, 1898. Two species under the genus *Scyllarides* have been reported from Indian coast. The subfamily Ibacinae is comprised of three genera (*Evibacus, Ibacus* and *Parribacus*) with a total of 15 species, represented by 1 genus and 1 species in Indian waters. Prasad and Tampi (1957, 1960) described the phyllosoma larvae of scyllarids collected off Mandapam, Cochin and Lakshadweep Sea. The subfamily scyllarinae includes 13 genera, of which 5 genera and 5 species are distributed in Indian waters. One species (total five species) coming under the one genus *Thenus* of the subfamily Theninae is of commercial importance in the Indian context.

3.4.1.5.1 Subfamily Arctidinae Holthuis, 1985

The subfamily Arctidinae has 2 genera and 17 species. A highly vaulted carapace, a three-segmented mandibular palp, and a shallow cervical incision along the lateral margin of the carapace are major characteristics of the species under the genera *Arctides* and *Scyllarides*. Species of *Scyllarides* are not sculptured and lack a transverse groove on the first abdominal somite (Lavalli and Spanier 2007).

3.4.1.5.1.1 Genus *Scyllarides* Gill, 1898

Fourteen species have been so far reported from this genus and two species are known from Indian waters.

Diagnosis: No transverse groove dorsally on the first abdominal somite; postorbital spine on carapace; distinct sculpturation on either side of the median line on abdominal somites.

3.4.1.5.1.1.1 *Scyllarides elisabethae* (Ortmann, 1894) [*Scyllarus elisabethae*]

English name: Cape Slipper lobster

Diagnosis: Distinct cervical and postcervical incisions on lateral margin of carapace; anterior margin of the carapace between the eye and the anterolateral angle evenly concave (Holthuis 1991).

Habitat and ecology: Depth ranges from 37 to 380 m (mostly less than 100 m) on substrate of fine sediments, mud or fine sand. The animals seem to dig into the mud.

Range distribution: Indo-west Pacific region from Mozambique to southeast Africa (Cape Province), Western Indian Ocean and southeast Atlantic.

A single female specimen measuring 120 mm CL, 330 mm TL and weight 740 g was caught by trammel nets off the Vizhinjam coast on the south west coast of India from a depth of 50 m (Thiyagarajan et al. 1998) (Fig. 3.3b).

Status: Rare

3.4.1.5.1.1.2 *Scyllarides tridacnophaga* Holthuis, 1967 (Fig. 3.11)

English name: Clamkiller slipper lobster

Diagnosis: Cervical groove narrow and shallow in its median area with the cardiac knob little pronounced; distinctly two-topped pregastric tooth. Median ridges on second to fourth abdominal somite sharp and distinctly set off from the rest of the surface. Three sharply defined orange spots on the first abdominal somite with the central spot most prominent and as distinct as the laterals (Holthuis1991).

Habitat and ecology: Depth ranges from 5 to 112 m; the species has been observed to open *Tridacna* shells.

Biology: Total body length up to 30 cm; carapace lengths vary between 6 and 12 cm.

Range distribution: Indo-West Pacific region (Red Sea, East Africa (Somalia, Kenya), Gulf of Aden, Pakistan, India and west coast of Thailand).

Along the Indian coast, the species was recorded from Mandapam, Gulf of Mannar (Anon 2010; Radhakrishnan et al. 1995) and Chennai on the southeast coast (Kizhakudan et al. 2012b) and Kollam and Khadiyapatanam on the southwest coast of India (Radhakrishnan personal communication) (Fig. 3.3b).

Status: Rare and recent occurrence of few numbers in trawl fishery.

Fig. 3.11 *Scyllarides tridacnophaga* (Photo by Tin Yam Chan, National Taiwan University

3.4.1.5.2 Subfamily: Ibacinae Holthuis, 1985

Diagnosis: Strongly dorsoventrally compressed carapace with a deep cervical incision along the lateral margin. The mandibular palp is simple or two-segmented (Holthuis 1991).

3.4.1.5.2.1 Genus *Parribacus* Dana, 1852

The genus includes six species. One species occurs in Indian waters.

Diagnosis: Coarsely squamose-tuberculate carapace, with no post-rostral or branchial carinae. Distance between the orbits more than twice as long as the distance between each orbit and the anterolateral angle of the carapace without any postero-median spine on the fifth abdominal somite. Mandibular palp two segmented (Holthuis 1991).

3.4.1.5.2.1.1 *Parribacus antarcticus* (Lund, 1793) [*Scyllarus antarcticus*] (Fig. 3.12)

English name: Sculptured Mitten lobster

Diagnosis: The transverse groove in the first abdominal somite is wide and naked, bearing at most a few hairs and tubercles in the median area; distinct tubercles seen on the anterior part of the second to third abdominal somites and the median carinae of these somites are elevated. The lateral margin of the fourth segment of

Fig. 3.12 *Parribacus antarcticus* (Photo by Tin Yam Chan, National Taiwan University)

the antenna bears six teeth. The two lateral teeth before the cervical incision are of almost equal size (Holthuis 1991).

Habitat and ecology: *Parribacus antarcticus* is an inhabitant of shallow tropical waters, preferably coral or stone reefs with a sandy bottom and taken at depths from 0 to 20 m. The species is nocturnal and in the daytime hides in crevices, sometimes in small groups.

Biology: Carapace lengths between 2 and 9 cm; maximum total length 20 cm. Tastes very good. It is sold as fresh or cooked.

Range distribution: The species is known from both western Atlantic (Florida to northeast Brazil), including the West Indian Islands and from the Indo-West Pacific region, extending from the east coast of south Africa to Japan, Taiwan, Indonesia and India (Holthuis 1985).

Along the Indian coast, the species is recorded from Minicoy and Suheli Par (Meiyappan and Kathirvel 1978; Rao et al. 1989) in Lakshadweep Islands and Gulf of Mannar, southeast coast of India (Radhakrishnan et al. 1995) (Fig. 3.3b).

Status: Not so common in Indian waters.

3.4.1.5.3 Subfamily: Scyllarinae Latreille, 1825

There are 13 genera under this subfamily with 55 recognized species (Chan, Chap. 2). Most species are small and are of no economic value. The subfamily is represented by five genera in India. No flagellum on the exopod of the first and third maxillipeds. The orbit is away from the anterolateral edge of the carapace. The carapace is vaulted and covered in tubercles.

3.4.1.5.3.1 Genus *Biarctus* Holthuis, 2002

The genus includes four species, and one species is known from Indian waters.

Diagnosis: Two median teeth on carapace before the cervical groove. No median carinae on abdomen. Narrow and deep grooves on somites 2 to 5. Abdominal pleurae rounded at the top and directed down. Thoracic sternum deeply V-shaped incised, without additional tubercles in the anterior margin. Pregastric tooth absent. A dark rounded median spot on first abdominal somite (Holthuis 2002).

One species is reported from India. Earlier they were included in the genus *Scyllarus* (Holthuis 1991).

3.4.1.5.3.1.1 *Biarctus sordidus* (Stimpson, 1860) [*Arctus sordidus*]

English name: Pigmy slipper lobster.

Scyllarus tutiensis Srikrishnadhas, Rahman and Anandasekaran, 1991 (Junior synonym).
Holthuis (2002) considers the species reported from Tuticorin as *S. tutiensis* (Srikrishnadhas et al. 1991) has close resemblance to *B. sordidus* in colour and general characteristics. Therefore, the species is considered as a junior synonym of *Arctus sordidus* Stimpson, 1860.

Diagnosis: Cardiac tooth well-developed and triangular. A complete transverse groove on the first abdominal somite, sometimes interrupted in the middle. A deep, triangular V-shaped incision with a median suture and a narrow groove on the anterior end of the thoracic sternum. Any one of the first four thoracic sternites has no median tubercle. Size varies between 10 and 20 mm CL and total length about 56 mm. A prominent circular or oval central dark spot on the first abdominal somite, which is surrounded by a pale ring (Holthuis 2002).
Range distribution: Distributed in Persian Gulf, Bay of Bengal (India, Sri Lanka), Malaysia, Indonesia, Gulf of Thailand, South China Sea, Singapore, Philippines, west coast of Australia (Holthuis 2002).

In India, the species has been recorded near Mumbai coast (Chhapgar and Deshmukh 1964; Sankolli and Shenoy 1974), near Mandapam in Gulf of Mannar (Prasad and Tampi 1961; 1968) and near Tuticorin (Srikrishnadhas et al. 1991) (Fig. 3.3b).

Habitat: Depth mostly between 10 and 20 m. For details refer Holthuis (2002).
Status: Not commonly found in lobster catches from India.

3.4.1.5.3.2 Genus *Bathyarctus* Holthuis, 2002

The genus has six species. One species is known from the Indian coast.

Diagnosis: Pregastric, gastric and cardiac teeth in the midline of carapace. No sharp rostral tooth. Abdominal somites with prominent median carinae, that of the fourth somite slightly to distinctly higher than that of somite III; dorsal surface of abdomen without a clear and sharp arrangement of narrow grooves. Anterior margin of the thoracic sternum truncate or convex, sometimes with a median tubercle, but without median incision; this margin situated on about the same level with anterolateral teeth of the rostrum. Propodus of P1 to P4 often with ventral setae (Holthuis 2002).

3.4.1.5.3.2.1 *Bathyarctus rubens* (Alcock and Anderson, 1894) [*Arctus rubens*]

Scyllarus rubens (Alcock and Anderson 1894) [different generic combination]

English name: Deep sea brown lobster

Diagnosis: Deep sea form. Anterior margin of the sternum truncate or slightly concave. Upper surface of posterior half of abdominal somites 1–4 practically smooth on either side of the transverse groove or with flattened indistinct tubercles. Median carina of abdominal somite IV distinctly higher than that of somite III. Small species. Carapace length 20 mm or less (Holthuis 2002).

Habitat: Depth of occurrence, 183–732m

Range distribution: This species has a wide distribution in the Indo-West Pacific region from Mozambique to the Philippines and New Caledonia including Madagascar, Sri Lanka, Indonesia, Australia, the Chesterfield and Loyalty Islands, Vanuatu and Fiji (Holthuis 2002).

In India, the species has been recorded from the Gulf of Mannar (Prasad and Tampi 1968) and off Cochin, the Arabian Sea (George 1967) (Fig. 3.3b).

Status: Not a regular component of the trawl catches from India.

3.4.1.5.3.3 Genus *Petrarctus* Holthuis, 2002

The genus includes six species and two species are reported from Indian waters.

Diagnosis: Carapace with gastric and cardiac teeth. Tooth absent on rostrum but with small dorsal tubercles. Pregastric tooth absent or replaced by a weak transverse carina or transverse row of tubercles. Longitudinal median carinae on abdominal somites 2–5. A wide transverse median groove on dorsal surface of somites 2 to 5 (Holthuis 2002).

Fig. 3.13 *Petrarctus rugosus* (Photo by Joe K Kizhakudan, CMFRI)

3.4.1.5.3.3.1 *Petrarctus rugosus* (H. Milne Edwards, 1837) [*Scyllarus rugosus*] (Fig. 3.13)

English name: Hunchback locust lobster

Diagnosis: The carapace has the median teeth before the cervical groove blunt and inconspicuous; the rostral tooth is reduced to a tubercle; the gastric tooth is most conspicuous. High tubercles on surface of the carapace. Median longitudinal carina on abdominal somites 2–5 and that of somite 3 is the highest; there is a wide transverse groove in each somite. The dorsal surface of the body is greyish or purplish brown with darker spots. Dark blue colour on the dorsal surface of the first abdominal somite (Holthuis 2002).

Habitat and ecology: Inhabits depths from 20 to 60 m.

Biology: Total body length reported is 2.5–6 cm.

Range distribution: This species is known from the Red Sea to Mozambique, including northern Madagascar, Comoros and Socotra. It is also found in the South China Sea, from Taiwan to the Philippines and Indonesia, Malaysia, Thailand, Vietnam, China, off northern Queensland, Australia (Holthuis 1991). The type locality of the specimen (*Scyllarus rugosus*) is from Pondicherry, southeast coast of India (Holthuis 2002).

In India, the species is distributed along the Chennai and Pondicherry coasts on the southeast coast. Prasad and Tampi (1968) described what they thought to be the

Fig. 3.14 *Petrarctus jeppiari* (Photo by Tin Yam Chan, National Taiwan University)

nisto stage of this species from Lakshadweep Archipelago. Tampi and George (1975), Prasad et al. (1980) and Prasad (1983) described various stages of the phyllosoma presumed to be of this species (Fig. 3.3b).

Status: Often found in trawl catches, probably of ornamental value.

3.4.1.5.3.3.2 *Petarctus jeppiaari* Yang, Kumar and Chan, 2017 (Fig. 3.14)
This is a new species recently recorded from Muttom, Kanyakumari district on the southwest coast of India (Yang et al. 2017).

Diagnosis: High cardiac tooth on carapace and the abdominal somites 3 and 4 with dorsal carinae of equal height (Yang et al. 2008). Morphologically resembles *P. veliger*. However, the colouration is different from *P. veliger* in that the articulated part of the abdominal tergite I lacks a large dark circular median spot. In that place, it bears a thin bright blue stripe somewhat similar to that of *P. rugosus* and *P. brevicornis*. After preservation, the blue line fades quickly with only the dark coloured inverted triangular spot still remaining. The identification of the specimen was confirmed by molecular genetic analysis (Yang et al. 2017).

Range distribution: Presently known only from Muttom, southwest coast of India (Yang et al. 2017) (Fig. 3.3b).

Habitat and ecology: Inhabits 150–200 m.

Biology: The Holotype ovigerous female measures 24.3 mm CL and Paratypes, males, 17.7–25.4 mm CL and females, 19.0–26.2 mm CL.

Status: Recorded from deepsea trawler catches at Jeppiaar Fishing Harbour, Muttom, Kanyakumari district, Tamil Nadu. Commercially not important due to the small size; probably of ornamental value.

3.4.1.5.3.4 Genus *Eduarctus* Holthuis, 2002

The genus has eight species, and one species is known from Indian waters.

Diagnosis: Short and rounded rostrum. Pregastric, gastric and cardiac teeth on carapace; rostral tooth present or absent. With numerous squamiform tubercles on dorsal surface of carapace. A median longitudinal carina on second, third and fourth abdominal somites, but often very high. A median, transverse groove on first abdominal somite. Pleura of the abdominal somites 2 to 5 bluntly rounded. Legs smooth, no hairs on the lower surface of the propodus of any of them (Holthuis 2002).

3.4.1.5.3.4.1 *Eduarctus martensii* (Pfeffer, 1881) [*Scyllarus martensii*] (Fig. 3.15)

English name: Striated locust lobster

Diagnosis: Carapace with two distinct teeth in the median line before the cervical groove, the rostral tooth is absent and replaced by an inconspicuous tubercle. The region between the postrostral and branchial carinae shows many tubercles; the abdomen has a conspicuously elevated longitudinal median carina on somites 2–5, that of somite 2 shows as an inverted V-shaped ridge when looked at dorsally; somite 1 shows a complete transverse groove behind which there are about 16 straight, parallel longitudinal unbranched grooves, which are quite characteristic for the species. The body is yellowish or reddish-brown. A darker brown transverse band may be present on the third abdominal somite. For details, refer Holthuis (2002).

Range distribution: This species is widely distributed in the Indo-West Pacific region. Known from East Africa, Mozambique, the Comoros Islands, northern Madagascar and Seychelles. In the west Pacific, it is known from Thailand, Malaysia, Cambodia, Vietnam, Indonesia, the Philippines, Taiwan and the south of Japan. It is also found in northern Australia and New Caledonia (Holthuis 1991).

Fig. 3.15 *Eduarctus martensi* (Photo by Tin Yam Chan, National Taiwan University)

The species is known from east and west coasts of India, Lakshadweep and Andaman and Nicobar Islands (Prasad and Tampi 1968; Tampi and George 1975; Prasad 1983) (Fig. 3.3b).

Habitat and ecology: The species has been found in depths between 6 and 79 m, mostly between 10 and 50 m (Holthuis 1991). The substrate that it inhabits is smooth, sometimes with shells. The total body length is 2–5 cm.

Status: Not commonly found in commercial trawl catches from India. Since the species is too small and have no commercial value, it is likely that the presence of the specimens in trawl catches may be overlooked.

3.4.1.5.3.5 Genus *Scammarctus* Holthuis, 2002

The genus includes two species, and one species is recorded from Indian coastal waters.

Diagnosis: Carapace with only pregastric, gastric and cardiac teeth in the median line; rostral tooth absent. Abdomen with a slightly elevated and median carina without narrow arborescent grooves. Anterior part of thoracic sternum gutter-like sunken and prolonged forward between the bases of third maxillipeds. No median tubercles on the sternum and no posterior teeth (Holthuis 2002).

Fig. 3.16 *Scyllarus batei batei* (Photo by Tin Yam Chan, National Taiwan University)

3.4.1.5.3.5.1 *Scammarctus batei batei* (Holthuis, 1946) [*Scyllarus batei*] (Fig. 3.16)
English name: Soft locust lobster.

Diagnosis: Carapace with two distinct teeth in the median line before the cervical groove; the rostral tooth is absent. Abdomen with a distinct sharp median carina on somites 1–5. Somite 1 with the transverse groove interrupted in the middle by the median carina; the fourth segment of the antenna has a single, distinct oblique median carina; no median tubercles on the sternites; dactyl of legs 3–5 with dorsal fringes of hair; body pale brown with the ridges and tubercles pale purple or reddish; first abdominal somite brick red in the anteromedian area (Holthuis 2002).

Habitat and biology: Depth ranges from 160 to 484 m on sandy and muddy substrates. Maximum body length about 7 cm.

Range distribution: This species is found in the western Indian Ocean, the western central Pacific and the Coral Sea; it is reported off Africa from Madagascar and Mozambique to the Gulf of Aden, and in the South China Sea from Taiwan to the Philippines and Indonesia (Holthuis 1991). It has also been reported from northern Australia, Fiji and New Caledonia (Holthuis 2002).

In India, the species was recorded from Arabian Sea (southwest coast of India) (George 1967) and Tuticorin, Gulf of Mannar, Bay of Bengal (Vaitheeswaran 2015) (Fig. 3.3b).

Status: Rare.

Fig. 3.17 *Thenus unimaculatus* (Photo by Abhilash, CMFRI)

3.4.1.5.4 Subfamily: Theninae Holthuis, 1985

This monotypic family was revised by Burton and Davie (2007). There is only one genus *Thenus* in the subfamily. Five species have been identified using both morphology and molecular methods. The species so far described as *Thenus orientalis* from the Indian coast is *Thenus unimaculatus* (Radhakrishnan et al. 2013). *Thenus indicus* is present along the Andhra Pradesh coast of India (Jeena 2013).

3.4.1.5.4.1 Genus: *Thenus* Leach, 1816

The genus includes five species, and two species are known from Indian waters.

Diagnosis: Orbits on the anterolateral angle of the carapace. Body strongly depressed. Lateral margin of the carapace with only the cervical incision. No teeth on the lateral margin of the carapace apart from the anterolateral and postcervical. Fifth leg of female without a chela.

3.4.1.5.4.1.1 *Thenus unimaculatus* Burton and Davie, 2007 (Fig. 3.17)

English name: Slipper lobster/Sand lobster

Diagnosis: Purple to black pigmentation blotch on inner surface of merus of second and sometimes third legs, usually large but variable in extent and may be reduced to a narrow streak; purple pigmentation occasionally surrounding eye socket on carapace; outer phase of propodus of P2 having upper-most longitudinal groove bearing obvious setae over atleast proximal half. Merus of third maxilliped with a small spine proximally on inner ventral margin; inner margin of ischium prominently dentate along the entire length. No single morphometric ratios that falls outside the following maximum and minimum values; carapace width (CM1) greater than 1.29 times carapace length (CL); length of propodus of pereiopod 1 (PL1) less than 0.23 times carapace length (CL); length of propodus of pereopod 2 (PL2) greater than 0.39 times carapace length (CL); width of propodus of pereopod 1 (PW1) greater than 0.35 times length (PL1) (Burton and Davie 2007).

Habitat and ecology: Depth ranges from 8 to 70 m, usually between 10 and 50 m; on soft substrate, sand or mud.

Biology: Maximum total body length about 25 cm; often appears as bycatch in trawls; also caught in gillnets. At Kollam, Kerala, peak fishery was observed from November to February. Total length varied between 61 and 230 mm in males and 46 and 250 mm in females. Length at recruitment (Lr) was 48 mm. Absolute fecundity varied from 14,750 to 33,250 mature eggs (Radhakrishnan et al. 2013).

Range distribution: This species is known from a few specimens collected from various locations in the Indo-West Pacific, near Thailand, the United Arab Emirates and Mozambique (Burton and Davie 2007; Lamsuwansuk et al. 2012).

In India, the species is distributed along the entire coast: Saurashtra region, Gujarat (Kizhakudan 2014), Mumbai (Kabli and Kagwade 1996a, b; Radhakrishnan et al. 2005), Kollam (Radhakrishnan et al. 2013), Cochin, Kerala, Kanyakumari district (Vijayanand et al. 2007), Nagapattinam (Murugan and Durgekar 2008), Mudusalodai, Tuticorin (Saha et al. 2009), Chennai in Tamil Nadu (Subramanian 2004; Kizhakudan et al. 2004), Visakhapatnam, Andhra Pradesh (Hossain 1975, 1978, Hossain 1975) and Andaman and Nicobar Islands (Anuraj et al. 2017).

Status: Commercially exploited in India from both east and west coasts of India (Kagwade et al. 1991; Radhakrishnan et al. 2005; Jeena et al. 2015), Andaman and Nicobar Islands (Anuraj et al. 2017).

3.4.1.5.4.1.2 *Thenus indicus* Leach, 1816 (Fig. 3.18)

Diagnosis: No spots on the pereopods and telson; pereopods slender; dorsal profile slightly concave; rostral processes sharp and directed anteriorly and upward; second antennal segment has five teeth; third maxilliped with a spine on the merus and dentition on ischium; morphometric ratio ML3/CL - > 0.45; MW1/CL- < 0.07 (Burton and Davie,2007).

Range distribution: Thailand (Lamsuwansuk et al. 2012), Queensland, Australia (Courtney et al. 2001; Jones 2007).

The species is known from Visakhapatnam, India (Fig. 3.3b) (Jeena 2013).

Fig. 3.18 *Thenus indicus* (Photo by Tin Yam Chan, National Taiwan University)

Habitat and ecology: Inhabitant of relatively shallow, inshore waters with muddy sediment; most abundant between 10 and 30 m.

Habitat: Depth of occurrence, inshore sea close to the coast (Lamsuwansuk et al. 2012).Status: Commercially exploited in Thailand and Australia.

3.5 Infraorder: Polychelida Scholtz and Richter, 1995

Family: Polychelidae Wood-Mason, 1875

The polychelids are large, primitive decapods that inhabit the depths of the world oceans down to 5000 m between latitudes 50°N and 55° S (Galil 2000).

Diagnosis: Carapace with dorsal orbits shallow or deeply incised; U- or V-shaped; eyes reduced, fused to anterior margin of carapace, directed laterally; with distinct median carina anterior to cervical groove; postorbital carinae non-aligned with branchial carinae but terminating distinctly mesial to branchial carinae; cervical and branchiocardiac grooves distinct across carapace, indicated at lateral margins by notches. Abdominal pleuron 2 distinctly larger than, and overlapping, pleura 1 and 3. Uropodal exopod entire, without diaeresis. Telson triangular. Pereopod 1 dactylus tapering distally, as long as pollex (Based on Ahyong 2009).

There are 6 extant genera (*Cardus, Homeryon,* **Pentacheles, Polycheles, Stereomastis** and **Willemoesia**) and 38 species under this family. No commercially important species. The family is represented by four genera (bold) and eight species in Indian waters.

3.5.1 Genus *Stereomastis* Bate, 1888

The genus includes 17 species, of which 11 are known from Indo-Pacific regions (Chan 2010). A new species from Turkey was added (Artuz et al. 2014). However, based on morphology, this species was synonymized with *Polycheles typhlops* for the time being (Chan, see Chap. 2). Three species are reported from Indian waters.

Diagnosis: Carapace distinctly longer than wide. Dorsal orbital sinuses U-shaped. Pollex of major chela without perpendicular spine on inner margin. Anterolateral margin of basal antennular segment rounded, with 1 or 2 anterolateral spines. Maxilliped 3 epipod vestigial. Pereopods 1–5 epipod vestigial. Dactylus and pollex of pereopods 3–4 relatively straight, weakly curved (Ahyong 2009).

3.5.1.1 *Stereomastis phosphorus* (Alcock, 1894) [*Pentacheles phosphorus*]

Polycheles phosphorus (Alcock, 1894) under different generic combination

English name: Pink blind lobster
Diagnosis: Outer proximal margin of basal antennular segment with one spine; median carina of abdominal tergites 1–4 with antrorse spine; dorsum of carapace between branchial and median post-cervical carinae with antrorse spine; branchial carina indicated by row of spines; branchial groove with row of spines. In most specimens, the mid-dorsal carina of the carapace (excluding the rostral spines) has a spinal formula of 1,1,2,1 in front of the cervical groove and 2,2,2 behind the groove. Colour uniformly rose pink except for some grey patches on branchial regions (extending up to the gastric regions) of the carapace (Griffin and Stoddart 1995).
Habitat: Depth of occurrence, 101–1479 m
Range distribution: This species is known from Indo-Pacific; Arabian Sea, Gulf of Mannar, the Bay of Bengal to Sri Lanka, the Andaman Sea and the Laccadive Sea (Griffin and Stoddart 1995; Ahyong and Brown 2002) (Fig. 3.3b).
Status: Moderately abundant.

3.5.1.2 *Stereomastis nana* (Smith, 1884) [*Pentacheles nanus*]

Polycheles nanus (Smith, 1884) [different generic combination]
Polycheles andamanensis (Alcock, 1894) [*Pentacheles andamanensis*]
Stereomastis andamanensis (Alcock, 1894) [*Pentacheles andamanensis*]
Stereomasts grimaldii (Bouvier, 1905) [*Polycheles grimaldii*]

Diagnosis: The mid-dorsal carina of the carapace, behind the rostral spines, has a spinal formula 1,1,2,1 before the cervical groove and 2,2,2 behind it. The spinal formula of the lateral edge of the carapace is 5–6:3–4:6–8. Median carina on abdominal tergite 5 with antrorse spine; antrorse spine on abdominal tergite 3 largest; lyre-shaped carina on abdominal tergite 6 prominently denticulate; basal tubercle on telson pointed (Griffin and Stoddart, 1995).

Habitat: Depth of occurrence, 724–2000 m

Range distribution: This species is found across the globe, from as northerly as Greenland to South Africa (Galil 2000). It is also known from Iceland, the Irish Sea, the Bay of Biscay across the Atlantic to Canada and the USA (Galil 2000). It has also been recorded from West Africa and the Azores, the East China Sea, Japan, Philippines, Indonesia, New Caledonia, Wallis and Fortuna Islands, Vanuatu, New Zealand, Solomon Islands and Australia (Galil 2000; Ahyong and Galil 2006). Additionally this species is known from the southwestern Atlantic, Gulf of Aden and the Arabian Sea (Griffin and Stoddart 1995). The species has also been recorded from Chile (Alvarenga and Cardoso 2014).

Along the Indian coast, the occurrence of the species is reported from the Bay of Bengal and the Andaman and Nicobar Islands (Galil 2000) (Fig. 3.3b).

Status: Moderately abundant.

3.5.1.3 *Stereomastis sculpta* (Smith, 1880) [*Polycheles sculptus*]

Diagnosis: The rostrum is bifid; the orbital notches broad, U-shaped with a single spine on the inner orbital angle. There are two spines on the basal antennular segment. The mid-dorsal carina of the carapace has the spinal formula, 1,2,1 before the cervical groove and 2,2,2 behind it. The spinal formula of the lateral margins of the carapace is 6:3:7–8. There are five spines on the slightly sinuous sub-lateral ridge of the branchial region; the last spine is larger than the rest. Median carina of abdominal segment 1–5 is produced into a spine. There is no spine on the anterior mid-point of the second abdominal pleuron (Griffin and Stoddart 1995).

Habitat: Depth of occurrence, 457–2836 m.

Range distribution: Indo-West Pacific, Indonesia, eastern Australia.

In Indian waters, the species is known from the Andaman Sea.

Status: Rare

3.5.2 Genus: *Polycheles* Heller, 1862

There are nine species under this genus, and one species is reported from Indian waters.

Fig. 3.19 *Polycheles typhlops* (Photo by Tin Yam Chan, National Taiwan University)

Diagnosis: Carapace distinctly longer than wide. Dorsal orbital sinuses V-shaped. Poleex of major chela without perpendicular spine on inner margin. Anterolateral margin of basal antennular segment rounded, with anterolateral spines. Maxilliped 3 epipod vestigial. Pereopod 15 epipod well developed. Dactylus and pollex of pereopods 34 relatively straight, weakly curved (Ahyong 2009).

3.5.2.1 *Polycheles typhlops typhlops* Heller, 1862 (Fig. 3.19)
Polycheles hextii (Alcock, 1894) [*Pentacheles hextii*]
Diagnosis: The carapace is long and narrow. The spinal formula of lateral edges of the carapace is 7–9:5–6:24–26. Colour bright orange, except in parts of carapace, abdominal terga and pleura, which is white. Cheliped with proximal portion of merus white, distal third of merus and carpus orange (Griffin and Stoddart 1995). One (rarely two) rostral spine. Inner basal margin of dorsal orbit spinose; abdominal pleuron 2 trianguloid anteriorly with rounded apex; uropodal exopod ventrally bicarinate (Ahyong 2009).
Habitat: Depth of occurrence 400–600 m
Range distribution: This species is known from the Pacific, Atlantic Ocean, the Mediterranean, Caribbean and Indian Ocean (Galil 2000; Ahyong and Poore 2004). It is present in the Gulf of Mexico, the USA and Bermuda. It is also found around the coast of Africa, including West Africa, Cape Verde Islands, Madagascar, Comoros Islands, the Gulf of Aden and South Africa. In the East China Sea, this species is present off Japan and Taiwan (Galil 2000). It is also

recorded off Australia, Philippines, Indonesia and the Maldives (Galil 2000). Recently this species has also been discovered off the coast of Brazil, Solomon Islands, New Caledonia, Fiji and Tonga (Ahyong and Chan 2004).

This species is also recorded from Arabian Sea, southwest coast of India and Bay of Bengal (Radhakrishnan, personal communication) (Fig. 3.3b).

Status: Rare

3.5.3 Genus *Pentacheles* Bate, 1878 [*Pentacheles gibba*]

Five species are known from world oceans and three species recorded from Indian waters.

Diagnosis: Carapace ovate or subrectangular; postorbital and postcervical carinae well defined or obsolescent; never swollen; dactyl of P5 simple or subchelate in male; chelate in female. Epipod of third maxilliped longer than ischium; basal antennular segment proximally quadrate, lamellar (Galil 2000).

3.5.3.1 *Pentacheles gibbus* Alcock, 1894
Polycheles gibbus (Alcock, 1894) [different generic combination]
Diagnosis: Carapace convex, gibbous; lateral margins convergent anteriorly and posteriorly. Dorsal surface of carapace thickly tomentose, minutely granulose. Anterior margin of carapace sinuous, bearing small, bifid rostral spine. Internal and external angles of orbital sinus unarmed. Eyestalk with small spine directed upward. Basal antennular segment foliaceous; mesial margin raised above rostrum, irregularly granulate; single spine on antero-external angle. Spine formula of lateral margins of carapace 5–6:3:16–19. Internal angle of orbital notch unarmed; carapace convex; abdominal tergite, pleura set with conical tubercles. Carinae on tergites 2–5 bicuspid. Sixth tergite set with prominent conic granules on each side of short crest. Tergites 2–5 bearing well-defined, transversely oblique submedian grooves. Pleura set with prominent conic granules. Second pleuron anteriorly produced, ovate. Telson with granulate tubercle medially, two smooth submedian carinae posteriorly, margins minutely spinulate (Galil 2000).
Habitat: Depth of occurrence, 1668–1703 m
Range distribution: Arabian Sea, Andaman Sea
Status: Rare.

3.5.3.2 *Pentacheles obscurus* Bate, 1878 [*Pentacheles obscura*]
Polycheles obscurus (Bate, 1878) [different generic composition]
Polycheles carpenteri (Alcock, 1894) [*Pentacheles carpenteri*]
Diagnosis: Carapace ovate, lateral margins convergent anteriorly and posteriorly. Dorsal surface of carapace thickly tomentose, minutely granulose laterally. Anterior margin of carapace sinuous, bearing small, bifid rostral spine. Internal

and external angles of orbital sinus unarmed. Eyestalk with small spine directed upward. Basal antennular segment foliaceous, mesial margin raised above rostrum, irregularly granulate; single spine on antero-external angle. Spine formula of lateral edges of carapace 5:3:27–28; spines diminish considerably in size posteriorly, posterior-most no more than granules. Median postrostral carina irregularly granulate anteriorly, median postcervical carina bearing two rows of antrorse granules. Gastro-orbital region unarmed. Branchial carina sinuous, obsolescent. Posterior margin of carapace smooth.

Abdominal tergites medially carinate; carinae on tergites 2–5 bicuspid; sixth tergite with short crest posteriorly. Tergites 2–5 bearing well-defined, transversely oblique, submedian grooves. Pleura nearly smooth; second pleuron anteriorly produced, ovate. Telson with granulate tubercle medially, two smooth submedian carinae posteriorly (Galil 2000).

Habitat: Depth of occurrence, 1100–3080 m
Range distribution: New Caledonia, Gulf of Eden, Madagascar

In Indian waters, the species is known from the Bay of Bengal (Carpenter's Ridge)

Status: Rare

3.5.3.3 *Pentacheles laevis* Bate, 1878
Polycheles laevis (Bate, 1878) [different generic combination]
Polycheles gracilis (Bate, 1878) [*Pentacheles gracilis*]
Polycheles granulatus Faxon, 1893
Polycheles beaumontii, (Alcock, 1894) [*Pentacheles beaumontii*]
Diagnosis: Carapace ovate. Dorsal surface of carapace granulose, sparsely setose. Anterior margin of carapace slightly concave, two long rostral spines adjoining at base. Internal and external angles of orbital sinus prominently produced, spinose. Eyestalk bearing spine, curved distad. Basal antennular segment lamellar, produced anteriorly to an acute point; mesial margin prominently spinose; single spine on antero-external angle. Lateral margins of carapace lined with upcurved spines, anterior-most spine largest, diminishing gradually in size posteriorly; spine formula 7–9:3–4:14–15. Median postrostral carina bearing irregular number of spines followed by close-set granules; median postcervical carina with two rows of antrorse, mammilate tubercles. Gastro-orbital region unarmed. Posterior margin of cervical groove unarmed. Branchial carina sinuous, spinulate; spinules increasing in size posteriorly. Posterior margin of carapace smooth. Abdominal tergites medially carinate; carinae on tergites 1–3 bearing antrorse beak; tergites 4–5 with blunt carina; sixth tergite non-carinate; tergites 2–5 with well defined, oblique sub-median grooves. Surface of pleura smooth, minutely punctate; anterior pleuron with mesial groove proximally; Telson with rounded tubercle anteriorly, two smooth, confluent carinae posteriorly (Galil 2000).

Habitat: Depth of occurrence, 347–2505 m.
Range distribution: Wide distribution. Atlantic and Pacific Oceans; Andaman Sea
Status: Rare.

3.5.3.4 Genus: *Willemoesia* Grote, 1873

Four extant species are known and of which one species is reported from Indian waters.

Diagnosis: Carapace dorsoventrally flattened, subrectangular or ovate. Anterolateral angle of carapace produced, spiniform. Lateral margins well defined, spinose; cervical and postcervical incisions dividing margin into three parts. Front bearing single rostral spine, lacking well defined orbits. Eyestalks globose, fixed beneath and parallel with anterior margin of carapace. Abdomen somewhat laterally depressed. First abdominal tergite narrow, pleura fused, abbreviated. Abdominal tergites 2–5 smooth or sculptured, medially carinate, carinae blunt or anteriorly spinose; sixth tergite smooth or sculptured. Frontal margin of carapace without orbital sinus (Galil 2000).

3.5.3.4.1 *Willemoesia leptodactyla* (Thomson, 1873)
[*Deidamia leptodactyla*]

Willemoesia indica Alcock 1901
Willemoesia secunda Sund, 1920

Diagnosis: Carapace oblong, lateral margins subparallel, anteriorly convergent. Dorsal surface of carapace densely covered with antrorse spinules. Rostral spine upcurved. Frontal margin of carapace transverse. Internal angle of orbital sinus spinose. Eyestalk bulbous, bearing short curved spine; Lateral margins of carapace spinose, spines increasing in size anteriorly, anteriormost largest; spine formula 8–10:5–7:15–25. Median postrostral carina prominent, granulate; spine formula variable; spines on median postcervical carina larger posteriorly. Gastro-orbital carina sinuous, sparsely spinulose. Branchial carina sinuous, prominent, spinose. Posterior margin of carapace smooth.

First abdominal tergite smooth, bearing tubercle medially on anterior margin. Tergites 2–5 sculptured, with deep submedian grooves, medially carinate, carinae culminating in antrorse spine anteriorly, spines increasing in size posteriorly. Sixth tergite with swollen margins, submedian crescents and lyre-shaped median carina; median carina more prominent posteriorly. Second pleuron heart shaped; pleura 3–5 with blunt rib mesially, indented posteriorly. Telson with T-shaped hump anteriorly, smooth convergent carinae posteriorly, unarmed margins, tip rounded. Posterior margin of uropodal endopod rounded (Galil 2000).

Habitat: Depth of occurrence, 2396–5124 m
Range distribution: Wide distribution. West and east Atlantic. Indo-West Pacific

The species was recorded from Bay of Bengal, Andaman Sea.

Status: Rare

3.6 Infraorder: Astacidea Latreille, 1802

Superfamily: Enoplometopoidea Saint Laurent, 1988

3.6.1 Family: Enoplometopidae Saint Laurent, 1988

3.6.1.1 Genus: *Enoplometopus* A. Milne Edwards, 1862

There are 12 species and 10 are from Indo-West Pacific region; two species are known from Indian coast. They are shelf dwellers that inhabit rocky seabed. *Enoplometopus* occurs in all three ocean basins and best known from the western and central Pacific (Wahle et al. 2012). Many species are morphologically extremely similar. Each species has very distinct and diagnostic colourations and can be easily separated by colourations. Special attention should be given to the spines on large the chela and the posterior margin of the abdominal tergite VI (Chan and Ng 2008).

3.6.1.1.1 *Enoplometopus occidentalis* (Randall, 1840)
[*Nephrops occidentalis*] (Fig. 3.20)

English name: Red reef lobster, Hawaiian reef lobster

Diagnosis: Rostrum with two to four spines on lateral margin. Carapace armed with five median (anterior most blunt), one postcervical, two intermediate, one supraocular, and three or four lateral spines. Chela broad and compressed, upper and lower faces of palm with longitudinal rows of tubercles; outer margin of dactyl with two or three spines on distal part and unarmed or with low blunt tubercles on proximal part. Second pereopod with dactyl 0.3 times as long as propodus; carpus and merus with distoventral spine. Pleura of abdominal somites 2–5 rounded or bluntly pointed. Telson with one lateral and three distolateral spines, distalmost spine, the largest. Several variations have been observed in the number of spines on the rostrum. Lateral margin of rostrum usually armed with two or three spines but can have up to four spines on each margin. Most common species found in the Indo-West Pacific (Poupin, 2003).

Colour: Body orange-red with white tip of spines. Lateral face of carapace with median white spot circled in dark orange. Similar spots on dorsal and lateral faces of abdomen. Chela orange with tubercles darker; fingers banded in light and dark orange. Ambulatory legs orange with narrow white or pale orange bands.

Habitat: Depth of occurrence, 0–100 m on hard bottoms (Poupin 2003).

Range distribution: West Indian Ocean; west Pacific, central Pacific, Hawaii.

The only record of the species in India is from Mandapam, Gulf of Mannar, southeast coast of India (Radhakrishnan and Jayasankar 2014) (Fig. 3.3c).

Status: Rare

Fig. 3.20 *Enoplometopus occidentalis* (Photo by Tin Yam Chan, National Taiwan University)

3.6.1.1.2 *Enoplometopus macrodontus* Chan and Ng, 2008 (Fig. 3.21)

Diagnosis: Body pubescent, with numerous long stiff setae. Rostrum elongated, distinctly overreaching antennular peduncle, armed with three or four pairs (rarely two or five on one side) of lateral teeth. Carapace with large supraocular spine and one large intermediate, five median, two lateral and one postcervical teeth; all teeth well developed; antennal spine large, strongly bent inwards. Small but distinct branchiostegal spine generally present. Dorsal surface of rostrum and carapace with scattered long stiff setae. Eye well developed, subspherical. Scaphocerite, including distolateral teeth, more or less reaching tip of antennular peduncle. Abdomen with some long stiff setae. Telson rectangular, slightly longer than wide, bearing two pairs of movable lateral spines and two pairs of movable posterolateral spines. Two posterolateral spines always next to each other, with inner one longer (Chan and Ng 2008).

Colour: Body generally reddish-orange with few white spots and lines. Carapace mostly uniformly orange, with one moderately large white spot behind branchiostegal spine, small white spots also distributed along antero-lateral margin behind orbit and branchiostegal spine; rostrum with three white transverse bands; tip of postcervical teeth always white. Antennular and antennal flagella not banded; abdomen with tergum orange except median keel alternated with red and white; pleura reddish and with large white spots at the junctions

Fig. 3.21 *Enoplometopus macrodontus* (Photo by E. V. Radhakrishnan)

between tergites and pleura, except for somite 2 which has a large white spot at anterior pleuron. Large cheliped reddish except for palm and carpus mostly orange. Pereopods II–V reddish and each covered with three thin white bands on merus and ischium.

Range distribution: Philippines, western Pacific and Arabian Sea, South west coast of India (Chan and Ng 2008, Radhakrishnan et al. 2011).

The species has been recorded from Ponnani on the southwest coast of India (Radhakrishnan et al. 2011) (Fig. 3.3c).

Habitat: Depth of occurrence, 90–200 m (Chan and Ng 2008)
Status: Rare
Superfamily: Nephropoidea Dana, 1852

3.6.2 Family: Nephropidae Dana, 1851 [Nephropinae]

There are 14 genera under this family. This family is represented by three genera and seven species along the Indian coast.

3.6.2.1 Genus *Acanthacaris* Bate, 1888

The genus is represented by two species. Only one species is known from Indian waters.

3.6.2.1.1 *Acanthacaris tenuimana* Bate, 1888

English name: Prickly deepsea lobster

Diagnosis: Body cylindrical, completely covered with small spines and sharp tubercles; carapace with a well-developed median rostrum which is laterally compressed with dorsal and ventral, but no lateral teeth. Eyes very small, lacking pigment; antennae long and whip-like; antennal scales well developed. Tail powerful, with a well-developed tail fan. First three pairs of pereopods ending in true chelae. The first pair equal, very slender, longer than the body, covered with sharp spinules and ending in elongate and slender fingers with long teeth on cutting edges, but without hairs. Fingers of first cheliped 1.5 to twice as long as palm. Second pair of pereopods very much longer and less spiny than third pair (Holthuis 1991).

Range distribution: Indo-West Pacific Ocean (Natal, Mozambique, Madagascar, Philippines, Indonesia, eastern Australia, China Sea) and Sea of Japan (Griffin and Stoddart 1995).

In India, the species is recorded from the Lakshadweep Islands (Bate 1888) (Fig. 3.3b).

Habitat: The depth of distribution is from 160 to 2161 m.
Status: Rare

3.6.2.2 Genus *Nephropsis* Wood-Mason, 1872

There are 15 species under this genus. Five species are known from Indian waters, of which *N. stewarti* is occasionally landed in commercial quantities (Kizhakudan and Thirumilu 2006; Dineshbabu 2008).

Diagnosis: Eye not pigmented. Body granular and hairy, but not covered with evenly placed large pearly tubercles. Pleura of second abdominal somite ending in a long sharp point (Holthuis 1991).

3.6.2.2.1 *Nephropsis carpenteri* Wood- Mason, 1885

English name: Ridgeback lobsterette

Diagnosis: Median dorsal carinae on third to sixth abdominal somites but not on second (Holthuis1991).

Biology: Alcock (1901) reported the size of a type specimen collected from southeast of Chennai as 33.5 mm CL (101.5.m TL). George (1967) recorded the size of 113 mm TL for a female specimen from southwest coast of India. The maximum size recorded by Holthuis (1991) was 120 mm TL. The largest size recorded

Fig. 3.22 *Nephropsis stewarti* (Photo by A.P. Dineshbabu)

was 154 mm TL (48 mm CL) for a specimen collected off Chennai (Thirumilu 2011).

Range distribution: This species is found in the Indo-West Pacific region and around Japan (Holthuis 1991).

Along the Indian coast, reported from the Arabian Sea and the Bay of Bengal (Fig. 3.3c) (George and Rao 1965b; Kizhakudan and Thirumilu 2006; Thirumilu 2011).

Habitat: Depth of occurrence 200–500 m
Status: Rare

3.6.2.2.2 *Nephropsis stewarti* Wood-Mason, 1872 (Fig. 3.22)

English name: Indian Ocean lobsterette

Diagnosis: Abdominal somites without any trace of a mid-dorsal carina. No postsupraorbital spine on carapace. The distance between the supraorbital spines and the gastric tubercle is less than half the distance between the gastric tubercle and the cervical groove. Exopod of uropod with a diaresis (Holthuis 1991).

Habitat and ecology: Depth 170–1060 m; forms small-scale fishery at Mangalore. During 2000–2006, the average annual landing of the species was estimated at

23.3 t with the highest landing in 2001 (51 t) and the lowest in 2005 (9 t) (Dineshbabu 2008).

Biology: Fishery was constituted by the length range 58–158 mm. Females <80 mm (total length) were found to be immature. Highest percentage of immature females (33%) was found during November (Dineshbabu 2008).

Range distribution: The type locality for this species is 25 miles off Ross Island on the eastern coast of the Andaman Islands, India, from a depth of 476–550 m (Holthuis 1991). Indo-West Pacific from Eastern Africa to Japan, the Philippines, Indonesia and Northwestern Australia (Chan 1998).

In Indian waters, reported from Tamil Nadu and Odisha coasts, off Mangalore, and Anadaman and Nicobar Islands (Fig. 3.3c) (Wood-Mason 1872; Dineshbabu et al. 2005; Jayaprakash et al. 2006; Kizhakudan and Thirumilu 2006; Dineshbabu 2008; Sreedhar et al. 2010).

Status: Not a regular species in the deepsea fishery but has moderate fishery potential.

3.6.2.2.3 *Nephropsis sulcata* Macpherson, 1990

English name: Grooved lobsterette

Diagnosis: Median groove of rostrum reaching distinctly beyond anterior part of lateral rostral teeth. Distance between supraorbital spine and gastric tubercle is half the distance between gastric tubercle and post cervical groove (Holthuis 1991).

Habitat: Depth of occurrence 415–1115 m

Biology: The maximum length is unknown, but the carapace length including the rostrum is between 1.5–3.0 cm in males and 1.8–3.4 cm in females (Holthuis 1991).

Range distribution: Indo-West Pacific Ocean; Natal (southern Africa), Madagascar, the Philippines, South China Sea, north-western and eastern Australia, Coral Sea, New Caledonia, Chesterfield Islands.

In Indian waters, the species is known from Lakshadweep Sea (Fig. 3.3c) (Griffin and Stoddart 1995).

Status: Rare

3.6.2.2.4 *Nephropsis ensirostris* Alcock, 1901

English name: Gladiator lobsterette

Diagnosis: Rostrum without lateral teeth; a strong postsupraorbital spine present behind the supraorbital spine. Abdominal somites 3–6 with a median dorsal carina. Anterior margin of pleura of second abdominal somite without spines. Telson without medio-dorsal spine (Holthuis 1991)

Habitat: Depth of occurrence 580–1160 m

Range distribution: The species is distributed in the Indo-West Pacific region including the Gulf of Aden, south of Sri Lanka, the Philippines, India and Indonesia.

The type locality of the species is Arabian Sea, north of the Lakshadweep Islands at Lat 13° 47'49' Long 73° 7'E (Fig. 3.3c) (Holthuis 1991).

Status: Rare

3.6.2.2.5 *Nephropsis suhmi* Bate, 1888

English name: Red and White lobsterette

Diagnosis: There are two post-orbital spines on each side of the carapace. A well-developed spine on the anterior margin of pleura of abdominal segments 2–4 but not on segment 5 Abdomen without medio-dorsal carina. Exopod of uropod without diaresis (Holthuis 1991).

Habitat: Depth of occurrence 786–2029 m

Range distribution: Indo-West Pacific Ocean; Madagascar, Maldives islands, northwestern and northeastern Australia, western Tasman Sea, New Caledonia and Indonesia

In Indian waters, the species is known from Arabian Sea including the Lakshadweep Sea and the Andaman and Nicobar Islands (Fig. 3.3c) (Griffin and Stoddart 1995).

Status: Rare

3.6.2.3 Genus: *Metanephrops* Jenkins, 1972

The genus is represented by eighteen extant and three fossil species (Late Cretaceous to Pliocene). They are found at depths of about 50–1000 m, mostly below 150 m (Jenkins 1972; Chan 1997; Tshudy 2003).

Diagnosis: Eyes large with black pigment; antennal scale present; body not uniformly spinulose (Holthuis 1991). The genus is represented by 18 species and 1 species is reported from Indian waters.

3.6.2.3.1 *Metanephrops andamanicus* (Wood-Mason, 1894)

English name: Andaman lobster

Diagnosis: Rostrum without lateral teeth; Second to fifth abdominal segments with a marked dorso-median carina, flanked by a pair of conspicuous longitudinal groove; 200 mm in total length (Holthuis 1991).

Habitat: Depth of distribution 250–750 m

Range distribution: East Africa, South China Sea, the Philippines, Indonesia and Andaman Sea

Status: Rare

3.7 Conclusion

Lobsters are economically important crustaceans comprised of four infraorders (Astacidea, Glypheidea, Achelata and Polychelida) under the suborder Macrura Reptantia with a rich fossil record dating back 360 Ma (Schram and Dixon 2004). They are presumed to include 9 extant families and 17 fossil families (De Grave et al. 2009). Chan (2010), and in Chap. 2 of this book, based on molecular evidence, compiled 6 extant families, 54 genera and 260 valid species (including 4 subspecies). Feldmann et al. (2012) included three new genera and species based upon examination of fossil material from the Triassic of China. Bracken-Grissom et al. (2014) examined phylogenetic relationships among major groups of all lobster families and 94% of genera using six genes (mitochondrial and nuclear) and 195 morphological characters across 173 species of lobsters. The combined molecular + morphology analysis revealed that lobsters are a non-monophyletic assemblage of crustaceans.

The lobster fauna of India is so diverse that representatives from the major three infraorders (Astacidea, Achelata, Polychelida) are reported to occur in the Arabian Sea, Bay of Bengal and the two Island systems, Lakshadweep in the Arabian Sea and Andaman and Nicobar Islands in the Bay of Bengal. The southern part of the Indian coast (Arabian Sea and Bay of Bengal) is the region represented by a large number of species. The Mudspiny lobster, *P. polyphagus*, exhibits a latitudinal distribution with commercial quantities harvested along the northwest and northeast coasts. The northern latitudes are area of lower species diversity. Although *T. unimaculatus* has a wider distribution throughout the Indian coast, *T. indicus* is restricted to the Andhra Pradesh coast. Some species are endemic to a particular region (*E. macrodontus, E. occidentalis, P. jeppiari*). With the extension of the trawling grounds to deeper areas, more new species and records can be expected.

References

Ahyong, S. T. (2009). The polychelidan lobsters. Phylogeny and systematics (Polychelida, Polychelidae). In J. W. Martin, K. A. Crandall, & D. L. Felder (Eds.), *Crustacean issues 18- decapod crustacean Phylogenetics* (pp. 369–378). Boca Raton: Taylor & Francis/CRC Press.

Ahyong, S. T., & Brown, D. E. (2002). New species and new records of Polychelidae from Australia (Decapoda: Crustacea) from Australia. *Raffles Bulletin of Zoology, 50*(1), 53–79.

Ahyong, S. T., & Chan, T.-Y. (2004). Polychelid lobsters of Taiwan (Decapoda: Polychelidae). *Raffles Bulletin of Zoology, 52*, 171–182.

Ahyong, S. T., & Galil, B. S. (2006). Polychelidae from the southern and western Pacific (Decapoda: Polychelida). *Zoosystema, 28*(3), 757–767.

Ahyong, S. T., & Poore, G. C. B. (2004). Deep-water galatheidae (Crustacea: Decapoda: Anomura) from southern and eastern Australia. *Zootaxa, 472*, 1–76.

Alcock, A. (1901). *A Descriptive Catalogue of the Indian Deep-sea Crustacea Decapoda Macrura and Anomala in the Indian Museum* being a revised account of the deep sea species collected by the Royal Indian Marine Survey Ship Investigator. Calcutta. 286 pp.

Alcock, A. (1906). *Catalogue of the India decapod crustacean in the collection of Indian Museum*, Part III, Macrura: 1–55.

Alcock, A. & Anderson, A.R.S. 1894. Natural history notes from H.M. Indian marine survey steamer "investigator", commander C.D. Oldham, R.N. commanding, series II, no 14. An account of a recent collection of deep sea Crustacea from the bay of Bengal and the Laccadive Sea. *Journal of the Asiatic Society of Bengal*, 63, pt 2 (3): 141–185, pl. 9.

Ali, D.M., Pandian, P.P., Somavanshi, V.S., John, M.E., & Reddy, K.S.N. (1994). *Spear lobster, Linuparus somniosus Berry & George, 1972 (Fam. Palinuridae) in the Andaman Sea.* Occasional paper, Fishery Survey of India, Mumbai, 6: 13pp.

Anon. (2010). Occurrence of the *Scyllarides tridacnophaga* Holthuis, 1967, a new record from the west coast of India. *CMFRI Newsletter, 125*, 15.

Anrose, A., Selvaraj, P., Dhas, J. C., Prasad, G. V. A., & Babu, C. (2010). Distribution and abundance of deep sea spiny lobster *Puerulus sewelli* in the Indian exclusive economic zone. *Journal of the Marine Biological Association of India, 52*(2), 162–165.

Anuraj, A., Kirubasankar, R., Kaliyamoorthy, M., & Roy, D. (2017). Genetic evidence and morphometry for shovel-nosed lobster, *Thenus unimaculatus* from Andaman and Nicobar Islands, India. *Turkish Journal of Fisheries and Aquatic Sciences, 17*, 209–215.

Artuz, M. L., Kubanc, C., & Kubanc, S. N. (2014). *Stereomastis artuzi* sp. Nov., a new species of Polychelidae (Decapoda, Polychelida) described from the Sea of Marmara, Turkey. *Crustaceana, 87*(10), 1243–1257.

Balasubramanyan, R. (1967). On the occurrence of palinurid spiny lobster in the Cochin backwaters. *Journal of the Marine Biological Association of India, 9*(2), 425–438.

Balasubramanyan, R., Satyanarayana, A. V., & Sadanandan, K. A. (1960). A preliminary account of the experimental rock lobster fishing conducted along the south-west coast of India, with bottom set gillnets. *Indian Journal of Fisheries, 7*, 405–422.

Balasubramanyan, R., Satyanarayana, A. V., & Sadanandan, K. A. (1961). A further account of the rock lobster fishing experiments with the bottom-set gillnets. *Indian Journal of Fisheries, 8*, 269–290.

Balss, H. (1925). Palinura, Astacura and Thalassinidea. *Wissenschaftliche Ergebnisse der Deutschen Tiefsee-Expedition auf dem Dampfer "Validivia" 1898–1899* 20:185–216, pls 18–19.

Bate, C. S. (1888). Report on the Crustacea Macrura collected by H.M.S. Challenger during the years 1873–76. *Report on the Scientific Results of the Voyage of H.M.S. Challenger during the years 1873–76*, 24: i-xc, 1–942, Figs. 1–76, Pls. 1–150.

Borradaile, L. A. (1906). Marine crustaceans XIII. The Hippidae, Thalassinidea and Scyllaridea. In J. S. Gardiner (Ed.), *The fauna and geography of the Maldives and Laccadive Archipelago* (pp. 750–754). Cambridge: Cambridge University Press.

Bracken-Grissom, H. D., Toon, A., Felder, D. L., Martin, J. W., Finley, M., Rasmussen, J., Palero, F., & Crandall, K. A. (2009). The decapod tree of life: Compiling the data and moving toward a consensus of decapod evolution. *Arthropod Systematics & Phylogeny, 67*(1), 99–116.

Bracken-Grissom, H. D., Ahyong, S. T., Wilkinson, R. D., Feldmann, R. M., Schweitzer, C. E., Breinholt, J. W., & Robles, R. (2014). The emergence of lobsters: Phylogenetic relationships, morphological evolution and divergence time comparisons of an ancient group (Decapoda: Achelata, Astacidea, Glypheidea, Polychelida). *Systematic Biology, 63*(4), 457–479.

Burton, T. E., & Davie, P. J. F. (2007). A revision of the shovel-nosed lobsters of the genus *Thenus* (Crustacea: Decapoda: Scyllaridae) with descriptions of three new species. *Zootaxa, 1429*, 1–38.

Chakraborty, R. D., Maheswarudu, G., Radhakrishnan, E. V., Purushothaman, P., Kuberan, P., Jomon, G., Sebastian, & Thangaraja, R. (2014). Rare occurrence of blunthorn lobster *Palinustus waguensis* Kubo, 1963 from the southwest coast of India. *Marine Fisheries Information Service T&E Series, 219*, 25–26.

Chan, T. Y. (1997). Crustacea Decapoda: Palinuridae, Scyllaridae and Nephropidae collected in Indonesia by the KARUBAR cruise, with an identification key for the species of Metanephrops. In A. Crosnier & P. Bouchet (Eds.), *Résultats des Campagnes MUSORSTOM, volume 16, 172* (pp. 409–431). Paris: Mémoires du Muséum National d'Histoire Naturelle Editions du Muse'um.

Chan, T.-Y. (1998). Shrimps and prawns, lobsters. In K. E. Carpenter & V. H. Niem (Eds.), *FAO Species identification guide for fisheries purposes. The living marine resources of the Western Central Pacific. Volume 2. Cephalopods, crustaceans, holothurians and sharks* (pp. 851–1043). Rome: Food and Agriculture Organization.

Chan, T.-Y. (2010). Annotated checklist of World's marine lobsters (Crustacea, Decapoda: Astacidea, Glypheida, Achelata, Polychelida). *The Raffles Bulletin of Zoology, 23*, 153–181.

Chan, T.-Y., & Ng, P. K. L. (2001). On the nomenclature of the commercially important spiny lobsters *Panulirus longipes femoristriga* (von Martens, 1872), *Panulirus bispinosus* Borradaile, 1899 and *Panulirus albiflagellum* Chan& Chu, 1996 (Decapoda: Palinuridae). *Crustaceana, 74*(1), 123–127.

Chan, T.-Y., & Ng, P. K. L. (2008). *Enoplometopus* A. Milne-Edwards, 1862 (Crustacea: Decapoda: Nephropidae) from the Philippines with description of one new species and a revised key to the genus. *Bulletin of Marine Science, 83*(2), 347–365.

Chan, T.-Y., & Yu, H.-P. (1995). The rare lobster genus *Palinustus* A. Milne Edwards, 1880 (Decapoda: Palinuridae), with description of a new species. *Journal of Crustacean Biology, 15*(2), 376–394.

Chan, T. -Y., Ma, K. Y., & Chu, K. H (2014). The deep sea spiny lobster genus *Puerulus* Ortmann, 1897 (Crustacea, Decapoda, Palinuridae) with description of five new species. In: Ahyong, S.T., T.-Y. Chan., L. Corbari and P.K.L. Ng (Eds.) Tropical Deep Sea Benthos. *Memoires du Museum national d'Histoire naturelle*, 204.

Chang, Y. J., Sun, C. L., Chan, Y., Yeh, S. Z., & Chiang, W. C. (2007). Reproductive biology of the spiny lobster, *Panulirus penicillatus*, in the southeastern coastal waters off Taiwan. *Marine Biology, 151*, 555–564.

Chekunova, V. I. (1971). Distribution of commercial invertebrates on the shelf of India, the northeastern part of Bay of Bengal and the Andaman Sea. In S. A. Bogdanov (Ed.), *Soviet Fisheries Investigations in the Indian Ocean* (pp. 68–83). Jerusalem: Israel Programme for Scientific Translations.

Chhapgar, B. F., & Deshmukh, S. K. (1961). On the occurrence of the spiny lobster *Panulirus dasypus* (H. Milne Edwards) in Bombay waters with a note on the systematics of Bombay lobsters. *Journal of the Bombay Natural History Society, 58*(3), 632–638.

Chhapgar, B. F., & Deshmukh, S. K. (1964). Further records of lobsters from Bombay. *Journal of the Bombay Natural History Society, 61*, 203–207.

Chopra, B. N. (1939). Some food prawns and crabs of India and their fisheries. *Journal of the Bombay Natural History Society, 41*(2), 221–234.

Chhapgar, B. F., & Deshmukh, S. K. (1971). Lobster fishery of Maharashtra. *Journal of the Indian Fisheries Association, 1*(1), 74–86.

Courtney, A. J., Cosgrove, M. G., & Die, D. J. (2001). Population dynamics of Scyllarid lobsters of the genus *Thenus* spp. on the Queensland (Australia) east coast. I. Assessing the effects of tagging. *Fisheries Research, 53*, 251–261.

de Alvarenga, F. M., & Cardoso, I. A. (2014). First record of *Stereomastis nana* (Polychelidae: Crustacean: Decapoda) from the South-Western Atlantic. *Marine Biodiversity Records, 7*, 4.

De Grave, S., Pentcheff, N. D., Ahyong, S. T., Chan, T.-Y., Crandall, K. A., Dworschek, P. C., Felder, D. L., Feldman, R. M., Fransen, C. H. J. M., Goulding, L. Y. D., Lemaitre, R., Low, M. E. Y., Martin, J. W., Ng, P. K. L., Schweitzer, C. E., Tan, D., Tshudy, S. H., & Wetzer, R. (2009). A classification of living and fossil genera of decapod crustaceans. *Raffles Bulletin of Zoology, 21*, 1–109.

De Man, J. G. (1916). The Decapoda of the Siboga expedition Part III. Families Eryonidae, Palinuridae, Scyllaridae and Nephropidae. *Siboga Expedition Monograph, 39A*(2), 1–122.

Deshmukh, S. (1966). The puerulus of the spiny lobster *Panulirus polyphagus* (Herbst) and its metamorphosis into post-puerulus. *Crustaceana, 10*, 137–150.

Dineshbabu, A. P. (2008). Morphometric relationship and fishery of Indian Ocean lobsterette, *Nephropsis stewarti* Wood_Mason, 1873 along the southwest coast of India. *Journal of the Marine Biological Association of India, 50*(1), 113–116.

Dineshbabu, A. P., Durgekar, N. R., & Zacharia, P. U. (2011). Estuarine and Marine Decapods of Karnataka Inventory. *Fishing Chimes, 30*(10 & 1), 20–24.

Dineshbabu, A. P., Sreedhara, A. P., & Muniyappa, Y. (2005). Report on the fishery of Indian Ocean Lobsterette, *Nephropsis stewarti* along Mangalore. *Marine Fisheries Information Service T&E Series, 184*, 18.

Farhadi, A., Jeffs, A. G., Farahmand, H., Rejiniemon, T. S., Smith, G., & Lavery, S. D. (2017). Mechanisms of peripheral phylogeographic divergence in the indo-Pacific: Lessons from the spiny lobster *Panulirus homarus*. *BMC Evolutionary Biology, 17*, 195. https://doi.org/10.1186/s12862-017-1050-8. 18.08.2017.

Feldmann, R. M., Schweitzer, C. E., Hu, S., Zhang, Q., Zhou, C., Xie, T., Huang, J., & Wen, W. (2012). Decapoda from the Luoping biota (Middle Triassic) of China. *Journal of Paleontology, 86*, 425–441.

Frisch, A. J. (2007). Short and long-term movements of painted lobster (*Panulrus versicolor*) on a coral reef at Northwest Island, Australia. *Coral Reefs, 26*, 311–317.

Galil, B. S. (2000). Crustacea Decapoda: Review of the genera and species of the family Polychelidae Wood-Mason, 1874. *Resultats des Compagnes MUSORSTOM, 21. Memoires du Museum national d'Historie naturelle, 184*, 285–387.

George, M. J. (1965). Observations on the biology and fishery of the spiny lobster *Panulirus homarus* (Linnaeus). *Proceedings of Symposium Crustacea*, Part IV. Marine biological Association of India, Mandapam camp, India: 1308–1316.

George, M. J. (1967). Two new records of scyllarid lobsters from the Arabian Sea. *Journal of the Marine Biological Association of India, 9*(2), 433–435.

George, M. J. (1973). The lobster fishery resources of India. In: *Proceedings of Symposium on Living Resources of the seas around India*, Cochin, December 1968, Special Publication, Central Marine Fisheries Research Institute: 570–580.

George, M. J., & George, K. C. (1965). *Palinustus mossambicus* Barnard (Palinuridae: Decapoda), a rare spiny lobster from Indian waters. *Journal of the Marine Biological Association of India, 7*(2), 463–464.

George, R. W., & Main, A. R. (1967). The evolution of spiny lobsters (Palinuridae): A study of evolution in the marine environment. *Evolution, 21*, 803–820.

George, M. J., & Rao, P. V. (1965a). A new record of *Panulirus longipes* (Milne Edwards) from the southwest coast of India. *Journal of the Marine Biological Association of India, 7*(2), 461–462.

George, M. J. & Rao, P. V. (1965b). On some decapod crustaceans from south west coast of India. *Proc. Symp. Crustacea. Part 1. Series 2*. Marine Biological Association of India. Mandapam Camp, India: 327–336.

George, K. C., Thomas, P. A., Appukuttan, K. K., & Gopakumar, G. (1986). Ancillary living marine resources of Lakshadweep. *Marine Fisheries Information Service T&E Series, 68*, 46–50.

Giraldes, B. W., & Smyth, D. M. (2016). Recognizing *Panulirus meripurpuratus* sp. nov. (Decapoda: Palinuridae) in Brazil- systematic and biogeographic overview of *Panulirus* species in Atlantic Ocean. *Zootaxa, 4107*(3), 353–366.

Goswami, B. C. B. (1992). Marine fauna of Digha coast. *Journal of the Marine Biological Association of India, 34*(1&2), 115–137.

Gravely, T. H. (1927). The littoral fauna of Krusadai Island in the Gulf of Mannar. Order: Decapoda (except Paguridae and Stomatopoda). *Bulletin of the Madras Government Museum, (Natural History), 1*(1), 135–155.

Gravely, T. H. (1929). Report on a systematic survey of deep sea grounds by "Lady Goschen" for 1927–28. *Madras Fisheries Department Bulletin, 21*(1), 153–187.

Griffin, D. J. G., & Stoddart, H. E. (1995). Deep-water decapod crustacea from eastern Australia: Lobsters of the families nephropidae, palinuridae, polychelidae and scyllaridae. *Records of the Australian Museum, 47*(3), 231–263.

Heller, G. (1865). Eie crusteen Reisi der oesterreichidchen Fregatte "Novara" un die Erde in den, Jehren 1857-1859 unter den Bertehlen des Commodore B. Von Wullerstorf- Urbair. *(Zoology) Series, 2*(5), 325–458.

Holthuis, L. B. (1985). A revision of the family Scyllaridae (Crustacea: Decapoa: Macrura).1. Subfamily Ibacinae. *Zoologische Verhandelingen, Leiden, 218*: 1-30, figs. 1-27.

Holthuis, L. B. (1991). Marine lobsters of the world. FAO species catalogue, Vol. 13. *FAO Fisheries Synopsis*, Food and Agriculture Organization, Rome, 125 (13): 1–292.

Holthuis, L. B. (2002). The Indo-Pacific scyllarinae lobsters (Crustacea, Decapoda, Scyllaridae). *Zoosystema, 24*(3), 499–683.

Hossain, M. A. (1975). On the squat lobster, *Thenus orientalis* off Visakhapatnam (Bay of Bengal). *Current Science, 44*(5), 161–162.

Hossain, M. A. (1978). Few words about the sand lobster, *Thenus orientalis* (Lund) Decapoda: Scyllaridae from Andhra coast. *Seafood Export Journal, 10*, 43–46.

Jawahar, P., Sundaramoorthy, B., & Chidambaram, P. (2014). Studies on breeding biology of *Panulirus homarus* (Linnaeus, 1758) Thoothukudi, southeastern coast of India. *Journal of Experimental Zoology India, 17*(1), 175–181.

Jayaprakash, A. A., Kurup, B. M., Venu, S., Thankappan, D., Pachu, A. V., Manjebrayakath, H., Thampy, P., & Sudhakar, S. (2006). Distribution, diversity, length-weight relationship and recruitment pattern of deep sea finfishes and shellfishes in the shelf-break area off southwest coast of Indian EEZ. *Journal of the Marine Biological Association of India, 48*(1), 56–61.

Jeena, N. S. (2013). Genetic divergence in lobsters (Crustacea: Palinuridae and Scyllaridae) from the Indian EEZ. *Ph.D Thesis*, Cochin University of Science and Technology, Kochi, India, May 2013, pp. 153.

Jeena, N. S., Gopalakrishnan, A., Kizhakudan, J. K., Radhakrishnan, E. V., Kumar, R., & Asokan, P. K. (2015). Population genetic structure of the shovel-nosed lobster *Thenus unimaculatus* (Decapoda, Scyllaridae) in Indian waters based on RAPD and mitochondrial gene sequences. *Hydrobiologia, 766*, 225. https://doi.org/10.1007/s10750-015-2458-z.

Jenkins, R. J. F. (1972). Metanephrops, a new genus of late Pliocene to recent lobsters (Decapoda, Nephropidae). *Crustaceana, 22*, 161–177.

John, C. C., & Kurian, C. V. (1959). A preliminary note on the occurrence of deep water prawn and spiny lobster off the Kerala coast. *Bulletin of Central Research Institute Trivandrum, Series C, 7*(1), 155–162.

Jones, S. (1965). The crustacean fishery resources of India. *Proceedings of Symposium Crustacea, Part IV, Series, 2*, 1328–1341.

Jones, C. M. (2007). Biology and fishery of the Bay lobster *Thenus* spp. In K. L. Lavalli & E. Spanier (Eds.), *The biology and fisheries of the slipper lobster* (pp. 325–358). Boca Raton: CRC Press/Taylor & Francis Group.

Jha, D. K., Kumar, T. S., Nazar, A. K. A., Venkatesan, R., & Saravanan, N. (2007). Spiny lobster resources of north Andaman Sea: Preliminary observations. *Fishing Chimes, 27*(1), 138–141.

Kabli, L. M., & Kagwade, P. V. (1996a). Reproductive biology of the sand lobster *Thenus orientalis* (Lund) from Bombay waters. *Indian Journal of Fisheries, 43*(1), 13–25.

Kabli, L. M., & Kagwade, P. V. (1996b). Age and growth of the sand lobster *Thenus orientalis* (Lund) from Bombay waters. *Indian Journal of Fisheries, 43*(3), 241–247.

Kagwade, P. V. (1987). Age and growth of the spiny lobster *Panulirus polyphagus* (Herbst). *Indian Journal of Fisheries, 34*(4), 389–398.

Kagwade, P. V. (1988). Reproduction in the spiny lobster *Panulirus polyphagus* (Herbst). *Journal of the Marine Biological Association of India, 30*(1&2), 37–46.

Kagwade, P. V., Manickaraja, M., Deshmukh, V. D., Rajamani, M., Radhakrishnan, E. V., Suresh, V., Kathirvel, M., & Rao, G. S. (1991). Magnitude of lobster resources of India. *Journal of Marine Biological Association of India, 33*(1&2), 150–158.

Kathirvel, M. (1975). On the occurrence of the puerulus larvae of the Indian spiny lobster, *Panulirus homarus* (Linn.), in Cochin backwaters. *Indian Journal of Fisheries, 22*(1&2), 287–290.

Kathirvel, M. (1998). A pictorial guide for identification of Indian spiny lobsters. Part 1. *Fish and Fisheries, 18*(4).

Kathirvel, M & James, D. B. (1990). The phyllosoma larvae from Andaman and Nicobar Islands. *Proceedings of First Workshop on Scientific results of FORV Sagar Sampada*, DOD, CMFRI, CIFT, Cochin: 147–150.

Kathirvel, M., & Nair, K. R. (2002). A new record of scyllarid lobster from southwest coast of India. *Fish and Fisheries, 32*, 4.

Kathirvel, M., Suseelan, C., & Rao, P. V. (1989). Biology, population and exploitation of the Indian deep sea lobster, *Puerlus sewelli*. *Fishing Chimes, February, 1989*, 16–25.

Kathirvel, M., Thirumilu, P., & Gokul, A. (2007). Biodiversity and economical value of Indian lobsters. In: *Zoological Survey of India. National Symposium on conservation and valuation of marine biodiversity* (Ed.). The director, zoological survey of India: 177-200.

Kizhakudan, J. K. (2002). First report of the spiny lobster, *Panulirus versicolor* (Latreille, 1804) from trawl landings at Veraval. *Marine Fisheries Information Service T&E Series, 172*, 6–7.

Kizhakudan, J. K., & Thirumilu, P. (2006). A note on the blunthorn lobsters from Chennai. *Journal of the Marine Biological Association of India, 48*(2), 260–262.

Kizhakudan, J. K., Thirumilu, P., & Manibal, C. (2004). Fishery of the sand lobster *Thenus orientalis* (Lund) by bottomset gillnets along Tamil Nadu. *Marine Fisheries Information Service, Technical and Extension Series, 181*, 6–7.

Kizhakudan, J. K., Krishnamoorthi, S., & Thiyagu, R. (2012a). Unusual landing of deep sea lobster *Palinustus waguensis* at Cuddalore. *Marine Fisheries Information Service T&E Series, 211*, 19.

Kizhakudan, J. K., Krishnamoorthi, S., & Thiyagu, R. (2012b). First record of the scyllarid lobster *Scyllarides tridacnophaga* from Chennai coast. *Marine Fisheries Information Service T&E Series, 211*, 13.

Kizhakudan, Joe. K. (2014). Reproductive biology of the female shovel-nosed lobster *Thenus unimaculatus* (Burton and Davie, 2007) from north-west coast of India. *Indian Journal of Geo-Marine Sciences, 43*(6), 927–935.

Kumar, T. S., Jha, D. K., Syed Jahan, S., Dharani, G., Abdul Nazar, A. K., Sakthivel, M., Alagarraaja, K., Vijayakumaran, M., & Kirubagaran, R. (2010). Fishery resources of spiny lobsters in the Andaman Island, India. *Journal of the Marine Biological Association of India, 52*(2), 166–169.

Kurian, C. V. (1963). Further observations on the deep water lobster *Puerulus sewelli* Ramadan off the Kerala coast. *Bulletin of the Department of Marine Biology and Oceanography University of Kerala, 1*, 122–127.

Kurian, C. V. (1965). Deep water prawns and lobsters off the Kerala coast. *Fishery Technology, 2*(1), 51–53.

Kurian, C. V. (1967). Further observations on deep water lobsters in the collections of R.V. Conch. *Bulletin of the Department of Marine Biology and Oceanography University of Kerala, 3*, 131–135.

Kuthalingam, M. D. K., Luther, G., & Lazarus, S. (1980). Rearing of early juveniles of spiny lobster *Panulirus versicolor* (Latreille) with notes on lobster fishery in Vizhinjam area. *Indian Journal of Fisheries, 27*(1&2), 17–23.

Lakshmi Pillai, S., & Thirumilu, P. (2006). Unusual landing of *Palinustus waguensis* at Chennai fishing harbour by indigenous gear. *Marine Fisheries Information Service T&E Series, 190*, 25–26.

Lakshmi Pillai, S., & Thirumilu, P. (2007). Extension in the distributional range of long-legged spiny lobster, *Panulirus longipes longipes* (A. Milne Edwards, 1868) along the southeast coast of India. *Journal of the Marine Biological Association of India, 49*(1), 95–96.

Lalithadevi, S. (1981). The occurrence of different stages of spiny lobsters in Kakinada region. *Indian Journal of Fisheries, 28*(1&2), 298–300.

Lamsuwansuk, A., Denduangiboripant, J., & Davie, P. J. F. (2012). Molecular and morphological investigations of shovel-nosed lobsters *Thenus* spp. (Crustacea: Decapoda: Scyllaridae) in Thailand. *Zoological Studies, 51*(1), 108–117.

Lavalli, K. L., & Spanier, E. (2007). Introduction to the biology and fisheries of slipper lobsters. In K. L. Lavalli & E. Spanier (Eds.), *The biology and fisheries of slipper lobster* (pp. 3–24). Boca Raton/London/New York: CRC Press/Taylor & Francis Group.

Lavery, S. D., Farhadi, A., Farahmand, H., Chan, T.-Y., Azhdehakoshpour, A., Thakur, V., & Jeffs, A. G. (2014). Evolutionary divergence of geographic subspecies within the scalloped spiny lobster *Panulirus homarus* (Linnaeus, 1758). *PLoS One, 9*(6), e97247. https://doi.org/10.1371/journal.pone.0097247.

Llyod, R. E. (1907). Contributions to the fauna of Arabian Sea with descriptions of new fishes and crustaceans. *Records of the Indian Museum, 1*, 1–12.

Meiyappan, M. M., & Kathirvel, M. (1978). On some new records of crabs and lobsters from Minicoy, Lakshadweep (Laccadives). *Journal of the Marine Biological Association of India, 20*(1&2), 116–119.

Murugan, A., & Durgekar, R. (2008). *Beyond the tsunami: Status of fisheries in Tamil Nadu, India: A snapshot of present and long-term trends* (p. 75). Bangalore, India: UNDP/UNTRS, Chennai and ATREE.

Mustafa, A. M. (1990). *Linuparus andamanensis*, a new spear lobster from Andaman. *Andaman Science Association, 6*(2), 177–180.

Nair, R. V., Soundararajan, R., & Dorairaj, K. (1973). On the occurrence of *Panulirus longipes longipes*, *Panulirus penicillatus* and *Panulirus polyphagus* in the Gulf of Mannar with notes on the lobster fishery around Mandapam. *Indian Journal of Fisheries, 20*(2), 333–350.

Oommen, V. P., & Philip, K. P. (1974). Observations on the fishery and biology of the deep sea spiny lobster *Puerulus Sewelli* Ramadan. *Indian Journal of Fisheries, 21*(2), 369–385.

Palero, F., Crandall, K. A., Abello, P., Macpherson, E., & Pascual, M. (2009). Phylogenetic relationships between spiny, slipper and coral lobsters (Crustacea, Decapoda, Achelata). *Molecular Phylogenetics and Evolution, 50*, 152–162.

Parulekar, A. H. (1981). Marine fauna of Malawan, central west coast of India. *Mahasagar, 14*, 33–34.

Patak, S.N., Feldman, R.M., Porter, M., & Tshudy, D. 2006. Phylogeny and evolution. In: Phillips, B.F. (Ed.). *Lobsters: Biology, management, aquaculture and fisheries*. Blackwell Scientific Publications, Oxford, p. 113-145.

Patil, A. M. (1953). Study of the marine fauna of Karwar coast and the neighbouring islands. Part IV. Echinodermata and other groups. *Journal of the Bombay Natural History Society, 51*, 429–434.

Phillips, B. F., Cobb, J. S., & George, R. W. (1980). General biology. In J. S. Cobb & B. F. Phillips (Eds.), *The biology and management of lobsters. I. Physiology and behavior* (pp. 1–82). New York: Academic Press.

Pillai, C. G., Mohan, M., & Kunhikoya, K. K. (1985). Observations on the lobsters of Minicoy atoll. *Indian Journal of Fisheries, 32*(1), 112–122.

Poupin, J. (2003). Reef lobsters Enoplometopidae A. Milne Edwards, 1862 from French Polynesia with a brief revision of the genus (Crustacea, Decapoda, Enoplometopidae). *Zoosystema, 25*(4), 643–664.

Powell, A. (1908). *Panulirus* or 'the spiny lobster' of Bombay. *Journal of the Bombay Natural History Society, 18*, 360–369. pls A,B,11 text figs.

Prasad, R. R. (1983). Distribution and growth: Studies on the phyllosoma larvae from the Indian Ocean: I. *Journal of the Marine Biological Association of India, 20*, 143–156.

Prasad, R. R. (1986). Distribution, habits and habitats of palinurid lobsters and their larvae. *Marine Fisheries Information Service T&E Series, 70*, 8–15.

Prasad, R. R., & Tampi, P. R. S. (1957). Phyllosoma of Mandapam. *Proceedings of the National Academy of Sciences, India Section B, 23*, 48–67.

Prasad, R. R., & Tampi, P. R. S. (1959). On a collection of palinurid phyllosomas from the Laccadive seas. *Journal of the Marine Biological Association of India, 1*(2), 143–164.

Prasad, R. R., & Tampi, P. R. S. (1960). Phyllosoma of scyllarid lobsters from the Arabian sea. *Journal of the Marine Biological Association of India, 2*(2), 241–249.

Prasad, R. R., & Tampi, P. R. S. (1961). On the newly hatched phyllosoma of *Scyllarus sordidus* (Stimpson). *Journal of the Marine Biological Association of India, 2*, 250–252.

Prasad, R. R., & Tampi, P. R. S. (1968). Distribution of palinurid and scyllarid lobsters in the Indian Ocean. *Journal of the Marine Biological Association of India, 10*(1), 78–87.

Premkumar, V. K., & Daniel, A. (1975). Distribution pattern of the economically important spiny lobsters of the Genus *Panulirus* white, in the Indian Ocean. *Journal of the Marine Biological Association of India, 17*, 36–40.

Ptacek, M. B., Sarver, S. K., Childress, M. J., & Herrnkind, W. F. (2001). Molecular phylogeny of the spiny lobster genus *Panulirus* (Decapoda:Palinuridae). *Marine and Freshwater Research, 52*, 1937–1047.

Purushottama, G. B., & Saravanan, R. (2014). A rare deep water Japanese Blunt-horn lobster *Palinustus waguensis* Kubo, 1963 (Decapoda: Palinuridae) found off Mangalore coast, central west coast of India. *Indian Journal of Geo-Marine Sciences, 43*(8), 1550–1553.

Radhakrishna, Y., & Ganapati, P. N. (1969). Fauna of Kakinada Bay. *Bulletin National Institiute Science India, 38*, 689–699.

Radhakrishnan, E. V., & Jayasankar, P. (2014). First record of the reef lobster *Enoplometopus occidentalis* (Randall, 1840) from Indian waters. *Journal of the Marine Biological Association of India, 56*(2), 88–91.

Radhakrishnan, E. V., & Thangaraja, R. (2008). *Sustainable exploitation and conservation of lobster resources in India, a participatory approach* (pp. 184–192). Kochi: Glimpses of Aquatic Biodiversity - Rajiv Gandhi Chair Special Publication. CUSAT.

Radhakrishnan, E. V., Menon, K. K., & Lakshmi, S. (2000). Small-scale traditional spiny lobster fishery at Tikkoti, Calicut. *Marine Fisheries Information Service, Technical and Extension Series, 164*, 5–8.

Radhakrishnan, E. V., Vijayakumaran, M., & Shahul Hameed, K. (1990). On a record of the spiny lobster *Panulirus penicillatus* (Olivier) from Madras. *Indian Journal of Fisheries, 37*(1), 73–75.

Radhakrishnan, E. V., Kasinathan, C., & Ramamoorthy, N. (1995). Two new records of scyllarids from the Indian coast. *The Lobster Newsletter, 8*(1), 9.

Radhakrishnan, E. V., Deshmukh, V. D., Manisseri, M. K., Rajamani, M., Kizhakudan, J. K., & Thangaraja, R. (2005). Status of the major lobster fisheries in India. *New Zeland Jouranl of Marine and Freshwater Research, 39*, 723–732.

Radhakrishnan, E. V., Pillai, S. L., Rajool Shanis, C. P., & Radhakrishnan, M. (2011). First record of the reef lobster Enoplometopus macrodontus Chan and ng, 2008 from Indian waters. *Journal of the Marine Biological Association of India, 53*(2), 264–267.

Radhakrishnan, E. V., Chakraborty, R. D., Baby, P. K., & Radhakrishnan, M. (2013). Fishery and population dynamics of the sand lobster *Thenus unimaculatus* Burton & Davie, 2007 landed by trawlers at Sakthikulangara fishing harbour on the southwest coast of India. *Indian Journal of Fisheries, 60*(2), 7–12.

Ramakrishna, S. J., & Alukdar, S. T. (2003). Marine invertebrates of Digha coast and some recommendations on their conservation. *Records of the Zoological Survey of India, 101*(3–4), 1–23.

Rajamani, M., & Manickaraja, M. (1997). The spiny lobster resources in the trawling grounds off Tuticorin. *Marine Fisheries Information Service T&E Series, 148*, 7–9.

Rao, G. S., Suseelan, C., & Kathirvel, M. (1989). Crustacean resources of the Lakshadweep Islands. In: Marine living resources of the Union Territory of Lakshadweep- An indicator survey with suggestions for development. *CMFRI Bulletin, 43*, 72–76.

Rao, P. V., & Kathirvel, M. (1971). On the seasonal occurrence of *Penaeus semisulcatus* de Haan, *Panulirus polyphagus* (Herbst) and *Portunus pelagicus* (Linnaeus) in Cochin backwater. *Indian Journal of Fisheries, 19*, 129–134.

Rao, P. V., & George, M. J. (1973). Deep sea spiny lobster, *Puerulus sewelli* Ramadan: Its commercial potentialities. In *Proceedings of symposium on living resources of the seas around India (CMFRI)* (pp. 634–640).

Richards, A. H., Bell, L. J., & Bell, J. D. (1994). Inshore fisheries resources of the Solomon Islands. *Marine Pollution Bulletin, 29*, 90–98.

Sankolli, K. N., & Shenoy, S. (1974). On the laboratory hatched six phyllosoma stages of *Scyllarus sordidus* (Stimpson). *Journal of the Marine Biological Association of India, 15*, 218–226.

Satyanarayana, A. V. V. (1961). A record of *Panulirus penicillatus* (Olivier) from the inshore waters off Quilon, Kerala. *Journal of the Marine Biological Association of India, 3*(1), 269–270.

Saha, S. N., Vijayanand, P., & Rajagopal, S. (2009). Length-weight relationship and relative condition factor in *Thenus orientalis* (Lund, 1793) along East Coast of India. *Current Research Journal of Biological Sciences, 1*(2), 11–14.

Scholtz, G., & Richter, S. (1995). Phylogenetic systematics of the raptantian Decapoda (Crustacea, Malacostraca). *Zoological Journal of the Linnean Society, 113*, 289–328.

Schram, F. R., & Dixon, C. J. (2004). Decapod phylogeny: Addition of fossil evidence to a robust morphological cladistic data set. *Bulletin Mitzunami Fossil Museum, 31*, 1–19.

Sewell, R. B. S. (1913). Notes on the biological work of the R.I.M.S "Investigator" during the survey season 1910–1911 and 1911-1912. I. *Asiatic Society of Bengal, 9*, 329–390.

Shanmugam, S., & Kathirvel, M. (1983). Lobster resources and culture potential. *CMFRI Bulletin, 34*, 61–65.

Silas, E. G. (1969). New findings on trawling grounds on the continental slope bordering the Wadge Bank and extending to the Gulf of Mannar. *Bulletin, Central Marine Fisheries Research Institute No., 12*, 34–37.

Singh, S. P., Groeneveld, J. C., Al-Marzouqi, A., & Wilows-Munro, S. (2017). A molecular phylogeny of the spiny lobster *Panulirus homarus* highlights a separately evolving lineage form the Southwest Indian Ocean. *PeerJ, 5*, e3356. https://doi.org/10.7717/peerj.3356.

Sreedhar, U., Raghu Prakash, R., & Raleswari, G. (2010). Present scenario of the coastal and Deepwater trawl resources of southeast coast of India. In Meenakumari et al. (Eds.), *Coastal fishery resources of india- conservation and sustainable utilisation* (pp. 77–89). Cochin: Society of Fisheries Technologists (India).

Srikrishnadhas, B., Rahman, M. K., & Anandasekharan, A. S. M. (1991). A new species of the scyllarid lobster *Scyllarus tutiensis* (Scyllaridae: Decapoda) from the Tuticorin Bay in the Gulf of Mannar. *Journal of the Marine Biological Association of India, 33*(1&2), 418–421.

Subramanian, V. T. (2004). Fishery of sand lobster *Thenus orientalis* (Lund) along Chennai coast. *Indian Journal of Fisheries, 51*(1), 111–115.

Tampi, P. R. S., & George, M. J. (1975). Phyllosoma larvae in the IIOE (1960–65) collections- systematics. *Mahasagar, 8*(1–2), 15–44.

Thangaraja, R., & Radhakrishnan, E. V. (2012). Fishery and ecology of the spiny lobster *Panulirus homarus* (Linnaeus, 1758) at Khadiyapatanam in the southwest coast of India. *Journal of the Marine Biological Association of India, 54*(2), 69–79.

Thiagarajan, R., Krishna Pillai, S., Jasmine, S., & Lipton, A. P. (1998). On the capture of a live south African cape locust lobster at Vizhinjam. *Marine Fisheries Information Service T&E Series, 158*, 18–19.

Thirumilu, P. (2011). Largest recorded ridge-back lobsterette, *Nephropsis carpenteri* Wood-Mason, 1885 from Chennai coast. *Marine Fisheries Information Service T&E Series, 209*, 13–14.

Thomas, M. M. (1979). On a collection of deep sea decapod crustaceans from the Gulf of Mannar. *Journal of the Marine Biological Association of India, 21*(1&2), 41–44.

Thurston, E. (1887). Preliminary report on the marine fauna of Rameshwaram and neighbouring islands, Madras (Madras government museum). *Science Series, 1*, 1–41.

Thurston, E. (1894). Rameswaram Island and fauna of the Gulf of Mannar, Madras (Madras Government Museum). *Science Series, 11*, 98–138.

Tikader, B. K., & Das, A. K. (1985). *Glimpses of animal life in Andaman and Nicobar Islands* (p. 170). Calcutta: Zoological Survey of India.

Tikader, B. K., Daniel, A., & Subba Rao, N. V. (1986). *Seashore animals of Andaman and Nicobar Islands* (p. 188). Calcutta: Zoological Survey of India.

Toon, A., Finley, M., Staples, J., & Crandall, K. A. (2009). Decapod phylogenetics and molecular evolution. In J. W. Martin, K. A. Crandall, & D. L. Felder (Eds.), *Crustacean issues 18: Decapod crustacean Phylogenetics* (pp. 15–29). Boca Raton: Taylor & Francis/CRC Press.

Tortell, P. (2004). *Thoughts on integrated coastal zone management (ICTM) in Saudi Arabia*. Wellington: The Regional Organization for the Conservation of the Environment of the Red Sea and Gulf of Aden – PERSGA.

Tsang, L. M., Ma, K. Y., Ahyong, S. T., Chan, T.-Y., & Chu, K. H. (2008). Phylogeny of Decapoda using two nuclear protein coding genes. Origin and Evolution of the Reptantia. *Molecular Phylogenetics and Evolution, 48*, 359–368.

Tsang, L. M., Chan, T.-Y., Cheung, M. K., & Chu, K. H. (2009). Molecular evidence for the Southern Hemisphere origin and deep sea diversification of spiny lobsters (Crustacea: Decapoda: Palinuridae). *Molecular Phylogenetics and Evolution, 51*, 304–311.

Tshudy, D. (2003). Clawed lobster (Nephropidae) diversity through time. *Journal of Crustacean Biology, 23*, 178–186.

Tsoi, K. H., Chan, T.-Y., & Chu, K. H. (2011). Phylogenetic and biogeographic analysis of the spear lobsters *Linuparus* (Decapoda: Palinuridae), with the description of a new species. *Zoologischer Anzeiger, 250*, 302–315.

Vijayanand, P., Murugan, A., Saravanakumar, K., Khan, S. A., & Rajagopal, S. (2007). Assessment of lobster resources along Kanyakumari (South East Coast of India). *Journal of Fisheries and Aquatic Science, 2*, 387–394.

Vaitheeswaran, T. (2015). A new record of Scyllarid lobster *Scyllarus batei batei* (Holthuis, 1946) (Family: Scyllaridae, Latreille, 1852) (Crustacea: Decapoda: Scyllaridae) off Thoothukudi coast of Gulf of Mannar, southeast coast of India 08°52.6'N 78°16'E and 08° 53.8'N 78°32'E (310 m). *International Journal of Marine Sciences, 5*(54), 1–2.

Vijayakumaran, M., Maharajan, A., Rajalakshmi, S., Jayagopal, P., Subramanian, M. S., & Remani, C. (2012). Fecundity and viability of eggs in wild breeders of spiny lobsters, *Panulirus homarus* (Linnaeus, 1758), *Panulirus versicolor* (latreille, 1804) and *Panulirus ornatus* (Fabricius, 1798). *Journal of the Marine Biological Association of India, 54*(2), 5–9.

Wahle, R. A., Tshudy, D., Cobb, J. S., Factor, J., & Jaini, M. (2012). Infraorder Astacidea Latreille, 1802 p.p.: The marine clawed lobsters. In F. R. Schram & J. C. von Vaupel Klein (Eds.), *Treatise on zoology- anatomy, taxonomy, biology, the Crustacea* (Vol. 9, pp. 3–108). Leiden: Brill.

Webber, W. R., & Booth, J. D. (2007). Taxonomy and evolution. In K. L. Lavalli & E. Spanier (Eds.), *The biology and fisheries of the slipper lobster* (pp. 25–52). Boca Raton: Taylor & Francis/CRC Press.

Wood-Mason, J. (1872). On *Nephropsis stewarti*, a new genus and species of macrurous crustaceans, dredged in deep water off the eastern coast of the Andaman Islands. *Proceedings of Asiatic Society of Bengal, 1872*, 151.

Yang, C.-H., Chen, I.-S., & Chan, T.-Y. (2008). A new slipper lobster of the genus *Petrarctus* (Crustacea: Decapoa: Scyllaridae) from the West Pacific. *The Raffles Bulletin of Zoology, 19*, 71–81.

Yang, C.-H., Bracken-Grissom, H., Kim, D., Crandall, K. A., & Chan, T.-Y. (2012). Phylogenetic relationships, character evolution, and taxonomic implications within the slipper lobsters (Crustacea: Decapoda: Scyllaridae). *Molecular Phylogenetics and Evolution, 62*(1), 237–250.

Yang, C.-H., Bijukumar, A., & Chan, T.-Y. (2017). A new slipper lobster of the genus *Petrarctus* Holthuis, 2002 (Crustacea: Decapoda: Scyllaridae) from the southwest coast of India. *Zootaxa, 4329*(5), 477–486.

Zacharia, P. U., Krishnakumar, P. K., Dineshbabu, A. P., Vijayakumaran, K., Rohit, P., Thomas, S., Sasikumar, G., Kaladharan, P., Durgekar, N. R., & Mohamed, K. S. (2008). Species assemblage in the coral reef ecosystem of Netrani Island off Karnataka along the southwest coast of India. *Journal of the Marine Biological Association of India, 50*(1), 87–97.

Applications of Molecular Tools in Systematics and Population Genetics of Lobsters

4

Jeena N. S, Gopalakrishnan A, E. V. Radhakrishnan, and Jena J. K

Abstract

Recent advances in molecular tools have facilitated the detection of new species, differentiation of cryptic species, revision in taxonomy, assessment of evolutionary relationships, phylogenetic analysis, phylogeography and stock delineation studies in marine lobsters. This chapter discusses the implications of these tools in lobster research through a comprehensive review of the works carried out globally across different species. The use of molecular markers for resolving conflicting issues in phylogeny and evolution of marine lobsters has been detailed at the beginning of this chapter. An account on the application of 'DNA barcoding' for species and larval identification of lobsters and use of multiple genes in their phylogenetic systematics as well as distinction of lineages is described afterwards. Mitochondrial genomes have been utilised as a major genetic marker in many areas of research. The complete mitochondrial genome of various lobster species characterised so far is given for reference. Better knowledge on the diet of planktonic phyllosoma larvae of lobsters is important for commercialisation of lobster mariculture. Molecular diet analysis of larvae utilising conventional markers and high-throughput DNA sequencing methods has been explained. Efforts to effectively manage a fishery require an understanding of population demographics and connectivity. By defining the scale of subpopulation structure shown by a species through genetic analysis, management units can be accurately identified. The stock delineation studies in lobsters using different types of molecular markers have been discussed in this chapter. DNA microsatellites are considered as a dominant class of markers in genomic analysis for assessing genetic structure of populations, linkage, parentage and relatedness.

Jeena N. S (✉) · Gopalakrishnan A · E. V. Radhakrishnan
ICAR-Central Marine Fisheries Research Institute, Cochin, Kerala, India
e-mail: jeenans@rediffmail.com

Jena J. K
Indian Council of Agricultural Research, New Delhi, India

The use of highly polymorphic microsatellite markers with their application in diverse fields like detection of fine-scale genetic structure over small spatial scales for better management and their utility in assigning marine-protected areas (MPAs), parentage analysis, etc., are narrated. The application of gene sequencing and next-generation sequencing (NGS) derived markers in lobster research is detailed, and it is anticipated that these markers due to their improved analytical power and high precision will soon replace the mtDNA and microsatellite markers used in this field since 2004. The present scenario of lobster genetic research in India is reviewed at the end.

Keywords

Molecular taxonomy · Stock delineation · Microsatellite markers · Gene sequencing

4.1 Introduction

Marine lobsters belong to the suborder Macrura Reptantia Bouvier, 1917, of Order Decapoda, Class Crustacea (Holthuis 1991). There are four infraorders under Macrura Reptantia: Astacidea (reef lobsters and clawed/true lobsters), Glypheidea (living fossils), Achelata (spiny, slipper and furry lobsters) and Polychelida (blind lobsters). Chan (2010) elucidated six extant families, 55 genera and 248 species of living marine lobsters. Lobsters form commercially valuable fisheries in many regions of the world with a total catch exceeding 279,000 mt in 2010 (FAO 2010). True lobsters (Nephropidae) form major share in world catch followed by spiny lobsters (Palinuridae) and slipper lobsters (Scyllaridae). Jeffs (2010) has opined that the global supply from wild fisheries appears to be at or close to its maximum.

Molecular genetic markers have been considered as powerful tools that can detect genetic identity of species, individuals or populations (Avise 1994). Employment of these markers have aided in the discovery of more species of marine lobsters and was found to be useful in revising the taxonomy as well as concepts of the evolutionary relationships between species and families. They have immense analytical power, an essential requirement to explore the genetic variability in populations. Tasks such as description of the genetic structure of populations, detecting admixture of subpopulations and stock assignment have been considered as main applications of genetic markers in fishery research (Cuéllar-Pinzón et al. 2016).

Recent advances in molecular tools have created huge impact on marine lobster phylogenetic studies. Their phylogeny had remained an argumentative issue on aspects like monophyly, acquirement or loss of stridulating organ, direction of evolution between shallow and deep waters, etc. A taxon (group of organisms) which forms a clade that consists of an ancestral species and all its descendants is considered as a monophyletic group. Though the concept of monophyletic position of marine lobsters differs between authors, many were of the view that marine lobsters do not comprise a monophyletic group (Ahyong and O'Meally 2004; Porter et al. 2005) and the relationships between taxonomic ranks differ from the earlier well-established scheme of Holthuis (1991). Significantly contrasting results have been noticed in various phylogenetic studies (Tsang et al. 2008; Toon et al. 2009). The latest phylogenetic analysis with molecular markers concluded that lobsters are indeed a monophyletic group (Chan 2010), and some members of Thalassinidea (e.g. squat lobster) which are not regarded as 'true lobsters' were excluded (Jeena 2013).

Palinuridae (spiny lobsters), Scyllaridae (slipper lobsters) and Synaxidae (furry or coral lobsters) are the major three families which come under the infraorder Achelata. They have a common unique larval phase called phyllosoma, and all of them lack chelae on their first pair of pereopods (Scholtz and Richter 1995; Dixon et al. 2003). There is considerable phenetic and genetic support for the divergence of Palinuridae and Scyllaridae, but there exists only weak support for the family Synaxidae. The latest phylogenetic analyses of Achelata and Palinuridae recommended Synaxidae to be a junior synonym of Palinuridae or under the super-family Palinuroidea (Palero et al. 2009a; Tsang et al. 2009; Chan 2010).

Palinuridae include 12 genera of spiny lobsters grouped into two lineages, Stridentes and Silentes, based on the presence or absence of an acoustic stridulating organ (Berry 1974; Pollock 1990). The Stridentes include nine genera: *Panulirus*, *Palinurus*, *Palinustus*, *Puerulus*, *Linuparus*, *Justitia*, *Nupalirus*, *Palinurellus* and *Palibythus*, while Silentes has only three genera: *Jasus*, *Projasus* and *Sagmariasus*. Using a set of nuclear molecular markers, Tsang et al. (2009) proved that Silentes is more primitive, constituting the basal lineage to Stridentes. They have also demonstrated that Stridentes were monophyletic, and the synamorphic stridulating organ appeared only once within their evolution.

Slipper lobsters of the family Scyllaridae, which are characterised by the flattened antennal flagellum, had been drawing lesser research focus perhaps due to the lower number of commercially important species (Lavalli and Spanier 2007). Scyllarid lobsters inhabit the coastal waters along the continental shelf and upper slope areas across the Equator, low latitudes and also temperate latitudes affected by warm water currents (Webber and Booth 2007). Large species of scyllarids are fished commercially (Duarte et al. 2010), and the smaller species are acclaimed for their taxonomic diversity (Chan 2010). Four subfamilies, namely, Arctidinae, Ibacinae, Scyllarinae and Theninae, were proposed based on the carapace shapes and morphology of the maxilliped, exopods and mandibular palp (Holthuis 1985, 1991; 2002). Yang et al. (2012) retrieved the phylogenetic relationships within Scyllaridae using molecular markers, and there are four subfamilies, 20 extant genera and 89 species (Arctidinae = 17, Ibacinae = 15, Theninae = 5 and Scyllarinae = 52 species respectively) at present.

There are two hypotheses that postulate the evolution of lobsters. The first one suggested the origin of lobsters from a deep-sea ancestral stock to the shallow-water genera and the second one, vice versa (Jeena 2013). The hypothesis of deep water to shallow water evolution was suggested earlier, analysing adult similarity and larval cladistics (e.g. George and Main 1967; Baisre 1994; George 2005; George 2006a, b). But recent genetic studies in Palinuridae and Scyllaridae supported the view of an onshore (shallow-water) reef origin, which then disseminated into offshore (deeper) reefs and ultimately adapted to typical soft deep-sea bottoms (Davie 1990; Tsang et al. 2009; Tsoi et al. 2011; Yang et al. 2012).

4.2 Molecular Identification and Phylogenetic Studies

The recent years witnessed numerous studies on the ecology, phylogeography, and molecular phylogeny of spiny lobsters. The Genus *Panulirus* is the largest group in the family Palinuridae (McWilliam 1995) which comprises species that are globally highly priced. On the other hand, slipper or shovel-nosed lobsters of the family Scyllaridae are often considered as a desirable incidental catch in commercial fishery (DiNardo and Moffitt 2007). The major species contributing to slipper lobster fishery were *Ibacus* and *Thenus* (FAO 2010). In India, *Thenus* species has become targeted fishery (Radhakrishnan et al. 2007).

Panulirus has been the most successful in terms of species diversity among all the genera in Palinuridae. There are 21 taxa (with three subspecies) at present (Chan 2010). Evolutionary divergence in *Panulirus homarus* was investigated by Lavery et al. (2014) which led to the conclusion that the two subspecies *P. homarus homarus* and *P. homarus megasculpta* are the same. They noted that the subspecies *P. homarus rubellus* was the most divergent one and the *P. homarus* "Brown" subspecies may be an allopatric population.

4.3 Identification of Species and Larvae

'Species' is the focal taxonomic unit of conservation biology. Molecular tools like 'DNA barcoding' have been extensively employed for taxonomy and biodiversity studies, serving efficiently to separate cryptic and very similar species. These tools are utilised in the vast and speciose group of decapod crustaceans also and comparing short fragments of DNA, some taxonomic impediments could be overcome by researchers which led to the discovery of more lobster species. Approximately 650 bp region from the 5′-end of the cytochrome *c* oxidase 1 (COI or COX) gene is designated as 'DNA barcode' or a standardised tool for molecular taxonomy and identification (Ratnasingham and Hebert 2007). The morphological variants of *Panulirus longipes* with partially overlapping distribution could be separated into two subspecies, *P. longipes longipes* and *P. l. bispinosus* (~3% divergent), using mitochondrial COI (Ravago and Juinio-Meñez 2002). A number of studies have pointed out that it is undesirable to rely on a single sequence for taxonomic identification (Mallet and Willmort 2003; Matz and Nielsen 2005). Other mitochondrial genes like 16S rDNA (16S) are also widely used in crustaceans. Burton and Davie (2007) investigated the genus *Thenus* using morphological and molecular tools to reassign five distinct species in the previous monotypic subfamily, Theninae. The DNA barcode was successfully employed to identify *Thenus* species in different countries (Jeena 2013; Iamsuwansuk et al. 2012).

Molecular tools have been in use for the authentic identification of phyllosoma larvae way back from 1990s. Silberman and Walsh (1992) demonstrated explicit discrimination of phyllosoma of two species of *Panulirus* inhabiting northwestern Atlantic using restriction fragment length polymorphism (RFLP) analysis of 28S rDNA. RFLP analysis of the mitochondrial COI gene to identify ten species of spiny lobsters from the Indo-Pacific was developed by Chow et al. (2006), which was later applied with modifications on mid- to final-stage phyllosoma collected from the northwestern Pacific. The phyllosoma of *Panulirus echinatus* in the central Atlantic, which was not noticed before in plankton samples, was identified using mitochondrial 16S rDNA sequence analysis (Konishi et al. 2006). Mitochondrial 16S rRNA gene analysis could reveal the identities of wild-caught phyllosoma of the slipper lobster *Eduarctus martensii* (Wakabayashi et al. 2017).

Molecular markers could also bring into light the cryptic speciation in some spiny lobsters of commercial importance. COI and 16S rRNA genes and nuclear gene sequences were used to evaluate the taxonomic status of *Panulirus argus* sampled from sites in the Caribbean Sea and Southwest Atlantic, which resulted in taxonomic ambiguity regarding this species (Tourinho et al. 2012). The genetic divergence found between specimens from the two areas urged the authors to hypothesise the existence of a cryptic species. Later, Giraldes and Smyth (2016), based on morphology, divided *Panulirus argus* into two different species: *P. argus* and *Panulirus meripurpuratus* sp. nov. The former species sensu stricto is distributed north of Amazon-Orinoco plume, a biogeographic barrier, and the latter to the south in Brazil. This information is supposed to aid in future management and conservation of *Panulirus* in the Atlantic Ocean and indigenous species from Brazilian waters. Similarly, analysis of the mitochondrial markers (16S and COI regions) revealed existence of at least six species in the morphologically close taxa of the genus *Palinurus*, resulting in the discovery of a new species, *Palinurus barbarae* (Groeneveld et al. 2006).

The molecular markers were found to be useful in detecting lineages also. Two very divergent allopatric lineages from the eastern Pacific and the central and western Pacific could be detected in *Panulirus penicillatus* (Chow et al. 2011; Abdullah et al. 2014; Iacchei et al. 2016). Analyses of molecular data could recover *P. homarus homarus* and *P. homarus rubellus* as separately evolving lineages in the Indo-West Pacific (Singh et al. 2017).

4.4 Implications of Molecular Tools in Taxonomy and Diet Analysis

Phylogenetic studies provide a sound basis for taxonomy. Researchers have presented an increased resolution when phylogenetic analyses were coalesced with multiple genes (Ahyong and O'Meally 2004; Porter et al. 2005). The majority of molecular phylogenetic research in spiny lobsters using mitochondrial markers focused on relationships between species within a genus. Mitochondrial gene sequences were used to deduce the phylogenetic relationships within *Palinuridae*

(Cannas et al. 2006). Phylogeny of the genus *Panulirus* (Ptacek et al. 2001), *Palinurus* (Groeneveld et al. 2007), *Linuparus* (Tsoi et al. 2011), *Jasus* (Ovenden et al. 1997) and the clawed lobster genera (Tshudy et al. 2009) were also reconstructed using sequence data derived from the mitochondrial genes. Phylogenetic relationship of Iranian lobsters was derived using mtDNA-COI RFLP method (Ardalan et al. 2010). Decapod phylogeny including lobster genera was also evaluated using mitochondrial and nuclear markers (Porter et al. 2005).

The first attempt on molecular phylogeny reconstruction of genus *Panulirus* using ribosomal DNA and morphological characters was by Patek and Oakley (2003). Evolutionary relationships and taxonomic discrepancies in infraorder Achelata were inferred from concatenated datasets of nuclear and mitochondrial genes (Palero et al. 2009a). Chan et al. (2009) attempted phylogenetic reconstruction of *Metanephrops* spp. using nuclear (histone H3) and mitochondrial genes for all 17 existing species. By means of mitochondrial and microsatellite markers, the monophyly patterns and phylogenetic relationships of *Palinurus* species were determined (Palero et al. 2009b). Phylogenetic reconstruction of Palinuridae was also undertaken using nuclear protein coding gene sequences (Tsang et al. 2009). Yang et al. (2012) ascertained the phylogeny of 54 scyllarid species using concatenated gene sequences.

The mitochondrial genomic fraction though relatively small, forms a critical component in eukaryotic DNA. Mitochondrial DNA, alternatively known as mtDNA or mitochondrial genome or mitogenome, is used as a fundamental genetic marker in a vast number of research areas spanning population genetics, evolutionary biology, phylogenetics, phylogeography, biodiversity and molecular ecology (Timbó et al. 2017). The complete mitochondrial genomes of many lobster species have been characterised (Table 4.1) and are available on the NCBI GenBank for reference.

For commercialisation of any lobster species, better understanding of their planktonic predatory phyllosoma larval diet is of paramount importance. If their diet were nutritionally modelled on the natural diet, there is higher probability that the mortalities experienced during larviculture might be reduced. Laboratory experiments indicated that the soft-bodied zooplanktons may be the item that form the major prey for mid- to late-stage phyllosoma larvae of spiny and slipper lobsters (Kittaka 1994). Molecular diet analysis with 18S rDNA could support this finding (Suzuki et al. 2006). The diets of phyllosoma, which are known to be opportunistic carnivores, correlate with the relative abundance of prey organisms in ambient water (Suzuki et al. 2008). A DNA-based diet approach has been proven to be a promising tool for analysing diets of phyllosoma of *Panulirus japonicus* (Suzuki et al. 2006; Chow et al. 2010) and *P. cygnus* (O'Rorke et al. 2012). A suite of group-specific primers suitable to amplify DNA fragment from specific loci of potential prey, but not from lobsters, was developed for *Jasus edwardsii* and *Scyllarus* sp. (Connell et al. 2014). High-throughput DNA sequencing techniques like 454 pyrosequencing is in use to identify prey species from phyllosoma midgut glands (O'Rorke et al. 2014).

Table 4.1 List of published complete mitogenomes of lobsters (in NCBI GenBank up to June 2018)

Sl. No	Organism	Common name	Mitogenome size (bp)	NCBI accession number	References
1	*Panulirus japonicus*	Japanese spiny Lobster	15,717	NC_004251.1	Yamauchi et al. (2002)
2	*Panulirus stimpsoni*	Chinese spiny lobster	15,677	NC_014339.1	Liu and Cui (2011)
3	*Panulirus ornatus*	Ornate rock lobster	15,680	HM_446347.1	Liang (2012)
4	*Homarus americanus*	American lobster	16,432	HQ_402925.1	Kim et al. (2011)
5	*Enoplometopus occidentalis*	Hawaiian red lobster	15,111	NC_020027.1	Shen et al. (2013)
6	*Panulirus versicolor*	Painted spiny lobster	15,767	KC_107808.1	Shen et al. (2013)
7	*Scyllarides latus*	Mediterranean slipper lobster	15,663	NC_020022.1	Shen et al. (2013)
8	*Homarus gammarus*	European lobster	14,316	NC_020020.1	Shen et al. (2013)
9	*Palinurellus wieneckii*	Indo-Pacific furry lobster	15,699	NC_021753.1	Yang et al. (2013)
10	*Thenus orientalis*	Flathead lobster, bay lobster	16,826	NC_024440.1	Tan et al. (2016)
11	*Sagmariasus verreauxi*	Green rock lobster	15,470	NC_022736.1	Doyle et al. (2015)
12	*Nephrops norvegicus*	Norway lobster	16,132	NC_025958.1	Gan et al. (2016)
13	*Enoplometopus debelius*	Debelius's dwarf reef lobster	15,641	NC_025592.1	Ahn et al. (2016a)
14	*Ibacus ciliatus*	Japanese fan lobster	15,696	NC_025581.1	Ahn et al. (2016b)
15	*Panulirus cygnus*	Australian spiny lobster	15,724	NC_028024.1	Kim et al. (2016)
16	*Metanephrops thomsoni*	Red-banded lobster	19,835	NC_027608.1	Ahn et al. (2016c)
17	*Panulirus homarus*	Scalloped/Indian spiny lobster	15,665	JN_542716.1	Xiao et al. (2017)

The karyological data in decapods are rare. Comparative cytogenetics in four species of Palinuridae with B chromosomes, ribosomal genes and telomeric sequences clearly indicated that there is a difference in situation in Scyllaridae and Palinuridae (Salvadori et al. 2012).

4.5 Population Genetic Studies

The information on stock structure has important implications for resource managers working to rebuild fisheries that have been overfished. Differences in stock structure of a species can be an indicative that there may be separate populations with distinct gene pools. Failure to detect the underlying genetic structure is a concern for commercially exploited species since it may result in overexploitation and depletion of localised subpopulations with corresponding thinning of genetic variation and ultimate loss of adaptive potential (Carvalho and Hauser 1994). New genetic techniques are powerful enough to detect much subtle genetic differences within species and to uncover previously unknown spatial genetic structures, ultimately providing essential information for future management of exploited marine species.

Evidence of spatial genetic structure can have significant implications for stock management. This may indicate restricted mixing of adults and larvae, question the idea of a single population and may suggest that regional regulation would be more appropriate. Also it provides a better idea regarding the genetic health and evolutionary potential of the fishery. Small effective population sizes and the associated declines in genetic diversity provide early warning signals of a fishery vulnerable to reduced productivity. For example, it has been accepted that to provide data on the maximum sustainable yield (MSY), it is first essential to reveal patterns of the population structure if it exist (Evans and Evans 1995). The knowledge of genetic structure would help fishery managers in estimating the contribution of new recruits from local and foreign sources (Diniz et al. 2010).

The arrival of fast and cheap genotypic screening and NGS-derived markers are fast replacing the classic marker types like allozymes, AFLPs, whole mitogenomes, RAPDs and RFLPs on mtDNA (Cuéllar-Pinzón et al. 2016). Though there are reports on the use of RAPD fingerprinting in population genetics in a wide array of crustaceans, relatively few reports are available with this marker in lobsters (Table 4.2). This technique was helpful in discriminating the tissues of American and European lobsters (Hughes and Beaumont 2004) and comparison of different species of lobsters (Park et al. 2005). Allozymes have been utilised to assess the stock structure in *Homarus americanus* (Shaklee 1983; Kornfield and Moran 1990), *H. gammarus* (Jørstad et al. 2004), *Panulirus marginatus* (Seeb et al. 1990) and *Nephrops norvegicus* (Stamatis et al. 2006).

Two class of markers, microsatellites and mitochondrial DNA-based ones, dominated the majority of the research in lobsters after 2004. There are many advantages for mitochondrial markers which make them ideal for use in population genetic studies for several reasons (Galtier et al. 2009). Mitochondrial DNA can generate signals about population history over short time frames due to its elevated mutation rate. Population structures of spiny lobsters such as *P. japonicus* (Inoue et al. 2007; Sekiguchi and Inoue 2010), *Palinurus elephas* (Cannas et al. 2006; Palero et al. 2008), *Panulirus regius* (Froufe et al. 2011), *P. argus* (Naro-Maciel et al. 2011), *Palinurus mauritanicus* (Palero et al. 2008) and achelate lobsters (Froufe et al. 2011) have been attempted with nucleotide sequence analysis of mitochondrial COI

Table 4.2 Genetic variability indices reported for different lobster species with RAPD and allozymes. Comments are based on the degree of genetic differentiation as suggested by Wright (1978)

Species	Category	Marker used	Nei's gene diversity 'h'	F_{ST} or G_{ST} or Φ_{ST}	Genetic similarity among populations	Genetic distance (GD) among populations	Comments on genetic differentiation	References
Homarus americanus	Lobster	RAPD	……	$F_{ST} < 0.000$ to 0.073	……	0.002–0.006	No genetic differentiation	Harding et al. (1997)
H. gammarus	Lobster	RAPD	……	0.1166	0.35–0.58 (BSI)	High	High	Ulrich et al. (2001)
N. norvegicus	Lobster	Allozyme	0.165–0.187	0.013	……	0.0013–0.0186	Shallow, but significant	Stamatis et al. (2006)
P. homarus homarus	Lobster	RAPD	0.1719	0.0136	0.95–0.96	0.5–0.6	No genetic differentiation	Jeena et al. (2016c)
T. unimaculatus	Lobster	RAPD	0.1446	0.0442	0.92–0.95	0.07–0.08	-do-	Jeena et al. (2016a)

gene. The cytochrome oxidase II (COII) gene was utilised to study genetic connectivity of the lobster *Jasus tristani* (Von der Heyden et al. 2007). The genetic diversity of *P. homarus* populations of southern Sri Lanka was compared with south India using COI gene showed no significant difference between populations (Senevirathna and Munasinghe 2014). Population structure studies employing mtDNA markers have been attempted worldwide in many lobster species (Table 4.3).

DNA microsatellites have proven to be one of the most powerful classes of markers in genomic analysis for deciphering population structure, linkage, parentage and relatedness. Microsatellites are highly variable short-tandem DNA repeats which can be developed for use in genetic studies. Their core-repeat units are short, ranging from two to six base pairs in length. Most common are the di-, tri- and tetranucleotide repeats. Microsatellite polymorphisms are detectable, and the results can be visualised using fluorescently labelled PCR primers with automated sequencing. Microsatellites are effective markers for detecting genetic subdivision even within marine species having high larval dispersal capabilities (Ruzzante et al. 1996). Its data can also be used for determining the effective population size as well as changes in population size.

Polymorphic microsatellite loci have been developed for commercially important spiny and slipper lobsters to conduct population genetic analysis. Microsatellite markers were characterised from partial genomic DNA libraries for *H. gammarus* (André and Knutsen 2010) and *Panulirus interruptus* (Ben-Horin et al. 2009) and also through next-generation sequencing (NGS) techniques in *Jasus edwardsii* (Thomas and Bell 2012), *Panulirus ornatus* (Delghandi et al. 2016), *P. argus* (Truelove et al. 2015b), *Panulirus guttatus* (Truelove et al. 2016), *P. echinatus* (Santos et al. 2018), *P. homarus* (Delghandi et al. 2015) and *Panulirus* spp. (Dao et al. 2013). Microsatellite primers have been developed in slipper lobsters as well (Rodríguez-Rey et al. 2013; Faria et al. 2014).

One of the critical parameters in connectivity models for optimising size and spacing of marine protected areas (MPAs) is the information on spatiotemporal patterns of population differentiation. Polymorphic microsatellite markers (msatDNA) are considered to be more efficient in resolving population structure at small spatial scales than mtDNA markers (Hellberg 2009). Primary results based on msatDNA proposed that a subregional population structure may exist in the Caribbean spiny lobster (*Panulirus argus*) among marine-protected areas (MPAs) in the Mesoamerican Barrier Reef (Truelove et al. 2015a). To enable management of locally based MPAs for lobster fishery, Truelove et al. (2015b) used powerful polymorphic microsatellite markers, which showed significant levels of differentiation among discrete size classes in population. Similarly, population genetic structure of the European lobster (*H. gammarus*) in the Irish Sea was investigated by Watson et al. (2016) using

Table 4.3 The population structure of different lobster species based on various molecular markers

Population	Region	Marker used	Genetic differences among demes	References
Spiny lobsters				
Panulirus japonicus	Japan	COI	No genetic differentiation	Inoue et al. (2007)
Palinurus mauritanicus	Northwestern Mediterranean and Atlantic	COI	No	Palero et al. (2008)
Panulirus regius	Cape Verde and southwestern Africa	COI	No	Froufe et al. (2011)
P. penicillatus	Western and Eastern Pacific	COI and 16S rDNA	High genetic differentiation	Chow et al. (2011)
Panulirus argus	N. Caribbean Sea	COI	No	Naro-Maciel et al. (2011)
	N. Caribbean Sea	Control region	No	
	Caribbean Sea and Brazil	Control region	High	Diniz et al. (2005)
Palinurus elephas	Mediterranean	COI	No	Cannas et al. (2006)
	Atlantic and Mediterranean	COI	Yes	Palero et al. (2008)
	Azores, Atlantic, Mediterranean, Sagres	COI	Yes, shallow	Froufe et al. (2011)
Panulirus interruptus	Pacific-Baja California	mtDNA RFLP	No	García-Rodríguez and Perez-Enriquez (2006)
Palinurus gilchristi	South Africa	Control region	No	Tolley et al. (2005)
Palinurus delegogae	SW Indian Ocean (Africa)	Control region	Shallow	Gopal et al. (2006)
Homarus gammarus	Subarctic	RFLP	Medium	Jørstad et al. (2004)
	Atlantic and Mediterranean	RFLP	Yes, significant	Triantafyllidis et al. (2005)
Jasus lalandii	South Africa	16S rRNA	Shallow	Matthee et al. (2008)
Jasus tristani	South Atlantic	CO II	Shallow	Von der Heyden et al. (2007)
N. norvegicus	Northeast Atlantic and Mediterranean	RFLP	Shallow	Stamatis et al. (2004)

(continued)

Table 4.3 (continued)

Population	Region	Marker used	Genetic differences among demes	References
P. homarus rubellus	SW Indian Ocean, African mainland (Mozambique and eastern South Africa) and southeast Madagascar	COI-like region	Yes, significant	Reddy et al. (2014)
P. homarus	West Indian Ocean (WIO), Arabian Sea and Tanzania	mtDNA control region	Yes, significant	Farhadi et al. (2013)
P. ornatus	Southeast Asian archipelago from Vietnam to Australia	mtDNA control region sequences and microsatellite loci	No	Dao et al. (2015)
Panulirus cygnus	Western Australian coastline	12S rRNA and microsatellites	No	Kennington et al. (2013)
Jasus edwardsii	New Zealand and Australia	Microsatellites and mathematical modelling	Yes, significant	Thomas and Bell (2012)
Panulirus interruptus	USA and Mexico	Control region and 16S rRNA	No	García-Rodríguez and Perez-Enriquez (2006)
Homarus gammarus	Irish Sea	Microsatellites	No	Watson et al. (2016)
Palinurus elephas	Atlantic Ocean and Mediterranean Sea	Mitochondrial control region and microsatellite loci	Yes	Babbucci et al. (2010)
P. homarus	Sri Lanka (Indian Ocean)	COI and Cytochrome B	No	Senevirathna et al. (2016)
Jasus frontalis	Southeastern Pacific Islands	COI	No	Porobić et al. (2013)
P. penicillatus	Pacific region	mtDNA control region	Yes	Abdullah et al. (2014)
P. ornatus	Indo-West Pacific	mtDNA control region	Yes	Yellapu et al. (2016)
N. norvegicus	Atlantic and Mediterranean	mtDNA control region	Yes	Gallagher et al. (2018)
Jasus edwardsii	Southern Tasmania, Australia, New Zealand	Microsatellites	Yes	Morgan et al. (2013)

(continued)

Table 4.3 (continued)

Population	Region	Marker used	Genetic differences among demes	References
N. norvegicus	Icelandic waters	Microsatellites	No	Pampoulie et al. (2011)
Panulirus interruptus	Gulf of California, coast of the Baja California Peninsula, United States	Mitochondrial control region and 16S rRNA	No	García-Rodríguez et al. (2017)
Homarus gammarus	Southwestern UK and Europe	Microsatellites	Yes	Ellis et al. (2017)
Panulirus homarus	West Indian Ocean	Mitochondrial control region, COI, microsatellite loci	Yes	Farhadi et al. (2017)
Panulirus echinatus	Central equatorial, south and mid-Atlantic Ocean	Microsatellite loci		Santos et al. (2018)
P. penicillatus	Japan and Taiwan	Microsatellite loci	No	Abdullah et al. (2017)
Slipper lobsters				
Scyllarides latus	Azores and Atlantic	COI	No	Froufe et al. (2011)
Scyllarides latus	Mediterranean and 13 locations from the NE Atlantic	Microsatellite markers	No	Froufe et al. (2011)
Scyllarides latus	NE Atlantic and western Mediterranean	Microsatellite markers developed using NGS	No	Faria et al. (2013)
T. unimaculatus	Andaman Sea (provinces of Thailand)	12S rRNA and COI	No	Wongruenpibool and Denduangboripant (2013)
Scyllarus arctus	NE Atlantic and selected locations of Iberian Peninsula	Microsatellites	No	Faria et al. (2014)
Scyllarides brasiliensis	Southwest Atlantic	COI and control region	No	Rodríguez-Rey et al. (2014)
T. unimaculatus	Indian Ocean	COI and CytB	No	Jeena et al. (2016a)

microsatellite markers to assess the effectiveness of management strategies in the first British marine-protected area, which revealed a panmictic population with no evidence of recent genetic bottlenecks and existence of large effective population sizes.

Decapod crustaceans display extensive variation in fertilisation strategies, spanning from single to multiple paternity, and knowledge of mating systems and behaviour are needed for the conversant management of commercially exploited lobster fisheries (Ellis et al. 2015). Multiple paternity in Norway lobster (*N. norvegicus*) was evaluated with microsatellite markers (Streiff et al. 2004). Paternity of individual broods in the European lobster *Homarus gammarus* from the Atlantic peninsula, assessed with microsatellites, indicated that multiple paternal fertilisations are either absent, unusual or highly skewed in favour of a single male (Ellis et al. 2015). Microsatellites were found to be successful in parentage assignment of *P. homarus* and in future, this may provide a practical tool for parentage analysis in hatchery production of juveniles in commercial breeding programmes of tropical spiny lobsters (Delghandi et al. 2017).

Due to the limited genetic resolution offered by traditional genotypic methods and weak genetic differentiation in marine species, decoding their genetic structure and deducing connectivity had been a challenging task. The significant growth of gene sequencing and NGS-derived markers after 2010 is being applied and expected to be implemented more in fishery research tasks in the near future (Ekblom and Galindo 2011). The hierarchical genetic structure of the American lobster (*Homarus americanus*) throughout the species' range could be delineated using restriction site-associated DNA sequencing (RAD sequencing), and individuals could be successfully assigned to their location of origin (Benestan et al. 2015). RAD sequencing was also useful to develop SNP markers in *H. gammarus* to detect genetic differentiation between lobsters of Mediterranean and Atlantic regions (Jenkins et al. 2018). In Australia and New Zealand, the population structure of the commercially important southern rock lobster *Jasus edwardsii* was investigated using double digest restriction site-associated DNA sequencing (ddRADseq). The aim was to identify a panel of SNP markers that could be used to trace the country of origin and assignment tests performed with the outlier SNP panel allocated 100% of the individuals to country of origin which demonstrated the usefulness of these markers for food traceability (Villacorta-Rath et al. 2016). DdRADseq technique was useful to investigate the chaotic genetic patchiness and postsettlement selection in *Jasus edwardsii* (Villacorta-Rath et al. 2018) in southeast Australia. Using ddRADseq libraries of two lobster species, *Jasus edwardsii* and *Sagmariasus verreauxi*, Souza et al. (2017) designed probes for population genomic studies in five other species of *Jasus*, and this data are expected to be useful to assess spatial–temporal genetic variation in *Jasus* species found in the Southern Hemisphere. The application of conventional marker types is expected to disappear or remain pictorial following the influx of fast and cheap next-generation sequencing (NGS)-derived markers (Cuéllar-Pinzón et al. 2016).

4.6 Indian Scenario

Lobsters are regarded as a low-volume, high-value resource from the Indian seas. Though it constitutes only 0.46% of the total marine crustacean landings (CMFRI, 2018), it had been an important export commodity comprising 0.25% in quantity and 1% in value (MPEDA 2009). Twenty-five species of lobsters have been reported so far from the Indian coast (Modayil and Pillai 2007) of which only four littoral and one deep sea forms are significant in commercial fishery.

Lobster fishery showed a declining average around 2200 mt for nearly 15 years from 1985, which again dropped to 1371 mt in 2004 (Radhakrishnan et al. 2005). The landing was estimated to be 2863 tonnes in 2017 (CMFRI 2018). The slipper lobster *Thenus unimaculatus* and the spiny lobster *Panulirus homarus* contributed the most to lobster fishery (CMFRI 2011). Despite the hardy nature and good growth rate of juveniles, prolonged larval duration extending over several months combined with limited success in hatchery production of seeds poses problem in large-scale lobster aquaculture (Kittaka and Booth 2000). The slipper lobster is fast evolving as a candidate species with advantages like shorter larval life, hardiness, high growth rates and good market price (Mikami and Kuballa 2007; Vijayakumaran and Radhakrishnan 2011). India has succeeded in completing the larval rearing of *Thenus* spp. (Kizhakudan et al. 2004).

The major focus of lobster research in India were centred on fishery, stock assessment, growth, culture and breeding with a few reported genetic works on larval identification (Dharani et al. 2009): 18S rDNA gene polymorphism (Rejinie Mon et al. 2011) and phylogeny reconstruction (Suresh et al. 2012). Molecular signatures of 11 commercially important species of lobsters from the Indian EEZ were generated with a set of mitochondrial as well as nuclear markers; genetic identities of widely distributed *Thenus unimaculatus* and *P. homarus homarus* were confirmed, and phylogeny was reconstructed to clarify the evolutionary relationships (Fig. 4.1; Jeena et al. 2016b). Mitochondrial and nuclear markers were utilised for identification of *Palinustus waguensis* from India (Chakraborty et al. 2016).

The identification of stock structure has been widely accepted as a sine qua non for sustainable management of marine fisheries (Reiss et al. 2009). The detection of genetic stocks (if any) with high precision markers is of extreme importance in the current context of alarming decline in lobster landings to formulate ideal conservation strategies. *Thenus orientalis*, which occur as by-catch in trawl fisheries, constitute the most important component of the lobster fishery on the northwest, southwest and southeast coasts of India (Radhakrishnan et al. 2007). Along Mumbai in the northwest, *Thenus* fishery collapsed in 1994 and is yet to recover, creating concern about the sustainability of the fishery (Deshmukh 2001). RAPD and mitochondrial DNA genes were used to investigate the genetic population structure of *T. unimaculatus* and *P. homarus* (Jeena et al. 2016a, c), which indicated low levels of genetic differentiation probably due to the high connectivity and resulting panmixia. The lengthy planktonic larval phase as well as transport and dispersal by monsoon currents of Northern Indian Ocean may be the reasons. Further studies integrating oceanic or coastal circulations with larval distribution pattern need to be carried out as

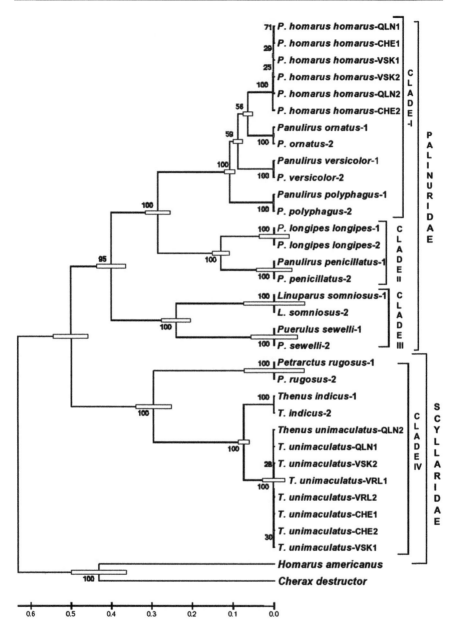

Fig. 4.1 Maximum likelihood tree of the 11 lobster species based on best-fitting nucleotide substitution model (TN93 + G + I) in MEGA5 inferred from haplotype sequence variation of the 1790 bp mtDNA region. Numbers at nodes indicate the bootstrap values. *H. americanus* (NC_015607.1) and *Cherax destructor* (NC_011243.1) from GenBank are included as outgroup species. VRL, CHE, QLN indicate collection locations Veraval (Gujarat), Chennai (Tamil Nadu) and Kollam (Kerala) in India for the species

attempted in lobsters like *Scyllarus* (Inoue and Sekiguchi 2005). High-resolution molecular markers derived from next-generation sequencing (NGS) technologies need to be applied for the evaluation of fine-scale genetic structure analysis of lobsters in Indian waters.

4.7 Conclusion

The use of molecular tools has contributed to the discovery of more species of lobsters and has thrown light on the evolution and systematics of marine lobsters. The diet analysis of phyllosoma may aid to improve our present level of knowledge regarding the exact feed requirement of larvae, which is one of the major bottle necks in lobster larviculture. The delineation of stock structure with various molecular markers will be helpful in identifying management units for ensuring sustainable harvest of this high value resource, which is under decline in many parts of the world due to indiscriminate exploitation. Since proven hatchery technology for commercial seed production is lacking for most of the species, more research focus may be given on conservation and management aspects. Genetic tools can provide information that may support the formulation of optimum management measures to save this depleting resource.

References

Abdullah, M. F., Chow, S., Sakai, M., Cheng, J. H., & Imai, H. (2014). Genetic diversity and population structure of pronghorn spiny lobster *Panulirus penicillatus* in the Pacific region. *Pacific Science, 68*(2), 197–211.

Abdullah, M. F., Cheng, J. H., Chen, T. I., & Imai, H. (2017). Development of compound polymorphic microsatellite markers for the pronghorn spiny lobster *Panulirus penicillatus* and comparison of microsatellite data with those of a previous mitochondrial DNA study performed in the northwestern Pacific. *Biogeography, 19*, 61–68.

Ahn, D. H., Min, G. S., Park, J. K., & Kim, S. (2016a). The complete mitochondrial genome of the Violet-spotted reef lobster *Enoplometopus debelius* (Crustacea, Astacidea, Enoplometopidae). *Mitochondrial DNA Part A, 27*(3), 1819–1820.

Ahn, D. H., Kim, S., Park, J. K., Shin, S., & Min, G. S. (2016b). The complete mitochondrial genome of the Japanese fan lobster *Ibacus ciliatus* (Crustacea, Achelata, Scyllaridae). *Mitochondrial DNA Part A, 27*(3), 1871–1873.

Ahn, D. H., Min, G. S., Park, J. K., & Kim, S. (2016c). The complete mitochondrial genome of the red-banded lobster *Metanephrops thomsoni* (Crustacea, Astacidea, Nephropidae): a novel gene order. *Mitochondrial DNA Part A, 27*(4), 2663–2664.

Ahyong, S. T., & O'Meally, D. (2004). Phylogeny of the Decapoda Reptantia: resolution using three molecular loci and morphology. *The Raffles Bulletin of Zoology, 52*(2), 673–693.

André, C., & Knutsen, H. (2010). Development of twelve novel microsatellite loci in the European lobster (*Homarus gammarus*). *Conservation Genetics Resources, 2*(1), 233–236.

Ardalan, M., Sari, A., Rezvani-Gilkolaei, S., & Pourkazemi, M. (2010). Phylogeny of Iranian coastal lobsters inferred from mitochondrial DNA Restriction Fragment Length Polymorphism. *Acta Zoologica Bulgarica, 62*(3), 331–338.

Avise, J. C. (1994). *Molecular Markers, Natural History and Evolution*. New York/London: Chapman and Hall.

Babbucci, M., Buccoli, S., Cau, A., Cannas, R., Goñi, R., Díaz, D., & Patarnello, T. (2010). Population structure, demographic history, and selective processes: Contrasting evidences from mitochondrial and nuclear markers in the European spiny lobster *Palinurus elephas* (Fabricius, 1787). *Molecular Phylogenetics and Evolution, 56*(3), 1040–1050.

Baisre, J. A. (1994). Phyllosoma larvae and the phylogeny of Palinuroidea (Crustacea: Decapoda): a review. *Marine and Freshwater Research, 45*(6), 925–944.

Benestan, L., Gosselin, T., Perrier, C., Sainte-Marie, B., Rochette, R., & Bernatchez, L. (2015). RAD genotyping reveals fine-scale genetic structuring and provides powerful population assignment in a widely distributed marine species, the American lobster (*Homarus americanus*). *Molecular Ecology, 24*(13), 3299–3315.

Ben-Horin, T., Iacchei, M., Selkoe, K. A., Mai, T. T., & Toonen, R. J. (2009). Characterization of eight polymorphic microsatellite loci for the California spiny lobster, *Panulirus interruptus* and cross-amplification in other achelate lobsters. *Conservation Genetics Resources, 1*(1), 193.

Berry, P. F. (1974). A revision of the *Panulirus homarus*- group of spiny lobsters (Decapoda, Palinuridae). *Crustaceana, 27*(1), 31–42.

Burton, T. E., & Davie, P. J. F. (2007). A revision of the shovel-nosed lobsters of the genus *Thenus* (Crustacea: Decapoda: Scyllaridae), with descriptions of three new species. *Zootaxa, 1429*, 1–38.

Cannas, R., Cau, A., Deiana, A. M., Salvadori, S., & Tagliavini, J. (2006). Discrimination between the Mediterranean spiny lobsters *Palinurus elephas* and *P. mauritanicus* (Crustacea: Decapoda) by mitochondrial sequence analysis. *Hydrobiologia, 557*(1), 1–4.

Carvalho, G. R., & Hauser, L. (1994). Molecular genetics and the stock concept in fisheries. *Reviews in Fish Biology and Fisheries, 4*(3), 326–350.

Chakraborty, R. D., Maheswarudu, G., Purushothaman, P., Kuberan, G., Sebastian, J., Radhakrishnan, E. V., & Thangaraja, R. (2016). Nuclear and mitochondrial DNA markers based identification of blunthorn lobster *Palinustus waguensi*s Kubo, 1963 from South-west coast of India. *Indian Journal of Biotechnology, 15*(2), 172–177.

Chan, T. Y. (2010). Annotated Checklist of the World's Marine Lobsters (Crustacea: Decapoda: Astacidea, Glypheidea, Achelata, Polychelida). *The Raffles Bulletin of Zoology, 23*, 153–181.

Chan, T. Y., Ho, K. C., Li, C. P., & Chu, K. H. (2009). Origin and diversification of the clawed lobster genus *Metanephrops* (Crustacea: Decapoda: Nephropidae). *Molecular Phylogenetics and Evolution, 50*(3), 411–422.

Chow, S., Suzuki, N., Imai, H., & Yoshimura, T. (2006). Molecular species identification of spiny lobster phyllosoma larvae of the genus *Panulirus* from the northwestern Pacific. *Marine Biotechnology, 8*(3), 260–267.

Chow, S., Suzuki, S., Matsunaga, T., Lavery, S., Jeffs, A., & Takeyama, H. (2010). Investigation on natural diets of larval marine animals using peptide nucleic acid-directed polymerase chain reaction clamping. *Marine Biotechnology, 13*, 305–313.

Chow, S., Jeffs, A., Miyake, Y., Konishi, K., Okazaki, M., Suzuki, N., & Sakai, M. (2011). Genetic Isolation between the Western and Eastern Pacific Populations of Pronghorn Spiny Lobster *Panulirus penicillatus*. *PLoS One, 6*(12), e29280. https://doi.org/10.1371/journal.pone.0029280.

CMFRI Annual report. (2010–2011). Central Marine Fisheries Research Institute, Cochin. pp: 163.

CMFRI Annual report. (2017–2018). Central Marine Fisheries Research Institute, Cochin. pp: 24.

Connell, S. C., O'Rorke, R., Jeffs, A. G., & Lavery, S. D. (2014). DNA identification of the phyllosoma diet of *Jasus edwardsii* and *Scyllarus* sp. *New Zealand Journal of Marine and Freshwater Research, 48*(3), 416–429.

Cuéllar-Pinzón, J., Presa, P., Hawkins, S. J., & Pita, A. (2016). Genetic markers in marine fisheries: Types, tasks and trends. *Fisheries Research, 173*, 194–205.

Dao, H. T., Todd, E. V. & Jerry, D. R. 2013. Characterization of polymorphic microsatellite loci for the spiny lobster *Panulirus* spp. and their utility to be applied to other Panulirus lobsters. *Conserv. Genet. Resour.*, 5 (1): 43-46.

Dao, H. T., Smith-Keune, C., Wolanski, E., Jones, C. M., & Jerry, D. R. (2015). Oceanographic currents and local ecological knowledge indicate, and genetics does not refute, a contemporary pattern of larval dispersal for the ornate spiny lobster, *Panulirus ornatus* in the South-East Asian Archipelago. *PloS one, 10*(5), e0124568.

Davie, P. J. F. (1990). A new genus and species of marine crayfish, *Palibythus magnificus*, and new records of *Palinurellus* (Decapoda: Palinuridae) from the Pacific Ocean. *Invertebrate Taxonomy, 4*(4), 685–695.

Delghandi, M., Goddard, S., Jerry, D. R., Dao, H. T., Afzal, H., & Al-Jardani, S. S. (2015). Isolation, characterization, and multiplexing of novel microsatellite markers for the tropical scalloped spiny lobster (*Panulirus homarus*). *Genetics and Molecular Research, 14*(4), 19066–19070.

Delghandi, M., Afzal, H., Al Hinai, M. S. N., Al-Breiki, R. D. G., Jerry, D. R., & Dao, H. T. (2016). Novel polymorphic microsatellite markers for *Panulirus ornatus* and their cross-species primer amplification in *Panulirus homarus*. *Animal Biotechnology, 27*(4), 310–314.

Delghandi, M., Saif Nasser Al Hinai, M., Afzal, H., & Khalfan Al-Wahaibi, M. (2017). Parentage analysis of tropical spiny lobster (*Panulirus homarus*) by microsatellite markers. *Aquaculture Research, 48*, 4718–4724. https://doi.org/10.1111/are.13293.

Deshmukh, V. D. (2001). Collapse of sand lobster fishery in Bombay waters. *Indian Journal of Fisheries, 48*(1), 71–76.

Dharani, G., Maitrayee, G. A., Karthikayalu, S., Kumar, T. S., Anbarasu, M., & Vijayakumaran, M. (2009). Identification of *Panulirus homarus* puerulus larvae by restriction fragment length polymorphism of mitochondrial cytochrome oxidase I gene. *Pakistan Journal of Biological Sciences (PJBS), 12*(3), 281.

DiNardo, G. T., & Moffitt, R. B. (2007). The Northwestern Hawaiian Islands lobster fishery: A targeted slipper lobster fishery. In *The biology and fisheries of the slipper lobster* (pp. 243–261). Boca Raton: CRC Press.

Diniz, F. M., Maclean, N., Ogawa, M., Paterson, I. G., & Bentzen, P. (2005). Microsatellites in the overexploited spiny lobster, *Panulirus argus*: Isolation, characterization of loci and potential for intraspecific variability studies. *Conservation Genetics, 6*(4), 637–641.

Diniz, F. M., Ogawa, M., Cintra, I. H. A., Maclean, N., & Bentzen, P. (2010). Genetic identification of fishing stocks: New tools for population studies of the spiny lobster *Panulirus argus* (Latreille, 1804). *Boletim Técnico-Científico do Cepnor, 10*(1), 95–111.

Dixon, C. J., Ahyong, S. T., & Schram, F. R. (2003). A new hypothesis of decapod phylogeny. *Crustaceana, 76*(8), 935–975.

Duarte, L. F. D. A., Severino-Rodrigues, E., & Gasalla, M. A. (2010). Slipper lobster (Crustacea, Decapoda, Scyllaridae) fisheries off the southeastern coast of Brazil: I. Exploitation patterns between 23°00′ and 29°65′S. *Fisheries Research, 102*(1), 141–151.

Ekblom, R., & Galindo, J. (2011). Applications of next generation sequencing in molecular ecology of non-model organisms. *Heredity, 107*, 1–15.

Ellis, C. D., Hodgson, D. J., André, C., Sørdalen, T. K., Knutsen, H., & Griffiths, A. G. (2015). Genotype reconstruction of paternity in European lobsters (*Homarus gammarus*). *PLoS One, 10*(11), e0139585.

Ellis, C. D., Hodgson, D. J., Daniels, C. L., Collins, M., & Griffiths, A. G. (2017). Population genetic structure in European lobsters: implications for connectivity, diversity and hatchery stocking. *Marine Ecology Progress Series, 563*, 123–137.

Evans, C. R., & Evans, A. J. (1995). Fisheries ecology of spiny lobsters *Panulirus argus* (Latreille) and *Panulirus guttatus* (Latreille) on the Bermuda Platform: estimates of sustainable yields and observations on trends in abundance. *Fisheries Research, 24*(2), 113–128.

FAO. (2010). Fishery statistical collections: global production. *Food and Agriculture Organization (FAO) of the UN*.

Farhadi, A., Farhamand, H., Nematollahi, M. A., Jeffs, A., & Lavery, S. D. (2013). Mitochondrial DNA population structure of the scalloped lobster *Panulirus homarus* (Linnaeus 1758) from the West Indian Ocean. *ICES Journal of Marine Science, 70*(7), 1491–1498.

Farhadi, A., Jeffs, A. G., Farahmand, H., Rejiniemon, T. S., Smith, G., & Lavery, S. D. (2017). Mechanisms of peripheral phylogeographic divergence in the indo-Pacific: lessons from the spiny lobster *Panulirus homarus*. *BMC Evolutionary Biology, 17*(1), 195.

Faria, J., Froufe, E., Tuya, F., Alexandrino, P., & Pérez-Losada, M. (2013). Panmixia in the endangered slipper lobster *Scyllarides latus* from the northeastern Atlantic and western Mediterranean. *Journal of Crustacean Biology, 33*(4), 557–566.

Faria, J., Pérez-Losada, M., Cabezas, P., Alexandrino, P., & Froufe, E. (2014). Multiplexing of novel microsatellite loci for the vulnerable slipper lobster *Scyllarus arctus* (Linnaeus, 1758). *Journal of Experimental Zoology Part A: Ecological Genetics and Physiology, 321*(2), 119–123.

Froufe, E., Cabezas, P., Alexandrino, P., & Pérez-Losada, M. (2011). *Comparative phylogeography of three achelata lobster species from Macaronesia (North East Atlantic)* (pp. 157–173). Boca Raton: CRC Press.

Gallagher, J., Finarelli, J. A., Jonasson, J. P., & Carlsson, J. (2018). Mitochondrial D-loop DNA analyses of Norway Lobster (*Nephrops norvegicus*) reveals genetic isolation between Atlantic and Mediterranean populations. *bioRxiv*, 258392.

Galtier, N., Nabholz, B., Glémin, S., & Hurst, G. D. D. (2009). Mitochondrial DNA as a marker of molecular diversity: a reappraisal. *Molecular Ecology, 18*(22), 4541–4550.

Gan, H. M., Tan, M. H., Gan, H. Y., Lee, Y. P., & Austin, C. M. (2016). The complete mitogenome of the Norway lobster *Nephrops norvegicus* (Linnaeus, 1758) (Crustacea: Decapoda: Nephropidae). *Mitochondrial DNA Part A, 27*(5), 3179–3180.

García-Rodríguez, F. J., & Perez-Enriquez, R. (2006). Genetic differentiation of the California spiny lobster *Panulirus interruptus* (Randall, 1840) along the west coast of the Baja California Peninsula. *Mexico Marine Biology, 148*(3), 621–629.

García-Rodríguez, F. J., Perez-Enriquez, R., Medina-Espinoza, A., & Vega-Velázquez, A. (2017). Genetic variability and historic stability of the California spiny lobster *Panulirus interruptus* in the Gulf of California. *Fisheries Research, 185*, 130–136.

George, R. W. (2005). Evolution of life cycles, including migration, in spiny lobsters (Palinuridae). *New Zealand Journal of Marine and Freshwater Research, 39*(3), 503–514.

George, R. W. (2006a). Tethys sea fragmentation and speciation of *Panulirus* spiny lobsters. *Crustaceana, 78*, 1281–1309.

George, R. W. (2006b). Tethys origin and subsequent radiation of the spiny lobsters (Palinuridae). *Crustaceana, 79*(4), 397–422.

George, R. W., & Main, A. R. (1967). The evolution of spiny lobsters (Palinuridae): a study of evolution in the marine environment. *Evolution, 21*, 803–820.

Giraldes, B. W., & Smyth, D. (2016). Recognizing *Panulirus meripurpuratus* sp. nov. (Decapoda: Palinuridae) in Brazil- Systematic and biogeographic overview of *Panulirus* species in the Atlantic Ocean. *Zootaxa, 4107*(3), 353–366.

Gopal, K., Tolley, K. A., Groeneveld, J. C., & Matthee, C. A. (2006). Mitochondrial DNA variation in spiny lobster *Palinurus delagoae* suggests genetically structured populations in the southwestern Indian Ocean. *Marine Ecology Progress Series, 319*, 191–198.

Groeneveld, J. C., Griffiths, C. L., & Van Dalsen, A. P. (2006). A new species of spiny lobster, *Palinurus barbarae* (Decapoda, Palinuridae) from Walters Shoals on the Madagascar Ridge. *Crustaceana, 79*(7), 821–833.

Groeneveld, J. C., Gopal, K., George, R. W., & Matthee, C. A. (2007). Molecular phylogeny of the spiny lobster genus *Palinurus* (Decapoda: Palinuridae) with hypotheses on speciation in the NE Atlantic/Mediterranean and SW Indian Ocean. *Molecular Phylogenetics and Evolution, 45*(1), 102–110.

Harding, G. C., Kenchington, E. L., Bird, C. J., Pezzack, D. S., & Landry, D. C. (1997). Genetic relationships among subpopulations of the American lobster (*Homarus americanus*) as revealed by random amplified polymorphic DNA. *Canadian Journal of Fisheries and Aquatic Sciences, 54*(8), 1762–1771.

Hellberg, M. E. (2009). Gene flow and isolation among populations of marine animals. *Annual Review of Ecology, Evolution, and Systematics, 40*, 291–310.

Holthuis, L. B. (1985). A revision of the family Scyllaridae (Crustacea: Decapoda: Macrura). I. Subfamily Ibacinae. *Zoologische Verhandelingen Leiden, 218*(1), 1–130.

Holthuis, L. B. (1991). Marine Lobsters of the world. An Annotated and illustrated catalogue of species of interest to fisheries known to date. *FAO Species Catalog, FAO Fisheries and Synopsis.*, 13:125. FAO-UN, Rome.

Holthuis, L. B. (2002). The Indo-Pacific scyllarine lobsters (Crustacea, Decapoda, Scyllaridae). *Zoosystema, 24*(3), 499–683.

Hughes, G., & Beaumont, A. R. (2004). A potential method for discriminating between tissue from the European Lobster (*Homarus gammarus*) and the American Lobster (*H. americanus*). *Crustaceana, 77*(3), 371–376.

Iacchei, M., Gaither, M. R., Bowen, B. W., & Toonen, R. J. (2016). Testing dispersal limits in the sea: Range-wide phylogeography of the pronghorn spiny lobster *Panulirus penicillatus*. *Journal of Biogeography, 43*(5), 1032–1044.

Iamsuwansuk, A., Denduangboripant, J., & Davie, P. J. (2012). Molecular and Morphological Investigations of Shovel-Nosed Lobsters *Thenus* spp. (Crustacea: Decapoda: Scyllaridae) in Thailand. *Zoological Studies, 51*(1), 108–117.

Inoue, N., & Sekiguchi, H. (2005). Distribution of Scyllarid phyllosoma larvae (Crustacea: Decapoda: Scyllaridae) in the Kuroshio Subgyre. *Journal of Oceanography, 61*(3), 389–398.

Inoue, N., Watanabe, H., Kojima, S., & Sekiguchi, H. (2007). Population structure of Japanese spiny lobster *Panulirus japonicus* inferred by nucleotide sequence analysis of mitochondrial COI gene. *Fisheries Science, 73*(3), 550–556.

Jeena, N. S. (2013). Genetic divergence in lobsters (Crustaceana: Palinuridae and Scyllaridae) from the Indian EEZ, Ph. D Thesis, Cochin University of Science and Technology, Cochin, pp 238.

Jeena, N. S., Gopalakrishnan, A., Kizhakudan, J. K., Radhakrishnan, E. V., Kumar, R., & Asokan, P. K. (2016a). Population genetic structure of the shovel-nosed lobster *Thenus unimaculatus* (Decapoda, Scyllaridae) in Indian waters based on RAPD and mitochondrial gene sequences. *Hydrobiologia, 766*(1), 225–236.

Jeena, N. S., Gopalakrishnan, A., Radhakrishnan, E. V., Kizhakudan, J. K., Basheer, V. S., Asokan, P. K., & Jena, J. K. (2016b). Molecular phylogeny of commercially important lobster species from Indian coast inferred from mitochondrial and nuclear DNA sequences. *Mitochondrial DNA Part A., 27*(4), 2700–2709.

Jeena, N. S., Gopalakrishnan, A., Radhakrishnan, E. V., Kizhakudan, J. K. & Sajeela K.A 2016c. Signs of panmixia in the scalloped spiny lobster *Panulirus homarus* (Linnaeus, 1758) along the Indian coast. In: Book of abstracts, Ist International Agrobiodiversity Congress, 06–09 November, New Delhi, p 104

Jeffs, A. (2010). Status and challenges for advancing lobster aquaculture. *Journal of the Marine Biological Association of India, 52*, 320–326.

Jenkins, T. L., Ellis, C. D., & Stevens, J. R. (2018). SNP discovery in European lobster (*Homarus gammarus*) using RAD sequencing. *Conservation Genetics Resources*. https://doi.org/10.1007/s12686-018-1001-8.

Jørstad, K. E., Prodöhl, P. A., Agnalt, A. L., Hughes, M., Apostolidis, A. P., Triantafyllidis, A., & Svåsand, T. (2004). Sub-arctic populations of European lobster, *Homarus gammarus*, in northern Norway. *Environmental Biology of Fishes, 69*(1), 223–231.

Kennington, W. J., Cadee, S. A., Berry, O., Groth, D. M., Johnson, M. S., & Melville-Smith, R. (2013). Maintenance of genetic variation and panmixia in the commercially exploited western rock lobster (*Panulirus cygnus*). *Conservation Genetics, 14*(1), 115–124.

Kim, S., Lee, S. H., Park, M. H., Choi, H. G., Park, J. K., & Min, G. S. (2011). The complete mitochondrial genome of the American lobster, *Homarus americanus* (Crustacea, Decapoda). *Mitochondrial DNA, 22*(3), 47–49.

Kim, G., Yoon, T. H., Park, W. G., Park, J. Y., Kang, J. H., Park, H., & Kim, H. W. (2016). Complete mitochondrial genome of Australian spiny lobster, *Panulirus cygnus* (George, 1962) (Crustacea: Decapoda: Palinuridae) from coast of Australia. *Mitochondrial DNA Part A., 27*(6), 4576–4577.

Kittaka, J. (1994). Culture of phyllosomas of spiny lobster and its application to studies of larval recruitment and aquaculture. *Crustaceana, 66*(3), 258–270.

Kittaka, J., & Booth, J. D. (2000). Prospectus for aquaculture. In B. F. Phillips & J. Kittaka (Eds.), *Spiny lobsters: Fisheries and culture* (2nd ed., pp. 465–473). London: Blackwell Science Ltd.

Kizhakudan, J. K., Radhakrishnan, E. V., George, R. M., Thirumilu, P., Rajapackiam, S., Manibal, C., & Xavier, J. (2004). Phyllosoma larvae of *Thenus orientalis* and *Scyllarus rugosus* reared to settlement. *The Lobster Newsletter, 17*(1).

Konishi, K., Suzuki, N., & Chow, S. (2006). A late-stage phyllosoma larva of the spiny lobster *Panulirus echinatus* Smith, 1869 (*Crustacea: Palinuridae*) identified by DNA analysis. *Journal of Plankton Research, 28*(9), 841–845.

Kornfield, I., & Moran, P. (1990). Genetics of population differentiation in lobsters. In I. Kornfield (Ed.), *Life history of the American Lobsters* (pp. 23–24). Orono: Lobster Institute.

Lavalli, K. L. & Spanier, E. 2007. In: Lavalli, K. L. & Spanier, E. (ed). *The biology and fisheries of the slipper lobster*. Vol. 17. CRC press.

Lavery, S. D., Farhadi, A., Farahmand, H., Chan, T. Y., Azhdehakoshpour, A., Thakur, V., et al. (2014). Evolutionary Divergence of Geographic Subspecies within the Scalloped Spiny Lobster *Panulirus homarus* (Linnaeus 1758). *PLoS ONE, 9*(6), e97247.

Liang, H. (2012). Complete mitochondrial genome of the ornate rock lobster *Panulirus ornatus* (Crustacea: Decapoda). *African Journal of Biotechnology, 11*(80), 14519–14528.

Liu, Y., & Cui, Z. (2011). Complete mitochondrial genome of the Chinese spiny lobster *Panulirus stimpsoni* (Crustacea: Decapoda): genome characterization and phylogenetic considerations. *Molecular Biology Reports, 38*(1), 403–410.

Mallet, J., & Willmort, K. (2003). Taxonomy: renaissance or Tower of Babel? *Trends Ecol. Evolution, 18*, 57–59.

Matthee, C. A., Cockcroft, A. C., Gopal, K., & von der Heyden, S. (2008). Mitochondrial DNA variation of the west-coast rock lobster, *Jasus lalandii*: Marked genetic diversity differences among sampling sites. *Marine and Freshwater Research, 58*(12), 1130–1135.

Matz, M. V., & Nielsen, R. (2005). A likelihood ratio test for species membership based on DNA sequence data. *Philosophical Transactions of the Royal Society of London. Series B, Biological Sciences, 360*(1462), 1969–1974.

McWilliam, P. S. (1995). Evolution in the phyllosoma and puerulus phases of the spiny lobster genus *Panulirus* White. *Journal of Crustacean Biology, 15*, 542–557.

Mikami, S., & Kuballa, A. V. (2007). Factors important in larval and postlarval molting, growth and rearing. In K. L. Lavalli & E. Spanier (Eds.), *The Biology and Fisheries of the Slipper Lobster* (pp. 91–110). Boca Raton: CRC Press.

Modayil, M. J. & Pillai, N. G. K. (2007). *Status and perspectives in marine fisheries research in India*. Central Marine Fisheries Research Institute.

Morgan, E. M., Green, B. S., Murphy, N. P., & Strugnell, J. M. (2013). Investigation of genetic structure between deep and shallow populations of the southern rock lobster, *Jasus edwardsii* in Tasmania, Australia. *PloS One, 8*(10), e77978.

MPEDA. (2009). Statistics of marine products 2009. The Marine Exports Development Authority (Government of India, Ministry of Commerce and Industry), Kochi. pp: 25, 59

Naro-Maciel, E., Reid, B., Holmes, K. E., Brumbaugh, D. R., Martin, M., & DeSalle, R. (2011). Mitochondrial DNA sequence variation in spiny lobsters: population expansion, panmixia, and divergence. *Marine Biology, 158*(9), 2027–2041.

O'Rorke, R., Lavery, S. D., Wang, M., Nodder, S. D., & Jeffs, A. G. (2014). Determining the diet of larvae of the red rock lobster (*Jasus edwardsii*) using high-throughput DNA sequencing techniques. *Marine Biology, 161*(3), 551–563.

O'Rorke, R., Lavery, S., Chow, S., Takeyama, H., Tsai, P., Beckley, L. E., & Jeffs, A. G. (2012). Determining the diet of larvae of western rock lobster (*Panulirus cygnus*) using high-throughput DNA sequencing techniques. *PLoS One, 7*(8), e42757.

Ovenden, J. R., Booth, J. D., & Smolenski, A. J. (1997). Mitochondrial DNA phylogeny of red and green rock lobsters (genus *Jasus*). *Marine and Freshwater Research, 48*(8), 1131–1136.

Palero, F., Abelló, P., Macpherson, E., Gristina, M., & Pascual, M. (2008). Phylogeography of the European spiny lobster (*Palinurus elephas*): Influence of current oceanographical features and historical processes. *Molecular Phylogenetics and Evolution, 48*(2), 708–717.

Palero, F., Crandall, K. A., Abelló, P., Macpherson, E., & Pascual, M. (2009a). Phylogenetic relationships between spiny, slipper and coral lobsters (Crustacea, Decapoda, Achelata). *Molecular Phylogenetics and Evolution, 50*(1), 152–162.

Palero, F., Lopes, J., Abelló, P., Macpherson, E., Pascual, M., & Beaumont, M. (2009b). Rapid radiation in spiny lobsters (*Palinurus* spp) as revealed by classic and ABC methods using mtDNA and microsatellite data. *BMC Evolutionary Biology, 9*(1), 263.

Pampoulie, C., Skirnisdottir, S., Hauksdottir, S., Olafsson, K., Eiríksson, H., Chosson, V., & Hjorleifsdottir, S. (2011). A pilot genetic study reveals the absence of spatial genetic structure in Norway lobster (*Nephrops norvegicus*) on fishing grounds in Icelandic waters. *ICES Journal of Marine Science, 68*(1), 20–25.

Park, S. Y., Park, J. S., & Yoon, J. M. (2005). Genetic differences and variations in slipper lobster (*Ibacus ciliatus*) and deep sea lobster (*Puerulus sewelli*) determined by RAPD analysis. *Gene and Genomics, 27*(4), 307–317.

Patek, S. N., & Oakley, T. H. (2003). Comparative tests of evolutionary trade-offs in a palinurid lobster acoustic system. *Evolution, 57*(9), 2082–2100.

Pollock, D. E. (1990). Palaeoceanography and speciation in the spiny lobster genus *Jasus*. *Bulletin of Marine Science, 46*(2), 387–405.

Porobić, J., Canales-Aguirre, C. B., Ernst, B., Galleguillos, R., & Hernández, C. E. (2013). Biogeography and historical demography of the juan fernández rock lobster, *Jasus frontalis* (Milne Edwards, 1837). *The Journal of Heredity, 104*(2), 223–233.

Porter, M. L., Pérez-Losada, M., & Crandall, K. A. (2005). Model-based multi-locus estimation of decapod phylogeny and divergence times. *Molecular Phylogenetics and Evolution, 37*(2), 355–369.

Ptacek, M. B., Sarver, S. K., Childress, M. J., & Herrnkind, W. F. (2001). Molecular phylogeny of the spiny lobster genus *Panulirus* (Decapoda: Palinuridae). *Marine and Freshwater Research, 52*(8), 1037–1047.

Radhakrishnan, E. V., Deshmukh, V. D., Manisseri, M. K., Rajamani, M., Kizhakudan, J. K., & Thangaraja, R. (2005). Status of the major lobster fisheries in India. *New Zealand Journal of Marine and Freshwater Research, 39*(3), 723–732.

Radhakrishnan, E. V., Manisseri, M. K., & Deshmukh, V. D. (2007). Biology and Fishery of the Slipper Lobster, *Thenus orientalis*, in India. In *The Biology and Fisheries of the Slipper Lobster* (pp. 309–324). Boca Raton: CRC press.

Ratnasingham, S., & Hebert, P. D. (2007). BOLD: The Barcode of Life Data System. *Molecular Ecology Notes, 7*(3), 355–364. (http://www.barcodinglife.org).

Ravago, R. G., & Juinio-Meñez, M. A. (2002). Phylogenetic position of the striped-legged forms of *Panulirus longipes* (A. Milne-Edwards, 1868) (Decapoda, Palinuridae) inferred from mitochondrial DNA sequences. *Crustaceana, 75*, 1047–1059.

Reddy, M. M., Macdonald, A. H., Groeneveld, J. C., & Schleyer, M. H. (2014). Phylogeography of the scalloped spiny-lobster *Panulirus homarus rubellus* in the southwest Indian Ocean. *Journal of Crustacean Biology, 34*(6), 773–781.

Reiss, H., Hoarau, G., Dickey-Collas, M., & Wolff, W. J. (2009). Genetic population structure of marine fish: mismatch between biological and fisheries management units. *Fish and Fisheries, 10*, 361–395.

Rejinie Mon, T. R., Joseph, M. V., & Huxley, V. A. J. (2011). 18S rRNA Gene Polymorphisms of *Panulirus homarus* Populations from Different Geographic Regions of Peninsular India. *Journal of Theoretical and Experimental Biology, 8*(1 & 2), 85–93.

Rodríguez-Rey, G. T., Cunha, H. A., Lazoski, C., & Solé-Cava, A. M. (2013). Polymorphic microsatellite loci from Brazilian and Hooded slipper lobsters (*Scyllarides brasiliensis* and *S. deceptor*), and cross-amplification in other scyllarids. *Conservation Genetics Resources, 5*(4), 985–988.

Rodríguez-Rey, G. T., Solé-Cava, A. M., & Lazoski, C. (2014). Genetic homogeneity and historical expansions of the slipper lobster, *Scyllarides brasiliensis*, in the south-west Atlantic. *Marine and Freshwater Research, 65*(1), 59–69.

Ruzzante, D. E., Taggart, C. T., Cook, D., & Goddard, S. (1996). Genetic differentiation between inshore and offshore Atlantic cod (*Gadus morhua*) off Newfoundland: microsatellite DNA variation and antifreeze level. *Canadian Journal of Fisheries and Aquatic Sciences, 53*(3), 634–645.

Salvadori, S., Coluccia, E., Deidda, F., Cau, A., Cannas, R., & Deiana, A. M. (2012). Comparative cytogenetics in four species of Palinuridae: B chromosomes, ribosomal genes and telomeric sequences. *Genetica, 140*(10–12), 429–437.

Santos, M. F., Souza, I. G., Gomes, S. O., Silva, G. R., Bentzen, P., & Diniz, F. M. (2018). Isolation and characterization of microsatellite markers in the spiny lobster, *Panulirus echinatus* Smith, 1869 (Decapoda: Palinuridae) by Illumina MiSeq sequencing. *Journal of Genetics* (97), Online Resources, pp e25–e30

Scholtz, G., & Richter, S. (1995). Phylogenetic systematics of the reptantian Decapoda (Crustacea, Malacostraca). *Zoological Journal of the Linnean Society, 113*(3), 289–328.

Seeb, L. W., Seeb, J. E., & Polovina, J. J. (1990). Genetic variation in highly exploited spiny lobster *Panulirus marginatus* populations from the Hawaiian Archipelago. *Fishery Bulletin, 88*(71), 3–18.

Sekiguchi, H., & Inoue, N. (2010). Larval recruitment and fisheries of the spiny lobster *Panulirus japonicus* coupling with the Kuroshio subgyre circulation in the western North Pacific: A review. *Journal of the Marine Biological Association of India, 52*, 195–207.

Senevirathna, J. & Munasinghe, D. (2014). Genetic diversity and population structure of *Panulirus homarus* populations of southern Sri Lanka and South India revealed by the mitochondrial COI gene region. International Conference on Food, Biological and Medical Sciences (FBMS-2014) Jan. 28–29, 2014 Bangkok (Thailand). https://doi.org/10.15242/IICBE.C0114541

Senevirathna, J. D. M., Munasinghe, D. H. N., & Mather, P. B. (2016). Assessment of Genetic Structure in Wild Populations of *Panulirus homarus* (Linnaeus, 1758) across the South Coast of Sri Lanka Inferred from Mitochondrial DNA Sequences. *International Journal of Marine Science, 6*(6), 1–9.

Shaklee, J. B. (1983). The utilization of isozymes as gene markers in fisheries management and conservation. *Isozymes: Current Topics in Biological and Medical Research II*. pp 213–247.

Shen, H., Braband, A., & Scholtz, G. (2013). Mitogenomic analysis of decapod crustacean phylogeny corroborates traditional views on their relationships. *Molecular Phylogenetics and Evolution, 66*(3), 776–789.

Silberman, J. D., & Walsh, P. J. (1992). Species identification of spiny lobster phyllosoma larvae via ribosomal DNA analysis. *Molecular Marine Biology and Biotechnology, 1*(3), 195–205.

Singh, S. P., Groeneveld, J. C., Al-Marzouqi, A., & Willows-Munro, S. (2017). A molecular phylogeny of the spiny lobster *Panulirus homarus* highlights a separately evolving lineage from the Southwest Indian Ocean. *PeerJ, 5*, e3356.

Souza, C. A., Murphy, N., Villacorta-Rath, C., Woodings, L. N., Ilyushkina, I., Hernandez, C. E., & Strugnell, J. M. (2017). Efficiency of ddRAD target enriched sequencing across spiny rock lobster species (Palinuridae: *Jasus*). *Scientific Reports, 7*(1), 6781.

Stamatis, C., Triantafyllidis, A., Moutou, K. A., & Mamuris, Z. (2004). Mitochondrial DNA variation in Northeast Atlantic and Mediterranean populations of Norway lobster, *Nephrops norvegicus*. *Molecular Ecology, 13*(6), 1377–1390.

Stamatis, C., Triantafyllidis, A., Moutou, K. A., & Mamuris, Z. (2006). Allozymic variation in Northeast Atlantic and Mediterranean populations of Norway lobster, *Nephrops norvegicus*. *ICES Journal of Marine Science, 63*(5), 875–882.

Streiff, R., Mira, S., Castro, M., & Cancela, M. L. (2004). Multiple paternity in Norway lobster (*Nephrops norvegicus* L.) assessed with microsatellite markers. *Marine Biotechnology, 6*(1), 60–66.

Suresh, P., Sasireka, G., & Karthikeyan, K. A. M. (2012). Molecular insights into the phylogenetics of spiny lobsters of Gulf of Mannar marine biosphere reserve based on 28 S rDNA. *Indian Journal of Biotechnology, 11*(2), 182–186.

Suzuki, N., Murakami, K., Takeyama, H., & Chow, S. (2006). Molecular attempt to identify prey organisms of lobster phyllosoma larvae. *Fisheries Science, 72*(2), 342–349.

Suzuki, N., Hoshino, K., Murakami, K., Takeyama, H., & Chow, S. (2008). Molecular diet analysis of phyllosoma larvae of the Japanese spiny lobster *Panulirus japonicus* (Decapoda: Crustacea). *Marine Biotechnology, 10*(1), 49–55.

Tan, M. H., Gan, H. M., Lee, Y. P., & Austin, C. M. (2016). The complete mitogenome of the Morton Bay bug *Thenus orientalis* (Lund, 1793) (Crustacea: Decapoda: Scyllaridae) from a cooked sample and a new mitogenome order for the Decapoda. *Mitochondrial DNA Part A, 27*(2), 1277–1278.

Thomas, L., & Bell, J. J. (2012). Characterization of polymorphic microsatellite markers for the red rock lobster, *Jasus edwardsii* (Hutton 1875). *Conservation Genetics Resources, 4*(2), 319–321.

Timbó, R. V., Togawa, R. C., Costa, M. M., Andow, D. A., & Paula, D. P. (2017). Mitogenome sequence accuracy using different elucidation methods. *PLoS One, 12*(6), e0179971.

Tolley, K. A., Groeneveld, J. C., Gopal, K., & Matthee, C. A. (2005). Mitochondrial DNA panmixia in spiny lobster *Palinurus gilchristi* suggests a population expansion. *Marine Ecology Progress Series, 297*, 225–231.

Toon, A., Finley, M., Staples, J., & Crandall, K. A. (2009). Decapod phylogenetics and molecular evolution. *Decapod Crustacean Phylogenetics. Crustacean Issues., 18*, 15–29.

Tourinho, J. L., Solé-Cava, A. M., & Lazoski, C. (2012). Cryptic species within the commercially most important lobster in the tropical Atlantic, the spiny lobster *Panulirus argus*. *Marine Biology, 159*(9), 1897–1906.

Triantafyllidis, A., Apostolidis, A. P., Katsares, V., Kelly, E., Mercer, J., Hughes, M., & Triantaphyllidis, C. (2005). Mitochondrial DNA variation in the European lobster (*Homarus gammarus*) throughout the range. *Marine Biology, 146*(2), 223–235.

Truelove, N. K., Griffiths, S., Ley-Cooper, K., Azueta, J., Majil, I., Box, S. J., & Preziosi, R. F. (2015a). Genetic evidence from the spiny lobster fishery supports international cooperation among Central American marine protected areas. *Conservation Genetics, 16*(2), 347–358.

Truelove, N. K., Ley-Cooper, K., Segura-García, I., Briones-Fourzán, P., Lozano-Álvarez, E., Phillips, B. F., Box, S. J., & Preziosi, R. F. (2015b). Genetic analysis reveals temporal population structure in Caribbean spiny lobster (*Panulirus argus*) within marine protected areas in Mexico. *Fisheries Research, 172*, 44–49.

Truelove, N., Behringer, D. C., Butler, M. J., IV, & Preziosi, R. F. (2016). Isolation and characterization of eight polymorphic microsatellites for the spotted spiny lobster, *Panulirus guttatus*. *PeerJ, 4*, e1467.

Tsang, L. M., Ma, K. Y., Ahyong, S. T., Chan, T. Y., & Chu, K. H. (2008). Phylogeny of Decapoda using two nuclear protein-coding genes: origin and evolution of the Reptantia. *Molecular Phylogenetics and Evolution, 48*(1), 359–368.

Tsang, L. M., Chan, T. Y., Cheung, M. K., & Chu, K. H. (2009). Molecular evidence for the Southern Hemisphere origin and deep sea diversification of spiny lobsters (Crustacea: Decapoda: Palinuridae). *Molecular Phylogenetics and Evolution, 51*(2), 304–311.

Tshudy, D., Robles, R., Chan, T. Y., Chu, K. H., Ho, K & Ahyong, S. (2009). Phylogeny of marine clawed lobsters (Families Nephropidae Dana, 1852 and Thaumastochelidae Bate, 1888) based on mitochondrial genes. In Martin J. W., Crandall K.A & Felder D.L. (ed.). *Decapod crustacean phylogenetics*, pp 357–368.

Tsoi, K. H., Chan, T. Y., & Chu, K. H. (2011). Phylogenetic and biogeographic analysis of the spear lobsters *Linuparus* (Decapoda: Palinuridae), with the description of a new species. *Zoologischer Anzeiger, 250*(4), 302–315.

Ulrich, I., Muller, J., Schutt, C., & Buchholz, F. (2001). A study of population genetics in the European lobster, *Homarus gammarus* (Decapoda, Nephropidae). *Crustaceana, 74*(9), 825–837.

Vijayakumaran, M., & Radhakrishnan, E. V. (2011). Slipper lobsters. In R. K. Fotedar & B. F. Phillips (Eds.), *Recent advances and new species in aquaculture* (pp. 85–114). Wiley-Blackwell.

Villacorta-Rath, C., Ilyushkina, I., Strugnell, J. M., et al. (2016). Outlier SNPs enable food traceability of the southern rock lobster, *Jasus edwardsii*. *Marine Biology, 163*(11), 223.

Villacorta-Rath, C., Souza, C. A., Murphy, N. P., Green, B. S., Gardner, C., & Strugnell, J. M. (2018). Temporal genetic patterns of diversity and structure evidence chaotic genetic patchiness in a spiny lobster. *Molecular Ecology, 27*(1), 54–65.

Von der Heyden, S., Groeneveld, J. C., & Matthee, C. A. (2007). Long current to nowhere? – Genetic connectivity of *Jasus tristani* populations in the southern Atlantic Ocean. *African Journal of Marine Science, 29*(3), 491–497.

Wakabayashi, K., Yang, C. H., Shy, J. Y., He, C. H., & Chan, T. Y. (2017). Correct identification and redescription of the larval stages and early juveniles of the slipper lobster *Eduarctus martensii* (Pfeffer, 1881) (Decapoda: Scyllaridae). *Journal of Crustacean Biology, 37*(2), 204–219.

Watson, H. V., McKeown, N. J., Coscia, I., Wootton, E., & Ironside, J. E. (2016). Population genetic structure of the European lobster (*Homarus gammarus*) in the Irish Sea and implications for the effectiveness of the first British marine protected area. *Fisheries Research, 183*, 287–293.

Webber, W. R., & Booth, J. D. (2007). Taxonomy and evolution. In K. L. Lavalli & E. Spanier (Eds.), *The biology and fisheries of the Slipper Lobster* (pp. 26–52). Boca Raton: CRC press.

Wongruenpibool, S., & Denduangboripant, J. (2013). Genetic diversity of purple-legged shovelnosed lobster *Thenus unimaculatus* in Thailand. *Genomics and Genetics, 6*(1), 64–70.

Xiao, B. H., Zhang, W., Yao, W., Liu, C. W., & Liu, L. (2017). Analysis of the complete mitochondrial genome sequence of *Palinura homarus*. *Mitochondrial DNA Part B, 2*(1), 60–61.

Yamauchi, M. M., Miya, M. U., & Nishida, M. (2002). Complete mitochondrial DNA sequence of the Japanese spiny lobster, *Panulirus japonicus* (Crustacea: Decapoda). *Gene, 295*(1), 89–96.

Yang, C. H., Bracken-Grissom, H., Kim, D., Crandall, K. A., & Chan, T. Y. (2012). Phylogenetic relationships, character evolution, and taxonomic implications within the slipper lobsters (Crustacea: Decapoda: Scyllaridae). *Molecular Phylogenetics and Evolution, 62*(1), 237–250.

Yellapu, B., Jeffs, A., Battaglene, S., Lavery, S. D., & Hauser, L. (2016). Population subdivision in the tropical spiny lobster *Panulirus ornatus* throughout its Indo-West Pacific distribution. *ICES Journal of Marine Science, 74*(3), 759–768.

Ecology and Global Distribution Pattern of Lobsters

5

E. V. Radhakrishnan, Bruce F. Phillips, Lakshmi Pillai S, and Shelton Padua

Abstract

This chapter presents a comprehensive overview of the global distribution of lobsters (nephropid, palinurid and scyllarid) in all oceans. Lobsters are found in tropical, subtropical and temperate regions, from the intertidal to great depths. Many species of the genus *Panulirus* prefer rocky or coral reefs and some are found in sandy/muddy substrates. The tropical zone has the largest number of species (174), followed by the subtropics with 71 species and the temperate region with 16 species. Among the 63 species of palinurids, 39 are distributed in the tropics and 19 species in the subtropical zone with many species overlapping in their distribution. Under the family Nephropidae, 32 species are found in the tropical belt, 15 in the subtropical zone and 10 species in the temperate zone. Three species under two genera, *Thymops* and *Thymopides* are distributed in the southern Atlantic and Indian Ocean region (50ºS). The Indo-West Pacific is the region with maximum diversity with the nephropids represented by 29 species, palinurids by 36 species and the scyllarids by 57 species. Among the total number of scyllarids, 63 species are distributed in the tropical zone, 26 in the subtropical and 3 species in the temperate region.

Rock lobsters use a range of different habitats at different phases of their life cycle from the water column during the pelagic larval phase to seagrass and algal meadows in puerulus stage and small holes in the reef in postpuerulus and juvenile stages to subadults (3–4 years of age) migrating across the deepwater regions of sand and reefs to settle on offshore, deepwater habitats as mature breeding lobsters. Many spiny lobster species exhibit ontogenetic habitat shift from the

E. V. Radhakrishnan (✉) · Lakshmi Pillai S · S. Padua
ICAR-Central Marine Fisheries Research Institute, Cochin, Kerala, India
e-mail: evrkrishnan@gmail.com

B. F. Phillips
School of Molecular and Life Sciences, Curtin University, Perth, WA, Australia
e-mail: B.Phillips@curtin.edu.au

© Springer Nature Singapore Pte Ltd. 2019
E. V. Radhakrishnan et al. (eds.), *Lobsters: Biology, Fisheries and Aquaculture*,
https://doi.org/10.1007/978-981-32-9094-5_5

postlarval settlement habitat of macroalgae, kelp or seagrass to benthic crevices sheltering concomitant with aggregation in crevices as larger juveniles, subadults and adults. Different species of lobsters may cohabit in the same region but may differ in their habitat selection. Lobsters cohabit with many species of sponges, sea urchins, echinoderms, fishes, decapods, seagrass and seaweeds in the coastal fishing grounds. Artisanal fishing using bottom-set gillnets in the coastal grounds removes these low trophic species regularly. The indiscriminate and constant removal of these low trophic species has the potential to cause serious ecological balance of the reef system.

Keywords
Ecology · Ontogenetic shift · Distribution · Settlement habitat

5.1 Introduction

Ecological systems are highly complex and dynamic, and discerning the interaction between the various components is essential for management aimed at achieving sustainability (Levin 1999; Partelow 2014). Despite a large volume of research data on lobsters, our knowledge on their role in ecosystem functions is scanty (Phillips et al. 2013). Lobsters are considered ecologically important due primarily to their large size, high abundance and prominent trophic functions (Cobb and Phillips 1980; Phillips and Kittaka 2000). An understanding of the structures and processes of the lobster stock within the ecosystem is needed in order to explain changes in lobster abundance. Lobsters are found in temperate, subtropical and tropical seas of the world (between 65°N and 60°S) with a large number of species inhabiting the tropical seas. They occupy habitats from the intertidal zone to the deep sea (3000 m) (Holthuis 1991). Many species of the genus *Panulirus* prefer rocky or coral reefs for shelter, whereas others are found on sandy/muddy substrates. The species of the genus *Jasus* are mainly found on rocky reefs, but can be found to inhabit various substrates from the intertidal zone to depths of 200–400 m (MacDiarmid and Booth 2003; Booth 2006). Adult slipper lobsters are found in all the world's oceans, temperate, subtropical and tropical, within the latitude 4°S to 40°N and at depths varying from 0 to at least 800 m (Holthuis 1991, 2002; Webber and Booth 2007). They are found in coastal waters, continental shelves and continental slopes and occupy a variety of habitats from soft substrates such as sand and mud to harder gravel surfaces and to rocky and reef crevices (Lavalli et al. 2007). Berry (1971) found that adult habitat preference varies among the Palinuridae from shallow intertidal surf zones to great ocean depths.

Knowledge on the life history processes as well as the habitat preference of lobsters can be helpful in predicting accurately the distribution pattern of lobster resources and in defining more meaningful management strategies for these fisheries (Rios-Lara et al. 2007). The hydrological environment and preferential food have been shown to influence habitat use in *Panulirus argus* in the Caribbean (Marx and Herrnkind 1985) and *Panulirus homarus* on the southwest coast of India (Thangaraja and Radhakrishnan 2012).

Examination of the relationship between environmental data and lobster catches show that catches of both the American lobster *Homarus americanus* and the European lobster *Homarus gammarus* are influenced by sea surface temperature (Dow 1977; Fogarty 1986; Campbell et al. 1991). Temperature is an important environmental variable that is generally accepted to have a pervasive influence on the behaviour of temperate lobsters (Herrnkind 1980; Lawton and Lavalli 1995). It may also be a critical factor in the case of catches of tropical species, as it may influence the availability of food in the vicinity (Thangaraja and Radhakrishnan 2012). Tegner and Levin (1983) and Edgar (1990) found the spiny lobsters implicated as predators in a variety of benthic habitats, and their selective predation was apparently responsible for the profound impact on species composition and size frequency distribution of sea urchins, mussels and gastropods (Griffiths and Seiderer 1980; Tegner and Levin 1983; Joll and Phillips 1984; Edgar 1990).

Lobsters coexist with several other vertebrate and invertebrate species in their natural environment. Sponges, gorgonians, sea urchins, starfishes, ornamental fishes, several decapods, seagrass and seaweeds have been reported to constitute the fauna in the lobster fishing grounds (Thangaraja and Radhakrishnan 2012). Lobster abundance in the fishing grounds has been directly correlated with these associated fauna and flora as some of them constitute the preferred food of lobsters. Artisanal fishing of lobsters with bottom-set gillnets and constant removal of the keystone low trophic species (crabs) may significantly impact the sensitive ecological balance of the reef ecosystem (Giraldes et al. 2015).

In this chapter we deal with the global distribution of lobsters and the related aspects of ecology in relation to the environment and the associated fauna in their natural habitats.

5.2 General Distribution

George and Main (1967) recognized two well-marked distribution patterns for lobsters belonging to the family Palinuridae: (i) a high latitude circumpolar distribution and (ii) a low latitude circumequatorial distribution. The genera which fall into the first group are *Jasus*, *Projasus* and *Palinurus*. The species under *Jasus* and *Projasus* are distributed in the Southern Ocean with each species endemic to a particular region. *Palinurus*, once might have had a circumpolar distribution in the Northern Hemisphere, is now limited to the Atlantic and southwestern Indian Ocean. The other members of palinurids which fall under equatorial distribution occur in an extended belt extending about 35° on either side of the equator and at depths from shallow water to depths over 1500 m. The genera which fall under deepwater equatorial distribution are *Puerulus*, *Linuparus*, *Palinustus* and *Justitia*. *Panulirus* is the only genus which falls under the category of shallow-water equatorial distribution. The members of the genus *Palinustus* live at shallower depths (80–400 m) compared to the nine species of *Puerulus* (250–1438 m) (George and Main 1967).

The global lobster fauna is represented by 260 species (including 4 subspecies) under 54 genera (Chap. 2). The family Nephropidae consists of 13 genera and 57 species and the Palinuridae, 12 genera, 59 species and 4 subspecies. The family Scyllaridae is represented by 19 genera under 4 subfamilies and includes 92 species (see Chap. 2 of this book). The tropical zone has the largest number of species (174), followed by the subtropical region with 71 species and temperate zone with 16 species. The tropical region referred to is the geographical zone between the Tropic of Cancer (23° 27'N and 23°27'N) and the Tropic of Capricorn (23° 27'S and 23°27'S). The subtropical region is the zone between the Tropic of Cancer and 35°N and the Tropic of Capricorn and 35°S.

Under the family Nephropidae, 32 species are distributed in the tropical belt and 15 in the subtropical zone, of which species under the genera *Nephropsis* and *Metanephrops* are represented maximum in the tropical zone. Three species under the two genera *Thymops* and *Thymopides* are distributed in the southern Atlantic and Indian Ocean region (Table 5.1). While palinurids have a wider distribution from 60°N to 50°S, the distribution of scyllarids is restricted between 45°N and 44°S (Lavalli and Spanier 2007). Among the 63 species of palinurids (including the 4 subspecies), 39 are tropical species (including the subspecies) and 19 species in the subtropical zone with the distribution overlapping considerably (Table 5.2). The distribution of *Jasus* is restricted to the subtropical (*J. caveorum, J. frontalis* and *J. lalandii*) and temperate latitudes (*J. edwardsii* and *J. paulensis*) of the Southern Hemisphere (Fig. 5.1a). The palinurid species *Justitia longimanus* is known in all the three major oceans (Fig. 5.1b). The genus *Panulirus* is currently represented by 24 species/subspecies, which are distributed across the tropical and subtropical belt (i.e. between 35°N and 35°S Lat.) of all the world's oceans (Fig. 5.1c, Fig. 5.1d), wherever suitable habitats for their existence are available in depths under about 100 m (Pitcher 1993). Among this, 16 are distributed in the tropical zone and 8 species in the subtropical latitudes (Table 5.3). Under the genus *Palinurus*, there are 6 species and majority of them (5 species) are distributed in the subtropical and temperate regions. The living fossil family Glypheidae includes 2 genera with 2 species and both are tropical. The family Enoplometopidae with a single genus *Enoplometopus* has 12 species and 9 species are distributed in the tropical region. The family Polychelidae is represented by 38 species under 6 genera and 29 species are distributed in tropical latitudes (Table.5.3).

The Indo-West Pacific (the region between 30°N and 30°S Lat. and 30°E and 120°W Long.), is the region with highest species diversity (131 species) with the nephropids represented by 29 species, palinurids by 36 species and the scyllarids by 57 species; 13 of the 24 species (including 4 subspecies) of *Panulirus* are distributed in this region (Tables 5. 1 & 5.2).

5.2.1 Global Distribution of Palinurid Lobsters

The palinurids normally inhabit at specified depths and latitude and at least a few of them overlap in their distribution. The 24 species of the genus *Panulirus*, which inhabit shallow areas down to 100 m depth, can be separated on the basis of substratum, depth, turbidity and temperature. Among the 39 species in tropical oceans, 28

Table 5.1 Distribution of species (first record) under each genus in major oceans

Family/subfamily/genera	East Pacific Ocean	Atlantic Ocean	Southern Indian/ Atlantic Ocean	Indo-West Pacific
Family Enoplometopidae				
Enoplometopus		3		6
Family Nephropidae				
Acanthacaris		1		1
Dinochelus				1
Eunephrops	1	3		
Homarus		2		
Homarinus		1		
Metanephrops	3	2		13
Nephropides		1		
Nephrops		1		
Nephropsis		6		9
Thaumastocheles	1	1		3
Thaumastochelopsis				2
Thymopides		1	1	
Thymops			2	
Thymopsis			1	
Family Glypheidae				
Laurentaeglyphea				1
Neoglyphea				1
Family Palinuridae				
Jasus	3	1	1	
©*Justitia*				1
Linuparus				4
Nupalirus	1			2
Palibythus				1
Palinurellus		1		1
Palinurus	1	2	2	1
Palinustus		1		4
Panulirus	5	6		13
Projasus	1			1
Puerulus				9
Sagmariasus				1
Family Scyllaridae				
Subfamily Arctidinae				
Arctides				3
Scyllarides	3	8		3
Subfamily Ibacinae				
Evibacus	1			
Ibacus	2			6
Parribacus	2	1		3
Subfamily Scyllarinae				
Acantharctus	1	1		1
Antarctus	1			

(continued)

Table 5.1 (continued)

Family/subfamily/genera	East Pacific Ocean	Atlantic Ocean	Southern Indian/ Atlantic Ocean	Indo-West Pacific
Bathyarctus		3		3
Biarctus	1			3
Chelarctus				5
Crenarctus				2
Eduarctus				8
Galearctus	1			6
Gibbularctus				1
Petrarctus	1			5
Remiarctus				1
Scammarctus				2
Scyllarus		9		
Subfamily Theninae				
Thenus				5
Family Polychelidae				
Cardus		1		
Homeryon	1			1
Pentacheles		2		4
Polycheles	2	2		5
Stereomastis	2	4		11
Willemoesia	1	3		

⁰Present in all the three oceans

species are distributed in the Indo-west Pacific region and 8 species in the Atlantic region (Table 5.2; Fig. 5.1c and 5.1d). One species *Justitia longimanus* is recorded from all the three major oceans.

In the genus *Jasus*, two species inhabit the temperate region (below 35° S) and the other three are in the subtropical latitudes. Among the five species, two are distributed in the Pacific Ocean and one species *J. edwardsii* in both southern Indian (southern Australia) and Pacific Ocean (Southern New Zealand). *J.lalandii* is the only species distributed in the Atlantic ocean. *J. paulensis* (*J. tristani*) is distributed in both southern Atlantic (Tristan da Cunha) and Indian Ocean (St. Paul Island). Population genetic studies and analysis of molecular variance revealed no significant population genetic differentiation between *J. tristani* and *J. paulensis,* though they are physically separated by thousands of miles, and, therefore, have been synonymized (Groeneveld et al. 2012).

Lobsters under the genus *Puerulus* are generally deepwater species. All the nine species of *Puerulus* are distributed in the Indo-west Pacific region (Table.5.1). Under the genus *Linuparus*, all species are distributed in the Indo-west Pacific region.

Among the eight subtropical species of *Panulirus,* five species are distributed in the Indo-Pacific region. One subtropical species under the genus *Palinurus* is distributed in SW Indian Ocean. Among the two temperate species of *Palinurus*, one species is distributed in the southern Indian Ocean (*Palinurus gichristi*) and another species (*Palinurus elephas*) is in North Atlantic Ocean (Table 5.2; Fig. 5.1c and 5.1d).

Table 5.2 Countries of occurrence and geographical distribution of spiny lobsters

Genus	Species	Common name	Countries	Geographic region
Tropical				
Panulirus				
	argus	Caribbean spiny lobster	USA, Bahamas, Cuba, Caribbean	West Atlantic
	meripurpuratus	Purple spiny lobster	Northeastern Brazil	West Atlantic
	gracilis	Blue lobster, langosta azul	Ecuador, Panama	Central East Pacific
	guttatus	Spotted spiny lobster	Bahamas, Belize	Caribbean, West Atlantic
	homarus homarus	Green Scalloped rock lobster	Kenya, Yemen, Somalia, Oman, Iran, India, Sri Lanka, Indonesia, Vietnam	Indo-west Pacific
	laevicauda	Langosta	Brazil	West Atlantic
	longipes longipes	Spotted-legged rock lobster	Okinawa Islands, Japan, South China Sea, India	Indo-West Pacific
	femoristriga	White-whiskered rock lobster	Japan, Indonesia, Philippines	Indo-West Pacific
	longipes bispinosus		Japan, Cook Islands, New Caledonia	Indo-West Pacific
	ornatus	Ornate rock lobster	Vietnam, Australia, Papua New Guinea Kenya, India, Indonesia	Indo-West Pacific
	penicillatus	Double-spined rock lobster	Reunion, Pacific Islands, Galapagos, India, Philippines	Indo-West Pacific
	polyphagus	Mud spiny lobster	India, Pakistan, Indonesia, Philippines, Vietnam, Western Australia	Indo-West Pacific
	regius	Longouste royale	Nigeria, Ghana, Congo, Cape Verde Islands	North East Atlantic
	versicolor	Painted rock lobster	India, South Africa, Southern Japan, Red Sea	Indo-West Pacific
	pascuensis	Longosta, crayfish	French Polynesia, Pitcairn Islands	East & Central Pacific
	echinatus	Spiny lobster	Cape Verde Islands, St. Helena Island, Tristan da Cunha Island	Southeast, Southwest and Central Atlantic

(continued)

Table 5.2 (continued)

Genus	Species	Common name	Countries	Geographic region
Palinurus				
	charlestoni	Cape Verde spiny lobster	Cape Verde Islands	East Central Atlantic
Puerulus				
	sewelli	Arabian whip lobster	Southwest and Southeast India, Pakistan, Somalia, Myanmar	Indo-West pacific
	angulatus	Banded whip lobster	Mozambique, Western Australia, Japan, Taiwan, Philippines, A&N Islands, Indonesia	Indo-West Pacific
	carinatus	Red whip lobster	Mozambique, Madagascar, South Africa	Indo-West Pacific
	velutinus	Velvet whip lobster	Indonesia, Philippines, Malaysia	Indo-West Pacific
	gibbosus		South Africa, Mozambique, Madagascar, Zanzibar, Somalia	Indo-West Pacific
	mesodontus		Japan to Fiji	Indo-West Pacific
	richeri		New Caledonia to Marquesas Islands	Indo-West Pacific
	sericus		New Caledonia	Indo-West Pacific
	quadridenticus		New Caledonia	Indo-West Pacific
Linuparus				
	somniosus	African spear lobster	South Africa, A&N Islands, Indonesia, Kenya, Tanzania, Thailand	Indo-West Pacific
	sordidus	Oriental spear lobster	East and west Australia, Indonesia, Vietnam, Philippines, Taiwan, Hong Kong	Indo-West Pacific
	trigonus	Japanese spear lobster	China, Hong Kong, Japan, East & western Australia, Philippines, Taiwan	Indo-West Pacific
	meridionalis		Australia, New Caledonia	Indo-West Pacific

Palinustus				
	mossambicus	Buffalo blunthorn lobster	Mozambique, Somalia	Indo-West Pacific
	truncatus	American blunthorn lobster	Brazil, French Guyana, Suriname, Venezuela	Western Atlantic Ocean
	unicornutus	Unicorn blunthorn lobster	Indonesia, Japan, Kenya, South Africa, Maldives, New Caledonia	Indo-West Pacific
	waguensis	Japanese blunthorn lobster	India, Indonesia, Japan, Taiwan Philippines, Madagascar	Indo-West Pacific
Justitia				
	longimanus[a]	West Indian furrow lobster	Bermuda, Bahamas, Cuba, Brazil, Madagascar, French Polynesia	Atlantic, Indo-West Pacific
Nupalirus				
	vericeli		Tuamotu, French Polynesia	Indo-West Pacific
Palibythus				
	magnificus		Western Samoa	Indo-West Pacific
Palinurellus				
	gundlachi	Caribbean furry lobster	Bahamas, Cuba, Southeast USA, West Indies	West Central Atlantic Ocean
	wieneckii	Indo-Pacific furry lobster	Indonesia, Japan, Mauritius, Saudi Arabia, Thailand	Indo-West Pacific
Subtropical				
Palinurus				
	delagoae	Natal spiny lobster	Southeast Africa	Indo-west Pacific
	mauritanicus	Pink spiny lobster	Mauritania, Algeria, France, UK, Spain, Portugal, Italy	Northeast Atlantic, Mediterranean
	barbarae	Giant spiny lobster	Walters Shoals, south Madagascar	SW Indian Ocean

(continued)

Table 5.2 (continued)

Genus	Species	Common name	Countries	Geographic region
Panulirus				
	cygnus	Western rock lobster	Western Australia	Indo-West Pacific
	inflatus	Langosta	Mexico, Guatemala	Eastern Central Pacific
	interruptus	California spiny lobster	USA (California), Mexico (Baja California)	Eastern Central Pacific
	japonicus	Ise-ebi Japanese spiny lobster	Japan, Korea, Taiwan	North-West Pacific
	marginatus	Hawaiian lobster	USA (Hawaii)	Indo-West Pacific
	brunneiflagellum		Ogasawara Islands, Southern Japan	Indo-West Pacific
	homarus rubellus	Red scalloped rock lobster	South Africa, Mozambique	Indo-West Pacific
	stimpsoni	Hong Kong rock lobster	Hong Kong, Taiwan, Vietnam, China	Indo-West Pacific
Palinustus				
	holthuisi		Japan, Taiwan, China	Indo-West Pacific
Nupalirus				
	chani		Loyalty Islands, Japan, Taiwan, New Caledonia	Indo-West Pacific
	japonicus		Japan, New Caledonia, Madagascar	North-West Pacific
Jasus				
	caveorum		Southeast Pitcairn Island, Chile	Southeast Pacific
	frontalis	Juan Fernandez rock lobster	Juan Fernandez Island	Southeast Pacific
	lalandii	Cape rock lobster	Southwest Africa	Southeast Atlantic
Projasus				
	bahamondei	Chilean jagged lobster	Chile, San Ambrosio Islands	Southeast Pacific
	parkeri	Cape jagged lobster	Natal, South Africa, Australia, New Zealand	Indo-West Pacific

Temperate			
Jasus			
	edwardsii	Red rock lobster	New Zealand, Eastern Australia	Southern Indian Ocean, South-West Pacific
	paulensis	St. Paul rock lobster	St. Pauls and New Amsterdam Islands	Southern Indian & Atlantic Ocean
Sagmariasus			
	verreauxii	Green rock lobster	Australia, New Zealand	Indo-West Pacific
Palinurus			
	elephas	Common spiny lobster, crawfish	UK, France, Spain, Italy	NE Atlantic
	gilchristi	Southern spiny lobster	South Africa	Southwest Indian Ocean

ᵃ*Tropical/subtropical*

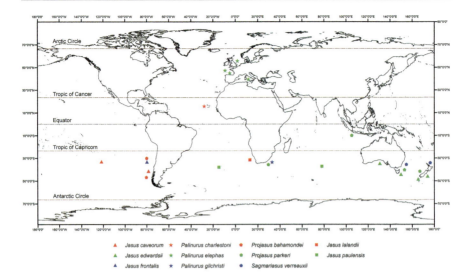

Fig. 5.1a Distribution of palinurid lobsters in world oceans. a. *Jasus* sp., *Palinurus* sp., *Projasus* sp., *Sagmariasus* sp.

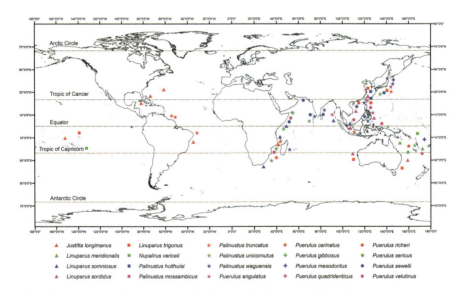

Fig. 5.1b *Justitia* sp., *Linuparus* sp., *Nupalirus* sp., *Palinustus sp.*, *Puerulus* sp.

5 Ecology and Global Distribution Pattern of Lobsters 163

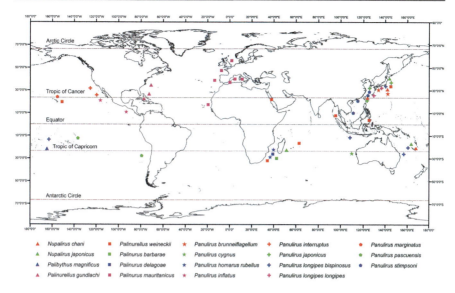

Fig. 5.1c *Panulirus* sp., *Palinurus* sp., *Palibythus* sp., *Nupalirus* sp., *Palinurellus* sp.

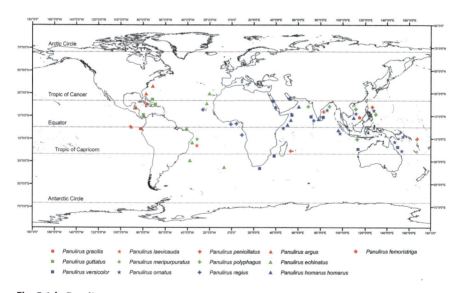

Fig. 5.1d *Panulirus* sp.

Some species of spiny lobsters are endemic to a particular region such as *P. cygnus* in western Australia, *P. gilchristi* and *P. delagoae* in South Africa, *P. interruptus* in southwest coast of USA, *P. marginatus* in Hawaii, *P. echinatus* in Tristan da Cunha and *P. homarus rubellus* in Southeastern Africa. The tropical species such as *P. homarus homarus*, *P. penicillatus*, *P. longipes longipes* and *P. versicolor* have a wider distribution in the Indo-Pacific region (Fig. 5.1c and 5.1d). While the distribution of some species is restricted to the tropical belt, some other species overlap in their distribution to both tropical and subtropical regions.

5.2.2 Global Distribution of Scyllarids

Among the scyllarids, 63 species are distributed in the tropical zone, 26 in the subtropical and 3 species in the temperate region (Table 5.3). The distribution of scyllarids is restricted to within the latitudes of 45°N and 44°S with most genera distributed between 30°N and 30°S (Lavalli and Spanier 2007). The Indo-West Pacific region is represented by 16 genera and 57 species, 13 species are exclusively in the Pacific Ocean, 22 species in the Atlantic Ocean and 1 species common in all the three oceans (Table 5.1). One species each in the genera *Scammarctus* (*S. batei arabicus*) and *Petrarctus* (*P. jeppiari*) are endemic to the Indian Ocean with the former found in the Gulf of Aden and the latter at Muttom, Kanyakumari district, on the southwest coast of India (Chan, Chap. 2 of this book). The genus *Evibacus* is represented by a single species. *E. princeps*, a subtropical species, is distributed along the west coast of Mexico in the Baja California region. Among the eight species under the genus *Ibacus*, six species are represented in the Indo-West Pacific region. The genera, *Arctides, Acantharctus, Scyllarides Bathyarctus* have representation in the Atlantic and Indo-West Pacific region whereas the eight species under the genus *Scyllarus* are exclusive to the Atlantic and the Mediterranean regions. Distribution of the five species in the genus *Thenus* is restricted to the Indo-West Pacific region. Two genera (*Antarctus* and *Ibacus*) represented by three species are distributed in the southern temperate zone.

Three species of *Parribacus* are represented in the Indo-West Pacific region.

5.3 Ecology

Lobsters are found in temperate, subtropical and tropical seas of the world (between 65°N and 60°S) with a large number of species from tropical seas. They occur from inter- and shallow subtidal habitats to the deep sea (3000 m) (Holthuis 1991). The American lobster, *Homarus americanus*, lives in a wide variety of habitats. They are most commonly found at 0–50 m depth on rocky reefs and also inhabit muddy areas, seagrass beds and sandy depressions (Lawton and Lavalli 1995). The European lobster, *H. gammarus*, inhabits depths between 0 and 150 m but usually not deeper than 50 m (Holthuis 1991). They are typically found on rocky substrates but may also burrow into hard mud or form depressions in sand. The type of habitat,

Table 5.3 Distribution of species in the three geographical regions

Family/genera	Tropical	Subtropical	Temperate	
Enoplometopidae				
Enoplometopus	9	3		
Nephropidae				
Thaumastocheles	3	2		
Thaumastochelopsis	2			
Thymopides	1		1	
Thymops			2	
Thymopsis			1	
Nephropsis	11	3	1	
Nephrops			1	
Nephropides	1			
Metanephrops	10	7	1	
Homarus			2	
Homarinus			1	
Eunephrops	1	3		
Dinochelus	1			
Acanthacaris	2			
	32	**15**	**10**	
Glypheidae				
Laurentoglyhea	1			
Neoglyphea	1			
	2			
Palinuridae				
Jasus		3	2	
Justitia	1			
Linuparus	4			
Nupalirus	1	2		
Palibythus	1			
Palinurellus	2			
Palinurus	1	3	2	
Palinustus	4	1		
Panulirus	16	8		
Projasus		2		
Puerulus	9			
Sagmariasus			1	
	39	**19**	**5**	
Scyllaridae				
Arctides	2		1	
Scyllarides	7	7		
Evibacus		1		
Ibacus	5	2	1	
Parribacus	4	2		
Acantharctus	2	1		
Antarctus			1	

(continued)

Table 5.3 (continued)

Family/genera	Tropical	Subtropical	Temperate	
Bathyarctus	5	1		
Biarctus	2	2		
Chelarctus	4	1		
Crenarctus	1	1		
Eduarctus	8			
Galearctus	4	3		
Gibbularctus	1			
Petrarctus	5	1		
Remiarctus		1		
Scammarctus	2			
Scyllarus	6	3		
Thenus	5			
	63	**26**	**3**	
Polychelidae				
Cardus	1			
Homeryon		2		
Polycheles	7	2		
Pentacheles	4	1		
Stereomastis	14	2	1	
Willemoesia	3	1		
	29	**8**	**1**	

substrate and the shelter characteristics chosen by both *H. americanus* and *H. gammarus* are similar (Cooper and Uzmann 1980). The Norway lobster, *Nephrops norvegicus*, is found sublittorally in soft muddy substrata, commonly at depths between 200 and 800 m. However, substantial populations are also recorded from depths <20 m in Scottish Sea Lochs. Many species of the genus *Panulirus* prefer rocky or coral reef for shelter whereas others are found on sandy/muddy substrates. *Jasus* species of lobsters are mainly found on rocky reefs, but can be found to inhabit various substrates from intertidal to 200–400 m (MacDiarmid and Booth 2003; Booth 2006). Adult slipper lobsters are found in all world oceans, temperate, subtropical and tropical oceans within the latitudes 4°S to 40°N and at depths varying from 0 to at least 800 m (Holthuis 1991, 2002; Webber and Booth 2007). They are found on a variety of habitats from soft substrates such as sand and mud to harder gravel surfaces and to rocky and reef crevices (Lavalli et al. 2007).

5.3.1 Substrate Preference

Predicting species distribution and habitat suitability modelling has considerable application in implementation of integrated coastal zone management and ecosystem-based fisheries management (Galparsoro et al. 2009). Information on distribution and particularly the abundance of lobsters in relation to the nature of the substrate characteristics of the benthic habitat is vital for the development of stock-recruitment relationship. In the laboratory, early stage juvenile American lobsters, *Homarus americanus*, showed a preference for shelters instead of gravel, when given a choice between silt-clay deposited over gravel versus gravel substrates (Pottle and Elner 1982). Juvenile lobsters of *H. gammarus* appear to have wide habitat tolerances, enabling them to inhabit a variety of substrate types whereas adults preferred rock crevices widely (Table 5.4). *H. gammarus* juveniles (~8 mm carapace length) selected coarse substrates in the laboratory settings, which offered suitable crevices, or burrowed extensively in fine, cohesive mud (Howard and Bennett 1979). Rock lobsters may use a range of different habitats at different phases of their life cycle from the water column during the pelagic larval phase to seagrass and algal meadows in puerulus and small holes in the reef in postpuerulus and juvenile stages to subadults (3–4 years of age), migrating across the deepwater regions of sand and reefs to settle on offshore, deepwater habitats as mature breeding lobsters (Bellchambers et al. 2012). The Western rock lobster, *P. cygnus*, is largely confined to the limestone reef systems, both surrounding the offshore islands and the deep waters fringing the central coast of Western Australia (Bellchambers et al. 2012). *P. argus* pueruli prefer complex structures regardless of food abundance whereas juveniles choose complex, food-rich habitats (Herrnkind and Butler IV 1986). Newly settled early benthic juveniles of *P. argus* prefer to occupy complex strands of macroalgae. However, no positive correlation between macroalgae and abundance of lobsters <35 mm CL could be established, although a predictable relationship between abundance of larger lobsters (>35 mm CL) with salinity and the density of octocorals, corals and solution holes could be established (Field and Butler IV 1994). Jones (1993) reported maximum abundance of *Thenus indicus* on fine-grain sediments (mud) compared to *T. orientalis* with maximum abundance on coarse-grain sediments. Laboratory studies on substrate preference of these two species corroborate this conclusion (Jones 1988). In the laboratory, *Ibacus peronii* was observed to prefer fine sandy substrates over coarse ones (Faulkes 2006).

5.3.2 Ontogenetic Habitat Shifts

In shallow marine environments, do marine animals, which attain large size in their adulthood, settle on shelter providing habitat in their early benthic life to escape from predators? Wahle and Steneck (1992) based on field experiments and video observations present strong evidence to show that *H. americanus* and other macruran decapods are strongly associated with shelter-providing habitats until they outgrow their vulnerable size. Ontogenetic habitat shifts are characteristic of certain aquatic organisms to maximize growth and to minimize predation risk (Snover 2008). Many spiny lobster species exhibit ontogenetic habitat shift from the postlarval settlement habitat of macroalgae, kelp or seagrass (at 6–15 mm CL for *P. argus*) to benthic crevice sheltering concomitant with aggregation in crevices as larger juveniles, subadults and adults (>15 mm CL) (Butler and Herrnkind 2000; Butler et al. 2006; Childress and Herrnkind 2002; Childress and Jury 2006). Some species demonstrate

Table 5.4 Habitat types and depth ranges of nephropid and palinurid lobsters (compiled from IUCN)

Species	Habitat	Depth
Homarus americanus	Postlarvae settle on cobble/boulder (<30 m); deepwater population on muddy substrates, peat reefs, seagrass beds	0–700 m; most common 0–50 m
Homarus gammarus	Rocky substrates, but may also burrow into cohesive mud or form depressions in sand	Within the continental shelf at depths of 150 m more common > 50m
Nephrops norvegicus	Burrows constructed in muddy substrates	20–800 m
Jasus edwardsii	Rocky reefs	1–250 m
Jasus lalandii	Rocky bottoms and in deep crevices	5–200 m
Jasus paulensis	Rocky substrates; most commonly found on kelp zones	0–60 m; more common 20–40 m
Jasus frontalis	Rocky and sandy substrates	2–200 m
Palinurus elephas	Rocky and coralligenous habitats	5–260 m
Palinurus gilchristi	Rocky areas	55–360 m
Palinurus delagoae	Muddy and sandy substrates as well as coral fragments and rocky substrates	100–600 m
Panulirus homarus rubellus	Rocky substrates	1–36 m
Panulirus homarus homarus	Rocky substrates	1–90 m
Panulirus polyphagus	Muddy and rocky substrates and is often seen near river mouths (Holthuis 1991).	3–90 m
Panulirus ornatus	Slightly turbid coastal waters, sandy and muddy substrates, rocky and coral reefs	8–50 m
Panulirus versicolor	Coral reef, most often on the seaward edge of reef plateaus	1–15 m
Panulirus longipes longipes	Rocky and coral reefs in shallow waters	1–20 m
Panulirus penicillatus	Rocky substratum; outer reef slopes, subtidal zone, small islands	1–16 m
Panulirus argus	Rocky reefs, coral reefs and seagrass beds, which are utilized for shelter (Holthuis 1991).	1–90 m
Panulirus echinatus	Offshore regions with rocky substrates, can be found in caves and crevices	0–35 m
Panulirus laevicauda	Shallow water down to 50 m on rocky reef and coral reef. It is found coexisting with *Panulirus argus*	0–50 m
Puerulus sewelli	Substrates of coarse sand, hard muds and shells (Holthuis 1991).	180–1300 m; common between 18–300 m
Linuparus somniosus	Rough substrates with sand and mud	216–375 m

(continued)

Table 5.4 (continued)

Species	Habitat	Depth
Palinustus waguensis	Rocky habitats and deep reef slopes	10–180 m
Projasus parkeri	Muddy substrates with coral and rock; most abundant on steep-sided volcanic seamounts	370–841 m
Justitia longimanus	Outer part of coral reef slopes and rocky reefs	0–300 m

ontogenetic shifts as they grow larger and larger, even sharing the den with conspecifics (Eggleston and Lipcius 1992; Mintz et al. 1994; MacDiarmid 1994). Childress and Herrnkind (2002) observed juveniles of *P. argus* raised with conspecifics undergoing transition at a smaller size than the solitary lobsters under the same condition. In the absence of conspecifics, bigger lobsters wait until they are large enough to change from algal dwelling to crevice sheltering. Ontogenetic habitat shifts for gregarious animals are probably influenced by the proximity of conspecifics.

5.3.3 Associated Fauna in Lobster Fishing Grounds

Lobsters cohabit with other vertebrate and nonvertebrate organisms in their natural habitats. The Kanyakumari coastline on the southwest coast of India has rocky patches in the inshore sea and coral reefs in the deeper areas, providing ideal conditions for the settlement of lobsters. The partially and fully submerged rocky outcrops with rocky crevices covered by oysters and macroalgae provide favourable shelter for the spiny lobsters. The high abundance of spiny lobsters along the Kanyakumari coast during October–January months is associated with the presence of the Brown mussel, *Perna indica*, which is the primary food of spiny lobsters (Thangaraja and Radhakrishnan 2012). *P. indica* constitutes about 10.5% of the fauna on the intertidal and subtidal rocks (Fig. 5.2). The associated fauna entangled in the bottom-set gillnets (BSGN) operating in the fishing grounds at Khadiyapatnam, a major lobster fishing centre in Kanyakumari district, belong to 10 genera of fishes, 5 species of crustaceans, 3 species of molluscs, 10 species of sponges, 3 genera of echinoderms and 14 species of seaweeds (Table 5.5). Echinoderms are represented by the starfishes and brittle stars (Fig. 5.3). The composition differs between the peak fishing (October–January) and non-fishing months (April–May) with sponges and echinoderms dominating during the fishing season (Fig. 5.4). The macroalgae, seagrass, sponges, gorgonians and the rocky crevices provide significant protection to juvenile lobsters, reducing predation risks (Childress and Herrnkind 1994).

An estimated 25 t of the associated fauna and flora is landed as bycatch by an average 70 gillnets operated per day during the 3 peak fishing months at Khadiyapatnam (Thangaraja and Radhakrishnan 2012). Giraldes et al. (2015) report that 80% of the gillnet bycatch in the artisanal fishing of *P. echinatus* of Brazil is constituted by crabs from the family Mithracinae, which are considered to be algae

Fig. 5.2 Composition of fauna and flora attached on the rocks at Khadiyapatnam lobster fishing ground. (Reproduced from Thangaraja and Radhakrishnan (2012), Fig. 6)

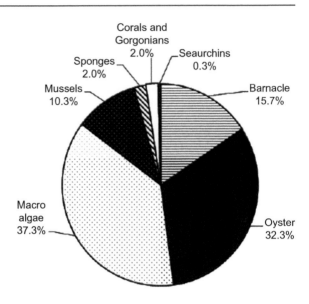

Table 5.5 List of associated fauna and flora entangled in bottom-set gillnets operated at lobster fishing grounds at Khadiyapatnam, Kanyakumari district, southwest coast of India

Group	Genus/species/common name
Fishes	*Lethrinus* sp., *Lutjanus* sp., *Ambassis* sp., *Pampus* sp., *Scarus* sp., *Trichiurus lepturus*, *Arothron* sp., *Epinephelus* sp., *Anguilla bicolor*, *Gymnothorax fimbriatus* and *Odonus niger*
Crustaceans	*Scylla* sp., *Portunus pelagicus*, *Charybdis* sp., *Fenneropenaeus indicus* and spider crabs
Molluscs	*Sepia* sp., *Loligo* sp., *Turbinella pyrum*
Echinoderms	*Stomoponeustes variolaris*, star fishes, brittle stars
Sponges	*Clarthria indica*, *C. pocera*, *C. arborescens*, *Psammoclema* sp., *Axinella ceylonensis*, *Dragmacidon* sp., *Acanthella cavemosa*,
Seaweeds	*Ahnfeltia plicata*, *Caulerpa racemosa*, *Chaetomorpha linoides*, *Chaetomorpha media*, *Gracilaria verrucosa*, *Hypnea musciformis*, *Padina commersonii*, *Sargassam wightii*, *Spatoglossum asperum*, *Ulva reticulata*, *Ulva fasciata* and *Enteromorpha sp.*

Reproduced from Thangaraja and Radhakrishnan (2012), Table 5

cleaners (*algae cleaning crews*) in the reef system. Studies show 75% increase in algal biomass in the reefs from where these crabs are removed, and the authors have expressed concern that constant removal of the keystone low-trophic species such as crabs and other invertebrates can create an ecological imbalance in the coral reef ecosystem.

Fig. 5.3 Associated fauna cohabiting with *Panulirus homarus homarus* at Khadiyapatnam, Kanyakumari district, southwest coast of India. (**a, b, c**): star fishes; (**d, e, f**): sponges; (**g, h**): gorgonians; (**i**): seagrass; (**j**): sea cucumber; (**k**): crab; (**l**): ornamental fish

5.3.4 Trophic Interaction and Cascades

Globally, anthropogenic and climate-induced interventions pose the greatest challenge to the structure and functioning of marine ecosystems (Ling et al. 2009). There is a need to understand how these multiple stressors influence ecosystem dynamics to suppress trends of ecosystem changes. Spiny lobsters coexist with

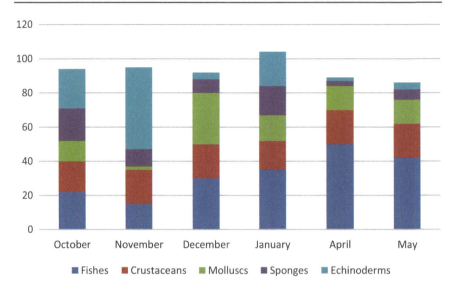

Fig. 5.4 Average monthly percentage composition of associated fauna landed by bottom-set gillnets (BSGN) during lobster fishing (October–January) and non-fishing season (April–May) at Kadiyapattanam. (Reproduced from Thangaraja and Radhakrishnan (2012), Fig. 4)

several other species and may act both as prey and predator. Lobsters are predators of molluscs and sea urchins worldwide, and their predation on sea urchins has reported to result in trophic cascades (Tegner and Dayton 1981; Robles 1987; Mayfield and Branch 2000; Shears and Babcock 2002). They may act as keystone species in the coastal multispecies ecosystems. By conducting *insitu* and *invitro* studies inside marine-protected areas in Tasmania, Ling et al. (2009) show that by indiscriminately removing large predators like *Jasus edwardsii*, the resilience of kelp beds could be drastically reduced against the proliferating sea urchin population. The spiny lobster *P. interruptus* distributed off the southern California coast prefers to feed on the sea urchin *Strongylocentrotus purpuratus* over *S. fransiscanus*. When offered different sizes of *S. fransiscanus* in the laboratory, lobsters preferred smaller sizes though they are capable of feeding on large sizes probably due to handling difficulty. Largescale exploitation of lobsters resulted in population explosion of sea urchins, which contributed to episodes of destructive grazing of kelp forests in the 1950s (Tegner and Dayton 1981). Therefore, there is an urgent need to adopt more integrated resilience-based approaches in the face of rapidly warming climate and unprecedented removal of predators from the world ocean systems (Ling et al. 2009). Guenther et al. (2012) are of the view that trophic cascades due to predator removal from an ecosystem may be due to various other factors. Macroalgal abundance also may depend upon nutrient availability, wave intensity, sedimentation and interaction of all these factors (Menge et al. 2003; Reed et al. 2011). No indirect correlation between lobster fishing measured as fishing effort and decline in macroalgal populations through increases in the abundance of sea urchins could be established, and no conclusive evidence was obtained to show that lobster fishing triggered trophic cascades in the southern California kelp forests (Guenther et al. 2012).

5.4 Summary

Lobsters inhabit all the major oceans, from shallow coastal waters to deeper waters in the tropical, subtropical, temperate and polar zones. While the distribution of temperate species of palinurids are restricted to the southern ocean, nephropids, especially the species coming under the genera *Homarus* and *Nephrops* are limited to the northern latitudes. While some species are endemic to a particular region, few others have a wider distribution in all the major oceans. There is rarely any overlapping in latitudinal or depth distribution within a genus. The tropics and subtropics have the maximum number of species with Indo-West Pacific region having the highest diversity. Distribution of species is governed not only by the surrounding aquatic environment, but by the different habitats they occupy, such as the soft muddy bottom by *Panulirus polyphagus* and the hard and rocky crevices by *P. homarus homarus*. While species which have wide tolerance to salinity *(P. polyphagus)* are mostly distributed near river mouths, there are species with limited capacity to tolerate wide salinity fluctuations *(P. penicillatus)*. Species within the genus *Palinustus* have a wide depth range (vertical) in distribution, occupying 20–200 m as in *P. waguensis*.

Marine organisms are sensitive to their physical environment, including the water temperature. Climate changes could result in animals moving away from their historical distributional ranges, and such shifts may have far-reaching implications for interspecies competition, the economics of fishing and ocean ecosystem health. Shallow-water species may be the most vulnerable to such changes.

References

Bellchambers, L. M., Mantel, P., Chandrapavan, A., Pember, M. B., & Evans, S. E. (2012). *Western rock lobster ecology – The state of knowledge marine stewardship council principle 2: Maintenance of ecosystem fisheries research report, no. 236*. Perth: Department of Fisheries, Western Australia.

Berry, P. F. (1971). *The spiny lobster (Palinuridae) of the east coast of Southern Africa: distribution and ecological notes. Investl. Rep. No. 27* (pp. 1–23). Durban: Oceanographic Research Institute.

Booth, J. D. (2006). *Jasus* species. In B. F. Phillips (Ed.), *Lobsters: Biology, management, aquaculture and fisheries* (pp. 340–358). Oxford: Blackwell.

Butler, M. J. I. V., & Herrnkind, W. F. (2000). Puerulus and juvenile ecology. In B. F. Phillips, J. S. Cobb, & J. Kittaka (Eds.), *Spiny lobster management* (pp. 276–301). Oxford: Blackwell Publishing.

Butler, M. J. I. V., Steneck, R. S., & Herrnkind, W. F. (2006). Juvenile and adult ecology. In B. F. Phillips (Ed.), *Lobsters: Biology and management* (pp. 263–309). Oxford: Blackwell Publishing.

Campbell, A., Noakes, D. J., & Elner, R. W. (1991). Temperature and lobster *Homarus americanus*, yield relationships. *Canadian Journal of Fisheries and Aquatic Sciences, 48*, 2073–2082.

Childress, M. J., & Herrnkind, W. F. (1994). The behavior of juvenile Caribbean spiny lobster in florida bay: seasonality, ontogeny and sociality. *Bulletin of Marine Science, 54*(3), 819–827.

Childress, M. J., & Herrnkind, W. F. (2002). Influence of conspecifics on the ontogenetic habitat shift of juvenile Caribbean spiny lobsters. *Marine and Freshwater Research, 52*(8), 1077–1084.

Childress, M. J., & Jury, S. H. (2006). Behaviour. In B. F. Phillips (Ed.), *Lobsters: Biology, management, aquaculture and fisheries* (pp. 78–112). Oxford: Blackwell Publishing.

Cooper, R. A., & Usmann, J. R. (1980). Ecology of juvenile and adult Homarus. In J. S. Cobb & B. F. Phillips (Eds.), *The biology and Management of Lobsters. Vol.!!. Ecology and management* (pp. 97–139). New York: Academic Press.

Cobb, J. S., & Phillips, B. F. (1980). Preface. In J. S. Cobb & B. F. Phillips (Eds.), *The biology and management of lobsters* (Vol. 1, pp. xi–xiii). New York: Academic Press.

Dow, R. L. (1977). Relationship of sea surface temperature to American and European lobster landings. *ICES Journal of Marine Science, 37*(2), 186–191.

Edgar, G. J. (1990). Predator-prey interactions in seagrass beds: The influence of macro-faunal abundance and size structure on the diet and growth of the western rock lobster *Panulirus cygnus* George. *Journal of Experimental Marine Biology and Ecology, 139*, 122.

Eggleston, D. B., & Lipcius, R. N. (1992). Shelter selection by spiny lobster under variable predation risk, social conditions, and shelter size. *Ecology, 73*, 992–1011.

Faulkes, Z. (2006). Digging mechanisms and substrate preferences of shovel nosed lobsters *Ibacus peronii* (Decapoda: Scyllaridae). *Journal of Crustacean Biology, 26*, 69–72.

Field, J. M., & Butler, M. J., IV. (1994). The influence of temperature, salinity, and postlarval transport on the distribution of juvenile spiny lobsters, *Panulirus argus* (Latreille, 1804) in Florida Bay. *Crustaceana, 67*, 26–45.

Fogarty, M. J. (1986). *Population dynamics of the American lobster Homarus americanus* (Ph D dissertation), University of Rhode Island, Kingston.

Galparsoro, I., Borja, A., Bald, J., Liria, P., & Chust, G. (2009). Predicting suitable habitat for the European lobster (*Homarus gammarus*), on the Basque continental shelf (Bay of Biscay), using Ecological-Niche Factor Analysis. *Ecological Modelling, 220*, 556–567.

George, R. W., & Main, A. R. (1967). The evolution of spiny lobsters (Palinuridae): A study of evolution in the marine environment. *Evolution, 21*, 803–820.

Giraldes, B. W., Zacaron Silva, A., Corrêa, F. M., & Smyt, D. M. (2015). Artisanal fishing of spiny lobsters with gillnets — A significant anthropic impact on tropical reef ecosystem. *Global Ecology and Conservation, 4*, 572–580.

Griffiths, C. L., & Seiderer, J. L. (1980). Rock-lobsters and mussels -limitations and preferences in a predator-prey interaction. *Journal of Experimental Marine Biology and Ecology, 44*, 95–109.

Groeneveld, J. C., Von der Heyden, S., & Matthee, C. A. (2012). High connectivity and lack of mtDNA differentiation among two previously recognized spiny lobster species in the southern Atlantic and Indian Oceans. *Marine Biology Research, 8*(8), 764–770. https://doi.org/10.1080/17451000.2012.676185.

Guenther, C. M., Lenihan, H. S., Grant, L. E., Lopez-Carr, D., & Reed, D. C. (2012). Trophic Cascades Induced by Lobster Fishing Are Not Ubiquitous in Southern California Kelp Forests. *PLoS One, 7*(11), e49396. https://doi.org/10.1371/journal.pone.0049396.

Herrnkind, W. F. (1980). Spiny lobsters: Patterns of movement. In J. S. Cobb & B. F. Phillips (Eds.), *The biology and management of lobsters* (Vol. II, pp. 349–401). New York: Academic Press.

Herrnkind, W. F., & Butler, M. J., IV. (1986). Factors regulating postlarval settlement and juvenile microhabitat use by spiny lobsters *Panulirus argus*. *Marine Ecology Progress Series, 34*, 23–30.

Holthuis, L. B. (1991). *Marine lobsters of the World. FAO species catalogue* (FAO Fisheries Synopsis) (Vol. 13). Rome: Food and Agriculture Organization. 125 (13): 1–292.

Holthuis, L. B. (2002). The Indo-Pacific scyllarinae lobsters (Crustacea, Decapoda, Scyllaridae). *Zoosystema, 24*(3), 499–683.

Howard, A. E., & Bennett, D. B. (1979). The substrate preference and burrowing behaviour of juvenile lobsters (*Homarus gammarus* (L.)). *Journal of Natural History, 13*(4), 433–438.

Joll, L. M., & Phillips, B. F. (1984). Natural diet and growth of juvenile western rock lobsters *Panulirus cygnus* George. *Journal of Experimental Marine Biology and Ecology, 7*(5), 145–169.

Jones, C. M. (1988). *The Biology and Behaviour of Bay lobsters, Thenus spp. (Decapoda: Scyllaridae) in Northern Queensland, Australia*. PhD. Dissertation (p. 190). Bisbane: University of Queensland.

Jones, C. M. (1993). Population structure of two species of *Thenus* (Decapoda: Scyllaridae) in northeastern Australia. *Marine Ecology Progress Series, 97*, 143–155.

Lavalli, K. L., & Spanier, E. (2007). Introduction to the biology and fisheries of slipper lobsters. In K. L. Lavalli & E. Spanier (Eds.), *The biology and fisheries of the slipper lobster*. New York: CRC Press, Taylor & Francis Group.

Lavalli, K. L., Spanier, E., & Grasso, F. (2007). Behavior and sensory biology of slipper lobsters. In K. L. Lavalli & E. Spanier (Eds.), *The biology and fisheries of slipper lobsters* (Crustacean issues) (Vol. 17, pp. 133–181). New York: CRC Press (Taylor & Francis Group).

Lawton, P., & Lavalli, K. L. (1995). Postlarval, juvenile, and adult ecology. In J. R. Factor (Ed.), *Biology of the lobster, Homarus americanus* (pp. 47–88). San Diego: Academic Press.

Levin, S. A. (1999). *Fragile dominion. Complexity and the commons*. Cambridge: Perseus Books Group.

Ling, S. D., Johnson, C. R., Frusher, S. D., & Ridgway, K. R. (2009). *Overfishing reduces resilience of kelp beds to climate-driven catastrophic phase shift*. www.pnas.org. https://doi.org/10.1073pnas.0907529106

MacDiarmid, A. B. (1994). Cohabitation in the spiny lobster *Jasus edwardsii* (Hutton, 1875). *Crustaceana, 66*, 341–355.

MacDiarmid, A. B., & Booth, J. (2003). Crayfish. In N. Andrew & M. Francis (Eds.), *The living reef. The ecology of New Zealand's rocky reefs* (pp. 120–127). Nelson: Craig Potton Publishing.

Marx, J. M., & Herrnkind, W. F. (1985). Factors regulating microhabitat use by young juvenile spiny lobsters, Panulirus argus: food and shelter. *Journal of Crustacean Biology, 5*, 650–657.

Mayfield, S., & Branch, G. M. (2000). Interrelations among rock lobsters, sea urchins, and juvenile abalone: implications for community management. *Canadian Journal of Fisheries and Aquatic Sciences, 57*, 2175–2185.

Menge, B. A., Lubchenco, J., Bracken, M. E. S., Chan, F., Foley, M. M., et al. (2003). Coastal oceanography sets. the pace of rocky intertidal community dynamics. *Proceedings of the National Academy of Sciences of the United States of America, 100*, 12229–12234.

Mintz, J. D., Lipcius, R. N., Eggleston, D. B., & Seebo, M. S. (1994). Survival of juvenile Caribbean spiny lobster: effects of shelter size, geographic location and conspecific abundance. *Marine Ecology Progress Series, 112*, 255–266.

Partelow, S. (2014). *Assessing sustainability in lobster fisheries as social-ecological systems: A framework and research protocol*. Lund: Lund University.

Phillips, B. F., & Kittaka, J. (2000). Preface. In B. F. Phillips & J. Kittaka (Eds.), *Spiny lobsters: Fisheries and culture* (2nd ed.). New York: Wiley-Blackwell.

Phillips, B. F., Wahle, R. A., & Ward, T. J. (2013). Lobsters as part of marine ecosystems – A review. In B. F. Phillips (Ed.), *Lobsters: Biology, management, aquaculture and fisheries* (2nd ed.). New York: Wiley-Blackwell.

Pitcher, R. (1993). Spiny lobster. In A. Wright & L. Hill (Eds.), *Nearshore marine resources of the South Pacific* (pp. 539–608). Suva: Institute of Pacific Studies, Honiara: Forum Fisheries Agency and Halifax: International Centre for Ocean Development.

Pottle, R. A., & Elner, R. W. (1982). Substrate preference behavior of juvenile American lobsters, *Homarus americanus*, in gravel and silt–clay sediments. *Canadian Journal of Fisheries and Aquatic Sciences, 39*(6), 928–932.

Reed, D. C., Rassweiler, A., Carr, M. H., Cavanaugh, K. C., Malone, D. P., et al. (2011). Wave disturbance overwhelms top-down and bottom-up control of primary production in California kelp forests. *Ecology, 92*, 2108–2116.

Rios-Lara, V., Salas, S., Bello-Pineda, J., & Irene-Ayora, P. (2007). Distribution patterns of spiny lobster (Panulirus argus) at Alacranes reef, Yucatan: Spatial analysis and inference of preferential habitat. *Fisheries Research, 87*, 35–45.

Robles, C. (1987). Predator foraging characteristics and prey population structure on a sheltered shore. *Ecology, 68*, 1502–1514.

Shears, N. T., & Babcock, R. C. (2002). Marine reserves demonstrate top-down control of community structure on temperate reefs. *Oecologia, 132*, 131–142.

Snover, M. L. (2008). Ontogenetic habitat shifts in marine organisms: influencing factors and the impact of climate variability. *Bulletin of Marine Science, 83*(1), 1–15.

Tegner, M. J., & Dayton, P. K. (1981). Population structure, recruitment, and mortality of two sea urchins (*Strongylocentrotus fransciscanus* and *S. purpuratus*) in a kelp forest near San Diego, California. *Marine Ecology Progress Series, 77*, 49–63.

Tegner, M. J., & Levin, L. A. (1983). Spiny lobsters and sea urchins: Analysis of a predator-prey interaction. *Journal of Experimental Marine Biology and Ecology, 73*, 125–150.

Thangaraja, R., & Radhakrishnan, E. V. (2012). Fishery and ecology of the spiny lobster Panulirus homarus (Linnaeus, 1758) at Khadiapatanam in the southwest coast of India. *Journal of the Marine Biological Association, 54*(2), 69–79.

Wahle, R. A., & Steneck, R. S. (1992). Habitat restrictions in early benthic life: experiments on habitat selection and *in situ* predation with the American lobster. *Journal of Experimental Marine Biology and Ecology, 157*(1), 91–114.

Webber, W. R., & Booth, J. D. (2007). Taxonomy and evolution. In K. L. Lavalli & E. Spanier (Eds.), *The biology and fisheries of the slipper lobster* (pp. 25–52). Boca Raton: Taylor & Francis/CRC Press.

6

Food, Feeding Behaviour, Growth and Neuroendocrine Control of Moulting and Reproduction

E. V. Radhakrishnan and Joe K. Kizhakudan

Abstract

This chapter encapsulates information on food, feeding behaviour, moulting, growth and neuroendocrine regulation of moulting and reproduction in lobsters. A lot of research has been carried out worldwide on growth and the processes controlling it in lobsters, mostly on homarid and palinurid lobsters. This chapter, through a review of documented research, describes the moult cycle in lobsters, which is similar to that of other crustaceans, and the process of growth that occurs with moulting. It also discusses differences in growth rates with age, sex and varying external factors, including food. The feeding behaviour in lobsters is also discussed. The relationship between growth, moulting and reproduction and the internal factors that regulate these processes are also dealt with. A detailed account of neuroendocrine hormones and mechanisms that play an important role in the lobster physiology is presented. An insight into the aspects presented here will help in developing rearing protocols for commercially important cultivable lobsters.

Keywords

Moulting · Growth · Food · Neuroendocrine hormones · Physiology · Reproduction

6.1 Introduction

Growth is an important life history process in our understanding of population ecology and stock assessment (Chang et al. 2012). Growth is also a factor of population abundance and production over time (Ricker 1975). Growth in crustaceans occurs at ecdysis (Aiken 1980) or moulting and is a cyclical event in the life

E. V. Radhakrishnan (✉) · J. K. Kizhakudan
ICAR-Central Marine Fisheries Research Institute, Cochin, Kerala, India
e-mail: evrkrishnan@gmail.com

of a crustacean. Information on moult cycle is key to improving aquaculture techniques (Mikami 2005). Moulting is a complex and energy-demanding process, and the life history of most crustaceans is synchronized with the moult cycle (Chang 1985). The interval between two moults, the intermoult, is the period when water absorbed during the moulting process is replaced by tissues. In adult females, the entire reproductive processes, mating, ovarian maturation, yolk synthesis, egg extrusion, fertilization and parental care of the brood, are completed in the intermoult period. Many environmental, physiological, behavioural and nutritional factors are believed to influence moulting and reproduction, and these two metabolic processes are regulated by several multifunctional neuroendocrine hormones located in the nervous system and mandibular regions. Advancements in crustacean endocrinology during the past 30 years, including development of more robust analytical methods to isolate and characterize the hormones and the identification of mRNA genes responsible for production of these hormones, paved way for better understanding of the interrelationships between moulting and reproduction in crustaceans (Adiyodi and Adiyodi 1970; Kleijn and van Herp 1995; Chang et al. 2001; Raviv et al. 2006).

The rate of growth in different species and similar species in different geographical areas is dependent on several biotic and abiotic factors. Among the biotic factors, food plays a major role in the growth process in crustaceans. In lobsters, the quality of food, its availability and the density of the lobster population in relation to food supplies decide the growth rate (Newman and Pollock 1974; Barkai and Branch 1988). Temperature is the major abiotic factor that influences growth in temperate and semitropical species. Several other abiotic factors such as salinity, oxygen and photoperiod also influence growth and reproduction in lobsters.

Food is a major factor that determines growth. Decapods are mostly omnivorous; some are strictly carnivorous and some show saprophagus characteristics (Joll and Phillips 1984: Sainte-Marie and Chabot 2002). Knowledge of an animal's natural diet is essential for studies on its nutritional requirements, its interactions with other organisms and its potential for culture (Williams 1981). Palinurid lobsters in general have been characterized as omnivorous with a diet consisting of crustaceans, gastropods, fish remains, asteroid echinoderms, porifera and marine plants (Lindberg 1955; Engle 1979; Kanciruk 1980; Colinas-Sánchez and Briones-Fourzán 1990; Díaz-Arredondo & Guzmán del Próo 1995; Briones-Fourzán et al. 2003), and they are benthic carnivores according to some researchers (Berry 1977; Davis 1977). They are prey for many marine species such as sharks and octopuses. Spiny lobsters exhibit ontogenetic shifts in habitat use during juvenile/adult phases for different behavioural purposes (Windell 2015) and show progressive dietary shift with increasing size (Sainte-Marie and Chabot 2002). They affect a range of prey types and habitat over their life cycle and enact multiple roles as prey, predators and keystone species and exhibit non-consumptive interactions (Boudreau and Worm 2012).

The chapter summarizes the scientific data gathered over the past many years on physiological processes and hormonal control of moulting, growth and reproduction and the food and feeding behaviour of lobsters.

6.2 Moult Cycle

Moulting (ecdysis) is a complex process which includes all the morphological, physiological and biochemical changes that an animal undergoes from the state of preparation for and up to the state of recovery from ecdysis (Waddy et al. 1995). There is a tendency to use moult in synonymous with ecdysis. Waddy et al. (1995) suggest that ecdysis is the most appropriate term than moulting to describe the moulting process. The most appropriate and precise terms used in moult parlour are proecdysis, metecdysis and postecdysis rather than premoult, intermoult and postmoult (Waddy et al. 1995).

6.2.1 Moulting (Ecdysis)

Spiny lobsters, like other decapods, shed their old exoskeleton periodically to grow, i.e. increase in length and weight. Ecdysis is not just casting off the outer body shell but also peeling off the chitinous lining of the oesophagus and foregut and the surface of the gills. At the end of the larval phase, a complete transformation of the flat leaf-like phyllosoma larva to an adult-shaped puerulus takes place. The process of ecdysis in both homarid and palinurid lobsters is similar, though there are differences in the time taken to complete each stage. As the lobster enters in the proecdysis phase, i.e. preparation for the moult, several morphological, anatomical and physiological changes take place. The proecdysis activities include resorption and storage of cuticular components in the haemolymph, deposition of new cuticle, limb regeneration, shifts in biochemical pathways and selective ion and water absorption (Waddy et al. 1995). For convenience, ecdysis has been divided into two phases: a passive and an active phase. During the passive phase (Stage D_4, which is part of the proecdysis), due to absorption of water, the carapace separates from the branchiostegal decalcified ecdysial sutures above the bases of the legs and thoracoabdominal membrane bulges outward at the junction of the carapace and the abdomen. If the lobster is disturbed or if the conditions are unfavourable, the lobster may prolong the phase for several hours. As the lobster enters the active ecdysial phase (irreversible), the thoracoabdominal membrane ruptures, and on losing the grip, the lobster either falls sideways or keeps its upright position and wriggles and jumps out of the old exoskeleton. Radhakrishnan and Vijayakumaran (1998) have provided photographic evidence of *P. homarus homarus* undergoing ecdysis in the upright position during the final process. The postecdysial phase begins with the lobster continuing to absorb large quantities of water to stretch the limbs and body and assumes its new length and volume. The metecdysis (intermoult) phase starts with the lobster actively feeding during which the new cuticle is secreted, mineralization occurs and deposition of new tissues commences replacing the water absorbed during ecdysis. The lobster is capable of maintaining a high level of coordinated activity, agility and neuromuscular control of the body functions until the rupture of the thoracoabdominal membrane and splitting of the carapace from the abdomen (active phase of ecdysis), but becomes defenceless and vulnerable during the final phase. *Panulirus homarus homarus* was reported to complete the active phase in 3–4 min (Radhakrishnan and Vijayakumaran 1998) whereas *P. homarus rubellus* takes approximately 3–7 min for the whole process from rupturing of the arthrodial membrane to shedding of the old exoskeleton (Berry 1971). However, it is not known whether the duration is shorter in juveniles as was found in *J. lalandii* (Paterson 1969) and in *P. argus* (Travis 1954). Kizhakudan et al. (2013) reported that the juvenile mudspiny lobster, *P. polyphagus*, exhibited higher moulting frequency, with the intermoult period increasing steadily. Increments in carapace length (CL) and weight at each moult showed significant correlation with increasing CL in both males and females. Spiny lobsters in general take lesser time (3–5 min) to complete the final ecdysial process compared to homarid lobsters (15–20 min). Passing this vulnerable phase rapidly may be necessary for survival since spiny lobsters moult in the presence of other lobsters and in at least a few species, cannibalism is prevalent. Captive newly moulted lobsters avoid other lobsters by quickly moving to secure places. Following Wahle and Fogarty (2006), Kizhakudan (2007) estimated the probability of ecdysis for lobsters of different size classes within specific time frames. In males and females of both species, the moult probability for shorter intermoult periods was highest for smaller animals. As growth progresses, larger animals tend to have larger intermoult periods.

Immediately after ecdysis, normal lobsters devour portions of the shed exoskeleton, which is probably a mechanism to recover calcium and minerals lost through their exuviae. No instance of cannibalism was recorded in the captive spiny lobster *P. homarus* (Berry 1971; Radhakrishnan and Vijayakumaran 1984a) unlike other species such as *J. lalandii* (Paterson 1969). However, Kizhakudan (2007) observed that *P. polyphagus* exhibited a strong tendency to consume moult exuviae, and whole moult recovery from *P. polyphagus* rearing tanks was always difficult. On the otherhand, *T. orientalis* (=*unimaculatus*) did not exhibit any feeding affinity for the moult exuviae, and it was possible to almost always collect the moults intact. The newly moulted lobsters avoid other lobsters and hide in the shelter. Eyestalk-ablated (EA) lobsters, on the other hand, become easy prey to others if EA lobsters and normal ones are kept together (Radhakrishnan, personal observation).

6.2.2 Moult Staging

Moult staging in crustaceans is usually done by examining the hardness of exoskeleton or observing the transparent edge of a pleopod under a microscope to see changes in setal morphology. Drach (1939) was the first to introduce a staging system for crustaceans that recognized the morphological and physiological changes associated with moulting. Drach and Tchernigoutzeff (1967) have divided crustacean moult cycle into four basic stages: A, immediate postmoult; B, postmoult; C, intermoult; and D, premoult. Knowles and Carlisle (1956) introduced stage E, ecdysis. These stages have been again subdivided into 13 stages. The scheme has been adopted by many researchers for moult staging in several crustaceans including crayfish (Stevenson et al. 1968; Mills and Lake 1975; Peebles 1977; Van Herp and

Bellon-Humbert 1978), lobsters (Aiken 1973; Dall and Barclay 1977; Lyle and MacDonald 1983) and prawns (Smith and Dall 1985). Aiken (1973, 1980), and Waddy et al. (1995) provide a detailed description of moult staging and setal staging in the American lobster *Homarus americanus*. Aiken (1973) was the first to develop a moult staging system based on setal development in pleopods of the American lobster *Homarus americanus*. The stages were assigned only after correlating setal changes with histological evidence, as the integumentary process varies from one area of the body to another (Aiken 1980; Waddy et al. 1995). Drach's (1939) intermoult stages are based on changes in the integument. Lyle and MacDonald (1983) reported a simple method of pleopod moult staging in wild populations of *P. marginatus*. Radhakrishnan (1989) followed the method of Lyle and MacDonald (1983) to moult stage the spiny lobster *P. homarus homarus* with modifications. Turnbull (1989) adopted a modified scheme of Lyle and MacDonald (1983) to moult stage the spiny lobster *P. ornatus*. However, there were several differences between *P. marginatus* and *P. ornatus* in setal development and duration of each stage. The moult staging referred to here is based on moult staging of the pleopod and changes associated with shell hardening (Table 6.2).

Stage A: The postecdysial stage begins soon after the lobster has completed ecdysis and constitutes 2% of the moult cycle (Drach 1944). *P. homarus homarus* spends 1.7% of the total moult cycle duration in this stage (Radhakrishnan 1989). The postecdysis stage is divided into A and B, and stage A is further separated into A_1 and A_2 (Table 6.1). Stage A_1 lasts for 8–10 h after ecdysis, depending upon the lobster size. The body of the lobster is soft and flaccid, but water absorption and ingestion processes continue and the animal acquires its new volume. The new setal wall is thin and wrinkled with the setal lumen full with the matrix. The setal bases are poorly developed with a thin cuticle between them (Turnbull 1989). Pigmentation is extended up to the edge of the thin cuticle. The animal is fully agile and can swim away from advancing co-habitants.

Stage B: The lobster is in the 'paper shell' stage. This stage occupies 2% of the moult cycle in *H. americanus*, but may vary in different species (Aiken 1980). The shell partially hardens, mineralization of new cuticle continues and the branchiostegal area continues to be soft. The cuticle between the setal bases is still thin. Setal walls thicken, and the matrix is restricted to the centre of the seta. The lobster begins feeding.

Stage C: The shell is fully hardened, but the branchiostegal area is still flexible and can be pressed by a finger. The stage can be arbitrarily divided into four substages (C_1 to C_4) on the basis of rigidity (Waddy et al. 1995). Although stage C is generally considered as metecdysis or intermoult, the actual metecdysial stage begins from stage C_4 onwards when the process of hardening is fully completed, and the lobster starts active feeding and accumulates organic reserves (Table 6.1). The setal wall is thickened with the lumen narrowing towards the centre. The cuticle becomes thick and extends below the setal bases. The pigmentation looks a little withdrawn at the bases of the setae. Stage C occupies 65% of the moult cycle in some species (Drach 1944). *P. homarus homarus* spends 40.6% of the total moult cycle duration in stage C (Radhakrishnan 1989). This stage C and may be

Table 6.1 Criteria for moult staging of *Panulirus homarus homarus* (size of the lobster, CL 45–50 mm; weight, 75–100 g) (Radhakrishnan & Vijayakumaran, unpublished)

Moult (ecdysis) stage	Time before (−) or after (+) ecdysis (h/days)	Distinguishing features
Ecdysis (E)	0 h	Final (irreversible) phase of ecdysis, lasts 3–4 min
Postecdysis or A_1 Postmoult (A) A_2 (B)	+ 8–10 h +10–14 h +1–4 days	Integument very soft; matrix of setae extends up to the tip Integument soft; cuticular nodes appear at the base of the setae Integument flexible; matrix within the setae tapers towards the centre
Metecdysis (intermoult) (C) (C_1-C_3) C_4	+5–7 days +7–15 days	Anterio-dorsal region of the carapace could be depressed by applying slight pressure. Branchiostegite region soft Hardening of exoskeleton complete, including the branchiostegite region
Proecdysis (premoult) (D) (D_0-D_3) D_4 (early) D_4 (late)	−15 to −20 days −48 h −6 h	Same as described by Aiken (1973) Longitudinal decalcified line appears on the branchiostegite. Region below the line very soft and will break under slight pressure Dorsal distention of the arthrodial membrane at the junction of carapace and abdomen

influenced by both intrinsic factors such as size, age, physical condition (Aiken 1973, 1980; Chittleborough 1975) and extrinsic factors such as temperature and availability of food (Travis 1954; Chittleborough 1975).

Stage D: The premoult or the proecdysis is the period when the lobster starts preparing for the next moult. Stage D is divided into the substages $D_0–D_4$, which are further subdivided by addition of *superscript characters* (D_1', D_1'', D_1''') (Charniaux-Cotton 1957; Drach 1939, 1944; Drach and Tchernigovtzeff 1967; Waddy et al. 1995). Stage D occupies 52.9% of the moult cycle in *P. homarus homarus* (Radhakrishnan 1989).

Substage D_0: Apolysis or retraction of the pigmented epidermis from the cuticle begins at this stage, leaving a transparent gap between the cuticle and the epidermis (Turnbull 1989). Stage D_0 also appears to be variable, and the lobster may spend weeks or months in this plateau. Waddy et al. (1995) consider D_0 as a resting stage and suggest that this period may be considered as a transitional phase when the lobster is neither in intermoult nor premoult stage. *P. homarus homarus* spends 34.9% of the total moult cycle duration in this stage (Radhakrishnan 1989).

Substage D_1: This substage is further divided into two or three substages (indicated with superscripts) D_1', D_1'' and D_1'''. Maximum retraction and scalloping of the epidermis at the tip of the pleopod are the first visible indications of the lobster entering D_1' stage. Invagination papillae form at the site of future setae. The distal part of the newly forming seta is visible as flaccid wisps of tissue in the retracted zone in *P. ornatus* (Turnbull 1989), whereas in *H. americanus* the shafts of setae are not well defined at this point (Waddy et al. 1995). No visible change in setal development between stages D_1' and D_1'" was evident in *P. ornatus*, and, therefore, the substage D_1" was omitted (Turnbull 1989). On entering stage D1'", the setae develop barbules, which are fully visible. In *H. americanus*, the barbules on setae become visible only in the D_2' stage (Aiken 1980).

Substage D_2: On entering the D_2 stage, a new epicuticle is visible on the pleopod epidermis. A new exocuticle is present in the gastric region of the carapace in *P. ornatus* and *H. americanus* (Turnbull 1989; Waddy et al. 1995). A transparent layer that corresponds to the new cuticle is visible above the epidermis in *P. ornatus*. The carapace below the ecdysial line starts softening as the lobster enters stage D_2 in *P. ornatus* (Turnbull 1989). The invagination in the epidermis at the site of formation of new setae in the pleopod of *P. ornatus* becomes more pronounced than in stage D_1'". In *H. americanus*, shafts of developing setae are visible but the proximal ends are not clearly defined. Since the pigmentation obscures the view of further development of the new setae, moult staging of the pleopod beyond this point becomes difficult in spiny lobsters (Lyle and MacDonald 1983; Turnbull 1989; Radhakrishnan 1989). However, in *H. americanus*, full development of setae is visible through stages D_2–D_3.

In *H. americanus*, pleopod stage D_2 is subdivided further into two substages (indicated with superscripts) D_2' and D_2". In stage D_2', shafts of setae are visible full length, but the proximal ends are bifurcate instead of blunt whereas in stage D_2". The proximal end of the new setae becomes blunt (Aiken 1980; Waddy et al. 1995). In stage D_3, the shafts of setae become thick and dark, and if folds or ripples are visible in the cuticle on the upper surface of pleopods, it is classified as D_3 (Waddy et al. 1995).

Stage D_4: The stage can be easily identified by the presence of a decalcified line on the branchiostegite (Radhakrishnan 1989). In *H. americanus*, the line appears in stage D_3 and indicates that moult will occur within a week at temperatures of 12–15 °C (Aiken 1980). In *P. homarus homarus*, the lobster can be predicted to moult within 48 hours after appearance of the decalcified line (Radhakrishnan 1989). In *P. ornatus*, softening of the exoskeleton below the ecdysial line occurs 3 days prior to ecdysis (Turnbull 1989).

Stage E: Ecdysis includes a passive (often classified as stage D_4 in *H. americanus*) and an active phase. During the passive phase, water is ingested, absorbed and redistributed within the body (Waddy et al. 1995). The actual process of ecdysis takes place during the active phase, when the lobster starts lifting the old carapace from the posterior margin, allowing the head to be raised and the cephalic appendages to be withdrawn. Finally, the lobster wriggles out of the old exoskeleton by giving a push to the abdominal region (Waddy et al. 1995; Radhakrishnan and Vijayakumaran 1998).

6.2.3 Water Uptake at Ecdysis

A spiny lobster approaching ecdysis can easily be identified by the appearance of a decalcified ecdysial line along the branchiostegite. In *P. argus*, this line is clearly visible 3–4 days before ecdysis during summer months (Travis 1954). In *P. homarus homarus*, this resorptive line is visible 48 h before ecdysis (Radhakrishnan and Vijayakumaran 1998). If not disturbed, *P. homarus* the lobster completes ecdysis within 8–10 h after the initiation of the process. Radhakrishnan and Vijayakumaran (unpublished) observed weight changes during ecdysis in normal and bilaterally ablated lobster *P. homarus homarus*. The weight change was represented as percentage of the preecdysial weight (Fig. 6.1). While the percentage total water uptake by a normal lobster during the whole process of ecdysis was an average 39.3 ± 4.95, the percentage of water uptake in bilaterally eyestalk-ablated lobster measured an average 54.54 ± 11.67 (Table 6.2). These authors also observed difference in

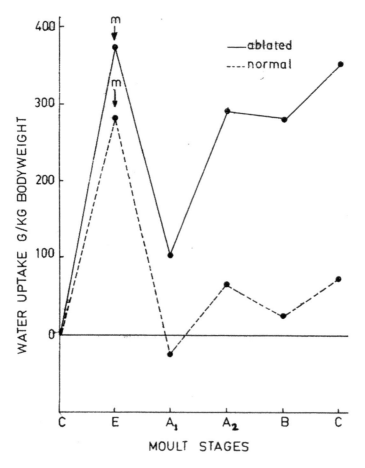

Fig. 6.1 Water uptake during ecdysis by normal and ablated *Panulirus homarus homarus*

Table 6.2 Weight changes and water uptake in control and eyestalk-ablated *Panulirus homarus homarus* during ecdysis (Radhakrishnan and Vijayakumaran, unpublished)

Treatment/ no. of animals	Weight (g)	*Percent increase in weight		*Percent water uptake			Total water uptake (%)
		Immediately after moult	After initial water uptake	Up to ecdysis (measured)	Up to ecdysis (calculated from animal wt. + moult wt.	Postmoult	
Control/5	104 ± 12.4	1.06 (−3.0 to 7.32)	11.09 ± 8.72	28.2 ± 4.03	29.6 ± 2.25	10.18 ± 3.69	39.3 ± 4.95
Ablated/7	75.8 ± 18.47	8.77 ± 5.39	29.85 ± 9.93	32.59 ± 5.83	36.04 ± 6.93	18.61 ± 5.95	54.54 ± 11.67

*Based on premoult weight

Table 6.3 Percent increase (of premoult CL and weight) in carapace length and weight after first moult in eyestalk-ablated *Panulirus homarus homarus* at different moult stages (weight taken after 15 days of moulting) (Radhakrishnan and Vijayakumaran, unpublished)

Moult stage	Weight	Carapace length
C	41.4 ± 14.8	11.6 ± 1.9
D_1–D_2	34.9 ± 10.3	11.2 ± 2.7
D_4	30.2 ± 3.7	10.2 ± 1.8

postecdysial weight if the ablation is carried out in different moult stages. If ablated in intermoult stage C, the percentage postmoult weight increase after the first moult was 41.4 ± 14.8 compared to 30.2 ± 3.7 in lobsters ablated in moult stage D_4 (Table 6.3). The percentage of water in the haemolymph of normal and eyestalk-ablated lobster was significantly different with ablated lobsters having higher water content in the haemolymph and in all tissues (Table 6.4).

6.3 Growth

Growth can be expressed as an increase in length, volume or weight with time (Hartnoll 1982, 2001). In marine organisms especially in finfishes, growth is considered to be continuous with fluctuations in its rates, according to life stages and environmental parameters. Growth in crustaceans also is a continuous process, but that can only be expressed in size increase during moulting, due to the rigid nature of the hard exoskeleton (Phillips et al. 1980). Growth in crustaceans is manifested by two processes: the moulting frequency and the size increase per moult. The interval between two successive moults, the intermoult period, increases with size from a few days in the larval and early juvenile stages to a year or more in adult lobsters, whereas the moult increment, expressed as a proportion of the size of the lobster, decreases as the lobsters grow. A linear relationship exists between the logarithm of the moult increment and the carapace length, and between the logarithm of moult interval and the carapace length (Mauchline 1977). Lobsters under the genus *Panulirus* continue to moult and grow even after attaining sexual maturity. However, the percent increase in growth decreases with increase in size (Aiken 1980).

6.3.1 Estimating Growth

Growth estimates in spiny lobsters can be determined by direct and indirect methods (Wahle and Fogarty 2006). Direct measure of growth is possible in captive specimens by measuring length and weight after each moult or by tag-recapture studies in the wild populations. Growth and age can be indirectly estimated by analysis of size frequency distribution of samples collected from wild population of lobsters. The quantification of lipofuscin, a metabolic byproduct that accumulates with age, has been used to estimate age in lobsters and crayfish (Wahle et al. 2001). The direct method provides both intermoult duration and size increment at each

Table 6.4 Percent water content in haemolymph of normal and ablated *Panulirus homarus homarus* in different moult stages (Radhakrishnan and Vijayakumaran, unpublished)

Treatment	Percent water content in different moult stages									
	D_4	E	A_1	B	C_1	C_4	D_0	D_1	D_2	D_3
Normal	83.7 ± 1.3	89.5 ± 0.4	89.7 ± 0.7	90.9 ± 0.9	90.2 ± 1.5	85.3 ± 1.1	83.7 ± 0.8	83.4 ± 0.8	83.7 ± 0.2	84.4 ± 0.6
Ablated	85.3 ± 1.4	–	93.4 ± 0.4	92.5 ± 0.8	93.6 ± 1.1	92.4 ± 0.8	90.9 ± 0.8	92.4 ± 0.9	–	–

moult, the two parameters by which growth can be estimated. The Hiatt growth diagram developed by Hiatt (1948) by plotting pre-exuvial size against post-exuvial size yielded straight-line relationships for increase in weight as well as length in a variety of crustaceans (Kurata 1962; Mauchline 1976). Mauchline (1976) suggested that instead of plotting postmoult to premoult length, more useful linear relationships can be obtained by plotting the log percent length increase against the body length or moult number. By adopting this method, a constant called *moult slope factor* can be generated, the factor by which percent size increase decreases at successive moults (Aiken 1980). Similarly, the *intermoult period slope factor* also can be generated by plotting log intermoult period against length or moult number, which defines the increase in intermoult period with successive moults. Mauchline (1977) also obtained a linear relationship for log carapace length (log L^3) cubed against log age (days).

In the absence of information on moult increment and intermoult duration, several methods are available for estimating growth. Most of the methods estimate various parameters of the von Bertalanffy (1938) growth function with the following formula:

$$L_t = L_\infty \left[1 - e^{K(t-t0)} \right]$$

where L_t is the length at age t, L_∞ is the asymptotic carapace length, K is the growth coefficient and t_0 is the parameter that positions the function on the age axis so that length and age (from hatching) of recruited size classes correspond (i.e. the age when the function intersects the age axis). The parameters L_∞ and K may be estimated from tag recapture data using methods such as the Fabens (1965) equation or by the Gulland and Holt (1959) plot (e.g. Prescott 1988; Pitcher 1993). Size frequency data also may be used to estimate L and K by means of Wetherall (1986) plots and the ELEFAN (Pauly et al. 1980) computer programme (Munro 1982) or modal progression (Uchida and Tagami 1984). These methods are explained in detail in a number of stock assessment manuals (Ricker 1975; Gulland 1983; Pauly 1983; Pauly and Morgan 1987; Sparre et al. 1989; Majowski et al. 1987); Morgan 1980.

George (1967) studied growth of *Panulirus homarus* from Colachel-Muttom zone of the Kanyakumari coast, on the southwest coast of India, using modal progression derived from length frequency data and found that the fishery is represented by six year-groups. A year-group which enters into the fishery at 131–140 mm total length (TL) (third year class) attains 300 mm (ninth year class) in 5 years. *P. homarus* (Linnaeus) attains an average annual growth rate of 28 mm per year. Mohammed and George (1968), based on recoveries of nearly 70 tagged *P. homarus homarus* along the southwest coast of India, estimated sizes attained at successive ages by fitting von Bertalanffy's growth equation. Based on these estimates they suggest fast growth in the first year of life after settlement as puerulus and considerably slower growth rate thereafter. Mohammed and George (1968) also found that males grew faster than females, and this conclusion was supported by the higher modal sizes of males in the commercial catch. Berry (1971), however, found similar growth for both sexes until maturity, and Smale (1978) observed faster growth in females in the subspecies *P. homarus rubellus*. Mohammed and George (1968) suggest that *P. homarus* fishery along the Kanyakumari coast of southwest coast of India is supported largely by 1 and 2 year-groups, and the calculated maximum size (L_∞) was 312.1 mm and 302.5 mm TL for males and females, respectively. These lobsters have been estimated to be 10 years old at the maximum size. Thomas (1972) observed growth per moult to be 4–9 mm in carapace length in *P. homarus*. Kathirvel (1973) observed in aquarium-held *Panulirus polyphagus* measuring 88 mm an 11 mm increase in CL in two successive moults. Nair et al. (1981) in a similar study observed that the growth increment per moult was 6.5–9.6 mm in *P. homarus*, 11.3–13.8 mm in *P. ornatus* and 5.5 mm in *P. penicillatus*. Cooper and Uzmann (1980) estimated both the annual moulting frequency and the average increment as a function of size in the American lobster *Homarus americanus*. Among the spiny lobsters of the genus *Panulirus*, the growth coefficient 'K' has been reported to vary from as less as 0.00026 in *P. cygnus* off Western Australia (Phillips et al. 1992) to as high as 0.58 in *P. penicillatus* in the Marshall Islands (Ebert and Ford 1986).

Kagwade (1987) used the modal progression method to estimate growth parameters in *P. polyphagus* and reported very high L_∞ values of 537 mm TL and 443 mm TL and annual $K = 0.2$ and 0.2231 for male and female *P. polyphagus* off Mumbai, Maharashtra, respectively, on the northwest coast of India. Juveniles of both sexes attained 85 mm in the first year, 145 mm in the second year and 205 mm in the third year and, thereafter, males grow faster. Kizhakudan et al. (2013) compared growth rates in the mudspiny lobster, *P. polyphagus*, held in captivity with growth estimates derived from the length composition of these lobsters in trawl landings at Veraval, northwest coast of India. They estimated von Bertalanffy's growth parameters L_∞ and K for males as 135 mm CL and 0.46 year^{-1} in the wild and 144.8 mm CL and 0.51 year^{-1} in captivity and for females as 124.7 mm (CL) and 0.38 year^{-1} in the wild and 119 mm (CL) and 0.43 year^{-1} in captivity. They reported a clear difference in the growth curves of males and females in captivity and in the natural habitat, with the VBGF curves indicating that males had a better growth output than females, both in the wild and in captivity.

Kagwade and Kabli (1996) carried out growth studies in *Thenus orientalis* (=*unimaculatus*), collected from trawl landings in Mumbai, on the northwest coast of India, using length frequency scatter diagram method. They estimated the von Bertalanffy growth parameters for *T. orientalis*, showing that its growth pattern was retrogressive and geometric throughout life. The $L\infty$, K and t_0 parameters derived from males were 368 mm, 0.1279, and −0.74, respectively, and for females the parameters were 300 mm, 0.1690 and − 0.70, respectively. Kizhakudan (2007) reported growth parameters L_∞ and K for *T. orientalis* (=*unimaculatus*) males as 125.8 mm CL and 0.56 year^{-1} in the wild and 119.2 mm CL and 0.74 year^{-1} in captivity and for females as 114 mm (CL) and 0.56 y^{-1} in the wild and 120.1 mm (CL) and 0.6 year^{-1} in captivity. Radhakrishnan et al. (2013) estimated the growth parameters for *T. unimaculatus* from the southwest coast of India as $L_\infty = 240$ mm, K (monthly) = 0.042 for males and $L_\infty = 260$ mm, K (monthly) = 0.05 for females.

Although von Bertalanffy growth model is the most popular model used for estimating growth, the greatest concern with application of this model to crustacean growth is that it cannot visualize the distinctive discontinuous growth of these taxa. Further, most estimates of the von Bertalanffy growth model for crustaceans involve certain assumptions concerning the age of the smallest individual (Wahle and Fogarty 2006). The inconsistency in reporting t_0 (as a negative value) with age of puerulus as 0 (CL ~ 9.0 mm), instead of hatching, leads to inappropriately high L_∞ values. The actual age of the postlarval lobsters of at least a few species is now known from the recent success in rearing them from egg to postlarvae. Pitcher (1993) suggested that growth comparisons should be restricted to the size range of the exploited size classes as different methods of estimating L_∞ are likely to give different results. The estimate of growth parameters of different species of spiny lobsters from different geographic locations by various methods are presented in Table 6.5 for a comparative view of lobster growth.

Indirect measures of growth and size at age are provided by the analysis of size frequency distribution of samples of the wild population. The so-called *age pigment*, lipofuscin, which accumulate in the brains, has been used with varying degrees of success in crustaceans, including lobsters, as a proxy for age (Nicol 1987; Belchier et al. 1994; Sheehy 1990, 1992; Wahle et al. 1996).

6.3.2 Influence of Environmental Factors on Growth

There are several environmental factors that can influence growth rate. Temperature may be the important factor for temperate and semitropical species. Laboratory-held *P. cygnus* at a constant elevated temperature year-round showed an increase in the growth rate due to shortening of intermoult duration with percentage increase in size remaining almost the same (Chittleborough 1975). The nutritional state also influences growth rate to a great extent by increasing the time between moults. Animals held in groups show faster growth rate compared to those kept individually (Cobb and Tamm 1974; Chittleborough 1975; Radhakrishnan 1989). Other factors that may influence growth rate are oxygen concentration (Chittleborough 1975), space (Sastry and French 1977) and light intensity. Loss of limbs and their subsequent regeneration causes reduction in growth (Chittleborough 1975; Kulmiye and Mavuti 2005), though it induces precocious moulting. Eyestalk ablation induces precocious moulting and higher weight gain in spiny lobsters due to the removal of the moult-inhibiting hormone (MIH), crustacean hyperglycaemic hormone (CHH) and several other endocrine factors (Quackenbush and Herrnkind 1981; Radhakrishnan and Vijayakumaran 1984a).

6.3.3 Sexual Differences in Growth

Differential growth between sexes is reported in many species of spiny lobsters. *P. ornatus* males grow faster than other species of the genus. *P. penicillatus* males

Table. 6.5 The von Bertalanffy growth parameters of tropical spiny lobsters from different geographic regions

Species/Location	Sex	L∞	K/year	Method	Sources
Panulirus homarus homarus Sri Lanka	Pooled	287 mm TL	0.43	ELEFAN 1B	Jayawickrama (1991)
Panulirus homarus homarus Sri Lanka	Male Female	127.0 mm CL 121.0 mm CL	0.41 0.39	Wetherall plot	Jayakody (1991)
Panulirus homarus homarus Somalia	Male Female	127.0 mm CL 110.0 mm CL	0.46 0.43	ELEFAN 1	Fielding and Mann (1999)
Panulirus homarus homarus India	Male Female	122.0 mm CL 118.0 mm CL	0.72 0.62	Tag recovery	Mohammed and George (1968)
Panulirus homarus homarus India	Pooled	119.4 mm CL	0.30	ELEFAN 1	Thangaraja et al. (2015)
Panulirus homarus homarus Oman	Male Female Pooled	144.5 mm CL 134.7 mm CL 143.16 mm CL	0.75 0.81 0.72	ELEFAN 1 Ford-Walford plot	Mehanna et al. (2012)
Panulirus homarus homarus Oman	Pooled Pooled	136.0 mm CL 128.9 mm CL	0.41 0.33	ELEFAN 1 Wetherall plot (mean)	Al-Marzouqi et al. (2007)
Panulirus homarus homarus Yemen	Male Female	136.0 mm CL 118.0 mm CL	0.46 0.44	LFDA method	Sanders and Bouhlel (1984)
Panulirus homarus homarus Kenya (captivity)	Male Female	115.0 mm CL 105.0 mm CL	0.26 0.33	Fabens method	Kulmiye and Mavuti (2005)
Panulirus homaru rubellus South Africa	Male Female	120.0 mm CL 94.2 mm CL	0.17700 0.33700	Tag recapture	Smale (1978)

(continued)

Table. 6.5 (continued)

Species/Location	Sex	L∞	K/year	Method	Sources
Panulirus polyphagus India	Male Female	537.0 mm TL 443.0 mm TL	0.20 0.22	Modal progression	Kagwade (1987)
Panulirus ornatus Torres Strait, Australia	Male Female	150.7 mm CL 163.5 mm CL	0.00157 0.00157	ML method (Palmer et al. 1991)	Phillips et al. (1992); Baisre and Cruz (1994)
Panulirus penicillatus Red Sea (captivity)	Male Female	141.3 mm CL 84.7 mm CL	0.04970 0.10660	Gulland-Holt	Plaut and Fishelson (1991)
Panulirus penicillatus Marshall Islands	Male Female	146.5 mm CL 96.5 mm CL	0.211 0.588	Fabens method (1965)	Ebert and Ford (1986)
Panulirus argus Cuba	Male Female	250.3 mm CL 170.9 mm CL	0.27 0.39	ML method (Palmer et al. 1991)	Phillips et al. (1992); Baisre and Cruz (1994)
Panulirus cygnus Western Australia	Male Female	165.6 mm CL 163.5 mm CL	0.00026 0.00026	ML method (Palmer et al., 1991)	Phillips et al. (1992)
Panulirus cygnus Seven Mile Beach, Western Australia	Male Female	158.8 mm CL 125.0 mm CL	0.17000 0.19000	Lipofuscin concentration	Sheehy et al. (1998)
Panulirus longipes Tonga	Male Female	133.0 mm CL 118.0 mm CL	0.31 0.42	Wetherall plot and ELEFAN	Munro (1988)

grow faster than *P. longipes* and *P. marginatus*. *P. cygnus* males and females grow relatively quick initially, but slow down substantially, following a similar pattern as that of *P. penicillatus* females. Female *P. homarus homarus* from India was reported to grow faster than males (Radhakrishnan et al. 2015). In the Sri Lankan population of *P. homarus*, females show negative allometric growth ('b value' less than 3) compared to isometric growth in males (Senevirathna et al. 2014). Higher b values (<3) were reported for the Sri Lankan population of male *P. longipes* and *P. penicillatus* (Sanders and Liynage 2009).

P. homarus reared under captivity also show differential growth between sexes. Jong (1993) observed faster growth of males when males and females are grown separately and also when both sexes are held together. Kizhakudan et al. (2013) observed that male *P. polyphagus* exhibited higher growth increments at moult than females. However, in the sand lobster, *T. unimaculatus*, females were found to grow with higher growth increments at moult than males (Kizhakudan 2007).

6.3.4 Growth of Injured Lobsters

Loss of antennae, limbs and other external injuries can impact growth negatively in lobsters. Negative or zero growth is not unusual in lobsters and can be due to several factors (Aiken 1980). Under captive conditions, mean lower moult increment after a moult and longer intermoult duration were reported in injured *P. homarus homarus* (Kulmiye and Mavuti 2005). Tag injury reduces growth by 65% in *P. homarus homarus*. Smale (1978) reported lower growth rate in tagged and recaptured *P. homarus rubellus* with necrotic wounds compared to growth in laboratory-held specimens. Longer intermoult duration by 5 weeks and reduced moult increment by 0.5 mm CL were observed in injured *P. argus* compared to uninjured specimens (Davis 1981). Injury resulted in 39% reduction in the growth of smaller *P. argus* (<65 mm CL) released back into the sea in Florida (Hunt and Lyons 1986).

6.4 Food Selection and Feeding Strategies

Identification of food consumed by spiny lobsters is difficult as they are nocturnal feeders and the stomach contents are in a semi-digested condition during daytime (Berry 1971). Therefore, information on the natural diet and feeding habit of spiny lobsters is scarce. The diet of juveniles of the spiny lobster *P. homarus rubellus* consists mostly of barnacles, while adults eat more of mussels (Berry 1971). The juveniles probably are not able to open larger mussels and, therefore, mostly subsist on smaller mussels and barnacles attached to rocks. Bivalves appear to be the main food of the spiny lobster *P. homarus megasculpta* from the southeast coast of Iran. Crabs, gastropods, barnacles and algae constitute the secondary food item. Polychaetes, fish, echinoderms and members of Ascidiacea are considered as incidentally ingested food (Mashaii et al. 2011). Fish is the most preferred diet of *P. echinatus* from Brazil (Goes and Lins-Oliveira 2009). Under captive conditions, Radhakrishnan (1989) observed that *P. homarus* was incapable of breaking open the hard-lipped clamshells, whereas they could feed on oysters. They are opportunistic feeders and wait for the oysters to open their shell for respiration.

Kizhakudan (2007) observed from the analysis of gut contents of *P. polyphagus* and *T. orientalis* (=*unimaculatus*) from the northwest coast of India that they preferred gastropods in the natural habitat, presumably due to the abundance of gastropods along that coast in comparison with bivalves. In captivity, however, the sand lobster showed a strong preference for bivalves. Experiments on dietary impacts on growth in these two species of lobsters indicated that *P. polyphagus* fed with the gastropod *Turbo* sp. and *T. orientalis* (=*unimaculatus*) fed with the clam *Mercia opima* showed better growth performance (Kizhakudan and Patel 2011). The average daily growth of male and female *P. polyphagus* fed on *Turbo* sp. was 0.14 mm CL (0.47 g) and 0.12 mm CL (0.33 g), respectively. Sand lobsters fed on fresh clam showed the highest average daily growth rate of 0.17 mm CL and 0.42 g.

There are several reports of spiny lobsters feeding on algae (Heydorn 1969; Munro 1974). The cellulose fibres probably stimulate growth and the assimilation of nitrogen in lobsters, especially when they are on high-protein diet (Joll and Phillips 1984). However, when they are feeding on a low-protein diet, the plant material probably acts as an extender, making a low-protein diet adequate for normal growth and survival (Castañeda-Fernández de Lara et al. 2005). Coralline algae seem to be an important diet for the spiny lobsters *P. cygnus* (Joll and Phillips 1984; Mayfield et al. 2000) and *P. echinatus* (Goes and Lins-Oliveira 2009) as they probably derive their calcium requirement from these algae (Kanciruk 1980). *P. cygnus* could achieve 35% absorption efficiency on feeding with coralline alga *Corallina cuvieri* (Joll and Phillips 1984).

6.4.1 Feeding Behaviour

George et al. (1955) provided an elaborate description of the digestive system of *P. polyphagus*. The feeding mechanism and behaviour of the spiny lobster *Panulirus homarus rubellus* have been studied in both wild and captive populations (Berry 1971). In the wild, the lobster usually targets a mussel separated from, or on the edge of a group and detach it by severing the byssus threads with the dactyls of its stout third maxilliped. The flat edge of the mussel is brought between the strong mandibles using the maxillipeds and the first walking leg, and the edge of the shell is crushed until the mantle is exposed. The lobster then severs the adductor muscle by inserting the dactyls of third maxillipeds or the first walking legs through the aperture and opens the mussel and feeds on the meat. Kizhakudan (2007) observed that when entire live gastropods are introduced into a *P. polyphagus* rearing tank, the lobster, holding the gastropod with the II and III pereiopods, would grind a hole in the shell with its hard dentition and suck the meat out. The entire process would take at least an hour for completion. When fed with bivalves, the lobster would use its mandibles to bite the edges of the bivalve and dig out the flesh. The sand lobster *T. orientalis* (=*unimaculatus*), when fed with live bivalves, would hold the shell with its I, III and IV pairs of pereiopods and probe the edges with the dactyli of the II pair of pereiopods, slowly wedging the tip of the dactyli into the shell and opening the shell to expose and cut the adductor muscle with its II pair of pereiopods. The meat is then scraped out and passed to the III maxilliped (Kizhakudan 2007).

Though spiny lobsters are considered a generalist species, they appear to be selective of the foods they eat. The proportion of food in their stomach is different from their relative abundance in the lobster habitat. Radhakrishnan (1989) observed captive *P. homarus* preferring mussel meat when a mixed diet consisting of mussel, clam and fish is provided in laboratory experiments. Mashaii et al. (2011) found that size, sex, moult stage or reproductive status have no influence on the frequency of occurrence of prey items in the stomach of feral population of *P. homarus* from southeast Iranian waters. Lipcius and Herrnkind (1982) observed lobsters in the intermoult stage foraging more actively than lobsters in the premoult condition, whereas postmoult juveniles prefer to remain in their shelter and therefore are less likely to be captured by baited traps (Herrera et al. 1991; Jernakoff et al. 1993).

Spiny lobsters detect the food by chemoreception using the antennules and tips of the pereiopods (Kanciruk 1980). Captive *P. homarus homarus* show unusual antennular movements, especially pulling the antennular tips with the third maxillipeds and first walking legs, when the chemical stimuli from food reach them. The lobster becomes restless and starts quick movements towards the food (Radhakrishnan, personal observation). The role of vision in locating the food seems to be little in majority of the Palinuridae (Kanciruk 1980). Eyestalk-ablated lobsters find their food by way of chemoreception. They exhibit some abnormal feeding behaviour by grabbing all the food in contact and keeping it near the mouth using the third maxillipeds and first pereiopods, probably due to hyperphagia (Radhakrishnan and Vijayakumaran 1998).

6.4.2 Feeding Ecology

Spiny lobsters are considered as key predators in various benthic habitats and their selective predation can have important effects on the structure of benthic communities (Phillips et al. 1980). Feeding ecology, food selection and food preferences have been studied in many species of *Panulirus* spiny lobsters, e.g., *P. longipes* (Dall 1974, 1975), *P. homarus* from South Africa (Smale 1978), *P. argus* (Andrée 1981; Briones-Fourzán et al. 2003), *P. cygnus* (Joll 1982, 1984), *P. homarus* from southeast coast of India (Radhakrishnan and Vivekanandan 2004), *P. interruptus* (Castañeda-Fernández de Lara et al. 2005; Díaz-Arredondo and Guzmán-del-Próo 1995), *P. elephas* (Goñi et al. 2001), *P. echinatus* (Góes and Lins-Oliviera 2009) and *P. homarus* from Iranian waters (Mashaii et al. 2011). The most extensive evidence on the cascading effects of predation comes from the studies on the spiny lobster *P. interruptus* preying upon two species of sea urchins, *Strongylocentrotus fransiscanus* and *S. purpuratus*, the defining species in California kelp forests (Tegner and Dayton 1981). The grazing of these sea urchins, in turn, altered the abundance of the giant kelp *Macrocystis pyrifera*. Similar studies have also been conducted in New Zealand on the spiny lobster *Jasus edwardsii* and the sea urchin *Evechinus chloroticus* (Andrew and MacDiarmid 1991; Shears and Babcock 2002). In South Africa, predation of *J. lalandii* and *P. homarus* alters abundance and size structure of mussel, urchin and gastropods on which they feed (Newman and Pollock 1974; Pollock 1979; Berry and Smale 1980; Barkai and McQuaid 1988; Mayfield et al. 2000). Thangaraja and Radhakrishnan (2012) observed close association between *P. homarus homarus* and the availability of the brown mussel (*Perna indica*) on inshore rocks along the southwest coast of India. Correlation between distribution of *P. homarus rubellus* and occurrence of the brown mussel *Perna perna* on the southeast coast of Africa has been reported by Berry (1971).

Castaneda-Fernandez et al. (2005) identified 26 taxa and unidentifiable crustaceans in the stomach of *P. interruptus* from the western Mexican coast. Combining the stomach content results with the availability of food resources in the environment, these authors compared the selectivity of lobsters for various prey types using the Manly selection index. Although some groups were estimated as being preferred, preferences varied inconsistently among sites and seasons. Studies on the spiny lobsters *P. cygnus* (Edgar 1990) and *J. lalandii* (Barkai and Branch 1988; Mayfield et al. 2000) show that lobsters may shift their food preference to some uncommon food sources in their location if the preferred food is unavailable. Such shifting is common in polyphagous species because this maximizes foraging efficiency when alternative food species become more abundant than the preferred food (Murdoch and Oaten 1975).

6.4.3 Ontogenetic Shifts in Diet

In lobsters, dietary preference and requirement undergo significant alterations as the lobsters grow and become more mobile with morphological changes in their feeding apparatus. In the American lobster *H. americanus*, some studies find no difference in identity or frequency of food items ingested by different size groups (12–125 mm CL) (Weiss 1970; Ennis 1973), and other studies found notable differences, especially in frequency of food items. Carter and Steele (1982) have suggested that *H. americanus* in the size range of 12–73 mm CL consume sea urchins, ophiuroids and mussels more frequently than the adult lobsters. Scarratt (1980) found lobsters consuming more of mussels, fish and crabs and less of echinoderms as they approach maturity. Sainte-Marie and Chabot (2002) have clear evidence of a progressive dietary shift with increasing lobster size with smaller lobsters relying more on soft or easily acquired food items such as small juvenile bivalves, macroalgae, meiobenthic crustaceans or foraminiferans, compared to larger lobsters feeding on bigger, more nutritious prey such as mussels, sea urchins and crabs. Ontogenetic shifts also lead to difference in growth and survival and also may minimize competition through reduced overlap in resource use, both within and between species (Werner and Gilliam 1984). Sea urchins, in addition to smaller crustaceans, are important for smaller *J. lalandii*, with more preference towards mussels in larger lobsters (Barkai and Branch 1988). The ontogenetic shift in dietary preferences also implies that different sizes may have different roles in community structure and function. Seasonal shifts in the diet of *P. cygnus* were reported with smaller lobsters less than 40 mm CL feeding more on molluscs and crustaceans than larger lobsters in summer and consuming more coralline algae in winter (Edgar 1990; Jernakoff et al. 1993).

Lobsters harbour a diverse intestinal microbial flora, which enter the alimentary system from the environment and through the food they consume. Feeding habits and the chemical nature of the environment probably decide the survival of microorganisms in the gut and formation of resident microflora in the intestine (Ganeshkumar et al. 2010). In wild *Panulirus versicolor* from the Andaman and Nicobar Islands, *Enterobacteriaceae* constitute 67% of the gut microflora, whereas in laboratory-reared animals *Vibrionaceae* was dominant (71%). The highest concentration of bacteria was observed in the midgut, where proteolytic and cellulolytic bacteria were dominant in the foregut region.

6.5 Neuroendocrine Systems

6.5.1 X-Organ-Sinus Gland Complex

The two important physiological functions, moulting and reproduction, are entwined in lobsters and are controlled by antagonistic hormones secreted and stored in the eyestalk, nervous system and several non-neural glands present in the cephalic region. The eyestalk is the pivotal organ housing several neuropeptides. The optic peduncle within the stalked eye consists of four ganglionic structures such as the lamina ganglionaris (LG), the medulla externa (ME), the medulla interna (MI) and the medulla terminalis (MT). The medulla terminalis tapers to form the optic nerve. The medulla terminalis contains a group of neurosecretory cells called medulla terminalis ganglionic X, the X-organ-sinus gland complex or the X-organ (Hanstrom 1937). A neurohaemal organ associated with the X-organ is the sinus gland (SG), located at the peripheral junction between the medulla externa and medulla interna. The axonal terminals originating from the perikarya of the neurosecretory cells of the X-organ end in the sinus gland, which is essentially a storage gland. In the spiny lobster *P. homarus homarus*, SG is visible to the naked eye as a very small bluish white opalescent body (Fernandez and Radhakrishnan 2010). The SG has an internal blood sinus with different axonic terminals containing the neurosecretory products. Fernandez and Radhakrishnan (2010) mapped the neurosecretory cells (NSCs) in the eyestalk, brain and thoracic ganglia of the spiny lobster *P. homarus homarus*. NSCs are distributed only in MT, MI and ME and maximum numbers are in MT. Six types of NSCs have been identified in the optic ganglia. Based on the neurosecretory activity, the secretory cycle of the NSCs has been classified into two (Durand 1956; Matsumoto 1962), three (Mohamed 1989) or four (Fernandez and Radhakrishnan 2010) secretory phases viz., vacuolar, secretory, quiescent and synthetic phases. While many consider the presence of peripheral vacuoles as the initial phase, Fernandez and Radhakrishnan (2010) considered synthetic phase as the first phase in *P. homarus homarus*, followed by the vacuolar phase during which the neurosecretory material gets filled up in the vacuoles. The X-organ-sinus gland complex of the eyestalk is considered as the site of synthesis and release of moult-inhibiting, vitellogenin-inhibiting and crustacean hyperglycaemic hormones (Keller 1992). All the hormones have been isolated, purified and characterized.

6.5.2 Endocrine Control of Moulting and Growth

Growth in crustaceans is under the control of two antagonistic hormones, the moult-inhibiting hormone (MIH), a neuropeptide produced and released from the X-organ in the eyestalks and stored in the sinus gland (Bliss 1951, 1953, 1956; Carlisle 1953; Hanstrom 1937, 1939, 1947; Passano 1951a, b, 1953), and the moulting hormone (MH), a steroid hormone produced by a pair of epithelial glands known as the Y-organs located in the cephalothorax (Gabe 1953, 1954, 1956; Echalier 1955, 1959). It has long been known that bilateral eyestalk ablation accelerates moulting in malacostracan crustaceans (Zeleny 1905; Smith 1940), which has led to the hypothesis that moulting is negatively regulated by an inhibiting hormone. Further studies have clearly shown that eyestalk removal leads to an increase in haemolymph ecdysteroids consequent upon increased ecdysteroid synthesis by the Y-organ (Chang et al. 1976; Keller and Schmid 1979). On the other hand, eyestalk or sinus gland extract injections have been shown to reduce haemolymph ecdysteroid titres (Hopkins 1982; Bruce and Chang 1984; Nakatsuji and Sonobe 2004). Although there have been indications of a third hormone in the central nervous system (Scudamore 1947; Carlisle 1953; McWhinnie and Mohrherr 1970), the classical bihormonal hypothesis of decapod moult control based on these two antagonistic hormones remained popular (Aiken 1980).

Neither the moult-inhibiting hormone (MIH) nor the moulting hormone (MH) had been isolated from any of the crustaceans during the 1950s and the early 1960s when the bihormonal hypothesis was developed. The presence of moult inhibition factors in the eyestalk was deduced from indirect evidence obtained with classical endocrinological techniques. Bilateral eyestalk ablation induced a variety of physiological changes, and it was not clear whether these functions were controlled by a single hormone or a hormone with multiple physiological effects (Kleinholz 1975). During the late 1960s, the focus of research began to shift from classical endocrinological methodologies to study the origin, site of synthesis and regulation of the hormones. The development of sophisticated analytical techniques by the early 1990s paved way for the isolation, purification and characterization of certain neuropeptides from the eyestalks. Probably the most important progress has been made in the elucidation of a novel family of large peptides from the X-organ-sinus gland system which includes the crustacean hyperglycaemic hormone (CHH), putative moult-inhibiting hormone (MIH) and vitellogenin (= gonad)-inhibiting hormone (VIH), which are believed to be involved in growth and reproduction (Kegel et al. 1991; Keller 1992). These peptides have so far only been found in crustaceans. Renewed interest in the neurohaemal pericardial organs has led to the identification of a number of cardioactive/myotropic neuropeptides, some of them unique to crustaceans. Recent studies reveal the existence of multiple forms of MIH/MIH-like molecules among penaeids and raise the possibility that molecular polymorphism may exist more generally among MIH (type II) peptides. The haemolymphatic MIH titre has been determined for two species, a crayfish (*Procambarus clarkii*) and a crab (*Carcinus maenas*). Important contributions have been made by immunocytochemical mapping of peptidergic neurons in the nervous system, which has provided evidence for a multiple role of several neuropeptides as neurohormones on the one hand and as local transmitters or modulators on the other.

6.5.3 Moult Inhibition

Growth in the spiny lobster is achieved by moulting, which is regulated by neuropeptides (moult-inhibiting hormone) produced and released by the X-organ-sinus gland complex in the eyestalk (Bliss 1951, 1953; Carlisle 1953, Hanstrom 1937,

1939, 1947; Passano 1951a, b, 1953) and a moulting hormone (steroid hormone) produced by a pair of epithelial glands known as Y-organs located in the cephalothorax (Gabe 1953, 1954, 1956). The MIH produced in the neurosecretory cells of the X-organ is transported via axons to the sinus gland for storage or release. The high titres of MIH suppress the synthesis of ecdysteroids and only when the Y-organ is relieved from the suppressive action of MIH that ecdysteroids are produced to reach a critical level so that moulting is initiated (Webster 1998). Dall and Barclay (1977) was sceptical of the role of MIH in palinurid lobsters as eyestalk ablation did not initiate precocious ecdysis. However, Quackenbush and Herrnkind (1981) reported precocious moulting in eyestalk-ablated *P. argus*. Radhakrishnan and Vijayakumaran (1984a) and Radhakrishnan (1994) demonstrated increased moulting frequency and higher weight gain after bilateral eyestalk ablation in *P. homarus, P. polyphagus and P. ornatus*. Most studies on isolation, purification and characterization of moult hormones were carried out in the American lobster *H. americanus*. Until early 1980s there was little knowledge of the chemical nature of MIH. Preliminary observations are that MIH is either a protein or a peptide and is trypsin sensitive (Rangarao 1965). Freeman and Bartell (1976) estimated the molecular weight of MIH isolated from shrimp between 1000 and 5000. Quackenbush and Herrnkind (1983) have reported partial purification of MIH from the spiny lobster *P. argus*. However, these isolates are obtained from whole eyestalks, and Chang (1985) cautions that care must be exercised during testing inhibitory activity of hormones that nonspecific toxicity of the extracts is not interfering with the normal physiology of the bioassay system. Bruce and Chang (1984) developed a bioassay and purification methods (Chang et al. 1987) for MIH and further sequenced the complete amino acid profile of Hoa-MIH of *H. americanus* (Chang et al. 1990). The eyestalk homogenates were able to significantly decrease the titres of circulating ecdysteroids when injected into ablated lobsters and also significantly delay induced moults in eyestalk-ablated juvenile lobsters, showing the presence of a moult-inhibiting factor in the eyestalks. These authors have isolated another closely related peptide from the eyestalk sinus gland, the crustacean hyperglycaemic hormone (CHH), the amino acid sequence of which has 96% identity with Hoa-MIH. These authors also reported high hyperglycaemic activity for Hoa-MIH, but low MIH activity for Hoa-CHH. Since some of these peptides have multiple biological activities (Hoa-MIH has both MIH and CHH activity), any one of these peptides cannot be designated as the moult-inhibiting hormone or the crustacean hyperglycaemic hormone (Chang et al. 1993). Despite the reported similarity between MIH and CHH, the two hormones are produced by different neurosecretory cells, which have been confirmed by in situ hybridization studies using MIH-encoding cDNA (Dircksen et al. 1988; Waddy et al. 1995).

There has been speculation about the process by which MIH prevents moulting in *H. americanus*. The MIH may directly inhibit or decrease secretion of ecdysone by Y-organs (Bruce and Chang 1984) or it may antagonize ecdysteroids at the level of the target tissues (Freeman and Bartell 1976; Freeman and Costlow 1979). The mean change in circulating ecdysteroids was a decrease of 29.2 pg/µl following injection of sinus gland extracts in a 5-month-old *H. americanus* compared to the control animals that had an increase of 28.4 pg/μl of serum ecdysteroids after the injection (Chang 1985). Sinus gland implants have been shown to delay moulting in crayfish, showing the presence of a moult-inhibiting principle in the eyestalk (Couch et al. 1976). Observations on the effect of eyestalk ablation (Spaziani et al. 1982; Radhakrishnan and Vijayakumaran 1984a, b; Juinio-Menez and Ruinata 1996) or sinus gland extracts (Soumoff and O'Connor 1982) on the incorporation of radiolabelled precursors by Y-organ cells also lend support to the hypothesis of a direct action of MIH upon the moulting gland.

6.5.4 Other Moult-Inhibiting Factors

Chang (1984) speculates that there could be even other sources of moult-inhibiting factors, may be the pleopods, in an ovigerous female that decreases circulating titres of ecdysteroids. An ovigerous female and a non-ovigerous adult female both had basal level of circulating ecdysteroids during the intermoult, though the intermoult period of ovigerous female was 130 days more than the non-ovigerous female, obviously, to facilitate the ovigerous female to release the larvae. In a study conducted on the effect of pleopod gland extract upon moult interval and ecdysteroid titres in eyestalk-ablated juvenile *H. americanus*, Chang (1985) observed a longer intermoult period (19.4 + 1.4 days) and a decrease in the concentration of circulating ecdysteroids (32.3 ng/ml) in pleopod gland extract-injected animals compared to shorter intermoult duration (17.7 + 1.6 days) and an increase in haemolymph ecdysteroids in the control group (28.4 ng/ml). An even more intriguing observation was that the ablated juveniles of *H. americanus* displayed rapid decrease in the level of ecdysteroids just prior to ecdysis, showing probably an alternative mechanism for the regulation of ecdysteroids in an induced moult. The rapid decline in circulating ecdysteroids may be necessary for the initiation of behaviour related to exuviations (Chang and Bruce 1980). Another hypothesis is that the catabolism pathways are activated as in insects (Chen et al. 1994) or that the ecdysteroid production by the Y-organs is under negative feedback control by circulating ecdysteroids as shown in the crayfish *O. limosus* (Dell et al. 1999).

6.5.5 Moult Gland and Moulting Hormone

Eyestalk ablation accelerates the moulting process in the absence of the moult-inhibiting hormone (MIH) produced and released by the X-organ-sinus gland complex in the eyestalk and activation of the moult gland or the Y-organ, which produces the moulting hormone. The steroid arthropod moulting hormone was first isolated from insects and was called ecdysone (Butenandt and Karlson 1954). The active form was subsequently determined to be 20-hydroxyecdysone (20-HE). The first isolation of 20-HE from a crustacean was from the spiny lobster (Hampshire and Horn 1966; Horn et al. 1966). Ecdysone and 20-HE, collectively known as the ecdysteroids, are the two most predominant hormones that possess moulting hormone activity (Lafont 1997). Ecdysteroids may also play a gonadotropic role in crustaceans (Chang et al. 2001).

During early premoult, Y-organs secrete ecdysteroids that circulate in the haemolymph in free form and are taken up by target tissues (Waddy et al. 1995). The concentration of ecdysterones in the haemolymph of *H. americanus* during the moult cycle follows a predictable pattern. They are barely detectable immediately after ecdysis, and the major peak occurs during stages D_1 and D_2. In stage D_3 when the old cuticle is being resorbed, haemolymph ecdysteroid levels drop rapidly to low levels characteristic of stage A (reviewed by Aiken and Waddy 1992). Though injections of the steroid 20-HE can shorten the intermoult period in *H. americanus*, when the injections are made prior to premoult (Rao et al. 1973; Chang 1989), elevated concentrations of exogenous 20-HE during premoult result in a consequent delay in ecdysis (Cheng and Chang 1991). The 20-HE-injected (1.0 μg/g wet weight) lobsters moulted 120 + 45.1 hr. after the injection, whereas the control group injected with saline moulted 56.4 + 28.1 hr. after the injections. From these studies, it is evident that normal rapid decline in circulating ecdysteroids at late premoult may be necessary for the initiation of behaviour related to exuviations (Chang and Bruce 1980).

6.5.6 Crustacean Hyperglycaemic Hormone (CHH)

Nearly 75 years ago, Abramowitz et al. (1944) observed that a factor in the crustacean eyestalk was able to elevate concentration of haemolymph glucose, and the hormone responsible for the diabetogenic effect is the crustacean hyperglycaemic hormone (for reviews see Chang et al. 2001). Chang et al. (1990) found two closely related neuropeptides in the sinus gland of *H. americanus*, the MIH and CHH. The complete amino acid sequence of both these peptides was elucidated, and the amino acid sequence of Hoa-CHH has 96% identity with Hoa-MIH.

Chang et al. (1998) have conducted assays for CHH in embryos, larvae and adults of *H. americanus* to determine if CHH was present in all life stages. They used an ELISA for detection of CHH immunoactivity. Surprisingly there was no detectable CHH during the first 50 days of embryonic life. However, the levels rapidly increased after this time. In the larvae, the CHH levels peaked at end of stages I and II. Most of the whole-body CHH was detected in the eyestalks and the balance in circulation in the haemolymph. Though no significant difference in circulating levels of CHH levels during the moult cycle was evident, on a finer scale, Keller and Chang (unpublished) observed a dramatic increase in CHH just prior to completion of ecdysis in *H. americanus*. Radhakrishnan and Vijayakumaran (personal observation) observed higher water content in the haemolymph and abdominal muscles of eyestalk-ablated *P. homarus* and presumed that higher influx of water in ablated lobsters may be due to a factor in the eyestalk. During ecdysis, a large quantity of water is absorbed by the body at various sites for expansion of the new exoskeleton, and Chung et al. (1999), based on their study in *C. maenas*, attributed this to a rapid increase in CHH at ecdysis. Apart from the X-organ-sinus gland, evidences show that there are other areas in the central nervous system, such as the suboesophageal ganglion, which also show CHH immunoactivity (Chang et al. 1999, 2001).

6.5.7 Methyl Farnesoate

Methyl farnesoate (MF) is a sesquiterpene that is related to the insect hormone (Chang et al. 2001). MF is secreted by the mandibular organ (Laufer et al. 1987), and relatively low amounts (0.1 ng/g) were detected in *H. americanus* larvae (Borst et al. 1987). When added to the water in which the larvae were developing, MF caused a significant increase in the time from hatching to stage IV (22 + 1.3 days) compared to the control (20.8 + 1.5 days) (Borst et al. 1987).

MF synthesis is negatively regulated by eyestalk factors (Laufer et al. 1986; Tsukimura and Borst 1992). Eyestalk ablation causes hypertrophy of the mandibular organ (Byard et al. 1975) and leads to increase in MF levels in the haemolymph, while injection of sinus gland extracts decreases MF to undetectable levels (Borst et al. 1994). MF stimulates ecdysteroid production in the crab Y-organs in vitro, suggesting that MF may be a positive stimulator of Y-organs (Chang et al. 1993; Tamone and Chang 1993). In adults, MF may have a reproductive role. In lobsters, the relative size of the mandibular organ increases much more in adult males than females (Chang et al. 2001). MF titres in females may be lower than the males because the mandibular organs are smaller and less active or MF is sequestered by gonads (Chang et al. 2001). However, the specific role of MF in lobster reproduction is ambiguous. In *H. americanus*, Byard et al. (1975) were not able to observe any apparent effect in late phases of female reproduction due to MF ablation, including secondary vitellogenesis and oviposition.

6.6 Eyestalk Ablation

Eyestalk ablation increases moulting frequency, higher weight gain and thereby accelerated growth in crustaceans. The relationship between eyestalk ablation and accelerated moulting and growth in the American lobster *H. americanus* was established many years ago (Sochasky 1973). No response to eyestalk ablation in *P. cygnus* prompted Dall (1977) to question the significance of MIH in spiny lobsters. However, Quackenbush and Herrnkind (1981) reported accelerated moulting and gonad development in bilateral eyestalk-ablated *P. argus*. Radhakrishnan and Vijayakumaran (1984a) reported faster moulting and higher weight gain in bilateral eyestalk-ablated juvenile, subadult and adult *P. homarus*, irrespective of the reproductive status and season. In subadult and adult females, when their eyestalks are removed, both moulting and gonad development are accelerated simultaneously. In the absence of MIH and GIH, lobsters moult more frequently and increase more in size at each moult. For example, eyestalk-ablated juvenile *P. homarus* weighing 20–25 g moulted five times in 108 days compared to three in the control group. The weight gain was an average 1.02 g/day in the experimental group compared to

0.35 g/day in the control group. In adult ablated lobsters (mean initial weight, 257 g), the second moult after eyestalk ablation was completed in 30.8 + 3.6 days, compared to 51.3 + 18.6 days by the control group (mean initial weight, 250 g). The weight gain by an eyestalk-ablated *P. homarus* was 455 g in 249 days (1.83 g/day) (Radhakrishnan and Vijayakumaran 1984a). Kizhakudan (2013) found that eyestalk ablation resulted in increased frequency of moulting in juvenile and subadult male and female *P. polyphagus* with ablated lobsters showing a higher rate of growth at each moult compared to the unablated ones. Growth increment, in terms of both carapace length and weight, was particularly high among the juvenile males. Ablated juvenile males showed an average increase in CL by almost 74% in about 114 days (0.7% increment per day), while the unablated ones showed an increase in CL of about 45% in 115 days (0.4% increment per day); the weight gain was 370% in 114 days (3.3% per day) in ablated juvenile males, while control males showed an increase of 174% in 115 days (1.5% per day). Based on this study, Kizhakudan et al. (2013) suggested the possibility of obtaining better growth in captive rearing of *P. polyphagus*, particularly in the juvenile grow-out phase.

The moult stage at which ablation was carried out has a bearing on the dominant activity, the moulting or the gonad development (Fernandez 2002). *P. homarus homarus* bilaterally ablated in early intermoult takes more time to enter into the premoult compared to those ablated in late intermoult. In those ablated in the premoult stage, the moulting process is rapidly completed. Subadult or adult lobsters ablated in the early premoult stage (D_0–D_1) resorb the fully developed ovary before moulting, whereas in lobsters ablated in the intermoult stage, the ovaries undergo precocious development and the unfertilized ova get oviposited on to the endopods of the pleopods even in the absence of mating. The lobster may moult even when the eggs/ova are still on the pleopods (Fernandez and Radhakrishnan 2016). Eyestalk removal eliminates the physiological blocks, and the loss of MIH and GIH simultaneously accelerates both the moult and the ovarian development processes in *P. homarus* unlike the American lobster in which the dominant activity at the time of eyestalk ablation, growth or reproduction, is accelerated. The resorbed ovary never regains development in the succeeding intermoult period and is permanently damaged. Vitellogenesis is suppressed throughout the succeeding moult cycles with the moulting activity predominating over reproduction.

Studies on eyestalk ablation in scyllarid lobsters are limited. Kizhakudan (2007) reported that eyestalk ablation resulted in increased frequency of moulting in both juvenile and subadult male and female *Thenus orientalis* (now *Thenus unimaculatus*). The intermoult period increased with increasing size in both ablated and control lobsters of either sex, and the relative duration was found to be lesser in the ablated lobsters. Females, both juveniles and subadults, were found to have slightly shorter intermoult duration than the males. This difference was reflected in both ablated and control lobsters. Male and female control lobsters moulted three times within 100 days, while most of the ablated lobsters moulted four times during the period.

6.7 Neuroendocrine Control of Reproduction

Precocious maturation of ovary after eyestalk ablation in the shrimp *Palaemon serratus* prompted Panouse (1943, 1947) to suggest that there are some gonadal inhibitory principles (GIH) in the eyestalk. The neurohormone (also called VIH, vitellogenin-inhibiting hormone) inhibits the synthesis of yolk proteins in vitro (Fingerman 1987; Quackenbush and Keeley 1988). GIH from the American lobster *H. americanus* has been isolated and sequenced (Soyez et al. 1991). The VIH occurs in two isoforms, Hoa-VIH-1 and Hoa-VIH-11, and only one has VIH activity. VIH is a 78-residue peptide with a molecular weight of 9135 Da having an amidated C-terminus and free N-terminus in *H. americanus* (Subramoniam and Kirubagaran 2010). A recombinant peptide related to the vitellogenin-inhibiting hormone of *H. americanus* was produced, and on testing the biologically active principle it was found that the amidated C-terminus is responsible for vitellogenesis inhibition (Ohira et al. 2006). Quackenbush and Herrnkind (1983) have partially purified the GIH from the spiny lobster *P. argus*. Radhakrishnan and Vijayakumaran (1984b) demonstrated precocious ovarian maturation in the bilaterally ablated juvenile and adult spiny lobster *Panulirus homarus homarus*. In juvenile lobsters, the ovary in the previtellogenic phase precociously entered into the vitellogenic phase after eyestalk ablation, whereas in subadults oviposition took place even in the absence of mating (Radhakrishnan and Vijayakumaran 1984b; Fernandez and Radhakrishnan 2016).

6.7.1 Immunocytochemical Localization of VIH in X-Organ-Sinus Gland

The neurosecretory cells in the medulla terminalis region of the X-organ are primarily responsible for the synthesis of VIH. Immunocytochemical studies have revealed the co-localization of VIH with another peptide, the crustacean hyperglycaemic hormone (CHH), in the same X-organ neurosecretory cells. In situ hybridization using RNA probes has revealed co-localization of VIH and CHH in the same neurosecretory cells of the X-organ. These results suggest that CHH and VIH may be synthesized in one cell group of the X-organ. The lobster VIH mRNA is only found in the eyestalk ganglia, indicating VIH is not produced in any other sites in the nervous system. The VIH in haemolymph increases significantly just before spawning when vitellogenin levels are declining. The VIH levels in the haemolymph, however, remain high after spawning and during primary vitellogenesis (Subramoniam 1999).

Recent microanalytical investigations show a close relationship between the three eyestalk neurohormones: VIH, CHH and MIH (CHH family peptides). The structural relationship among these hormones was elucidated from the cloning and sequencing of their DNAs, and this also reflected in their overlapping physiological functions (Subramoniam 1999).

6.7.2 Gonad Stimulatory Hormone (GSH)

The supraoesophageal ganglion and the thoracic ganglion in *P. homarus* possess unique neurosecretory cells, which are involved in the synthesis and production of gonad stimulatory hormones (Otsu 1960; Eastman-Reks and Fingerman 1984; Fernandez 2002; Fernandez and Radhakrishnan 2010). There are eight distinct types of NSCs in the thoracic ganglia, and though lesser in number compared to the optic and supraoesophageal ganglion, most of the NSCs are much larger (Fernandez 2002).

Injection of thoracic ganglion extract into subadult lobsters had lesser effect on ovarian maturation compared to the supraoesophageal ganglion extract. The ovarian stimulatory effect of these injections was evident only when ganglia from bilateral eyestalk-ablated females in which vitellogenesis is already initiated are used. This shows probably higher secretory activity of vitellogenesis-stimulatory hormones during the process of vitellogenesis (Fernandez 2002).

Reproduction is under the control of a vitellogenesis (gonad)-inhibiting hormone (VIH/GIH) produced by the X-organ in the eyestalk and a vitellogenesis (gonad)-stimulating hormone (VSH/GSH) believed to be secreted by neurosecretory cells located on the brain and thoracic ganglion. Other than the neurosecretory hormones, another hormone, methyl farnesoate, secreted by the mandibular organ has been implicated in moulting and reproduction. Experimental evidence shows that it has a positive role in reproduction of crayfish and crabs, but its involvement in lobster reproduction is not conclusive (Subramoniam and Kirubagaran 2010). The role of biogenic amines as neuroregulators in stimulating the secretion of neurosecretory hormones involved in vitellogenesis has also been suggested (Fingerman 1997; Kirubagaran et al. 2005).

6.7.3 Biogenic Amines

Biogenic amines function as neurotransmitters and neuromodulators in crustaceans. Some of them also subserve as neuroregulators to control the release of neurohormones (Subramoniam 1999). 5-hydroxytryptamine (5-HT) stimulates the release of several eyestalk neuropeptides. The opioid met-enkephalin has been shown to inhibit ovarian maturation by specific stimulation of VIH from the X-organ and concomitant inhibition of GSH release from the brain/thoracic ganglia (Sarojini et al. 1995). Dopamine inhibits 5-HT-stimulated ovarian maturation under in vivo conditions in the crayfish *Procambarus clarkii*.

The bihormonal concept of ovarian maturation by inhibitory and stimulatory neuropeptides originating from the central nervous system is still valid, though the methyl farnesoate released by the mandibular organ has a major role in ovarian maturation in many decapods. The functional role of neurohormones in controlling moulting and reproduction has been elucidated from identification, isolation, synthesis and characterization of several hormones. The genes responsible for production of these hormones have been identified and probably further advances open the possibility of synthesizing some novel neuropeptides to enhance growth and egg production in the near future.

6.8 Conclusion

Studies on the food, feeding behaviour and growth of commercially important lobsters find practical applications in developing rearing protocols for maximizing lobster production through aquaculture. The effects of external factors on moulting and growth in these animals cannot be ignored. Similarly, it is a well-established fact that the neuroendocrine system regulates major physiological functions in lobsters, particularly growth and reproduction. The strong linkage between the neuroendocrine system and external factors, such as food and environment, impacts the physiological processes of growth and reproduction, and a deep understanding of these impacts is essential to improve the candidature of the species for aquaculture.

References

Abramowitz, A. A., Hisaw, F. L., & Papandrea, D. N. (1944). The occurrence of a diabetogenic factor in the eyestalks of crustaceans. *The Biological Bulletin, 86*, 1–5.

Adiyodi, K. G., & Adiyodi, R. G. (1970). Endocrine control of reproduction in decapod crustacea. *Biological Reviews of the Cambridge Philosophical Society, 45*(2), 121–164.

Aiken, D. E. (1973). Proecdysis, setal development and molt prediction in the American lobster (*Homarus americanus*). *Journal of the Fisheries Research Board of Canada, 30*, 1337–1344.

Aiken, D. E. (1980). Molting and growth. In J. C. Cobb & B. F. Phillips (Eds.), *The biology and management of lobsters* (Vol. 1, pp. 91–163). New York: Academic.

Aiken, D. E., & Waddy, S. L. (1992). The growth process in crayfish. *Reviews in Aquatic Sciences, 6*, 335–381.

Al-Marzouqi, A., Al-Nahdi, A., Jayabalan, N., & Groeneveld, J. C. (2007). An assessment of the spiny lobster *Panulirus homarus* fishery in Oman – Another decline in the western Indian Ocean? *Western Indian Ocean Journal of Marine Science, 6*(2), 159–174.

Andree, S. W. (1981). *Locomotor activity patterns and food items of benthic postlarval spiny lobsters, Panulirus argus*. MS thesis, Florida State University, Miami.

Andrew, N. L., & MacDiarmid, A. B. (1991). Interrelation between sea urchins and spiny lobsters in northeastern New Zealand. *Marine Ecology Progress Series, 70*, 211–222.

Baisre, J. A., & Cruz, R. (1994). The Cuban spiny lobster fishery. In B. F. Phillips, J. S. Cobb, & J. Kittaka (Eds.), *Spiny lobster management* (pp. 119–132). London: Fishing News Books.

Barkai, A., & Branch, G. M. (1988). Energy requirement for a dense population of rock lobsters *Jasus lalandii*: novel importance of unorthodox and orthodox food sources. *Marine Ecology Progress Series, 50*, 83–96.

Barkai, A., & McQuaid, C. (1988). Predator-prey role reversal in a marine benthic ecosystem. *Science, 242*, 62–64.

Belchier, M., Shelton, P. M. J., & Chapman, C. J. (1994). The identification and measurement of fluorescent age-pigment abundance in the brain of a crustacean (*Nephrops norvegicus*) by confocal microscopy. *Comparative Biochemistry and Physiology, 108B*(2), 157–164.

Berry, P. F. (1971). The biology of the spiny lobster *Panulirus homarus* (Linnaeus) off the east coast of southern Africa. *Investl Representative Oceanography Research Institute, 28*, 75.

Berry, P. F. (1977). A preliminary account of a study of biomass and energy flow in a shallow subtidal reef community on the east coast of South Africa, involving the rock lobster *Panulirus homarus*. *Circular-CSIRO Division of Fisheries Ocenography (Australia), 7*, 24.

Berry, P. F., & Smale, M. J. (1980). An estimate of production and consumption rates in the spiny lobster *Panulirus homarus* on a shallow littoral reef off the Natal coast, South Africa. *Marine Ecology Progress Series, 2*(4), 337–343.

Bliss, D. E. (1951). Metabolic effect of sinus gland or eyestalk removal in the land crab, *Gecarcinus lateralis*. *The Anatomical Record, 111*, 502–503.

Bliss, D. E. (1953). Neurosecretion and crab metabolism in the land crab, *Gecarcinus lateralis* (Freminville). I. Differences in the respiratory metabolism of sinus glandless and eyestalkless crabs. *Biological Bulletin (Woods Hole, Mass), 104*, 275–296.

Bliss, D. E. (1956, November 20). Neurosecretion and the control of growth in a decapod crustacean. In K. G. Wingstrand (Ed.), *Bertil Hanstrom, Zoological papers in Honour of his Sixty-fifth Birthday* (pp. 56–75). Zoological Institute, Lund.

Borst, D. W., Laufer, H., Landau, M., Chang, E. S., & Hertz, W. A. (1987). Methyl Farnesoate (MF) and its role in crustacean reproduction and development. *Insect Biochemistry, 17*, 1123–1127.

Borst, D. W., Tsukimura, B., Laufer, H., & Couch, E. F. (1994). Regional differences in methyl farnesoate production by the lobster mandibular organ. *The Biological Bulletin, 186*(1), 9–16.

Boudreau, S. A., & Worm, B. (2012). Ecological role of large benthic decapods in marine ecosystems: a review. *Marine Ecology Progress Series, 469*, 195–213.

Briones-Fourzan, P., Lara, V. C. F., de Lozano Alvarez, E., & Estrada-Olivo, J. (2003). Feeding ecology of the three juvenile phases of the spiny lobster *Panulirus argus* in a tropical reef lagoon. *Marine Biology, 142*, 855–865.

Bruce, M. J., & Chang, E. S. (1984). Demonstration of a molt inhibiting hormone from the sinus gland of the lobster, *Homarus americanus*. *Comparative Biochemistry Physiology Part A, 79*(3), 421–424.

Butenandt, A., & Karlson, P. (1954). ljber die Isolierung eines Metamorphosehormons der Insekten in kristellisierter Form. *Zeitschrift für Naturforschung, 96*, 389–391.

Byard, E. H., Shivers, R. R., & Aiken, D. E. (1975). The mandibular organ of the lobster, *Homarus americanus*. *Cell and Tissue Research, 162*, 13–22.

Carlisle, D. B. (1953). Preliminary note on the structure of the neurosecretory system of the eyestalk of *Lysmata seticaudata* Risso (Crustacea). *C R Hebd Seances Academic Science, 236*, 2541–2542.

Carter, J. A., & Steele, D. H. (1982). Attraction to and selection of prey by immature lobsters (*Homarus americanus*). *Canadian Journal of Zoology, 60*, 326–336.

Castaneda-Fernandez, V., Butler, M. J., Hernandez-Vazquez, S., del Proo, S. G., & Serviere-Zaragoza, E. (2005). Determination of preferred habitats of early benthic juvenile California spiny lobster *Panulirus interruptus* on the Pacific coast of Baja California Sur, Mexico. *New Zealand .Journal of Marine Freshwater Research, 56*, 1–9.

Chang, E. S. (1984). Ecdysteroids in Crustacea: Role in reproduction, molting, and larval development. In W. Engels, W. H. Clark, A. Fischer, & D. F. Went (Eds.), *Advances in invertebrate reproduction* (Vol. 3, pp. 223–230). Amsterdam: Elsevier Science Publishers.

Chang, E. S. (1985). Hormonal Control of Molting in Decapod Crustacea. *American Zoologist, 25*(1), 179–185.

Chang, E. S. (1989). Endocrine regulation of molting in decapod crustacean. *Reviews in Aquatic Sciences, 1*, 131–157.

Chang, E. S. (2001). Crustacean hyperglycemic hormone and family old paradigms and new perspectives. *American Zoologist, 41*, 380–388.

Chang, E. S., & Bruce, M. (1980). Ecdysone titers of juvenile lobsters following molt induction. *The Journal of Experimental Zoology, 214*, 157–160.

Chang, E. S., Sage, B. A., & O'Connor, J. D. (1976). The qualitative and quantitative determination of ecdysones in tissues of the crab *Pachygrapsus crassipes* following molt induction. *General and Comparative Endocrinology, 30*, 21–33.

Chang, E. S., Bruce, M. J., & Newcomb, R. W. (1987). Purification and aminoacid composition of a peptide with molt-inhibiting activity from the lobster, *Homarus americanus*. *General and Comparative Endocrinology, 65*, 56–64.

Chang, E. S., Prestwich, G. D., & Bruce, M. J. (1990). Aminoacid sequence of a peptide with both molt-inhibiting and hyperglycemic activities in the lobster, *Homarus americanus*. *Biochemical and Biophysical Research Communications, 171*, 818–826.

Chang, E. S., Bruce, M. J., & Tamone, S. L. (1993). Regulation of crustacean molting: A multihormonal system. *American Zoologist, 33*, 324–329.

Chang, E. S., Keller, R., & Chang, S. A. (1998). Quantification of crustacean hyperglycemic hormone by ELISA in hemolymph of the lobster, *Homarus americanus*, following various stresses. *General and Comparative Endocrinology, 111*, 359–366.

Chang, E. S., Chang, S. A., Beltz, B. S., & Kravitz, E. A. (1999). Crustacean hyperglycemic hormone in the lobster nervous system: localization and release from cells in the subesophageal ganglion and thoracic second roots. *The Journal of Comparative Neurology, 414*, 50–56.

Chang, E. S., Chang, S. A., & Mulder, E. P. (2001). Hormones in the lives of crustaceans: an overview. *American Zoologist, 41*, 380–388.

Chang, Y.-J., Sun, C.-L., Chen, Y., & Yeh, S.-Z. (2012). Modelling the growth of crustacean species. *Reviews in Fish Biology and Fisheries, 22*(1), 157–187.

Charniaux-Cotton, H. (1957). Croissance regeneration et determinisme endocrinien des caracteres sexuels *d'Orchestia gammarella* (Pallas) Crustace Amphipode. *Annales des Sciences Naturelles – Zoologie et Biologie Animale, 19*, 411–560.

Chen, J. H., Kabbouh, M., Fisher, M. J., & Rees, H. H. (1994). Induction of an inactivation pathway for ecdysteroids in larvae of the cotton leafworm, *Spodoptera littoralis*. *The Biochemical Journal, 301*, 89–95.

Cheng, J. H., & Chang, E. S. (1991). Ecdysteroid treatment delays ecdysis in the lobster, *Homarus americanus*. *The Biological Bulletin, 181*(1), 169–174.

Chittleborough, R. G. (1975). Environmental factors affecting growth and survival of juvenile western rock lobsters. *Australian Journal of Marine & Freshwater Research, 26*, 177–196.

Chung, J. S., Dircksen, H., & Webster, S. G. (1999). A remarkable, precisely timed release of hyperglycemic hormone from endocrine cells in the gut is associated with ecdysis in the crab *Carcinus maenas*. *Proceedings of the National Academy of Sciences of the United States of America, 96*, 13103–13107.

Cobb, J. S., & Tamm, G. R. (1974). Social conditions increase intermolt period in juvenile lobsters *Homarus americanus*. *Journal of the Fisheries Research Board of Canada, 32*, 1941–1943.

Colinas-Sánchez, F., & Briones-Fourzán, P. (1990). Alimentación de las langostas Panulirus argus y P. guttatus (Latreille 1804) en el Caribe Mexicano. *Anales del Instituto de Ciencias del Mar y Limnología, Universidad Nacional Autónoma de México, 17*, 89–109.

Cooper, R. A., & Uzmann, J. R. (1980). Ecology of juvenile and adult Homarus. In J. S. Cobb & B. F. Phillips (Eds.), *The biology and management of lobsters* (Ecology and management) (Vol. 2, pp. 97–178). New York: Academic.

Couch, E. F., Daily, C. S., & Smith, W. B. (1976). Evidence for the Presence of Molt-Inhibiting Hormone in the sinus Gland of the Lobster *Homarus americanus*. *Journal of the Fisheries Research Board of Canada, 33*(7), 1623–1627.

Dall, W. (1974). Indices of nutritional state in the western rock lobster, *Panulirus longipes* (Milne Edwards). I. Blood and tissue constituents and water content. *Journal of Experimental Marine Biology and Ecology, 16*, 167–180.

Dall, W. (1975). Indices of nutritional state in the western rock lobster, *Panulirus longipes*, II. Gastric fluid constituents. *Journal of Experimental Marine Biology and Ecology, 18*, 1–18.

Dall, W. (1977). Review of the physiology of growth and moulting in rock lobsters. *Circular, CSIRO Division of Fisheries & Oceanography. (Aust.), 7*, 75–81.

Dall, W., & Barclay, M. C. (1977). Induction of viable ecdysis in the western rock lobster by 20-hydroxyecdysone. *General and Comparative Endocrinology, 31*(3), 323–334.

Davis, G. E. (1977). Effects of recreational harvest on a spiny lobster, *Panulirus argus*, population. *Bulletin of Marine Science, 27*, 223–236.

Davis, G. E. (1981). Effect of injuries on spiny lobster, *Panulirus argus*, and implications for fishery management. *Fishery Bulletin, 78*, 979–984.

Dell, S., Sedlmieier, D., Bocking, D., & Dauphlin-Villemant, C. (1999). Ecdysteroid biosynthesis in crayfish Y-organs: Feedback regulation by circulating ecdysteroids. *Archives of Insect Biochemistry and Physiology, 41*(3), 148–155.

Díaz-Arredondo, M. A., & Guzmán-del-Próo, S. A. (1995). Feeding habits of the spiny lobster (*Panulirus interruptus* Randall, 1840) in Bahía Tortugas, Baja California Sur. *Ciencias Marinas, 21*, 439–462.

Dircksen, H., Webster, S. G., & Keller, R. (1988). Immunocytochemical demonstration of the neurosecretory systems containing putative moult-inhibiting hormone and hyperglycemic hormone in the eyestalk of brachyuran crustaceans. *Cell and Tissue Research, 251*(1), 3–12.

Drach, P. (1939). Mue et cycle d'intermue chez les Crustacés Décapodes. *Annales de l'Institut Océano-graphique, Monaco, 18*, 103–391.

Drach, P. (1944). Étude préliminaire sur le cycle d,intermue et son conditionnement hormonal chez Leander serratus (Pennant). *Bulletin biologique de la France et de la Belgique, Paris, 78*, 40–62.

Drach, P., & Tchernigovtzeff, C. (1967). Sur la method de determination des stades d'intermue et son application general aux Crustaces. *Vie et Milieu (A), 18*, 597–607.

Durand, J. B. (1956). Neurosecretory cell types and their secretory activity in the crayfish. *The Biological Bulletin, 14*, 62–76.

Eastman-Reks, S., & Fingerman, M. (1984). Effects of neuroendocrine tissue and cyclic AMP on ovarian growth in vivo and in vitro in the fiddler crab, *Uca pugilator*. *Comparative Biochemistry and Physiology Part A, 79*, 679–668.

Ebert, T. A., & Ford, R. F. (1986). Population ecology and fishery potential of the spiny lobster *Panulirus penicillatus* at Enewetak Atoll, Marshall Islands. *Bulletin of Marine Science, 38*, 56–67.

Echalier, G. (1955). Role de l'organe Y dans la d6terminisme de la mue de Carcinides (*Carcinus*) moenas L. (Crustaces D6capodes); experiences d'implantation. *CR Hebd Seances Academic Science, 240*, 1581–1583.

Echalier, G. (1959). L'organe Y et ledeterminisme de la croissance et de la mue chez Carcinus maenas (L.), Crustace Decapode. *Annales des Sciences Naturelles-Zoologie et Biologie Animale, 12*, 1–59.

Edgar, G. J. (1990). Predator-prey interactions in seagrass beds. I. The influence of macrofaunal abundance and size-structure on the diet and growth of the western rock lobster *Panulirus cygnus* George. *Journal of Experimental Marine Biology and Ecology, 139*, 1–22.

Engle, J. M. (1979). Ecology and growth of juvenile California spiny lobster, *Panulirus interruptus* (Randall). PhD thesis, University of Southern California.

Ennis, G. P. (1973). Food, feeding and condition of lobsters, Homarus americanus, throughout the seasonal cycle in Bonavista Bay, Newfoundland. *Journal of the Fisheries Research Board of Canada, 30*, 1905–1909.

Fabens, A. J. (1965). Properties and fitting of the von Bertalanffy growth curve. *Growth, 29*, 265–289.

Fernandez, R. (2002). Neuroendocrine control of vitellogenesis in the spiny lobster *Panulirus homarus* (Linnaeus, 1758). PhD thesis, Central Institute of Fisheries Education (Deemed University), Mumbai, India, pp 189

Fernandez, R., & Radhakrishnan, E. V. (2010). Classification and mapping of neurosecretory cells in the optic, supraoesophageal and thoracic ganglia of the female spiny lobster *Panulirus homarus* (Linnaeus, 1758) and their secretory activity during vitellogenesis. *Journal of the Marine Biological Association of India, 52*(2), 237–248.

Fernandez, R., & Radhakrishnan, E. V. (2016). Effect of bilateral eyestalk ablation on ovarian development and moulting in early and late intermoult stages of female spiny lobster *Panulirus homarus* (Linnaeus, 1758). *Invertebrate Reproduction and Development, 60*(3), 238–242.

Fielding, P. J., & Mann, B. Q. (1999). The Somalia inshore lobster resource. *A survey of the lobster fishery of the northeastern region (Puntland) between FOAR and EYL during November 1998.* IUCN Eastern Africa Programme, pp. 1–37.

Fingerman, M. (1987). Endocrine mechanisms in crustaceans. *Journal of Crustacean Biology, 7*, 1–24.

Fingerman, M. (1997). Crustacean endocrinology: a retrospective, prospective and introspective analysis. *Physiological Zoology, 70*, 257–269.

Freeman, J. A., & Bartell, C. K. (1976). Some effects of molt-inhibiting hormone and 20-hydroxyecdysone upon molting in the grass shrimp, *Palaemonetes pugio*. *General and Comparative Endocrinology, 28*, 131–142.

Freeman, J. A., & Costlow, J. D. (1979). Hormonal control of apolysis in barnacle mantle tissue epidermis, in vitro. *The Journal of Experimental Zoology, 270*, 333–346.

Gabe, M. (1953). Sur l'existence chez quelques crustaces malacostraces d' un organe comparate a la glande de la mue des insects. *C Hebd Seances Seances Sciences, 237*, 1111–1113.

Gabe, M. (1954). Partcularites morphologiques de l'organe Y (glande de lamae) des crustaceas Malacostraces. *Bulletin de la Société zoologique de France, 79*, 166 (abtsr.).

Gabe, M. (1956). Histologie compare de la glande de mue (organe Y) des crustaceas Malacostraces. *Annual Science Natural Zoological Biological Animal* [11] *18*, 145–152 (Fisheries Research Board of Canada Traslated. Series, 1586).

Ganeshkumar, A., Baskar, B., Santhanakumar, J., Vinithkumar, N. V., Vijayakumaran, M., & Kirubagaran, R. (2010). Diversity and functional properties of intestinal microbial flora of the spiny *lobster Panulirus versicolor* (Latreille, 1804). *Journal of the Marine Biological Association India, 52*(2), 282–285.

George, M. J. (1967). Observations on the biology and fishery of the spiny lobster *Panulirus homarus* (Linnaeus). *Proceedings of the Symposuim on Crustacea, Part IV*. Marine biological Association of India, Mandapam Camp, India, 1308–1316.

George, C. J., Reuben, N., & Muthe, P. T. (1955). The digestive system of *Panulirus polyphagus* (Herbst). *Journal of Animal Morphology and Physiology, 2*, 14–27.

Goes, C. A., & Lins-Oliveira. (2009). Natural diet of the spiny lobster, *Panulirus echinatus* Smith, 1869 (Crustacea: Decapoda: Palinuridae), from São Pedro and São Paulo Archipelago, Brazil. *Brazilian Journal of Biology, 69*(1), 143–148.

Goñi, R., Quetglas, A., & Reñones, O. (2001). Diet of the spiny lobster *Panulirus elephas* (Decapoda: Palinuridae) from the Columbretes Islands Marine Reserve (north-western Mediterranean). *Journal of the Marine Biological Association of the United Kingdom, 81*, 347–348.

Gulland, J. A. (1983). *Fish stock assessment: a manual of basic methods* (FAO/Wiley series on food and agriculture). Chichester: Wiley.

Gulland, J. A., & Holt, S. J. (1959). Estimation of growth parameters for data at unequal time intervals. *Journal du Conseil CIEM, 44*, 200–209.

Hampshire, F., & Horn, D. H. S. (1966). Structure of crustecdysone, a crustacean moulting hormone. *Chemical Communications, 1966*, 37–38.

Hanstrom, B. (1937). Incretory organs and hormonal functions in invertebrates. VIII. Neurosecretory organs of unknown function. I. The X organ of crustaceans. *Ergebnisse der Biologie, 14*, 214–219 (*Fisheries Research Board of Canada*, Translational. Series, 2731).

Hanstrom, B. (1939). *Hormones in Invertebrates*. London/New York: Oxford University Press.

Hanstrom, B. 1947. The brain, the sense organs and the incretory organs of the head in the crustacean Malacostraca. *Acta Universitatis Lundensis* [N.S.], *58*, 1–44.

Hartnoll, R. G. (1982). Growth. In: L. G. Abele (Ed.), *The biology of Crustacea: 2. Embryology, morphology and genetics. The biology of Crustacea* (pp. 111–196). New York: Academic.

Hartnoll, R. G. (2001). Growth in Crustacea-twenty years on. *Hydrobiologia, 449*, 111–122.

Herrera, A., Ibarza'bal, D., Foyo, J., & Espinosa, J. (1991). Alimentacio'nnatural de la langosta *Panulirus argus* en la regio'n de Los Indios (plataforma SW de Cuba) y su relacio'n con el bentos. *Review of Investigation Marine, 12*, 172–182.

Heydorn, A. E. F. (1969). Notes on the biology of *Panulirus homarus* and on length weight relationships of *Jasus lalandii*. *South African Division of Sea Fish Investigation Report, 69*.

Hiatt, R. W. (1948). The biology of the lined shore crab, *Pachygrapsus crassipes* Randall. *Pacific Sciences, 2*, 135–213.

Hopkins, P. M. (1982). Growth and regeneration patterns in the fiddler crab, *Uca pugilator*. *Biological Bulletin (Woods Hole), 163*, 301–319.

Horn, D. H. S., Middleton, E. J., Wunderlich, J. A., & Hampshire, F. (1966). Identity of the moulting hormones of insects and crustaceans. *Chemical Communications, 1966*, 339–340.

Hunt, J. H., & Lyons, W. G. (1986). Factors affecting growth and maturation of spiny lobsters, *Panulirus argus*, in the Florida Keys. *Canadian Journal of Fisheries and Aquatic Sciences, 43*, 2243–2247.

Jayakody, D. S. (1991). On the growth, mortality and recruitment of the spiny lobster (Panulirus homarus) in Sri Lankan waters. *NAGA, ICLARM Quarterly*, 38–42.

Jayawickrama, S. J. C. (1991). Fishery and population dynamics of *Panulirus homarus* (Linnaeus) from Mutwal, Sri Lanka. *Journal of the National Science Council of Sri Lanka, 9*(1), 53–62.

Jernakoff, P., Phillips, B. F., & Fitzpatrick, J. J. (1993). The diet of post-puerulus western rock lobster *Panulirus cygnus* George at Seven Mile Beach, Western Australia. *Marine and Freshwater Research, 44*, 649–655.

Joll, L. M. (1982). Foregut evacuation of four foods by the Western rock lobster, *Panulirus cygnus* in Aquaria. *Australian Journal of Marine & Freshwater Research, 33*, 939–943.

Joll, L.M. (1984). Natural diet and growth of juvenile Western rock lobster *Panulirus cygnus*. PhD thesis. Perth: University of Western Australia.

Joll, L. M., & Phillips, B. F. (1984). Natural diet and growth of juvenile Western rock lobster *Panulirus cygnus*. *Journal of Experimental Marine Biology and Ecology, 75*, 145–169.

Jong, K. (1993). Growth of the spiny lobster Panulirus homarus (Linnaeus, 1758), depending on sex and influenced by reproduction (Decapoda, Palinuridae). *Crustaceana, 64*(1), 18–23.

Juinio-Menez, M. A., & Ruinata, J. (1996). Survival, growth and food conversion efficiency of *Panulirus ornatus* following eyestalk ablation. *Aquaculture, 146*(3&4), 225–235.

Kagwade, P. V. (1987). Age and growth of the spiny lobster *Panulirus polyphagus* (Herbst) of Bombay waters. *Indian Journal of Fisheries, 34*(4), 389–398.

Kagwade, P. V., & Kabli, L. M. (1996). Age and growth of the sand lobster *Thenus orientalis* (Lund) from Bombay waters. *Indian Journal of Fisheries, 43*(3), 241–247.

Kanciruk, P. (1980). Ecology of juvenile and adult Palinuridae (spiny lobsters). In J. S. Cobb & B. F. Phillips (Eds.), *The biology and management of lobsters* (pp. 59–96). New York: Academic.

Kathirvel, M. (1973). The growth and regeneration of an aquarium held spiny lobster, *Panulirus polyphagus* (Herbst). *Indian Journal of Fisheries, 20*(1), 219–221.

Kegel, G., Reichwein, B., Tensen, C. P., & Keller, R. (1991). Aminoacid sequence of crustacean hyperglycemic hormone (CHH) from the crayfish, *Orconectes limosus*. Emergence of a Novel neuropeptide. *Peptides, 12*(5), 909–913.

Keller, R. (1992). Crustacean neuropeptides: structures, functions and comparative aspects. *Experientia, 48*, 439–448.

Keller, R., & Schmid, E. (1979). In vitro secretion of ecdysteroids by Y-organ and lack of secretion by mandibular organs of the crayfish following molt induction. *Journal of Comparative Physiology, 130*, 347–353.

Kirubagaran, R., Peter, S. M., Dharani, G., Vinithkumar, N. V., Sreeraj, G., & Ravindran, R. (2005). Changes in vertebrate type steroids and 5-hydroxytryptamine during ovarian recrudescence in the Indian spiny lobster *Panulirus homarus*. *New Zealand Journal of Marine Freshwater Research, 39*, 527–537.

Kizhakudan, J. K. (2007). Reproductive biology, ecophysiology and growth in the mudspiny lobster *Panulirus polyphagus* and the sand lobster *Thenus orientalis*. PhD thesis, Bhavnagar University, Gujarat, 169 p.

Kizhakudan, J. K. (2013). Effect of eyestalk ablation on moulting and growth in the mudspiny lobster *Panulirus polyphagus* (Herbst, 1793) held in captivity. *Indian Journal of Fisheries, 60*(1), 77–81.

Kizhakudan, J. K., & Patel, S. K. (2011). Effect of diet on growth of the mud spiny lobster *Panulirus polyphagus* (Herbst, 1793) and the sand lobster *Thenus orientalis* (Lund, 1793) held in captivity. *Journal of the Marine Biological Association of India, 53*(2), 167–171.

Kizhakudan, J. K., Kizhakudan, S. J., & Patel, S. K. (2013). Growth and moulting in the mud spiny lobster, *Panulirus polyphagus* (Herbst, 1793). *Indian Journal of Fisheries, 60*(2), 79–86.

Kleijn, D. P. V., & van Herp, F. (1995). Molecular biology of neurohormone precursors in the eyestalk of Crustacea. *Comparative Biochemistry and Physiology Part B, 112*, 573–579.

Kleinholz, L. H. (1975). Purified hormones from the crustacean eyestalks and their physiological specificity. *Nature (London), 258*, 256–257.

Knowles, F. G. W., & Carlisle, D. B. (1956). Endocrine control in the Crustacea. *Biological reviews of the Cambridge Philosophical Society, 31*, 396–473.

Kulmiye, A. J., & Mavuti, K. M. (2005). Growth and moulting of captive *Panulirus homarus homarus* in Kenya, western Indian Ocean. *New Zealand Journal of Marine and Freshwater Research, 39*(3), 539–549.

Kurata, H. (1962). Studies of age and growth of crustacean. *Bulletin Hokkaido Reg Fisheries of Research Laboratory, 24*, 1–114.

Lafont, R. (1997). Ecdysteroid related molecules in animals and plants. *Archives of Insect Biochemistry and Physiology, 35*, 3–20.

Laufer, H. M., Landau, M., Borst, D., & Homola, H. (1986). The synthesis and regulation of methyl farnesoate, a new juvenile hormone for crustacean reproduction. In M. Porchet, J. C. Andries, & A. Dhainaut (Eds.), *Advances in invertebrate reproduction* (pp. 135–143). Amsterdam: Elsevier Publications.

Laufer, H., Borst, D., Baker, F. C., Carrasco, C., & Sinkus, M. (1987). Identification of a juvenile hormone-like compound in a crustacean. *Science, 235*, 202–205.

Le Soyez, D., Caer, J. P., Noel, P. Y., & Rossier, J. (1991). Primary structure of two isoforms of the vitellogenesis inhibiting hormone from the lobster Homarusamericanus. *Neuropeptides, 20*, 25–32.

Lindberg, R. G. (1955). Growth, population dynamics, and field behaviour in the spiny lobster, *Panulirus interruptus* (Randall). *University of California, Publication Zoological, 59*, 157–248.

Lipcius, R. N., & Herrnkind, W. F. (1982). Molt cycle alterations in behavior, feeding and diel rhythms of a decapod crustacean, the spiny lobster *Panulirus argus*. *Marine Biology, 68*, 241–252.

Lyle, W. G., & MacDonald, C. D. (1983). Molt stage determination in the Hawaiian spiny lobster *Panulirus marginatus*. *Journal of Crustacean Biology, 3*(2), 208–216.

Majowski, J., Hampton, J., Jones, R., Laurec, A., & Rosenberg, A. A. (1987). Sensitivity of length-based methods for stock assessment. Report of Working group III. *ICLARM Conference Proceedings, 13*, 363–372.

Mashaii, N., Rajabipour, F., & Shakouri, A. (2011). Feeding habits of the scalloped spiny lobster *Panulirus homarus* (Linnaeus, 1758) (Decapoda: Palinuridae) from the southeast coast of Iran. *Turkish Journal of Fisheries and Aquatic Sciences, 11*, 45–54.

Matsumoto, K. (1962). Experimental studies of the neurosecretory activities of the thoracic ganglion of a crab, *Hemigrapsus* sp. *General and Comparative Endocrinology, 2*, 4–11.

Mauchline, J. (1976). The Hiatt growth diagram for crustacean. *Marine Biology, 35*, 79–84.

Mauchline, J. (1977). Growth of shrimps, crabs and lobsters – An assessment. *Journal du Conseil/ Conseil Permanent International pour l'Exploration de la Mer, 37*(2), 162–169.

Mayfield, S., Branch, G. M., & Cockcroft, A. C. (2000). Relationships among diet, growth rate, and food availability for the South African rock lobster, *Jasus lalandii* (Decapoda: Palinuridea). *Crustaceana, 73*, 815–834.

McWhinnie, M. A., & Mohrherr, C. J. (1970). Influence of eyestalk factors, intermolt cycle and season upon C-leucine incorporation into protein in the crayfish (*Orconectes virilis*). *Comparative Biochemistry and Physiology, 34*, 415–437.

Palmer, M. J., Phillips, B. & Smith, G. T. (1991). Application of Nonlinear Models with Random Coefficients to Growth Data. Biometrics, 47, 623.

Mehanna, S., Al-Shijibi, S., Al-Jafary, J., & Al-Senaidi, R. (2012). Population dynamics and management of scalloped spiny lobster *Panulirus homarus* in Oman coastal waters. *J. Biol. Agri. Healthcare, 2*(10), 184–194.

Mikami, S. (2005). Moulting behaviour responses of Bay lobster, *Thenus orientalis*, to environmental manipulation. *New Zealand .Journal of Marine Freshwataer Research, 39*, 297–302.

Mills, B. J., & Lake, P. S. (1975). Setal development and moult staging in the crayfish *Parastacoides tasmanicus* (Erichson) (Decapoda, Parastacidae). *Australian Journal of Marine & Freshwater Research, 26*(1), 103–107.

Mohamed, K. S. (1989). Studies on the reproductive endocrinology of the penaeid prawn *Penaeus indicus* H. Milne Edwards. PhD thesis, Cochin University of Science and Technology.

Mohammed, K. H., & George, M. J. (1968). Results of the tagging experiments on the Indian spiny lobster, *Panulirus homarus* (Linnaeus) – Movement and growth. *Indian Journal of Fisheries, 15*, 15–26.

Morgan, G. R. (1980). Population dynamics of spiny lobsters. In J. S. Cobb & B. F. Phillips (Eds.), *The biology and management of lobsters. Vol. 2. Ecology and management* (pp. 189–218). New York: Academic.

Munro, J. L. (1974). The biology, ecology, exploitation and management of Caribbean reef fishes, Part V.I. The biology, ecology and bioeconomics of Caribbean reef fishes: Crustaceans (spiny lobsters and crabs). Research Report Zoological Department, University of West Indies, 3, 57 pp.

Munro, J. L. (1982). Estimation of the parameters of the von Bertalanffy growth equation from recapture data at variable time intervals. *ICES Journal of Marine Science, 40*, 199–200.

Munro, J. L. (1988). Growth and mortality rates and state of exploitation of spiny lobsters in Tonga. *South Pacific Commission Workshop on Pacific Inshore Fisheries Resources, New Caledonia* (Background Paper, Vol. 51, 34pp).

Murdoch, W. W., & Oaten, A. (1975). Predation and population stability. *Advances in Ecological Research, 9*, 1–131.

Nair, R. V., Soundararajan, R., & Nandakumar, G. (1981). Observations on growth and moulting of spiny lobsters *Panulirus homarus* (Linnaeus), *P. ornatus* (Fabricius) and *P. penicillatus* (Olivier) in captivity. *Indian Journal of Fisheries, 28*(1&2), 25–35.

Nakatsuji, T., & Sonobe, H. (2004). Regulation of ecdysteroid secretion from the Y-organ by molt-inhibiting hormone in the American crayfish *Procambarus clarkii*. *General and Comparative Endocrinology, 135*, 358–364.

Newman, G. G., & Pollock, D. E. (1974). Growth of the rock lobster *Jasus lalandii* and its relationship to benthos. *Marine Biology, 24*, 339–346.

Nicol, S. (1987). Some limitations on the use of the lipofuscin ageing technique. *Marine Biology, 93*, 609–614.

Ohira, T., Okumura, T., Suzuki, M., Yajima, Y., Tsusui, N., Wilder, M., & Nagasawa, H. (2006). Production and characterization of recombinant vitellogenesis-inhibiting hormone from the American lobster *Homarus americanus*. *Peptides, 27*, 1251–1258.

Otsu, T. (1960). Precocious development of the ovaries in the crab, *Potamon dehaani*, following implantation of the thoracic ganglion. *Annotationes Zoologicae Japonenses, 33*, 90–96.

Palmer, M. J., Phillips, B., & Smith, G. T. (1991). Application of nonlinear models with random coefficients to growth data. *Biometrics, 47*, 623.

Panouse, J. B. (1943). Influence de Ì ablation du pédoncle oculaire sur la croissance dè Ì ovaire chèz la crevette Leander serratus. *Comptes rendus de l'Académie des Sciences, Paris, 217*, 553–555.

Panouse, J.B. 1947. La glande du sinus et la maturation des produits génitaux chez les crevettes. The *Bulletin of Biology* Fr. Belg (Suppl)

Passano, L. M. (1951a). The X-organ sinus gland neurosecretory system in crab. *The Anatomical Record, 111*, 462–665.

Passano, L. M. (1951b). The X-organ, a neurosecretory gland controlling molting in crabs. *The Anatomical Record, 111*, 559.
Passano, L. M. (1953). Neurosecretory control of molting in crabs by the X-organ sinus gland complex. *Physiologiia Comparata et Oecologia, 3*, 155–189.
Paterson, N. F. (1969). The behavior of captive rock lobster *Jasus lalandii* (H. Mile-Edwards). *Annals. South African Museum, 52*(10), 225–264.
Pauly, D. 1983. Some simple methods for the assessment of tropical fish stocks. *FAO Fish Tech. pap*.234: 32 p.
Pauly, D., & Morgan, G. R. (Eds.). (1987). Length-based methods in fisheries research. *ICLARM Corif Proceedings, 13*, 468 pp.
Pauly, D., David, N., & Ingles, J. (1980). ELEFAN I. User's instructions and program listings. Mime O. Report
Peebles, J. B. (1977). A rapid technique for molt staging in live *Macrobrachium rosenbergii*. *Aquaculture, 12*(2), 173–180.
Phillips, B. F., Cobb, J. S., & George, R. W. (1980). General biology. In J. S. Cobb & B. F. Phillips (Eds.), *The biology and management of lobsters* (Vol. II, pp. 1–82). New York: Academic.
Phillips, B. F., Palmer, M. J., Cruz, R., & Trendall, J. T. (1992). Estimating growth of the spiny lobsters *Panulirus cygnus, P. argus* and *P. ornatus*. *Australian Journal of Marine & Freshwater Research, 43*, 1177–1188.
Pitcher, R. (1993). Spiny lobster: In: A. Wright, & L. Hill (Eds.), *Nearshore marine resources of the South Pacific, Suva: Institute of Pacific Studies* (pp. 539–608). Honiara: Forum Fisheries Agency and Halifax: International Centre for Ocean Development.
Plaut, I., & Fishelson, L. (1991). Population structure and growth in captivity of the spiny lobster *Panulirus penicillatus* from Dahab, Gulf of Aqaba, Red sea. *Marine Biology, 111*, 467–472.
Pollock, D. E. (1979). Predator-Prey Relationships Between the Rock Lobster *Jasus lalandii* and the Mussel *Aulacomya ater* at Robben Island on the Cape West Coast of Africa. *Marine Biology, 52*, 347–356.
Prescott, J. (1988). Tropical spiny lobsters: An overview of their biology, the fisheries and the economics with particular reference to the double-spined rock lobster *P. penicillatus* (SPC Workshop on Pacific Inshore Fisheries Research), New Caledonia, WP 12, 36 pp.
Quackenbush, L. S., & Herrnkind, W. F. (1981). Regulation of molt and gonadal development in the spiny lobster, *Panulirus argus* (Crustacea: Palinuridae): Effect of eyestalk ablation. *Comparative Biochemistry and Physiology, 69*, 523–527.
Quackenbush, L. S., & Herrnkind, W. F. (1983). Partial characterization of eyestalk hormones controlling molt and gonadal development in the spiny lobster, *Panulirus argus*. *Journal of Crustacean Biology, 3*(1), 34–44.
Quackenbush, L. S., & Keeley, L. L. (1988). Regulation of vitellogenesis in the fiddler crab *Uca pugilator*. *The Biological Bulletin, 5*, 321–331.
Radhakrishnan, E. V. (1989). Physiological and biochemical studies on the spiny lobster *Panulirus homarus*. PhD thesis, University of Madras, Chennai.
Radhakrishnan, E. V. (1994, August 29–31). Commercial prospects for farming spiny lobsters. *Aquaculture towards the 21st century: Proceedings of the INFOfish-Aquatech'94 Conference*, Colombo, Sri Lanka, pp 96–102.
Radhakrishnan, E. V., & Vijayakumaran, M. (1984a). Effect of eyestalk ablation in the spiny lobster *Panulirus homarus* (Linnaeus) 1. On moulting and growth. *Indian Journal Fisheries, 31*, 130–147.
Radhakrishnan, E. V., & Vijayakumaran, M. (1984b). Effect of eyestalk ablation in the spiny lobster *Panulirus homarus* (linnaeus): 3. on gonadal maturity. *Indian Journal Fisheries, 31*, 209–216.
Radhakrishnan, E. V., & Vijayakumaran, M. (1998). Observations on the moulting behaviour of the spiny lobster *Panulirus homarus* (Linnaeus). *Indian Journal Fisheries, 45*(3), 331–338.
Radhakrishnan, E. V., & Vivekanandan, E. (2004). Prey preference and feeding strategies of the spiny Lobster *Panulirus homarus* (Linnaeus) predating on the green mussel *Perna viridis*

(Linnaeus). *Program and Abstracts of the 7th international conference and workshop on lobster biology and management*, 40 pp.

Radhakrishnan, E. V., Chakraborty, R. D., Baby, P. K., & Radhakrishnan, M. (2013). Fishery and population dynamics of the sand lobster *Thenus unimaculatus* (Burton & Davie, 2007) landed by trawlers at Sakthikulangara Fishing Harbour in the south-west coast of India. *Indian Journal of Fisheries, 60*(2), 7–12.

Radhakrishnan, E. V., Thangaraja, R., & Vijayakumaran, M. (2015). Ontogenetic changes in morphometry of the spiny lobster, *Panulirus homarus homarus* (Linnaeus, 1758) from southern Indian coast. *Journal of the Marine Biological Association of India, 57*(1), 5–13.

Rangarao, K. (1965). Isolation and partial characterization of the molt-inhibiting hormone of the crustacean eyestalk. *Experientia, 21*, 593–594.

Rao, K. R., Fingerman, S. W., & Fingerman, M. (1973). Effects of exogenous ecdysones on the molt cycles of fourth and fifth stage American lobsters, *Homarus americanus*. *Comparative Biochemistry & Physiology Part A, 44*(4), 1105–1120.

Raviv, S., Parnes, S., Segall, C., Davis, C., & Sagi, A. (2006). Complete sequence of *Litopenaeus vannamei* (Crustacea: Decapoda) vitellogenin cDNA and its expression in endocrinologically induced sub-adult females. *General and Comparative Endocrinology, 145*, 39–50.

Ricker, W. E. (1975). Computation and interpretation of biological statistics of fish populations. *Journal of the Fisheries Research Board of Canada, 191*, 1–382.

Sainte-Marie, B., & Chabot, D. (2002). Ontogenic shifts in natural diet during benthic stages of American lobster (*Homarus americanus*), off the Magdalen Islands. *Fishery Bulletin, 100*(1), 106–116.

Sanders, M. J., & Bouhlel, M. (1984). Stock assessment of the rock lobster (*Panulirus homarus*) inhabiting the coastal waters of the People's Democratic Republic of Yemen. FAO, Roame, RAB/81/002/21: 67p.

Sanders, M., & Liyanage, U. (2009). *Preliminary assessment for the spiny lobster fishery of the south coast (Sri Lanka)* (pp. 1–44). Sri Lanka: NARA.

Sarojini, R., Nagabhushanam, R., & Fingerman, M. (1995). In vitro inhibition by dopamine of 5-hydroxytryptamine-stimulated ovarian maturation in the red swamp crayfish, *Procambarus clarkii*. *Experientia, 52*(7), 707–709.

Sastry, A. N. & French, D.P. (1977). Growth of American lobster, *Homarus americanus* Milne-Edwards, under controlled conditions. *Circular-CSIRO Division of Fisheries and Oceanography (Aust.). 7*, 11 (abstr.).

Scarratt, D. J. (1980). The food of the lobster. In J. D. Pringle, G. J. Sharp, & J. F. Caddy (Eds.), *Proceedings of the workshop on the relationship between sea urchin grazing and commercial plant/animal harvesting* (Canadian technical report of fisheries and aquatic sciences, pp. 66–91), 954.

Scudamore, H. (1947). The influence of the sinus glands in the crayfish. *Physiological Zoology, 20*, 187–208.

Senevirathna, J. D. M., Thushari, G. G. N., & Munasinghe, D. H. N. (2014). Length-weight relationship of spiny lobster *Panulirus homarus* population inhabiting southern coastal region of Sri Lanka. *Journal of Environmental Sciences, 3*(2), 607–614.

Shears, N. I., & Babcock, R. I. (2002). Marine reserves demonstrate top-down control of community structure on temperature reefs. *Oecologia, 132*(1), 131–142.

Sheehy, M. R. J. (1990). Potential of morphological lipofuscin age pigment as an index of crustacean age. *Marine Biology, 107*, 439–442.

Sheehy, M. R. J. (1992). Lipofuscin age pigment accumulation in the brains of aging field and laboratory reared crayfish *Cherax quadricannatus* (von Martens) (Decapoda: Parastacidae). *Journal of Experimental Marine Biology and Ecology, 161*, 79–89.

Sheehy, M., Caputi, N., Chubb, C., & Belchie, M. (1998). Use of lipofuscin for resolving cohorts of western rock lobster (*Panulirus cygnus*). *Canadian Journal of Fisheries and Aquatic Sciences, 55*, 925–936.

Smale, J. M. (1978). Migration, growth and feeding in the Natal rock lobster *Panulirus homarus* (Linnaeus). *South African Association for Marine Biological Research Investigation Report*, Vol. 47.

Smith, R. I. (1940). Studies on the effect of eyestalk removal upon young crayfish (*Cambarus clarkii* Girard). *The Biological Bulletin, 79*, 145–152.

Smith, D. M. & Dall, W. (1985). Moult staging in the tiger prawn *Penaeus esculentus*. In: *Proceedings of the Second Australian Prawn Seminar*, Kooralbyn, Queensland, Australia, pp 85–93.

Sochasky, J. B. (1973). Failure to accelerate moulting following eyestalk ablation in decapod crustaceans: a review of the literature. *Fisheries Research Bd Canadian Technical Report, 431*, 1–127.

Soumoff, C., & O'Connor, J. (1982). Response of Y-organ secretory activity by molt inhibiting hormone in the crab *Pachygrapsus crassipes*. *General and Comparative Endocrinology, 48*, 432–439.

Soyez, D. Le Caer, J. P., Noel, P. Y. & Rossier, J. (1991). Primary structure of two isoforms of the vitellogenesis inhibiting hormone from the lobster Homarus americanus. Neuropeptides, 20, 25–32

Sparre, P., Ursin, E., & Venema S. C. (1989). Introduction to tropical fish stock assessment. 1. *Manual* (Fisheries technical Paper, 3M/I: xii + 337 pp). Rome: FAO.

Spaziani, E., Ostedgaard, L. S., Vensel, W. H., & Hegman, J. P. (1982). Effects of eyestalk removal in crabs: Relation to normal premolt. *Journal of Experimental Zoology Part A, 221*(3), 323–327.

Stevenson, J. R., Guckert, R. C., & Cohen, J. D. (1968). Lack of correlation of some proecdysal growth and developmental processes in the crayfish. *Biological Bulletin (Woods Hole. MA), 134*, 160–175.

Subramoniam, T. (1999). Egg production of economically important crustaceans. *Current Science, 76*, 350–360.

Subramoniam, T., & Kirubagaran, R. (2010). Endocrine regulation of vitellogenesis in lobsters. *Journal of the Marine Biological Association of India, 52*(2), 229–236.

Tamone, S. L., & Chang, E. S. (1993). Methyl farnesoate stimulates ecdysteroid secretion from crab Y-organs in vitro. *General and Comparative Endocrinology, 89*, 425–432.

Tegner, M. J., & Dayton, P. K. (1981). Population structure, recruitment and mortality of two sea urchins *Strongylocentrotus fransiscanus* and *S. purpuratus* in a kelp forest. *Marine Ecology Progress Series, 5*, 155–168.

Thangaraja, R., & Radhakrishnan, E. V. (2012). Fishery and ecology of the spiny lobster Panulirus homarus (Linnaeus, 1758) at Khadiyapatanam in the southwest coast of India. *Journal of the Marine Biological Association of India, 54*(2), 69–79.

Thangaraja, R., Radhakrishnan, E. V., & Chakraborthy, R. D. (2015). Stock and population characteristics of the Indian rock lobster *Panulirus homarus* homarus (Linnaeus, 1758) from Kanyakumari, Tamilnadu, on the southern coast of India. *Indian Journal of Fisheries, 62*(3), 21–27.

Thomas, M. M. (1972). Growth of the spiny lobster, *Panulirus homarus* (Linnaeus) in captivity. *Indian Journal of Fisheries, 19*, 125–129.

Travis, D. F. (1954). The molting cycle of the spiny lobster, *Panulirus argus* Latreille.1. Molting and growth in laboratory-maintained individuals. *The Biological Bulletin, 107*, 433–450.

Tsukimura, B., & Borst, D. W. (1992). Regulation of methyl farnesoate in the hemolymph and mandibular organ of the lobster *Homarus americanus*. *General and Comparative Endocrinology, 86*, 297–303.

Turnbull, C. T. (1989). Pleopod cuticular morphology as an index of moult stage in the Ornate Rock Lobster *Panulirus ornatus* (Fabricius). *Australian Journal of Marine & Freshwater Research, 40*, 285–293.

Uchida, R. N., & Tagami, D. T. (1984). Biology, distribution, population structure, and pre-exploitation abundance of spiny lobster, *Panulirus marginatus* (Quoy and Gaimard), in the Northwestern Hawaiian Islands. In: Grigg, R. W. & Tanoue, K. T. (Eds.), *Proceedings of the*

Symposuim on Resourch Investigation in the Northwestern Hawaiian Islands, Vol. 1, April 24–25, 1980, University of Hawaii, Honolulu, Hawaii, pp. 157–198. UNIHI-SEAGRANT-MR-84-01.

Van Herp, F., & Bellon-Humbert, C. (1978). Setal development and molt prediction in the larvae and adults of the crayfish *Astacus leptodactylus* (Nordmann, 1842). *Aquaculture, 14*(4), 289–301.

Von Bertalanffy, L. (1938). A quantitative theory of organic growth. *Human Biology, 10*(2), 181–213.

Waddy, S. L., Aiken, D. E., & de Kleijn, D. P. V. (1995). Control of growth and reproduction. In J. F. Factor (Ed.), *Biology of the lobster, Homarus americanus* (pp. 217–266). San Diego: Academic.

Wahle, R. A., & Fogarty, M. J. (2006). Growth and development: Understanding and modeling growth variability in lobsters. In B. F. Phillips (Ed.), *Lobsters: biology, management aquaculture and fisheries* (pp. 1–44). Oxford: Blackwell Publishing.

Wahle, R. A., Tully, O., & O'Donovan, V. (1996). Lipofuscin as an indicator of age in crustaceans: analysis of the pigment in the American lobster *Homarus americanus*. *Marine Ecology Progress Series, 138*, 117–123.

Wahle, R. A., Tully, O., & O'Donovan, V. (2001). Environmentally mediated crowding effects on growth, survival and metabolic rate of juvenile American lobsters (*Homarus americanus*). *Marine and Freshwater Research, 52*, 1157–1166.

Webster, S. G. (1998). Neuropeptides inhibiting growth and reproduction in crustaceans. In G. M. Coast & S. G. Webster (Eds.), *Recent advances in arthropod endocrinology* (pp. 33–52). Cambridge: Cambridge University.

Weiss, H. M. (1970). The diet and feeding behavior of the lobster, *Homarus americanus*, in Long Island Sound. PhD dissertation, University, Connecticut, Storrs, CT, 80 p.

Werner, E. E., & Gilliam, J. F. (1984). The ontogenetic niche and species interactions in size-structured populations. *Annual Review of Ecology, Evolution, and Systematics IS*, 393–425.

Wetherall, J. A. (1986). A new method for estimating growth and mortality parameters from length-frequency data. *ICLARM Fishbyte, 4*, 12–14.

Williams, M. J. (1981). Methods for analysis of natural diet in portunid crabs (Crustacea: Decapoda: Portunidae). *Journal of Experimental Marine Biology and Ecology, 52*(1), 103–113.

Windell, S. C. (2015). Spiny Lobster (*Panulirus interruptus*) Use of the Intertidal Zone at a Santa Catalina Island MPA in Southern California. Capstone Projects and theses (Paper 579).

Zeleny, C. (1905). Compensatory regulation. *The Journal of Experimental Zoology, 2*, 1–102.

Lobster Fisheries and Management in India and Indian Ocean Rim Countries

7

E. V. Radhakrishnan, Joe K. Kizhakudan, Saleela A, Dineshbabu A. P, and Lakshmi Pillai S

Abstract

The chapter summarises overview of the global lobster fisheries with special focus on status and challenges faced by lobster fisheries of India. Annual production by the states in India, gears used, species composition, biology of major species and the fishery management in comparison to the fishing regulations in the best managed fisheries in the world are discussed. The world total capture fisheries production in 2014 was 93.4 million tonnes, including output from inland waters. Production from world lobster capture fisheries was an average 269,604 t (2003-2016). Annual landing crossed 0.3 million t in 2014 and was 314, 806 t in 2016. Although lobster constitutes just 0.3% of the total global capture fisheries production, the resource is a commercially valuable seafood with high unit value. Canada stands first with an annual production of 0.09 million tonnes (2015) followed by USA (0.069 m t), with both these countries together contributing about 51.8% of the world production. Between 2000 and 2015, the fisheries production of the American clawed lobster, *Homarus americanus*, increased over 85% (1,57,064 t). All the palinurids together form 30.2% of the total global production. Australia with an average annual production of 11,392 t (2008-2014) contributes maximum to the total spiny lobster production. The most productive regions in the northern hemisphere from the lobster fisheries point of view are FAO fishing area 21 (USA and Canada) followed by the fishing area 27. In the southern hemisphere, the FAO fishing area 57 (Australia and Indonesia) lands an average 13,485 t (2008–2012) with an increase in landing during 2014 (14,017 t) and 2015 (15,029 t). The single most important spiny lobster fishery in the world is that of the Caribbean or Florida spiny lobster, *Panulirus argus*.

E. V. Radhakrishnan (✉) · J. K. Kizhakudan · Saleela A · Dineshbabu A. P · Lakshmi Pillai S
ICAR-Central Marine Fisheries Research Institute, Cochin, Kerala, India
e-mail: evrkrishnan@gmail.com

© Springer Nature Singapore Pte Ltd. 2019
E. V. Radhakrishnan et al. (eds.), *Lobsters: Biology, Fisheries and Aquaculture*,
https://doi.org/10.1007/978-981-32-9094-5_7

Lobster is a valuable crustacean resource harvested by India and the countries bordering the Indian Ocean. Although lobster constitutes only 0.05% of the annual marine fish catch of India (2010–2015), export fetches an average Rs. 196 crores (USD 32 million) in foreign exchange annually. Commercially exploited lobsters in India belong to the families Palinuridae and Scyllaridae. The multispecies resource is intensively exploited by both artisanal and mechanised sectors, with the latter contributing around 67% of the total catch. Among the 38 species constituting the lobster fauna of India, only 4 species of palinurids (*Panulirus polyphagus*, *P. homarus homarus*, *P. ornatus*, *Puerulus sewelli*) and 1 species of scyllarid (*Thenus unimaculatus*) significantly contribute to the fishery.

Though lobsters are distributed almost throughout the Indian coast, major fisheries are located on the northwest, northeast and the southern Indian region. The two northwestern states Gujarat and Maharashtra and the northeast state of West Bengal together contribute an average 65% of the total annual country catch (2007–2014). *P. polyphagus* forms 61% of the total annual catch and is the dominant species along the northwest and northeast coasts. *P. homarus homarus*, *P. ornatus* and *T. unimaculatus* are the major species constituting the fishery along the southern region. The major share of the lobster landing in India is held by the trawlers, forming 67% of the total annual catch. The share of the state of Gujarat to the total annual lobster landing is 28%, followed by Maharashtra, 27%; Tamil Nadu, 15%; Kerala, 13%; West Bengal, 10%; and Andhra Pradesh, 5% (2007–2014). In Gujarat, *P. polyphagus* forms 67% of the total catch and *T. unimaculatus* the rest. The annual total catch for 2014 and 2015 shows an upward trend with the production almost nearing 3000 t in 2015. In Maharashtra, the current fishery is almost completely dominated by *P. polyphagus* (94.2%), though during 1978–1985, the percentage contribution of *P. polyphagus* was 54%, with *T. unimaculatus* contributing the rest. Total collapse of the *T. unimaculatus* fishery in 1995 is a classic example of recruitment overfishing, which is not precluded by growth overfishing. In Tamil Nadu, the major species contributing to the fishery are *P. homarus homarus* (40%) and *T. unimaculatus* (38%). The Kanyakumari district in Tamil Nadu on the southwest coast of India was once the major lobster landing region of the country (313 t in 1965). However, intensive exploitation and absence of effective fishing regulations resulted in gradual decline, with the average catch reaching at 9 t (2003–2010). In Kerala, *T. unimaculatus*, *P. homarus homarus* and *P. sewelli* are the three major species supporting fisheries. The deep-sea lobster fishery is centred around the southern region with meagre landing in recent years.

Stock assessment study conducted on *P. polyphagus* fishery of Mumbai concluded that the estimated combined average MSY of males and females (168.42 t) is very low against the yield (243.5 t) and therefore effort reduction to the tune of 40% is necessary for the sustainability of the stock. Management of the spiny and slipper lobster trawl fishery is complicated by failure to enforce regulations due to the practical difficulties of implementing fishing restrictions, as the resource is a bycatch in trawlers. Mesh regulation and closed season for lobster fishing during the peak breeding months are management options that can be considered for implementation in the artisanal fishery. The Central Government has notified minimum legal size (MLS) for lobster export; however, the state governments have not enforced MLS for fishing, except the state of Kerala. Co-management initiatives to regulate artisanal fishing using gillnets are being considered by constituting Fishery Management Councils with representatives of the fishermen, scientists, administrators, fishery professionals and NGOs. If the fishermen can agree to a minimum legal size for fishing, ban on using gillnet and trammel net for lobster fishing and take a decision for a fishing holiday during the peak breeding season, the resource may be prevented from further depletion.

The lobster fisheries of the countries bordering Indian Ocean and the resource management in these countries are discussed in the light of unorganised lobster fishery management in India. Globally, so far, 259 marine fisheries have been certified by the MSC, of which 13 are lobster fisheries and 5 are exclusively spiny lobster fisheries. The first lobster fishery certified by MSC was the Western rock lobster, *Panulirus cygnus* trap fishery, which has been recertified for the fourth time.

Keywords

Annual production · States · Fishing gears · Fishery · Resource management · Certification · IORC

7.1 Introduction

Lobsters are a low-volume, high-value crustacean resource with high export demand and are commercially exploited by many countries around the world. The high unit value together with their simple harvest methods and potential for live transport have made them attractive to fishermen as a source of revenue and as a food source for the humans. The resource is intensively exploited both by artisanal and industrial sectors, and most stocks around the world are either fully exploited or even overexploited (Pitcher 1993; Phillips et al. 2013). Further decline in landings is likely due to overfishing, habitat destruction, climate change and coastal pollution (Jeffs 2010).

The lobster fishery, though not a high-volume resource, represents a significant economic resource for India with an annual export earning of Rs. 196 crores (US$32 million) annually. Lobster stocks are exploited by artisanal fishermen by traditional traps and gillnets and by diving in shallow waters (5–10 m), whereas commercial fishing is carried out by baited traps and by trawling in deeper waters deployed at a depth of 30–50 m. Deep-sea lobsters are fished by trawlers with specialised trawl nets at depths of 200–300 m. In Australia and South Africa, recreational fishing also contributes significantly to the total catch. Recreational fishing is regulated by permits or fishing licenses and by catch quotas. While some countries have enforced strict management regulations, others may not have realistically implemented or closely enforced sustained management programmes, resulting in catch decline. For example, the Western rock lobster fishery in Australia, valued at USD150–300 million annually, is regulated both by input (fishing capacity and effort control and biological control measures) and output (TAE and TAC) controls with significant economic benefits (Penn et al. 2015). In some tropical countries in Asia and Africa, the fishery is open access and has limited capacity to control gear selection, fishing effort and other biological control measures. Strict enforcement of regulations, limited entry of new boats into the fishery, the regulation of fishing gears, catch quotas and a minimum legal size for capture have made lobster fisheries in many countries sustainable, whereas in poorly managed fisheries, the resource is under heavy fishing pressure and is facing overexploitation and stock collapse. Being a low-volume resource compared to other marine fish resources and prawns, no specific management measures to regulate lobster fishing are in existence in India, except the minimum legal size for export notified by the Ministry of Commerce and Industry, Government of India (Radhakrishnan et al. 2005). Co-management or participatory management of the resource involving the stakeholders in fishing, trading and export and the government representatives probably may be promising, as it may address many issues related to resource sustainability, equity and efficiency in coastal resource management (Radhakrishnan et al. 2005; Mohamed et al. 2017).

Ecolabelling of marine fisheries products has the potential to influence the fishing sector to follow certain principles and bring about certain changes in fishing practices. This gives an opportunity to the producers to show to the consumer that the fish is harvested in an ecologically sustainable manner. Ecolables are, thus, 'seals of approval' given to products that are deemed to have fewer negative impact on the environment than other competitively similar products (Deere 1999). Several lobster fisheries are now certified by MSC (Marine Stewardship Council), an international certifying agency, and even some fishery, the Western rock lobster fishery of Australia, has been recertified for the fourth time.

The Indian Ocean rim countries contribute nearly 7.3% to the total global lobster fisheries production with significant landings by Australia and Indonesia (FAO 2015). The palinurid lobsters of the genera *Panulirus* and *Jasus* and the slipper lobsters form the major share of the total landings.

This chapter of fishery, biology and management of lobster fisheries in India has focused on commercially exploited fisheries of major maritime states in India and the biological information available on major species. The production and management of lobster fisheries in Indian Ocean rim countries is presented based on FAO catch data and research papers and reports from these countries. The analysis on lobster fishery in India is entirely based on the landing and biological data generated by the Fisheries Resources Assessment Division and Crustacean Fisheries Division of the Central Marine Fisheries Research Institute and the research papers published by researchers in the institute and other organisations.

Table 7.1 World capture fisheries production of lobsters (FAO 2016)

	Annual catch (t)
2003	225,766
2004	233,791
2005	231,209
2006	253,899
2007	237,621
2008	253,939
2009	251,188
2010	283,334
2011	284,294
2012	297,334
2013	291,802
2014	306,548
2015	308,926
2016	314,806

7.2 Global Fisheries Production

The total global annual average capture fisheries production of lobsters is 269,604 t (2003–2016) (FAO-Fishstat Database 2016) (Table 7.1). The production in 2016 was an estimated 314,806 t. The annual production presented here is less the production by ghost shrimps, which are not considered as lobsters in the revised taxonomy (Chan 2010). The production has steadily increased from 0.23 million t in 2003 to 0.31 million t in 2014, and this has been mainly due to the contribution by homarid lobsters which have shown an increase of over 70% from 0.084 million t in 2003 to 0.17 million t in 2014 (Table 7.2). The American clawed lobster *Homarus americanus* and the European clawed lobster *H. gammarus* are commercially significant and form the bulk of the total global lobster catch, with an average total annual landing of 1,13,705 t (2003–2014), constituting 43.3% of the global landings. On the other hand, annual global production from spiny lobster fisheries (*Panulirus*, *Jasus* and *Palinurus*) has fluctuated between 68,638 t and 87,940 t since 2003 (2003–2014). All the palinurids together contribute 30.2% (79,341 t) to the total global landings (Table 7.2). The Caribbean or Florida spiny lobster *Panulirus argus* alone contributes 9.6% to the total global landings (2014–2015) and 39% to the *Panulirus* production (Table 7.3). In a recent publication, the *P. argus* distributed in Brazilian waters has been reassigned as a new species, *P. meripurpuratus* (Giraldes & Smyth, 2016). *Nephrops norvegicus*, the single species under the genus *Nephrops*, contributes an average 62,993 t (23.9%) annually to the total world production of lobsters (Table 7.2). The global catch of scyllarid species has exhibited a downward trend in landings since 2003, touching the lowest landing of 1689 t in 2009. However, the landings attained a phenomenal increase of over 10,000 t during 2010 and 2011 but again plummeted to 1716 t in 2014 (FAO-Fishstat Database 2016). Total annual average landing of scyllarids is less than 5000 t, with a contribution of 1.8% to the total annual catch.

Table 7.2 Global group-wise landing of lobsters (FAO 2016)

	2003	2004	2005	2006	2007	2008	2009	2010	2011	2012	2013	2014	Average	%
Panulirus	65,510	72,809	70,936	64,147	56,204	65,215	57,870	67,136	70,217	75,961	76,850	70,695	67,796	25.8
Jasus	10,724	10,684	10,912	10,278	10,736	10,006	9136	10,493	8793	8263	8790	8560	9781	3.7
Palinurus	1099	1259	1772	1574	1698	1469	1646	2034	1935	2048	2300	2334	1764	0.7
Metanephrops	974	757	1069	965	1090	748	810	1042	923	876	911	1012	931	0.4
Nephrops	56,319	57,533	58,151	70,389	75,999	72,524	72,164	66,451	60,981	59,227	51,409	54,763	62,993	23.9
Homarus	84,133	84,551	86,117	101,126	87,218	100,510	106,809	124,857	129,927	147,209	146,989	165,008	113,705	43.3
Scyllarids	4907	4652	4277	4181	3616	2809	1689	10,383	10,376	2503	3624	1716	4561	1.7
Lobsters nei	2095	2089	1437	1196	1042	678	1127	953	919	910	917	2460	1319	0.5
Total	**225,761**	**234,334**	**234,671**	**253,856**	**237,603**	**253,959**	**251,251**	**283,349**	**284,071**	**296,997**	**291,790**	**306,548**	**262,849**	100

Table 7.3 World catch (species-wise) during 2014–2015 (estimated from FAO catch data)

Species	Quantity (t)	
	2014	2015
Homarus americanus	159,814	157,064
H. gammarus	5194	4875
Nephrops norvegicus	54,763	49,402
Panulirus argus	27,543	31,320
P. meripurpuratus	6787	6100
P. cygnus	7022	7166
Jasus lalandii	1781	1541
Palinurus elephas	369	387
P. gilchristi	774	652
Sagmariasus verreauxi	188	193
J. edwardsii	5800	5627
P. longipes	3077	2548
P. ornatus	2554	2089
P. homarus	6116	9669
P. gracilis	127	73
P. rubellus	244	272
J. paulensis	701	628
P. polyphagus	919	104
Palinurus spp.	1181	1658
Thenus spp.	1018	1297
Ibacus	469	91
Scyllarids	184	176
Metanephrops challengeri	789	875

The FAO statistical data on world capture fisheries production of lobsters is likely to be an underestimate, as lobster landings from many countries are not reflected in the data. The annual lobster landing data from India is not included until 2013, and from 2014 onwards landings estimated by FAO have been shown as Lobsters nei.

7.3 Major Fishing Areas and Countries

The major lobster fishing countries in the world are Canada, the USA, the UK, Indonesia, Australia, Bahamas, Mexico, Cuba and Brazil. Canada stands first with an annual production of 90,875 t in 2015 followed by the USA (69,203 t). The USA and Canada together land about 51.8% of the total global production (Table 7.4). When countries producing spiny lobsters are considered, Australia stands first with an annual production of 11,392 t followed by Indonesia (10,582 t) (2008–2014) (Table 7.5). However, Indonesia surpassed Australia in 2015 with an annual production of 16,750 t (FAO 2016)). The other countries in the order of maximum production of spiny lobsters are Bahamas, Brazil, Cuba and Nicaragua. The spiny lobster landing of Brazil shown in FAO statistics as *P.*

Table 7.4 Annual lobster capture fisheries production by major countries (FAO 2016)

Major lobster fishing countries			Production (t)					
	1980	1990	2000	2010	2011	2012	2013	2014	2015
Canada	20,089	47,857	45,331	67,277	66,978	74,790	74,686	92,799	90,875
USA	19,873	30,906	40,662	55,253	60,186	70,020	70,535	69,263	69,203
Indonesia	216	826	3596	7651	10,541	13,549	16,482	10,062	16,750
Australia	14,456	15,266	19,837	11,462	10,807	9756	11,301	11,231	10,629
Brazil	8023	9223	6469	6866	6976	7386	6726	6787	6100
Bahamas	2894	5808	9023	9692	8505	9761	6088	6569	6526
Cuba	10,567	7957	7478	4458	5010	4467	4621	4371	4220
Nigeria	0	2600	1939	4398	3697	4289	4586	5001	5216
Nicaragua	1849	783	6534	3800	4102	4427	4494	4724	6473
Mexico	2530	2358	2799	3260	3228	3041	3535	4459	4529
UK		26,983	29,578	41,406	37,699	35,812	31,477	33,849	28,965
New Zealand	4615	3122	2824	2906	3458	2699	2820	3577	3649
Dominican Republic	166	750F	1286	1001	2568	2505	2542	2454	1282
South Africa	6841	4856	2006	4121	2518	2507	2514	2398	2076
Zanzibar	0	0	306	396	912	1682	1695	2190	1929
Honduras	2199	4012	2470	3151	4313	1556	1658	4503	6157
Pakistan	48	470	807	1029	1080	1246	1356	1107	1257
Malaysia	2	691	1103	730	725	794	857	819	345
Others	13,772	14,537	10,386	11,441	50,990	10,575	10,466	40,385	42,745

argus is shown as *P. meripurpuratus*, as the species is now recognised as *P. meripurpuratus* (Giraldes & Smyth, 2015). India is in the 14th position based on total landings (CMFRI data), though the lobster production from India has not been figured in the FAO Statistical Database until 2013.

The most productive regions in the northern hemisphere from the lobster fisheries point of view are FAO fishing area 21 (USA and Canada) (Fig. 7.1) with a mean annual landing of 1, 16,987 t (*Homarus americanus*) (2008–2012) followed by fishing area 27 with a total annual production of 66,271 t (Table 7.6). The European lobster, *H. gammarus*, the Norway lobster, *Nephrops norvegicus*, and the palinurid, *Palinurus elephas*, are the three constituent species. While the total production shows remarkable increase for fishing area 21 during 2014 and 2015, the landings were lower for fishing area 27 for both 2014 and 2015. The fishing areas 34 and 31 are the other regions with high production. Maximum production of the Florida spiny lobster (Caribbean) *P. argus* is from the FAO fishing area 31 (28,287 t, 2008–2014), and the countries contributing to the production are Cuba, Bahamas, Mexico, the USA, Honduras and Nicaragua. The African countries bordering the East Atlantic (fishing area 34) with the constituent species *P. homarus*, *P. ornatus* and *P. regius* contribute 4522 t (2008–2012) with an increase in production by

Table 7.5 World ranking of spiny lobster-producing countries (FAO 2016 data) 2008–2014

Country	Landings (t)	Rank
Australia	11,392	1
Indonesia	10,582	2
Bahamas	7807	3
Brazil	6923	4
Cuba	4718	5
Nicaragua	4235	6
Nigeria	3912	7
Mexico	3199	8
South Africa	2701	9
New Zealand	2739	10
Honduras	2630	11
USA	2518	12
Dominican	1751	13
India©	1568	14
Japan	1253	15
Zanzibar	1095	16
Korea	1042	17
Pakistan	990	18
Malaysia	834	19

©CMFRI data

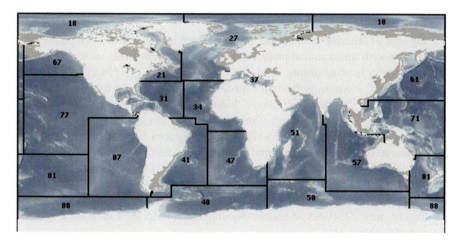

Fig. 7.1 FAO Major Fishing areas (Fisheries and Aquaculture Department, FAO, Rome)

2015 (8608 t). The highest productive regions in the southern latitude are fishing areas 47, 51, 57 and 81. The various cold water species under the genera *Jasus* and *Sagmariasus* (*J. paulensis, J. edwardsii, J. lalandi, S. verreauxi*) are landed by the countries South Africa, Australia and New Zealand (fishing areas 47, 57 and 81). The fishing area 57 (Australia and Indonesia) lands an average 13,485 t (2008–2012) with an increase in landing during 2014 (14,017 t) and 2015 (15,029 t). The

Table 7.6 Estimated annual lobster production (t) by FAO fishing areas (FAO 2016)

FAO fishing area	Species	Mean (2008–2012)	2014	2015
61	*Panulirus longipes*	2450	1906	1730
51	*P. delagoae, P. ornatus, P. homarus homarus*	3483	6751	7032
71	*P. ornatus, P. h. homarus, T. orientalis*	8688	11,184	15,828
57	*P. cygnus, J. edwardsii, P. h. homarus*	13,485	14,017	15,929
31	*P. argus*	27,221	27,565	31,342
41	*P. meripurpuratus*	7003	6787	6100
77	*P. gracilis, P. interruptus, P. argus*	3027	4326	4242
34	*P. regius, P. h. homarus, P. ornatus*	4522	6312	8608
47	*Jasus lalandii, J. paulensis, P. gilchristi*	3271	3759	2765
81	*J. edwardsii, S. verreauxi*	2705	3723	3803
87	*J. frontalis*	79	90	91
27	*Palinurus elephas, N. norvegicus, H. gammarus*	66,206	57,256	51,347
37	*P. elephas, N. norvegicus*	5250	3607	4806
34	*P. mauritanicus, H. gammarus, Palinurus* sp., *Panulirus* spp.	5265	6312	8608
21	*H. americanus*	116,987	159,814	157,064

fishing area 71, sharing both northern and southern hemispheres, is a high production region with mean annual production of 8688 t (2008–2012) with a phenomenal increase in landing during 2014 (11,184 t) and 2015 (15,828 t). The tropical spiny lobster, *P. ornatus*, is the major species contributing to the total landing from fishing areas 71 (Australia, Vietnam, Indonesia, Papua New Guinea and New Caledonia) and 51 (481 t) (Kenya, Tanzania). *Panulirus cygnus* is an endemic species fished from Western Australia (7279 t) (fishing area 57). *P. homarus homarus* is the major contributor to the total landing from fishing area 51 (East Africa, Kenya, Somalia, Yemen, Oman, Iran), and *P. longipes* is a major fishery (2, 450 t) in fishing area 61 (Taiwan, Japan, Korea, China). *Panulirus interruptus, P. marginatus, P. penicillatus* and *P. inflatus* are the major species forming a fishery (3027 t) in fishing area 77 (USA, Mexico, Nicaragua, Honduras, Costa Rica and El Salvador) (Table 7.6).

7.4 Fishing Gears and Methods in India

Fishing gears of various kinds are used for lobster fishing in India. They differ from place to place in design, fabrication and mode of operation (Mohan Rajan et al. 1981). Some gears are specifically employed to catch lobsters, and in others, lobsters are incidentally caught while targeting other commercially important demersal groups. Some gears used a few decades ago are not currently used as the demand now is for live, intact lobsters. In some regions, traditional gears used by the fishermen during those earlier years are also used currently without any modifications. Mohan Rajan et al. (1981) provide a detailed account of various types of gears and fishing methods adopted by lobster fishermen worldwide. The various types of gears used for lobster fishing in India are traps, tangle nets (gillnets and trammel nets), anchor hooks, scoop nets, trawl nets, bully nets, stake nets, dol nets and umbrella nets, of which some of them are not currently in operation. Apart from these gears, lobsters are also caught by spearing and handpicking by divers.

7.4.1 Traps

Although there are different types of traps used for fishing, the traditional lobster traps are designed to exclusively catch lobsters. The traps employed for lobster fishing along the southwest coast of India from Vizhinjam (south of Trivandrum) to Kanyakumari are region-specific and are generally called 'Colachel type', the shape of which is described as 'heart-shaped' or 'arrow heads' by different authors (Mohan Rajan et al. 1981). The traps made of palmyrah stalk and fibres are fabricated by the local lobster fishermen themselves (Fig. 7.2). The trap size, fabrication and operation are explained in detail by Miyamoto and Shariff (1961). The operation of the traps is mostly confined to 8–12 m depth near rocky shores. Generally, the traps are baited with mussels, a food relished by the lobsters. The lobster traps are operated by fishermen who are skin divers. While skin diving for pearl and sacred chank

Fig. 7.2 Traditional lobster trap made of palmyrah stalk used along the southwest coast of India (Photo by A. Saleela, CMFRI)

fishing by Tuticorin fishermen have been known for centuries, the existence of a diving tradition to fish lobsters in Kanyakumari is not that well known. The spiny lobster *Panulirus homarus homarus* readily enters the traps, but not *P. ornatus* and *P. versicolor* (Presscot 1988). Average catch obtained at Colachel-Muttom in Kanyakumari district during 1959–1960 was about 2.4 lobsters/trap/day (Miyamoto and Shariff 1961). These traps have a short life span of 2–3 weeks, as they are made of biodegradable material and are subjected to mechanical damage due to which the catch is often lost. The lobster traps are still used by the lobster fishermen at Colachel and Enayam in Kanyakumari district, Tamil Nadu and Vizhinjam and Mulloor in south Kerala.

The Central Institute of Fisheries Technology, Cochin, developed improved durable lobster traps as a substitute for the traditional traps of short life span and low efficiency. The improved traps have a mild steel rod frame mounted with 25 mm square-welded mesh, plastic coated for corrosion prevention (Fig. 7.3). The size of each trap is 70X55X40 cm (Srinivasa Gopal and Edwin 2013). The trap is a single entry type with a trunk-shaped funnel of 350 mm in length located at one end (Mohan Rajan 1991). An escape gap of 150x35 mm is provided on one side, 10 cm above the base of the trap for juvenile lobsters to escape if trapped inside. The traps baited with mussels are kept near the mouth of rock crevices by a diver fisherman at dusk and hauled in the morning. The lobsters in a trap will be alive and intact and fetch better price than that caught by anchor hooks.

The major disadvantages of trap fishing are that fishermen can set a maximum of 10 or 15 traps only at a time. Further, only diver fisherman can operate the traps which needs skill and physical strength.

Fig. 7.3 Rectangular traps with single funnel-shaped entry in operation along the southwest coast of India (Photo by R. Thangaraja)

7.4.2 Dol Nets

Dol nets are stationary nets fixed at about 20–40 m depth within sight of the shore, which take advantage of the strong tidal currents. The details of the structure of the net, fabrication, fixing and their operation are given by Raje and Deshmukh (1989). Lobsters moving along with the strong current are caught in the bag net and collected by the fishermen daily. The net is mainly used for fishing Bombay duck (*Harpodon nehereus*) and non-penaeid prawns abundant along the north Maharashtra and Gujarat coasts. In Gujarat, 29% of the lobster catch is landed by the *dol nets*.

7.4.3 Bully Nets or 'Gadas'

The bully net or *Gadas* is mainly used by the artisanal lobster fishermen of Mumbai (Jones 1965). It is a small hoop net of about 45 cm in diameter and depth with 3 cm mesh. The hoop is fastened at right angles to a long pole. A pole carrying a stiff wire probe is used to force the lobsters hiding in crevices to come out, and they are caught by the hoop net (Smith 1958). The net is operated from either a non-mechanised boat or from outboard engine boats in inshore areas.

7.4.4 Scoop Nets

In India, fishing by scoop nets was most prevalent along the Kanyakumari coast during the 1950s and 1960s when the lobster population was dense in the fishing grounds in nearshore rocky areas (Miyamoto and Shariff 1961; George 1967a). Here, a lure line either baited or non-baited is lowered to the area populated by lobsters and slowly pulled up until the lobsters follow the lure line and reach very close to the fishing craft. Once they are very close to the surface, the lobster is scooped up by the net. The fabrication and operation of the scoop net are explained in detail by Miyamoto and Shariff (1961). This gear is no longer used by the lobster fishermen of Kanyakumari district.

7.4.5 Anchor Hooks

Lobster fishing by anchor hooks was mainly carried out by the Muttom-Colachel lobster fishermen of Kanyakumari district (George 1967a) during the 1950s and early 1960s, the details of which are explained by Miyamoto and Shariff (1961). The hook has a shape of a grapnel anchor with six sharp arms of 10–12 cm made of cast iron. The anchor hook with mussel as bait is slowly lowered until it reaches the bottom. When the bite of the lobster on the mussel is felt, the anchor hook is pulled up with great force so that the lobster is hooked by its fleshy lower abdominal part.

The lobsters caught are wounded or mutilated and fetch low price. The fishermen discontinued this method of fishing when the demand was for intact lobsters on the export market.

7.4.6 Bottom-Set Gillnets

The bottom-set gillnets are considered as an effective gear to exploit demersal resources from a commercial point of view (Melville-Smith et al. 1999). Tangle nets or gillnets are used for lobster fishing by Gujarat, Maharashtra, Kerala, Tamil Nadu and Andhra Pradesh fishermen to catch lobsters inhabiting the shallow inshore rocky areas. During the 1950s and early 1960s, fishermen were using gillnets made of hemp twine for lobster fishing (George 1967a). The nets used at Muttom, Kanyakumari district of Tamil Nadu, were locally called *Kantativala*. Each piece of net has a length of 42′ (12.8 m), a breadth of 12′ (3.7 m) and a mesh size of 8 cm. One unit consists of 8–12 pieces of such nets (Mohan Rajan et al. 1981). In seventies, the hemp twine was replaced by cotton threads, and in recent times polyethylene monofilament yarn is widely used. The fishermen usually rerig the old webbing to use as lobster fishing gillnets. Five to nine floats are tied to each piece of the head rope and lead or stone sinkers at the bottom line to keep the net like a wall. The nets are lowered and set at the bottom with a marker float and a long buoy line is attached at the beginning of the first piece and at the end of the last piece. The nets are set at dusk in such a way that the entire net circles a rocky patch. Hauling of the net is done at dawn and after removing the catch, the net is reset again in the same place. The polyethylene monofilament nets currently used for lobster fishing vary in sizes, but usually each piece of the net is 40′ long and 12′ wide (12.8X3.7 m) with a mesh size of 7.5 to 8.0 cm. Eight to 12 such pieces are joined together to form a unit (Saleela, personal communication).

In Tuticorin, fishermen uses a modified bottom-set gillnet, 'plastic valai' instead of the 'monofilament nylon valai' with larger mesh size (130 mm instead of 85 mm), which could catch larger-size lobsters (Manickaraja 2004). These nets with a length of 160 m instead of the usual 90 m are operated at a depth of 60 m (usual operation at depths of 4–6 m) from boats with inboard engines of 10 to 15 HP.

Experimental lobster fishing was carried out during 1958–1959 and 1959–1960 at Quilon, Varkala and Vizhinjam fishing grounds in south Kerala and Colachel-Muttom grounds on the western coast of Kanyakumari district to test the operational efficiency of an improved form of bottom-set gillnet, which was found to be successful and adopted by the fishermen of these regions (Balasubramanyan et al. 1960, 1961).

Gillnets are operated by non-mechanised boats (very few now) or by an outboard engine boat or by a mechanised boat. The percentage use of these boats for lobster fishing differs within each state. For example, in Tamil Nadu, 90% of the lobster catch by gillnets is by outboard gillnetters (OBGN), whereas in Maharashtra, 88% of the catch by gillnets is landed by mechanised gillnetters (MGN). In Gujarat, 46% of the gillnet catch is landed by OBGN and 24% is by MGN. The motorised or mechanised boats help the fishermen to reach the fishing grounds faster and return with the catch quicker, thereby saving a lot of time. The percentage of catch landed by non-mechanised boats are negligible in all the states, except in Kerala, where fishermen in some landing centres still use non-mechanised crafts and 80% of the gillnet catch is landed by these boats.

7.4.7 Trammel Nets

The trammel nets (multifilament) and disco nets (monofilament) are also gillnets. Trammel nets (multifilament) have an inner and outer mesh size of 120 mm and a middle layer with 50 mm mesh size, whereas disco nets have a 75 mm mesh size (monofilament). The retention period for trammel net varies from 8 to 12 hours, but in the case of disco net, it varies from 6 to 8 hours (Murugan amd Durgekar 2008). Though not widely used, Tamil Nadu fishermen in Kanyakumari district and Chennai were using the net for lobster fishing in certain areas. The trammel net (*Kalralvalai* in Tamil) with its three-layered mesh is a nonselective gear and catches lobsters of all sizes, and juveniles and undersized lobsters form a major percentage of the catch (Fig. 7.4). Radhakrishnan and Vijayakumaran (2003) reported that 50% of the lobster catch by trammel net at Kovalam, south of Chennai, consists of juveniles, compared to 25% by the gillnets.

7.4.8 Cast Net

George (1967a) reports two types of cast net operated by lobster fishermen of Tikkoti area (north of Calicut, Kerala) for lobster fishing in nearshore rocky areas. These cast nets, locally called 'Muruvala' and 'Kara vala', are made of hemp twines and operated during daytime at a depth of 2 to 3 fathoms (3.7 to 5.5 m) from a

Fig. 7.4 Trammel net used for fishing lobsters (Photo by R. Thangaraja)

dug-out canoe. The only difference between these two types of cast nets is the mesh size, which is 4.5 to 5 cm in the former and 4 cm in the latter. These nets are not currently used for lobster fishing at this part of the Indian coast.

7.4.9 Stake Net

The stake net, which takes the advantage of high tidal amplitude and strong currents, is a stationary net fixed either in backwaters or in the inshore sea to catch fishes, prawns or lobsters moving along with the current. In Bhavnagar, Gujarat, long and short stake nets fixed in intertidal areas, locally called 'Bandhans' or 'Wada net', catch juvenile lobsters, which are used for pit culture. The net made of synthetic twines has a length of 225–400 m and a mesh size of 1.5–2.5 cm (Philipose 1994). The lobsters caught in the net are collected during low tide in live condition.

7.4.10 Umbrella Nets

The umbrella nets are circular-shaped nets in which some bait such as ribbonfish or 'Bombay duck' fish is kept to lure the lobsters and lowered to reef patches in near-shore areas. This type of net is used in Gulf of Kachch and Bhavnagar zones of Gujarat (Kizhakudan and Patel 2009).

7.4.11 Gunjajal (Bag Net)

'Gunjajal' is used by traditional fishermen in Kachchh and Bhavnagar areas of Gujarat to catch lobsters. The fishermen take the advantage of the season, the strong tidal currents and the lunar periodicity to catch lobsters.

7.4.12 Shore Seine

In north Andaman Islands, lobsters are also incidentally caught in shore seines, which are mainly operated for catching fin fishes (Kumar et al. 2010).

7.4.13 Trawl Nets

Trawling is the most effective gear to catch lobsters inhabiting soft substrates. The mud spiny lobster *P. polyphagus* and the slipper lobster *T. unimaculatus* are mainly landed by bottom trawlers. The deep-sea lobsters are exclusively landed by trawl nets. There is no specific trawl net for lobster fishing and they are often caught as a bycatch in fish or shrimp trawl nets. The rock-dwelling species such as *P. homarus homarus* and *P. ornatus* are either incidentally caught by trawl nets when they leave their rocky habitat for foraging during night or during the spawning migration (Moore and MacFarlane 1984). In Tamil Nadu, major share of the lobster catch of these species is by trawl nets. Since trawl net is a nonselective gear, lobsters of all sizes are caught, which makes it difficult to enforce size restriction on fishing. The trawl nets are operated by mechanised boats, which operate on a single-day basis and also engage in multiday fishing for 10–15 days.

7.4.14 Handpicking

Handpicking of lobsters from inshore rocky areas is commonly practised along the southwest and southeast coast of India, especially during certain seasons when the seawater clarity is good. The lobsters caught by hand may be either intact or would have lost their antennae or the legs. In Tuticorin, skin divers catch *P. ornatus* and *P. homarus* during November/December to March/April, if they find them hiding in the isolated rocks near the chank (*Xancus pyrum*) fishing grounds (Rajamani and Manickaraja 1991). At Vizhinjam and in the north Andaman Islands, fishermen use torch lights to locate the lobsters hiding in rocky crevices (Saleela, personal communication; Jha et al. 2007). Handpicking of lobsters during low tide from shallow reefs is also practised by fishermen in the Gulf of Kachchh and Bhavnagar in Gujarat (Kizhakudan and Patel 2009).

7.4.15 Spearing

Spearing is a very common method of lobster fishing in many Pacific Island nations. In India, this method of fishing is practised by lobster fishermen in Andaman and Nicobar Islands to catch lobsters from inshore rocky areas, especially during low tide. Nearly 75% of the lobster catch in Andaman Islands is by spearing (locally called bow and arrow method). This method of lobster fishing was mainly practised by fishermen from south Andaman Islands (Kumar et al. 2010), Gulf of Kachchh and Gulf of Khambhat in Gujarat during earlier periods, but not now, as these lobsters will be mutilated and not accepted by the export trade (Kizhakudan and Patel 2009).

7.5 Fisheries Production in India

Commercial fishing of lobsters in India began from 1950 onwards. The lobster production estimates are based on the landing data collected by the Central Marine Fisheries Research Institute, Cochin, which is available from 1968 onwards. The estimated annual average landing of lobsters from capture fisheries (1968–2016) fluctuated between a minimum of 679 t in 1980 and a maximum of 4082 t in 1985 with an average annual catch of 1921 t. The annual landing was 2003 t in 2014 and 2976 t in 2015, showing a remarkable increase in production over the previous years (CMFRI Annual Report, 2016–2017). The landing shows wide fluctuation year to

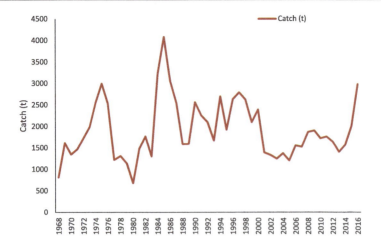

Fig. 7.5 Annual lobster landing in India (1968–2015)

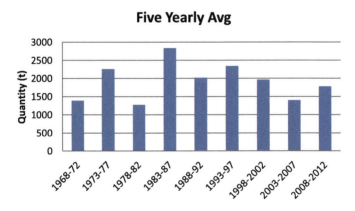

Fig. 7.6 Five yearly average landing of lobsters in India

year. The production increased from 812 t in 1968 to a peak of 4082 t in 1985 with an unprecedented fall in landing in 1980 (679 t) (Fig. 7.5). The average landing during the following 15 years (1986–2000) was 2229 t and 1534 t during 2001 to 2014. The decadal trend shows an increase in annual average landing from 1823 t during 1968–1977 to 2057 t (1978–1987) and 2179 t (1988–1997) followed by a decline (1671 t) during 1998–2007. The annual average production during the following 9 years (2008–2016) was significantly different from the previous decade with a significant increase in production of 200 t (1, 871 t). The annual lobster catch appears to be cyclic in nature with a period of low catch followed by a high catch in every 5-year period after which the same pattern occurs (Fig. 7.6). Such variations have been reported in lobster fisheries across the globe and were correlated with several environmental factors (Dow 1977; Campbell et al. 1991; Fogarty 1986; Harris et al. 1988; Polovina and Mitchum 1992; Phillips et al. 1994).

7.6 State-wise Production

The states along the west coast of India bordering the Arabian Sea are Gujarat, Maharashtra, Goa, Karnataka and Kerala, and on the east coast, the states bordering the Bay of Bengal are Tamil Nadu, Andhra Pradesh, Odisha and West Bengal. Although lobsters are distributed almost throughout the coast, major landing centres are on the northwest, northeast and southern Indian coast (Fig. 7.7) The states contributing significantly to the all India production are Gujarat, Maharashtra, Kerala, Tamil Nadu and West Bengal. The three east coast states which were not contributing much to the production (West Bengal, Odisha and Andhra Pradesh) showed an increase in landings from 2005 onwards. On the contrary, drastic decline in landings was visible in Kerala from 2010 onwards (Table 7.7). The annual landing in the states shows wide year-to-year fluctuations. The state-wise annual average landing of lobsters (1989–2014) shows that maximum landing is in Gujarat (726 t) forming 38% of the total landing (1908 t), followed by Maharashtra with a catch of 483 t (25%). These two states together from the northwest coast contributed 63% to the total annual landing

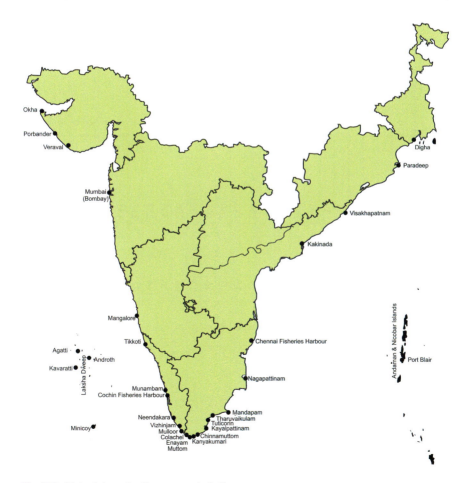

Fig. 7.7 Major lobster landing centres in India

Table 7.7 Annual lobster landings in states of India (1989–2012)

	West Bengal	Odisha	Andhra Pradesh	Tamilnadu	Pondicherry	Kerala	Karnataka	Goa	Maharashtra	Gujarat	Total
1989	0	1	1	164	9	74	2	6	386	946	1589
1990	0	0	4	365	11	123		5	787	538	1833
1991	0	0	10	362	20	496	2	4	735	924	2553
1992	0	0	9	502	3	206	0	1	453	920	2094
1993	0	0	3	411	4	40	13	93	237	958	1759
1994	1	0	4	559	11	443	0	4	405	1262	2689
1995	11	2	5	294	0	97	0	0	288	1225	1922
1996	0	0	5	252	0	112	0	0	1132	1130	2631
1997	6	0	0	346	0	265	47	0	818	1306	2788
1998	89	0	12	998	0	64	0	0	442	1090	2695
1999	25	9	24	254	0	513	2	0	291	976	2094
2000	0	0	13	142	0	535	49	1	611	1080	2431
2001	0	0	3	160	7	264	46	0	506	403	1389
2002	29	0	16	195	0	395	78	0	402	217	1332
2003	21	0	13	202	0	386	55	1	385	182	1245
2004	52	0	38	226	5	264	2	0	599	185	1371
2005	27	102	26	305	0	45	6	0	417	273	1201
2006	17	3	114	257	0	163	10	0	665	322	1551
2007	97	20	21	217	6	149	1	0	684	422	1617
2008	190	8	55	433	2	224	0	1	382	699	1994
2009	372	42	128	361	3	159	0	0	333	504	1902
2010	302	15	145	188	6	52	0	0	249	764	1720
2011	789	9	121	194	0	99	0	0	175	373	1761
2012	162	15	281	205	8	38	2	0	199	729	1640
2013	222	26	49	299	8	85	0	15	297	410	1411
2014	10	68	24	206	14	99	0	2	420	737	1580

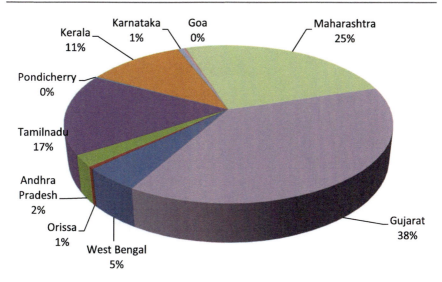

Fig. 7.8 Mean percentage of total lobster landings from states (1989–2012)

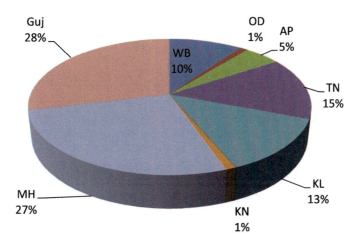

Fig. 7.9 Mean percentage of total landings from states (2007–2014) (Guj- Gujarat, MH-Maharashtra, TN-Tamilnadu, KL-Kerala, AP-Andhrapradesh,OD-Odisha, WB-West Bengal)

of lobsters in India (Fig. 7.8). Tamil Nadu contributed 17% (316 t), and Kerala with an annual landing of 217 t formed 11% of the total production. West Bengal recorded an average catch of 91 t (7%) and during 2008–2012, landed an average 316 t. Andhra Pradesh with a landing of 44 t (2%) also has a small contribution to the total production. The state of Karnataka landed an average 13 t and Odisha 9 t annually during this period. The annual average total landing shows a lower figure (1626 t) during 2000–2012. The percentage contribution by the states during 2007–2014 shows a different scenario compared to that during 1989–2012, with the state of Gujarat landing 28% of the total annual landing followed by Maharashtra, 27%; Tamil Nadu, 15%; Kerala, 13%; West Bengal, 10%; and Andhra Pradesh, 5% (Fig. 7.9).

Fig. 7.10 Landing of *Panulirus polyphagus* at New Ferry Wharf, Mumbai (Photo by V. D. Deshmukh, CMFRI)

7.7 Species-Wise Production

The lobster fishery in India mainly consists of five species of spiny lobsters, *Panulirus polyphagus, P. homarus, P. ornatus, P. versicolor* and *Puerulus sewelli*, and the slipper lobster *Thenus unimaculatus*. The spiny lobster *P. polyphagus* is the principal species (Fig. 7.10) followed by the slipper lobster *T. unimaculatus* (Fig. 7.11) and the deep-sea lobster *P. sewelli* (2%) (Fig. 7.12). The mud spiny lobster, *P. polyphagus*, forms 61% of the national annual production, and *T. unimaculatus* constitutes 26%. *P. homarus homarus* and *P. ornatus* contribute 8% and 2%, respectively (Fig. 7.13). The species that are commercially not significant are *P. penicillatus* and *P. longipes longipes*. *P. versicolor* is landed in small volumes along the southern coast.

In Gujarat, *Panulirus polyphagus* constitutes 67% of the total landings (2007–2012), followed by *T. unimaculatus* (33%). In Maharashtra, the fishery is almost completely dominated by *P. polyphagus* (94.2%), and the rest is contributed by *P. ornatus* (5%) and *T. unimaculatus* (0.7%). The species composition of the fishery in these northwestern states has changed over the years. During 1980–1985, *P. polyphagus* formed only 45% of the total landing at Veraval and the rest was contributed by *T. unimaculatus*. In Mumbai, though *P. polyphagus* was the major species (54%) in the fishery during 1978–1985, *T. unimaculatus* contributed the balance (Kagwade et al. 1991). However, the collapse of fishery for *T. unimaculatus* in 1995 changed the species composition during the rest of the period.

In Tamil Nadu, *P. homarus* (40%) and *T. unimaculatus* (38%) are the two major species representing the commercial fishery. The other species supporting the fishery

Fig. 7.11 *Thenus unimaculatus* from Chennai Fisheries Harbour (Photo by E. V. Radhakrishnan)

Fig. 7.12 The deep sea lobster *Puerulus sewelli* from Cochin Fisheries Harbour (Photo by E. V. Radhakrishnan)

are *P. polyphagus* (8.9%), *P. ornatus* (6.7%) and *P. versicolor* (5.6%). *Puerulus sewelli* forms less than 1% of the total annual catch. The lobster fishery in Kerala is mainly composed of three species, of which the slipper lobster *T. unimaculatus* is the dominant species forming 47.4% of the total annual catch, followed by *P. homarus* (23.7%) and *P. sewelli* (22.2%). The contribution of *P. ornatus* is just 6%.

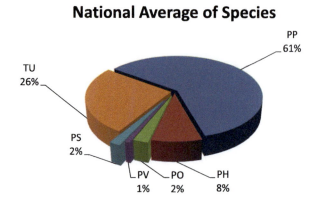

Fig. 7.13 National average of commercially important species of lobsters. PP- *Panulirus polyphagus*, TU- *Thenus unimaculatus*, PS- *Puerulus sewelli*, PH- *Panulirus homarus*, PO- *Panulirus ornatus*, PV- *Panulirus versicolor*

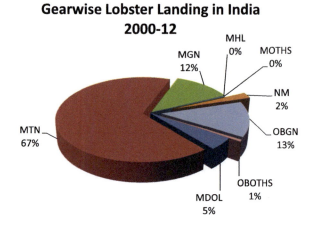

Fig. 7.14 Landing of lobsters by different gears in India. MTN-Mechanised trawl net, MGN- Mechanised gillnet, MDOL- Mechanised dolnet, NM- Non-mechanised gillnet, OBGN-Outboard gillnet

In West Bengal, 99.4% of the landing is represented by *P. polyphagus* and the rest by *T. unimaculatus*. In Andhra Pradesh, *T. unimaculatus* (71.3%), *P. polyphagus* (26.9%) and *P. homarus* (1.8%) constituted the bulk of the landing. In Odisha, *T. unimaculatus* and *P. polyphagus* together formed 95% of the total landing.

7.8 Gear-wise Production

Lobster fishing is carried out by a variety of gears which differs from state to state. The multiday and single-day fishing trawlers together land 67% of the total annual catch of lobsters, whereas 28% of the catch is landed by gillnets (based on 2000–2012 landing data). The catch brought by the trawlers is mostly in chilled condition. The gillnets are operated from different types of fishing crafts, of which 13% of the catch is landed by the outboard gillnetters and 12% by the mechanised (inboard) gillnetters. Landing by the non-mechanised crafts is just 2% and by the dol nets, 5% (Fig. 7.14).

In Gujarat, an average 51.5% of the lobster catch (471 t) is landed by trawlers and 34.4% by gillnetters (2000–2012). Among the gillnetters, 65.4% of the catch is landed by outboard gillnetters (OBGN) and the rest by mechanised gillnetters

(MGN). Lobster landing by the *dol* nets is, on an average, 14.2% of the total catch. In Maharashtra, trawlers land an average 62.6% of the total catch (431 t) and the gillnetters, 35.7%. Among the gillnetters 92.3% of the catch is landed by MGN. *Dol* nets land only 1.7% of the total catch. In Tamil Nadu, trawlers land 54.9% and gillnetters, the rest of the total catch (237 t). The OBGN landed 90.5% of the gillnet catch and MGN, 6%. In Kerala, 91% of the lobster catch (213 t) is by trawlers and the rest by gillnetters, of which almost the entire catch is landed by non-mechanised boats. In Andhra Pradesh, 71.5% of the lobster catch (75 t) is by trawlers, and in West Bengal, trawlers land the entire catch (158 t).

7.9 Lobster Fisheries of Gujarat

7.9.1 Introduction

The state of Gujarat bordering the Arabian Sea on the northwest coast of India has a coastline of 1600 km, which is the longest among the maritime states of the country. Three zones are recognised along coastal Gujarat – south Gujarat, Saurashtra and Kachchh. The continental shelf is relatively wider along south Gujarat coast compared to Saurashtra coast. Coastal Gujarat is endowed with diverse marine habitats such as coral reefs, wetlands, salt marshes, lagoons and mangroves. There are 15 coastal districts, of which Valsad is the southernmost and Kachchh, the northernmost. The state is influenced by the southwest monsoon, which occurs from June to September. The coastal waters are influenced by the West India Coastal Current (WICC) which flows from north to south during June to September. The current reverses during the northeast monsoon, and the East India Coastal Current (EICC) flows from south to north during November to February (Shetye 1998). The coastal waters of Gujarat are highly productive due to the influence of monsoon-modulated currents and upward Ekman pumping promoted by positive wind stress curl induced by the Findlater jet (Qasim 1982; Madhupratap et al. 1996; Luis and Kawamura 2004). Fishing is a perennial activity along this coast, except for sectoral breaks during the seasonal trawl ban period.

Since the initiation of commercial trawling along the Gujarat coast in the 1960s and the introduction of shrimp trawls, several significant changes have taken place in this operation, which include diversion of fishing activities towards exploitation of cephalopods; migration of trawl operations into deeper waters in the depth range of 80–100 m for exploiting cephalopods and demersal fishes such as threadfin breams, lizardfishes and bull's eye; conversion of short-trip operations (1–2 days) to long-trip operations (5–8 days); reduction in cod-end mesh size; and multiple-gear operation by trawl boats.

The thriving fishery for lobsters along the Gujarat coast may be due to the natural conditions that provide lobsters with suitable habitat on the continental shelf and excellent conditions for lobster feeding and growth. The chief area of commercial fish production is the Saurashtra coast. The two major fishing harbours are Veraval and Porbandar from where the majority of trawl boats operate. Lobster landings are maximum at Veraval and the fishing harbour of Jakhau in Kachchh. There are several smaller fishing harbours and several fishing villages, where small quantities of lobsters are landed. Lobster fishing probably received an impetus with export demand and the establishment of freezing plants.

7.9.2 History of Lobster Fisheries

Commonly called 'Titan' or 'Jinga', spiny lobsters were exploited in Gujarat even before the introduction of mechanisation in the fishing sector. However, lobster fishing was almost restricted to select communities of traditional fishermen in remote pockets along the coast. Lobster fishing was then a sustenance fishery of low intensity, and hence production statistics seldom reflect the existence of these fisheries. Interaction with local fishermen along the Saurashtra and Kachchh coasts revealed that lobsters were fished and locally consumed during lean fishery periods. Lobster fishing was also treated as a sport in earlier days, when the sizes of the lobster caught were much bigger than what is now available to the fishery. Traditional knowledge imparted down generations over the years within the fishing communities are still applied by the non-mechanised and artisanal sector to exploit lobsters.

In 1962, with the export of frozen lobster tails to the USA, lobsters became an important component of the fishery. The lobster landings of the state gradually became significant to the extent that it began to appear as a separate item in the CMFRI Annual Report from 1976 onwards; earlier it used to be included under the group *other crustaceans*. The Regional Centre of CMFRI at Veraval initiated intensive studies on fishery and biology of lobsters from 1975 onwards. However, scientific publications regarding the early years of lobster fishing along the Gujarat coast are limited.

Kagwade et al. (1991) provided a comprehensive report on the lobster resources along the Indian coast (1975–1984) and concluded that maximum landings of lobsters among the Indian states were by Gujarat, contributing 46% to the total lobster landings of the country. Bottom trawling by FORV *Sagar Sampada* indicated the occurrence of good abundance (2–85 kg/hr) of *Thenus orientalis* (= *T. unimaculatus*) between 40 and 75 m depth off Veraval to Dwaraka (Suseelan et al. 1990). The status of crustacean resources in general and lobsters in particular along the Indian coast, a comprehensive description of the fishery and biology of major species of lobsters and the various issues in management have been published by Suseelan and Pillai (1993), Radhakrishnan and Vijayakumaran (2003), Radhakrishnan et al. (2005, 2007a, b), Radhakrishnan and Thangaraja (2008), Kizhakudan and Patel (2010) and Kizhakudan (2014). A list of crustacean fauna of Gujarat coast, including lobsters, has been published by Trivedi et al. (2015).

7.9.3 Species Composition

The lobster fishery of Gujarat is predominantly supported by the mud spiny lobster *Panulirus polyphagus* and the slipper lobster *Thenus unimaculatus*. The slipper lobsters are relatively new entrants into the fishery, and until about four decades ago,

Fig. 7.15 Species composition of lobsters in Gujarat fishery. PP- *Panulirus polyphagus*, PH- *Panulirus homarus*, TU- *Thenus unimaculatus*

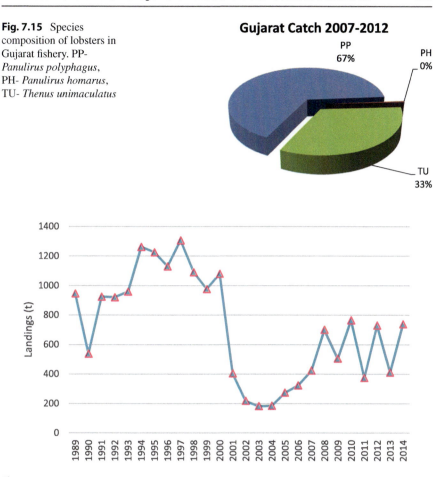

Fig. 7.16 Annual lobster landing in Gujarat (1989–2014)

P. polyphagus formed the mainstay of Gujarat's lobster trade forming 67% of the total landings followed by *T. unimaculatus* (33%) (Fig. 7.15). Other lobster species that occur in the fishery are the rock spiny lobster, *P. homarus homarus*, and the painted spiny lobster, *P. versicolor*. The predominant gears used for lobster fishing are trawl net, trap net, *dol* net, stake net and umbrella net. The major landing centres are Veraval, Porbandar, Mangrol, Jakhau, Okha and Jaffrabad.

7.9.4 Fishery

The major lobster landing centres in Gujarat are Veraval, Mangrol, Porbandar and Kachchh (Fig. 7.7). Gujarat stands first in terms of lobster landings in India with an annual average landing of 714 t (1989–2014) (Fig. 7.16). The annual average landing during 1975–1984 was 813 t forming 46% of the total annual average catch

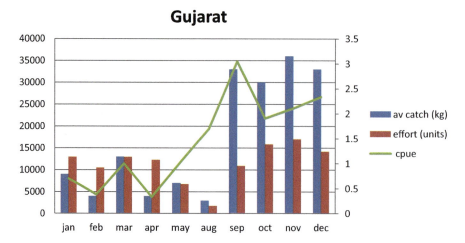

Fig. 7.17 Monthly average landing, effort and CPUE of lobsters in Gujarat

(Kagwade et al. 1991). The contribution of lobsters to the crustacean landings has been marginal and has decreased from 8.2% in 1971–1975 to less than 1% in 1996–2000, clearly indicating the fishing pressure on this resource.

Landings of the spiny lobster *P. polyphagus* decreased considerably, and by 2000, trawl landings of lobsters were dominated by the slipper lobster, *T. unimaculatus* (Kizhakudan and Thumber 2003). The catch shows high annual fluctuation with the lowest landings in 2003 (182 t) and the highest in 1997 (1306 t) (Fig. 7.16). The period 1994–1998 may be considered as the golden period for lobster landings of the state with an average annual landing of 1203 t. The landing was lowest during 2002–2005 with an annual average landing of 212 t. The decadal average landing shows vast differences, with the 1990–1999 period recording an annual average landing of 1033 t and the 2000–2009 period an average landing of 451 t, which is just 44% of the landing during the previous decade.

Lobsters appear in the trawl catches almost throughout the year except during July. However, from 2010 onwards there was no landing during June and July. The lobsters start appearing in the trawl catch from September onwards and peak landing is during October to December (Fig. 7.17). Kizhakudan and Thumber (2003) identified the peak fishing period of spiny and slipper lobsters along the coast as October to January, coinciding with the peak breeding periods of December to March for spiny lobsters and October to December for slipper lobsters.

7.9.4.1 Fishing Methods

Different fishing methods are used to capture lobsters along the Gujarat coast. The lobster fishery of Gujarat is essentially a trawl-based fishery where they appear as a bycatch in shrimp trawls. The single-day and multiday trawlers together land an average 70% (514 t) of the total annual landing (736 t), and the bottom-set gillnets together contribute 23% (168 t) to the total landing (1989–2012) (Fig. 7.18). The

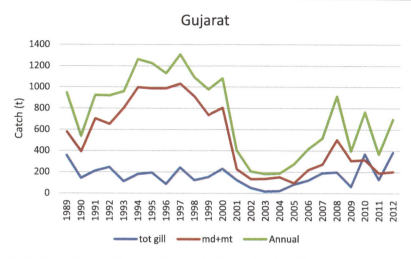

Fig. 7.18 Gear-wise annual landing of lobsters in Gujarat. Tot gill- Gillnets, Md+mt- mechanized trawlers

rest is landed by mechanised *dol* net boats. The artisanal fishermen use a traditional bully net called 'gudwa' for lobster fishing in the shallow waters. However, a change in landing pattern by different gears was evident during 2007–2012, when the contribution by mechanised trawlers to the total landing declined from 70 to 44%, whereas percentage landings by the gillnets and the *dol* nets increased to 41 and 15%, respectively. Though the size composition in the gillnet landing is not available, the catch was likely to have had a higher percentage of juveniles, as the fishing used to be carried out in inshore nursery areas. Mean catch rate was 0.05 kg/hr. for trawlers, 0.08 for dol nets and 0.04 kg/hr. for gillnets (2007–2012) (Fig. 7.19a–c).

7.9.4.2 Fishing Communities

The fishing communities engaged in lobster fishery were the Machis and Muslim fishers and some tribal communities who were involved in coastal activities. The traditional fishing villages are usually situated near a reef-lined coast (Jaleshwar, Hirakot, Sutrapada, Muldwaraka, Dhamlej, Okha, Jakhau, Ummergaon, Mangrol Bara, Rupen, Veraval, Chorwad, Shil, Salaya, Dwaraka Bet, Belapur), and most of them are adjacent to river openings or bar mouths (Muldwaraka, Veraval, Dhamlej, Jaleshwar, Sutrapada, Chorwad, Navibandar, Shil, Mahuva, Bhavnagar, Diu, Mangrol Bara).

7.9.4.3 Traditional Knowledge in Lobster Fishing

Based on the habitat, the fishermen used to classify the lobster population into three groups – the calcite reef population, the muddy bottom population and the deeper sandy bottom population. The aggregations found in muddy bottoms were classified as the 'safed titan' or 'jinga' (white pale-grey) and the ones collected from the rock/

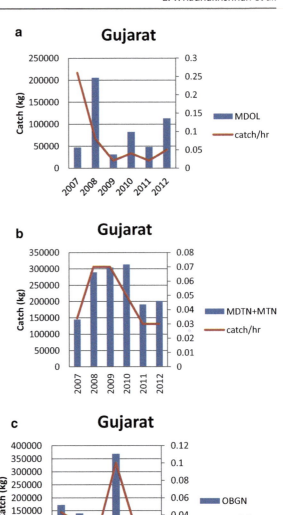

Fig. 7.19 (a) Lobster catch and CPUE in dolnet fishery in Gujarat. (b) Lobster catch and CPUE in trawl net in Gujarat. (c) Lobster catch and CPUE in gillnet in Gujarat

reef areas as the 'kalo titan' (black, dark green). The reef populations also had the 'pattovala titan' (*P. versicolor*) and 'kantavala titan' (*P. homarus*). One of the earliest methods of hunting lobsters was spear them out from hiding places on shallow water reefs, practised by the fishermen in the Gulf of Kachchh and Gulf of Khambhat. The lobsters would often be caught along with reef crabs (*Carcinus* sp.) and mud crabs (*Scylla* sp.). Although spearing is no longer used, handpicking is still practised in reef patches beyond Mundra (in Kachchh) and Mahuva (in Bhavnagar).

Adjacent to shallow water reefs, framed nets (circular nets or umbrella nets) were used to catch lobsters. The nets would be lowered with some fish (ribbonfish or Bombay duck) to lure the lobsters. These operations were carried out in the shallow water reef surfaces in the Kachch and Bhavnagar zones. Other gears predominantly used were the long and short tidal stake net and Gunjajal (bag net) in Kachch and Bhavnagar.

The fishermen had a thorough understanding of the behaviour of spiny lobster aggregations and used the knowledge to their advantage. For example, the fishermen of Jaleshwar and Sutrapada villages, while employing trap gillnet on the reefs, were aware that large spiny bodied 'Titan' assembled during Aug-Oct at the edge of the reefs for breeding, and these animals, when baited, are easily trapped. Another information that has been passed down the generations is that 'large-bodied mud spiny lobster (pale greenish coloured) is available more in muddy areas – the ones caught in the nearshore/reef areas are darker and usually 'juveniles'. The fishermen also relate lobster availability to season, lunar periodicity, water current, etc.

7.9.4.4 Lobster Fishery at Different Centres in Gujarat

The major fishing harbours where lobsters are landed by trawl net or by trawl net and gillnet are Veraval, Jakhau, Porbandar, Okha and Manupul, in descending order of importance. During the 10-year period from 1985–1986 to 1994–1995 (April to March), Veraval accounted for 35.4% of the total lobster landings among these five centres, followed by Jakhau (31.3%) and Porbandar (16%) (Fig. 7.20). Among the smaller fishing centres and villages where lobster fishing and landings occur, Rupan and Salaya contribute maximum to the landings (Fig. 7.21).

7.9.4.4.1 Fishery at Veraval

The estimated total annual average landing of lobsters at Veraval during 1989–2012 was 103 t. A decadal analysis revealed that the average landing during 1990–1999 was 187 t, contributing 18% to the total landing of the state (1033 t). A drastic decline in catches after 2000 resulted in the average landing during 2000–2009 falling to 32 t, contributing just 7% to the total landing. Earlier, Kagwade et al. (1991) reported that the average annual catch of lobsters at Veraval during 1980–1985 was 270 t at a rate of 3.02 kg/U. The average trawl catch and catch rate reduced to 32 t and 0.55 kg/U, respectively, by 2007–2012. *P. polyphagus* constituted 85% of the total catch and the rest by *T. unimaculatus* (Fig. 7.22).

Kagwade et al. (1991) reported that the average annual trawl catch of *P. polyphagus* for the period 1980–1985 was 121.7 t at a catch rate of 1.36 kg/U and the catch

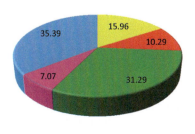

Fig. 7.20 Percentage contribution in lobster landings by major landing centres in Gujarat

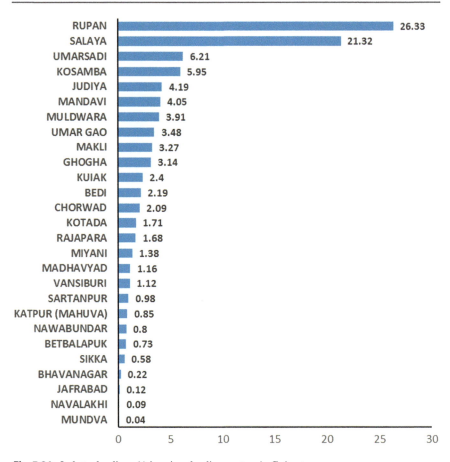

Fig. 7.21 Lobster landings (t) in minor landing centres in Gujarat

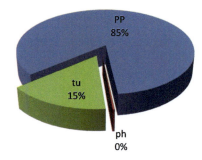

Fig. 7.22 Percentage composition of species landed in Veraval fishing harbour. PP-*Panulirus polyphagus*, tu-*Thenus unimaculatus*, ph-*Panulirus homarus homarus*

of *T. unimaculatus* was 148.3 t at a catch rate of 1.66 kg/U. The average landing of *P. polyphagus* for the period 1989–2012 was 45 t, forming 44% of the total landing, while *T. unimaculatus* formed 56% of the total landing, with an average landing of 58 t during the same period. The average catch of *P. polyphagus* for the period

2007–2012 improved marginally and stood at 48 t with a catch rate of 0.4 kg/U, whereas the catch of *T. unimaculatus* further declined to 9 t (15%) with a catch rate of 0.14 kg/U. The 10-year average catch of *P. polyphagus* declined drastically from 74 t in 1990–1999 to 16 t in 2000–2009 and that of *T. unimaculatus* from 113 t in 1990–1999 to 15 t in 2000–2009. Both species being a bycatch in trawls, species-specific fishing regulations were not in force.

Radhakrishnan et al. (2005), in describing the status of the lobster fishery of Veraval, reported that the annual mean carapace length of *P. polyphagus* decreased from 68.5 mm in males and 74.9 mm in females in 1997 to 41.5 mm and 47.9 mm, respectively, in 2001. This implied that the fishery was being subject to high level of exploitation and was dependent on the entry of new recruits to the fishery. This observation possibly also indicated a contracting size structure in the population as a whole because of overfishing and natural geographic variation in size due to the oceanographic and ecological conditions (Mohan 1997). The trawl catches were mostly comprised of adults, whereas the catch in gillnets and other traditional gears were more of juveniles, as these gears are operated on the shallow reefs near the coast. Peak landings of *P. polyphagus* also coincide with the breeding season (October–December), catching a large percentage of the breeding stock. Maximum landings by gillnets were during September to October.

7.9.4.4.2 Fishery at Porbandar

The average total landing of lobsters at Porbandar fishing harbour on the northern Gujarat during 2007–2012 was 24 t. *Panulirus polyphagus* constituted 62% of the total landing (15 t), while *T. unimaculatus* formed 37% (9 t) (Fig. 7.23). Small quantities of *P. homarus* (0.9%) and *P. ornatus* were landed occasionally. The major portion of the landing at Porbandar is by trawlers (95%) and the rest by gillnets. The average trawler catch and catch rates during 2007–2012 were 23 t and 1.09 kg/U, respectively.

7.9.4.4.3 Trawl Centres in Kachchh

Ghosh et al. (2008) reported heavy landing of *P. polyphagus* (900 t in 2 months) by trawlers and outboard mechanised gillnetters in September and October 2007 at Okha and Rupenbander. While the trawling grounds are off the Jakhau coast,

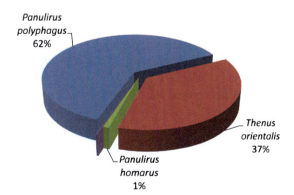

Fig. 7.23 Percentage composition of species landed in Porbandar, Gujarat

gillnetters operate in the rocky grounds off the Dwaraka-Okha coast. The depth of operation by the multiday fishing trawlers was 15–20 m and the gillnetters, 12–30 m. Unprecedented catch of spiny lobsters is not unusual though congregation of adult lobsters in shallow areas is quite intriguing. The lobsters landed measured an average 128–312 mm in total length with 26.5% of females carrying eggs, indicating a breeding congregation.

7.10 Lobster Fishery of Maharashtra

7.10.1 Introduction

The state of Maharashtra has a lengthy coastline stretching 720 km along the Arabian Sea and has one of the most important *large marine ecosystems*. The coast is influenced by the southwest monsoon and the West India Coastal Current (WICC), which flows southward during April–September. The current reverses during the northeast monsoon (East India Coastal Current) and starts flowing northward during November–February (Shetye 1998). The continental shelf is wide along the entire coastline with a muddy bottom. The five coastal districts are Sindhudurg, Ratnagiri, Raigad, Greater Mumbai and Thane, of which Greater Mumbai is important from the lobster fisheries point of view.

7.10.2 History of Lobster Fisheries

Information on lobster fisheries of Maharashtra goes back to the 1930s. Rai (1933) mentioned the occurrence of *Panulirus ornatus* and *P. fasciatus* (*P. polyphagus*) in commercial fish landings of Bombay Presidency. Chopra (1939) provides a description of the lobsters present in Bombay waters and states that the species of lobster along the east coast is *P. polyphagus*, while *P. ornatus* is common in the fishery of the Bombay (Mumbai) coast. Both these authors have misidentified *P. versicolor* as *P. ornatus* and have also reported it as the most common species, although *P. polyphagus* is the dominant species along the Bombay coast (Chhapgar and Deshmukh 1961, 1964, 1971). Despite records of lobsters in the fisheries of Bombay being available, there was no concerted effort to develop the fishery, probably because of the limited domestic demand (Kagwade 1993). Records of lobster landings from Maharashtra are available from 1978 onwards.

The fishery of lobsters along the Mumbai coast is a trawl-based fishery with lobsters appearing as a bycatch in shrimp trawls. There are no exclusive regulatory measures on fishing for lobsters. Other than some reports on the fishery and biology of lobsters from Mumbai, there have been few biological studies of the principal species until the late 1980s. The first-stage phyllosoma of *P. versicolor* in Bombay waters and other palinurid lobsters from Indian waters was described by Deshmukh (1968). Kagwade (1987a, b, 1988a, b) undertook a substantial biological study of *P. polyphagus* off Mumbai. The length frequency and the catch and effort data were analysed to derive estimates of age and growth. Kagwade (1993) estimated the stock of *P. polyphagus* off Mumbai and suggested certain management measures. Kagwade et al. (1991) published a detailed report of the magnitude of lobster resources of India based on the fishery data (1975–1984) collected from major lobster landing centres along the Indian coast by Fishery Resource Assessment Division of CMFRI.

The slipper lobster *T. unimaculatus* had contributed significantly to the lobster fishery of the Mumbai coast during the 1970s and 1980s. The lobster appeared in good quantities in the fish landings as a bycatch until the early 1990s. Kagwade and Kabli (1996a, b) present comprehensive accounts of the reproductive biology and age and growth of *T. unimaculatus* (*T. orientalis*) off Bombay. The morphometry of different body parts and their relationships have been estimated and conversion factors derived from these relationships. Deshmukh (2001) analysed the fishery data for *T. orientalis* and provides a critical account of the reasons for the collapse of slipper lobster fishery of the Mumbai coast.

7.10.3 Fishery

The lobster fishery of the Maharashtra coast is a year-round fishery, which has most important commercial and artisanal sectors. Commercial fishing of lobsters was presumed to have initiated off the Mumbai coast in the early 1960s. However, the statistical data on landings is available from 1978 onwards. The main fishing season is from September to March, though in some years good landings were reported in April also.

The estimated annual average landing of lobsters in Maharashtra was 473 t during 1989–2014 (Fig. 7.24). The catches during 2013 and 2014 were 297 t and 420 t, respectively. Between 1981 and 2014, landings varied between a minimum of 175 t in 2011 and a maximum of 2504 t in 1985, showing wide annual fluctuation in landings, which is a common phenomenon (Kagwade 1993). The first 10-year developmental stage of the fishery (1981–1990) had an annual average landing of 938 t, which was reduced to 534 t during the following decade (1991–2000), and then the fishery almost stabilised at an average landing of 465 t during the next 10-year cycle (2001–2010). During the 10-year period from 1975 to 1984, annual average landing was 484 t, forming 27.5% of the total annual all India landings (Kagwade et al. 1991). Maharashtra had the highest ever landing of lobsters in 1985 (2504 t).

7.10.3.1 Fishing Season and Methods
Fishing commences in August after the trawl ban and continues till May (Fig. 7.25). Lobsters are exploited by trawlers, gillnets and dol nets. The bulk of the landing is by mechanised trawlers with an average catch of 316 t (1989–2012), forming 65.7% of the total annual landings, followed by mechanised gillnets (143 t) contributing 29.8% to the total catch (Fig. 7.26). The *dol* nets land an average 4.6 t and outboard

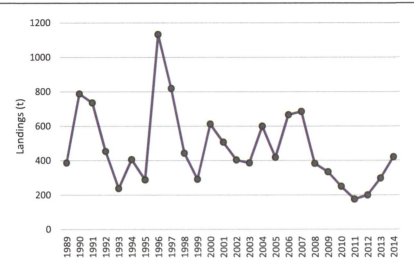

Fig. 7.24 Annual average lobster landings in Maharashtra

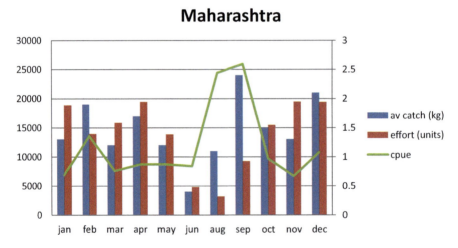

Fig. 7.25 Average monthly catch, effort and CPUE landed by trawl net in Maharashtra (2007-2012)

gillnets 9.4 t, and, from 2009 onwards, multiday trawlers account for the bulk of the landings (average 149 t). The average catch and catch rate (catch/hr) in trawl net during 2007–2012 were 209 t and 1.2 kg/hr, respectively. In gillnets, the catch and catch rates were 165 t and 0.9 kg/hr and in dol nets, 15 t and 0.04 kg/hr, respectively (Fig. 7.27a–c). In Maharashtra, the biggest two trawl landing centres are the Sassoon Docks and the New Ferry Wharf in Mumbai. The trawlers operating from Dahanu in the north and Ratnagiri in the south land the catch at these two landing centres (Deshmukh 2001).

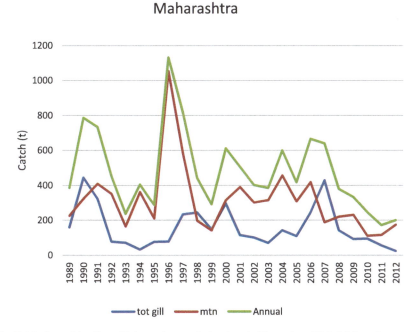

Fig. 7.26 Annual landing of lobsters by trawls (mtn) and gillnets (tot gill) in Maharashtra

7.10.3.2 Species Composition

The two major species constituting the fishery are the mud spiny lobster *Panulirus polyphagus* and the slipper lobster *Thenus unimaculatus*. During 1978–2012, *P. polyphagus* constituted 72% of the catch and the rest was by *T. unimaculatus* (Fig. 7.28). *P. ornatus* formed about 5% of the landings. The fishery of *T. unimaculatus* was collapsed in 1995 (Deshmukh 2001), and no recovery of the fishery is reported till now, except landing of some stray numbers. During 2007–2012, the fishery was almost comprised of *P. polyphagus* with *P. ornatus* contributing 5% to the total catch (Fig. 7.29).

7.10.3.3 Lobster Fishery in Mumbai

The Mumbai Research Centre of CMFRI has carried out detailed investigations on the biology, fishery, stock assessment and reproductive biology of spiny and slipper lobsters. The estimated annual average landing of lobsters at Mumbai during 1978–1985 was 402 t at the rate of 9.46 kg/U (Kagwade et al. 1991). The annual total landings increased from 200 t in 1978 to a peak of 1043 t in 1986 and recorded the lowest landing of 82 t in 1995 (Fig. 7.30). The estimated annual average landing during 1978 to 2012 was 250 t, of which spiny lobsters formed 72% (180 t). During the developmental stages of the fishery (1978–1987), the average landing of lobsters was 264 t of which 59% of the catch was *P. polyphagus* and the rest *T. unimaculatus*. In the following 10 years (1988–1997), annual average landing of lobsters was 132 t, and in the next succeeding 10-year period (1998–2007), the average annual

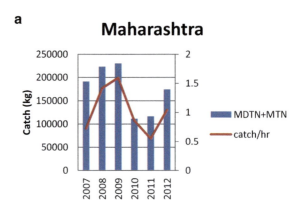
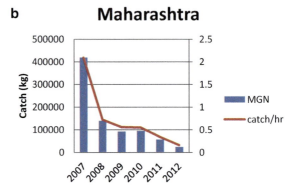
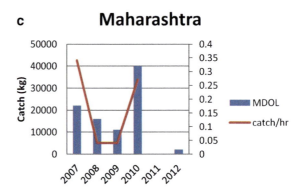

Fig. 7.27 (a) Catch and CPUE of lobsters landed by trawl net in Maharashtra (2007–2012). (b) Catch and CPUE of lobsters landed by Gillnets in Maharashtra (2007–2012). (c). Catch and CPUE of lobsters landed by dolnets in Maharashtra (2007–2012)

catch was marginally improved to 149 t. The slipper lobster *T. unimaculatus* disappeared from the fishery in 1995.

The lobster fishery of the Mumbai coast is predominantly a trawl-based fishery, with the mechanised trawlers landing 93% of the total catch (2007–2012). The gillnets contribute only 6.4% of the total catch. Gillnets are operated in the shallow waters and the catch mostly consists of juveniles which are recruiting into the fishery.

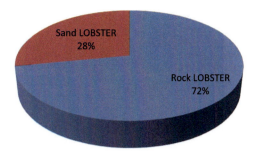

Fig. 7.28 Percentage composition of rock lobster (*Panulirus polyphagus*) and sand lobster (*Thenus unimaculatus*) landed in Maharashtra (1978–2012)

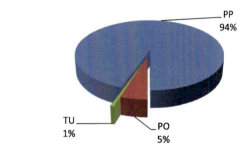

Fig. 7.29 Percent composition of lobsters landed in Maharashtra (2007–12). PP *Panulirus polyphagus*, TU *Thenus unimaculatus*, PO *Panulirus ornatus*

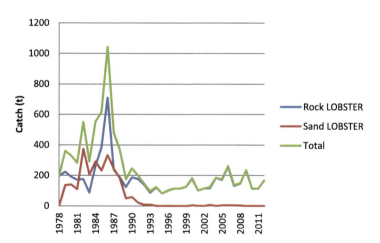

Fig. 7.30 Annual lobster landings in Mumbai

7.10.3.3.1 Species Composition and Biological Features of *Panulirus polyphagus*

Two species of lobsters are found in the fishery: the Mud spiny lobster, *P. polyphagus*, and the slipper lobster, *T. unimaculatus*. The relative contribution of the species in terms of the catch was 59% and 41%, respectively, until the collapse of the slipper lobster fishery in 1995. Kagwade et al. (1991) report average catch of *P. polyphagus* at Bombay (Mumbai) as 218 t with a catch rate of 5.11 kg/U during 1978–1985.

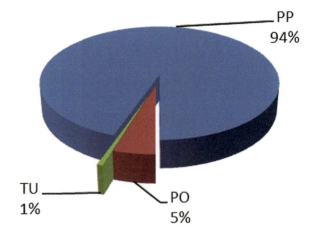

Fig. 7.31 Species composition of lobsters in Mumbai (2007–2012). PP-*Panulirus polyphagus*, TU-*Thenus unimaculatus*, PO-*Panulirus ornatus*

During the same period, average annual catch of *T. orientalis* (*T. unimaculatus*) was 185 t at the rate of 4.35 kg/U, which means contribution to the total catch by the slipper lobster was 46%. The composition of the fishery has changed over the years with *P. polyphagus* forming almost a single species fishery (94%) along the Mumbai coast (Fig. 7.31). The yearly landings show high fluctuations with a trend of decline and revival. The production peaked in 1986 with a landing of 709 t (Fig. 7.30). When the landings were partitioned into 10-year cohorts, the average landing was 264 t during the developmental stage of the fishery (1978–1987). However, during the second 10-year period (1988–1997), the annual average landing was 132 t. During the third phase (1998–2007), the fishery showed slight improvement with an annual average landing of 149 t. *Panulirus ornatus* contributes about 5% to the catch. The slipper lobster *T. orientalis* (*T. unimaculatus*) has almost disappeared from the fishery since 1995 and made an appearance again in certain years, with negligible annual landings of 2 to 3 t. Though *P. versicolor* has been reported to occur along the Mumbai coast, it hardly contributes to the fishery.

Panulirus polyphagus, though available round the year, displays seasonal abundance with peak fishery during October to March (Kagwade et al. 1991; Kagwade 1993). The major percentage of the landings (93%) is by single-day and multiday trawlers and the rest by gillnets (6.4%) and *dol* nets. Kagwade (1987a) studied the age and growth of *P. polyphagus* of lobsters landed at Bombay. The size of both males and females in the fishery ranged from 75 to 385 mm TL, the size between 160 mm and 230 mm forming the mainstay of the fishery. Recruitment of juveniles, ranging from 70 mm to 120 mm TL, was generally observed during December to February, in shallow nearshore waters, and hence it follows that the fishery, which is at its peak between October and March, is mostly dominated by the new recruits (1 to 2 year olds), just entering into the fishery. Any fishery subjected to high exploitation rates (0.81 for males and 0.68 for females in the case of *P. polyphagus* of Mumbai) (Kagwade 1994) is mostly made up of new recruits suggesting that annual landings are reasonably good indicators of recruitment into the fishery (Baisre and Cruz 1994). The drastic fall in catches in the second phase of the fishery as pointed out earlier (263 t in the first phase and 132 t in the second phase) is an indicator of the high fishing pressure the *P. polyphagus* fishery of Mumbai is subjected to.

7.10.3.3.2 Stock Status

Kagwade (1994) used the catch and length frequency data of *P. polyphagus* landed at Mumbai for the period 1976–1986 for stock assessment studies. The study concluded that the estimated combined average MSY of males and females (168.42 t) is very low against the yield (243.5 t) and therefore effort reduction to the tune of 40% is necessary for the sustainability of the stock. With the assumption that natural mortality remained constant, the maximum yield of *P. polyphagus* could have been obtained at the exploitation ratio of only 0.46 for males and 0.53 for females. However, it was as high as 0.66–0.80 for males and 0.62–0.71 for females, and with such exploitation ratios, the biomass of the stock is reduced to less than 50%, indicating overfishing of the stock. The future sustainability of the stock is doubtful at such high exploitation ratios (Radhakrishnan et al. 2005).

7.10.3.3.3 Size at Maturity and Breeding Season

The size at which 50% of females of *P. polyphagus* matures was estimated at 175 mm TL. However, majority of the females mature at 205 mm, with the males maturing at 265 mm, and the annual production of eggs was as high as 0.14 and 4.72 million in specimens of 180 mm and 353 mm TL, respectively (Kagwade 1988a). The predominance of females was observed throughout the year (1:1.16). Though egg-bearing lobsters were observed round the year, major breeding activity was in January and September, with a minor one in March and June (Kagwade 1988a). However, Radhakrishnan et al. (2005) reported September as the spawning peak for *P. polyphagus* in Mumbai waters (Fig. 7.32). The coincidence of peak fishing and breeding always results in high percentage of breeding females in the catch (Fig. 7.33). In the trawler catches, percentage of females declined from 40.7% in 1978 (Kagwade 1988a)

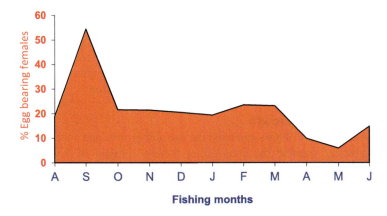

Fig. 7.32 Spawning season of *P. polyphagus* along the Mumbai coast

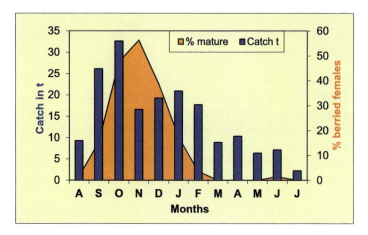

Fig. 7.33 Monthly landing of *P. polyphagus* and percentage of egg bearing females in lobster fishery of Mumbai

to 24.06% in 1988 due to heavy exploitation of spawners. Such heavy exploitation of breeding females is detrimental to the stock, and either a fishing ban during peak breeding season or returning of egg-bearing females back into the sea to protect the breeding population is suggested (Kagwade 1993).

7.10.4 Slipper Lobster Fishery

Commercial fishery for *Thenus unimaculatus* was initiated in 1978, with an annual catch of 2 t. This is a single species fishery exclusively landed by trawlers along the Mumbai coast. The total catch reached a maximum of 375 t in 1982 and subsequently declined to an average of 243 t (1983–1985) before reaching another peak (334 t) in 1986. Thereafter, the catch declined sharply, landing only 2.2 t in 1994 (Deshmukh 2001; Radhakrishnan et al. 2005). Thus, the slipper lobster fishery along the Mumbai coast lasted for just 17 years (Deshmukh 2001). Catch has not improved even after 15 years and the mean landing during 2007–2012 was only 2.2 t. The average annual catch during 1978–1985 was 185 t with a catch rate of 4.35 kg/U (Kagwade et al. 1991). *T. unimaculatus* appears throughout the year in the fishery with peak landings in June followed by September.

7.10.4.1 Age and Growth

Kagwade and Kabli (1996b) estimated the age and growth and concluded that it is a slow-growing species; the commercial fishery consists of 3-year-old males (210 mm TL) and 4.5-year-old females (230 mm TL). Though the estimated asymptotic lengths for males and females were 368 mm TL and 300 mm TL, respectively, the maximum size of males recorded in the fishery was 292 mm TL and females 297 mm TL. Growth rates were reported to be similar for the first 2 years, and thereafter, males recorded faster growth rate compared to females.

7.10.4.2 Reproductive Biology

Kagwade and Kabli (1996a) carried out a comprehensive study of reproductive biology of *T. unimaculatus* landed at Mumbai. Sexes are separate with the genital opening of the males on the coxopodite of the fifth walking leg and the oviduct opening of the females on the coxopodite of the third walking leg. Out of the four pairs of pleopods, the first three pairs of pleopods in males have narrow and curved endopods, whereas in females all the four pairs of pleopods possess endopods, which carry setae to hold the fertilised eggs. The ovarian structure and development is typical of the lobster family. Egg-bearing females appear in the catch from September to April with maximum percentage during November to January (24.5 to 57.2%).

7.10.4.2.1 Size at Maturity

The size at female maturity of *T. unimaculatus* along the Mumbai coast was estimated at 107 mm total length based on both the ovary and percentage of egg-bearing females in the fishery (Kagwade and Kabli 1996a). Subramaniam (2004) estimated the size at maturity for the Chennai population as 105.5 mm TL, whereas size at maturity for the Veraval population was estimated at 61–65 mm CL (150 mm TL) (Kizhakudan 2014). Kagwade and Kabli (1996a) estimated the size at maturity by combining the percentage of females (50%) with ovaries in early maturing to spent stages along with percentage of egg-bearing females.

7.10.4.2.2 Spawning Season

Breeding periods are defined as those containing seasonal or monthly presence of egg-bearing females (Groeneveld and Rossouw 1995). In tropical spiny lobsters with a large percentage of egg-bearing females throughout the year, defining spawning period is rather difficult. Unlike the spiny lobsters, there is a well-defined season for spawning in the slipper lobster *T. unimaculatus*. They may produce one or two broods of eggs within the same spawning season or in a year. Kagwade and Kabli (1996a) report that the breeding season is between September and April in Mumbai population of *T. unimaculatus*, with a major peak in November and a minor one in March, based on the percentage of egg-bearing females in the fishery. Hossain (1978) and Subramanian (2004) also suggested that *T. unimaculatus* from the east coast of India spawned more than once a year. Deshmukh (2001) re-analysed the 2-year data on the percentage of egg-bearing females (Kagwade and Kabli 1996a) from Mumbai and suggested a single well-defined spawning period between October and January. Radhakrishnan et al. (2007a, b) suggested a single spawning period for the species from Mumbai waters (October to March) based on higher incidence of mature females during October to December and larger percentage of berried females during October to March. The fecundity ranged from 20,050 to 53,280 for 240 mm-sized females. The fecundity showed a linear relationship with total length (157 to 238 mm TL). The linear regression equation for fecundity-length relationship was $Y = -46.66 + 0.4164\,x$, where y is the number of eggs in thousands and x is the total length of the lobster.

7.10.5 Collapse of the Slipper Lobster Fishery Along the Mumbai Coast

Deshmukh (2001) critically evaluated the biological reasons for the collapse of *T. unimaculatus* fishery along the Mumbai coast. Large-scale exploitation of the spawning females, which formed 60% of the total catch, probably impacted the recruitment process, resulting in rapid decline of the fishery in subsequent years. The exploitation of breeding stock when continued further during the declining phase of the fishery resulted in total collapse of the stock. This is a classic example of recruitment overfishing which is not precluded by growth overfishing (Radhakrishnan et al. 2005).

7.11 Kerala Fisheries

7.11.1 Introduction

The state of Kerala is a narrow strip of land situated in the southwest corner of the Indian Peninsula between 8°18' and 12°48'N and 74°52' and 77°22'E and bounded on the west by the Arabian Sea. With a coastline 590 kilometres long and a continental shelf off this coast covering about 40,000 km^2, the coastal waters of the state are considered to be among the most productive in the Indian Ocean. The coast is influenced by the West India Coastal Current flowing from north to south during the southwest monsoon (June–September) and the East India Coastal Currents flowing from south to north during November to January (Shetye 1998). The upwelling during the monsoon season and the runoff from the nearly 43 rivers flowing into the Arabian Sea make this ecosystem highly productive. The inshore sea is rocky in some places and muddy in most parts and has reefs in the southern region. The state also has extensive backwater systems. The estimated annual fish production by the state is 0.6 million tonnes (2014), and Kerala stands fourth among Indian states in terms of total annual landing of lobsters.

7.11.2 History of Lobster Fisheries

There is not much information on the early years of lobster fishing along the Kerala coast. However, it is believed that lobster fishing commenced as early as 1958 on the southern part of the state (Balasubramanyan et al. 1961). Artisanal fishing using traps appears to have been in operation for at least 50 years, and shallow water lobsters have probably been caught along the Vizhinjam and Quilon coast during this period. Artisanal lobster fishermen continue to fish lobsters using the traditional lobster traps (Fig. 7.1). In some areas along the Thiruvananthapuram coast, fishermen use torch to locate the lobsters hiding inside rocky crevices and handpick them. Artisanal lobster fishing has a fairly long tradition along the Calicut coast (George 1967a). Although the availability of the deep-sea lobsters off the Quilon district (*Quilon Bank*) was known from the deep-sea trawling operations carried out during

1959–1961 by the Kerala University Marine Science Department Research Vessel *Conch* and by R/V *Kalava* of Indo-Norwegian Project (Alcock 1901; John and Kurian 1959; Kurian 1965), no systematic survey of the deep-sea fishery resources was carried out until 1967, and the potential commercial concentration of the deepwater spiny lobster, *Puerulus sewelli*, was brought to light during the exploratory survey conducted by the Indo-Norwegian Project fishing trawlers (Tholasilingam et al. 1968; Silas 1969; Rao and George 1973; Mohammed and Suseelan 1973; Suseelan 1974; Oommen and Philip 1976) and the Fishery Survey of India vessels. The Research Vessel *FORV Sagar Sampada* continued the scientific exploration of the offshore fishing grounds hitherto unexploited along the east and west coasts of India between 1985 and 1988, and Suseelan et al. (1990) analysed the crustacean components of the trawl catches and provided information on catch, catch rates, spatial and temporal distribution and abundance of edible prawns, lobsters and crabs. Though there was an initial upsurge in deep-sea fishing during the 1970s and 1980s, poor catch and lower demand for deep-sea prawns resulted in a lull. However, from 1999 onwards, mechanised multiday fishing boats based at Kochi and Quilon have been regularly fishing for deep-sea prawns and lobsters in the Quilon Bank.

7.11.3 Fishing Grounds and Fishing Methods

A variety of lobsters are distributed along the Kerala coast. However, the resource constitutes a fishery only at certain pockets, of which the fishery at Tikkoti near Kozhikode assumes considerable importance. Lobsters are captured by a variety of fishing gears, of which 88% is by the mechanised trawlers and the rest by the gillnets. During the 1950s, the traditional fishermen of Tikkoti and nearby villages used a local gillnet made of hemp twine, called *Muruvala*, with a mesh size of 4.5–5.0 cm for fishing lobsters. A type of cast net called *Kara vala* was also used by some fishermen (George 1967a). The other important lobster landing centres in Kerala are the Cochin Fisheries Harbour and Munambam in Ernakulam district, Neendakara-Sakthikulangara Fishing Harbour in Kollam district and Vizhinjam and Mulloor in Thiruvananthapuram district (Fig. 7.7). Apart from a small-scale fishery in and around Ernakulam district, deep-sea trawlers fishing in 'Quilon Bank' land the catch at Cochin Fisheries Harbour. At Neendakara and Sakthikulangara fishing harbours, the slipper lobster *Thenus unimaculatus* is landed as a bycatch by single-day and multiday fishing trawlers (Radhakrishnan et al. 2013). Vizhinjam and Mulloor are essentially artisanal lobster fishing centres, where lobsters are mainly captured by traps and bottom-set gillnets. Occasional catches by diver fishermen have also been reported (Saleela, personal communication).

7.11.4 Species Distribution

Six species in the genus *Panulirus* occur in the shallow waters along the Kerala coast: *P. homarus homarus*, *P. ornatus*, *P. polyphagus*, *P. versicolor*, *P. penicillatus* and *P. longipes longipes*. Among the shallow water palinurids, *P. homarus homarus* is the predominant species captured along the Kerala coast, although small numbers of *P. ornatus* and *P. versicolor* also form occasional catches. *P. polyphagus* is an important constituent of the fishery along the Calicut coast (Radhakrishnan et al. 2000). The occurrence of *P. penicillatus* along the Quilon coast was reported by Satyanarayana (1961). The deep-sea lobster *P. sewelli* is fished at depths of 150–400 m by trawlers operating from Kochi and Neendakara fishing harbours. The major areas of occurrence are Quilon Bank, Alleppy and Ponnani of which the Quilon Bank is the most important fishing ground for this species. Kathirvel et al. (1989) provide a detailed description on the biology, population parameters and exploitation of *P. sewelli* along the Indian coast. Other deep-sea lobsters reported to occur along the Kerala coast are *P. waguensis* (Chakraborty et al. 2014), *Nephropsis stewarti* (Holthuis 1991), *N. carpenteri* (George and Rao 1965), *Enoplometopus macrodontus* (Radhakrishnan et al. 2011), *Scyllarides tridacnophaga* (Radhakrishnan, personal communication), *S. elisabethae* (Thiagarajan et al. 1998) and *P. typhlops*. *Nephrops stewarti* is occasionally found in small quantities in deep-sea catches. Among the scyllarid lobsters, *T. unimaculatus* is the only species which supports a commercial fishery (Radhakrishnan et al. 2013).

7.11.5 Fishery

The annual average landing of lobsters in Kerala has been estimated at 207 t (1989–2014) (Fig. 7.34). The major percentage (88%) of the catch is landed by trawlers and the rest by bottom-set gillnets (Fig. 7.35). Total landings show wide fluctuations with the highest landing of 535 t in 2000 and the lowest in 2010 (18 t). Decadal landings were on an average 192 t during the 10-year period of 1989–1998; however, a rapid increase in landings (337 t) was observed between 1999 and 2008,

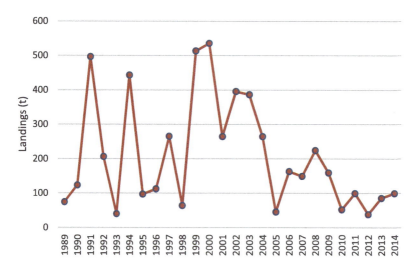

Fig. 7.34 Annual landing of lobsters in Kerala

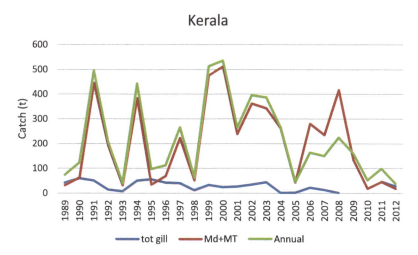

Fig. 7.35 Annual landing of lobsters by major gears in Kerala. Tot gill- total gillnets, Md+mt- mechanized trawl

which coincides with the expansion of fishing operations in deeper waters. The average annual landing of the deep-sea lobster *P. sewelli* from the Quilon Bank during 1999–2002 was 340 t (Radhakrishnan et al. 2005). However, annual landing during 2010–2014 has been extremely poor with an average catch of 66 t. The total annual average catch for the period 2007–2014 was 108 t, and that landed by the trawlers was on an average 104 t, with a catch rate of 0.25 kg/U. Shallow water lobsters are landed as a bycatch by the single-day and multiday fishing trawlers, whereas gillnets operating in shallow rocky areas exclusively target lobsters. The slipper lobster *T. unimaculatus* has contributed the maximum (53.3%) to the total catch of the state. Of the six species of spiny lobsters, *P. homarus homarus*, *P. ornatus* and *P. sewelli* are the three species which form the mainstay of the commercial fishery, contributing 28%, 6% and 13%, respectively, to the total landings of the state (1989–2012) (Fig. 7.36). *Panulirus versicolor* is poorly represented in the catch, with its presence mostly restricted to the Thiruvananthapuram coast. Similarly, *P. polyphagus* catch is restricted to the Tikkoti coast of Kozhikode (Calicut) district.

Lobsters are landed throughout the year except during June-July when the trawl ban is in force. Peak landing of lobsters by trawlers is from November to March, which coincides with the fair season for fishing along the Kerala coast (Fig. 7.37). This is also the period when the fishery for the deep-sea lobster *P. sewelli* and the slipper lobster *T. unimaculatus* is at its peak. During this period, 74% of the total annual catch by the state is landed. The highest catch rate occurs in September just after the fishing season starts after the trawl ban, and then there is a gradual decline until December. The catch rate improves in January, February and March, and then there is a progressive decline leading to the period when the catches are the lowest. Gillnet fishery for lobsters is at its peak from September to December.

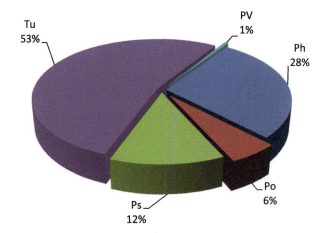

Fig. 7.36 Species composition of lobsters along the Kerala coast. TU *Thenus unimaculatus*, Ph *Panulirus homarus homarus*, Po *Panulirus ornatus*, PV *Panulirus versicolor*, Ps *Puerulus sewelli*

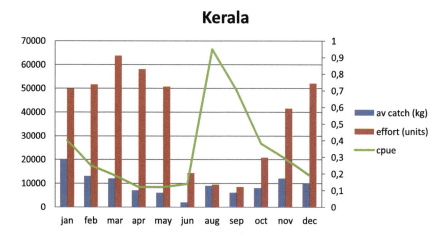

Fig. 7.37 Monthly average lobster catch, effort and CPUE in Kerala

7.11.5.1 Lobster Fishery Along the Thiruvananthapuram Coast

The shallow water fishery at Vizhinjam and Mulloor in Thiruvananthapuram district is by traps and gillnets. Small quantities of slipper lobsters are landed at Poovar, south of Thiruvananthapuram. Trap fishing at Vizhinjam is carried out by the traditional fishermen and constitutes 70% of the total catch. The fishermen set the traps, made of palmyrah stalks and baited with mussel or dead fish, near the rocky crevices during daytime and lift the traps early in the morning on the subsequent day to collect the lobsters. The traps are left permanently in the sea. The polyethylene monofilament bottom-set gillnets of 75–80 mm mesh size are set on rocky areas near the coast and hauled on the following day morning. The spiny lobster *P. homarus homarus* constitutes 99% of the lobster catch in these centres. The peak period of fishing in both the centres is from end of July to December. The annual average landing at Vizhinjam was 1 t (2000–2012) with the peak landing of 2.6 t in 2006 and the minimum of 0.3 t in 2012. The CPUE ranged from 0.06 kg/U in January to 0.79 kg/U in December. Though *P. homarus homarus* is landed throughout the year, peak season for the fishery is from August to November, when 62% of the total catch is landed. At Mulloor, annual average landing (2009–2014) was 0.6 t. The major contribution to the catch is by traps (70%) and the rest is by bottom-set gillnets.

Domestic market for lobsters in Thiruvananthapuram district is focused on beach resorts and restaurants associated with the tourism industry. The traders stock the lobsters in live condition until they get a good price. Lobsters weighing >200 g is considered as the first grade, which fetches Rs. 500–2000/kg, depending on the demand. The second grade (100–200 g) fetches Rs. 300–1500/kg and the third grade weighing <100 g is sold at Rs. 100–400/kg.

7.11.5.2 Small-Scale Lobster Fishery at Tikkoti, Calicut

There is a lack of reliable information regarding the early years of lobster fishing along the north Kerala coast. The earliest report on the lobster fishery of the coast is by George (1967a). The lobster fishery at Tikkoti and nearby fishing villages situated around 40 km north of Calicut is entirely an artisanal fishery. Fishing is by bottom-set gillnets operating near the rocky areas of the coast, though the traditional fishermen in earlier years had used a type of cast net called *Kara vala*, apart from the gillnet called *Muruvala* (George 1967a).

The annual average catches at Tikkoti landed by the cast net during 1963 and 1964 were 4625 numbers (1.4 t) and 6046 numbers (1.8 t), respectively (George 1973). Kagwade et al. (1991) reported that the total annual average catch (1982–1986) at Calicut was 8 t at the CPUE of 4.5 kg/U with 5.9 t of *P. homarus*, at the rate of 3.51 kg/U, and 1.7 t of *P. polyphagus*, with a CPUE of 1.03 kg/U. Radhakrishnan et al. (2000) observed that three species, *P. homarus* (74%), *P. polyphagus* (20%) and *P. ornatus* (6%), constitute the lobster fishery at Tikkoti. For the first time (1995–1996), *P. versicolor* was reported from this part of the coast, though they did not contribute to the fishery. During 1994–1996, the estimated annual landing was an average 2.3 t. However, the catch was drastically declined to 0.5 t in 1996–1997.

The fishery is seasonal which starts immediately after the monsoon in September. *Panulirus homarus* appears first in the fishery followed by *P. polyphagus*. *P. ornatus* enters into the fishery in December or January. However, this pattern is not consistent in all the years. *Panulirus polyphagus* had a progressive decline over the years, from 0.6 t in 1995–1996 to 0.09 t in 1996–1997. There is no data to substantiate that the stock is unique to this area. In *P. homarus*, the size of lobsters in the catch ranged from 41 to 100 mm CL in both females and males. Considering the 3-year period of observation, 37% of males and 47% of females were in the size range of 61–70 mm CL. In *P. polyphagus*, the size of males ranged between 51 and 110 mm CL with maximum numbers in the size range of 71–80 mm CL and 81–90 mm CL. In females, the size varied from 61 to 100 mm CL with the dominant sizes in the length range of 81–90 mm CL, showing that juveniles were less than 10% in the catches. The breeding season for *P. homarus* was in February (44%) and March (91%), unlike the peak season for breeding of the species (November–December) distributed along the Kanyakumari coast. The price of lobsters ranged from Rs. 150 to 300/kg depending upon the weight. George (1967b) estimated the rate of exploitation of *P. homarus* at Tikkoti as 28.4%. Over the years there was no substantial increase in effort, as fishing is carried out by a few traditional fishermen in that area. The sudden fluctuation in landings is characteristic of the spiny lobster fishery and is more likely due to the result of differences in oceanographic and ecological conditions.

7.11.5.3 Lobster Landings at Ernakulam District

The lobster landing centres of Ernakulam district include the two fishing harbours, Cochin Fisheries Harbour and Munambam, and a minor centre, Vypeen. The total annual average landing for the period 2007–2012 was an estimated 20.3 t. Two commercially important species are landed by the trawlers; the deep-sea lobster *P. sewelli* and the scyllarid *T. unimaculatus*, which is the major species landed in these centres, constituting 97.5% of the total landing. The total annual average landing of *P. sewelli* and *T. unimaculatus* by the multiday fishing trawlers in the three centres together was an estimated 1.2 t and 18.8 t, respectively. Maximum landing of *T. unimaculatus* was in 2008 (7.3 t) and the minimum in 2012 (0.4 t). In 2010, an estimated 3.7 t of *Nephropsis stewarti* was landed at Cochin Fisheries Harbour. Such unusual catches of lobsters have been reported also from Okha, Gujarat and Mumbai (Ghosh et al. 2008). The occurrence of the deep-sea reef lobster *Enoplometopus macrodontus* was reported from Ponnani, south of Calicut (Radhakrishnan et al. 2011). This is the first global record of the species apart from its occurrence in the Philippines (Chan and Ng 2008).

7.11.5.4 Lobster Fishery of Neendakara-Sakthikulangara

Neendakara-Sakthikulangara Fishing Harbour is an important lobster landing centre along the south Kerala coast. The catch is landed as a bycatch by both single-day and multiday fishing trawlers. The total annual average landing for the period 2007–2012 was an estimated 72 t. Four species contribute to the commercial catch; three species of palinurids, *P. homarus homarus*, *P. ornatus* and the deep-sea lobster *P. sewelli*, and the scyllarid lobster *T. unimaculatus*. The annual average landing of *P. homarus homarus* was 18 t forming 25% of the total catch. *Panulirus ornatus* constituted 5 t (7%) and *P. sewelli* 14 t, forming 19%. The landing of *P. sewelli* was highest during 2007 (32.6 t) and 2008 (19.8 t). The annual average landing of *T. unimaculatus* in these two landing centres together was 35 t forming 49% of the total catch. The highest landing of the species (67.4 t) was in 2008. The other palinurid species reported from this part of the coast and not contributing to the fishery are *P. versicolor*, *P. penicillatus*, *P. polyphagus* and *P. waguensis*. Few numbers of the scyllarid *S. tridacnophaga* are also landed occasionally.

Though CMFRI has been monitoring the crustacean landings from these centres for about five decades, there are no focused studies on the biology and fisheries of the principal species of lobsters landed here, probably due to that the fishery is insignificant compared to other high-volume crustacean resources. The earliest report on the lobster fishery of Quilon was by Balasubramanian et al. (1961). During 1959–1960, experimental fishing of lobsters was conducted off Quilon to test the operational efficiency of a newly designed bottom-set gillnet. Catch (138 kg for 8 days of operation) and catch rates (0.97 kg/unit) were reasonably high though it was a lean season for the fishery. The size range of lobsters captured varied from 120 to 370 mm TL with a modal length at 220–270 mm TL (400–800 g). The data showed that the fishery was in its nascent stage and had good potential for expansion.

Although Neendakara-Sakthikulangara landing centres are important from the deep-sea fisheries point of view, this low-volume resource did not get much attention in biological and population studies, in spite of its high demand in the export market and high unit value compared to other seafood.

Radhakrishnan et al. (2013) report the fishery and stock parameters of the slipper lobster *T. unimaculatus* landed at Sakthikulangara during 2005–2010 (Table 7.8). The annual average landing was 19 t at a mean catch rate of 0.09 kg/hr. forming 0.48 to 1.16% of the annual trawl landings of the centre. The peak season for the fishery is from November to February. The catchable stock of *T. unimaculatus* is in the range of 46–250 mm TL, and the length distribution of both males and females is nearly normal. The annual average landing of the species during 1982–1999 varied from 6.7 t to 114.6 t (Subramanian 2004).

7.11.5.5 Biology and Stock Assessment of *Thenus unimaculatus*

The length-weight relationship of *T. unimaculatus* from Sakthikulangara and from Mumbai and Tuticorin is shown in Table 7.9 for comparison. The relationships show similarity and are negatively allometric in all the three populations. Females are heavier (total weight) than males of equivalent carapace length, with the difference increasing as the animals grow. The growth rates are identical in both sexes till they attained 2 years, and thereafter the females grew faster than males. The commercial catch at Sakthikulangara is maintained by 1 and 2 year classes, whereas in Mumbai population, it was 3-year-old males and 4.5-year-old females contributing to the fishery (Kagwade and Kabli 1996b). This implies that the fishery is supported by one or two intensively fished early year classes which are entering into the fishery. During the 6-year period of the fishery, the exploitation rates for male and female lobsters were above 0.5 (E_{opt}) indicating that the species is overexploited off the Quilon coast (Table 7.8). The current level of fishing is very high and catches appear to be declining from a peak 32 t in 2008 to 12 t in 2010.

While estimating size at maturity, it was assumed that female lobsters that are carrying eggs on the ovigerous setae are mature. Unlike palinurids, the spermatophore in *T. unimaculatus* was rarely found sticking on to the sternal plate of females (Kagwade and Kabli 1996a). The spermatophore probably falls off soon after fertilisation and oviposition. The smallest female in berried condition measured 61 mm in CL, but substantial numbers above 72 mm CL were found to be mature (Radhakrishnan et al. 2013). However, the percentage of females carrying eggs

Table 7.8 Biological characteristics and stock parameters of *Thenus unimaculatus* off Quilon, Kerala (2005–2010)

Biological characteristics	Population parameters
Size range (mm TL)	
Males	61–230
Females	46–250
N (numbers)	
Males	3292
Females	2775
Modal length (mm TL)	
Males	143
Females	153 and 168
Asymptotic length (mm TL)	
Males	240
Females	260
Asymptotic weight (g)	
Male	800
Female	806
Sex ratio	
Male/female	1:0.9
Length at 50% capture (mm TL)	
Male	147 (at age 1.9 years)
Female	153 (at age 1.6 years)
Fecundity 61–84 mm CL	14,750 to 33,250
Dominant size in the fishery (year classes)	1 and 2
Growth coefficient (K) monthly (mm)	
Male	0.04
Female	0.05
Natural mortality (M)	
Male	0.64
Female	0.7
Fishing mortality (F)	
Male	3.89
Female	2.58
Total mortality (Z)	
Male	4.52
Female	3.29
Exploitation rate (E)	
Male	0.86
Female	0.78
Catch (tonnes) annual mean	19
Mean catch rate (kg/hr)	0.09

could never reach the 50% level by which the size at maturity could be arrived at. In Veraval population of *T. unimaculatus*, the percentage of berried females never exceeded more than 26% in any adult size groups (Kizhakudan 2014). Kagwade and Kabli (1996a) estimated the size at maturity of Mumbai population at 107 mm TL,

Table 7.9 Length-weight relationships for *Thenus unimaculatus* along the coasts of India

Relationship	Sex	Regression	N	r	Area	References
Carapace length (mm) vs total weight (g)	Female Male	Log W = -4.6322 + 2.9758 log CL Log W = −4.4826 + 2.9161 log C L	281 343	0.9918 0.9931	Tuticorin, Tamil Nadu	Saha et al. (2009)
Total length (mm) vs total weight (g)	Female Male	Log TW = -3.9393 + 2.7938 log TL Log TW = −3.7069 + 2.7013 log TL	332 444	0.9578 0.9189	Mumbai, Maharashtra	Kabli and Kagwade (1996)
Carapace length (mm) vs total weight (g)	Female Male	Log W = 0.001558 + 2.76 log CL Log W = 0.00237 + 2.65 log CL	2775 3292	0.9600 0.9500	Sakthikulangara, Kerala	Radhakrishnan et al. (2013)

and the SOM in the Veraval population of *T. unimaculatus* was estimated at 61–65 mm CL (Kizhakudan 2014). The spawning season was reported to be from September to April with peak abundance of berried females during November off Sakthikulangara and Mumbai. Unlike the west coast population, the southeast coast stock had two spawning peaks, one in February–March and another in June–August (Subramanian 2004).

7.12 Lobster Fisheries of Tamil Nadu

7.12.1 Introduction

The coastline of Tamil Nadu has a length of 1076 km, which constitutes about 15% of the total coastal length of India, and forms a part of the Coromandel Coast of Bay of Bengal as well as the Indian Ocean and the Arabian Sea. The coast is divided into two major coastal areas, the southwest coast comprising the coastal district of Kanyakumari and the southeast coast encompassing 12 districts. The continental shelf along the southwestern part of the Kanyakumari district is wider which joins with the Wadge Bank, an offshore area in the Indian Ocean, south of Kanyakumari. The continental shelf on the east coast of Tamil Nadu is relatively narrow and rarely exceeds 35 km in width. Along the west coast, the current flows equatorward (southerly) when the southwest monsoon winds are active (April–September) and becomes stronger when the winds peak during July–August (Shetye 1998). The West India Coastal Current (WICC) is strongest along the southwest coast of India and is hardly noticeable off the northwest coast. Strong upwelling influences the coastal areas resulting in high productivity. During November–January the current reverses, and the East India Coastal Current (EICC) turns around Sri Lanka and flows northward along the west coast against the weak winds of the season (Shetye et al. 1991).

The sea becomes calm, and this is the period (October–April) when maximum fishing activity takes place along the west coast. The currents and the countercurrents play a major role in distribution, spawning, larval dispersal, post-larval settlement and the fishery of lobsters. The topography of the east coast continental shelf is different from the west coast with rocks, coral reefs and sandy and muddy bottoms providing a variety of habitats for different species of lobsters.

Miyamoto and Shariff (1961) provide a detailed description of the topography of the Kanyakumari coast from the southern tip to Colachel. The sea bottom along the 60 km nearshore coastline of Kanyakumari district on the west coast is rocky with sandy patches in between. Natural reefs parallel to the coast exist in deeper waters. The 5 fathom line is well within about a mile from the shore (Miyamoto and Shariff 1961). The rocks with seaweeds, sea urchins and rich bivalve fauna offer excellent conditions for feeding and growth and are an ideal habitat for lobsters. Kanyakumari has perhaps the largest marine species diversity in India, due to the rich and diversified coastal ecosystem.

7.12.2 History of the Fisheries

The history of lobster fisheries in Tamil Nadu dates back to the early 1950s. Fishing for lobsters probably existed during this period but may have been of a subsistence scale as there was no demand for lobsters in the domestic or the export market. There are no reliable statistics of lobster landings available for this period. Lobsters did not appear as a separate item in the estimated fish landings of CMFRI until 1975. Since 1955, spiny lobster fishing has become more attractive due to export demand. The records show that the USA imported 0.020 million pound (9 t) lobster tails from India in 1956, valued at USD 0.009 million (Smith 1958). In 1957, the Government of India engaged Dr. Miyamoto, a Fishery Officer from FAO, to conduct a review of the lobster fishing methods adopted by Kanyakumari fishermen and to suggest improvements in the existing fishing methods and to estimate the potential of the resource along the southwest coast of India. During December 1957 to February 1958, Miyamoto and Shariff (1961) conducted a survey on the existing craft, gear and the fishing methods adopted by lobster fishermen along the Kanyakumari coast and submitted a report. The major fishing centres are Cape Comorin (Kanyakumari), Puthanthurai, Pallam, Muttam (Muttom), Kadiampatnam (Khadiyapatanam) and Kolachel (Colachel). The four kinds of fishing gears in use along the coast during this period were anchor hooks, scoop nets, traps and hemp bottom-set gillnets. The gears, anchor hooks and scoop nets were the predominant fishing gears which were gradually replaced by gill (monofilament) and trammel nets. Lobster traps were in use in certain villages in Kanyakumari district. When the water clarity is good, fishermen dive and catch lobsters in certain villages, which constitute only a small portion of the commercial landing. The bottom-set gillnets are now widely used by the Tamil Nadu fishermen all along the coast, with traps restricted to only two fishing villages in Kanyakumari district.

Though the lobster landings were not separately shown in the CMFRI Annual Report for 1959, there was a report on lobster fishing and landings on the Kanyakumari coast in the Fishery Biology section. Balasubramanyan et al. (1960, 1961) conducted experimental lobster fishing in 1958 and 1959 using bottom-set gillnets along the Kollam-Kanyakumari coast to study the operational efficiency of the nets for lobster fishing. The pioneering studies on biology and fishery of *Panulirus homarus* in Kanyakumari district were carried out by Chacko and Balakrishnan Nair (1963) and George (1967a), which provide information on landings by different gears, the fishing season, size and sex composition as well as the first age and growth estimates based on length frequency data. Mohammed and George (1968) using mark-recovery techniques estimated growth and movement of the spiny lobster *P. homarus* in the fishing grounds of Muttom-Colachel in Kanyakumari district. Nair et al. (1973) provide a detailed account of spiny lobsters constituting the lobster fishery in Gulf of Mannar. During the developmental years of the fishery (1960–1970), there were no controls on fishing, and since there was an export demand, the number of fishing villages and fishermen engaged in lobster fishing grew rapidly. In 1962, 2.3 t frozen lobster tails valued at 0.23 million rupees (USD7600) were exported, and 4 years later (1966), the exports had grown to 81 t valued at Rs. 1.47 million (USD 35000) (George 1965). The 10-year (1964–1973) average catch was around 111 t. The largest catch of 246 t was taken in 1972–1973. Though George (1967a) suggested enforcement of minimum legal size of 130–140 mm TL for fishing for *P. homarus* in Kanyakumari district, no controls were enforced by the State Fisheries Department. By the late 1970s, the fishing effort was still rising rapidly, whereas the annual catch had stabilised at around 80 t (Kagwade et al. 1991). The declining trend in catch continued, and by the beginning of the twenty-first century (2004), the catch from Kanyakumari district was around 9 t.

7.12.3 Species and Distribution

Though spiny lobsters have been recorded both from the east and west coasts of the state of Tamil Nadu, the resource has a discontinuous distribution along the coast. The major fishing areas are the Kanyakumari district on the west coast and the Tirunelveli coast, Tuticorin, Mandapam, Nagapattinam, Cuddalore and Chennai on the east coast (Fig. 7.7). Six species in the genus *Panulirus* have been reported to occur in the shallow waters of Tamil Nadu, of which the species forming commercial fishery are the spiny lobsters *P. homarus homarus* and *P. ornatus*. The scyllarid species, *T. unimaculatus*, also contributes significantly to the total landing. Other species constituting small localised fishery are *P. versicolor* and *P. polyphagus*. *Panulirus penicillatus* and *P. longipes longipes*. Occasional catch of the scyllarid, *Scyllarides tridacnophaga*, is of recent origin from Kanyakumari, Mandapam and Chennai. The deep-sea lobsters *P. sewelli* and *P. waguensis* and the nephropid lobster *N. stewarti* appear in the fishery in certain months, but they are of minor volumes. Other prominent species recorded along the Tamil Nadu coast are *Parribacus antarcticus* from Gulf of Mannar landed at Pamban, *Scyllarus tutiensis* (*Biarctus sordidus*) from Tuticorin, *Petrarctus rugosus* from Chennai and Puducherry and the reef lobster *Enoplometopus occidentalis* from Gulf of Mannar landed at Mandapam.

7.12.4 Fishing Methods

A variety of fishing methods are used to capture spiny lobsters. Though spiny and scyllarid lobsters are not targeted by trawlers (multiday and single-day fishing), maximum landing was by the trawlers (55%) with an annual average landing of 180 t (1989–2012). While spiny lobsters are incidentally caught by the trawlers, the scyllarid lobster *Thenus unimaculatus* is caught while trawling for prawns. The contribution by the outboard gillnetters (OBGN) is 26% with an annual average catch of 87 t, followed by non-mechanised crafts (14.0%) with a landing of 47 t (Fig. 7.38). The gears used by the non-mechanised fishermen are generally traps, bottom-set gillnets and trammel nets. In certain fishing villages, fishermen divers make a small contribution to the catches. During the 1950s and early 1960s, the main gears for lobster fishing along the Kanyakumari coast were anchor hooks, traps and scoop nets (Miyamoto and Shariff 1961; George 1967a). The catches landed by multiday trawlers are dead lobsters, whereas the gillnet and trap fishermen bring the lobsters in live condition. While it is easier to remove the lobster from traps and gillnets without any damage to the antennae or walking legs, the catch from trammel nets has to be carefully taken out to prevent physical damage to the lobsters, which will fetch lesser price compared to the intact live lobsters. The diver fishermen bring the hand-caught lobsters to the shore and bury them in the moist sand by which the lobster can be kept alive for 6–7 hours.

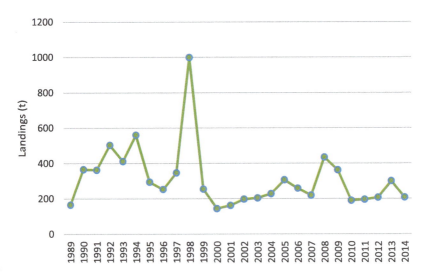

Fig. 7.38 Annual landing of lobsters in Tamilnadu

7.12.5 Catch and CPUE Trends

Tamil Nadu stands third in terms of total annual landing of lobsters by the maritime states with an average annual estimated landing of 311 t (1989–2014) (Fig. 7.39) forming 17% of the total all India landing. The landings show wide fluctuations from year to year. The peak landing of 998 t was in 1998. When the catch was partitioned into 10-year periods, average annual catch for the years 1989–1998 was 425 t and for the next 10-year period (1999–2008), 238 t, which is just 56% of the first 10-year period. The average annual catch for the period 1975–1984 was an estimated 343 t, forming 19.5% of the total catch (Kagwade et al. 1991). The average annual landing for 2003–2012 was 259 t which is only 69% of the catch landed during 1975–1984. The deficit was mainly due to the reduction in landings from Kanyakumari (8.9 t on average from 2003–2010, based on the market figures of Kanyakumari District Fishermen Sangams Federation (KDFSF), and 9.4 t based on CMFRI average annual catch data (2001–2010) compared to the average annual catch (80 t) during 1978–1980 (Kagwade et al. 1991). The average annual landing was even higher (109 t) during the earlier period (1964–1978), with the highest landing of 246 t in 1972–1973.

Mechanised trawlers landed an average 180 t (1989–2012) annually followed by gillnetters (152 t) (Fig. 7.40). In trawlers, lobsters appear as a bycatch. The dominant species are *P. homarus homarus*, *P. ornatus*, *P. polyphagus* and *T. unimaculatus*. The total catch and catch/hr in trawl (multiday and single day combined) and outboard gillnet for the period 2007–2012 are shown in Fig. 7.41a, b, respectively. Catch and catch rate in trawls increased from 114 t and 0.28 kg/hr in 2007 to 347 t and 238 t with a catch rate of 0.875 and 0.816 kg/hr, respectively, in 2008 and 2009. The average annual catch then declined to 125 t with a catch rate of 0.34 kg/hr in 2012. In the gillnet, the catch and catch rate decreased from 100 t and 0.046 kg/hr in 2007 to 67 t and 0.031 kg/hr in 2012. The catch rates available for the total catch

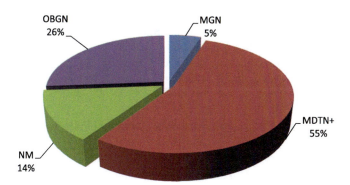

Fig. 7.39 Percentage landing of lobsters by different crafts and gears in Tamilnadu. MGN-Mechanised Gillnetter, OBGN-Outboard Gillnetter, NM-Non-mechanised Gillnetter, MDTN+-Mechanised trawler

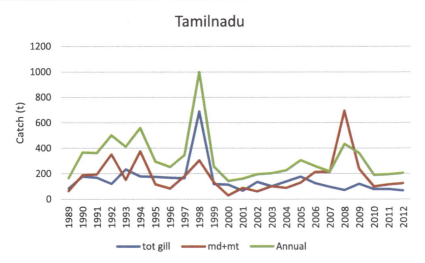

Fig. 7.40 Lobster landing by different gears in Tamilnadu. tot gill- all types of Gillnets, md+mt- mechanized trawlers

for the state may not yield adequate information upon which management measures could be based. Further, lobster being a bycatch in trawls, estimation of catch/hr. based on total effort spent by the trawler may provide only an overall trend.

7.12.6 Species Composition and Seasonal Trends

While the species composition of major fishing areas may be different, five species mainly constitute the fishery of the state. If all the major species landed during 1989–2012 are considered, the scyllarid *Thenus unimaculatus* accounts for 35% of the total landings, followed by *P. homarus homarus* (34%), *P. ornatus* (16.4%), *P. versicolor* (*10.1%*) and the deep-sea lobster *P. sewelli* (4.5%) (Fig. 7.42). If only palinurid lobsters are considered, *P. homarus* accounts for 52.3%, followed by *P. ornatus* (25.2%), *P. versicolor* (15.6%) and *P. sewelli* (6.9%) (Fig. 7.43). Landings have been reported throughout the year except March (trawl ban period), with no specific seasonal trend. The highest trawl landing was in September (Fig. 7.44).

7.12.7 Southwest Coast (Kanyakumari District) Fishery

Kanyakumari, the southernmost district in the state of Tamil Nadu, has a coastline of 70 km, of which 10 km is on the east coast of India. It is located between 77° 15′ and 77° 36′E longitude and 8° 03′ and 8° 35′N latitude. The west coast of Kanyakumari district was the most important lobster fishing area in India during the 1950s and 1960s. There is no record of the exact year when lobster fishing was started along the Kanyakumari coast. The fishery would have started in the early

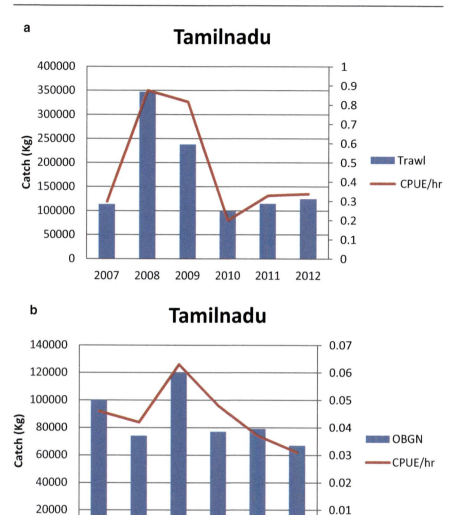

Fig. 7.41 (**a**) Lobster catch and catch/hr in trawlnets operated by mechanized boats in Tamilnadu. (**b**) Lobster catch and catch/hr in gillnets operated by outboard engine boats (OBGN) in Tamilnadu

1950s probably on a subsistence scale initially. With the advent of refrigeration technology and establishment of freezing plants to process and store the large quantity of prawns and lobsters landed along the southwest coast, fishing for lobsters was intensified. However, there is no reliable statistics of lobster landing in the state during this period. Though there are records of landing figures of prawns, shrimps and crustaceans for the years 1950–1954 in the Annual Report of the Central Marine Fisheries Research Station, Mandapam Camp (later known as CMFRI), published

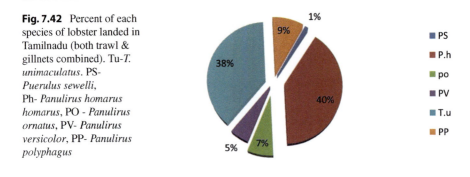

Fig. 7.42 Percent of each species of lobster landed in Tamilnadu (both trawl & gillnets combined). Tu-*T. unimaculatus*. PS- *Puerulus sewelli*, Ph- *Panulirus homarus homarus*, PO - *Panulirus ornatus*, PV- *Panulirus versicolor*, PP- *Panulirus polyphagus*

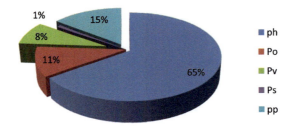

Fig. 7.43 Percentage of palinurid species (excluding *Thenus unimaculatus*) landed in Tamilnadu. Ph: *Panulirus homarus homarus*, PO: *Panulirus ornatus*, PV: *Panulirus versicolor*, PS: *Puerulus sewelli*, PP: *Panulirus polyphagus*

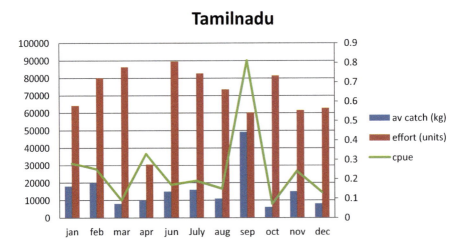

Fig. 7.44 Monthly catch, effort and CPUE of lobsters landed by trawlersalong the Tamilnadu coast

on 31 March 1955, there was no separate entry for lobster landings (Miyamoto and Shariff 1961). The impetus for lobster fishing would have been due to the establishment of refrigeration facilities and opening of export markets for lobsters. For the first time in 1956, India exported 9 t of frozen lobster tails to USA, valued at US$ 0.009 million (Smith 1958).

The artisanal lobster fishery has evolved from the use of anchor hooks, traps and scoop nets, as the major gear for fishing in the 1950s, to a fishery today primarily using the bottom-set gillnet. Traps made of palmyrah stalks which were also used for lobster fishing during that period exist now in some villages (Fig. 7.1). Miyamoto and Shariff (1961) in their Status report submitted to the Government of India on the craft, fishing gears used and the method of fishing adopted by Kanyakumari fishermen for lobster fishing, evaluated the merits and demerits of the fishing methods. Balasubramanyan et al. (1960, 1961) examined the operational efficiency of bottom-set gillnets by conducting experimental fishing along the Quilon-Kanyakumari coast. Mohammed and George (1968) conducted tag-recovery studies on *P. homarus* along the Colachel-Muttom coast during 1964–1965 fishing season and estimated age and growth of the spiny lobster *P. homarus*. Complete descriptions of the early fishery, landings by different gears and biology of the dominant species, *P. homarus*, are provided by George (1967a, 1973). After that there was a wide gap in biological research data on the spiny lobster fishery of Kanyakumari coast, though catch data were regularly collected by CMFRI.

7.12.7.1 Fishing Villages

Vijayanand et al. (2007) listed 38 fishing villages in Kanyakumari district where fishermen are engaged in lobster fishing, of which 36 are on the west coast. However, KDFSF/SIFFS (South Indian Federation of Fishermen Society) survey report indicated 44 fishing villages in Kanyakumari district, of which only 14 villages are actively engaged in lobster fishing. The villages with more than 1000 artisanal fishers are Keezmanakudy, Manakudy, Pallam, Muttom, Khadiyapatnam, Colachel, Kurumbanai and Enayam. However, only 10% of the total number of fishermen (13780) are involved in lobster fishing, and 2% alone depend entirely on lobster fishing for their livelihood. Among the lobster fishermen, less than 10% are the traditional trap fishermen, and 26% of the traps available in these villages are used for lobster fishing. The fishing village of Enayam is the home of more than half of the trap fishermen of Kanyakumari. The vast majority of the fishing grounds are within 1 km from the shore and therefore belong entirely to the artisanal fishermen.

7.12.7.2 Fishing Gear and Fishing

The four types of fishing gears used by the lobster fishermen in the 1950s were anchor hooks, scoop nets, traps and hemp bottom-set gillnets. Traps made of palmyrah stalks are the oldest gear used mainly by the Muttom and Colachel fishermen (Fig. 7.2). Anchor hooks and scoop nets were subsequently introduced and hemp bottom-set gillnets still later. Miyamoto and Shariff (1961) provide an elaborate description of the gear and the method of fishing operation adopted by the lobster fishermen in Kanyakumari. The gear-wise and month-wise landing of lobsters (in numbers) in the two major fishing villages, Muttom and Colachel, for the period 1958–1959 to 1961–1962 were provided by George (1973). While maximum numbers of lobsters were landed by anchor hooks, followed by traps at Muttom, more numbers of lobsters were landed by traps at Colachel. Traditional fishermen still use traps made of local palmyrah stalks, which are set at the rocky crevices by the divers in the evening. Most of the fishermen currently use traps made of powder-coated steel frames which are more durable than the palmyrah traps (Fig. 7.3). The Central Institute of Fisheries Technology (CIFT) introduced, for the first time, semicircular steel traps with a flat base and escape vents at the sides at Khadiyapatanam and Enayam lobster fishing centres and conducted experimental fishing during 1980–1982 fishing season (Mohanrajan et al. 1981, 1984, 1988; Meenakumari and Mohanrajan 1985). In the anchor hook fishery, anchor-shaped hooks with six sharp arms made of cast iron are let down to the bottom by means of baited lines. When the fisherman feels the lobster biting the bait (mussel meat), the line is pulled suddenly, thereby hooking the lobster on its abdomen. This fishing is done during night. In the scoop net fishery, baited lines are slowly pulled up when the lobster bites the bait and while nearing the surface scooped up by a round scoop net (George 1967b). The bottom-set gillnets with floats at the top and sinkers at the bottom are set around the rocks in the evening and hauled in the early morning. During night when the lobsters come out of rock crevices foraging for food, they are entangled in the net. During peak fishing season, the nets are hauled every day, and in the lean season, the nets are left at sea for 3–4 days before they are hauled. At some centres, fishermen have been using trammel nets, which catch large quantity of juvenile lobsters. The three-layered, trammel nets with the inner mesh 4 cm across catch even lobsters measuring 20 mm in carapace length.

Mechanised trawling has been a part of Kanyakumari fisheries since the late 1960s. There are two fishing ports, Colachel on the west coast and Chinnamuttom on the east coast. The Colachel boats mostly fish along the coast of Kerala, while the Chinnamuttom trawlers go for fishing south of Cape Comorin towards the Wadge Bank. Since the lobster fishing grounds are rocky, no trawling operations are carried out in this area. Occasionally, deep-sea lobsters are landed at Chinnamuttom.

7.12.7.3 Fishing Season

The lobster fishing season along the Kanyakumari coast starts normally on retreat of the southwest monsoon in September and continues until April. However, small quantities are also landed throughout the year. George (1973) reported November to April as the peak season for lobster fishing at Colachel-Muttom area, whereas Radhakrishnan and Thangaraja (2008) report that the duration of fishery is from October to April with peak fishing during November to January. However, the 8-year average monthly landing data (2003–2004 to 2010–2011) presented by KDFSF shows two peaks in lobster landings, August–September and April. The lobster abundance in the fishing ground seems to be greatly influenced by the environmental conditions and the food availability (Thangaraja and Radhakrishnan 2012).

7.12.7.4 Species Composition

Five species in the genus *Panulirus* occur in the shallow waters of Kanyakumari coast (George 1973). *P. homarus homarus* is the dominant species forming nearly 92% of the total catch, while *P. ornatus* and *P. versicolor* contribute to the rest of the catch (Radhakrishnan et al. 2005). Traps and gillnets are the two major gears operated along the Kanyakumari coast. Among the eight major lobster fishing villages, Khadiyapatnam, Colachel, Muttom and Enayam are the most important traditional lobster fishing centres. In Khadiyapatnam, *P. homarus homarus* forms about 85% of the total catch followed by *P. ornatus* (Fig. 7.45), whereas in Colachel, north of Khadiyapatnam, *P. homarus homarus* constitutes 95% of the catch (Fig. 7.46). At Muttom, an adjacent village to Khadiyapatnam, *P. homarus homarus* form 79%, followed by *P. ornatus* (18%) and *P. versicolor* (Fig. 7.47). At Chinnamuttom, on the southeast coast, major species is *P. homarus homarus* (94%) (Fig. 7.48). However, there may be marginal year-to-year variations in catch percentage.

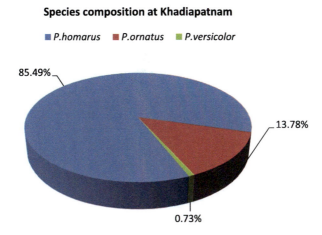

Fig. 7.45 Species composition of lobsters landed at Khadiyapatnam

Fig. 7.46 Species composition of lobsters landed at Colachel

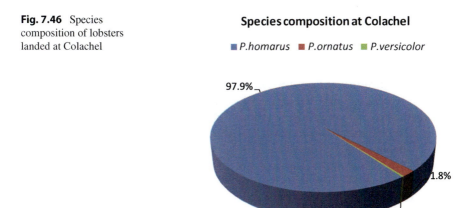

Fig. 7.47 Species composition of lobsters landed at Muttom

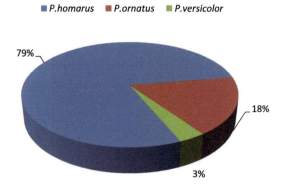

Fig. 7.48 Species composition of lobsters landed at Chinnamuttom

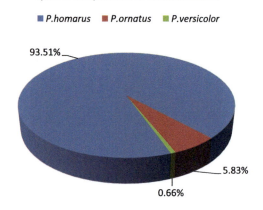

P. penicillatus and *P. longipes longipes* have also been recorded from the coast. *P. homarus homarus* is the only species captured by traps. De Bruin (1969) reported that *P. versicolor* avoid entering traps, no matter whatever is the design of the trap, but *P. penicillatus* could be captured in traps. The scyllarid, *T. unimaculatus*, is rarely landed at Colachel by trawlers. Recently, *Scyllarides tridacnophaga* has been occasionally appearing in the fishery but is generally rare.

7.12.7.5 Biology of *Panulirus homarus homarus*

Studies on biology and general ecology of *P. homarus homarus* along the Kanyakumari coast are relatively scarce. There is a big gap in research between early studies by George (1967a, b; 1973) and the recent investigations (Radhakrishnan and Thangaraja 2008; Thangaraja 2011; Thangaraja and Radhakrishnan 2012; Radhakrishnan et al. 2015; Thangaraja et al. 2015; Thangaraja and Radhakrishnan 2017; Saleela 2015). This provides an opportunity to evaluate how the fishery evolved through the 45 years of its existence and to draw some general exploitation patterns which may be of value when considering the current status of the spiny lobster fishery along the Kanyakumari coast.

7.12.7.5.1 Catch and CPUE Trends

George (1967a) conducted a detailed study of the lobster fishery from two major landing centres, Colachel and Muttom, in Kanyakumari district for the period 1959–1962 and collected data on landings by different gears and fishing season and studied the biology of the most dominant species, *P. homarus*. The length frequency data of *P. homarus* collected was analysed to derive estimates of age and growth. The total catch declined from a high of 24 t (numbers converted to weight assuming individual weight of lobster as 300g) in 1959 to 12.4 t in 1962. Estimated landings for the southwest coast (Kanyakumari district in Tamil Nadu state) were 213 t with a CPUE of 16.7 Kg and 301 t with a CPUE of 52.5 Kg in 1965 and 1966, respectively. However, the catch and CPUE was declined to 51 t and 15.7 Kg in 1967 and further down to 32.4 t and 10 Kg in 1968 (George 1973). The rapid increase in landings in 1965 and 1966 probably coincided with the expansion of fishing along the Kanyakumari coast as a consequence of increased demand for spiny lobster tails in the US market (George 1973). Export of frozen lobster tails increased from 40 t in 1962 valued at Rs. 0.23 million (USD 46,000) to 80 t in 1966 valued at Rs. 1.47 million (USD 2,09,400) (George 1967b). Kagwade et al. (1991) estimated the total annual average landing in Kanyakumari region as 80 t during 1978–1980, showing an improvement compared to 1968 landings. However, the declining trend of the fishery continued, and the average annual catch (2003–2010) was an estimated 9 t. Catch and effort is highest during the beginning of the season (November) and declines gradually in the following months until April, the end of the fishing season. The first 3 months of the season produced 72.4%, 83% and 70% of the total landings in 1964–1965, 1965–1966 and 1966–1967 seasons (lobster fishing season is from November to April). The average effort spent in the first 2 months (November and December) during these 3 years formed nearly 75% of the total effort spent in the entire season. Annual CPUE (kg/unit/night) follows the same trend with highest values in November (20 kg/unit) and lowest in April (8.7 kg/unit). The fishery showed an increasing trend with the effective season extending from November–December to November–January.

George (1967a) estimated the rate of exploitation of lobsters at 41.5% from the population estimates calculated from the catch and effort data and from the tag-recovery studies (Mohammed and George 1968) and suggested exploitation levels could be enhanced to 60% without affecting the stock. The catch improved with increasing effort but with wide annual fluctuations (67 to 246 t) until 1975 and thereafter started a declining trend. Powers and Bannerot (1984) have observed that at high levels of exploitation, observed fluctuations in landings in the Florida spiny lobster fishery corresponded to fluctuations in recruitment, as landings consist primarily of new recruits. The exploitation in the Kanyakumari fishery is very high, and the wide fluctuations in landings may be due to similar recruitment variability.

The catch trends in lobster fishery of three major lobster landing villages show that Muttom is a major lobster fishing village with an average annual catch of 8.7 t (2000–2004). The total annual catch by trap fishing is 7 t with a CPUE of 1.2 Kg/unit. Annual landing by the bottom-set gillnets is 1.9 t with a CPUE of 0.68 Kg/unit. The traps contributed nearly 78% of the total catch and bottom-set gillnets, the rest

(Saleela, personal communication). Peak trap fishing is during August to January when 70% of the total annual catch is landed. Fishing by bottom-set gillnets is mainly during October to January when 73% of the annual catch is caught.

The annual average landing of lobsters at Khadiyapatnam, the village adjacent to Muttom (2004–2010), was 1.7 t with a CPUE of 0.99 Kg/unit. The annual average landing at Colachel (2008–2014) was 3.4 t with a CPUE of 0.75 Kg/unit.

7.12.7.5.2 Length Distribution

Based on the length frequency data of *P. homarus* collected during 1958–1964, length at successive ages was calculated (George 1967a). George (1967a) reported the modal size groups of lobsters in the fishery were between 131–140 mm and 291–300 mm in total length in different months of the season. Among these size groups, the most predominant groups were 171–180 mm, 181–190 mm, 191–200 mm and 201–210 mm. The fishery, therefore, is largely supported by 1 and 2 year classes. From the modal progression of length, George (1967a) identified 9 year classes in the fishery and concluded that lobsters measuring 300–310 mm in total length are 10-year-olds. Estimates from Mohammed and George (1968) based on tag-recovery studies show significant difference in growth between males and females, with males showing faster growth rate. According to these authors, *P. homarus* after settlement reaches about 200 mm TL in 2 years, and thereafter the growth rate becomes slow. The study also revealed that *P. homarus* has restricted movement in the fishing ground, as the recoveries of tagged lobsters were within 12 km from site of release.

The biological characteristics of *P. homarus homarus* exploited in the artisanal fishery in Kanyakumari coast during 1958–1959 and 2007 are shown in Table 7.10. The sizes of *P. homarus* caught in the traps at Muttom (2000–2004) ranged from 91 to 260 mm TL in males and 102 to 320 mm TL in females, with a mean size of 151 mm and 164 mm in males and females, respectively. The sizes of *P. homarus* in the gillnet fishery at Khadiyapatnam varied from 92 to 300 mm TL in males and 102 to 289 mm TL in females with a mean size of 160 mm and 105 mm in males and females, respectively. The sizes of *P. homarus* observed in the commercial fishery at Colachel were 90–250 mm TL (males) and 85–280 mm TL (females) with a mean size of 148 mm TL and 176 mm TL in males and females, respectively.

The stock of *P. homarus homarus* at Khadiyapatanam adjacent to the Muttom fishing village during 2007 was in the size range of 22–115 mm CL (Thangaraja et al. 2015). The size of females in the gillnet and trammel net fishery ranges from 31 to 120 mm CL, whereas the males are in the size range of 21–100 mm CL. Nearly 52.3% of males caught are in the size range of 41–60 mm CL and 32.3% of females in the size range of 31–60 mm CL. The overall sex ratio was 1:1.4 in favour of females. The size frequency distribution of males and females in the fishery indicates that nearly 41% of *P. homarus homarus* caught are below maturity. On the other hand, the modal size groups of *P. homarus* in the anchor hook-trap fishery at Colachel-Muttom during 1958–1965 were in the length range of 171–210 mm TL (60–75 mm CL), showing that majority of lobsters caught were above the size at maturity (George 1967a). The predominant length range of *P. homarus* caught at

Table 7.10 Biological characteristics of the spiny lobster *Panulirus homarus homarus* population off Kanyakumari coast

Characteristics	Parameters	Authors
Size range (TL mm)	91–320	George (1967a)
Mean total length (TL)		
Males (TL mm)	151	
Females (TL mm)	164	
Size range (mm CL)	22–115	Thangaraja et al. (2015)
Mean carapace length males (mm)	61.66	
Females (mm)	64.53	
Mean body weight males (g)	179.2	
Females (g)	191.0	
Growth coefficient (K)		Mohammed and George (1968)
Males	0.717	
Females	0.601	
Males+females	0.30	Thangaraja et al. (2015)
Asymptotic length (L)		Mohammed and George (1968)
Males mm (TL)	312.37	
Females mm (TL)	303.17	
Males+females mm CL	119.4 (315.5 TL)	Thangaraja et al. (2015)
Total mortality (Z)	2.51	
Natural mortality (M)	0.67	
Fishing mortality (F)	1.84	
Exploitation rate (E) 1965	0.42	George (1967a)
Exploitation rate (E) 2007	0.73	Thangaraja et al. (2015)
Probability of capture L_c 50 (mm CL)	46.75	
Length at female maturity (mm CL)		
Setal method	53.0	
Percentage of ovigerous females	61.0	
Length at male maturity (mm CL)	63.0	
Yield/recruit (Y/R)	0.39	
Biomass/recruit (B/R)	2.23	
Estimated annual average catch (t) (2003–2010)	8.9	
Estimated annual average catch (t) (1958–1962)	90	George (1967a)
Maximum estimated catch (t) (1966)	301	

Colachel-Muttom during experimental fishing using bottom-set gillnet in 1959–1960 was 220–270 mm TL (75–85 mm CL) (Balasubramanyan et al. 1961), indicating the contracting size structure in the population as a whole because of intensive fishing as was observed in the Oman and Somalia lobster fishery (Fielding and Mann 1999). The size structure of the fishable population has shifted towards an increasing proportion of smaller lobsters. Therefore, the commercial fishery is mostly represented by new recruits in each fishing year as in the Florida fishery (Powers and Sutherland 1989). George (1967a) had cautioned that indiscriminate

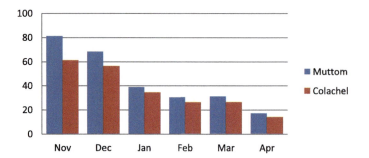

Fig. 7.49 *Panulirus homarus homarus*: Monthly percentage of berried females in total gillnet catch at Muttom and Colachel, Kanyakumari district (George 1967a)

fishing of lobsters including a significant proportion of juvenile lobsters measuring 90–140 mm TL (35–50 mm CL) is likely to be detrimental to sustainability of the resource, which has become true.

7.12.7.5.3 Breeding Season

Indirect evidence of percentage occurrence of females with eggs on their pleopods has been taken as the breeding season of *P. homarus homarus*. Though a small percentage of females was observed throughout the season at Muttom and Colachel, higher percentage of egg-bearing females during November and December (Fig. 7.49) shows that these months are the active breeding season. Nearly 80% of the reproductive potential is produced in the size ranges 51–80 mm CL. Thus, removal of large fecund females by the fishery is not likely to influence significantly upon the reproductive potential of the population.

George (1973) proposed a potential estimate of the lobster stock along the Kanyakumari coast at 500 t based on the data available at that time. Assuming that 60% of the biomass could be harvested, the annual catchability is 300 t, which was realised in 1966. However, harvest levels could never attain the same volume in the following years, though annual landing of around 91 t was achieved until 1978–1979. He also suggested the need to fix a minimum harvestable size for fishing (130 or 140 mm TL for *P. homarus*) (George 1967a). Reasons for the decline in landings from 47–300 t per annum (1960–1980) to the current 10–15 t per annum may be linked to indiscriminate and uncontrolled fishing in the absence of regulatory measures and influence of some non-fishery factors such as monsoonal upwelling and related oceanographic and ecological conditions.

7.12.7.5.4 Size at Maturity

Thangaraja and Radhakrishnan (2017) determined the size at the onset of maturity (SOM) of female *P. homarus homarus* from external secondary sexual characteristics, the macroscopic examination of the ovary and from the percentage of females carrying eggs. The size at which 50% of the female population reaches sexual maturity based on females carrying eggs was 61.0 mm CL, whereas that based on 'setose' method was 54.0 mm CL, 'windows' method 55.0 mm CL and the length at which they attain physiological maturity (ovary colour and weight) 55.0 mm CL. The smallest berried female was 45.5 mm CL. The gonadosomatic index (GSI) of females (30–49 mm CL) ranged from 0.15 + 0.27 (30–49 mm CL) to 0.95 + 0.3 (70–79 mm CL). Month-wise, GSI of females increased from 0.6 ± 0.2 in October (beginning of breeding) to 0.9 ± 0.1 in November (peak breeding) and 0.7 ± 0.52 in December.

The calculated size at maturity for females from the Kanyakumari stock of *P. homarus homarus* is close to the Sri Lankan population of the same subspecies (55–59 mm CL) (De Bruin 1962), while Jayakody (1989) estimated the size at maturity based on percentage of egg-bearing lobsters as 59.5 mm CL. Size at maturity was estimated as 63.4 mm CL in Kenyan population of *P. homarus homarus* based on the same criteria (Kulmiye et al. 2006), while Thangaraja and Radhakrishnan (2017) reported 61.0 mm CL as the size at maturity for the Kanyakumari population (Table 7.11). *P. homarus rubellus* females of east coast of South Africa attain maturity at 54.0 mm CL (Berry 1971) and the Somalian population of *P. homarus homarus* at 58.0 mm CL (Fielding and Mann 1999). Size at maturity for the subspecies *P. homarus homarus* in Oman waters varied from 69.2 mm CL to 75.9 mm at various regions (Mohan 1997). The size at maturity estimated from secondary sexual characteristics such as presence of pleopodal ovigerous setae and 'window' on sternal

Table 7.11 Size at female maturity of *Panulirus homarus homarus* estimated by different methods at various geographical locations

Authors		Setal method	Ovary (GSI)	Egg bearing	Windows method	Country/location
Al Marzouqi et al. 2007		–	–	63.9 mm CL	–	Oman (Al-wasta)
		–	–	65.6 mm CL	–	Oman (Dhofar)
Kulmiye et al. 2006		52.6 mm CL	52.0 mm CL	63.4 mm CL	–	Kenya (Mambrui)
Mohan	1997	–	–	59.5 mm CL	–	Oman (Dhofar)
		–	–	69.2 mm CL	–	(Oman) Shuwaymiyah
		–	–	75.9 mm CL	–	(Oman) Sudh
				72.8 mm CL	–	(Oman) Mugsyl
Berry	1971	–	–	54.0 mm CL		South Africa (Natal)
Jayakody	1989	–	–	59.5 mm CL	–	Sri Lanka (South coast)
De Bruin	1962	–	–	55–59 mm CL		Sri Lanka (West coast)
George	1963			60–70 mm CL		East Aden
Heydorn	1969			50.0 mm CL		South Africa
Thangaraja and Radhakrishnan	2017	54.0 mm CL	55.0 mm CL	61.0 mm CL	55.0 mm CL	Southwest coast of India (Khadiyapatnam)

plate is lower than that calculated from percentage of egg-bearing females. Presence of ovigerous setae and window is an indicator of functional maturity in female *P. homarus homarus*, i.e. they are capable of mating and carry eggs on the pleopods.

7.12.7.5.5 Morphometric Relationships

The morphometric relationships were calculated for the southern Indian population of *P. homarus homarus* for the carapace length from 20.4 mm to 101.1 mm (Radhakrishnan et al. 2015), with the differences increasing as the animals increase in size. Females were heavier than males of an equivalent carapace length. For a given carapace length, females were also slightly longer (total length) than males and have a wider second abdominal segment. The growth rate is faster in females than in males. The third walking leg in males grows allometrically after a certain carapace length which marks the beginning of attaining sexual maturity in males. The size at sexual maturity of males was estimated as 63.0 mm CL, which was derived from the CL-III walking leg relationship.

The length-weight relationship of *P. homarus* subspecies from different geographical regions is shown in Table 7.12. The CL/weight relationship of the Kanyakumari *P. homarus homarus* stock is very similar to those of other subspecies along the South African coast (*P. homarus rubellus*) and Somalia and Oman coast (*P. homarus homarus*). However, Sri Lankan female *P. homarus homarus* is 13% lower in weight (Sanders and Liyanagae 2009) for an equivalent CL than the Kanyakumari female lobsters. The male lobsters from Kanyakumari for a given carapace length are comparatively heavier than other subspecies and that of the same subspecies from Sri Lanka.

7.12.8 Southeast Coast Fishery

The southeast coast lobster fishing region encompasses Kanyakumari in the south to Chennai in the north. The major areas of lobster landing by trawlers along the east coast of Tamil Nadu are Tuticorin, Mandapam, Nagapattinam, Cuddalore and Chennai. A significant percentage of lobsters are also landed by gillnets and trammel nets in several artisanal lobster fishing villages. Small quantities are also landed by skin divers engaged in chank fishing. Palinurid and scyllarid lobsters are incidentally caught by the trawlers targeting mainly for prawns. Nair et al. (1973) provide description of an artisanal fishery for lobsters at Vedalai, near Mandapam. Rajamani and Manickaraja (1997a, b) provide descriptions of the Tuticorin gillnet and trawl fishery, Subramaniam (2004) on the *Thenus orientalis* (*T. unimaculatus*) fishery of Chennai coast, Kizhakudan et al. (2004) on the bottom-set gillnet fishery for *Thenus orientalis* (*T. unimaculatus*) and Kagwade et al. (1991), Radhakrishnan and Vijayakumaran (2003), Radhakrishnan et al. (2005) and Radhakrishnan and Thangaraja (2008) on the status, management and conservation of the lobster fishery along the Indian coast.

Table 7.12 Length-weight relationships in *Panulirus homarus* subspecies from different geographic regions

Country	Subspecies	Sex	Length-weight equation	References
South Africa	*Panulirus homarus rubellus*	Male	$Y = 0.00243 CL(mm)^{2.7767}$	Berry, P.F. 1971. *Investl Rep. Oceanogr.Res.Inst. S. Afr.*28:1–75
		Female	$Y = 0.00177 CL(mm)^{2.8590}$	
Somalia	*Panulirus homarus homarus (megasculpta)*	Male	$Y = 0.0022 \; CL(mm)^{2.8031}$	Fielding, P.J and B.Q. Mann. 1999. *IUCN*, Nairobi, 35 pp.
		Female	$Y = 0.0014 CL \; (mm)^{2.9388}$	
Sri Lanka	*Panulirus homarus homarus*	Male	$Y = 0.0015 CL \; (mm)^{2.71}$	Sanders, M.J and Liyanage, U. 2009.*The project report of National Aquatic Resources Research and Development Agency (NARA), Sri Lanka*
		Female	$Y = 0.0022 \; CL \; 9 \; mm)^{2.54}$	
Oman	*Panulirus homarus homarus (megasculpta)*	Male	$Y = 0.003 CL \; (mm)^{2.629}$	Al-Marzouqi, Al-Nahdi, A., Jayabalan, A. N., and J. C. Groeneveld, 2007.*J. Mar. Sci.*, 6(2): 159–174
		Female	$Y = 0.0024 CL \; (mm)^{2.717}$	
		Male	$Y = 0.0018 CL \; (mm)^{2.8597}$	Al-Marzouqi, A, Chesalin, M, Al-Shajibi, S, Al-Hadabi, A and Al-Senaidi, R. 2015.*J Aquac Mar Biol* 3(1):00056. DOI: https://doi.org/10.15406/jamb.2015.03.00056
		Female	$Y = 0.0016 CL \; (mm)^{2.9124}$	
India	*Panulirus homarus homarus*	Male	$Y = 0013 CL \; (mm)^{2.9059}$	Radhakrishnan, E.V. Thangaraja, R and M. Vijayakumaran.2015. *J. Mar. Biol. Ass. India*, 57(1): 5–13
		Female	$Y = 0.0013 CL \; (mm)^{2.9205}$	

7.12.8.1 Lobster Fishery of Tuticorin

Tuticorin is one of the major lobster landing centres of the state. Lobsters are landed both by trawls and gillnets. Lobsters appear as a bycatch in shrimp trawls almost throughout the year, whereas gillnets are operated in shallow rocky areas exclusively for lobsters.

7.12.8.1.1 Trawl Fishery

The mechanised trawlers based at Tuticorin Fisheries Harbour carry out trawling operations in the Gulf of Mannar, off Manapad in the south and off Erwadi on the north mainly for fishing for prawns. During such operations, occasionally good quantities of lobsters are also captured and landed as a bycatch in shrimp trawls operating at depths ranging from 20 to 60 m (Rajamani and Manickaraja 1997b). The total average annual catch for Tuticorin during 1978–85 was reported as 12 t with a catch rate of 4.39 kg/unit (Kagwade et al. 1991). The estimated annual landings of lobsters at Tuticorin Fisheries Harbour during 1991–1992 and 1992–1993 were 32.5 t and 42.6 t, respectively. The estimated average monthly landings fluctuated from 1.3 t in March to 10.7 t in October (33% of total catch) during 1991–1992 and from 0.1 t in September to 23.7 t in May (55.6% of total catch) during 1992–1993. The average catch rates for the 2 years were 1.4 kg/U and 1.9 kg/U, respectively (Rajamani and Manickaraja 1997b). *Panulirus ornatus* dominated the fishery, constituting an average 86.5% of the total catch. The average annual effort continued to rise almost 14-fold from 2083 units in 1991–1992 to 31,871 units in 2007–2012, while the average annual catch decreased to 0.7 t with a CPUE of 0.02 kg/U.

7.12.8.1.2 Gillnet Fishery

Kayalpattinam and Tharuvaikulam are the two major gillnet landing centres near Tuticorin (Rajamani and Manickaraja 1997a). At Kayalpattinam, bottom-set gillnets landed 17.5 t of lobsters during the year 1976. Although lobsters were caught throughout the year, larger quantities were landed during August–January and April–May; the peak season was in November. In 1993, the lobster catches increased from 42.2 t with a catch rate of 6.5 kg/unit to the peak catch of 50.6 t with a catch rate of 5.5 kg/unit in 1994. However, the landings declined to 4.4 t with a catch rate of 1.1 kg/unit in 2002 (Radhakrishnan et al. 2005). At Tharuvaikulam, landings gradually decreased from 11 t with a catch rate of 1.1 kg/unit in 1993 to 1.1 t with a catch rate of 0.6 kg/unit in 2002. During 2007–2012, the average annual landing was 4.7 t, an improvement over the 2002 landings. The total annual average landing at Tuticorin district by gillnets (includes non-mechanised, motorised gillnetters and outboard gillnetters) during 2007–2012 was 22.6 t with a catch rate of 0.07 kg/U. In spite of the increased effort, the gillnet fishery for the spiny lobsters seems to sustain itself under the present rate of exploitation.

Though lobsters were landed throughout the year, the peak fishery was from October to December and April to May (Rajamani and Manickaraja 1997b). Manickaraja (2004) reported operation of a modified bottom-set gillnet with larger mesh size by lobster fishermen of Kayalpattinam. The fishermen shifted from usage of monofilament nylon net to the plastic valai (gillnet), by increasing the mesh size from 85 to 130 mm. The operational depth was 60 m instead of 4 to 6 m. The yield was of larger lobsters 160–390 mm TL compared to 135 to 290 mm TL in monofilament net. *Panulirus ornatus* formed 71% of the catch (54.5% in monofilament net) and the rest by *P. versicolor*.

7.12.8.1.3 Species Composition

The commercial lobster fishery at Tuticorin is currently represented by *P. ornatus* (66%), *P. homarus homarus* (25%) and *P. versicolor* (9%). The deep-sea lobster *P. sewelli* is landed occasionally. The slipper lobster *T. unimaculatus* is not a regular species supporting the fishery. Recently, a single specimen of *Scyllarus batei batei* was reported from Tuticorin coast (Vaitheeswaran 2015). Kagwade et al. (1991) reported an average annual landing of 4.4 t of *P. ornatus* at the catch rate of 1.55 kg/unit for the 8-year period from 1978 to 1985. The fishery for the species was present all round the year showing only marginal fluctuations ranging between 0.3 t in January and 0.5 t in November. The average annual landing of *P. homarus* during the same period was 8 t with a catch rate of 2.84 kg/unit. The fishery for the species was throughout the year at Tuticorin. During 1976 at Kayalpattinam, *P. homarus* was the dominant species (70%) and *P. ornatus* the rest of the landings. During 1978–1989, *P. homarus* continued to be the major species forming 72.3% of the total catch (Radhakrishnan et al. 2005).

7.12.8.1.4 Change in Relative Abundance of Species

A change in the relative abundance of species in both trawl and gillnet fishery of Tuticorin was evident during 1991–1993 period with *P. ornatus* dominating the fishery, forming an average of 86.5% of the total trawl landings. The monthly landing of *P. ornatus* ranged from 1.2 t to 8 t in 1991–1992 and from 0.1 t to 20.5 t in 1992–1993. *Panulirus homarus* catch ranged from a minimum of 0.06 t to 2.7 t in 1991–1992 and 0.03 t to 3.1 t in 1992–1993 (Rajamani and Manickaraja 1997a). Peak landing of *P. ornatus* was in October (8 t) in 1991–1992 and May (20.5 t) in 1992–1993. *Panulirus homarus* has also followed the same pattern in landings in both the years (2.6 t in October during 1991–1992 and 3.1 t in May during 1992–1993). At Tharuvaikulam, *P. ornatus* was the major species constituting 90% of the total landings (Rajamani and Manickaraja 1997a). During 2007–2012, *P. ornatus* continued to be the major species in the gillnet fishery at Tuticorin with a total annual average landing of 14.9 t (66%) followed by *P. homarus homarus* 5.7 t (25%) and *P. versicolor* 2 t (9%), whereas in trawls, *P. ornatus* forms 86% of the total catch and the rest by *P. homarus* (1991–1993) (Rajamani and Manickaraja 1997b).

7.12.8.1.5 Length Composition

The total length of *P. ornatus* in trawl fishery ranged from 113 to 233 mm in males with maximum numbers in the size range of 111–140 mm (40.0%). In females, the size ranged from 128 to 452 mm and the predominant size ranged from 181 to 190 mm (40.9%). In *P. homarus* the size ranged from 89 to 209 mm in males with nearly 85% of the catch in the range of 101–140 mm. In females, the size ranged from 107 to 194 mm with 46.2% of the catch in the size range of 101–130 mm. The sex ratio of *P. ornatus* was in favour of females (59.5%), whereas in *P. homarus* males were dominant (66.7%) (Rajamani and Manickaraja 1997a). In the gillnet fishery of Tharuvaikulam (1990–1992), the length range of males varied from 120 to 342 mm TL and females from 117 to 363 mm TL. The average mean sizes recorded for males and females were 199.6 and 210.2 mm, respectively. Large-sized lobsters measuring more than 300 mm TL were encountered in the fishery during June and March in the case of males. Large-sized females were observed in May, June, September, October and March. Smaller-sized lobsters measuring less than 130 mm TL were encountered in the catch during July, October and November. The length range of *P. homarus* in the fishery ranged from 100 to 192 mm TL with mean size at 149.4 mm for males and from 110 to 271 mm TL in females with mean size at 162 mm. In *P. ornatus*, predominance of females was observed in most of the months constituting an average 58.4% (Rajamani and Manickaraja 1997b). Ovigerous females were never encountered either in trawl or the gillnet fishery, and they may probably be migrating to deeper waters for breeding (Rajamani and Manickaraja 1997b; Kagwade et al. 1991; Radhakrishnan et al. 2005). However, egg-bearing females were landed by gillnetters operating in deep waters off Nagapattinam in small quantities (Radhakrishnan, personal observation). At Kayalpattinam, the modal length of male *P. homarus* occurring in the fishery decreased from 245 mm TL during 1978 to 145 mm TL during 2002. The modal length of females also decreased from 195 mm TL to 165 mm TL during this period. Meanwhile, the modal length of both males and females of *P. ornatus* showed an increase from 175 mm to 195 mm TL (Radhakrishnan et al. 2005).

No information on stock and exploitation rate of lobsters from Tuticorin is available. *Panulirus homarus* was reported to breed round the year at Tuticorin with a major peak from December to March. Fecundity of a single brood in *P. homarus* (carapace length, 47–94 mm; weight, 156.75–900 g) ranged from 95,530 to 480,590. The fecundity-carapace length relationship estimated was $F = 7.8405 \, CL^{283.827}$, and the carapace length and egg mass weight relationship was $0.8969 \, CL^{31.5772}$. The estimated length at first maturity of female was 53 mm CL (Jawahar et al. 2014).

7.12.8.1.6 Lobster Fishing by Skin Diving

Capturing lobsters by hand by fishermen divers has been reported from many countries. In Somalia, diver catches sometimes exceed that of the tangle net catches, especially when water clarity is good (Fielding and Mann 1999). At Tuticorin, skin divers engaged in chank fishing also catch lobsters by hand from November–December to March–April, when water clarity is good (Rajamani and Manickaraja 1991). Two species of spiny lobsters *P. ornatus* and *P. homarus* are captured with the former being the dominant one. During January–February 1991, an estimated 1.6 t of lobsters of *P. ornatus* were captured by the chank fishermen, of which 87% were landed in January. The total number of divers involved in chank fishing during January was 28,440. The length of *P. ornatus* ranged from 154 to 450 mm total length (TL) in males and 169 to 423 mm TL in females. However, the predominant size range was 211 to 280 mm TL in both the sexes. The sea bottom is sandy with rocks in small patches. The divers catch the lobsters from the crevices among the rocks and bring them to the boat, where they are kept alive, until sold to traders.

7.12.8.2 Lobster Fishery of Mandapam

Spiny lobsters are landed both by trawlers and gillnets at Mandapam. In spite of its importance as a resource, very little information on the fishery and rate of exploitation is available. Trawlers landed an average 2.7 t of lobsters at Mandapam during 2007–2012. An average 5.4 t of slipper lobsters were landed during 1976 by trawl nets and gillnets. *Thenus orientalis* (*T. unimaculatus*) was caught in trawl nets throughout the year. The size ranged from 90 to 200 mm in total length. Gillnet fishing is mainly carried out by fishermen from Vedalai, a fishing village adjacent to Mandapam.

7.12.8.2.1 Gillnet Fishery

During 1976, the bottom-set gillnets landed 8.9 t of lobsters. Among the spiny lobsters, *P. ornatus* (70%) was the dominant species ranging in size from 240 to 344 mm TL. Peak landings were observed in October–December. Nair et al. (1973) provided a detailed report of lobster fishing off Vedalai, which commences from October and extends up to May. Good catches are obtained from December to March. The main fishing areas are Manauli Island, New Island and Hare Island located off Vedalai. During the 4 months of fishing from December to March, 1971–1972, an average 1.4 t of lobsters were landed by bottom-set gillnets. *Panulirus ornatus* is the most dominant species forming 68.7% of the total catch, followed by *P. homarus*, 21.2%; *P. versicolor*, 7%; *P. longipes longipes*, 2.8%; and *P. penicillatus*, 0.4%. *Panulirus homarus* is reported to be more common on the northern side of the islands at 5–8 m depth and *P. ornatus* on the southern side at 9–16 m depth. *Thenus unimaculatus* was landed by gillnets during June–August.

7.12.8.2.2 Species Composition

Six species of spiny lobsters are recorded from Mandapam region of the Gulf of Mannar (Nair et al. 1973). *Panulirus ornatus* is the dominant species landed by trawlers at Mandapam and Pamban. *Panulirus homarus homarus* forms a fairly good fishery at Valinokkam, south of Mandapam. Other species reported from Mandapam but contributing to the lobster fishery of the region on a smaller scale are *P. versicolor*, *P. longipes longipes* and *P. sewelli*. During 2007, trawlers based at Mandapam (Gulf of Mannar) landed 2 t of *P. sewelli*. *Panulirus penicillatus* and *P. polyphagus* are rarely caught from Gulf of Mannar. Nair et al. (1973) reported the occurrence of *P. penicillatus*, *P. longipes longipes* and *P. polyphagus* from Mandapam. The occurrence of *P. polyphagus* is reported for the first time from the southeast coast of India. A detailed description of the habitat, distribution and fishery of these species was given by these authors. *Panulirus longipes longipes* is fished from the coral reefs around the Manauli Island, New Island and Hare Island in the Gulf of Mannar. This subspecies prefers reefs constituted by a particular genus of coral, viz. *Acropora* (four species, *A. formosa*, *A. nobilis*, *A. surculosa* and *A. erythraea*, are very common around these islands), and is absent in the other reef beds. Four specimens of *P. penicillatus* were collected from the lobster landings during February–March 1972. They were caught in bottom-set nylon wall nets at a depth of 7–16 m from the windward surf zone of the coral reefs off Vedalai. One large male specimen of *P. polyphagus* measuring 117 mm CL was caught from the same ground at 9 m depth. The other noncommercial species of lobsters recorded from Mandapam are the scyllarid *Scyllarides tridacnophaga* (Radhakrishnan et al. 1995) and the reef lobster, *Enoplometopus occidentalis* (Radhakrishnan and Jayasankar 2014).

7.12.8.3 Lobster Fishery of Chennai (Madras) Coast

The lobster fishery of Chennai coast is constituted both by spiny and slipper lobsters. While spiny lobsters are landed by trawlers as well as gillnets and trammel nets operating in inshore fishing grounds, slipper lobster fishery is mainly trawl based. However, along the Chennai coast, slipper lobsters are also landed by bottom-set gillnets operating at depths of 20–40 m (Kizhakudan et al. 2004). Gillnets and trammel nets are operated in rocky fishing grounds. There is no targeted trawl fishery for lobsters along the Tamil Nadu coast, and they are landed as bycatch in shrimp and cephalopod trawls. The single-day and multiday fishing trawlers based at fisheries harbour conduct the fishing operations mostly north of Chennai (south Andhra Pradesh coast).

7.12.8.3.1 Species

The major species constituting the trawl fishery of spiny lobsters are *P. homarus homarus*, *P. ornatus*, *P. versicolor*, *P. polyphagus*, *P. sewelli*, *N. stewarti* and *P. waguensis*. The slipper lobster *T. unimaculatus* is the only commercial species of scyllarid in the fishery. Kizhakudan and Thirumilu (2006) reported landing of *P. waguensis* along with *Nephropsis stewarti* and *N. carpenteri* at Madras Fisheries Harbour in 2004.

7.12.8.3.2 Trawl Fishery

Kagwade et al. (1991) report annual average landings of lobsters at Madras as 10.9 t (1978–1984). The spiny lobster *P. polyphagus* was poorly represented in the fishery with an annual estimated average landing of 0.4 t forming 3.6% of the lobster landings. The average annual catch of *P. homarus* was 4.7 t forming 43% of the total lobster landings and *P. ornatus*, 0.2 t forming just 1.9% of the total catch. The average annual catch of *T. orientalis* (*T. unimaculatus*) was 5.6 t forming 52% of the total landing (Kagwade et al. 1991).

The total estimated annual average trawl landing of spiny lobsters at Madras Fisheries Harbour for the period 1980–2008 was 83.3 t (Fig. 7.50). The annual

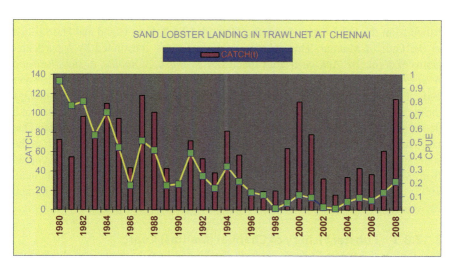

Fig. 7.50 Catch and CPUE of *Thenus unimaculatus* landed by trawlnets at Chennai Fishing Harbour

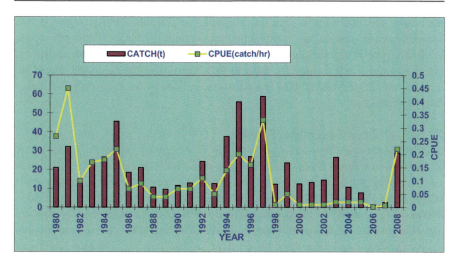

Fig. 7.51 Catch and CPUE of lobsters landed by trawlnet at Chennai Fishing Harbour

average landing of spiny lobsters was 21.3 t with a catch rate of 0.11 kg/hr. When the landings were partitioned into 10-year periods, average landing of spiny lobsters for the period 1980–1989 was 22.3 t with a catch rate of 0.16 kg/hr. Average landing for the succeeding 10-year period (1990–1999) was 24% higher (27.6 t), but the catch rate declined to 0.12 kg/hr. The average catch was further reduced to 13.2 t (2000–2008) with a catch rate of 0.04 kg/hr. The estimated annual average landing of slipper lobster for the period 1980–2008 was 62 t with a catch rate of 0.28 kg/hr. (Fig. 7.51). The average landing declined during 2007–2012 (45 t and CPUE 1.72 Kg). The catch on partitioning to 10-year periods showed 81.3 t with a catch rate of 0.56 kg/hr. during 1980–1989, 45.5 t with a catch rate of 0.19 kg/hr. for 1990–1999 and 58.8 t with a catch rate of 0.09 kg/hr. for 2000–2008. During 2007–2012, an average 0.35 t of lobsters was landed by gillnets at the Madras Fisheries Harbour.

In Nagapattinam, a major trawl landing centre, an average 47.6 t of *T. unimaculatus* with a CPUE of 3.32 Kg was landed during 2007–2012.

7.12.8.3.3 Biology of *T. unimaculatus*

Subramanian (2004) provides a description of the fishery and biology of *T. orientalis* (*T. unimaculatus*) along the Chennai coast. The main fishery was from September to February with peak landing from October to November. Females were larger with a mean size of 156.9 mm TL, against 153.6 mm for male. The asymptotic length calculated by von Bertalanffy growth equation for females was 298.91 mm with K as 0.2967/year and 288.91 for males with K as 0.3192/year, the initial age (t_0) being considered as 0 for both sexes. The size at maturity of female based on 50% of egg-bearing females was 105.5 mm TL. The two peak spawning spells were February–March and June–August and peak recruitment during January–February and

May–July. On the northwest coast, Kagwade and Kabli (1996a) reported two spawning peaks, a major one in November and a minor in March. The length-weight relationship of *T. unimaculatus* collected from Mudusal Odai (Cuddalore district) landing centre was provided by Saha et al. (2009). The 'b' value for males and females was 2.9161 and 2.9758, respectively, indicating faster growth in females than the males.

7.12.8.3.4 Deep-Sea Lobster Fishery

The deepwater spiny lobsters *P. sewelli* and *P. waguensis* support a small fishery along the north Tamil Nadu coast. Except for the four specimens of *P. waguensis* caught during 1965, there were no reports of the species from the Indian coast (George and George 1965). Large trawlers operating in the depth range of 300–450 m off Puducherry to Nagapattinam landed *P. waguensis* along with *N. stewarti* and *N. carpenteri* at Chennai Fisheries Harbour during 2004. There is no regular fishery for deep-sea lobsters and these are occasional catches. The total landing of *P. waguensis* was 0.53 t in February, April, August and September with latter 2 months landing bulk of the catch. The specimens landed were in the size range of 32–75 mm in carapace length. Kizhakudan and Thirumilu (2006) were the first to report a fishery of this magnitude from the Indian coast, though there were reports of stray landing of the species by trawls on the southwest coast (Chakraborty et al. 2014) and by gillnets on the southeast coast (Pillai andThirumilu 2007; Kizhakudan et al. 2012). The presence of this deep-sea lobster in gillnets operating in inshore waters is intriguing.

7.12.8.3.5 Gillnet Fishery

The bottom-set- gillnet landing of lobsters at Chennai Fisheries Harbour for the period 2000–2005 was an estimated 22.5 t with a catch rate of 9.5 kg/unit. The species composition and percentage of each species representing the fishery were *P. homarus*, 67.5%; *P. ornatus*, 7%; *P. versicolor*, 7.5% and *P. polyphagus*, 18%. The species composition and percentage of each species representing the fishery were *P. homarus*, 67.5%; *P. ornatus*, 7%; *P. versicolor*, 7.5% and *P. polyphagus*, 18%.

At Kovalam, a gillnet lobster landing centre south of Chennai, the fishery starts after the northeast monsoon. The dominant species in the fishery is *P. homarus* followed by *P. ornatus*. Peak landing is during December–January months. Fishermen use both bottom-set gillnets and trammel nets for lobster fishing. Trammel nets catch lobsters as small as 23 mm in carapace length, whereas the smallest size in the gillnet was 38 mm CL. An analysis of the catch data shows 50.8% of *P. homarus* in the trammel net belonging to the length range of 23–53 mm CL, compared to 17.6% in the gillnet. On the contrary, the percentage of larger lobsters was higher in the gillnets. The size composition shows 25% of lobsters in the size range of 73–98 mm CL in trammel nets compared to 43.2% in the gillnet (Fig. 7.52) (Radhakrishnan et al. 2005). The study shows that operating gillnets is economically more beneficial than fishing with trammel nets.

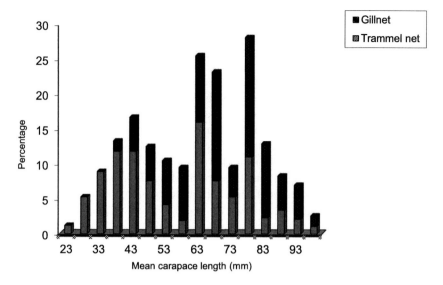

Fig. 7.52 Size frequency of *Panulirus homarus homarus* in trammel net and gillnet catches at Kovalam, Chennai

7.13 West Bengal Fisheries

7.13.1 Introduction

West Bengal is the northeastern-most maritime state of India bordering the Bay of Bengal. The length of the state coastline is 158 km, and, if tidally influenced creeks around the islands are considered, the total length of coastline is estimated to be 900 km (Dutta et al. 2016). The coastline of West Bengal spreads along the southern edge of its two maritime districts, 24 Parganas (south) and Purba Medinipur. The Purba Medinipur district has 38 marine fish landing centres, and 24 Parganas (south), 21 centres. The continental shelf is wide, extending up to 150 Km, shallow, having a muddy bottom, and its configuration is affected by the large river systems and tidal currents (Dutta et al. 2016). The coast is influenced both by the West India Coastal Current and the East India Coastal Currents (Shetye 1998). The West Bengal coastal ecosystem is part of the Large Marine Ecosystem of Bay of Bengal. The major marine fish landing centres are Digha and Contai. The annual average landing of marine fishes in 2014 by the state was 0.07 million t (87% less than 2013 landings of 0.26 million t), contributing only 2.1% to the total marine fish landings in India of 3.74 million t (CMFRI Annual report 2013–2014).

7.13.2 History of Lobster Fisheries

There is no information on the early years of lobster fishing along the West Bengal coast. The earliest reports on distribution of lobsters along the West Bengal coasts are by George (1967a) and Jones (1965). Lobsters may be part of the trawl catch ever since trawling was in operation along the West Bengal coast. However, fishery was not fully developed, as there was no market for lobsters during the 1950s and early 1960s. The spiny lobsters *P. polyphagus* and *P. ornatus* and the slipper lobster *T. unimaculatus* constitute the lobster fauna of the West Bengal coast (Ramakrishna et al. 2003). Chatterjee et al. (2007) reported two species of lobsters along the Digha coast of West Bengal: *P. ornatus* and the slipper lobster *T. unimaculatus*. *Panulirus polyphagus* is the major species of spiny lobster contributing to the lobster fishery along the northeast coast.

7.13.3 Fishery

The lobster landing data for West Bengal is available from 1994 onwards (CMFRI Annual Report). The annual average trawl landing of lobsters is (1994–2013) 121 t. The catch has fluctuated between a minimum of 1 t in 1994 and a maximum of 789 t in 2011. During the 10-year period from 1994 to 2003, the annual average landing was 19 t, and during the following 6 years (2004–2013), the average landing was 203 t, showing a phenomenal increase in recent years (Fig. 7.53). This increase in catch was also reflected in the total annual landing of lobsters in India for this period. The landing was almost entirely by the single-day and multiday fishing trawlers. Although there are three species of lobsters distributed along the

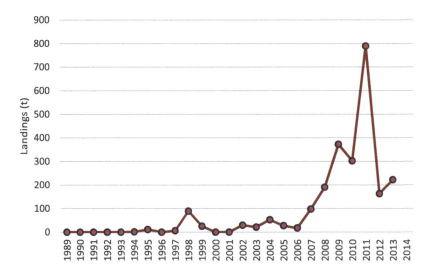

Fig. 7.53 Annual landing of lobsters in West Bengal (1989–2013)

coast, only two are of commercial importance. *P. polyphagus* constitutes 99.4% of the yearly catch and the rest is by *T. unimaculatus*. *P. polyphagus* also exhibits latitudinal distribution along the Indian coast, the species being the most dominant along the Maharashtra-Gujarat (northwest coast) and the West Bengal coast (northeast coast).

Size frequency data on the commercial catch and biological information on the principal species are currently not available. This information needs to be collected to understand the dynamics of the fishery and for future management.

7.14 Lobster Fisheries of Andhra Pradesh

7.14.1 Introduction

Lobster landing in Andhra Pradesh has never been large in volume until 2000. The landings gradually increased from 2001, probably with expansion of the fishing grounds and increasing demand for lobsters in the export market. The landing was not consistent initially and showed a highly fluctuating trend.

7.14.2 Fishery

Lobster landing data for the state of Andhra Pradesh is available from 1989 onwards at CMFRI. The total annual average landing (1989–2014) was an estimated 43 t (Fig. 7.54). Analysis of decadal landings shows the first 10-year period of the fishery (1989–1998) with only an average landing of 8 t and the following decade

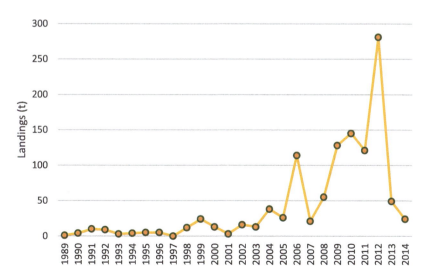

Fig. 7.54 Annual landing of lobsters in Andhra Pradesh (1989–2014)

(1999–2008) with an average landing of 43 t. During 2009–2012, there was a substantial increase in catch with an average of 167 t. However, the catch showed significant decline in 2013 and 2014.

The major part of the landings was by mechanised single-day and multiday fishing trawlers (95%). The gillnets (both non-mechanised and mechanised together) landed the rest of the catch. Four species in the genus *Panulirus*, *P. homarus homarus*, *P. polyphagus*, *P. ornatus* and *P. versicolor*; the deep-sea lobster, *P. sewelli*; and the slipper lobster, *T. unimaculatus*, constitute the lobster fauna of the Andhra Pradesh coast. Species-wise landings are available from 2007 onwards, and the annual average landing during 2007–2012 was an estimated 117 t. The species supporting the commercial fishery are *T. unimaculatus*, with an average landing of 83 t (71%); *P. polyphagus*, 31 t (27%); and *P. homarus homarus*, 2.5 t (2%). *Panulirus ornatus*, *P. versicolor* and *P. sewelli* have appeared in the fishery in certain years but not in commercial quantities, except *P. sewelli*.

7.14.3 CPUE and Catch Trends

At Visakhapatnam Fishing Harbour, the catch and CPUE of *P. polyphagus* landed by trawlers during 2007–2012 were an average 33.4 t and 0.3 kg/U (0.003 kg/hr), respectively. The catch and CPUE of *T. unimaculatus* were 43 t and 0.39 kg/U (0.003 kg/hr), respectively. Though the catch of *P. polyphagus* has fluctuated around 4 t, the landings have improved to an average 7 t during 2011 and 2012. The increase in unit effort from 2007 to 2012 was 42%, whereas the trawling hours increased twofold.

Research on lobsters from the region is very limited. Lalithadevi (1981) reported occurrence of juveniles of *P. homarus* and puerulii of *P. polyphagus* from the bay and estuary of Kakinada. Hossain (1978) conducted a detailed study on the biology of *T. orientalis* (*T. unimaculatus*). Molecular studies revealed the presence of *T. indicus* apart from *T. unimaculatus* along the Andhra Pradesh coast (Jeena 2013).

7.15 Odisha Lobster Fisheries

7.15.1 Fishery

The contribution of Odisha to the lobster catch of India is very meagre (Fig. 7.55). The catch fluctuated between a minimum of 3 t and a maximum of 102 t. The entire catch is landed by mechanised trawlers. The average landing between 1989 and 2005 was less than 10 t.

The slipper lobster *T. unimaculatus* contributed 78.4% to the total catch, followed by *P. polyphagus* (16.5%) and *P. homarus homarus* (5%). Biological information on the major species constituting the fishery is not available.

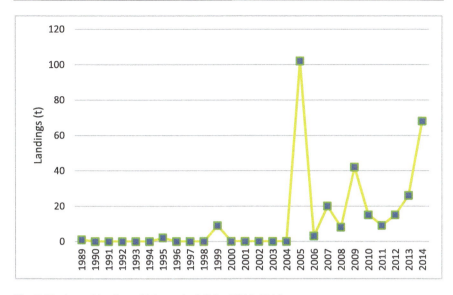

Fig. 7.55 Annual landing of lobsters in Odisha (1989–2014)

7.16 Karnataka Lobster Fisheries

7.16.1 Fishery

Karnataka also contributes a smaller volume to the lobster landings of India with an annual average landing of 19 t (2000–2012). Though the landings were initially promising (an average 57 t from 2000 to 2004), the catch was less than a tonne in the succeeding years. Although there are nine species of lobsters along the Karnataka coast, none of them are landed in commercial volumes consistently. The lobsters belonging to the genus *Panulirus* are *P. homarus homarus*, *P. polyphagus*, *P. ornatus*, *P. versicolor*, *P. penicillatus* and *P. longipes longipes*. The deep-sea lobsters *P. sewelli* and *N. stewarti* and the slipper lobster *T. unimaculatus* are sporadically reported in the commercial catch (Zacharia et al. 2008; Dineshbabu et al. 2011).

A comprehensive description of the fishery and morphometry of *N. stewarti* was published by Dineshbabu (2008). During 2000–2006, an average 23.3 t was landed annually at Mangalore by deep-sea trawlers operating off north Kerala and South Karnataka coasts at depths of 250–500 m. However, the fishery survived only for a brief period with sporadic landings in the following years.

An average 2 t of lobsters are also landed annually at the union territories of Puducherry and Daman and Diu.

7.17 Island Fisheries

7.17.1 Introduction

There are two major Oceanic Island groups located within the EEZ of India, the Andaman and Nicobar Islands in the Bay of Bengal and the Lakshadweep Islands in the Arabian Sea. The Andaman and Nicobar (A&N) Islands, a union territory of India, is an archipelago of 572 islands and islets, of which 38 are inhabited, located in the southeast Bay of Bengal, between latitudes 6°45'N and 13°41'N and longitudes 97°57' and 93°57' E. The north Andaman Groups of Islands are believed to be an extension of mountain systems along the Indo-Myanmar border region and are volcanic islands with surrounding reefs, whereas the Nicobar Group of Islands are mostly coral islands surrounded by coral reefs. These Islands are part of the Bay of Bengal Large Marine Ecosystem (BOBLME) and are known for their rich marine coastal and deep-sea biodiversity. During the southwest monsoon, the Islands get heavy rainfall.

The Lakshadweep Group of Islands, situated in the Arabian Sea, are an archipelago of 12 atolls, 3 reefs and 5 submerged banks with 32 sq.km. of geographical area. There are 36 islands of which 10 are inhabited and 16 islets are uninhabited (Devaraj and Pillai 2001). In some islands, the lagoons are on the western side and in others the lagoons have a north-south orientation; the southwest monsoon brings heavy rainfall to the islands. These islands are well known for the tuna fishery.

The systematic exploration of deep-sea marine fauna in the Bay of Bengal and seas around the Andaman and Nicobar Islands and Lakshadweep was conducted by the R.I.M.S. *Investigator* launched in 1881. Heller (1865), Alcock and Anderson (1894), Alcock (1901, 1906), Borradaile (1906), Lloyd (1907), Sewell (1913) and De Man (1916) were the first to enlist the lobster fauna of Andaman and Lakshadweep seas. The species composition, distribution and fishery of lobsters in these two island systems are discussed.

7.17.2 Andaman and Nicobar Island Lobster Fishery

Alcock (1901) listed three species of deepwater lobsters belonging to the family Nephropidae and seven species of Eryonidae in Andaman and Nicobar waters. Two species belonging to Palinuridae were also recorded (Balss 1925; Chekunova 1971). The occurrence of the deepwater nephropid lobster *Nephrops stewarti* on the east coast of the Andaman Islands was reported by Wood-Mason (1873). Six species of spiny lobsters in the genus *Panulirus* occur in shallow waters around the A&N Islands (Silas and Alagarswamy 1983). Shanmugham and Kathirvel (1983), Jha et al. (2007) and Kumar et al. (2010) also recorded six species of spiny lobsters in the A&N Islands; *Panulirus homarus homarus, P. ornatus, P. polyphagus, P. versicolor, P. penicillatus* and *P. longipes longipes* are the common species. The slipper lobster *Thenus orientalis* (*T. unimaculatus*) is distributed in South Andaman (Shanmugham and Kathirvel 1983). In addition to these shallow water species, the deep-sea lobsters *Puerulus sewelli* (Radhakrishnan et al. 2013; Anrose et al. 2010), the spear lobster *Linuparus somniosus* (Mustafa 1990; Ali et al. 1991) and *Puerulus angulatus* (Balss, 1905) were also reported from A&N waters.

7.17.2.1 Fishing Methods

In the South Andaman Islands, traditional tribal fishermen catch spiny lobsters from shallow areas by spearing, and 75% of the catch is landed by this method of fishing. Diver fishermen handpick lobsters, using torchlight, from inshore areas of North Andaman Islands during low tide. The bottom-set gillnets are mainly used by middle Andaman fishermen and 18% of the catch is by this gear. The deep-sea lobsters and the slipper lobster *T. unimaculatus* are landed by trawl nets. The major lobster landing centres are Wandoor (South Andaman) and Aerial Bay (North Andaman). Lobsters are also landed at some minor landing centres. Campbell Bay is a major landing centre in the Nicobar Group of Islands (Kumar et al. 2010).

7.17.2.2 Fishery

Though the A&N Islands are considered to be a potentially important region from a lobster fisheries point of view, there is no organised fishing for lobsters (Kumar et al. 2010). Commercial fishing for lobsters probably started as early as the 1950s. Reliable lobster landing data for the Islands is not available. The estimated total landing for the years 1999–2000 was 3.16 t and 73% was landed at Port Blair. Peak landing is in January (2.3 t). The dominant species in the fishery in A&N Islands is *P. penicillatus* (67%). *Panulirus versicolor* formed 26%, while *P. ornatus* and *P. homarus homarus* contributed 2% each to the catch. Jha et al. (2007) reported *P. versicolor* as the major species (85%) in the fishery in North Andamans. At Port Blair, 90% of the catch is constituted by *P. penicillatus*, followed by *P. versicolor* (5%), *P. longipes longipes*, *P. polyphagus* and *P. homarus homarus* (Kumar et al. 2010). The potential of the deep-sea lobster, *P. sewelli*, fishery in Andaman waters was evident from the results of the exploratory survey conducted by Fishery Survey of India vessels. During March 1998, a CPUE of 60 kg/hr. was obtained by the trawlers (Anrose et al. 2010). However, the resource was not exploited on a regular basis though small quantities are landed during the fishing season.

7.17.2.3 Export

An average 6.8 t of frozen lobster was exported from the A&N Islands during 2010–2011 (Gopal et al. 2013). Annually, an average 4 t of frozen lobster is exported by Andaman-based exporters (Kumar et al. 2010). Small quantities of live lobsters are also sent to the mainland from where they are exported to Southeast Asian countries. Live lobsters are also supplied to the local hotels and tourist resorts.

7.17.3 Lakshadweep Island Lobster Fishery

The earliest record of lobsters in the Lakshadweep Sea was by Alcock (1901, 1906) and Borradaile (1903, 1906). The Central Marine Fisheries Research Institute,

Table 7.13 Main islands of Lakshadweep and the lobster fauna

Islands	Lobster species
Minicoy	*Panulirus homarus homarus, Panulirus versicolor, Panulirus penicillatus, Parribacus antarcticus*
Kadmat	*Panulirus versicolor*
Kiltan	*Panulirus versicolor*
Chetlat	*Panulirus versicolor*
Agatti	*Panulirus versicolor, Panulirus penicillatus*
Kalpeni	*Panulirus versicolor, Panulirus penicillatus*
Bitra	*Panulirus versicolor*
Kavaratti	*Panulirus versicolor, Panulirus penicillatus*
Cheriya Kare (Suheli Par)	*Panulirus penicillatus, Parribacus antarcticus*
Androth	*Panulirus penicillatus*
Lakshadweep Sea	*Nephropsis ensirostris, Nephropsis sulcata, Acanthacaris tenuimana, Eduarctus martensii, Petrarctus rugosus, Stereomastis phosphorus*

Cochin, conducted a comprehensive survey of the Lakshadweep Islands during January to March 1987 to assess the fishery potential of the Islands with suggestions for development (Anon 1989). The occurrence of three species of *Panulirus* has been reported: *Panulirus versicolor* is the most common species in all the islands, except Androth and Cheriya Kare islands (Rao et al. 1989; Pillai et al. 1985). *Panulirus penicillatus* is the second most common (Table 7.13). *Panulirus homarus homarus* is recorded only from the southern island of Minicoy (Meiyappan and Kathirvel 1978). The scyllarid lobster *P. antarcticus* has been reported from Minicoy and Kiltan Islands. The lobsters recorded in different islands are listed in the table below (Meiyappan and Kathirvel 1978; Pillai et al. 1985; Rao et al. 1989). In the Minicoy Island fishery, *P. versicolor* constituted 93% of the landing and *P. penicillatus* and *P. homarus*, 5% and 2%, respectively (Pillai et al. 1984).

There is no organised fishery for lobsters in Lakshadweep Islands. Local fishermen use various methods for fishing: hand catching and bottom-set gillnets. Incidental catches in fish traps are reported occasionally. The catch is mainly consumed by local people and some catch is sold to hotels and tourist resorts. Attempts to send live and frozen lobsters to the mainland were not successful as traders could not ensure regular supply, probably due to transportation problems. Currently, the resource is underutilised, and if proper marketing channels and transportation facilities could be developed, lobsters could be shipped to the mainland for further processing and export.

7.18 Participatory Lobster Conservation and Management

Almost all lobster fishing countries have management regulations designed to conserve the resource for future and enhance production. The only management measure to regulate lobster fishing in India is the Minimum legal size fixed for export by the Government of India (Radhakrishnan et al. 2005). The commercial lobster fishery of Kanyakumari is probably the oldest in the country and is almost entirely an artisanal fishery with no competition from trawlers. The entire inshore sea of Kanyakumari on the west coast is rocky with a sandy bottom in between. Thus the lobster stock appears to be located in very shallow water and is accessible by all types of non-mechanised fishing gear. The reefs in the deeper areas probably shelter the new recruits. There is no closed season during the breeding period or a ban on capturing and marketing egg-bearing lobsters. Since there are no controls on the quantity or size of lobsters captured, the ever-increasing fishing pressure has resulted in a decline in total catch and CPUE. The current production from Kanyakumari is 10% of the average landings during 1966–1978 (91 t) and just 3% of the highest landings recorded (304 t) in 1966 (George 1973). Though George (1967a) suggested minimum legal size for fishing lobsters in Kanyakumari district based on scientific observations, no regulatory measures were implemented. Since *P. homarus homarus* is an inshore species with restricted movement (Mohamed and George 1968), the species is highly vulnerable to fishing. Furthermore, peak breeding coincides with active fishing, and consequently more than 60% of females caught during November–December months are egg bearing (Radhakrishnan and Thangaraja 2008). Fishermen in general are using gillnet for lobster fishing, which captures juveniles and subadults, which are newly recruited into the fishery. Apart from the information available on stock during the late 1950s and 1960s (George 1973), there was no information on stock status and exploitation rate of spiny lobster fishery until 2014 (Thangaraja et al. 2015).

Fishing pressure on Kanyakumari lobster stock is unlikely to be reduced in the short term. The entire stock is accessible to the artisanal fishermen. Since fishing and peak breeding season are concurrent, a great percentage of breeding females are likely to be captured and unlikely to be returned. Lobsters below the size at maturity, especially males, form almost 50% of the total catch. It is suggested that a minimum size limit of 60 mm CL on the capture of female lobsters would greatly reduce the chances of recruitment overfishing (Thangaraja and Radhakrishnan 2017). In most lobster fisheries, the ideal target levels for egg production per recruit are 30% of the unexploited levels (Fielding and Mann 1999).

With this in background, a project on community participation in lobster resource management was initiated in 2002 with the support of Marine Export Development Authority (MPEDA) at Khadiyapatanam and Enayam, the major lobster fishing villages in Kanyakumari district and trawl landing centres in Mumbai and Gujarat (Radhakrishnan et al. 2005). Fishermen are aware that the lobster fishery is gradually becoming depleted and some regulatory measures are necessary, which will benefit them in the long run. Village level meetings, distribution of educative pamphlets and posters, V-notching of egg-bearing lobsters and their release back to sea and distribution of lobster traps to wean the fishermen away from using the gillnets and trammel nets were some of the activities implemented under this programme (Figs. 7.56, 7.57, 7.58, 7.59, and 7.60). Implementation of a minimum legal size for fishing, ban on trammel and gillnets and closure of fishery during peak breeding season (November) are some of the regulatory measures suggested to the state government. Establishment of fishery management councils with representatives from

Fig. 7.56 Participatory management of lobster fishery: Distribution of CIFT designed lobster traps to fishermen at Kanyakumari, Tamilnadu, and Southwest coast of India (Photo by E. V. Radhakrishnan, CMFRI)

Fig. 7.57 Participatory management of lobster fishery: School children participating in a rally organized at a Veraval lobster fishing village, Gujarat, northwest coast of India (Photo by Joe K. Kizhakudan, CMFRI)

Fig. 7.58 Participatory management of lobster fishery: Stakeholders participating in a Workshop to discuss sustainable lobster fishing practices at Veraval, Gujarat (Photo by Joe K. Kizhakudan, CMFRI)

Fig. 7.59 Participatory management of lobster fishery: V- marking egg bearing lobsters for release back to sea at Kanyakumari, Tamilnadu, and Southwest coast of India (Photo by E. V. Radhakrishnan, CMFRI)

fishermen organisations, NGOs, fisheries department officials, scientists and district administration was proposed. The fishermen were not agreeable to the voluntary release of egg-bearing lobsters as their daily catch is only a few numbers of lobsters and lobster is a highly priced resource, which they cannot afford to release without any incentive from the government. A systematic approach to develop a strong community management system may be the appropriate step towards effective management of this valuable resource.

Fig. 7.60 Participatory management of lobster fishery: Communication tools developed for creation of awareness among the stakeholders (Photo by E. V. Radhakrishnan, CMFRI)

7.19 Deep-Sea Lobster Fishery

7.19.1 Introduction

Though the existence of the deep-sea palinurid lobster, *Puerulus sewelli*, along the west coast of India (Alcock 1901) and Gulf of Mannar (Alcock and Anderson 1894) was known as early as the turn of the last century, the availability of the species, in commercial concentrations, at fishing grounds along the Indian coast was known only after its rediscovery by John and Kurian (1959) and subsequent exploratory surveys and experimental fishing carried out along the west coast of India for over a period of two decades, from 1967, by R.V. Varuna of the Indo-Norwegian Project and by other vessels, viz. Tuna, Velameen and Klaus Sunnana (Kathirvel et al. 1989). Complete descriptions of the distribution, early fishery, landings, effort and biology of the species are provided by John and Kurian (1959), Kurian (1965), Tholasilingam et al. (1968), Silas (1969), Joseph (1972, 1974, 1986), Pillai and Ramachandran (1972), Chekunova (1973), Rao and George (1973), Oommen (1974, 1985), Oommen and Philip (1976), Somavanshi and Bhar (1984), Ninan et al. (1984), Philip et al. (1984), Joseph and John (1987), Kathirvel (1998), Sulochanan and John (1988), Kathirvel et al. (1989), Suseelan et al. (1990) and Anrose et al. (2010).

7.19.2 Distribution

P. sewelli is distributed in the Western Indian Ocean, including the coasts of Somalia and Pakistan, the southwest coast of India, the Gulf of Aden, the Gulf of Mannar (Holthuis 1991), the Andaman and Nicobar Islands and along the coast of Myanmar (Nakken and Aung 1980). It is distributed along the southwest coast in the areas lying between latitudes 7 °00' N and 18 °00' N at 150–450 m depth. In the Gulf of Mannar, it occurs between 7 ° 00'N and 9°00'N latitudes at 146–400 m depth. In the Andaman Sea, the distribution of the species is between 8°00' N and 12°00' N latitudes at 150–380 m depth (Kathirvel et al. 1989). The fishery is concentrated between Quilon and Kanyakumari on the southwest coast and between Mandapam and Tuticorin in the Gulf of Mannar on the southeast coast of India (Fig. 7.5).

7.19.3 Seasonal and Area-Wise Abundance

On the southwest coast of India, *P. sewelli* is found to be abundant at depths varying between 150 and 200 m during January to April. The greatest population density, with a catch rate of 200–300 kg/hr., was recorded off Mandapam (725 sq.km) on the southeast coast. In April 1970, the deep-sea trawler of the Integrated Fisheries Project (IFP), *Klaus Sunnana*, operated off the Gulf of Mannar at a depth of 146–366 m landed 16.88 t of *P. sewelli* with a catch rate of 137.2 kg/hr (Kathirvel et al. 1989). The areas next in the order of abundance were Kanyakumari, Colachel, Alleppey and Cochin (100–150 kg/hr) and off Quilon and Thiruvananthapuram (150–200 kg/hr). The 'Quilon Bank', the most productive fishing ground on the southwest coast of India, covers an area of about 3300 sq.km (Kathirvel et al. 1989). Overall catch rates of *P. sewelli* are higher on the west coast than the east coast of India.

Exploratory surveys conducted off the southwest coast of India by IFP and CIFNET vessels found greater abundance of lobsters (187–522 kg/hr) during January to May in the depth zone of 151–200 m, whereas the resource was found to move to deeper waters (201–300 m) as the season advances (Oommen and Philip

1976; Kathirvel et al. 1989). During 1984 to 2006, Fishery Survey of India conducted deep-sea fishing operations in the depth contour of 200–500 m between 8° and 15°N latitude. The highest catch rate of 29.18 kg/hr. was obtained on the west coast at 11°N latitude. On the east coast, the maximum catch was from 11°N latitude, registering a CPUE of 3.42 Kg/hr. The mean catch rate on the west coast in 1984 was high (4.71 kg/hr), which increased to 16.66 kg/hr. in 1987, but declined drastically to 0.03 kg/hr. in 1992. After a lull of more than a decade, in 2006, the catch improved marginally (0.41 kg/hr) (Anrose et al. 2010). The lobster *P. sewelli* was often landed along with a variety of deep-sea prawns. The production of prawns in general was relatively greater between 275 and 375 m depth zone.

7.19.4 Potential Estimates

Several estimates on the potential of the *P. sewelli* fishery from different grounds have been made, based on the exploratory surveys conducted since 1967. Joseph (1972) estimated the potential yield at 2000 t and the standing crop at 6000 t based on the catch landed during 1969–1971. Oommen and Philip (1976) estimated the standing stock at 12941 t from the southwest coast and 1869 t from the Gulf of Mannar. James et al. (1987) estimated the sustainable potential for the deep-sea lobster as 8000 t for the southwest coast and 1200 t for the southeast coast.

7.19.5 Commercial Exploitation

Though the occurrence of deep-sea lobsters along the Indian coast was known since the late 1950s, commercial fishing operation by some entrepreneurs was conducted only during 1979–1980. However, organised fishing was initiated only in 1988, when some trawlers from Visakhapatnam conducted operations during the pre-monsoon months between Cochin and Kanyakumari fishing grounds. The fishing was discontinued after 5 years as the economic return was not satisfactory. In 1999, Cochin- and Neendakara-based medium fishing trawlers restarted deep-sea fishing in the Quilon Bank as the prawn catch from shallow waters was not economically attractive. The deep-sea lobster *P. sewelli* was landed by trawlers as a bycatch along with deep-sea shrimps. The fishery was seasonal, commencing by September–October and extending until February–March. The total landing of *P. sewelli* during 1999–2000 was 574 t, with a peak monthly landing of 180 t in December. The landing decreased to 297 t and 236 t during the years 2000–2001 and 2001–2002, respectively. Maximum monthly landing was recorded in December (110 t) during 2000–2001 and March (49 t) in 2001–2002.

7.19.5.1 Length Composition
The size of *P. sewelli* caught off the southwest coast of India ranges from 36 to 207 mm in total length (Kathirvel et al. 1989). The dominant size groups of males and females in the catch measured 151–160 mm and 161–170 mm, respectively.

The relationship between the carapace length and total length is expressed by the formula $Y = 4.0721 + 0.4647\ X$ for males and $Y = 2.6160 + 0.4410\ X$ for females, where Y is the carapace length and X the total length. The length-weight relationship was found to be $\text{Log}\ W = -4.9525 + 3.1705\ \text{Log}\ L$ for males and $\text{Log}\ W = -4.8332 + 3.1130\ \text{Log}\ L$ for females.

Sexes were more or less equally distributed, with females comprising 47% of the catch during 2000–2001 and 56% during 2001–2002 (Radhakrishnan et al. 2005).

7.19.5.2 Reproductive Biology and Spawning

The reproductive biology of *P. sewelli* has been studied in detail, and the morphology of the sexual organs is typical of the palinurid lobsters (Kathirvel et al. 1989). It is heterosexual and sexually dimorphic. The female attains sexual maturity at a total length of 120–129 mm. Egg-bearing lobsters are encountered throughout the year. However, two spawning peaks have been noticed, the major one during January–April and another in October. The percentage of berried lobsters during the major peak breeding season ranged from 51 to 83. The spermatophore seems to be temporarily attached to the sternum, and falls off after fertilisation, unlike in the lobsters of the genus *Panulirus*. The eggs are attached to the pleopods until they hatch. The fecundity ranged from 10,170 to 36,400 in specimens measuring 136 mm and 196 mm in total length, respectively.

7.19.5.3 Breeding Migration

Puerulus sewelli exhibits vertical migratory movement from deeper waters to 150–200 m depth zone in January for breeding and remains there until April–May. Since peak fishing (January–April) and breeding coincide, large numbers of breeding females are caught. Though the exact impact of such a removal of the breeding population is not known, it may result in recruitment overfishing and consequent lower recruitment of young lobsters into the fishery. Occurrence of smaller size classes during December–January indicated entry of young ones into the fishery during these months. Maximum numbers of immature lobsters were recorded in January.

In the Andaman and Nicobar Islands, the occurrence of another deep-sea lobster *Linuparus somniosus* was reported (Ali et al. 1994). However, the resource was not available in commercial quantities for exploitation.

During 2000–2006, an average 23.3 t of the deep-sea lobster *Nephropsis stewarti* was landed at Mangalore. The highest annual landing was in 2001 (51 t) (Dineshbabu 2008). The specimens were in the size range of 58–158 mm TL.

7.20 Management

Although lobster is not a large marine resource in India, it is a high-value seafood, with almost the entire catch exported, fetching a substantial foreign exchange. The maritime states manage their fisheries and have their own regulations, but within the policy framework of the fisheries policy of the Central Government. Little attention has been paid to lobsters by the state governments, being a minor fishery resource, and this neglect in framing and enforcing fishing regulations has led to heavy fishing pressure on this vulnerable resource, leading to decline in landings in majority of the states. Since 2001, the annual catch has been stabilised around an average 1500 t. In 2015, the production has crossed 2000 t and in 2016 the landing was a phenomenal 2976 t. Although the reasons for such a sudden spurt in catch are not understood, increase in production is always a positive sign.

Lobster fisheries in India are characterised by the absence of or inadequate regulations in fishing or non-enforcement of existing regulations. The Central Government has notified minimum legal size (MLS) for major spiny and slipper lobster species for export (Radhakrishnan et al. 2005). However, fishing within the territorial limit (20 nautical miles from shore) is under the jurisdiction of the state governments. Lobster fishing, whether by the trawlers or the artisanal gears, is within this region. Multiday and single-day trawlers land 67% of the annual production and 28% is by the gillnets.

Lobster landing trend in the state of Gujarat shows that there has been a significant decline in landings of both *P. polyphagus* and *T. unimaculatus* along the coast during 2001–2004, but has recovered and stabilised at around 700 t. However, trawl landings were much lower than that during the 1990s. Lobster being a small-volume bycatch in shrimp trawls, no specific fishing regulations are enforced. The artisanal fishermen in Kachchh and Bhavnagar operate trap gillnet, long and short tidal stake net and bag net on shallow reefs, and the catch predominantly consists of juvenile lobsters. Enforcing a ban on taking egg-bearing lobsters may be one management option that the state government can consider. Co-management programmes implemented by Central Marine Fisheries Research Institute had a marginal impact on the exploitation of egg-bearing lobsters by the trawl fishermen.

In Maharashtra, intensive fishing of the spiny lobster *P. polyphagus* has resulted in overexploitation of the resource, as is the case in many global lobster fisheries. The high value of whole cooked frozen lobster in the international market has led fishermen to fish the species intensively up to the point of overexploitation. The collapse of *T. unimaculatus* fishery shall be an eye opener for the Maharashtra State Fisheries Department. The lobster fishermen have realised that resource management and some practical solutions to prevent large-scale harvesting of egg-bearing lobsters are necessary to conserve the resource. Management of the spiny and slipper lobster fishery of Maharashtra coast is complicated by failure to enforce regulations due to the practical difficulties of implementing fishing restrictions, as the resource is a bycatch in trawlers. Mesh regulation and closed season for lobster fishing during the peak breeding months are not practical management measures, as the state has already been enforcing a fishing holiday during 10 June to 15 August. Though lobster is a bycatch in trawlers, the fishermen target the resource at specific depths where they congregate for breeding. Maximum landings of *P. polyphagus* and *T. unimaculatus* are during their breeding season (September to March), and a prohibition on bringing the ovigerous females to the shore and marketing it may be one regulatory measure that can be enforced (Fig. 7.61). This means that the fishermen will have to release back the egg-bearing females into the sea during the peak breeding months (Kagwade et al. 1991; Radhakrishnan et al. 2005). Further,

Fig. 7.61 Egg bearing *Panulirus polyphagus* landed at Mumbai Fisheries Harbour (Photo by: V.D. Deshmukh, CMFRI)

capturing juvenile lobsters by the gillnet fishermen from inshore areas during their recruitment also should be regulated by enforcing size restrictions in fishing. The co-management workshops organised by CMFRI have brought out the issues before the stakeholders and called upon the fishermen and the exporters to find an amicable solution so that the spiny lobster fishery could be prevented from collapse like the slipper lobster fishery. Although lobsters have been recorded almost along the entire Kerala coast, the fishery is mostly restricted to certain regions. Major portion of the landing is by trawlers operating in the inshore and deeper fishing grounds. *T. unimaculatus* followed by the shallow water spiny lobster *P. homarus homarus* and the deep-sea species *P. sewelli* mainly constitute the fishery. Though low in volume, the slipper lobster *T. unimaculatus* is a valuable resource, which needs strategic management for future sustainability. Size restriction or catch regulation may not be a practical measure for a species landed as a bycatch in trawls. Though the state enacts a closed season for trawling from 15 June to 31 July, this period not being the breeding season for the species may not help in protecting the breeding stock. Releasing the egg-bearing lobsters during the peak breeding season (November to January) through participation of fishermen into the management decision-making process may be a viable option (Radhakrishnan et al. 2013).

The spiny lobster *P. homarus homarus* is a shallow water species, and their proximity to the shore has made them highly vulnerable to fishing. The high value for the live lobster has led fishermen to fish the species extensively in the absence of enforcement of any regulatory measures, except the minimum legal size for export notified by the Ministry of Commerce and Industry, Government of India (Radhakrishnan et al. 2005). The operation of traditional traps is restricted to only at certain pockets along the coast and bottom-set gillnets are widely used. The number of species recorded along the southern Kerala coast has been reduced from five to three, and the annual production has declined from an average 3.8 t to the current landing of an average 1.0 t. The fishermen shall be encouraged to use more of the eco-friendly traps than the gillnets for lobster fishing. Recently, a minimum legal size of 200 g for fishing of *P. homarus homarus* has been recommended by the expert committee constituted by the Government of Kerala (Mohammed et al. 2014). Fishermen are beginning to realise that resource management by introducing closed season during peak breeding period and a minimum size for fishing is necessary for the sustainability of the stocks.

Tamil Nadu stands third in lobster landings of the country with an average annual production of 332 t (1989–2012). Lobster catch is just 2% of the total marine fish landings of the state, and therefore, lobster is not considered as a major resource. However, considering the high unit value of lobsters in the international market, the resource has to be carefully managed for long-term sustainability.

The lobster fishery on the western coast of the state is mainly from the artisanal sector, whereas on the east coast, harvesting is both by trawlers and gillnets, with the major share of the landing by the trawlers (54.3%). Lobsters are a bycatch in trawls and therefore no specific regulatory measures for lobster fishing are implemented. The lobster fishery by the artisanal sector has been subjected to heavy fishing pressure over the past six decades, and this trend is likely to continue due to attractive price for live lobsters in the export market. There is no scope for increasing the effort, as catch rates have significantly declined and the mean size in the fishery has substantially reduced over the years. Further, there is no restriction on capturing and selling undersized and egg-bearing lobsters. The *P. ornatus* breeders along the east coast are naturally protected as they breed in deeper waters along the Sri Lankan coast and probably inaccessible to the Indian fishermen during the peak spawning season. Releasing the egg-bearing lobsters and fixing the minimum legal size for capture for *P. homarus* at 60 mm carapace length may have considerable benefits. Such restrictions are likely to lead to catch reductions which may not be acceptable to fishermen.

Lobster fishing along the Kanyakumari coast is essentially a coastal activity, and the fishing grounds are notionally under local community control. The lobster fishing is carried out by the artisanal fishermen who reside in the villages overlooking the fishing grounds. Each ground may not belong to a particular village, and fishermen from two or three villages may be fishing in the same ground. Therefore, effective management may require cooperation and coordination from these villages. The local religious institution has an overall control of the welfare of the fishermen community. Therefore, community-based co-management, as Pomeroy and Rivera-Guieb (2006) suggested, may be the ideal option for management of lobster fishery of Kanyakumari. The management decision process should follow a consultative approach involving the fishermen and the government, as participatory governance and community-based management approaches are an effective way of sustainably managing a natural resource. There is a plan to establish Lobster Fishery Management Councils to co-manage the lobster fishery in Kanyakumari.

Deep-sea fishing along the southwest coast of India, especially in the Quilon Bank, for shrimps has resumed since 1999. The fishing efficiency, owing to new technology, has increased significantly, with good market demand for deep-sea shrimps. Fishermen are now able to precisely target the resource they want in the deep-sea fishing grounds. Slow growth, low fecundity and the probability of a long larval phase pose stock recruitment problems, unless the breeding stock is maintained at optimal levels. Investigations to estimate the breeding stock sufficient enough to maintain a safe level of recruitment into the fishery have to be carried out. The fishing effort and the overall exploitation levels have to be monitored so that steps can be taken to maintain the breeding stock at or above the current level.

The deep-sea lobster *P. sewelli* is a valuable crustacean resource and if exploited judiciously may contribute to a sustainable fishery. Since fishing coincides with the breeding migration, the catch consists of a significant proportion of breeding females. Juveniles also constitute a fairly good portion of the catch during April–May and in October. Kathirvel et al. (1989) suggested the need for regulatory measures for better management and improvement of the stock. However, there are practical difficulties for enforcing regulatory measures for a low-volume resource such as *P. sewelli*, which is a bycatch in the deep-sea shrimp trawls.

While the lobster fisheries in most of the developed economies have right-based management in place with some biological and effort controls, there have been limited controls or non-enforcement of existing regulations in many developing countries (Table 7.14). In India, where maritime states manage their own fisheries, fishing regulations have to be formulated and implemented by the states (Radhakrishnan et al. 2007a, b). Resource management should deviate from the top-down approach to bottom-up approach involving the fishermen, local traders and exporters, and management will be successful only when the fishermen feel that they are the owners of the resource. Managing a resource landed as a bycatch by the trawlers in a multispecies tropical fishery is difficult. Trawler fishermen may be motivated to release back the egg-bearing lobsters through educational and extension programmes, and if convinced, the fishermen may cooperate with such voluntary measures. As lobster fishing being a socioeconomic activity, implementation of any legal measures to regulate fishing and marketing shall consider the livelihood issues of the various stakeholders to make it successful and sustainable.

7.21 Lobster Fisheries and Management in Indian Ocean Rim Countries

7.21.1 Introduction

The countries bordering the Indian Ocean (both FAO fishing areas 51 and 57) contribute significantly to the world lobster capture fisheries production. The total landing of palinurid lobsters from the Indian Ocean during 2012 was 16,695 t, with the fishing area 51 contributing 5363 t and fishing area 57, 11,332 t (FAO 2012). Countries with high production are Australia, Indonesia, Malaysia and South Africa. A brief description of lobster fisheries and management of Indian Ocean rim countries, South Africa, Madagascar, Tanzania, Kenya, Somalia, Yemen, Saudi Arabia, Sultanate of Oman, Iran, Sri Lanka, Malaysia, Indonesia and Australia (Fig. 7.62) is presented.

Table 7.14 Management measures implemented by major lobster fishing countries

Management strategies	India (all species)	Western Australia (P. cygnus)	South Africa (J. lalandii)	New Zealand (J. edwardsii)	USA California (P. interruptus)	Mexico (Baja California) (P. interruptus)	Gulf of Mexico (P. argus)	CARICOM countries (P. argus)
Input controls								
Limited entry		X			X	X	X	X
Vessel size		X	X					
Gear restrictions		X			X		X	
Pot limits		X					X	X
Escapement gaps		X	X	X	X			X
Closed season	X	X	X	X	X	X	X	X
Closed areas		X	X			X	X	
Minimum legal size	X	X	X	X	X	X	X	X
Maximum legal size		X						
Minimum tail weight							X	X
Restriction on catch of egg-bearing lobsters	X	X	X	X	X	X	X	X
Prohibition on destructive gear							X	
Output controls								
Total allowable catch (TAC)		X	X	X	X		X	
Individual transferrable quota (ITQ)		X	X	X				
Bag limits (recreational)		X		X			X	
Reporting of landing	X	X			X		X	

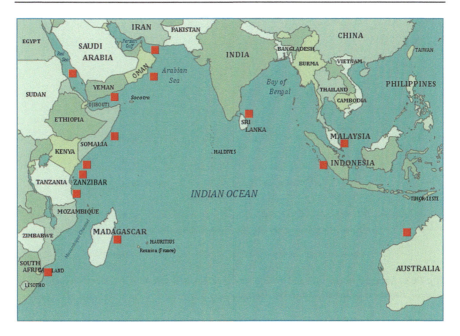

Fig. 7.62 Indian Ocean Rim countries

7.21.2 Lobster Fisheries of South Africa

7.21.2.1 Introduction

The South African coastline is approximately 3000 km long. While the east coast is influenced by the warm, nutrient-poor Agulhas current which flows southward from tropical latitudes of Mozambique and Madagascar, the west coast is bathed in cold waters of the Benguela current system. More than 11,000 marine species have been recorded around South Africa, 5% of total world species, indicating high biodiversity.

The rock lobster fishery of South Africa mainly targets the west coast rock lobster *Jasus lalandii* and the deepwater south coast lobster *Palinurus gilchristi*. A smaller fishery for the east coast deepwater lobster *Palinurus delagoae* also contributes to the annual total average landing of 2940 t (2003–2014) (Fig. 7.63) (FAO 2016). A shallow water artisanal fishery for *Panulirus homarus rubellus* also exists along the KwaZulu-Natal coast, which is not reflected in the FAO landing figure. The contribution of lobster fishery to the total fisheries catch is just 0.4%. However, in terms of value, it is 9% (R 290 million per annum).

7.21.2.2 The Trap Fishery for the West Coast Rock Lobster *Jasus lalandii*

The West coast rock lobster, *J. lalandii*, is harvested by trap or hoop net, in waters shallower than 100 m. The commercial fishing commenced in the late 1800s was peaked in the early 1950s yielding an annual catch of 18,000 t. The catch declined

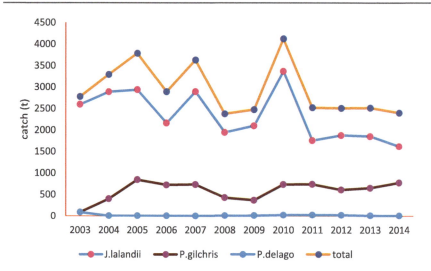

Fig. 7.63 Annual species-wise and total lobster landings in South Africa (FAO catch data). *Jasus lalandii, Palinurus gilchristi, Palinurus delagoae*

to 10,000 t during the 1960s and continued to decline to 2000 t in recent years. The annual average landing for the period 2003–2012 is an estimated 2333 t (FAO 2016) (Fig. 7.63). For 2010, the total allowable catch (TAC) for west coast rock lobster has been set at 2393 t that includes 1685 t and 457 t for direct commercial exploitation (offshore and inshore, respectively) and 251 t for recreational fishers, respectively.

7.21.2.2.1 Management

Management of the fishery is through TAC (total allowable catch), total allowable effort (TAE), minimum legal size (MLS) and protection of egg-bearing and soft-shelled lobsters. The MLS of 89 mm CL fixed in 1933 was lowered to 76 mm CL in 1959 in order to promote the catch, which was again increased to 89 mm CL in 1979 and later to 75 mm CL in 2000.

7.21.2.3 Industrial Longline Trap Fishery for the Deepwater Southern Spiny Lobster *Palinurus gilchristi*

South coast rock lobster *Palinurus gilchristi* is endemic to the southern coast of South Africa. The well-managed and sustainable lobster fishery for *P. gilchristi* is under operation along the southern coast of South Africa since 1974. This is the second largest lobster fishery in South Africa with an annual average yield of 591 t (2003–2012) (Fig. 7.63) (FAO 2016). The fishing grounds are on the Agulhas Bank and further eastwards at a depth of 50–200 m. The fishery is capital intensive and restricted to local fishing companies, requiring large ocean-going vessels and specialised equipment. Fishing is year round. The management strategy is a combination of TAC and TAE. The current TAC limit is 475 t tail mass. The total export value in 2012 was approximately R190 million.

7.21.2.4 Trap Fishery for the Deepwater East Coast Spiny Lobster *Palinurus delagoae*

The third species contributing to the total lobster production of South Africa is the Natal deepwater spiny lobster *P. delagoae*. The fishing operation is confined to the province of KwaZulu-Natal with an average annual landing of 18 t (2003–2012) (FAO 2016).

The other three deepwater species landed as a bycatch in multispecies trawl fishery in small quantities are *Metanephrops andamanicus*, *Scyllarides elisabethae* and *Nephropsis stewarti*. The deep-sea lobsters *Projasus parkeri*, *Puerulus angulatus*, *P. carinatus* and *Linuparus somniosus* are other palinurids recorded off the Southeast African coast (Berry 1971). *Jasus lalandii* and *P. delagoae* are straddling stocks shared with South Africa's northwestern northeastern neighbours, but there is little cooperative management (Pollock et al. 2000).

7.21.2.5 Experimental Longline Trap Fishery

An experimental longline trap fishery was operated along the KwaZulu-Natal coast between 1994 and 1997 and again between 2004 and 2007. The fishery targeted the deepwater spiny lobster *Palinurus delagoae*, the slipper lobster *Scyllarides elisabethae* and the deep-sea crab *Chaceon macphersoni*. Fishing was conducted in deep water (100–450 m), on rocky and gravel substrata between the Mozambique border in the north and about Port Edward in the south. The vessels set bottom longlines with strings of 100–200 traps attached; traps were barrel-shaped (c. 0.8 m long), made of plastic with a funnelled top entry, and baited with hake heads. Traps were set about 10 m apart on branch lines tied to an anchored bottom longline. Catch rates declined sharply during both periods (Groeneveld and Cockcroft 1997; Groeneveld et al. 2013a, b).

Catches of both lobster species combined were 120 t in 1994, 59.8 t in 1995 and 38.1 t in 1996. *Palinurus delagoae* catches declined from 89.5 t (1994) to 7.8 t (1997). A combined catch of 103 t of target species was landed between 2004 and 2007 (65% *P. delagoae*, 28% *S. elisabethae* and 7% *C. macphersoni*). Catch rates of *P. delagoae* were declined by 42% over the 4 years, and the average CL declined by 4%. Slipper lobster catches were almost constant. Due to operational difficulties, the longline trap fishery was suspended; however, the multispecies trawl fishery is continuing.

7.21.2.6 The East Coast Rock Lobster *Panulirus homarus rubellus*

The east coast rock lobster *P. homarus rubellus* occurs from central Mozambique and Madagascar to East London. In South Africa, the two subspecies of *P. homarus*, *P. homarus rubellus* and *P. homarus homarus* coexist. Berry (1974) found that *P. homarus rubellus* constitutes 98.3% in southern KwaZulu-Natal (KZN) coast, 88.3% in northern KZN, 79.7% in southern Mozambique and 100% in southeast Madagascar. The occurrence of *P. homarus homarus* varied from 1.6% in southern KZN to 9.2% in northern KZN with a few numbers of 'hybrids' in these areas (2.5% in northern KZN and 5.1% in south Mozambique).

In South Africa, the fishery is restricted to KwaZulu-Natal district on the east coast (Berry 1971). They inhabit rocky reefs in the surf zone at depths of 1–36 m. The fishery was carried out until 1965 mainly by artisanal local fishermen (Heydorn 1969), and in 1969 a company was formed to exploit it commercially, which was discontinued later. Presently, the resource is exploited by permitted recreational/subsistence fishers only with an annual landing ranging from 130 to 450 t (www.kznwildcoast.co.za). Steyn and Schleyer (2014) estimated the catch from recreational fishery for invertebrates in KwaZulu-Natal coast as 92.8 t in 2012, of which the rock lobster formed 43% (40.3 t). *P. homarus rubellus* constituted 95%; *P. versicolor*, 1%; *P. longipes longipes*, 1%; *P. ornatus*, 1%; and *P. penicillatus*, 1% (Steyn et al. 2008). Out of the total catch, 93% is by divers, 4% by traps and 3% by hook and line (Steyn and Schleyer 2014).

7.21.2.7 Management

The fishery is managed by a 4-month closed season (November to February), a minimum size limit of 65 mm carapace length, gear restrictions (no boats or SCUBA), fishing restrictions in certain areas, daily bag limits for recreational fishers and prohibition on catching egg-bearing and soft-shelled lobsters (Steyn and Schleyer 2011). However, illegal fishing and catching of undersized lobsters have led to a decline in catch.

7.21.3 Lobster Fisheries of Tanzania, Zanzibar and Madagascar

The three other African countries contributing significantly to world lobster production are Mozambique, Zanzibar and Madagascar. Zanzibar, the semiautonomous archipelago under Tanzania, is also a major lobster landing country. *Panulirus ornatus*, *P. versicolor*, *P. longipes*, *P. homarus rubellus* and *P. penicillatus* constitute the lobster fishery in Tanzania and Zanzibar. *Panulirus ornatus* is the dominant species followed by *P. versicolor* and *P. longipes* (Mutagyera 1975). Annual average landing of *P. ornatus* by Tanzania is 481 t (2010–2012) (FAO 2016). The annual lobster landing by Zanzibar in 2003–2014 was 713 t (FAO 2016) (Fig. 7.64), and more than 50% was constituted by *P. ornatus*. The annual average lobster landing by Madagascar was 424 t during 2003–2014 (FAO 2016) (Fig. 7.65). Mozambique landed an average 335 t during the same period (Fig. 7.66). The Tolagnaro region is the most important centre for lobster fishing, contributing 70% of the total lobster production of Madagascar. Five species of spiny lobsters, *P. homarus rubellus*, *P. ornatus*, *P. versicolor*, *P. longipes* and *P. penicillatus*, occur along the Madagascar coast. *Panulirus homarus rubellus* constitutes 70% of the catch (Rabarison 2000; Bautil 2002). The scyllarids reported from the Tanzanian coast are *Thenus orientalis*, *Scyllarides squamosus* and *Parribacus antarcticus* (Kyomo 1999).

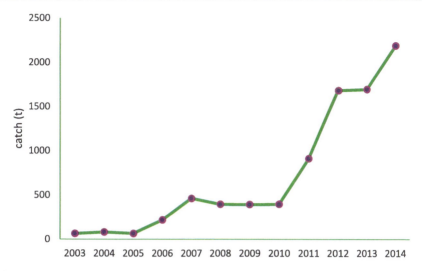

Fig. 7.64 Annual lobster landings Zanzibar (FAO catch data)

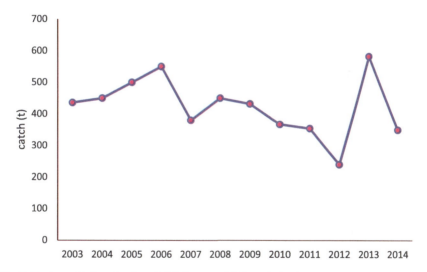

Fig. 7.65 Annual lobster landings in Madagascar (FAO catch data)

7.21.4 Lobster Fisheries of Kenya

7.21.4.1 Fishery

Lobster fisheries in Kenya are mainly located in the Lamu Archipelago on the northern coast of Kenya, and it contributes 60% of the total annual landings (Maina and Samoilys 2011). Landings are from artisanal fishermen and fishing is mainly by

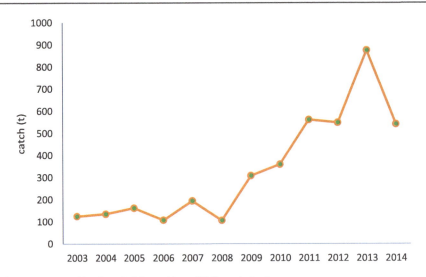

Fig. 7.66 Annual landings in Mozambique (FAO catch data)

diving and capturing with spear guns and hand nets. Lobster traps were found to be ineffective as the major species is *P. ornatus*, which does not enter the traps.

The fishery is mainly constituted by five species, *P. ornatus*, *P. longipes longipes*, *P. homarus homarus*, *P. versicolor* and *P. penicillatus*. *Panulirus ornatus* constitutes 80% of the catch by weight (Mutaguyera 1978). Lobsters are sold live, and *P. ornatus* is the most sought-after species due to its high value and ability to live outside water for long. The deepwater *Puerulus angulatus* and *P. sewelli* are landed in small quantities by trawlers.

The annual average landing is 111 t (2003–2014) (Fig. 7.67) (FAO 2016). However, the catch showed an upward trend with landing of 367 t in 2015. The catch is mostly exported to Singapore, Hong Kong, UAE, Italy, Portugal and Cyprus. The fishermen get an average KSH 500–600 during the southeast monsoon (SEM) and KSH 600–1000 during the northeast monsoon (NEM). Lobsters are caught predominantly during the calm NEM season (September to April) with peak fishing during October–March. Since most of the coastal region is covered by sea grasses and mangroves, lobster fishermen have to travel long distances (locally referred to as *Kwendago*) to reach the rocky substrata and coral reef to catch lobsters. Okechi and Polovina (1995) conducted a field study to test the feasibility of artificial shelters to catch *P. ornatus* at Gazi Bay.

The fishery has been showing signs of decline from 28 kg/trip/fisherman to 2.5 kg, which suggests that current stock is only 15% of that found in the early 1980s (Maina and Samoilys 2011).

7.21.4.2 Management

The current management regulations include a minimum legal size of 65.5 mm carapace length (250 g equivalent), prohibition on taking egg-bearing lobsters and a ban on using spear guns, stakes, crowbars and SCUBA gear for diving to catch

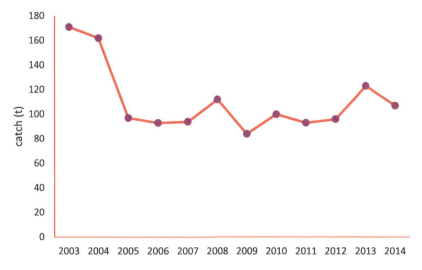

Fig. 7.67 Annual lobster landings in Kenya (FAO catch data)

lobsters. The stakeholders rejected the proposal to raise the MLS from 65.5 mm CL to 84 mm CL and compromised for 70 mm CL. Mutagyera (1978, 1983) suggested a minimum size limit of 90 mm CL for *P. ornatus*, though the mean size at maturity is 98.9 mm CL. None of the mature females were found below 80 mm. They proposed minimum catchable weight of 300 g and rejected introduction of limiting licenses. Various options for seasonal closures and area restrictions on fishing were discussed.

7.21.5 Lobster Fisheries of Somalia

7.21.5.1 Fishery

The Somali maritime zone is one of the most important large marine ecosystems (the Somali Current Marine Ecosystem) in the Indian Ocean. The ecosystem is characterised by the seasonal upwelling and consequent high productivity resulting from the Somali current. The Somali coastline is approximately 3300 km long and is divided into two major coastal areas; the north coast is bordered by the Gulf of Aden and is about 1300 km, while the 2000 km east coast forms the western edge of the Indian Ocean (Fielding and Mann 1999).

Several species of the genus *Panulirus* are found along the coast of Somalia, as well as two species of deepwater lobster (*P. sewelli* and *P. carinatus*), which are fished at depths of 150–400 m by trawlers. The other deepwater species landed occasionally are *Metanephrops andamanicus* and the slipper lobster *T. orientalis*. The predominant lobster species is *Panulirus homarus homarus* (megasculpta form), although very small numbers of other tropical lobster species such as *Panulirus ornatus* and *Panulirus longipes* also occur. Lobsters such as *P. ornatus*, *P. versicolor* and *P. penicillatus* are found on the northern part of the coast and

P. homarus is rare. The north coast species do not contribute to lobster fishery. It is quite surprising that *P. homarus homarus* forms a fairly good fishery along the coast of Yemen (George 1963; Sanders and Bouhlel 1984) and the Sultanate of Oman on the north (Johnson and Al-Abdulsalaam 1991), while 250 km across the Gulf of Eden on the northern part of the Somalia coast, the species is uncommon (Fielding and Mann 1999). Fielding and Mann (1999) conducted a field survey to assess lobster densities in the fishing grounds and collect biological and gear type and CPUE in the fishery so that appropriate management measures could be suggested.

Shallow water lobsters are captured by a variety of fishing gears along the east coast of Somalia, from tangle nets, traps and spearing to hand capture by diving. No reliable data on lobster landings are available. Mean landing (1997–2006) was an estimated 453 t valued at USD 4.39 million (Anon 2016). Haakonsen (1983) proposed a potential catch of 500 t of shallow water lobsters for the east coast of Somalia. The annual landing in Somalia by FAO (2016) estimate is 500 t. The resource is probably overfished according to anecdotal information from traders. The fishermen felt that reasons for overfishing are increasing fishing effort and the capture of lobsters by foreign trawlers.

7.21.5.2 Management

Though there are several management measures to regulate marine fisheries, non-compliance by industrial fisheries and the artisanal sector has led to overexploitation of lobster resources leading to decline in landings over the years. The suggested management measures are an MLS of 60 mm CL for *P. homarus*, prohibition on capturing egg-bearing lobsters, escape gap for traps and 100 mm mesh size for tangle nets (Fielding and Mann 1999).

7.21.6 Lobster Fishery of Yemen

The length of coastline of Yemen is 1906 Km along the Arabian Sea, Gulf of Aden and the Red Sea. Yemen borders Saudi Arabia on the north and Sultanate of Oman on the northeast.

7.21.6.1 Fishery

In Yemen, rock lobsters are known to exist since 1962. However, commercial exploitation of lobsters began in 1975. The potential yield varies between 300 and 1000 t and was re-estimated to be 700 t covering the entire fishery (Valle et al. 1993). It had a lucrative lobster fishery during 1988–1998 (833 t) (FAO 1998). The fishery virtually collapsed in the early 2000s from peaks of around 400 t in the late 1990s. The mean landing during 2003–2014 was 168 t (FAO 2016) (Fig.7.68). The collapse was attributed to the widespread use of tangle nets than traps to capture lobsters. Artisanal fishing accounts for well over 90% of total production (Alabsi and Komatsu 2014). Although landings have recovered a little, they remain at 70–80% of the earlier peak catch (Morgan 2006a).

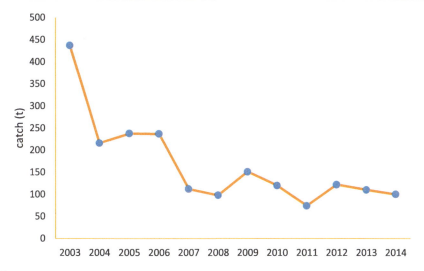

Fig. 7.68 Annual lobster landings in Yemen (FAO catch data)

The lobster fishery is mainly constituted by the spiny lobsters *Panulirus homarus homarus* (megasculpta form) and *P. versicolor* and the deep-sea lobster *P. sewelli*. Fishing is mainly by traps operating from small boats locally called *smbuks* or *huri*. Main fishing season is from October to December and the closed season for fishing is from June to September. The deep-sea lobster *P. sewelli* is exploited by trawlers operating at 200–600 m depth. Deep-sea lobster catches reached a peak of 1500 t in 1976–1977 off Yemen's Gulf of Aden coast. The landings have since been very small on the order of 100 t. In 1995, the recorded catch was just a couple of tonnes, which may be due to a lower effort and abundance.

7.21.6.2 Management

The Minimum Legal Size (MLS) for capture is limited to 19 cm total length. Gear restrictions such as use of traps only for fishing and prohibition of taking egg-bearing lobsters are other management measures enforced. The traps per boat is restricted to 60 numbers. However, the enforcement is weak, resulting in illegal fishing and export (Morgan 2004).

7.21.7 Lobster Fisheries of Saudi Arabia

Saudi Arabia occupies 80% of the Arabian Peninsula and is bordered on the west by the Red Sea and on the east by the Gulf that lies between Iran and the Arabian Peninsula (Morgan 2006b). The Red Sea coast represents about 79% of total length of the coast. Traditional or artisanal fisheries and industrial fisheries operate both the Red Sea and Gulf coasts. Artisanal fisheries in the Red Sea coast and trawl fisheries in the Arabian Gulf together land an average 12 t of lobsters annually (1998–2002) (FAO 2002). Annual average landing during 2003–2014 is 22 t (Fig. 7.69).

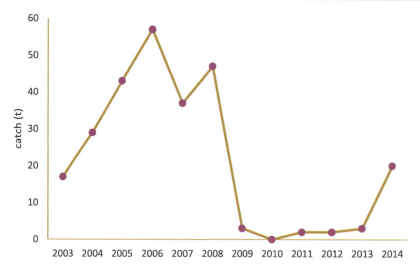

Fig. 7.69 Annual lobster landings in Saudi Arabia (FAO catch data)

Major species are *P. homarus homarus* (megasculpta form) and *P. versicolor*. *P. penicillatus* supports a semicommercial fishery north of Jeddah to the Gulf of Aqaba (Tortell 2004). The scyllarid lobster *T. orientalis* (probably *T. unimaculatus*) and *Scyllarides* are landed as bycatch by trawlers from Arabian Gulf.

Saudi Arabia banned lobster fishing from 1 August 2010 for a period of 3 years following a sharp decline in landings. Management is through seasonal ban on fishing, gear restrictions, prohibition on catching egg-bearing lobsters and minimum size of capture.

7.21.8 Lobster Fisheries of Sultanate of Oman

7.21.8.1 Introduction

The existence of lobsters and its value were not known until the 1970s when an incidental catch of about 10 t was landed at Masirah Island. Thereafter, lobsters became one of the most valuable and highly priced crustaceans in Oman, as well as an important export commodity. Though it is distributed along the entire coast of Oman, major fisheries are located in the area between Ras Al-Hadd and Dalkut (a distance of approximately 1100 km) (Mehanna et al. 2012).

7.21.8.2 Fishery

The lobster fishery of Oman is mainly supported by the palinurid spiny lobster *Panulirus homarus homarus* (megasculpta form), which represents 33% of the total lobster catch (Annual fishery statistics book 2011). The nearshore distribution of lobsters along the Arabian Sea coast of Oman makes them easily accessible to traditional fishers using traps and gillnets operating from their small (4–11 m length)

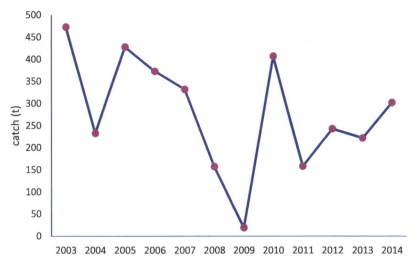

Fig. 7.70 Annual landing of lobsters in Sultanate of Oman (FAO catch data)

motorised fiberglass boats. Fishing season is restricted to 2 months (December to January) from an earlier 6 months. In general, lobster catches in Oman coastal waters have declined over the past several decades, and this has been attributed to the effects of improper fishing methods and overfishing (Al-Marzouqi et al. 2015). Annual average lobster landings were 799 t during 1988–2002, whereas average landings declined to 279 t (2003–2014) (Fig. 7.70) (FAO 2016). Historical data show that annual landings of *P. homarus* reached a peak of 2122 t in 1986 with a mean of 1500 t between 1985 and 1992 (Al-Marzouqui et al. 2015). The landings declined from about 2000 t in 1988 to only about 158 t in 2011 (Mehanna et al. 2012). Gross revenue from lobster fishery seriously decreased from up to six million OR (1 OR ≈ 2.6 $) (1988) to less than one million OR in 2011.

7.21.8.3 Management

The spiny lobster fishery is managed with a minimum size limit of 80 mm carapace length (CL) for all species along the coast of Oman, only 2-month fishing season (15 October–15 December until 2008, changed to March and April since 2010) and prohibition of fishing berried females (Articles 12 and 14 of the Marine Fishing Law). Only trap fishing is permitted; nets, spears and any other gears are banned. However, these regulations have not been strictly enforced. Small lobsters and berried females were common in the catch and the fishermen went to catch lobster in the closed months. In 2002, 379 t was officially landed with a value of 4.3 million USD, which constitutes 0.3% of total volume and 3.9% of total value of export. A 2002 estimate made as part of a report of the DGFR was that 850 t of illegally caught lobsters were exported (mainly to Dubai) annually. This is approximately three times the 'official' landings. An analysis of biological indicators such as size structure and sex composition shows that change in fishing season had a positive impact on the landings. However, more measures such as regional size limits, different closing periods and quota limits were suggested to improve the fishery (Al-Marzouqui et al. 2015).

7.21.9 Lobster Fisheries of Iran

7.21.9.1 Fishery

The total length of Iranian coast is 2700 km and the Oman Sea coastline is 1800 km. Three species of palinurids (*P. homarus homarus*, *P. versicolor* and *P. polyphagus*) and two species of scyllarids, *T. orientalis* (probably *T. unimaculatus*) and *Scyllarides squamosus*, constitute the lobster fishery in Iranian shallow waters of the Persian Gulf and the Gulf of Oman (Ardalan et al. 2010). Annual average lobster production (1988–1997) was 14 t (FAO 1998). Annual average landing during 2003–2014 is 17 t (Fig. 7.71) (FAO 2016). The continental shelf is narrow with an average width of 12 nautical miles. Fishing in Oman Sea is mainly by trawlers operated by industrial fishing fleets. Artisanal fishing of lobsters is by traps and gillnets.

There has been a 70% decline in lobster catches in Iranian waters in the Oman Sea, and in order to enhance the stock, 30 artificial shelters were deployed (Azhdari and Azhdari 2008). A small sport fishery for the lobsters exists in Iranian waters (Emmerson 2016).

Research on spiny lobsters distributed along the Iranian coast of Gulf of Oman was focused on (i) biosystematics and the appropriate tool (Sari 1991); (ii) population dynamics (Fatemi 1998); (iii) studies on growth characteristics (Rajabipour and Mashaii 2003), reproduction, feeding biology and ecology of the habitats (Mashaii 2003); and (iv) management of commercial catch (Mashaii and Rajabipour 2002, 2003).

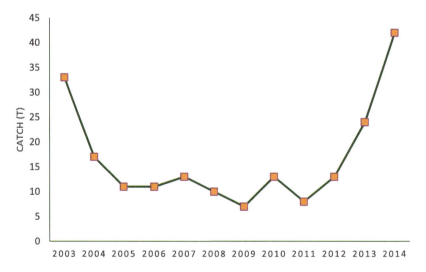

Fig. 7.71 Annual lobster landings in Iran (FAO catch data)

7.21.9.2 Management
Fisheries legislation is in place, but compliance is limited. Lobster fishery is governed by MLS and prohibition on taking egg-bearing lobsters.

7.21.10 Lobster Fisheries of Sri Lanka

7.21.10.1 Introduction
The length of Sri Lankan coastline is 1380 km. The coast supports lobster fishery because of its unique coastal features such as reefs, lagoons, bays and rocky areas. The important fishing grounds are located on the southern coast and the west coast. Lobster fishery is one of the major small-scale fisheries along the southern coast.

7.21.10.2 Species and Fishing Season
Five species of spiny lobsters, *P. homarus homarus*, *P. longipes*, *P. versicolor*, *P. penicillatus* and *P. polyphagus*, occur along the west, south and east coast of Sri Lanka (De Bruin 1962; Jayakody 1989, 1999). *Panulirus polyphagus* is restricted to the mud banks of Mullaitivu, and the species has almost disappeared from the fishing grounds due to overfishing (Liyanage and Long 2009). *Panulirus homarus homarus* is the predominant species on the west coast, *P. versicolor* on the east and *P. ornatus* on the north. The Sri Lankan lobster fishery is mainly constituted by *P. homarus homarus* (70–86%), followed by *P. ornatus* and *P. penicillatus* (3%) and *P. versicolor* (6%). The artisanal fishermen fish throughout the year except during the closed season, February, September and October. Lobster fishing season starts in August and peak fishing is during the calm period from November to March. The landing figures are not available in FAO catch statistics. Over 95% is exported mainly to Japan, Hong Kong, the UK and Singapore, and 5% is consumed locally and a good portion served in restaurants. An average 234 t, valued at 16.8 million, was exported between 2000 and 2008 of which 60% come from south coast fishing grounds. Export price varies from US$29 to 55/Kg and live *P. ornatus* even fetches the premium price of US$65/Kg.

Population studies of *P. homarus homarus* at Mutwal show that the stock is not subjected to high fishing pressure as lobster is not the main constituent of the bottom-set gillnet fishery (Jayawickrema 1991). Senevirathna et al. (2016) studied the population genetic structure of the spiny lobster *P. homarus homarus* from southern coast of Sri Lanka using mitochondrial DNA markers.

In Sri Lanka, four species of slipper lobsters, *Parribacus antarcticus*, *Scyllarus batei*, *Scyllarus* sp. and *Thenus orientalis*, support the fishery. Among them, only two *Parribacus antarcticus* and *Thenus orientalis* are exported (Koralagama et al. 2007).

7.21.10.3 Management
Several management measures are in force but not strictly followed resulting in overall decline in catches. The MLS for all species except *P. ornatus* is 60 mm CL and 100 mm CL for *P. ornatus*. Taking egg-bearing lobster and soft-shelled lobsters

is prohibited. Though lobster fishermen require a license for fishing, only 10% of the fishermen possess licenses.

The fishery is characterised by high exploitation levels and landing of undersized lobsters, getting poor remuneration to the fishermen. Catching undersized lobsters is to be prevented especially during the recruitment period. One of the major recommendations based on the preliminary assessment of the south coast lobster fishery (Sanders and Liyanage 2009) was an extension of the current closure period by another 1.5 months. The South Coast Lobster Fishery Management Committee agreed to extend the closure period to a total of 4.5 months, September, October, second half of December, January and February. Controlling the fishing effort, introduction of quota system (TAC, 12 t), introduction of new eco-friendly fishing gear and establishment of lobster sanctuaries are the other recommendations. No person other than the permit holder should possess, sell or export spiny or slipper lobster meat (except under the authority of permit issued by the Director) according to the Fisheries and Aquatic Resources Act No.2 of 1996.

7.21.11 Malaysian Lobster Fisheries

7.21.11.1 Introduction

The total length of Malaysian coastline is 4675 Km with Peninsular Malaysia on the west and East Malaysia on the east. East Malaysia covers the northern part of Borneo Island bordering the Pacific Ocean (FAO fishing area 71). The Sabah state on the northeast of Eastern Malaysia and eastern and southern parts of western Peninsular Malaysia accounts for 97.4% of the lobster landings of the country. The rest is landed along the northern Peninsular Malaysia coast (FAO fishing area 57). The potential of spiny lobsters as a valuable crustacean resource in Sabah was first realised in the 1960s, but it was during the late 1980s that spiny lobsters became a targeted fishery. The baited pots did not yield a good catch, and persistent poor catches forced the fishermen to adopt other methods such as diving and hand catching lobsters from reefs and trammel net fishing.

7.21.11.2 Fishery

Five species constituted the spiny lobster fishery; *P. longipes* is the most common followed by *P. versicolor* and *P. ornatus*. *P. ornatus* is the most sought-after species in the export market and fetched the export price of RM 115/kg (2004), whereas *P. versicolor* RM 80/kg and *P. longipes* RM 75/kg. *P. polyphagus* and *P. homarus* are of less demand and lower price.

Annual average landings were 30 t from fishing area 57 and 1087 t from fishing area 71 (2003–2014). Annual average total landing during 2003–2014 is 1113 t (Fig. 7.72) (FAO 2015). An average 114 t was exported between 1990 and 2001.

7.21.11.3 Management

Fishery management is enforced under the Fisheries Act 1985. Lobster fishing only by hand/diving/pot fishing is permitted. No take zones for lobsters in MPAs and

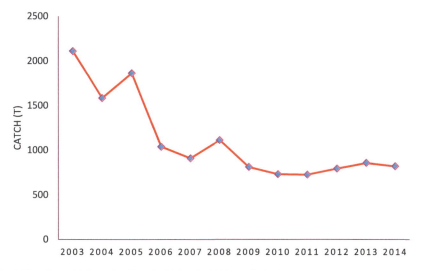

Fig. 7.72 Annual lobster landings in Malaysia (FAO catch data)

MMEs exist. Export of *P. ornatus*, *P. versicolor* and *P. longipes* is regulated by MLS. Annual export quota is limited to 70 t which is being reviewed. Export season is restricted to May and November. Taking egg-bearing lobsters is prohibited.

7.21.12 Lobster Fisheries of Indonesia

7.21.12.1 Introduction

The Indonesian archipelago lies between the Indian Ocean and the Pacific Ocean extending 5120 km from east to west and 1760 km from north to south. The tropical rock lobster fishery in Indonesia is a high-value, export-oriented, open-access artisanal fishery, constituted by six species of palinurids and few other scyllarids with the most important species varying regionally (Milton et al. 2014). On the southern Java coast, the fishery is by traps and gillnets.

7.21.12.2 Fishery

The dominant species in the fishery is *P. homarus homarus* followed by *P. penicillatus* (Milton et al. 2012). The trap fishery is mostly constituted by *P. penicillatus*. The gillnetters operate at deeper areas (25–100 m), and the catch is predominantly *P. homarus homarus* with some *P. versicolor* and *P. longipes* (Milton et al. 2014).

The lobster species reported from Indonesian waters are *Panulirus penicillatus* (Chow et al. 2011; Kalih 2012; Abdullah et al. 2014), *Linuparus somniosus* (Wowor 1999), *P. versicolor* (Ongkers et al. 2014), *P. homarus*, *P. longipes*, *P. ornatus* and *Parribacus antarcticus* (Kalih, 2012). *Metanephrops andamanicus*, *Puerulus mesodontus*, *Thenus indicus* and *Scyllarides haanii* from South Java also contribute to the lobster fauna of Indonesia (Wardiatno et al. 2016a, b, c). Kalih et al. (2013)

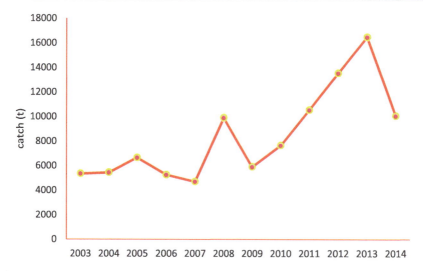

Fig. 7.73 Annual lobster landings in Indonesia (FAO catch data)

reported occurrence of seven species of palinurids from the coastal waters of Lombok Island including *P. longipes longipes* and *P. longipes femoristriga*. The Palabuhanratu Bay in West Java Island facing the Indian Ocean has a broad diversity of fauna (Wahiyudin et al. 2017). There are all together 12 species of lobsters, i.e. *Panulirus ornatus, P. versicolor, P. penicillatus, P. homarus, P. longipes longipes, P. polyphagus, Linuparus somniosus, Palinustus waguensis, Puerulus mesodontus, Parribacus antarcticus, Thenus indicus* and *Metanephrops andamanicus*. Three of the 12 species, i.e. *Panulirus ornatus, P. polyphagus, and Parribacus antarcticus*, are reported as new distribution record from south of Java, Indonesia.

A capture-based lobster farming industry is developing in several parts of Indonesia with technical and scientific support from ACIAR (Chap. 12).

The annual average total landing of lobsters in Indonesia is 8456 t (2003–2014). The catch from Indian Ocean region (FAO fishing area 57) is 2608 t, and 5848 t is from the Pacific Ocean side (fishing area 71) (Fig. 7.73). Annual average production from 1993 to 2002 were 1365 t and 1844 t from these two regions, respectively. However, annual production registered an upward trend with total landing reaching 10,062 t and 16,750 t in 2014 and 2015, respectively (FAO 2016).

7.21.12.3 Management

Management measures such as spatial closures adopted by lobster fisheries worldwide may not be practical in the South Java region, as fishermen can still fish from outside this area. Large-scale temporal closures and size restrictions may affect the livelihood of coastal fishermen as they depend on this resource. Further, these measures cannot be successfully implemented with community management alone (Milton et al. 2014). A uniform management code may not be a practical management measure for regulating lobster fisheries of Indonesia. Prohibition of catching egg-bearing lobsters, MLS and restrictions on gear are useful measures for fishery management. Large-scale exploitation of puerulus and juveniles is prevalent for the purpose of lobster farming.

7.21.13 Lobster Fisheries of Australia

Australia is the top ranking lobster-producing country in the world with an annual average production of 14,150 t (2003–2012) (FAO 2016). Major species constituting commercial fishery are the Western rock lobster *Panulirus cygnus*, the cold water species *Jasus edwardsii* and *J. verreauxi* and the tropical species *P. ornatus* and *P. homarus*.

7.21.13.1 West Coast Rock Lobster *Panulirus cygnus*
The west coast rock lobster, *Panulirus cygnus*, has historically been Australia's most valuable single species wild capture fishery (managed) with an annual landing of 5947 t and an estimated value of A$359 million in 2014. The fishery operates between Shark Bay and Cape Leeuwin using baited traps (pots), with 235 vessels operating in 2014. Fishing is by beehive or batten pots with three escape gaps. The lobster catch is exported as live, frozen, whole cooked or whole raw lobsters to Asia, the USA and Europe. There is also a small domestic market, mainly for the whole cooked rock lobster.

7.21.13.1.1 Management
The fishery is regulated by several legal instruments: the *Fish Resources Management Act 1994*, the *Fish Resources Management Regulations 1995*, the *West Coast Rock Lobster Management Plan 1993*, the West Coast Rock Lobster Managed Fishery Licence and the Commonwealth Government *Environment Protection and Biodiversity Conservation Act 1999*. There are a maximum size limit for fishing, 115 mm CL (females) for south and 105 mm CL for the northern stock, and a minimum size of 77 mm CL (November 15 to January 31) and 76 mm CL (1 February to 30 June). The fishery is managed by total allowable effort (TAE) and individual transferable efforts (ITE). The annual average landing is 8953 t (2003–2014) (Fig. 7.74) (FAO 2016).

7.21.13.2 Southern Red Rock Lobster *Jasus edwardsii*
The South Australia commercial and recreational rock lobster fishery is supported by *Jasus edwardsii*. Fishing is by steel mesh pots with a top entry and two escape gaps. There are currently 180 licenses in the southern zone fishery. Annual landing during 2002–2003 was 1766 t valued at A$63.8 million. The TAC limit was 1900 t. In the northern zone, annual catch during 2002–2003 was 595 t valued at A$18.8 million. There are 69 licenses and the TAC was limited to 525 t. Annual average landing of the species during 2003–2014 was 3668 t (Fig. 7.74) (FAO 2016).

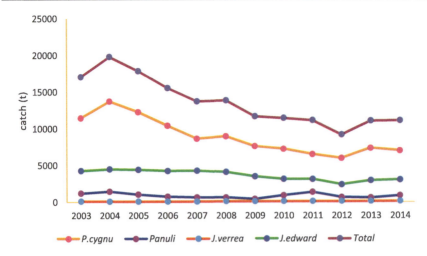

Fig. 7.74 Annual species-wise and total lobster landings in Australia (FAO catch data). *Panulirus cygnus, Panulirus* spp., *Jasus verreauxi, Jasus edwardsii*

7.21.13.2.1 Management

Both zones are currently governed under Fisheries Management Act 2007. Maximum legal size is 23.5 cm CL (male) and 18 cm CL (female). Minimum size limit is 10 cm CL (male) and 9 cm CL (female).

7.21.13.3 Eastern Rock Lobster *Jasus verreauxi*

The species is distributed along the southeastern coast of Australia. It reaches a maximum weight of 8 Kg (26 cm CL). The annual average landing during 2003–2014 was 122 t (FAO 2016) (Fig. 7.74).

7.21.13.4 *Panulirus ornatus*

The landing of other tropical palinurids, mainly *P. ornatus* in Torres Strait on the northern Australian coast, is an average 901 t. The MLS for *P. ornatus* is 75 mm CL.

More than 20 species of deep-sea lobsters under the genera *Acanthacaris, Metanephrops, Nephropsis, Linuparus, Projasus, Puerulus, Polycheles, Stereomastis* and *Willemoesia* have been recorded from the eastern Australian coast (Griffin and Stoddart 1995). The scyllarids distributed in the region come under the genera *Scyllarus, Ibacus* and *Thenus*.

7.22 Global Certified Lobster Fisheries

The demand for safe and sustainably certified seafood is increasing worldwide, concurrent with a growing population. We have been witnessing increasing global awareness and a concerted demand for sustainable lobster fishery management during the past 20 years, as global capture fisheries production has been showing signs of stagnation and even depletion of some fisheries. Most of the spiny lobster fisheries are vulnerable to human exploitation, and, as the resource is low in volume compared to other marine resources, management of the fishery is often neglected and fishery regulations are not strictly enforced. There is a tendency to fish more, as the resource fetches the highest unit price in the global seafood market. In India, though lobsters are fished all along the coast, commercial level fishery is restricted to only a few maritime states with an average annual production of 300–900 t. Annual average catch is just 10–30% of the production during the 1970s, and the current situation is due to overexploitation and ineffective management. However, there are some well-managed lobster fisheries elsewhere in the world, and these fisheries have been certified as sustainable, eco-friendly and well managed by the Marine Stewardship Council (MSC), an independent certifying agency.

7.22.1 Marine Stewardship Council (MSC)

The Marine Stewardship Council (MSC) is an international non-profit organisation established in 1996 as a joint initiative of the World Wildlife Fund for Nature (WWF) and Unilever, at the time the world's largest frozen fish buyer and processor, to address the problem of unsustainable fishing and safeguard seafood supplies for the future. The MSC runs the only certification and eco-labelling programme for wild capture fisheries that is consistent with the ISEAL Code of Good Practice for Setting Social and Environmental Standards and the United Nations Food and Agriculture Organization (FAO) Guidelines for the eco-labelling of fish and fishery products from marine capture fisheries. These guidelines are based on the FAO Code of Conduct for Responsible Fishing, which require credible fishery certification and eco-labelling schemes to include:

- Objective, third-party fishery assessment using scientific evidence.
- Transparent processes with built-in stakeholder consultation and objection procedures.
- Standards based on the sustainability of target species, ecosystems and management practices.

Each fishery must prove that it complies with the MSC's three overarching principles:

1. Sustainable fish stocks
   The fishing activity must be at a level which sustains the fish population. Certified fisheries must operate in a way that avoids overexploitation so that fishing can continue indefinitely.
2. Minimising environmental impact
   Fishing operations should be managed so that the structure, productivity, functions and diversity of the ecosystem on which the fishery depends is maintained.

3. Effective management

The fishery must meet all local, national and international laws and have a management system which can respond to changing circumstances while maintaining sustainability.

Thirty-one further criteria support these three main principles. The certification is achieved through a two-stage process: pre-assessment and full assessment. Once pre-assessments have been completed, a timetable and process for full assessments is developed. The pre-assessments are conducted by a number of appropriately experienced and qualified assessors, and full assessment is undertaken by independent, accredited MSC certification bodies. There are currently ten organisations worldwide accredited to undertake full MSC fisheries assessments.

7.22.2 Certification of Lobster Fisheries

Globally, so far, 259 marine fisheries have been certified by the MSC, of which 13 are lobster fisheries and 5 are exclusively spiny lobster fisheries. The first lobster fishery certified by MSC was the Western rock lobster, *Panulirus cygnus*, spot and trap fishery.

The MSC-certified lobster fisheries are:

1. The Maine lobster trap fishery.

    The Maine lobster trap fishery of the American clawed lobster *Homarus americanus* was independently certified to the Marine Stewardship Council environmental standard for sustainable fishing on 8 March 2013. Annual average landing (2000–2007) was 33,000 t.
2. The Gulf of Maine lobster fishery.

    The Gulf of Maine lobster fishery of *H. americanus* was independently certified on 10 May 2013. Annual average landing in 2013 was 57,000 t with a landed value of US$364 million. A parallel and separate assessment and certification of Maine lobster fishery in the same area is under consideration of the MSC on the basis of a request by another consortium of companies.
3. The Gaspesie lobster trap fishery.

    The lobster fishery (*H. americanus*) on the Gaspesie Peninsula in Quebec, Canada, along the shores of the Gulf of St. Lawrence on the Atlantic coast was independently certified on 5 March 2015. In 2013, the fishermen landed 1370 metric tonnes of lobster.
4. The Eastern Canada offshore lobster fishery.

    The Eastern Canada offshore lobster trap fishery of *H. americanus* was independently certified as sustainable on 2 June 2010 and recertified on 30 June 2015. The total allowable catch (TAC) is 720 t. The depth of fishing is 100–320 m. The lobsters are sold in the live market.
5. The Bay of Fundy, Scotian Shelf and Southern Gulf of St. Lawrence lobster trap fishery.

The Bay of Fundy, Scotian Shelf and Southern Gulf of St. Lawrence lobster trap fishery of *H. americanus* along the Atlantic coast of Canada (Nova Scotia and New Brunswick) was independently certified as sustainable on 22 July 2014. Nearly all lobsters from the Atlantic coast of Canada have been certified sustainable. In 2014, the landed value of all lobster fisheries in Canada was C$853 million, of which C$671 million (79%) was generated by independent harvesters in the fishery.

6. The Iles-de-la-Madeleine fishery.

The Iles-de-la-Madeleine lobster (*H. americanus*) trap fishery was independently certified as sustainable on 16 July 2013. The fishery is located in lobster fishing area (LFA) 22 surrounding the Magdalen Islands in the Gulf of St. Lawrence off the east coast of Canada. The landings during 2010 were worth C$26 million.

7. The Prince Edward Island lobster trap fishery.

The Prince Edward Island lobster trap fishery of *H. americanus* achieved MSC certification on 17 November 2014. The annual landing is approximately 14,000 t over the past few years.

8. The Normandy and Jersey lobster fishery.

The Normandy and Jersey lobster fishery of the European lobster *H. gammarus* was certified as sustainable on 14 June 2011. The fishery is shared between France and the UK. Annual landing is between 270 and 290 t.

9. The Juan Fernandez rock lobster fishery.

The Chilean Juan Fernandez rock lobster trap fishery of *Jasus frontalis* achieved MSC certification on 6 January 2015. The archipelago is situated about 400 miles off the Chilean coast with about 800 inhabitants. Annual landing is 100 t.

10. The Australian Western rock lobster fishery.

In March 2000, the Western rock lobster *Panulirus cygnus* fishery became the first in the world to attain MSC certification. The fishery was recertified in December 2006 and in 2017 it was recertified again for the fourth time. The lobsters are harvested by baited pots and traps. Annual landing was approximately 5500 t in 2010–2011. The fishery recently moved from an input (effort control) to an output (catch quota) management system. The consequence of the introduction of these quota management measures has been a significant reduction in the number of pots being used in the fishery. The estimated value is A$200 million per year.

11. The Tristan da Cunha lobster fishery.

The Tristan da Cunha lobster fishery of the spiny lobster *Jasus tristani* was certified as sustainable on 20 June 2011. Annual landing (2008–2009) is 435 t.

12. The Mexican Baja red rock lobster fishery.

The Mexican Baja red rock lobster wire trap fishery of the spiny lobster *Panulirus interruptus* was certified as sustainable in April 2004. The fishery was recertified in June 2011 for a second term. Annual average landing (2005–2010) is approximately 1400 t. The Baja California red rock lobster fishery was the first Latin American and first community-based developing world artisanal

fishery certified to the MSC standard. More than 500 artisanal fishermen from 10 cooperatives participate in the fishery.
13. The Sian Kaan and Banco Chinchorro Reserves spiny lobster fishery.

The Sian Kaan and Banco Chinchorro Reserves small-scale artisanal spiny lobster fishery of the spiny lobster *Panulirus argus* was certified as sustainable on 31 July 2012. The fishery operates in nearshore waters of the Sian Kaan and Banco Chinchorro Biosphere Reserves off the coast of the state of Quintana in the Yucatan Peninsula, Mexico. Annual landing is approximately 280 t since the mid-1990s.

First MSC-Certified Whole Lobster

The first whole lobster MSC certification has been granted to *Homarus americanus* to be sold in Lidl supermarkets across the UK. The MSC-certified lobsters come from the cold, clear waters off New Brunswick, Canada. The lobsters are cooked and frozen before being supplied to the UK.

7.23 Summary

World lobster production has been showing an upward trend since 2014, and this has been mainly due to higher landings of the American lobster *H. americanus* and improved production by Indonesia. However, it is unlikely that this trend will continue for long, as the resource has been the target of heavy exploitation due to premium price for the product and increased demand. Apart from anthropogenic intervention, habitat loss, environmental degradation, diseases and climate change have also been playing pivotal roles in reducing the productivity of many fisheries. Overexploitation due to poor enforcement of fishing regulations in many countries is also the likely cause of decline in global production. There are some practical difficulties in implementing management measures in certain fisheries like trawl fisheries, where the resource is appearing as a bycatch. In some fisheries, the peak breeding and fishing seasons coincide, and thereby egg-bearing lobsters form a major share of the catch.

Although lobster is not a large marine resource in India, it is a high-value seafood, with almost the entire catch exported fetching a substantial foreign exchange. The maritime states manage their fisheries but within the policy framework of the fisheries policy of the Central Government. Little attention has been paid to lobsters by the state governments, being a minor fishery resource, and this neglect in framing and enforcing fishing regulations has led to heavy fishing pressure on this vulnerable resource, leading to decline in landings in majority of the states. The stock of *T. unimaculatus* on the northwestern coast has been significantly depleted and so also the deep-sea lobster, *P. sewelli*, of the southern Indian coast. In the artisanal sector, the state governments have to ban the destructive fishing practices and encourage fishermen to use traps than the gillnets by subsidising the manufacturing cost of the traps. Implementation of a fishing holiday during the peak breeding season is another measure by which capturing of the egg-bearing lobsters can be minimised. Establishment of Fishery Management Councils to oversee overall fishing activities by the coastal fishing villages may be another proactive step for effective co-management of the resource.

In order to support the co-management system under formulation, more intensive biological research with new approaches to stock assessment are needed. A fishery data collection system should be part of the management system to develop input controls for exploitation of the resource on a sustainable basis. The future challenge of management will be to encourage fishermen to use traps with escape vents instead of gillnets, to make them agree for a temporary closure of fishing during the peak breeding season to minimise capture of breeding females and to set annual catch quotas, which are both biologically and economically beneficial. The fishermen on the southwest coast believe that there are two categories of lobsters, one a resident population and the other that are brought into the coastal fishing grounds by the coastal southerly currents during the southwest monsoon. The scientific evidence for such indigenous knowledge of the fishermen should be generated to explain the resilience of the lobster population along the Kanyakumari coast, even under decades of intensive fishing though in reduced volumes. There is absolutely no information on puerulus settlement and recruitment pattern of lobsters along the southwest coast. The future research must also focus on factors responsible for observed temporal and spatial variations in lobster abundance along the Kanyakumari coast. Establishing artificial reefs to provide additional shelter and a lobster sanctuary (no fishing zone) may benefit rejuvenation of the declining lobster stock along the southwest coast of India.

The countries bordering the Indian Ocean contribute significantly to the global lobster production. While countries such as Australia and South Africa have strict fishing regulations for both commercial and recreational fishing, many other countries in the region are not strictly enforcing the regulatory measures. Strict enforcement of fishing regulations with active participation of fishers and other stakeholders such as local traders and exporters is necessary for maintaining long-term sustainability of this vulnerable but valuable resource. Ecolabelling is probably an effective and practical solution that can ensure resource management on a sustainable platform.

References

Abdullah, M. F., Alimuddin, M. M., Salma, A. J., & Imai, S. H. (2014). Genetic isolation among the Northwestern, Southwestern and Central-Eastern Indian Ocean populations of the pronghorn spiny lobster *Panulirus penicillatus*. *International Journal of Molecular Sciences, 15*, 9242–9254.

Alabsi, N., & Komatsu, T. (2014). Characterization of fisheries management in Yemen: A case study of a developing country's management regime. *Marine Policy, 50*, 89–95.

Alcock, A. (1901). *A descriptive catalogue of the Indian deep-sea Crustacea, Decapoda, Macrura and Anomala in the Indian Museum, being a revised account of the deep-sea species collected by the Royal Marine Ship 'Investigator'*, Calcutta, p 286.

Alcock, A. (1906). Catalogue of the Indian decapod Crustacea in the collections of the Indian Museum. Part III. *Macrura*, 1–55.

Alcock, A., & Anderson, A.R.S. (1894). Natural history notes from H.M. Indian Marine Survey Steamer "Investigator", Commander C.D. Oldham, R.N. Commanding, Series II, no 14. An account of a Recent Collection of Deep sea Crustacea from the Bay of Bengal and the Laccadive sea. *Journal of the Asiatic Society of Bengal, 63*, pt 2 (3), 141–185, pl.9

Ali, D. M., Pandian, P. P., Somavanshi, V. S., John, M. E., & Reddy, K. S. N. (1994). Spear lobster, *Linuparus somniosus* Berry & George, 1972 (fam. Palinuridae) in the Andaman Sea. *Occasional Paper, Fishery Survey of India, Mumbai, 6*, 13.

Al-Marzouqi, A., Chesalin, M., Al-Shajibi, S., Al-Hadabi, A., & Al-Senaidi, R. (2015). Changes in the Scalloped Spiny Lobster, *Panulirus homarus* Biological Structure after a Shift of the Fishing Season. *Journal of Aquaculture & Marine Biology, 3*(1), 00056.

Anon. (1989). Marine living resources of the union territory of Lakshadweep: An indicative survey with suggestions for development. *CMFRI Bulletin, 43*, 375.

Anrose, A., Selvaraj, P., Dhas, J. C., Prasad, G. V. A., & Babu, C. (2010). Distribution and abundance of deep sea spiny lobster *Puerulus sewelli* in the Indian Exclusive Economic zone. *Journal of the Marine Biological Association of India, 52*(2), 162–165.

Ardalan, M., Sari, A., Rezvani-Gilkolaei, S., & Pourkazemi, M. (2010). Phylogeny of Iranian Coastal Lobsters inferred from mitochondrial DNA restriction fragment length polymorphism. *Acta Zoologica Bulgarica, 62*(3), 331–338.

Azhdari, H., & Azhdari, Z. (2008). Nestling of lobsters in three different new designed artificial reefs in the Oman Sea. *Iranian Journal of Fisheries Sciences, 17*, 11–22.

Baisre, J., & Cruz, R. (1994). The Cuban spiny lobster fishery. In B. F. Phillips, J. Cobb, & J. Kittaka (Eds.), *Spiny lobster management* (pp. 119–131). Oxford: Fishing News Books.

Balasubramanyan, R., Satyanarayana, A. V. V., & Sadanandan, K. A. (1960). A preliminary account of the experimental rock lobster fishing conducted along the southwest coast of India, with bottom-set-gillnets. *Indian Journal of Fisheries, 7*, 405–422.

Balasubramanyan, R., Satyanarayana, A. V. V., & Sadanandan, K. A. (1961). A further account of the rock lobster fishing experiments with the bottom-set-gillnets. *Indian Journal of Fisheries, 8*, 269–290.

Balss, H. (1925). Macrura d. Deutschen Tiefsee-Exp.2. Naturtha Tel A. *Wissenschaftliche Ergebnisse Der Deutschen Tiefsee-Expedition 'Valdivia', 20*(5), 221–315.

Bautil, B. (2002). *Conception d'un systeme d'exploitation de la peche langoustiere Rapport de mission du 8 au 21 decembre 2002* (Project PCT-FAO/MAG0170 (A): Decembre 2002) (43 p). Antananrivo: FAO.

Berry, P. F. (1971). *The biology of the spiny lobster Panulirus homarus (Linnaeus) off the east coast of southern Africa* (Investigation Report on Oceanographic Research Institute, Vol. 28, 75 pp).

Berry, P. F. (1974). A revision of the *Panulirus homarus*-group of spiny lobsters (Decapoda, Palinuridae). *Crustaceana, 27*, 31–42.

Borradaile, L. A. (1906). Marine crustaceans XIII. The Hippidae, Thalassinidea and Scyllaridae. In J. S. Gardiner (Ed.), *The fauna and geography of the Maldives and Laccadive Archipelago* (pp. 750–754). Cambridge: Cambridge University Press.

Campbell, A., Noakes, D. J., & Elner, R. W. (1991). Temperature and lobster, *Homarus americanus* yield relationships. *Canadian Journal of Fisheries and Aquatic Sciences, 48*, 2073–2082.

Chacko, P. I. & Balakrishnan Nair, N. (1963). Size and sex compositions of catches of the lobster *Panulirus dasypus* (Latreille) along Kanyakumari district coast in 1960-62. *Proceedings of the Indian Science Congress, 50*(3), 50.

Chakraborty, R. D., Maheswarudu, G., Radhakrishnan, E. V., Purushothaman, P., Kuberan, P., Jomon Sebastian, G., & Thangaraja, R. (2014). Rare occurrence of blunthorn lobster *Palinustus waguensis* Kubo, 1963 from the southwest coast of India. *Marine Fisheries Information Service, Technical and Extension Series, 219*, 25–26.

Chan, T. Y., & Ng, P. K. L. (2008). *Enoplometopus* A. Milne-Edwards, 1862 (Crustacea: Decapoda: Nephropidae) from the Philippines, with description of one new species and a revised key to the genus. *Bulletin of Marine Science, 83*(2), 347–365.

Chan, T.-Y. (2010). Annotated checklist of World's marine lobsters (Crustacea, Decapoda: Astacidea, Glypheida, Achelata, Polychelida). *The Raffles Bulletin of Zoology*, Supplement No. 23, 153–181.

Chatterjee, T. K., Hussain, A., & Mitra, S. (2007). Faunal diversity in prawns and crabs in Digha and adjacent coast in West Bengal with notes on the relationship of their abundance with physico-chemical parameters. *Journal of the Bombay Natural History Society, 104*(3), 311–315.

Chekunova, V. I. (1971). Distribution of commercial invertebrates on the shelf of India, the north-eastern part of Bay of Bengal and the Andaman Sea. In: S.A. Bogdanov (Ed.), *Soviet fisheries investigations in the Indian Ocean* (pp. 68–83). Jerusalem: Israel Programme for Scientific Translations.

Chhapgar, B. F., & Deshmukh, S. K. (1961). On the occurrence of the spiny lobster *Panulirus dasypus* (H. Milne Edwards) in Bombay waters with a note on the systematics of Bombay lobsters. *Journal of the Bombay Natural History Society, 58*(3), 632–638.

Chhapgar, B. F., & Deshmukh, S. K. (1964). Further records of lobsters from Bombay. *Journal of the Bombay Natural History Society, 61*, 203–207.

Chhapgar, B. F., & Deshmukh, S. K. (1971). Lobster fishery of Maharashtra. *Journal of Indian Fisheries Association, 1*(1), 74–86.

Chopra, B. N. (1939). Some food prawns and crabs of India and their fisheries. *Journal of the Bombay Natural History Society, 41*, 221–234.

Chow, S., Jeffs, A., Miyake, Y., Konishi, K., Okazaki, M., Suzuki, N., Abdullah, M. F., Imai, H., Wakabayasi, T., & Sakai, M. (2011). Genetic isolation between the Western and Eastern Pacific populations of Pronghorn Spiny Lobster *Panulirus penicillatus*. *PLoS One, 6*, e29280. https://doi.org/10.1371/journal.pone.0029280.

CMFRI. (2014). *CMFRI annual report, 2013–14*. Director, Central Marine Fisheries Research Institute, Cochin, Kerala, India.

De Bruin, G. H. P. (1962). Spiny lobsters of Ceylon. *Bulletin of Fisheries Research Statistics on Ceylon, 14*, 1–28.

De Bruin, G. H. P. (1969). The ecology of spiny lobsters *Panulirus* spp. of Ceylon waters. *Bulletin of Fisheries Research Statistics on Ceylon, 20*(2), 171–190.

De Man, J. G. (1916). The Decapoda of the Siboga expedition Part III. Families Eryonidae, Palinuridae, Scyllaridae and Nephropidae. *Siboga Expedition Monograpgh, 39A*(2), 1–122.

Deere, C.L. 1999. *Ecolabelling and sustainable fisheries*. Washington, DC/Rome: IUCN/FAO.

Deshmukh, S. (1968). On the first phyllosomae of the Bombay spiny lobsters (*Panulirus*) with a note on the unidentified first *Panulirus* phyllosoma from India (Palinuridae). *Crustaceana, 2*(Suppl), 47–58.

Deshmukh, V. D. (2001). Collapse of sand lobster fishery in Bombay waters. *Indian Journal of Fisheries, 48*(1), 71–76.

Devaraj, M., & Pillai, N. G. K. (2001). An overview of the marine fisheries research in the Lakshadweep. *Geological Survey of India, Special Publication, 56*, 83–94.

Dineshbabu, A. P. (2008). Morphometric relationship and fishery of Indian Ocean lobsterette, *Nephropsis stewarti* Wood_Mason, 1873 along the southwest coast of India. *Journal of the Marine Biological Association of India, 50*(1), 113–111.

Dineshbabu, A. P., Durgekar, R. N., & Zacharia, P. U. (2011). Estuarine and marine decapods of Karnataka inventory. *Fishing Chimes, 30*, 10–11.

Dow, R. L. (1977). Relationship of sea surface temperature and American and European lobster landings. *Journal of Conseil/International Council for the Exploration Mer, 37*, 186–190.

Dutta, S., Chakraborty, K., & Hazra, S. (2016). The status of the marine fisheries of West Bengal coast of the northern bay of Bengal and its management options: a review. *Proceedings of the Zoological Society, 69*(1), 1–8.

Emmerson, W. D. (2016). A guide to, and checklist for, the Decapoda of Namibia, South Africa and Mozambique, Vol.1, 524 p.

FAO. (1998). *Database of Global Fishery Statistics*. Rome: FAO.
FAO. (2002). *Fishstat Database Plus*. Rome: Fisheries and Aquaculture Department, Food and Agriculture Organization of the United Nations. www.fao.org/fishery/statistics/software/fishstat.
FAO. (2012). *FAO-Fishstat Database Plus*. Rome: Fisheries and Aquaculture Department, Food and Agriculture Organization of the United Nations. www.fao.org/fishery/statistics/software/fishstat. FAO Fisheries Circular No. 710. Rome Review of the state of world fishery. 47.
FAO. (2015). *FAO-Fishstat Database Plus*. Rome: Fisheries and Aquaculture Department, Food and Agriculture Organization of the United Nations. www.fao.org/fishery/statistics/software/fishstat.
FAO-Fishstat Database Plus. (2016). Fisheries and Aquaculture Department, Food and Agriculture Organization of the United Nations. Rome. www.fao.org/fishery/statistics/software/fishstat
Fatemi, M. R. (1998). *Population dynamics and stock assessment of the dominant lobster species in Chabahar*. PhD thesis, Tehran: Azad Islami University, Department of Science and Research.
Fielding, P. J. & Mann, B. Q. (1999). The Somalia inshore lobster resource. *A survey of the lobster fishery of the northeastern region (Puntland) between FOAR and EYL during November 1998*. IUCN Eastern Africa Programme, pp. 1–37.
Fogarty, M. J. (1986). *Population dynamics of the American lobster (Homarus americanus)*. PhD thesis, University of Rhode Island, Kingston, 227 pp.
George, R. W. (1963). Report to the Government of Aden on the craw-fish resources of the eastern Aden Protectorate. *Expanded programme of technical assistance* (Report 1696, 23pp). Rome: FAO.
George, M. J. (1967a). Observations on the biology and fishery of the spiny lobster *Panulirus homarus* (Linnaeus). InL *Proceedings of the Symposium on Crustacea*, Part IV. Marine biological Association of India, Mandapam camp, India, pp. 1308–1316.
George, M. J. (1967b, February 3). The Indian spiny lobster In: *Souvenir 20th Anniversary Central Marine Fisheries Research Institute,*, Mandapam, India.
George, M. J. (1973). The lobster fishery resources of India. In *Proceedings of the symposium on living resources seas around India* (Special Publication) (pp. 57–580). Cochin: CMFRI.
George, M. J., & George, K. C. (1965). *Palinustus mossambicus* Barnard (Palinuridae: Decapoda), a rare spiny lobster from Indian waters. *Journal of the Marine Biological Association of India, 7*(2), 463–464.
George, M. J., & Rao, P. V. (1965). On some decapod crustaceans from south west coast of India. In: Proceedings of the symposium on crustacea. Part 1. Series 2. Marine Biological Association of India, Mandapam Campus, India, pp. 327–336.
Ghosh, S., Mohanraj, G., Asokan, P. K., Dhokia, H. K., Zala, M. S., & Bhint, H. M. (2008). Bumper catch of spiny lobsters by trawlers and gillnetters at Okha, Gujarat. *The Marine Fisheries Information Service: Technical and Extension Series, 198*, 16–17.
Gopal, K., Sachithanandam, V., Dhivya, P., & Mohan, P. M. (2013). Current status of exported fishery resources in Andaman Islands. *Asian Journal of Marine Science, 1*(1), 12–16.
Griffin, D. J. G., & Stoddart, H. E. (1995). Deep-water Decapod Crustacea from Eastern Australia: Lobsters of the Families Nephropidae, Palinuridae, Polychelidae and Scyllaridae. *Records of the Australian Museum, 47*, 231–263.
Groeneveld, J. C., & Cockcroft, A. C. (1997). Potential of a trap-fishery for deep-water rock lobster *Palinurus delagoae* off South Africa. *Marine and Freshwater Research, 48*(8), 993–1000.
Groeneveld, J. C., & Rossouw, G. J. (1995). Breeding period and size in the southcoast rock lobster, *Palinurus gilchristi* (Decapoda: Palinuridae). *South African Journal of Marine Science, 15*, 17–23.
Groeneveld, J. C., Goñi, R., & Diaz, D. (2013a). *Palinurus* species: Chapter 11. In B. F. Phillips (Ed.), *Lobsters: Biology, management, aquaculture and fisheries* (2nd edn, pp. 326–356). Oxford: Wiley-Blackwell.
Groeneveld, J. C., Everett, B. I., Fennessy, S. T., Kirkman, S. P., Santos, J., & Robertson, W. R. (2013b). Spatial distribution patterns, abundance and population structure of deep-sea crab

Chaceon macphersoni based on complementary analyses of trap and trawl data. *Marine and Freshwater Research, 64*(6), 507–517.

Haakonsen, J. M. (1983). Somalia's fisheries case study (FAO report, pp. 170–184). Mogadiscio: Somalia Academy of Sciences and Arts.

Harris, G. P., Davies, P., Nunez, M., & Meyers, G. (1988). Interannual variability in climate and fisheries in Tasmania. *Nature, London, 333*, 754–757.

Heller, G. (1865). Eiecrusteen Reisi der osteerreichischenFregaatte 'Novara' undieErde in den, Jehren 1857–59 unter den Betehlen des Commodore B. von WullerstorfUrbair. *Zool, 2*(3), 1–280.

Heydorn, A. E. F. (1969). Notes on the biology of *Panulirus homarus* and on the length/weight relationships of *Jasus lalandii*. *Investigational Report of the Division of Sea Fisheries, South Africa, 69*, 1–26.

Holthuis, L. B. (1991). Marine lobsters of the World. FAO species catalogue, vol. 13. *FAO Fisheries Synopsis, Food and Agriculture Organization, Rome, 125*(13), 1–292.

Hossain, M. A. (1978). Few words about the sand lobster *Thenus orientalis* (Lund) (Decapoda: Scyllaridae). *Seafood Export Journal, 10*(1), 43–46.

Jawahar, P., Sundaramoorthy, B., & Chidambaram, P. (2014). Studies on breeding biology of *Panulirus homarus* (Linnaeus, 1758) Thoothukudi, southeastern coast of India. *Journal of Experimental Zoology, India, 17*(1), 175–181.

Jayakody, D. S. (1989). Size at onset of sexual maturity and onset of spawning in female *Panulirus homarus* (Crustacea: Decapoda: Palinuridae) in Sri Lanka. *Marine Ecology Progress Series, 57*, 83–87.

Jayakody, D. S. (1999). *South coast lobster fishery – 1999* (24 p). NARA: Lobster Fishery Assessment Report.

Jayawickrema, S. J. C. (1991). Fishery and population dynamics of *Panulirus homarus* (Linnaeus) from Mutwal, Sri-Lanka. *Journal of the National Science Council of Sri Lanka, 19*(1), 52–61.

Jeena, N. S. (2013, May). *Genetic divergence in lobsters (Crustacea: Palinuridae and Scyllaridae) from the Indian EEZ*. PhD thesis, Cochin University of Science and Technology, Kochi, India, p. 153.

Jeffs, A. G. (2010). Status and challenges for advancing lobster aquaculture. *Journal of the Marine Biological Association of India, 52*(2), 320–326.

Jha, D. K., Kumar, T. S., Nazar, A. K. A., Venkatesan, R., & Saravanan, N. (2007). Spiny lobster resources of North Andaman Sea: preliminary observations. *Fishing Chimes, 27*(1), 138–141.

John, C. C., & Kurian, C. V. (1959). A preliminary note on the occurrence of deep water prawn and spiny lobster off the Kerala coast. *Bulletin of Central Research Institute, Trivandrum, Series C, 7*(1), 155–162.

Johnson, W. D., & Al-Abdulsalaam, T. Z. (1991). The scalloped spiny lobster (*Panulirus homarus*) fishery in the Sultanate of Oman. *The Lobster Newsletter, 4*, 1–4.

Jones, S. (1965). The crustacean fishery resources of India. *Proceedings of the Symposium on Crustacea, Part IV, Series, 2*, 1328–1341.

Joseph, K. M. (1972). A profile of deep sea fishery resources off the southwest coast of India. *Seafood Export Journal, 4*(1), 97–104.

Joseph, K. M. (1974). Demersal fisheries resources off the North-west coast of India. *Bulletin Exploratory Fisheries Project, 1*, 1–45.

Joseph, K. M. (1986). Some observations on potential fishery resources from the Indian EEZ. *Fishery Survey of India Bulletin, 14*, 1–20.

Joseph, K. M., & John, M. E. (1987). Potential marine fishery resources. Seminar on Potential Marine Fishery Resources. *CMFRI Special Publications, 30*, 18–43.

Kabli, L. M., & Kagwade, P. V. (1996). Morphometry and conversion factors in the sand lobster *Thenus orientalis* (Lund) from Bombay waters. *Indian Journal of Fisheries, 43*(3), 259–254.

Kagwade, P. V. (1987a). Morphological relationships and conversion factors in spiny lobster *Panulirus polyphagus* (Herbst). *Indian Journal of Fisheries, 34*(3), 348–352.

Kagwade, P. V. (1987b). b. Age and growth of the spiny lobster *Panulirus polyphagus* (Herbst). *Indian Journal Fisheries, 34*(4), 389–398.

Kagwade, P. V. (1988a). Reproduction in the spiny lobster *Panulirus polyphagus* (Herbst). *Journal of Marine Biological Association of India, 30*(1&2), 37–46.

Kagwade, P. V. (1988b). Fecundity in the spiny lobster *Panulirus polyphagus* (Herbst). *Journal of Marine Biological Association of India, 30*(1&2), 114–120.

Kagwade, P. V. (1993). Stock assessment of the spiny lobster *Panulirus polyphagus* (Herbst) off north-west coast of India. *Indian Journal of Fisheries, 40*(1&2), 63–73.

Kagwade, P. V. (1994). Estimates of the stocks of the spiny lobster *Panulirus polyphagus* (Herbst) in the trawling grounds of Bombay. *Journal of the Marine Biological Association of India, 36*(1&2), 161–167.

Kagwade, P. V., & Kabli, L. M. (1996a). Reproductive biology of the sand lobster *Thenus orientalis* (Lund) from Bombay waters. *Indian Journal of Fisheries, 43*(1), 13–25.

Kagwade, P. V., & Kabli, L. M. (1996b). Age and growth of the sand lobster *Thenus orientalis* (Lund) from Bombay waters. *Indian Journal of Fisheries, 43*(3), 241–247.

Kagwade, P. V., Manickaraja, M., Deshmukh, V. D., Rajamani, M., Radhakrishnan, E. V., Suresh, V., Kathirvel, M., & Rao, G. S. (1991). Magnitude of lobster resources of India. *Journal of the Marine Biological Association of India, 33*(1&2), 150–158.

Kathirvel, M. (1998). A pictorial guide for identification of Indian spiny lobsters. Part 1. *Fish and Fisheries, 18*, 4.

Kathirvel, M., Suseelan, C., & Vedavyasa Rao, P. (1989). Biology, population and exploitation of the Indian deep sea spiny lobster, *Puerulus sewelli* Ramadan. *Fishing Chimes, 8*(11), 16–25.

Kizhakudan, J. K. (2014). Reproductive biology of the female shovel-nosed lobster *Thenus unimaculatus* (Burton & Davie, 2007) from north-west coast of India. *Indian Journal of Geo-Marine Sciences, 43*(6), 927–935.

Kizhakudan, J. K., & Patel, S. K. (2009, February 9–12). Traditional knowledge and methods in lobster fishing in Gujarat. In: E. Vivekanandan, T. M. Najmudeen, T.S. Naomi, A. Gopalakrishnan, K.V. Jayachandran, & M. Harikrishnan (Eds.), *Marine ecosystems: Challenges and opportunities* (Book of Abstracts, pp. 92–93). Cochin: Marine Biological Association of India.

Kizhakudan, J. K., & Patel, S. K. (2010). Size at maturity in the mud spiny lobster *Panulirus polyphagus* (Herbst, 1793). *Journal of the Marine Biological Association of India, 52*(2), 170–179.

Kizhakudan, J. K., & Thirumilu, P. (2006). A note on the bluntthorn lobsters from Chennai. *Journal of the Marine Biological Association of India, 48*(2), 260–262.

Kizhakudan, J.K. & Thumber, B.P.2003. Fishery of Marine crustaceans in Gujarat. In: M. R. Boopendranath, et al. (Eds.), *Sustainable fisheries development: Focus on Gujarat* (p. 207). Cochin: Society of Fisheries Technologists (India.

Kizhakudan, J. K., Thirumilu, P., & Manibal, C. (2004). Fishery of the sand lobster *Thenus orientalis* (Lund) by bottom-set gillnets along Tamilnadu coast. *Marine Fisheries Information Service Technical and Extension Series, 181*, 6–7.

Kizhakudan, J. K., Krishnamoorthi, S., & Thiyagu, R. (2012). First record of the scyllarid lobster *Scyllarides tridacnophaga* from Chennai coast. *Marine Fisheries Information Service Technical and Extension Series, 211*, 13.

Koralagama, D. N., Amaraslnghe, O., & Jayakody, S. (2007). Export market of lobsters in Sri Lanka, with special reference to slipper lobsters. In *Proceedings of the fourth academic sessions,* Technical Session III (NARA), pp. 84–90.

Kulmiye, A. J., Mavuti, K. M., & Groeneveld, J. C. (2006). Size at onset of maturity in spiny lobsters *Panulirus homarus homarus* from Mambrui, Kenya. *African Journal of Marine Science, 28*, 51–55.

Kumar, T. S., Jha, D. K., Syed Jahan, S., Dharani, G., Abdul Nazar, A. K., Sakthivel, M., Alagarraaja, K., Vijayakumaran, M., & Kirubagaran, R. (2010). Fishery resources of spiny lobsters in the Andaman Island, India. *Journal of the Marine Biological Association of India, 52*(2), 166–169.

Kurian, C. V. (1965). Deep water prawns and lobsters off the Kerala coast. *Fishery Technology, 2*(1), 51–53.

Kyomo, J. (1999). Distribution and abundance of crustaceans of commercial importance in Tanzania mainland coastal waters. *Bulletin of Marine Science, 65*(2), 321–335.

Lalithadevi, S. (1981). The occurrence of different stages of spiny lobsters in Kakinada region. *Indian Journal of Fisheries, 28*(1&2), 298–300.

Liyanage, U., & Long, B. (2009). *Status of the South coast lobster fishery 2009* (CENARA Project Reports on Lobsters, p. 43). Colombo: National Aquatic Resources Research and Development Agency.

Luis, A. J., & Kawamura, H. (2004). Air-sea interaction, coastal circulation and primary production in the eastern Arabian Sea: a review. *Journal of Oceanography, 60*(2), 205–218.

Lloyd, R. E. (1907). Contribution to the fauna of the Arabian Sea with description of new fishes and crustaceans. *Records of the Indian Museum, 1*, 1–12.

Madhupratap, M., Kumar, S. P., Bhattathiri, P. M. A., Kumar, M. D., Raghukumar, S., Nair, K. K. C., & Ramaiah, N. (1996). Mechanism of the biological response to winter cooling in the northeastern Arabian Sea. *Nature, 384*, 549–552.

Maina, G. W., & Samoilys, M. (2011). Lamu lobsters – a dwindling resource. *SWARA, 2011*(July–September), 34–35.

Manickaraja, M. (2004). Lobster fishery by a modified bottom-set-gillnet at Kayalapattanam. *The Marine Fisheries Information Service: Technical and Extension, 181*, 7–8.

Mashaii, N. (2003). *The management survey of the rock lobster, Panulirus homarus Linnaeus, 1758, in the coastal waters of Systan and Balouchestan province* (Final report of the project, 157 pp). Iranian Fisheries Research Organization (In Persian).

Mashaii, N., & Rajabipour, F. (2002). A survey about commercial catch of the rock lobster, *Panulirus homarus* Linnaeus, 1758, in the Iranian seashores of Oman Sea, at 2000. *Pajouhesh-va-Sazandegi, In Animal and Fisheries Sciences, 55*, 44–49. (In Persian).

Mashaii, N., & Rajabipour, F. (2003). Commercial catch management of spiny lobster, *Panulirus homarus* Linnaeus, 1758, in the coastal waters of Sistan and Balouchestan province. *Iranian Science of Fishery Journal, 12*(3), 175–192. (in Persian).

Meenakumari, B., & Mohanrajan, K. V. (1985). Studies on materials for traps for spiny lobsters. *Fisheries Research, 3*, 309–321.

Mehanna, S., Al-Shijibi, S., Al-Jafary, J., & Al-Senaidi, R. (2012) Population dynamics and management of scalloped spiny lobster *Panulirus homarus* in Oman coastal waters. *Journal of Biology, Agriculture and Healthcare, 2*(10), 184–194.

Meiyappan, M. M., & Kathirvel, M. (1978). On some new records of crabs and lobsters from Minicoy, Lakshadweep (Laccadives). *Journal of the Marine Biological Association of India, 20*(1&2), 116–119.

Melville-Smith, R., Cliff, M., & Anderton, S. (1999). *Catch, effort and the conversion from gill nets to traps in the Peel-Harvey and Cockburn Sound blue swimmer crab (Portunus pelagicus) fisheries* (Fisheries research report, Vol. 113, 24 pp). Perth: Western Australian Marine Research Laboratories.

Milton, D. A., Satria, F., Proctor, C. H., Prasetyo, A. P., Utama, A. A., & Fauzi, M. (2014). Environmental factors influencing recruitment and catch of tropical Panulirus lobsters in Southern Java, Indonesia. *Continental Shelf Res, 91*, 247–255.

Miyamoto, H., & Shariff, A. T. (1961). Lobster fishery off the southwest coast of India-anchor hook and trap fisheries. *Indian Journal Fisheries, 8*, 252–268.

Mohamed, K. S., Vijayakumaran, K., Zacharia, P. U., Sathianandan, T. V., Maheswarudu, G., Kripa, V., Narayanakumar, R., Rohit, P., Joshi, K. K., Sankar, T. V., Edwin, L., Ashok Kumar, K., Bindu, J., Gopal, N., & Puthra, P. (2017). Indian Marine Fisheries Code: Guidance on a Marine Fisheries Management Model for India. *CMFRI Marine Fisheries Policy Series, 4*, 120 p.

Mohammed, K. H., & George, M. J. (1968). Results of the tagging experiments on the Indian spiny lobster, *Panulirus homarus* (Linnaeus)- movement and growth. *Indian Journal of Fisheries, 15*, 15–26.

Mohammed, K. H., & Suseelan, C. (1973). Deep sea prawn resources off the southwest coast of India. In *Proceedings of the symposium on living resources around India, Special Publications,* C.M.F.R.I, India, pp. 614–633.

Mohan, R. (1997). Size structure and reproductive variation of the spiny lobster *Panulirus homarus* over a relatively small geographic range along the Dhofar coast in the Sultanate of Oman. *Marine and Freshwater Research, 48,* 1085–1091.

Mohanrajan, K. V. (1991). *Studies on spiny lobster fishery of southwest coast of India*. PhD Thesis (p. 207). Cochin, Kerala, India: Cochin University of Science and Technology.

Mohanrajan, K. V., Meenakumari, B., & Balasubramanyan, R. (1981). Spiny lobsters and their fishing techniques. *Fishery Technology, 18,* 1–11.

Mohanrajan, K. V., Meenakumari, B., Kandoran, M. K., & Balasubramanyan, R. (1984). Lobster fishing with modern traps – A viability report. *Fish Technology News Letter, 3*(12).

Mohanrajan, K. V., Meenakumari, B., & Kesavan Nair, A. K. (1988). Development of an efficient trap for lobster fishing. *Fishery Technology, 25,* 1–4.

Moore, R., & MacFarlane, W. (1984). Migration of the ornate rock lobster, *Panulirus ornatus* (Fabricius) in Papua New Guinea. *Australian Journal of Marine & Freshwater Research, 35,* 197–212.

Morgan, G. (2006a). Country review: Yemen. In C. De Young (Ed.), *Review of the state of world marine capture fisheries management: Indian Ocean* (FAO fisheries technical paper, 488) (pp. 337–348). Rome: FAO.

Morgan, G. (2006b). Country review: Saudi Arabia. In C. De Young (Ed.), *Review of the state of world marine capture fisheries management: Indian Ocean* (FAO fisheries technical paper, 488) (pp. 303–314). Rome: FAO.

Murugan, A., & Durgekar, R. (2008). India: A snapshot of present and long-term trends. In *Beyond the tsunami. Status of fisheries in Tamilnadu* (p. 75). Bangalore, India: UNDP/UNTRS, Chennai and ATREE.

Mustafa, A. M. (1990). *Linuparus andamanensis*, a new Spear lobster from Andaman. *Andaman Science Associations, 6*(2), 177–180.

Mutagyera, W. B. (1975). A preliminary report on the spiny lobster fishery of Zanzibar. *African Journal of Tropical Hydrobiology and Fisheries, 4*(1), 51–59.

Mutagyera, W. B. (1978). Some observations on the Kenya lobster fishery. *East African Agricultural and Forestry Journal, 43*(4), 401–407.

Mutagyera, W. B. (1983). Palinurid lobster biology and fishery in Kenya Waters. *Kenya Aquatica Bulletin, 1,* 26.

Nair, R. V., Soundararajan, R., & Dorairaj, K. (1973). On the occurrence of *Panulirus longipes longipes, Panulirus penicillatus* and *Panulirus polyphagus* in the Gulf of Mannar with notes on the lobster fishery around Mandapam. *Indian Journal of Fisheries, 20*(2), 333–350.

Ninan, T. V., Basu, P. P., & Verghese, P. K. (1984). Observations on the demersal fishery resources along the Andhra Pradesh coast. *Fishery Survey of India Bulletin, 13,* 13–22.

Okechi, J. K., & Polovina, D. E. (1995). An evaluation of artificial shelters in the artisanal spiny lobster fishery in Gazi Bay, Kenya. *South African Journal of Marine Science, 16,* 373–376.

Ongkers, O. T. S., Pattiasino, B. J., Tetelepta, J. M. S., Natan, Y., & Pattikawa, J. A. (2014). Some biological aspects of painted spiny lobster (*Panulirus versicolor*) in Latuhalat waters, Ambon Island, Indonesia. *Aquaculture Aquarium Conservation & Legislation, 7,* 469–474.

Oommen, P. V. (1985). Deep sea resources of the southwest coast of India. *Bulletin IFP, 11,* 1–85.

Oommen, P. V., & Philip, K. P. (1976). Observations on the fishery and biology of the deep sea spiny lobster *Puerulus sewelli* Ramadan. *Indian Journal of Fisheries, 21*(2), 369–385.

Penn, J. W., Caputi, N., & de Lestang, S. (2015). A review of lobster fishery management: the Western Australian fishery for *Panulirus cygnus*, a case study in the development and implementation of input and output-based management systems. *ICES Journal of Marine Science, 72,* 22–34.

Philip, K. P., Premchand, B., Avhad, G. K., & Joseph, P. S. (1984). A note on the deep sea demersal resources off Karnataka, North Kerala Coast. *Fishery Survey of India Bulletin, 13,* 23–29.

Phillips, B. F., Cruz, R., Brown, R. S., & Caputi, N. (1994). Predicting the catch of spiny lobster fisheries. In B. F. Phillips, J. S. Cobb, & J. Kittaka (Eds.), *Spiny Lobster Management* (pp. 285–300). Oxford/London: Blackwell Scientific Publications Ltd.

Phillips, B. F., Melville-Smith, R., Kay, M. C., & Vega-Velazquez, A. (2013). *Panulirus* species. In B. F. Phillips (Ed.), *Lobsters: biology, management, aquaculture and fisheries* (2nd ed., pp. 289–325). Oxford: Wiley.

Philipose, K. K. (1994). Lobster culture along the Bhavnagar coast. *Marine Fisheries Information Service Technical and Extension Series, 130*, 8–12.

Pillai, V. N., & Ramachandran, V. S. (1972). Some trends observed in the deep sea lobster catches of the vessel Blue Fin during the period Jan. 1969 to Dec. 1971. *Seafood Export Journal, 4*(6).

Pillai, S. L., & Thirumilu, P. (2007). Extension in the distributional range of long-legged spiny lobster *Panulirus longipes longipes* (A. Mile-Edwards, 1868) along the southeast coast of India. *Journal of the Marine Biological Association of India, 49*(1), 95–96.

Pillai, C. S. G., Mohan, M., & Kunhikoya, K. K. (1985). Observations on the lobsters of Minicoy Atoll. *Indian Journal of Fisheries, 32*(1), 112–122.

Pitcher, R. (1993). Spiny lobster: In: *Nearshore marine resources of the South Pacific*. In A. Wright & L. Hill (Eds.), *Suva: Institute of Pacific studies* (pp. 539–608). Honiara: Forum Fisheries Agency and Halifax: International Centre for Ocean Development.

Pollock, D. E., Cockcroft, A. C., Groeneveld, J. C., & Schoeman, D. S. (2000). The fisheries for *Jasus* species in the south-east Atlantic and for *Palinurus* species of the south-west Indian Ocean. In B. F. Phillips & J. Kittaka (Eds.), *Spiny lobsters: fisheries and culture* (2nd ed., p. 679). Oxford: Fishing News Books/Blackwell Science.

Polovina, J. J., & Mitchum, G. T. (1992). Variability in spiny lobster *Panulirus marginatus* recruitment and sea level in the Northwestern Hawaiian Islands. *Fishery Bulletin U.S, 90*, 483–493.

Pomeroy, R. S., & Rivera-Guieb, R. (2006). *Fishery co-management: A practical handbook* (288 p). Wallingford: CABI/IDRC Publications.

Powers, J. E., & Bannerot, S. P. (1984). *Assessment of spiny lobster resources of the Gulf of Mexico and southern United States* (p. 25). Miami: National Marine Fisheries Service, Southwest Fisheries Center.

Powers, J. E., & Sutherland, D. L. (1989). Spiny lobster assessment, cpue, size frequency, yield per recruit and escape gap analyses. *National Marine Fishery. Servey*, Southwest Fisheries Center. CRD-88/89-24. p. 75.

Presscot, J. (1988). *Tropical spiny lobster: An overview of their biology, the fisheries and the economics with particular reference to the double spined rock lobster P. penicillatus*. Workshop on pacific inshore fishery resources, SPC/Inshore Fishery Research/WP.18 17 March 1988.

Qasim, S. Z. (1982). Oceanography of the northern Arabian Sea. *Deep Sea Research, 29*, 1041–1068.

Rabarison, A. (2000). Rapport final du project de recherché sur l'etude des stocks de spiny lobster neritiques. *Ministere de la Recherche Scientifique, Convention 004/93 MADR/DG/ONE, Antananarivo*, 51 p.

Radhakrishnan, E. V., & Jayasankar, P. (2014). First record of the reef lobster *Enoplometopus occidentalis* (Randall, 1840) from Indian waters. *Journal of the Marine Biological Association of India, 56*(2), 88–91.

Radhakrishnan, E. V., & Thangaraja, R. (2008). Sustainable exploitation and conservation of lobster resources in India-a participatory approach. *Glimpses of Aquatic Biodiversity-Rajiv Gandhi Chair Special Publications, 7*, 184–192.

Radhakrishnan, E. V. & Vijayakumaran, M. (2003). The status of lobster fishery in India and options for sustainable management. In: V.S. Somavanshi (Ed.), *Large marine ecosystem: Exploration and exploitation for sustainable development and conservation of fish stocks* (pp. 294–311). Fishery Survey of India.

Radhakrishnan, E. V., Kasinathan, C., & Ramamoorthy, N. (1995). Two new records of scyllarids from the Indian coast. *The Lobster Newsletter, 8*(1), 9.

Radhakrishnan, E. V., Koumudi Menon, K., & Lakshmi, S. (2000). Small-scale traditional spiny lobster fishery at Tikkoti. *The Marine Fisheries Information Service: Technical and Extension Series, 164*, 5–8.

Radhakrishnan, E. V., Deshmukh, V. D., Manisseri, M. K., Rajamani, M., Kizhakudan, J. K., & Thangaraja, R. (2005). Status of the major lobster fisheries in India. *New Zealand Journal of Marine and Freshwater Research, 39*, 723–732.

Radhakrishnan, E. V., Manisseri, M. K., & Deshmukh, V. D. (2007a). Biology and fishery of the slipper lobster *Thenus orientalis* in India. In K. L. Lavalli & E. Spanier (Eds.), *Biology and fishery of the slipper lobster* (pp. 309–324). Boca Raton: CRC Press.

Radhakrishnan, E. V., Manisseri, M. K., & Nandakumar, G. (2007b). Status of research on crustacean fishery resources. In M. J. Modayil & N. G. K. Pillai (Eds.), *Status and perspectives in marine fisheries research in India* (pp. 135–172). Kochi: CMFRI.

Radhakrishnan, E. V., Lakshmi Pillai, S., Shanis, R., & Radhakrishnan, M. (2011). First record of the reef lobster *Enoplometopus macrodontus* Chan & Ng, 2008 from Indian waters. *Journal of the Marine Biological Association of India, 53*(2), 264–267.

Radhakrishnan, E. V., Chakraborty, R. D., Baby, P. K., & Radhakrishnan, M. (2013). Fishery and population dynamics of the sand lobster *Thenus unimaculatus* Burton & Davie, 2007 landed by trawlers at Sakthikulangara fishing harbour on the southwest coast of India. *Indian Journal of Fisheries, 60*(2), 7–12.

Radhakrishnan, E. V., Vijayakumaran, M., & Thangaraja, R. (2015). Ontogenetic changes in morphometry of the spiny lobster, *Panulirus homarus homarus* (Linnaeus, 1758) from southern Indian coast. *Journal of the Marine Biological Association of India, 57*(1), 5–13.

Rai, H. S. (1933). Shell fisheries of the Bombay presidency. *Journal of the Bombay Natural History Society, 36*, 884–897.

Rajabipour, F., & Mashaii, N. (2003). Length weight relationship of the spiny lobster, *Panulirus homarus* Linnaeus, 1758, from southeast of Iran. *7th international conference and workshop on lobster biology and management*, Hobart, Tasmania, 8–13th February 2003. Abstract Book.

Rajamani, M., & Manickaraja, M. (1991). On the collection of spiny lobsters by skin divers in the Gulf of Mannar off Tuticorin. *The Marine Fisheries Information Service: Technical and Extension Series, 113*, 17–18.

Rajamani, M., & Manickaraja, M. (1997a). On the fishery of the spiny lobster off Tharuvaikulam, Gulf of Mannar. *The Marine Fisheries Information Service: Technical and Extension Series, 146*, 7–8.

Rajamani, M., & Manickaraja, M. (1997b). The spiny lobster resources in the trawling grounds off Tuticorin. *The Marine Fisheries Information Service: Technical and Extension Series, 148*, 7–9.

Raje, S. G., & Deshmukh, V. D. (1989). On the dolnet operation at Versova, Bombay. *Indian Journal of Fisheries, 36*(3), 239–248.

Rao, P.V., & George, M. J. (1973). Deep-sea spiny lobster, *Puerulus sewelli* Ramadan: the commercial potentialities. In *Proceedings of the symposium on living resources of the seas around India, Special Publication, CMFRI*, pp. 634–640.

Rao, G. S., Suseelan, C., & Kathirvel, M. (1989). Crustacean resources of the Lakshadweep Islands. In Marine living resources of the Union Territory of Lakshadweep- An indicator survey with suggestions for development. *CMFRI Bulletin, 43*, 72–76.

Saha, S. N., Vijayanand, P., & Rajagopal, S. (2009). Length-weight relationship and relative condition factor in *Thenus orientalis* (Lund, 1793) along east coast of India. *Current Research Journal of Biological Sciences, 1*(2), 11–14.

Saleela, K. N. (2015). Nutritional studies on the spiny lobster *Panulirus homarus* from the south west coast of India, Manonmaniam Sundaranar University, Tirunelveli, Tamil Nadu, p. 199.

Sanders, M. J., & Bouhlel, M. (1984). Stock assessment for the rock lobster (*Panulirus homarus*) inhabiting the coastal waters of The People's Republic of Yemen. *FAO Project for Development of Fisheries in areas of the Red Sea and Gulf of Aden*, RABSI/002, 68 p.

Sanders, M., & Liyanage, U. (2009). *Preliminary assessment for the spiny lobster fishery of the south coast (Sri Lanka)* (pp. 1–44). Sri Lanka: NARA.

Sari, A. (1991). *A biosystematic survey on lobsters of Chabahar, Iran*. Unpublished MSc thesis. University of Tehran.

Satyanarayana, A. V. V. (1961). A record of *Panulirus penicillatus* (Olivier) from the inshore waters off Quilon, Kerala. *Journal of the Marine Biological Association of India, 3*(1), 269–270.

Senevirathna, J. D. M., Munasinghe, D. H. N., & Mather, P. B. (2016). Assessment of Genetic Structure in Wild Populations of *Panulirus homarus* (Linnaeus, 1758) across the South Coast of Sri Lanka Inferred from mitochondrial DNA sequences. *International Journal of Marine Science, 6*(6), 1–9.

Sewell, R. B. S. (1913). Notes on the biological work of the R.I.M.S "Investigator" during the survey season 1910–1911 and 1911–1912. I. *Journal of the Asiatic Society of Bengal, 9*, 329–390.

Shanmugham, S., & Kathirvel, M. (1983). Lobster resources and culture potential. *CMFRI Bulletin, 34*, 61–65.

Shetye, S. (1998). West India coastal current and Lakshadweep High/Low. *Sadhana, 23*(5&6), 637–651.

Shetye, S. R., Gouveia, A. D., Shenoi, S. S. C., Michael, G. S., Sundar, D., Almeida, A. M., & Santanam, K. (1991). The coastal current off Western India during the northeast monsoon. *Deep Sea Research, 12*, 1517–1529.

Silas, E. G. (1969). Exploratory fishing by R. V. Varuna. *Bulletin of Central Marine Fishery of Research Institute, 12*, 1–86.

Silas, E. G., & Alagarswamy, K. (1983). General considerations of mariculture potential of Andaman and Nicobar Islands. *CMFRI Bulletin, 34*, 104–107.

Smith, F. G. W. (1958. Florida St. Bd. Conservation of Education Service, *11*, 1136.

Somavanshi, V. S., & Bhar, P. K. (1984). A note on the demersal fishery resources of Gulf of Mannar. *Fishery Survey of India Bulletin, 13*, 12–17.

Srinivasa Gopal, T. K., & Edwin, L. (2013). Development of fishing industry in India. *Journal of Aquatic Biology Fisheries, 1*(1&2), 38–53.

Steyn, E., & Schleyer, M. H. (2011). Movement patterns of the East Coast rock lobster *Panulirus homarus rubellus* on the coast of KwaZulu-Natal, South Africa. *New Zealand Journal of Marine and Freshwater Research, 45*(1): 85–101.

Steyn, E, & Schleyer, M.H. (2014). Results of questionnaire surveys to obtain catch statistics of recreational invertebrate fisheries in KZN: *2008–2012 Oceanographic Research Institute Report Unpublished Report 310* (in preparation).

Steyn, E., Fielding, P. J., & Schleyer, M. H. (2008). The artisanal fishery for East Coast rock lobsters *Panulirus homarus* along the Wild Coast, South Africa. *African Journal of Marine Science, 30*, 497–506.

Subramaniam, V. T. (2004). Fishery of sand lobster *Thenus orientalis* (Lund) along Chennai coast. *Indian Journal of Fisheries, 51*(1), 111–115.

Sulochanan, P., & John, M. E. (1988). Offshore, deep sea and oceanic fishery resources off Kerala coast. *Bulletin of the Fishery Survey of India, 16*, 27–48.

Suseelan, C. (1974). Observations on the deep sea prawn fishery off the southwest coast of India with special reference to pandalids. *Journal of the Marine Biological Association of India, 16*(2), 491–511.

Suseelan, C., & Pillai, N. N. (1993). Crustacean fishery resources of India: a review. *Indian Journal of Fisheries, 40*(1&2), 104–111.

Suseelan, C., Muthu, M.S., Rajan, K.N., Nandakumar, G., Neelakanta Pillai, N., Surendranatha Kurup, N., & Chellappan, K. (1990). Results of an exclusive survey for deep sea crustaceans off southwest coast of India. In *Proceedings of the first workshop on scientific results of FORV Sagar Sampada*, 5–7 June 1989, pp. 347–359.

Thangaraja, R. (2011). Ecology, reproductive biology and hormonal control of reproduction in the female spiny lobster *Panulirus homarus* (Linnaeus, 1758). PhD thesis. Mangalore University, India, pp. 172.

Thangaraja, R., & Radhakrishnan, E. V. (2012). Fishery and ecology of the spiny lobster *Panulirus homarus* (Linnaeus, 1758) at Khadiyapatanam in the southwest coast of India. *Journal of the Marine Biological Association of India, 54*(2), 69–79.

Thangaraja, R., & Radhakrishnan, E. V. (2017). Reproductive biology and size at onset of sexual maturity of the spiny lobster *Panulirus homarus homarus* (Linnaeus, 1758) in Khadiyapatnam, southwest coast of India. *Journal of the Marine Biological Association of India, 59*(2), 19–28.

Thangaraja, R., Radhakrishnan, E. V., & Chakraborthy, R. D. (2015). Stock and population characteristics of the Indian rock lobster *Panulirus homarus homarus* (Linnaeus, 1758) from Kanyakumari, Tamilnadu, on the southern coast of India. *Indian Journal of Fisheries, 62*(3), 21–27.

Thiagarajan, R., Krishna Pillai, S., Jasmine, S., & Lipton, A. P. (1998). On the capture of a live South African Cape locust lobster at Vizhinjam. *The Marine Fisheries Information Service: Technical and. Extension Series, 158*, 18–19.

Tholasilingam, T., Venkatraman, G., Krishna Kartha, K. N., & Karunakaran Nair, P. (1968). Exploratory fishing off the southwest coast of India by M.F.V. Kalava. *Indian Journal of Fisheries, 11*, 547–558.

Trivedi, D. J., Trivedi, J. N., Soni, G. M., Purohit, B. D., & Vachhrajani, K. D. (2015). Crustacean fauna of Gujarat state of India: A review. *Electrical Journal of Environmental Science, 8*, 23–31.

Vaitheeswaran, T. (2015). A new record of Scyllarid lobster *Scyllarus batei batei* (Holthuis, 1946) (Family: Scyllaridae, Latreille, 852) (Crustacea: Decapoda: Scyllaridae) off Thoothukudi coast of Gulf of Mannar, southeast coast of India 08°52.6'N 78°16'E and 08°53.8'N 78°32'E (310 m). *International Journal of Marine Science, 5*(54), 1–2.

Valle, S. V. et al. (1993). *Actual situation and stock assessment for the Rock Lobster Panulirus homarus in the Gulf of Aden.* MSRRC.

Vijayanand, P., Murugan, A., Saravanakumar, K., Khan, S. A., & Rajagopal, S. (2007). Assessment of lobster resources along Kanyakumari (Southeast coast of India). *Journal of Fisheries and Aquatic Science, 2*, 387–394.

Wardiatno, Y., Hakim, A. A., Mashar, A., Butet, N., Adrianto, L., & Farajallah, A. (2016a). First record of *Puerulus mesodontus* Chan, Ma & Chu, 2013 (Crustacea, Decapoda, Achelata, Palinuridae) from south of Java, Indonesia. *Biodiversity Data Journal, 7*(4), e8069. https://doi.org/10.3897/BDJ.4.e8069.

Wardiatno, Y., Hakim, A. A., Mashar, A., Butet, N. A., & Adrianto, L. (2016b). Two newly recorded species of the lobster family Scyllaridae (*Thenus indicus* and *Scyllarides haanii*) from South of Java, Indonesia. *HAYATI Journal of Biosciences, 23*, 101–105.

Wardiatno, Y., Hakim, A. A., Mashar, A., Butet, N. A., Adrianto, L., & Farajallah, A. (2016c). On the presence of the Andaman lobster, *Metanephrops andamanicus* (Wood-Mason, 1891) (Crustacea Astacidea Nephropidae) in Palabuhanratu Bay (S-Java, Indonesia). *Biodiversity Journal, 7*(1), 17–20.

Wowor, D. (1999). The spear lobster, *Linuparus somniosus* Berry & George1972 (Decapoda, palinuridae) in Indonesia. *Crustaceana, 72*(7), 673–684.

Zacharia, P. U., Krishnakumar, P. K., Dineshbabu, A. P., Vijayakumaran, K., Rohit, P., Thomas, S., Sasikumar, G., Kaladharan, P., Durgekar, R. N., & Mohamed, K. S. (2008). Species assemblage in the coral reef ecosystem of Netrani Island off Karnataka along the southwest coast of India. *Journal of the Marine Biological Association of India, 50*(1), 87–97.

A Review of the Current Global Status and Future Challenges for Management of Lobster Fisheries

8

Bruce F. Phillips and Mónica Pérez-Ramírez

Abstract

The management of *Panulirus ornatus* in Torres Strait, *Panulirus argus* in the Bahamas and also in Florida, *Panulirus cygnus* in Western Australia and *Panulirus interruptus* in Baja California shared by the USA and Mexico is assessed. These major lobster fisheries have survived the last 7 years without major collapse, and Indonesia's catch is continuing to increase. The world of fisheries management is now at a much more complicated place than it was 10 or more years ago. Modelling of data from catches in the fishery, effort and sales prices of product and costs of fishing is now common. Other items also encountered by scientists, managers and industry managing the fisheries include ecosystem-based fishery management, management performance reviews, MSC reviews, productivity commission inquires, status of the stock assessments for permission to sell product overseas, bioeconomic modelling to improve harvest strategies, cost–benefit analysis and significantly the effects of climate change.

Keywords

Management · Fisheries · *Panulirus* · Challenges

B. F. Phillips (✉)
School of Molecular and Life Sciences, Curtin University, Perth, WA, Australia
e-mail: B.Phillips@curtin.edu.au

M. Pérez-Ramírez
FAO Representation in Mexico Farallón 130, Mexico City, Mexico
e-mail: Monica.PerezRamirez@fao.org

© Springer Nature Singapore Pte Ltd. 2019
E. V. Radhakrishnan et al. (eds.), *Lobsters: Biology, Fisheries and Aquaculture*,
https://doi.org/10.1007/978-981-32-9094-5_8

8.1 Introduction

In 2013, we published a detailed review of the commercial fishing for lobsters of the genera *Panulirus* (Phillips et al. 2013). This chapter includes some of that material but in a shortened form. For addition information, the reader is referred to the 2013 publication.

(a) **The Management of the Tropic Rock Lobster Fishery *Panulirus ornatus* in Torres Strait, Australia**

The artisanal fishery on *P. ornatus* is located in the Torres Strait and the east coast of Papua New Guinea (PNG). The fishery is an important source of livelihoods since the access is mainly restricted to the indigenous people (Phillips et al. 2010). The resource is shared by Australia's Torres Strait Protected Zone (TSPZ), PNG's area of the TSPZ and Gulf of Papua and the State of Queensland (south of TSPZ). Lobsters are caught by fishers while freediving to about 4 m or using hookah to around 20 m. Trawling is banned permanently in both countries (Moore and MacFarlane 1984). Within the Australian TSPZ, non-indigenous fishers may fish from dinghies associated to a mother vessel.

The TSPZ was officially created in 1984. Fisheries management in the TSPZ is undertaken under the Torres Strait Fisheries Act 1984 that put in place catch sharing arrangements. The fishery has effectively been managed by input controls. Catches have fluctuated inter-annually (Table 8.1). Australia endorses PNG boats to fish in Australian waters (Williams 2004). Most volume is exported to China.

A plan of management is being developed for the Fishery to transition the management arrangements to output controls through the allocation of TRL quota units to the two sectors. The plan pursuant to section 15A of the Torres Strait Fisheries Act 1984 will:

- Determine a total allowable catch (TAC) each season.
- The TAC will be determined by the PZJA in line with requirements of the industry (Patterson et al. 2017).

Many different fishery scenarios and management measures were tested through a management strategy evaluation (MSE) for the Torres Strait Tropical Rock Lobster Fishery (TSTRLF) (Plagányi et al. 2012). MSE included a bio-economic model to estimate subsector profits. This model and updates now provide a method of assessment and prediction on the effects of management arrangements and economic performance of the fishery.

The TSTRLF management objectives consider economic performance, but also social and cultural factors. They include the objectives of protecting the traditional way of life and livelihood of traditional inhabitants, particularly in relation to their traditional fishing rights and appropriate controls on fishing gear and fishing effort to minimize impacts on the environment (Patterson et al. 2017; Prescott and Steenbergen 2017). Fishery governance has progressively improved as a result of stakeholders'

Table 8.1 Landings (t) of *Panulirus* spp. for 2009–2015 (from FAO 2016 and "from some other sources")

Location	Species	2009	2010	2011	2012	2013	2014	2015	2016
Western Australian spiny lobster	P. cygnus	7634	7260	6327	6988	7379	7077	7156	6087
Caribbean spiny lobster	P. argus	23,521	27,432	28,668	30,116	27,325	26,664	30,239	
USA, Florida	P. argus	2063	2567	3350	3213	2557	3617	3208	3425
Bahamas	P. argus	7138	9482	8505	9761	6088	6509	6525	
Brazil	P. meripurpuratus	7268	6866	6776	7451	6726	6787	6100	
California spiny lobster (USA and Mexico)	P. interruptus	2139	3213	3335	2895	2952	4118	3093	
Torres Strait Australia ornate spiny lobster*	P. ornatus	444	940	951	678	613	913	476	
Indonesia Tropical spiny Lobster	nei	5892	7651	10,541	13,549	16,482	10,062	16,750	

*The FAO data are best considered indicative of the scale commercial catches of each species from each area rather than being accurate representations. Brazil is no longer *Panulirus argus* but now *Panulirus meripurpuratus* (Giraldes and Smyth 2010) engagement (Prescott and Steenbergen, 2017), advocacy and participation in advisory groups and decision-making in PZJA. Overall, steady progress in this fishery is being made as it moves towards a quota allocations system (Pascoe et al. 2017) incorporating social and economic objectives together with ecological sustainability.

(b) **The Bahamas Fishery for *Panulirus argus***

The spiny lobster fishery for *Panulirus argus* in the Bahamas is worth about US $90 million, employs about 9000 fishers and covers a massive 11,650 km^2 of ocean. Fishing is conducted using traps, condominiums (*casitas*) and diving.

In 2017, the Bahamian spiny lobster fishery has asked for assessment to the Marine Stewardship Councils Global standard for sustainable fishing. The MSC certification is a private governance company with a goal to create a market based on sustainable operations. It involves a third-party certification process and an environmental standard based on the following: (1) status of the target fish stock, (2) ecosystem impact of fishing and (3) governance system. The fishery performance is assessed against the MSC standard and conditions on improvements may rise as result of such process. Conditions are actions that the fishery must comply with within a set time. Certification has a limited 5-year duration; within this period, certified fisheries must submit to annual audits by evaluating conditions' progress. If the progress towards meeting conditions is not achieved, the certification may be suspended. The outcome of the MSC assessment will take about 18 months to be decided.

Of particular interest will be the assessment of the impacts of hurricanes including the recent effects of "Irma" in early 2017. Phillips et al. (2017) have determined that hurricanes have effects of decreasing the catch of *P. argus* in the Caribbean 2 years after they occur. The last few years have received a series of strong hurricanes in the Caribbean. Catches in the Bahamas (Table 8.1) have shown considerable reductions between 2013 and 2015, but the reasons are not yet explained.

An additional problem is the quality of the catch data on *P. argus* for the Bahamas. Smith and Zeller (2016) reported that "reconstructed total catches (e.g. reported catches and estimates of unreported catches) were 2.6 times the landings presented by the FAO for the Bahamas. This discrepancy was primarily due to unreported catches from the recreational and subsistence fisheries in the FAO data. The recreational fishing accounted for 55% of reconstructed total catches". There are obviously many problems to be overcome in this fishery.

(c) **An Updated Assessment of *P. argus* in Florida**

The Caribbean spiny lobster fishery allows both commercial and recreational fishing permits, providing income to coastal communities. Between 20% and 25% of the lobster catches come from recreational fishers.

In 2011, an amendment to the fishery management plan (FMP) for spiny lobster in the Gulf of Mexico and South Atlantic was approved. In addition to establishing annual catch limits, the amendment redefines biological reference points and removes several other species from the FMP (Gulf of Mexico Fisheries Management Council 2011, Southeast Fishery Bulletin FB11-98, 2011). The new management arrangement includes an annual catch limit of 7.63 million lbs. (3468 t) and an annual catch target of 6.59 million lbs. (2995 t). These new arrangements are aimed at achieving sustainability of the Florida spiny lobster fishery.

At the conclusion of the review of the *P. argus* fishery in Florida in 2013, Phillips et al. (2013) commented on the fact that declines in the catch of the fishery were not explained. However, more recent catches (Table 8.1) clearly indicted good catch levels since that time to 2016.

The latest review of the fishery is given in a paper published by the Gulf of Mexico Fishery Management Council, "Modifications to Management Benchmarks, Annual Catch Limit, Annual Catch Target, and Prohibition of Traps for Recreational Harvest in the South Atlantic Exclusive Economic Zone" (http://sero.nmfs.noaa.gov/sustainable_fisheries/gulf_sa/spiny_lobster/documents/pdfs/gulf_sa_spiny_lob_reg_am4.pdf published in 2017). This review includes comments on future management considerations of:

1. The Deepwater Horizon MC252 oil spill in 2010. The full consequences are unknown, but the impact on the physical environment of the Gulf of Mexico are expected to be significant.
2. Climate change. Global climate change could have significant effects on the Gulf of Mexico and South Atlantic fisheries; however, the extent of these effects cannot be quantified at time. The most recent review of its impacts on lobsters is in Phillips et al. (2017).
3. Hurricanes. These remain an even present threat. "Irma" in 2017 may have caused damage to the Florida area, but no details are available.
4. Source of recruiting puerulus. The management system currently accepts that most of the recruits to Florida are from sources other than Florida. A paper by Kough et al. (2013) estimated that between 10 and 40% of larvae are from Florida and retained in Florida waters.
5. Stock assessment. Lack of definition of stock size and source makes an assessment status impossible using standard methodology.

Despite all these problems, the fishery in Florida for *P. argus* is well managed and produces good catches.

(d) **Western Australian *Panulirus cygnus* Fishery**

The western rock lobster *P. cygnus* is found only in temperate and subtropical waters off the west coast of Australia (Phillips et al. 2010).

Several research and models to predict the future catches have been developed. The level of puerulus settlement is mainly related to sea surface temperature and ocean currents. The settlement levels are highly correlated with catches up to 4 years later (Phillips 1986; Caputi et al. 1995, 2003). The seasonal fluctuations in puerulus settlement have led to high variability in catches but enforcement authority has set properly management measures. Fishers also have understood seasonal fluctuations and take them into account in their fishing operations.

In 2006/2007, a significant decline in puerulus settlement started; the lowest level recorded since 1968 was shown in the 2008/2009 season (Department of Fisheries 2011). A risk assessment workshop was able to identify the changes in environmental conditions and the productivity in the eastern Indian Ocean as the most likely factors responsible for the low puerulus settlement (Brown 2009). Later, the level of puerulus settlement in 2010/2011 was above that of the previous three seasons (Department of Fisheries 2011).

The downturn in settlement may impact future landings, and with no action taken the downturn may also affect the state of the broodstock negatively. Effort reductions were applied into the fishery in several seasons (2007–2010). In the 2008/2009 season, a catch limit of 7800 t was set, and a TAC of 5500 t was introduced the next season. The fishery faced the transition to management by output control when individual catch limits (5500 t) were introduced. After 2010/2011, reference points focusing on maximum economic yield (MEY) rather than maximum sustainable yield (MSY) were adopted.

Melville-Smith (2011) reviewed factors affecting population resilience of temperate fisheries, including climate change, fishing activities, invasive species and coastal development. The conclusion was that it is tempting where fished populations fail to identify a single responsible factor. According to Melville-Smith (2011), several factors such as environmental, biological and economical factors and management are involved into the failure of fishing populations to recover. For the western rock lobster fishery, the combination of the factors mentioned earlier and the response of management by reducing effort at the time rebuilding broodstock have been positive in recovery fishery.

A major review by Penn et al. (2015) examined the development and implication of input- and output-based management systems using the *P. cygnus* fishery as a case study. Penn et al. (2015) concluded that linking biological controls with and evolutionary approach to management may allow sufficient fishery-based data for management decision to be effective. Price/earnings ratios could be used to analyse trends in license values and industry's economic viability over time under both output and input control management.

Another item of interest is an assessment of risk management for the Western Rock Lobster Council Inc. Risk is the effect of uncertainty on the ability of an organization to meet its objectives. A 2016 assessment for the council was undertaken by Peter Cooke of the company *Acknowledge*. A series of recommendation have been developed by Peter Cooke for the fishing industry to consider (Kim Colero, Western Rock Lobster Council Inc., Western Australia, personal communication).

The Western Rock Lobster Fishery was awarded MSC certification as a well-managed fishery in March 2000, the first fishery in the world to receive this imprimatur. It continues to maintain certification.

(e) **The Baja Red Rock Lobster Fishery (*Panulirus interruptus*)**

The fishery is operated in the North Pacific coast (28.6°N 115.5°W–26.6°N 113.2°W) by ten fishing cooperatives that hold exclusive access to benthic highly valued species in geographic adjacent areas defined by the federal government. Lobster is caught with baited traps, and 90% of the annual catch (~1300–1500 t) is exported to Asia. Market conditions, financial capacity of the cooperatives and sense of community made the lobster fishery vital for local livelihoods (McCay et al. 2014).

The cooperatives are vertically integrated in a regional federation (FEDECOOP) that has capacity to organization, management and marketing, bridging users with local and government levels. The fishery is defined as "multi-level co-managed" (Finkbeiner and Basurto 2015) since its management is carried out by both centralized (National Commission of Aquaculture and Fisheries, CONAPESCA) and decentralized (National Fisheries Institute, INAPESCA) governmental agencies, their respective regional delegations and the FEDECOOP. In addition, since 2003, there is a subcommittee functioning as an inclusive representation of fishery stakeholders. The subcommittee exposes the challenges faced by the fishery to design strategies and foster decision-making. Thus, some management measures are top-down fixed, and other measures come from bottom-up processes, through negotiation between FEDECOOP and CONAPESCA (see McCay et al. 2014). Two main instruments regulate the lobster fishery: the National Fisheries Law and the Mexican Official Standards (NOM-006-PESC-1993 modified in 2007).

As part of the exclusive access granted, the cooperatives are exhorted to get involved into the production of resource knowledge. FEDECOOP employs fish biologists to provide technical assistance to its members and record information. FEDECOOP members use transect surveys and are familiar with the importance of record feasible data to assess their fishing resource. Cooperatives actively participate in scientific monitoring with the regional delegation of INAPESCA and have funded scientific research in educational institutions. The remote location of exclusive fishing areas as well as the stringent internal vigilance rules set by the cooperatives may prevent poaching (Pérez-Ramírez et al. 2012; McCay et al. 2014).

In 2004, FEDECOOP was certified by the MSC. Addressing the conditions raised in the first certification (2004–2009) strengthened the cooperation between FEDECOOP and the local fishing agency but also the involvement of fishing scientists. Conditions pointed out to evaluate the ecosystem impacts of fishing and to develop appropriate harvest strategies. After celebrating a stakeholder workshop and a joint research project between public research institutions, conditions were achieved (see Bellchambers et al. 2015). After a long assessment process, recertification (2011–2016) conditions were placed on stock assessment and a reporting system for bycatch and bait was developed. Providing an accurate method for stock assessment remains open and behind target during the 2015 surveillance audit. The cooperatives and the local fishing agency introduced a logbook to quantify bycatch and bait to address the reporting system condition.

(f) **Future Research for Management**

The data in Table 8.1 suggests that the lobster fisheries worldwide have survived the last 7 years without major collapse, and Indonesia's catch is continuing to increase. We cannot find publications to support and describe the fisheries in Indonesia. However, we are advised that the catches shown in Table 8.1 do not include spiny lobster aquaculture data.

The world of fisheries management is now a much more complicated place than it was 10 or more years ago. Modelling of data from catches in the fishery, effort and also sales prices of product and costs of fishing are now common. Some of these important models are those of Plagányi et al. (2011, 2013, 2014a, 2014b). Other items now encountered by scientists, managers and industry are ecosystem-based fishery management, management performance reviews, MSC reviews, productivity commission inquires, status of the stock assessments for permission to sell product overseas, bioeconomic modelling to improve harvest strategies and cost–benefit analysis, to name just a few.

It may lead to better management of the fishery, but they all increase management costs and in some cases contradictive information. A constant problem is ensuring the quality of the basic catch and effort data and its accuracy over time. India does not provide catch statistics to FAO. However, in Chapter 7, there is a good review of catches and fishing operations for lobsters in India and the history of the lobster fisheries over time.

No one knows the full extent of changes that climate change may bring to the fisheries worldwide. However, in a recent book, Phillips and Pérez-Ramírez (2017) have reviewed the situation globally. A chapter is specifically devoted to lobsters. A number of *Panulirus* species have already been identified as being affected by climate change including *P. cygnus, P. argus, P. interruptus, P. marginatus* and *P. japonicas* (Phillips et al. 2017). Hobday and Cvitanovic (2017) have reviewed the reasons why there has been a delayed response to the implications of climate change in Australia. There conclusions were "implementation of management and policy responses (to climate change) have lagged because societal and fisher awareness of climate change have lagged" (Hobday and Cvitanovic 2017). This may also be the reason in many other countries.

Regarding certification, the process requires continual knowledge generation, collaborative problem-solving and learning. The certified lobster fisheries mentioned earlier have realized that certification may promote new collaborative partnerships to support the process and to address conditions (Bellchambers et al. 2015). These collaborative actions between stakeholders at different levels of organization are related to stakeholders' objectives, skills and capacities. Nevertheless, some developments are still required to facilitate coordination: (1) develop a strategic risk framework to address investment and time frame and (2) institutional building for improve assistance supporting the longevity of certification. Enabling public policy on fisheries certification may be also critical. As a learning process, it should attempt negotiation and reflection actions in such way that the stakeholders learn from each other and reflect what they have learned to improve resource management.

Challenges that some lobster fisheries in Australia are facing include an ongoing programme to identify and capture costs and benefits of certification, improvements in community acceptance and license to operate and the changing market dynamics including commencement of Australia/China free-trade agreement in 2019 (Kim Colero, personal communication*)*. An excellent review of opportunities and the outlook for the Australian spiny lobster fisheries has been made by Plagányi et al. (2017).

8.2 Conclusions

The management of the spiny lobsters *Panulirus ornatus* in Torres Strait, *Panulirus argus* in the Bahamas and also in Florida, *Panulirus cygnus* in Western Australia and *Panulirus interruptus* in Baja California shared by the USA and Mexico all appear to be adequate at this time, as measured by catch levels. The rapidly increasing catch in Indonesia remains to be explained. Many spiny lobster fisheries are now certified and this has led to a necessity for a significant increase in data gathering from the fisheries. Western Australia, which was the first fishery certified, is now undergoing a programme to identify and capture costs and benefits of certification, as well as improvements in community acceptance of the fishery, and license to operate as well as changing market dynamics. Some of these aspects, and others, are likely to affect other lobster fisheries worldwide.

Not discussed in this chapter but mentioned in some other chapters in this volume is the effect of rapid development of aquaculture of spiny lobsters. Achievements in Australia, which are not published because the research results are considered *commercial in confidence*, indicate that the studies are close to finality, which will result in the development of new spiny lobster industries in competition with wild lobster fisheries.

References

Bellchambers, L. M., Phillips, B. F., & Pérez-Ramírez, M. (2015). From certification to recertification the benefits and challenges of MSC: A case study using lobsters. *Fisheries Research, 182*, 88–97.

Brown, R. S. (2009). *Western rock lobster low puerulus settlement risk assessment: Draft report for public comment* (pp. 1–52). Perth: Western Australian Department of Fisheries.

Caputi, N., Chubb, C., & Brown, R. S. (1995). Relationships between spawning stock, environment, recruitment and fishing effort for the western rock lobster, *Panulirus cygnus*, in Western Australia. *Crustaceana, 68*, 213–226.

Caputi, N., Chubb, C., Melville-Smith, R., Pearce, D., & Griffin, D. (2003). Review of relationships between life history stages of the western rock lobster, *Panulirus cygnus*, in Western Australia. *Fisheries Research, 65*, 47–61.

Department of Fisheries. (2011). *Puerulus settlement index.* http://www.fish.wa.gov.au/Species/Rock-Lobster/Lobster-Management/Pages/Puerulus-Settlement-Index.aspx

Finkbeiner, E. M., & Basurto, X. (2015). Re-defining co-management to facilitate small-scale fisheries reform: An illustration from northwest Mexico. *Marine Policy, 51*, 433–441.

Giraldes, B. W., & Smyth, D. M. (2010). Recognizing *Panulirus meripurpuratus* sp. (Decapoda: Palinuridae) in Brazil—Systematic and biogeographic overview of *Panulirus* species in the Atlantic. *Zootaxa, 4107*(3), 353–366.

Gulf of Mexico Fisheries Management Council. (2011). Amendment 10 to the fishery management plan for spiny lobster in the Gulf of Mexico and South Atlantic with Draft Environmental Impact Statement. Gulf of Mexico Fishery Management Council, National Oceanic and Atmospheric Administration and South Atlantic Fishery Management Council. Available online.

Hobday, A. J., & Cvitanovic, C. (2017). Preparing Australian fisheries for the critical decade:insights from the past 25 years. *Marine and Freshwater Research, 68*(10), 1779–1787.

Kough, A. S., Paris, C. B., & Butler, M. J. (2013). Larval connectivity and the international management of fisheries. *PLoS ONE, 8*(6), e64970.

McCay, B. J., Micheli, F., Ponce-Díaz, G., Murray, G., Shester, G., Ramirez-Sanchez, S., & Weismang, W. (2014). Cooperatives, concessions, and co-management on the Pacific coast of Mexico. *Marine Policy, 44*, 49–59.

Melville-Smith, R. (2011). *Factors potentially affecting the resilience of temperate marine populations*. Report prepared for the Australian Government Department of Sustainability, Environment, Water, Population and Communities on behalf of the State of the Environment 2011 Committee. Canberra: DSEWPaC.

Moore, R., & MacFarlane, W. (1984). Migration of the ornate rock lobster, *Panulirus ornatus* (Fabricius), in Papua New Guinea. *Australian Journal of Marine and Freshwater Research, 35*, 197–212.

Pascoe, S. D., Plagányi, E. E., & Dichmont, C. M. (2017). Modelling multiple management objectives in fisheries: Australian experiences. *ICES Journal of Marine Sciences, 74*(2), 464–474.

Patterson, H., Georgeson, L., Noriega, R., Koduah, A., Helidoniotis, F., Larcombe, J., Nicol, S., & Williams, A. (2017). Chapter 1: Overview. In *Fishery Status Reports 2017*. Canberra: Australian Bureau of Agricultural and Resource Economics and Sciences. CC BY 4.0

Penn, J. W., Caputi, N., & de Lestang, S. (2015). A review of lobster fishery management: The Western Australian fishery for *Panulirus cygnus*, a case study in the development and implementation of input and output-based management systems. *ICES Journal of Marine Sciences, 72*, i22–i34.

Pérez-Ramírez, M., Ponce-Díaz, G., & Lluch-Cota, S. (2012). The role of MSC certification in the empowerment of fishing cooperatives in Mexico: The case of red rock lobster co-managed fishery. *Ocean & Coastal Management, 63*, 24–29.

Phillips, B. F. (1986). Prediction of commercial catches of the western rock lobster *Panulirus cygnus*. *Canadian Journal of Fisheries and Aquatic Sciences, 43*(11), 2126–2130.

Phillips, B. F., & Pérez-Ramírez, M. (Eds.). (2017). *Climate change impacts on fisheries and aquaculture: A global analysis*. Oxford: Wiley-Blackwell. 1048 pp.

Phillips, B. F., Melville-Smith, R., Linnane, A., Gardner, C., Walker, T. I., & Liggins, G. (2010). Are the spiny lobster fisheries in Australia sustainable? *Journal of the Marine Biological Association of India, 52*(2), 139–161.

Phillips, B. F., Melville-Smith, R., Kay, M. C., & Vega-Velazquez, A. (2013). *Panulirus* species. In B. F. Phillips (Ed.), *Lobsters: Biology, management, aquaculture and fisheries* (pp. 289–325). New York: Wiley.

Phillips, B. F., Pérez-Ramírez, M., & de Lestang, S. (2017). Lobsters in a changing climate. In B. F. Phillips & M. Pérez-Ramírez (Eds.), *Climate change impacts on fisheries and aquaculture: A global analysis* (pp. 815–850). Oxford: Wiley-Blackwell.

Plagányi, E. E., Weeks, S. J., Skewes, T. D., Gibbs, M. T., Poloczanska, E. S., Norman-López, A., Blamey, L. K., Soares, M., & Robinson, W. M. L. (2011). Assessing the adequacy of current fisheries management under changing climate: A southern synopsis. *ICES Journal of Marine Sciences, 68*(6), 1305–1317.

Plagányi, E. E., Deng, R., Dennis, D., Hutton, T., Pascoe, S., van Putten, I., & Skewes, T. (Eds.). (2012). *An integrated management strategy evaluation (MSE) for the Torres Strait Tropical Rock Lobster Panulirus ornatus Fishery* (AFMA & CSIRO draft final project report, AFMA project 2009/839). Cleveland: CSIRO.

Plagányi, E. E., van Putten, I., Hutton, T., Deng, R. A., Dennis, D., Pascoe, S., Skewes, T., & Campbell, R. A. (2013). Integrating indigenous livelihood and lifestyle objectives in managing a natural resource. *Proceedings of the National Academy of Sciences of the United States of America, 110*, 3639–3644.

Plagányi, E. E., Dennis, D., Campbell, R., Haywood, M., Pillans, R., Tonks, M., Murphy, N., & McLeod, I. (2014a). *Torres Strait rock lobster (TRL) fishery surveys and stock assessment: TRL fishery model, used to calculate the upcoming TAC updated using the 2014 survey data and the previous year's CPUE data*. AFMA Project 2013/803. June 2015 Milestone report.

Plagányi, E. E., van Putten, I., Thébaud, O., Hobday, A. J., Innes, J., Lim-Camacho, L., Norman-López, A., Bustamante, R. H., Farmery, A., Fleming, A., Frusher, S., Green, B., Hoshino, E., Jennings, S., Pecl, G., Pascoe, S., Schrobback, P., & Thomas, L. (2014b). A quantitative metric to identify critical elements within seafood supply networks. *PLoS ONE*. https://doi.org/10.1371/journal.pone.0091833.

Plagányi, E. E., McGarvey, R., Gardner, C., Caputi, N., Dennis, D., de Lestang, S., Hartmann, K., Liggins, G., Linnane, A., Ingrid, E., Arlidge, B., Green, B., & Villanueva, C. (2017). Overview, opportunities and outlook for Australian spiny lobster fisheries. *Reviews in Fish Biology and Fisheries*. https://doi.org/10.1007/s11160-017-9493-y.

Prescott, J., & Steenbergen, D. J. (2017). Laying foundations for ecosystem-based fisheries management with small-scale fisheries guidelines: Lessons from Australia and Southeast Asia. In S. Jentoft, R. Chuenpagdee, M. J. Barragán-Paladines, & N. Franz (Eds.), *The small-scale fisheries guidelines. Global implementation* (MARE Publication Series 14). Cham: Springer. https://doi.org/10.1007/978-3-319-55074-9_1.

Smith, N. S., & Zeller, D. (2016). Unreported catch and tourist demand on local fisheries of small island states: the case of The Bahamas, 1950-2010. *Fishery Bulletin, 114*, 117–131. Southeast Fishery Bulletin FB11-98, 2011.

Williams, G. (2004). Torres Strait lobster fishery. In A. Caton & K. McLoughlin (Eds.), *Fishery status reports 2004: Status of fish stocks managed by the Australian Government* (pp. 43–52). Canberra: Bureau of Rural Sciences.

Reproductive Biology of Spiny and Slipper Lobster

9

Joe K. Kizhakudan, E. V. Radhakrishnan, and Lakshmi Pillai S

Abstract

This chapter presents an account of the reproductive biology of lobsters through a review of exhaustive research carried out across the globe in different species. Lobsters are sexually dimorphic and show marked behavioural changes particularly during the breeding stage. Description of the male and female reproductive structures (primary and secondary), gonadal development, size at maturity, breeding behaviour, external indicators of maturity and morphometric ratios signalling onset of sexual maturity is presented. The structure and nature of the spermatophore and its deposition on the female for fertilization in palinurid and scyllarid lobsters are also discussed. While palinurid lobsters have been the subject of study on a large scale, globally, studies on scyllarid lobsters are relatively restricted. The reproductive cycle in different species of lobsters is described with discussions on breeding season, spawning migration, spawning periodicity, mating behaviour and breeding in captivity. The effects of eyestalk ablation on enhancing maturation in captivity are also discussed. Studies on the reproductive biology of commercially important lobsters are useful in conservation and management of the resource in their natural habitat and improving their aquaculture potential. Results of such studies can be collated to derive minimum fishing size limits and identify closed seasons for protecting spawning populations, as well as in developing successful husbandry practices for captive rearing and propagation.

Keywords

Reproductive biology · Spermatophore · Maturation · Mating behaviour · Breeding · Palinurids

---

J. K. Kizhakudan (✉) · E. V. Radhakrishnan · Lakshmi Pillai S
ICAR-Central Marine Fisheries Research Institute, Cochin, Kerala, India
e-mail: jkkizhakudan@gmail.com

© Springer Nature Singapore Pte Ltd. 2019
E. V. Radhakrishnan et al. (eds.), *Lobsters: Biology, Fisheries and Aquaculture*,
https://doi.org/10.1007/978-981-32-9094-5_9

9.1 Introduction

The science of reproduction in lobsters is an intriguing subject which has attracted the attention of a lot of researchers worldwide. Lobsters are sexually dimorphic and show marked behavioural changes, particularly during the breeding seasons. Secondary sexual characteristics form an important aspect of study in understanding the reproductive biology of lobsters.

Several studies have been directed worldwide on the propagation of lobsters, both in the wild and in captivity. Detailed information on the reproductive biology and behaviour of lobsters, particularly palinurid and homarid lobsters, have been documented over the last five decades. Studies on reproduction in *Jasus lalandii* by Fielder (1964) and Heydorn (1965), on fecundity in *Panulirus longipes* by Morgan (1972), on the comparison of spermatophoric masses and mechanisms of fertilization in the Southern African spiny lobsters by Berry and Heydorn (1970), on the biology of *P. homarus* by Berry (1971), on *Palinurus gilchristi* by Pollock and Augustyn (1982) and *Jasus edwardsii* and *J. tristani* by Pollock and Goosen (1991) are some of the most important early studies which provide primary insight into this complicated biological process. MacDiarmid (1987) and MacDiarmid et al. (1991) studied reproductive patterns and behaviour in a population of *J. lalandii* in a protected reserve in northern New Zealand. MacDiarmid (1987) demonstrated the impact of male aggressive behaviour on mate selection by female lobsters. Several field studies of tropical spiny lobsters indicate that rapid and repetitive brood cycles are common, up to five broods per annum (Briones-Fourzan and Lozano-Alvarez 1992; Juinio 1987; MacDiarmid 1987). Nelson et al. (1988a, b) and Waddy and Aiken (1992) monitored the effects of environmental variables on gonadal maturation in female lobsters and stressed the importance of monitoring the effects of environmental variables. Juinio (1987) carried out an extensive study of *P. penicillatus* (Oliver, 1791) in the Philippines. Lyons et al. (1981) summarized the reproductive biology of the Florida spiny lobster *P. argus* in populations under light, heavy and no fishing pressure. Lipcius (1985) and Herrnkind and Lipcius (1989) also studied reproductive behaviour in *P. argus*. Briones-Fourzan and Lozano-Alvarez (1992) studied two populations of closely related palinurids, *P. inflatus* and *P. gracilis*. Quackenbush and Herrnkind (1983) and Lipcius and Herrnkind (1987) studied the effects of photoperiod on the moult cycle and reproductive behaviour in *P. argus*. Gomez et al. (1994) determined the breeding period of *P. longipes longipes* in the Philippines by studying ovarian development and egg bearing in the fished population of females. George (1965) has recorded the breeding season of *P. homarus* along the south-west coast of India. The maturation cycle in female *P. japonicus* has been described in detail by Nakamura (1990) and Minagawa and Sano (1997). Pinheiro and Lins-Oliveira (2006) described the reproductive biology of the brown spiny lobster *P. echinatus* from Brazil. Hearn and Toral-Granda (2007) described the reproductive biology of the red slipper lobster *P. penicillatus* from Galapagos Islands. Chang et al. (2007) studied the reproductive biology of *P. penicillatus* from Taiwan waters. Freitas et al. (2007) described the reproductive biology of the spiny lobster *P. regius* from Cape Verde, West Africa. Melville-Smith et al.

(2009) give detailed insight into the reproductive biology of Australia's western rock lobster *P. cygnus*, identifying key issues relevant to the management of its fishery.

From the Indian subcontinent, Radhakrishnan (1977) reared juveniles of *P. homarus* to maturity in the laboratory and attained maturation and breeding. Radha and Subramoniam (1985) described spermatophore formation and biochemistry in *P. homarus*. Hussain and Amjad (1980) have studied the breeding and fecundity of *P. polyphagus* along the Pakistan coast. Kagwade (1988a, b) has described reproduction and fecundity in *P. polyphagus* from Maharashtra waters. Kizhakudan (2007) studied the reproductive biology of *P. polyphagus* from Gujarat and described its size at maturity (Kizhakudan and Patel 2010). Pillai et al. (2014) described the male reproductive biology of *P. homarus*. Thangaraja and Radhakrishnan (2017) studied the reproductive biology of *P. homarus homarus* from the Kanyakumari coast of Tamil Nadu and estimated the size at physiological and functional maturity of females.

Several studies on the histological structure and development of the gonads have also been carried out, particularly in palinurid lobsters (Fielder 1964; Berry and Heydorn 1970; Radha and Subramoniam 1985; Nakamura 1990; Minagawa and Sano 1997; Kizhakudan 2007; Pillai et al. 2014) and homarid lobsters (Schade and Shivers 1980; Kooda-Cisco and Talbot 1982).

In comparison with spiny lobsters, there is relatively very little information on the biology and reproductive behaviour of scyllarid lobsters. Fecundity and spawning seasons have been described in the Mediterranean locust lobster, *Scyllarides latus* (Martins 1985), and the slipper lobster, *S. nodifer*, from the northeastern Gulf of Mexico (Lyons 1970; Hardwick and Cline 1990). In Australia, the reproductive characteristics of *T. orientalis* have been relatively better documented (Kneipp 1974; Branford 1980; Jones 1988). Stewart et al. (1997) have described the size at first maturity and reproductive biology of *Ibacus peronii* and *Ibacus* sp. from Australian waters. Stewart and Kennelly (1997) have described the fecundity and egg size in *I. peronei*. Haddy et al. (2005) studied the reproductive biology of the Balmain bugs *I. chacei*, *I. brucei* and *I. alticrenatus*. De Martini and Williams (De Martini and Williams 2001) studied fecundity and egg size in *Scyllarides squammosus*. DeMartini et al. (2005) discussed indicators of sexual maturity in *S. squammosus* and the spiny lobster *P. marginatus*. Hearn and Toral-Granda (2007) have described the reproductive biology of the Galapagos slipper lobster *Scyllarides astori* from Galapagos Islands. Oliveira et al. (2008) assessed the size at sexual maturity and fecundity of the slipper lobster *Scyllarides deceptor* along the southern Brazilian coast. It has now been confirmed that the slipper lobster in Indian waters, identified earlier as *T. orientalis* (Chhapgar and Deshmukh 1964), is in fact *T. unimaculatus* (Burton and Davie 2007; Jeena et al. 2015). From Indian waters, Kagwade and Kabli (1996) recorded the breeding peaks, fecundity and size at first maturity of *Thenus unimaculatus* from the Mumbai coast. Hossain (1975, 1976, 1978a, 1978b, 1978c) has documented observations on the biology and sexual characteristics of *T. unimaculatus* from the Andhra Coast. Rahman et al. (1987) studied the egg development in *T. unimaculatus*. An account of the reproductive biology of female *T.*

unimaculatus from the south-east coast of India has been presented by Rahman (1989). Moulting in female *T. unimaculatus* has been described by Rahman and Subramoniam (1989). Kizhakudan (2007) studied the reproductive biology and ecophysiology of *T. orientalis* from the north-west coast of India. Kizhakudan (2014) described reproductive biology of the female slipper lobster *T. unimaculatus*.

9.2 Sexual Dimorphism

Morphological differences between sexes, sexual dimorphism, is a dominant state in which females often grow larger and reach bigger size and weight than males (Gopal et al. 2010). Studies on sexual dimorphism is particularly useful in taxonomic, biological, behavioural and evolutionary studies (Accioly et al. 2013). Morphological distinction between adult males and females is evident in lobsters, and many decapods crustaceans exhibit discontinuous growth in body shape. Often allometry is related to reproduction in crustaceans.

9.2.1 Sexual Dimorphism in Morphology and Secondary Sexual Characteristics

Ontogenetic sexual dimorphism has been observed in the spiny lobster *P. homarus homarus*, in which females attain larger length and weight than males of the same age (Radhakrishnan et al. 2015). Tidu et al. (2004) observed that females of *Palinurus elephas* grow faster than males, and the total length is more in females than males of the same age. The abdomen is obviously wider in females of *P. homarus homarus* compared to males, to accommodate the fertilized eggs on the abdominal pleopods until larvae are hatched out. In males, allometric elongation of the second and third walking legs was evident on attaining functional size at maturity (Berry 1970; Kulmiye et al. 2006; Radhakrishnan et al. 2015). Elongation of legs in adult males plays an important role during mating. Males have uniramous pleopodal expods, whereas pleopods are biramous in females, another sexually dimorphic morphological difference between males and females. Exopod length of females grows proportionally relative to carapace length at onset of maturity in females. Kulmiye et al. (2006) used this characteristic to estimate functional maturity in *P. homarus homarus* from Kenya.

9.3 Male Reproductive System

9.3.1 Morphology of Sexual Organs

The male reproductive system in tropical spiny lobsters consists of paired testes and a pair of vas deferens leading to a narrow ejaculatory duct (Fig. 9.1). The testis is a

Fig. 9.1 Testes with vasa deferentia in *Panulirus homarus homarus* male

semi-transparent light yellowish organ placed over the dorsal surface of the midgut gland, beneath the heart. The mature testis is an H-shaped structure with two anterior lobes and two posterior lobes, as described by Mathews (1951), Fielder (1964), Paterson (1969), Nakamura (1990), Pillai (2007) and Pillai et al. (2014) for *P. penicillatus*, *J. lalandii*, *J. novaehollandiae*, *P. japonicus* and *P. homarus*, respectively. The anterior lobes extend to the gut and the posterior lobes extend backwards up to the first abdominal segment. Paired vas deferens arise from the outer side of the posterior lobes and open through a chitinous sigmoid-shaped gonopore on the coxopodites of the fifth walking leg on either side. Each vas deferens arises as a slender tube with a highly convoluted proximal end leading into a thicker and dilated distal end which gradually thins down again to form the ejaculatory duct. The thicker intermediate portion shows a tendency to vary in size with the maturation process and contains a mucoid substance. The proximal coiled region and the intermediate thicker portion of the vas deferens are the sites of spermatophore production. The spermatophoric mass consists of a highly convoluted tube containing the sperm mass embedded in a gelatinous matrix secreted by the inner lining of the proximal vas deferens. The outer covering of the spermatophore is a layer of thick and hardened gelatinous material secreted in the distal vas deferens. The spermatophores are retained in the vas deferens until copulation. Following this, the spermatophores are ejaculated through the ejaculatory duct at the distal portion of the vas deferens.

In slipper lobsters, the testes, situated dorsal to the alimentary tract, are a pair of highly convoluted white tubular structures joined by a transverse bridge, giving it an H-shaped appearance (Fig. 9.2). The lobes extend backwards into the abdominal region. The vas deferens arises posterior to the transverse bridge. The proximal vas deferens is highly convoluted while the distal vas deferens is straight and opens through the genital pore on the coxa of the fifth pair of pereiopods. The vas deferens in a mature male holds the spermatophoric mass which is seen as a white gelatinous substance. The spermatophoric mass is expelled from the terminal tubular part of the vas deferens and through a pair of gonopores on the coxa of the fifth pair of pereiopods at the time of copulation.

Fig. 9.2 Maturing testis in *Thenus unimaculatus*

Immature testes appear as transparent or translucent white structures while mature testes are yellow-orange in colour. Three stages of follicular development have been described in the mature individuals of *Panulirus laevicauda*, indicated by (a) the predominance of spermatogonia I and II, (b) increasing numbers of spermatocytes I and II and (c) spermatocytes I and II (de Lima and Gesteira de Lima and Gesteira 2008).

9.3.2 Spermatophoric Mass Morphology and Biochemistry

Sperm transfer from male to female during mating in crustaceans is effected by means of a spermatophore, which is a specialized sperm packet serving as a vehicle for sperm transport. The spermatophore essentially contains the sperms surrounded by protective layers of cellular secretions produced in the vas deferens. The spermatophoric mass is generally composed of a spermatophoric tube, a basal adhesive matrix and a protective gelatinous matrix. Histochemical observations reveal that the wall of the spermatophore tube consists of neutral mucopolysaccharide, whereas the sperm mass and the gelatinous matrix are rich in acidic mucopolysaccharides (Radha and Subramoniam 1985). Pillai et al. (2014) studied the ultrastructure of germ cells in the testis of the spiny lobster *P. homarus homarus* from south-west coast of India. While the extruded spermatophore is sticky and thick in *P. homarus homarus* (Pillai et al. 2014), it is granular in *P. homarus rubellus* from South Africa (Berry 1970). The spermatozoa of *P. homarus rubellus* is roughly spherical in shape with a diameter of 10–15 μ (Berry 1970). Pillai et al. (2014) described the structure of spermatozoa of *P. homarus homarus* with a nucleus, lens-shaped acrosome and microtubules.

Berry and Heydorn (1970) present a vivid comparative account of the spermatophore structure in the South African spiny lobster *Panulirus homarus, Panulirus gilchristi, Jasus lalandii, Puerulus angulatus* and *Linuparus trigonus* and distinguished three basic types of spermatophoric mass. In *Jasus lalandii,* they reported a gelatinous thread-like spermatophoric mass, with no specific shape and which disintegrates in seawater. The highly convoluted spermatophore is randomly distributed throughout the protective matrix. In *Panulirus gilchristi, Puerulus angulatus* and *Linuparus trigonus*, they observed a gelatinous spermatophoric mass which does not disintegrate in seawater. It has a characteristic shape and also a definite orientation and distribution within the protective matrix. In *Panulirus homarus*, they found that the spermatophore has a granular, putty-like consistency, which hardens in seawater. This type of spermatophore, too, has a definite orientation and distribution within the protective matrix. George (2005) distinguishes two types of spermatophoric mass in Palinuridae, a paired spermatophoric mass that hardens quickly in seawater and spreads in the Stridentes subgroup of the Palinuridae and a soft gelatinous spermatophoric mass that rapidly breaks up in seawater in the Silentes subgroup.

The nature of the spermatophoric mass in spiny lobsters and scyllarid lobsters is completely different. The matrix of the spermatophoric mass in most tropical spiny lobsters changes from its gelatinous form to a hardened stiff form when exposed to seawater. Spermatophore extrusion is done continuously, and it is deposited as two layers over the window area, the outer layer being the gelatinous matrix which serves to cement the spermatophore to the female plastron and the inner core layer being the convoluted spermatophore chambers. The spermatophore soon hardens and turns blackish in colour, in which condition it is called a "tar spot" (Fig. 9.3).

In the absence of observable spermatophores on ovigerous *Scyllarides nodifer*, Lyons (1970) postulated that fertilization in this species is probably internal. DeMartini et al. (2005) did not observe external spermatophores in *Scyllarides squammosus*. However, Spanier and Lavalli (2006) suggested the possibility of different reproductive strategies within the genus *Scyllarides* since Martins (1985), Almog-Shtayer (1988) and Spanier and Lavalli (1998) reported the presence of

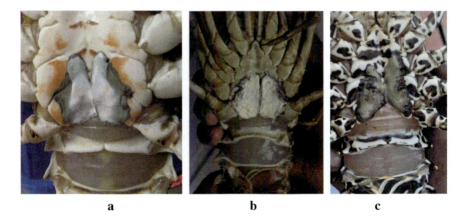

Fig. 9.3 Spermatophore in spiny lobsters. (**a**) *Panulirus ornatus* (**b**) *Panulirus polyphagus* (**c**) *Panulirus versicolor*

spermatophore on females of *Scyllarides latus* up to 10 days prior to egg extrusion. Sharp et al. (2007) suggested the spermatophores of some scyllarid species remain gelatinous and highly ephemeral, similar to that described by Berry and Heydorn (1970) for *Jasus lalandii*. While they did not find any spermatophore on egg-bearing individuals of *Scyllarides nodifer* and *Scyllarides aequinoctialis*, they reported the presence of eroded and fresh spermatophores on individuals of *Parribacus antarcticus*. Lyons (1970) too had reported spermatophore deposition in this species.

The spermatophoric mass of *T. unimaculatus* lacks an external gelatinous matrix (Kizhakudan 2014). The spermatophores remain embedded in a fibrillar mucoid matrix, which does not harden on exposure to seawater. It breaks open in a few hours after exposure to seawater. The freshly released spermatophores are milky white in colour, soft, delicate and embedded in a gelatinous matrix of mucoid fibrils. Impregnation is seen on the ventral side of the first abdominal segment as two parallel lines. The spermatophore packets are ovoid in shape and microscopic in size. They hold the active spermatozoa within. The gelatinous matrix in which the spermatophores are embedded holds a fibrillar network to which the spermatophores are attached (Fig. 9.4). The spermatophoric mass in scyllarid lobsters lasts for only a few hours and cannot be retained for consequent spawnings, unlike in the case of spiny lobsters.

MacDiarmid and Butler (1999) established that there exists a positive relation between male size and spermatophore area in the Caribbean spiny lobster *Panulirus argus* and the southern temperate rock lobster *Jasus edwardsii*. Butler et al. (2011) affirmed that male size has a significant impact on spermatophore weight and number of sperms transferred to the female in *P. argus*. MacDiarmid and Butler (1999) reported that larger males tend to apportion spermatophore area depending on the size of the female, depositing larger spermatophores on larger females. The males of *Panulirus ornatus* can spread a large spermatophoric mass over almost the entire sternum of large, fully mature females that develop four sets of softened windows to mark the prepared area (George 2005).

Fig. 9.4 Spermatophore in scyllarid lobsters. (**a**) *Thenus unimaculatus* (**b**) *Scyllarides tridacnophaga*

9.4 Female Reproductive System

9.4.1 Morphology of Female Reproductive Tract

In tropical lobsters, the ovary is situated dorsally, beneath the heart and over the midgut gland. The mature ovary is usually H- shaped, with two anterior lobes extending into the cephalic region and two posterior lobes extending up to the first abdominal segment. Nakamura (1990) reported that the left lobe is slightly longer than the right in *Panulirus japonicus*. Kizhakudan (2007, 2014) found that the ovaries in *P. polyphagus* and *T. unimaculatus* conform to the general H-shaped pattern described in other lobsters (Fig. 9.5). A pair of semi-transparent slender oviducts arising just below the transverse connection between the two lobes open to the exterior through a pore on the coxa of the third walking leg. Cau et al. (1988) reported the presence of two transversal commissures bridging the right and left lobes in the gonad of the slipper lobster *Scyllarus arctus*. The ovaries in *Petrarctus rugosus* have also been found to have two transverse commissures (Fig. 9.6) (Kizhakudan, pers. observ). The appearance, size, aspect ratio and colour of the ovaries change with progressive maturation. Nakamura (1990) described the immature ovary as white or weakly yellowish in colour, which turned reddish orange in the mature state. Berry (1971) reported the colour of the ripe ovary in *P. homarus* to be coral red. Silva and Landim (2006) reported that in the mature stage, the ovaries of *P. echinatus* and *P. laevicauda* are orange in colour, while those of *P. argus* are consistently red. The mature ovaries of astacuran lobsters on the other hand are usually dark green, as in *Homarus* spp., or bluish, as in *Nephrops* spp. (Aiken and Waddy 1980).

9.4.2 Ovarian Stages and Histology

Ovarian development is usually studied through macroscopic and microscopic staging. Mota-Alves and Tomé (1965, 1966) described five ovarian developmental

Fig. 9.5 Ripe ovary of *Thenus unimaculatus* (in situ view)

Fig. 9.6 Ripe ovary of *Petrarctus rugous*

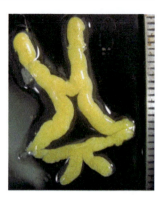

stages in females of *P. argus* and *P. laevicauda*, whereas six ovarian developmental stages were later recognized in both the species (Soares and Peret 1998a, 1998b; Soares et al. 1998). Pinheiro and Lins-Oliveira (2006) described the following six stages of gonadal development in female *Panulirus echinatus*: "immature", "initial maturation", "final maturation", "mature", "spawned" and "rematuration". They noted that the colour changed from brownish-white to light-yellow, orange, red and brownish grey from the immature to the spawned stages. Silva and Landim (2006) observed the following four stages of ovarian maturation in *P. argus*, *P. laevicauda* and *P. echinatus*: "immature", "prematuration", "mature" and "spawning/ resorption". Chang et al. (2007) described the following five stages of ovarian development in *Panulirus penicillatus*: "inactive", "developing/redeveloping", "ripe/re-ripe", "spent" and "recovery", with the colour transitioning from white to light orange, dark orange, white/light yellow and white. Fernandez (2002) classified the ovarian development of *P. homarus* into five stages based on morphological characteristics, colour, texture and gonadosomatic index (GSI): immature (stage V_1), primary vitellogenic (stage V_2), secondary vitellogenic (stage V_3), mature (stage V_4) and spent (stage V_5). The colour of the ovary changed from white to cream, bright coral red, deep coral red and light coral red with GSI ranging from 0.32 in stage V_1 to 4.9 in stage V_4 and 0.65 in stage V_5.

Stewart et al. (1997) described the progression of ovaries in *Ibacus peronii* and *Ibacus* sp. through four maturation stages, with the colour changing from translucent/white in stage 1 to creamy yellow in stage 2, yellow/orange in stage 3 and bright orange in stage 4. They reported the spent ovaries to be creamy yellow in colour with residual orange eggs seen through the ovary wall. Kagwade and Kabli (1996) recognized five stages of ovarian development in *T. orientalis* (=*T. unimaculatus*). Kizhakudan (2014) described the following five stages of ovarian maturation—"immature", "early maturing", "late maturing/mature", "ripe", "spawning" and "spent/recovery"—in the shovel-nosed lobster *Thenus unimaculatus* and observed that the colour of the developing ovary progresses from translucent to white to creamy or dark yellow and finally to dark orange in the ripe and spawning stage (Fig. 9.7).

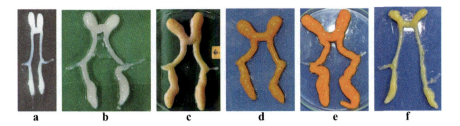

Fig. 9.7 Ovaries of *Thenus unimaculatus* in different stages of development. (**a**) Immature (**b**) early maturing (**c**) late maturing (**d**) ripe (**e**) spawning (**f**) spent

The ovarian wall consists of an outer epithelial layer, a middle layer of connective tissue supplied with blood vessels and an inner germinal epithelium. The thickness of the connective tissue changes with the ovarian maturation cycle. The ovaries in immature and maturing individuals lack a lumen; the ovaries of mature individuals have a central lumen. The germinal epithelium forms a series of inward folds as it runs through the length of the ovary. Ova are formed from the germinal epithelium and are initially surrounded by a single layer of flat follicular cells. Developed oogonia, primary and secondary oocytes and developing ova can be seen distributed from the central germinal epithelium to the peripheral wall of the ovary. Transverse sections of a spent ovary reveal resorbed oocytes and infiltration of connective tissue. The ovarian wall at this stage is relaxed.

Transverse sections of the oviduct reveal that the wall of the oviduct is made up of an outer thin layer of epithelium, a middle layer of connective tissue and an inner layer of columnar epithelium. The tubular oviduct in slipper lobsters is more compressed than round. Kizhakudan (2007) observed that unlike in the case of *P. polyphagus* which has a large central lumen in the oviduct, the lumen of the oviduct in *T. orientalis* (=*T. unimaculatus*) is smaller, as are the villi, and the lumen is placed towards the peripheral regions on either side of a central layer of connective and muscular tissues.

9.5 Ovarian Development (Vitellogenesis)

Ovarian maturation in lobsters progresses through three distinct phases depending on the extent of yolk deposition during oocyte formation and development, namely, pre-vitellogenesis phase, primary vitellogenesis phase and secondary vitellogenesis phase (Minagawa and Sano 1997; Nakamura 1990; Fernandez 2002; Kizhakudan 2014). In the pre-vitellogenesis phase, which is the immature phase in ovarian development, the outer wall of the ovary is thin, and the developing oocytes, mostly surrounded by follicle cells, have large and well-defined nuclei with prominent nucleoli, and there is no indication of yolk deposition. The cytoplasm is basophilic and takes haematoxylin stain.

As vitellogenesis sets in, the cytoplasm gradually becomes more eosinophilic. The primary vitellogenesis phase is usually completed through three stages. In stage 1, many oocytes have a few peripheral vacuoles scattered in the cytoplasm, which takes haematoxylin stain. In stage 2, peripheral vacuoles increase in number and tend to be distributed uniformly around the nucleus. The germinal zone is placed centrally and developing oocytes are seen near the peripheral region; yolk deposition begins in the cytoplasm of the developing oocytes. The cytoplasm turns slightly eosinophilic. In stage 3, the peripheral vacuoles are larger in size; yolk deposition becomes dense; eosinophilic granules increase in number among the peripheral vacuoles and the egg membrane is seen between the oocytes and the follicle cells.

The secondary vitellogenesis phase is the mature phase, where yolk granules accumulate abundantly in the developed oocytes. This phase is completed in two stages. In stage 1, mature oocytes are seen in the ovarian cavity. There is dense yolk deposition which masks the nuclei. The cytoplasm is highly eosin-positive. In stage 2, the size of the oocytes becomes maximum, while the nuclei decrease greatly in size and can hardly be observed; yolk deposition is complete. The ovarian cavity is packed with mature oocytes and the ovarian wall is strong and tense.

9.5.1 Resorption

Most authors recognize a "spawning/resorption" or a "spent/recovery" stage in the ovarian maturation cycle of lobsters. Following spawning, residual oocytes and other ovarian cells are usually resorbed into the ovaries. In this state, the ovary becomes shrunken, flaccid, with dispersed pigmented areas and empty spaces internally, and it often resembles an immature ovary. However, the gonad of a spawned lobster is never entirely restored to the condition of original immaturity (Silva and Landim 2006).

9.5.2 Cement Glands

Cement glands, also called tegumental glands, pleopodal glands or abdominal glands, are accessory structures developed on the pleopods of female lobsters prior to egg extrusion. They are milky white in appearance and develop from the edges of the pleopods towards the centre. They secrete a sticky substance that helps the eggs attach or cement to the pleopods and abdomen after extrusion. Cement glands are often used as an indicator of sexual maturity in female lobsters as their development has been found to show high correlation with ovarian development in the American lobster (Stephens 1952; Aiken and Waddy 1982). Cement gland development can be staged under a microscope, based on the extent of pleopod surface covered by the glands. The use of cement gland evaluation has been adopted in several later studies (Campbell and Robinson 1983; Miller and Watson 1991; Ugarte 1994; Comeau and

Savoie 2002; Gendron 2003; Waddy and Aiken 2005; Reeves et al. 2011; Silva et al. 2012; Watson et al. 2013).

Decapod cement glands were first recognized and described by Braun (1875) and later by Cano (1891). Herrick (1894) first described the cement glands in the pleopods of the American lobster and Yonge (1932, 1937) described the structures in the European lobster. While cement glands have been extensively studied in the American lobster, there are not too many descriptions of these structures and their functions in tropical spiny lobsters of the genus *Panulirus*. In scyllarid lobsters, Stewart et al. (1997) mention the presence of a sticky white substance covering the ovigerous setae on the pleopodal endopods of a female *Ibacus peronii* capture in the process of oviposition, when some eggs had already been deposited on the endopods of the first set of pleopods. Although they do not hypothesize on the origin of the substance, they suggest that it is used to attach the eggs to the ovigerous setae.

Stephens (1952) recognized four developmental stages in the cement glands on the pleopods of the crayfish *Orconectes*, and Aiken and Waddy (1982) adopted a modified version of this description for the American lobster, using only the endopodite for the purpose since they found that the cement glands were most accessible and most abundant on this appendage. They advocated staging based on cement gland activity and spread, which usually begin along the medial and lateral edges of the endopodite and progresses into the central and the distal regions. While the early stages (1 and 2) are difficult to detect with the naked eye, advanced stages (3 and 4) can be clearly discerned (Table 9.1).

Table. 9.1 Stages of cement gland development in the American lobster *Homarus americanus* (Aiken and Waddy, 1982)

Stage	Description
Stage 1	Tissue thickened between nodes along lateral and medial edges, but individual glands not visible to unaided eye. With transmitted light at x10 and x20, individual glands of ~100 μm may be seen as small indistinct spots, but no "rosettes" are formed
Stage 2	Some cement gland activity apparent in central region, especially proximal end. Some glands with rosette appearance when viewed with transmitted light at x10 to x20. Visible to unaided eye as small white dots medial to nodes
Stage 3	Gland rosettes well developed in central region of endopodite. Visible to unaided eye, as distinct white dots in central region and continuous white mass in medial and lateral regions. Glands 100–150 ¡lm diameter. Ratio of rosette to gland diameter less than 0.75
Stage 4	Glands engorged throughout, often appearing to be arranged in rows. Visible to unaided eye as a continuous white mass in lateral and medial regions and in proximal portion of central region. Diameter of glands greater than 200 ¡lm; ratio of rosette to gland diameter greater than 0.75
Degenerating	Variable number of large cement glands scattered throughout pleopod, but general cement gland development more typically stage 1. Condition associated with resorbing or spent ovaries

9.6 Fecundity/Brood Size

The number of eggs carried on the pleopods by a female is often defined as the fecundity. Pollock and Goosen (1991) used the term brood size, as the term is most appropriate for repetitively breeding females. Chubb (1994) redefined fecundity as the total number of eggs produced by an individual female in a breeding season. Tuck et al. (2000) classified types of fecundity as potential fecundity (number of oocytes in the ovary), actual fecundity (number of eggs on the pleopods at the time of capture) and effective fecundity (estimated number of eggs on the pleopods at the time of hatching). Spiny lobsters have smaller egg size, higher fecundity and prolonged larval phase compared to clawed lobsters (Pollock 1997).

While Berry (1971) reported a positive correlation between fecundity and lobster size in *P. homarus rubellus*. Linear relationship between carapace length and weight of eggs/number of eggs in was not evident in *P. homarus homarus* from Chennai, on the south-east coast of India (Vijayakumaran et al. 2005). The number of eggs in a brood of *P. homarus homarus* from Chennai (52.2 to 94.4 mm CL) ranged from 1,20,544 to 4,49,585 (Vijayakumaran et al. 2012). In *P. homarus rubellus*, the smaller size class (50–59 mm CL) was estimated to produce about 100,000 eggs and the largest class (90–99 mm CL) about 900,000 eggs with the 70–79 mm size class contributing maximum to the reproductive potential (Berry 1971). Marginally, lower egg weight and egg numbers were observed in laboratory-reared *P. homarus*, compared to the wild specimens. However, average number of eggs/gram body weight was similar in both captive and wild females. Maximum number of eggs per gram body weight in females from wild as well as under captivity was recorded in the size group of 61–70 mm CL, and the maximum number of eggs per female (100 mm CL) recorded was 6,28,930 (Vijayakumaran et al. 2012). The number of eggs per brood of *P. homarus homarus* (carapace length, 47–94 mm; weight, 156.75 to 900 g) from Tuticorin ranged from 95,530 to 4,80,590 (Jawahar et al. 2014). While constructing the CL-fecundity relationship curve, caution should be exercised, as wide variations in egg numbers and egg weight were recorded due to the smaller size of brood in the second spawning from a single mating (Vijayakumaran et al. 2012). In *P. ornatus* from southern coastal waters of Bulukumba, South Sulawesi, Indonesia, the number of eggs in a single brood varied from 88,000 in the smallest lobster (CL77.7 mm) to 1,546,226 (CL 188.2) in the biggest one (Musbir et al. 2018).

Kizhakudan (2007) estimated the fecundity of *P. polyphagus* as ranging from 1,58,000 eggs to 9,31,000 eggs, with an average fecundity of 4,55,000 eggs for reproductively active females in the size range of 64–119 mm CL and the fecundity of *T. orientalis* (=*unimaculatus*) as ranging from 19,600 eggs (60 mm CL) to 59,500 eggs (102 mm CL), with an average fecundity of 39,300 eggs for reproductively active females in the size range of 60–102 mm CL. A linear relationship was found in both species between carapace length and fecundity. Hossain (1979) has estimated the relationship between total length and fecundity as ($F = -24493.6 + 207.784$ TL, $r = 0.6015$) or total wet weight of the female ($F = +2507.73 + 86.985$TW, $r = 0.6243$), showing a positive and linear relationship between the two variants.

While the brood fecundity from one spawning is an average 17,905, the total annual reproductive potential from three spawnings in a year was an estimated 53,715 per one adult female. Kagwade and Kabli (1996) also found linear relationship between total length and fecundity of *T. orientalis* from Bombay waters, $y = -46.66 + 0.4164x$.

Chubb (1994) has expressed the opinion that clutch size from first spawning from fresh sperm deposition should only be considered while estimating fecundity. Clutch size may be variable in different subspecies of *P. homarus*. While the maximum number of eggs for the 90–99 mm CL size class recorded for *P. homarus homarus* from the Indian coast was 6,28,930 (Vijayakumaran et al. 2012), a maximum of 6,67,000 was recorded for the subspecies *P. homarus megasculpta* from Oman waters (Mohan 1997) and 9,00,000 for *P. homarus rubellus* from South African region (Berry 1971). Mohan (1997) has reported that the size group ranging from 70.1 to 75 mm CL and 80.1 to 85 mm CL contribute 44 and 35.8% of eggs produced by *P. homarus megasculpta* (=*P. homarus homarus*) from different regions of Dhofar coast in the Sultanate of Oman.

Egg loss commonly occurs during incubation in lobsters, which appears to be mainly due to a combination of disease and nemertean egg predators (Aiken and Waddy 1980; Chubb 2000). Prolonged exposure to air, handling stress at the primary and secondary holding centres, suboptimal conditions in broodstock holding water and long duration of transport are some of the reasons for egg loss in wild lobsters procured for breeding purposes (Vijayakumaran et al. 2012).

9.7 Maturity

9.7.1 Size at Sexual Maturity

Size at first maturity is usually defined as the minimum size at which 50% of the population has attained maturity. In decapods, the size at first maturity is described in terms of physiological maturation (the size at which the gonads attain maturity) and physical maturation (the size at which the animal is capable of mating and spawning). These processes may not occur at the same size (Jones 1988) and both must be known to derive a reliable estimate of the size at true sexual maturity (Stewart et al. 1997). MacDiarmid and Sainte-Marie (2006) recognized the following three indicators of maturity: morphological maturity (assessed from the development of external secondary sexual characteristics), physiological maturity (assessed from the development of gonads and accessory glands) and functional maturity (assessed from internal or external behavioural features indicating past or recent breeding activity).

Several factors limit the estimation of size at first maturity in lobsters. Most lobster species tend to change habitats with the onset of maturation, resulting in a high degree of variability in the catchability of mature and immature lobsters. Size at maturity estimates based on such fished populations of lobsters can thus be highly biased (Farmer 1974; Tully et al. 2001). Staging based on histological examination of gonads is perhaps the most reliable method for assessing the maturity state of the animal. Most lobster researchers resort to assessment of multiple criteria for estimating the size at first maturity. In female lobsters, these include:

- The functional criteria of egg-bearing (Kensler 1967; Heydorn 1969; Aiken and Waddy 1980).
- The presence of fresh or spent spermatophores and resorbing ova.
- Physiological criteria of ovary colour and size and oocyte size.
- Development of cement glands on the pleopods.
- Morphological criteria of abdomen and pleopod development (MacDiarmid and Sainte-Marie 2006).

Female spiny lobsters are considered sexually mature when they are capable of egg production and have the capacity to mate and spawn. Size at maturity and size at onset of breeding usually refer to functional maturity (Chubb 1994). The size of the smallest female carrying eggs is sometimes cited as the size at maturity. However, it has no relevance while formulating management measures. Majority of researchers working with *Panulirus* species have adopted incidences of mated but non-berried and berried females for estimating size at maturity (Chittleborough 1974, 1976; Kanciruk and Herrnkind 1976; Gregory et al. 1982; MacDonald 1982; Juinio 1987; Jayakody 1989; Chubb 1991; Kulmiye et al. 2006; Kizhakudan and Patel 2010). Size at maturity of the same species varies between different geographical regions (Table. 9.2). In the natal population of the spiny lobster *P. homarus rubellus*, 50% maturity based on percentage of females carrying eggs was estimated at a carapace length of 54 mm, although a few may breed at a carapace length of 50 mm (Berry 1971). However, SOM based on ovary development was estimated at 51 mm CL. In the Transkeian population of *P. homarus rubellus*, the estimated length at first maturity was 50 mm CL and the smallest female carrying eggs had a carapace length of 43 mm CL as opposed to 50 mm CL in the natal population (Heydorn 1969). De Bruin (1962) estimated the size at first maturity of the Sri Lankan west coast population of *P. homarus homarus* at a carapace length between 55 and 59 mm CL, whereas Jayakody (1989) reported 59.5 mm CL as the size at maturity based on egg-bearing lobsters from the southern Sri Lankan coast. Jawahar et al. (2014) estimated the size at functional maturity of *P. homarus homarus* from Tuticorin at 53 mm CL. Thangaraja (Thangaraja 2011) estimated the size at maturity for the same subspecies along the Kadiyapattanam coast on the south-west coast of India at 61.0 mm CL, based on 50% of females carrying eggs, and 53.0 mm based on 50% of females with ovigerous setae. However, size at maturity based on ovarian development was estimated at 55.0 mm CL (Thangaraja and Radhakrishnan 2017). Size at maturity for the same subspecies from Kenyan waters has been estimated at 63.4 mm CL based on the *berry* method (Kulmiye et al. 2006). George (1963) estimated the size at maturity of *P. homarus megasculpta* (=*P. homarus homarus*) females to be between 60 mm and 70 mm CL at Aden (Yemen). Johnson and Al-Abdulsalam (Johnson and Al-Abdulsalaam 1991) provided a higher estimate of 80–85 mm CL for the same subspecies off the coast of Oman. Mohan (1997) reported the size at maturity of *P. homarus megasculpta* (=*P. homarus homarus*)

Table. 9.2 Size at maturity of different species of spiny lobsters

Species	Size at maturity CL (mm)	Method of estimation	Geographical region	Authors
Panulirus guttatus	32.0	50% berry	Florida keys, USA	Robertson and Butler (2003)
Panulirus argus	79.0–81.0	50% berry	Cuba	Cruz and Bertelsen (2008)
	79.0	50% berry	Brazil	
	75.0	50% berry	Florida keys	
	85.0	50% berry	Dry Tortugas	
Panulirus polyphagus	66.0–70	50% ovary	Veraval, India	Kizhakudan and Patel (2010)
	71.0–75.0	50% berry		
	74.8 (deduced from TL)	50% berry	Mumbai, India	Kagwade (1988)
Panulirus ornatus	107.2	Berry (minimum)	Malaysia	Zakaria and Kassim (1999)
	78.6	Berry (minimum)	Papua New Guinea	MacFarlane and Moore (1986)
	67.4	Ovary (minimum)		
Panulirus versicolor	80.0–90.0	Berry (minimum)	Palau, New Caledonia	MacDonald (1982)
	80.0	Allometry (CL < III maxilliped)	Unspecified	George and Morgan (1979)
	98.0	Berry (minimum)	Great barrier reef	Frisch (2007)
Panulirus penicillatus	56.46	Ovary	Taiwan	Chang et al. (2007)
	66.63	Berry		
	40.0–50.0	Allometry (CL < IIIWL)	Red Sea (Saudi Arabia)	Hogarth and Barrat (1996)
	45.0–50.0	Berry	Philippines	Juinio (1987)
	75.0–79.0	Berry	Solomon Islands	Presscott (1988)
Panulirus marginatus	75.0–80.0	Berry	Hawaii	McGinnis (1972)
Panulirus cygnus	89.5	Berry	Freemantle, Australia	Melville-Smith and de Lestang (2006)
	65.0	Berry	Abrolhos Islands	
	81.4	Berry	Jurien Bay	
Panulirus inflatus	63.73	Berry	Mexican Pacific	Perez-Gonzalez et al. (2009)

(continued)

Table. 9.2 (continued)

Species	Size at maturity CL (mm)	Method of estimation	Geographical region	Authors
Panulirus homarus homarus	52.6	Setal	Mambrui, Kenya	Kulmiye et al. (2006)
	63.4	Berry		
	50.5	Exopod length		
	53.0	Setal	Kadiyapattanam, Kanyakumari, India	Thangaraja (2011)
	61.0	Berry		
	55.0	Ovary	Kadiyapattanam, Kanyakumari	Thangaraja and Radhakrishnan (2017)
	54.0	Setal		
	55.0	Window		
	53.0	Berry	Tuticorin, India	Jawahar et al. (2014)
	38.0–47.0	Ovary	Southern Sri Lanka	Jayakody (1989)
	59.5	Berry		
	55.0–59.0	Berry	Western Sri Lanka	De Bruin (1962)
	77.8	Berry	Dhofar, Oman	Al-Marzouqi et al. (2007)
	75.1	Berry	Al-Wusta	
	73.7	Berry	Al-Sharqiya	
	80.0–85.0		Oman	Johnson and Al-Abdulsalam (1991)
	69.0–76.0	Berry	Oman	Mohan (1997)
	60.0–70.0	Berry	Yemen	George (1963)
Panulirus homarus rubellus	54.0	Berry	Natal, South Africa	Berry (1971)
	51.0	Ovary		
	50.0	Berry	Transkei, South Africa	Heydorn (1969)

from three locations along the coast of Oman, ranging from 69 to 76 mm CL. The difference in size at first maturity within a subspecies from different geographic locations may be attributed to both extrinsic factors (e.g. water temperature, food availability, population densities) and some intrinsic factors such as age and growth rate, whereas variation between subspecies may be a combination of extrinsic factors and intrinsic genetic capacity (Kulmiye et al. 2006). Kizhakudan and Patel (2010) found that in *Panulirus polyphagus* from Gujarat coast in north-west India, the critical maturation phase extends between 56 and 65 mm CL for males and between 66 and 75 mm CL for females, and the size at onset of sexual maturity judged from the 25% success rate in development of different sexual characteristics is at 51–55 mm CL for males and 51–60 mm CL for females. The size at first maturity in male and female *T. unimaculatus* from the same coast was assessed as 51–55 mm CL and 61–65 mm CL, respectively, with very little time gap between the attainment of morphological, physiological and functional maturity (Kizhakudan 2007; Kizhakudan 2014).

The assessment of size at maturity in males is a more challenging exercise. Several early studies on spiny lobsters and clawed lobsters use the presence of mature spermatozoa in the vasa deferentia to assess male physiological maturity (Heydorn 1969; Berry 1970; Farmer 1974; MacDiarmid 1989a, b; Turner et al. 2002). Change in the size of the first cheliped in clawed lobsters or the second or third pereiopods in spiny lobsters, relative to CL, has often been used as an indicator of functional maturity. Radhakrishnan et al. (2015) determined the size at onset of sexual maturity in male *P. homarus homarus* as 63.0 mm CL based on allometric growth of the third pereiopod. Slopes of regression of log transformed data showed positive allometric growth of the third pereiopod for males ($b = 1.17$) compared to negative allometry in females ($b = 0.97$).

In male lobsters (MacDiarmid and Sainte-Marie 2006), the precocious development of the second and third pereiopods in male spiny lobsters, after they attain physiological maturity, is associated with their ability to extract females from their dens (Berry 1970; Lipcius et al. 1983; Bertelsen and Horn 2000). Robertson and Butler (2003) estimated the size at maturity for male *P. guttatus* using the onset of allometric growth of the second walking leg as an indicator of male maturity. From estimates of size at maturity for two species, *P. polyphagus* and *T. orientalis* (=*T. unimaculatus*), Kizhakudan (2007) concluded that in both species, males mature earlier than females; while smaller males mate with larger females, possibly of the same clutch, in *T. orientalis*, larger males mate with smaller females, possibly of the same clutch, in *P. polyphagus*.

9.7.2 Secondary Sexual Characteristics and External Indicators of Maturity

The onset and progress of maturity in lobsters are often discernible through external structures or characteristics, which make a preliminary assessment of maturity state close to accurate. All lobster species bear distinct secondary sexual characteristics which either develop or begin to undergo marked changes with the onset of maturity. The common external indicators of maturity in spiny lobsters include penile process, gonopore and II to V pair of pereiopods in males and setal brush on pereiopods, ovigerous setae on abdominal pleopods and mating windows on the ventral sternal plates in females. In scyllarid lobsters, the common external indicators of maturity include gonopore in males and ovigerous setae on abdominal pleopods and gonopore in females.

9.7.2.1 Gonopore and Penile Process in Male Lobsters

The gonopore at the end of the ejaculatory duct, on the coxa of the fifth pair of pereiopods, is a complex structure associated with a pointed and serrated penile process terminating in a distal tuft of setae (Fig. 9.8). In juvenile male *P. polyphagus*, the gonopores are visible as tiny openings on the coxa of the fifth pair of pereiopods. As the animal grows, a small knob-like protrusion develops on the lateral edges of the coxa of the leg. As the animal approaches sexual maturity, this structure

Fig. 9.8 Penile process and gonopore in male lobsters. (**a**) *Panulirus longipes* (**b**) *Panulirus homarus homarus* (**c**) *Panulirus polyphagus* (**d**) *Panulirus versicolor* (**e**) *Puerulus sewelli* (**f**) *Linuparus somniosus*

becomes a prominent pointed penile process with a curved tip bearing a tuft of setae. The active gonopore in a mature male is membranous, with very strong elastic folds and is highly sensitive. The pigmentation on the gonopore is also an indicator of sexual maturity. While the gonopore is translucent white initially, as the penile process develops, it develops a pinkish pigmentation, often tending to reddish pink at the time of breeding.

The gonopore, situated in the form of a simple aperture at the base of the fifth walking leg, is seen to increase in size as the animal grows. There are no other distinct secondary sexual characteristics marking the onset of sexual maturity in male *T. orientalis*.

9.7.2.2 Pereiopods in Male Lobsters

The II–V pairs of pereiopods grow relatively longer in the males as they approach sexual maturity. These legs play an important role in holding the female at the time of courting and mating.

9.7.3 Gonopore in Female Lobsters

The female gonopore, situated in the form of a simple aperture at the base of the third walking leg, is seen to increase in size as the animal grows, but is relatively much smaller than the male gonopore.

9.7.4 Pereiopods with Setal Egg Brush in Female Lobsters

The pereiopods in females play an important role during incubation. The dactyli at the tip of the pereiopods, particularly the fifth pair of pereiopods, develop long, dense setae which are used as an egg brush, to clean the eggs carried on the abdominal pleopods (Fig. 9.9).

9.7.5 Ovigerous Setae on Abdominal Pleopods in Female Lobsters

A commonly acknowledged external indicator of sexual maturity in female lobsters is the presence of fully developed ovigerous setae on the abdominal pleopods (Fielder 1964; Pollock 1982). The endopods of the abdominal pleopods in females bear ovigerous setae (Fig. 9.10) which are used for attaching spawned eggs until the end of the incubation period. In juvenile females, the pleopods are devoid of setae. As the female enters into the sub-adult phase, the pleopods enlarge and the endopods bifurcate and develop long ovigerous setae (Fig. 9.11) (Kizhakudan 2007). On attaining maturity, the ovigerous setae bear lots of seteoles on each seta for carrying eggs.

In *Jasus lalandii* (Paterson 1969) and several other palinurids, such as *P. japonicus* (Nakamura, 1940), *P. argus* (Sutcliffe 1953) and *P. longipes* (George 1958),

Fig. 9.9 Dactyli of fifth pair of walking legs in *Panulirus homarus homarus* (female) bearing setal egg brush

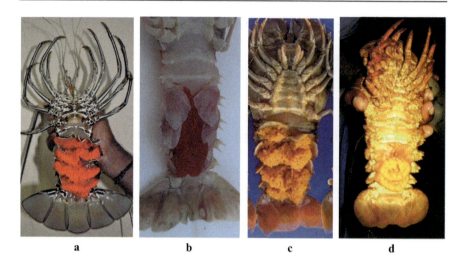

Fig. 9.10 Ovigerous setae bearing eggs. (**a**) *Panulirus versicolor* (**b**) *Puerulus carinatus* (**c**) *Thenus unimaculatus* (**d**) *Scyllarides tridacnophaga*

Fig. 9.11 Pleopods of an adult female *Thenus unimaculatus*

there is a regular annual cycle of formation and disappearance of ovigerous setae but not in *P. homarus* (Berry 1971) and many other spiny lobsters or any of the clawed lobsters (Aiken and Waddy 1980). Sometimes the setae are found to persist, after two or three breeding cycles, even when the animals enter a non-breeding phase. Therefore, while the setae may be indicators of sexual maturity, they may not always be indicative of the breeding cycle of the female. Chittleborough (1976) observed that the presence of setae on the endopodites is an unreliable indicator of sexual maturity. Kizhakudan (2007) observed that the appearance and development of the ovigerous setae for the first time coincides with the onset of sexual maturity in both *P. polyphagus* and *T. orientalis (= T. unimaculatus)*. The number and size of the setae were found to reach a maximum at the time of egg bearing.

9 Reproductive Biology of Spiny and Slipper Lobster

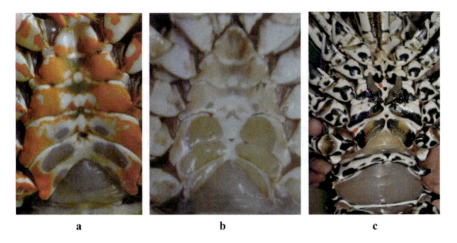

Fig. 9.12 Mating windows on sternal plate in female lobsters. (**a**) *Panulirus homarus homarus* (**b**) *Panulirus polyphagus* (**c**) *Panulirus versicolor*

9.7.6 Mating Windows on Ventral Sternal Plate in Female Lobsters

George (2005) mentions the role of decalcified windows on the female sternum in the spiny lobster genus *Panulirus* as an indicator of sexual maturity (Fig. 9.12). This feature has been highlighted by several researchers (Lindberg 1955; Velázquez 2003; George 2005; Kizhakudan and Patel 2010). As the female enters its breeding phase, a process of decalcification of the ventral sternal plates commences. The plates at the base of the fifth, fourth and third pairs of pereiopods develop soft mating windows which function as the sites of spermatophore deposition during mating. There are four pairs of windows in *P. polyphagus*: two at the base of the fifth ventral sternite and one each at the bases of the fourth and third ventral sternites (Kizhakudan and Patel 2010). The windows are separated by distinct calcified ridges. The process of decalcification begins from the fifth pair of pereiopods and progresses up to the third, indicating that the female lobster has reached functional sexual maturity and is ready for impregnation.

9.7.7 Increased Abdominal Width in Female Lobsters

Greater abdominal width in female lobsters is considered to be an adaptation for increasing reproductive output (Radhakrishnan et al. 2015). Female spiny lobsters have been found to exhibit positive allometric growth in abdomen width (Berry 1971; Minagawa and Higuchi 1997; Kizhakudan and Patel 2010). Radhakrishnan et al. (2015) observed positive allometric growth of second abdomen width in immature females of *P. homarus homarus* and isometric growth in mature females.

This ontogenetic transition from positive allometry to isometry could be the result of a change in energy allocation (Claverie and Smith 2009).

9.7.8 Morphometry: Relationship Between Carapace Length and Somatic Lengths as Indices of Sexual Maturity

Discontinuities (changes in slope) in the linear relationships between body size and certain externally visible features have often been used as indicators of physical maturity in lobsters. Female maturity has been associated with allometric changes in the length of pleopods, the length of the pleopodal setae, telson length or the width of the abdomen relative to carapace length, which all change in relation to preparation for first spawning (Street 1969; Hossain 1978b; Aiken and Waddy 1980; Pollock and Augustyn 1982; Jones 1988; Stewart et al. 1997; Lizárraga-Cubedo et al. 2003; Kulmiye 2004; DeMartini et al. 2005). MacDiarmid and Sainte-Marie (2006) recommend that easily measured appendage length to body size relations should be routinely applied to provide estimates of female SOM (Size at Onset of Maturity) in lobsters, but only after this approach has been validated by undertaking histological studies of gonadal maturation.

George and Morgan (1979) described a method of determining maturation-related changes in pereiopods and other body dimensions by plotting leg size against the CL, for immature and mature lobsters, and reading the size at maturity from the point of upward deflection or the point of intersection of the regression lines of immature and mature lobsters. In estimating the onset of sexual maturity in female *P. polyphagus* from 51 to 55 mm CL onwards, Kizhakudan (2007) found that the regression lines of VSL on CL had different slopes and there was a clear difference in the relationship between the two parameters before 50 mm CL and after 50 mm CL, signifying changes due to the animal's entry into sexual maturity. While females have a higher VSL to CL ratio initially, the males exceed them beyond 50 mm CL, indicating the changes required for the formation of mating windows in the females; the ventral sternite in females becomes broader and increase in length is suppressed. Comparing the regression lines of length of 3rd, fourth and fifth pereiopods against CL on male and female *P. polyphagus,* Kizhakudan (2007) found that while the third and the fifth pairs of pereiopods are relatively longer in females in the juvenile phase, the lengths take an upper deflection in males at about 45.1 mm CL and 44 mm CL, respectively. The fourth pair of pereiopods are almost similar in length in both the sexes initially, but the length in males takes an upward deflection at 36 mm CL. These are important indicators of the onset of sexual maturity in males, as the pereiopods play a major role in the act of copulation and impregnation. Similarly, in *T. unimaculatus*, it was observed that while the third and fourth pereiopods remain longer in females, the fifth pair of pereiopods is relatively longer in females in the juvenile phase, and its length takes an upper deflection in males at about 47 mm CL.

Kulmiye et al. (2006) observed that the length of the second pereiopod in male *P. homarus homarus* increased in relative size at the onset of maturity, confirming the usefulness of this measurement as an indicator of male functional maturity. The exopod length in females was found to increase proportionally, relative to CL, at onset of maturity. Radhakrishnan et al. (2015) estimated the size at onset of maturity in male *P. homarus homarus* from the south-east coast of India using changes in the allometric growth of the third pereiopod. Proportional changes in exopod and leg length have been described as useful indicators of sexual maturity in several other *Panulirus* species (Berry 1970; Grey 1979; George and Morgan 1979; Juinio 1987; Plaut 1993; Gomez et al. 1994; Evans et al. 1995; Hogarth and Baratt Hogarth and Barratt 1996; De Martini et al. DeMartini et al. 2005).

9.8 Pleopod Dimorphism

Pleopod dimorphism is a characteristic of spiny lobsters in which the male pleopod is uniramous, having leaf-like exopodite only, whereas female pleopod is biramous with an exopod and rod-like endopods on all except the first pleopod (Fig. 9.13). Proportional changes in exopod length in relation to maturity have been found for several *Panulirus* species (Berry 1971; Grey 1979; George and Morgan 1979; Juinio 1987; Plaut 1993; Gomez et al. 1994; Evans et al. 1995; Hogarth and Baratt Hogarth and Barratt 1996; De Martini et al. 2005). Kulmiye et al. (2006) observed significant difference in exopod size of immature and mature females in relation to carapace length of *P. homarus homarus* from Kenya. The regression lines crossed at a carapace length of 50.5 mm CL, the estimated functional size at maturity, which was found to correlate well with physiological maturity determined from female gonads (50.5 mm vs. 52 mm CL). The larger exopods in mature females help them to cover the fertilized eggs from abrasion during walking and continuous fanning to provide aeration.

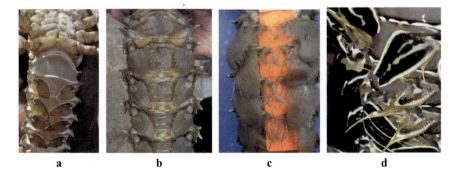

Fig. 9.13 Pleopod dimorphism in lobsters. (**a**) *Panulirus homarus homarus* (M) (**b**) *Linuparus somniosus* (M) (**c**) *Linuparus somniosus* (F) (**d**) *Panulirus versicolor* (F)

9.9 Reproductive Cycle

Temperate and subtropical species of *Panulirus* generally have a distinct annual reproductive cycle with a well-defined breeding season in spring-summer or summer-autumn with eggs hatching within 1–2 months after oviposition, whereas tropical populations of spiny lobsters may breed year-round with peak spawning activity in certain months (George 1965; Berry 1971: Chubb 1994). The peak spawning activity by different subspecies of the same species distributed in different geographical regions probably coincides with favourable environmental conditions and ocean currents for dispersal of phyllosoma larvae and availability of preferred food. In the natal population of *P. homarus rubellus*, Berry (1971) observed breeding frequency to be more than once per year, which is also true for *P. argus* (Sutcliffe, 1953) and *P. japonicus* (Ino, 1950). In the feral population of *P. homarus rubellus*, evidence of repetitive breeding was based on a combination of three factors: the degree of development of ovaries, the state of spermatophoric mass and the presence of eggs on the pleopods. Based on these criteria, Berry (1971) concluded that the intermoult period in larger sizes of *P. homarus rubellus* from South African coast is sufficiently long to accommodate more than two broods (Berry 1971).

On the south-west coast of India (Muttom), *P. homarus homarus* was reported to breed during November–December (George 1965), whereas at Tuticorin on the south-east coast of India, maximum breeding activity was observed to be in March and another during December to February (Jawahar et al. 2014). In Sri Lankan waters, spawning activity was the highest between August and January (Jayakody 1994).

9.10 Breeding Season

Breeding season is influenced by a host of environmental conditions, especially temperature and photoperiod, and annual variations in these conditions may influence the onset of maturation and breeding activity (Chittleborough 1976; Lipicus, 1985; Lipicus 1987). Although *P. homarus* breeds throughout the year, breeding activity reaches a peak in certain months. In *P. homarus rubellus*, peak breeding is during the summer months. The lowest incidence of egg bearing is during autumn and early winter, with a progressive increase in breeding activity in mid-winter followed by summer months (Berry 1971). In Sri Lankan population of *P. homarus homarus*, De Bruin (1962) indicated the breeding period from August to March with a peak in December, whereas in Kanyakumari on the south-west coast of India, the peak breeding months for the same species are from November to December (George 1965; Radhakrishnan and Manisseri 2003; Thangaraja and Radhakrishnan 2012). Two breeding peaks were reported in *P. homarus homarus* from Tuticorin, on the south-east coast of India, with a high percentage of egg-bearing lobsters during March and also from December to February (Jawahar et al. 2014).

Observations on the lobster fishery from coastal reefs and along the north Tamil Nadu coast of the Bay of Bengal indicate the following peak breeding months for spiny lobsters: October to January and March to May for *P. homarus homarus* and

October to November and February to May for *P. versicolor* (Kizhakudan, pers. observ.). Observations on the fishery of sand lobsters *T. unimaculatus* from coastal sandy beds along this coast indicate peak breeding months to be December to January and March to May. Second-year spawners begin spawning in early December, followed by virgin spawners in late December and January. Successive second spawning has been observed mostly during February and March. Peak breeding of the hunchback lobster *P. rugosus* has been observed to be August to October (Kizhakudan, pers. observ.).

9.10.1 Spawning Migration and Aggregation

Long-distance migration of spiny lobsters for spawning purposes has been reported in *P. ornatus*. In August to September each year, adult lobsters in the northern Torres Strait undertake a mass migration across the Gulf of Papua to breeding grounds of Yule Island in Papua New Guinea. Adult females walk across an open sand-mud seabed for 2–3 months at an average speed of 6.1 km/day. The maximum recorded distance was 511 km. No return migration was detected and probably there is a high mortality of lobsters due to combined physiological stress of migrating and breeding (Moore and MacFarlane 1984; MacFarlane and Moore 1986; Bell et al. 1987; Trendall and Prescott 1989; Pitcher et al. 1991; Dennis et al. 1992). Such long-distance breeding migrations have not been reported in any *P. homarus* subspecies, though local movement and spawning aggregations in deeper areas for egg hatching are known (Prescott and Pitcher 1991; Kulmiye 2004). The local movement of the ovigerous females to certain specific locations may be to take advantage of the prevailing currents for dispersion of the long-lived phyllosoma larvae to offshore areas.

Kizhakudan (pers. observ.) noted that although there exists a fishery for *P. polyhagus* and *P. ornatus* in the Bay of Bengal along the north Tamil Nadu coast, the occurrence of berried specimens of these species are very rare in collections from coastal reefs; however, they have been found to occur in considerable numbers in trawl landings, indicating migration of berried lobsters to deeper waters and non-reef areas. Adult mature specimens of *P. ornatus* have not been encountered in gill net catches from the coastal reefs, suggesting that they begin their migration to deeper waters well before mating. On the other hand, mature adults of *P. polyphagus* have been encountered in the reef fishery, with a skewed ratio of increased number of adult males in the population, indicating post-mating migration of this species to deeper waters. The sex ratio of adult populations of *P. homarus homarus* and *P. versicolor* in the reef areas are usually close to 1:1, suggesting that these species do not undertake breeding migrations to a great extent. In the case of *T. unimaculatus*, lowering of temperature during the north-east monsoon and change in current pattern during October to November trigger the spawning process, when a coastward movement of the breeding lobsters has been observed. The migrating population is usually dominated by berried, ready-to-spawn and spent females, with very few active adult males.

9.10.2 Frequency of Egg Bearing or Spawning Periodicity

Spiny lobsters may produce from one to four broods of eggs within the same breeding season (Chubb 1994). While smaller lobster size groups (50–59 mm CL) of *P. homarus rubellus* reproduce twice, larger females (>60 mm CL) may produce up to three or four broods per year (Berry 1971). Vijayakumaran et al. (2005) reported an average of four repetitive spawns per year with even five spawns in 33% of captive specimens of *P. homarus homarus*. This repetitive spawning in captive specimens may be due to the interaction of various factors such as abundant food, food quality, lower energy expenditure in foraging for food, favourable environmental conditions and population density (Chittleborough 1976). Repetitive spawning also will depend upon the viability of sperms in the remnant spermatophoric mass after the first spawning and the state of the ovary. Vijayakumaran et al. (2005) found only 20% of the captive *P. homarus homarus* spawning for the second time which took place within 2–3 days after first spawning. Though year-round breeding activity was observed in both captive and wild lobsters, no synchronization in peak spawning season between wild and captive groups was reported. While the peak breeding season for *P. homarus homarus* along the southern Indian coast is from November to December, no specific period for maximum spawning was noticed in captive population. For example, peak spawning activity in captive *P. homarus homarus* was observed in April and July to August in 2000–2002 and February and August in 2003 (Vijayakumaran et al. 2005).

Repeated spawning within the same season may affect brood size and egg quality in spiny lobsters (Creaser 1950; Ino 1950; Briones-Fourzan and Lozano-Alvarez 1992). However, Berry (1971) did not find reduction in either brood size or egg quality from repetitive broods in *P. homarus rubellus*.

9.10.3 Mating and Spawning Behaviour

Mating behaviour and cycles vary between different species of lobsters. Mating in *P. polyphagus* occurs 1–12 days prior to spawning (egg release) and not immediately after the premate moult (Kizhakudan 2007). Fertilization is external and occurs only at the time of spawning. Mating is seen to be associated with the development of orangish colour in the ventral sinus of the female, indicating the presence of vitellogenin. The time between the premate moult and mating, during which the pre-copulatory courtship takes place, ranged from 20 to 40 days. Males exhibit aggressive behaviour among themselves and separate dens should be provided for active males. Usually, the largest male monopolizes the attention of all the females in one tank, and a single male successfully impregnates up to four females. Reproductive success is maximum between males and females of the same age group wherein the males are larger than their female counterparts. Older males show poor response to young females breeding for the first time. Males of 70–80 mm CL were found to mate with females of 66–75 mm CL, and males of 80–100 mm

CL mated with females of 75–95 mm CL. However, males of 100–120 mm CL did not show any response to reproductively active females of 66–75 mm CL.

In the wild collection of breeders from Chennai as well as from the holding centre in Kanyakumari, Vijayakumaran et al. (2005) observed that the smallest breeder of *P. homarus* had a CL of 52.2 mm (156.8 g total weight), while the biggest one had a CL of 94.4 mm (850 g total weight). A smaller size at maturity was recorded for *P. homarus*. The smallest *P. ornatus* breeder measured 104.4 mm CL and 1049 g in total weight, and the biggest one measured 145.1 mm CL and 2800 g in total weight. The smallest *P. versicolor* breeder had 66.1 mm CL and 310 g total weight, while the biggest one measured 95.1 mm CL and 850 g.

Courtship begins immediately after the female completes the premate moult. The reproductively active female performs stridulation to attract a reproductively active male. Vision and antennal sensitivity also play a major role in detection of a mating partner by the male. The male tracks the female, usually from behind, continuously probing with the antennae. The male chases the female continuously and feeding activity reduces considerably for some days. Courtship behaviour is best observed in the early evening hours. The chasing activity continues during the day, while mating occurs between late night and early morning hours. There is no aggression between the sexes while mating.

Towards the final phase of courting, the male moves towards the female and copulation takes place frontally as the male vertically embraces the female, from head to tail. The duration of copulation is very short and lasts for only 2–3 min. The extensible cuticular membranes of the active male penile process stretch and open outwards. As the male moves over the female, the membranes of the gonopore stick on to the soft window areas on the sternum at the bases of the fifth, fourth and third pereiopods of the females. The male deposits the spermatophore along with its gelatinous matrix over the window areas. Spermatophore extrusion is done continuously, and it is deposited as two layers, the outer layer being the gelatinous matrix which serves to cement the spermatophore to the female plastron and the inner core layer being the convoluted spermatophore chambers. The implantation of the spermatophores of the male lobster on the plastron of the female is called impregnation and the spermatophore-bearing female is called an impregnated female. As soon as impregnation is complete, the lobsters separate immediately. There is no conspicuous post-mating cohabitation or guarding of the females by the males.

In spiny lobsters, the spermatophoric mass, when freshly deposited on the female, is white and soft. It soon darkens to a lichen-green colour and becomes hard. It eventually turns blackish in colour, the pigmentation proceeding inwards, from the margins. In this condition, it is also known as the "tar spot". The tar spot is butterfly-shaped, broader at the base (over the windows of the fifth walking leg) and tapering towards the top (over the windows of the third walking leg). It is quite smooth on the outside. The bigger the male, the bigger is the spermatophore attachment.

The impregnated female now prepares itself for spawning, which takes place 2–10 days after mating. Before egg extrusion, the female scrapes the surface of the tar spot with the chelate part of the fifth leg to release the sperms. The eggs from the oviduct pass over this sperm mass and fertilization takes place in the brood chamber. It was observed that in young females breeding for the first time, the spermatophore is almost completely scraped off and used up for fertilizing the first batch of eggs, in 90% of the cases, and only the margins of the attachment remain for a few weeks on the sterna. However, in older females, a considerable part of the spermatophore attachment is retained after the first egg extrusion. It remains throughout the incubation period and is used to fertilize successive batches of eggs, unless it is tampered with during frequent sampling and handling of the female held in captivity.

The fertilized eggs attach themselves to the ovigerous setae on the pleopods. This process is called oviposition. Three pairs of dark scars mark the remnants of the spermatophoric mass after it is completely scraped off. Spawning in the female is usually complete and the ovaries are very rarely seen to be in a partially spent state. A mature female in the late intermoult stage usually spawns twice, and sometimes thrice, during a single intermoult phase, the gap between a hatching and the next spawning ranging from 2 to 5 days. Rematuration takes place after a span of about 2–3 months, during which the animals undergo at least one moulting. When mating fails to occur or impregnation does not take place properly, the mature females release unfertilized eggs, which are pale pink in colour. These eggs are shed from the pleopods in about 2–5 days after spawning. Whenever mating occurs, the eggs released are fertilized and are bright orange in colour. A single impregnation usually suffices for two and sometimes even three successive spawnings within the same maturation cycle of the female.

The incubation period in *P. polyphagus* ranges from 25 to 32 days and is easily influenced by water temperature (Kizhakudan 2007). Radhakrishnan (1977) reported the incubation period of *P. homarus* as days under captive conditions. During this period, the females maintain their lower abdomen and tail slightly bent inwards and the setose exopods of pleopods and dactyli of the pereiopods are used to maintain constant fanning and grooming of the eggs. The female tends to remain in isolation and shows very little movement. Feed intake is also considerably reduced during the incubation period.

At the time of hatching, the female straightens its abdomen and extends the tip of the abdomen to the water surface, and the phyllosoma that hatch out are fanned away with the pleopods. Hatching takes place only during the early morning hours and is usually completed the same day. A considerable number of eggs are prone to be lost during the incubation period if water quality and tank bottom quality are not maintained well or if the animal is subjected to increased handling stress.

Mating has been seen to coincide with moulting in some species of *Jasus* (Kanciruk 1980). Evidence of premoult females evoking increased activity in males has been reported in *Jasus lalandii* (Silberbauer 1971). Mating between 2 hours and 63 days after moulting has been observed in three species of *Jasus* (Kittaka and MacDiarmid 1994). Kanciruk and Herrnkind (1976), using external fouling as an indicator of time since last moult in *P. argus*, determined that females do mate soon after moulting, whereas Chittleborough (1976) reported that moulting and mating may be separated by 2–97 days in *P. cygnus*. Vijayakumaran et al. (2005) reported that in captive broodstock of *P. homarus*, the two events were separated by 2–47 days, with an average of 19.4 ± 14.0 days. The interval was higher in females which spawned again immediately after hatching and in those which performed a second mating after hatching. The evidence suggests that mating need not occur immediately after moulting for successful spawning. In the bigger females above 69 mm CL which spawned more frequently, mating generally followed moulting, whereas in the smaller ones, two or three moults were recorded between successive mating. Kittaka and MacDiarmid (1994) reported that breeding in spiny lobsters occurs after the female moults, as the moult provides fresh ovigerous setae on the endopods of the pleopods for attachment of eggs.

The average days taken to mate after a moult (19.4), to spawn after mating (2.2), to incubate the eggs (26.0) and to moult after hatching (28.5) add up to 76 days to complete the whole cycle in *P. homarus*. This suggests that *P. homarus* can spawn on an average up to 4 times in a year (Vijayakumaran et al. 2005). In comparison, the cycle is completed in 96 days in *P. argus* (moulting to mating, 13–20 days; mating to spawning, 6–9 days; spawning to larval release, 20–30 days; and larval release to moulting, 38–56 days) (Kanciruk and Herrnkind 1976).

Male aggression towards females is rare as evidenced in only two matings: one with unsuccessful spermatophore deposition and the other where sperm deposition was successful but spawning did not occur; both of these females moulted before further mating, a clear evidence that they were not ready for egg production (Vijayakumaran et al. 2005). Up to four copulations have been reported by a single pair in *Jasus edwardsii*, *P. argus* and *P. japonicus* (Lipcius et al. 1983; Kittaka 1987; Deguchi 1988; Deguchi et al. 1991). Two matings within a period of 30 minutes in *P. homarus* indicate such a possibility in this species also. No fighting between males was found and the reported aggressive male embrace in *P. homarus* leading to loss of limbs as reported by Berry (1970) was not observed. On average, 2.20 ± 1.13 days were taken for spawning after mating with 15% spawning in 1 day and 66% in 2 days (range: 1–7 days) (Vijayakumaran et al. 2005).

In scyllarid lobsters, males are smaller than females and are generally more active. Courtship lasts for a few days prior to mating, when the males actively move around in the tank often flipping over while swimming. During the courtship period, the males are very active at night and are often seen swimming even during the day, chasing the females, which are less active. Mating usually takes place during late night hours.

There is no compulsory premate moult in this species. A female that is ready to mate has well-developed ovigerous setae and exhibits a colour change in the ventral sinus. The male uses its antennules to sense the presence of the female during the courtship chase, and as it catches up with the female, it climbs on to the female and turns it over, holding on to its tail end. Copulation takes place with the animals facing ventrally in reverse position so that the ventral sternal tip of the male is slightly in front of the female's sternum.

Copulation is slightly prolonged in this species and lasts for nearly 5 minutes. As the pair hold on to each other tightly with their legs, the female probes the integumental membrane on the last abdominal segments of the male and nibbles at the

Fig. 9.14 Spermatophore packets of *Petrarctus rugosus* in fibrillar gel matrix

Fig. 9.15 Empty spermatophore packets of *Thenus unimaculatus*

uropod. As a result, reproductively active males are often seen with damaged uropod fans and tail ulcers which invite bacterial invasion.

When copulation is over, the impregnated female curves its abdomen inwards and holds it at a slight elevation from the tank bottom. It is not very active and tends to crouch in some dark corner above the substratum. It carries the impregnation till the early morning hours. The freshly released spermatophores are milky white in colour, soft, delicate and embedded in a gelatinous matrix of mucoid fibrils (Figs. 9.14 and 9.15). The impregnation is seen on the ventral side of the first abdominal segment as two parallel lines, usually extending from the coxa of the fifth pair of pereiopods to the posterior tip of the second abdominal segment. The spermatophore packets are ovoid in shape and microscopic in size. They hold the active spermatozoa within. The gelatinous matrix in which the spermatophores are embedded holds a fibrillar network to which the spermatophores are attached. When the spermatozoa are released for fertilization, the empty spermatophore packets are left attached to the fibrils in the matrix.

Egg laying commences in about 2 h and the eggs released are guided from the sternal region to the abdominal brood chamber where they are attached to the ovigerous setae on the endopod of the pleopods. The abdomen remains completely curved inwards, forming a "U"-shaped cup-like brood chamber laterally sealed by the exopodites and the teeth-like extensions of the abdominal tergites. The endopodal branches and their setae spread like a floor on the ventral side of the chamber when all the eggs and spermatozoa mixed with water have been pushed in. Almost 90% of the eggs get attached to the pleopodal setae, irrespective of whether they are

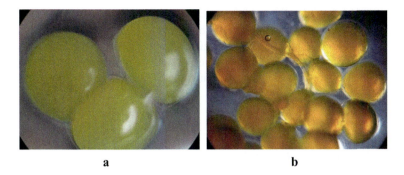

Fig. 9.16 Fresh eggs of *Thenus unimaculatus*. (**a**) unfertilized (**b**) fertilized

Fig. 9.17 Fertilized egg of *Scyllarides tridacnophaga*

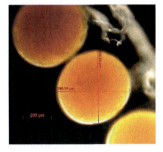

fertilized or not. The fertilized eggs are dark yellow or orange in colour, while the unfertilized eggs turn pale cream or pinkish. The unfertilized eggs are shed off in 3–5 days (Figs. 9.16 and 9.17).

The incubation period in *T. orientalis* (=*T. unimaculatus*) ranges from 32 to 37 days, during which embryonic development takes place inside the eggs (Kizhakudan et al. 2004; Kizhakudan 2005). The abdomen continues to be curved inwards and the eggs are constantly fanned with the exopods of the pleopods and cleaned with the setal brushes on the dactyli of the pereiopods. Locomotor activity and feed intake are very much reduced during the incubation period and the female tends to remain in isolation.

At the time of hatching, the female holds the inwardly curved abdomen at a slightly elevated angle, and the phyllosoma that hatch out are fanned away with the pleopods. Hatching takes place in batches only during the early morning hours and is usually completed in 1–3 days. After hatching, the empty egg capsules are seen attached to the ovigerous setae. The capsules are shed, along with a part of the setae, about 48 hours after hatching. Water quality, tank bottom quality and handling stress, particularly during the incubation period, greatly influence the success rate of hatching. Kizhakudan (pers. observ.) found that larval size and number of stages is related to the egg diameter. The egg diameter ranges from 395 to 470 µm in *P. rugosus*, 510 to 550 µm in *S. tridacnophaga* and 850 to 910 µm in *T. unimaculatus*.

During the breeding period, the intermoult phase is highly extended in the female. After the first brood of eggs has been hatched, the female begins preparing itself internally for the next breeding, within the same intermoult phase. Interestingly, the ovigerous setae that developed before the first breeding continue intact for the next breeding also, in a single intermoult phase. The entire process takes up to 150 days, and in such periods, there is a considerable delay of growth in these animals.

9.11 Breeding in Captivity

The number of spawnings, moults per year and overall fecundity for many palinurid species is the consequence of the interaction between important ecological factors, such as temperature, food availability and quality, substrate suitability, population density and water chemistry. Tropical spiny lobsters breed throughout the year and the number of broods per year is dependent on water quality and the quality and quantity of feed available, as proved in captive breeding of *P. homarus* and *P. ornatus*; however, the peak breeding in captive breeders of *P. homarus* was between July and August when more than 80% of the breeders were found to carry eggs compared to lean months when about 16% spawned (Vijayakumaran et al. 2005).

The number of phyllosomae released by captive breeders of *P. homarus* ranged from 0 to 5,29,180. In a few instances, the whole egg mass was shed with 0% hatching, although a maximum hatching of 99.0% was also recorded. Unlike the wild breeders, an average of 95.0% of eggs were fertilized with the minimum being 76.5% and the maximum 100% (Vijayakumaran et al. 2005).

In experiments conducted by Kizhakudan (2007) in Veraval, Gujarat, in about 250 days of culture, male *P. polyphagus* grew faster than females. All males above 70 mm CL were sexually active, with well-developed penile processes and pigmentation. Females above 66–70 mm CL, with full windows, became active for breeding after a premating moult. Breeding success was observed in 15 numbers of the 24 stocked, out of which 5 were unfertilized (improper spermatophore attachment). After an incubation period of 25–32 days, leading to hatching, seven lobsters bred again, of which 5 were fertilized and 2 were unfertilized. Post breeding, the females underwent a transition moult resulting in reduced ovigerous setae and reduced windows. These animals completed one more moult again, in a span of 150 days from the last spawning, to enter into a second breeding cycle. A total of 20 animals survived, 13 entered the breeding phase and 4 had unfertilized eggs. Of these, 10 rematured again and bred.

Unlike the palinurid lobsters, scyllarid lobsters do not require a premating moult (Kizhakudan, pers. observ). In experiments conducted on *T. unimaculatus*, it was found that mating generally occurred in the night and egg extrusion started a few hours after this. Hatchery-raised seed of *T. unimaculatus* raised in a recirculatory system with in situ sandy substrate matured in a period of 340 days from hatching. While maturation and spawning were achieved without any external interventions, unilateral eyestalk ablation was tried out in late juveniles and sub-adults. This was found to improve growth rate by 10–15%. Following mating and spawning, the larvae produced were viable with a hatching success of 87–90% (Kizhakudan, unpub.). In another set of experiments, sand lobster broodstock were raised over a period of 9 months, from seed stage with 75% survival rate in a high-density rearing system, with sand substrate and no water exchange (CMFRI 2010).

Juveniles of *P. homarus* collected from gillnet fishery from Kovalam, Tamil Nadu, and held in indoor cement tanks attained maturity and spawned viable eggs (Radhakrishnan 1977). Male and female juveniles of *P. ornatus* collected from the coastal waters of Chennai were raised in the field laboratory of CMFRI at Kovalam in recirculatory cement tank systems with reduced transmission of incident light, for about 2 years. Maturation and breeding in captivity was achieved without eye ablation or hormonal administration (CMFRI 2010). Spawning/egg extrusion was first observed in a female having a body weight of 1117 g and 115 mm CL. The berried female was isolated and held in a separated spawning tank. Fertilized eggs were incubated for a period of 25 days before hatching out. Hatching took place in a phased manner, with 90% of the eggs hatching out on the first day and the remaining 10% on the following day. The total count of hatched out larvae was approximately 1,50,000. The female was returned to the broodstock tanks after egg hatching was completed, following which it mated and spawned a second time (Kizhakudan, unpub.).

9.12 Eyestalk Ablation

A lot of research documents eyestalk ablation as one of the most effective methods to accelerate moulting frequency and hence growth in crustaceans reared in captivity (Abramovitz and Abramovitz 1940; Smith 1940; Scudamore 1948; Bauchau 1948). The eyestalk in crustaceans is home to a neuro-secretory complex called the X-organ, which secretes the moult-inhibiting hormone (MIH), which has a major role in controlling the moult process (Passano 1953, 1960). Removal of the eyestalk naturally suppresses the production of MIH, resulting in increased moulting activity. The effect of eyestalk ablation in acceleration of moulting and weight gain in the American lobster, *H. americanus*, has been well documented (Sochasky et al. 1973; Rao et al. 1973; Aiken and Waddy 1975; Mauviot and Castell 1976). On the other hand, delayed moulting in the same species following eyestalk ablation has also been reported (Donahue 1951; Flint 1972). Similarly, Travis (1951, 1954) and Dall (1977) did not obtain positive responses to eyestalk ablation in the spiny lobsters *P. argus* and *P. cygnus*, respectively. Following speculations concerning the role of the eyestalk in the production of MIH, Aiken (1980) suggested that since MIH functioned primarily to regulate seasonal moulting, it was probably not so significant in tropical lobsters and may not accelerate moulting in palinurid lobsters. While Quackenbush and Herrnkind (1981) obtained accelerated gonadal development in *P. argus* following eyestalk ablation, they could not obtain increase in growth or weight. However, Radhakrishnan and Vijayakumaran (1982, 1984) reported remarkable acceleration of moulting frequency and consequent weight gain following eyestalk ablation in *P. homarus*. They found that bilateral eyestalk ablation in *P. homarus* resulted in accelerated moulting and gonadal development in all size groups (from early juveniles to adults), irrespective of the reproductive status and the season, indicating the presence of MIH and gonad-inhibiting hormone (GIH) in the eyestalk.

Unlike in the case of *P. homarus* (Radhakrishnan and Vijayakumaran 1982, 1984) Kizhakudan (2013) observed that, although there was a general trend of improved growth increments in ablated lobsters, the effects of eyestalk ablation in *P. polyphagus* seem to change with sex and size and that juvenile males responded best to eyestalk ablation, in terms of growth. Ninety percent of eyestalk-ablated *P. polyphagus* bred and spawned in a week's time after ablation. Further growth progress and maturation in the ablated lobsters was slightly diminished. Forty percent of the lobsters entered a second breeding cycle with an immediate successive moult. There was no delay and therefore continuous egg production could be induced, compromising the growth. Fecundity and the quality of eggs however diminished.

9.13 Conclusion

Studies on the reproductive biology of commercially important lobsters are ultimately aimed at conservation and management of the resource in their natural habitat and improving their aquaculture potential. Prediction of minimum size limits to be observed for protecting spawning populations from being destroyed by fishing activities and awareness of spawning seasons are direct outcomes of such studies. Any information derived from observations of reproductive biology and behaviour in captivity will be of use in developing husbandry practices for the species and thus improve its candidature for aquaculture. Successful captive propagation has the added advantage of improving the status of natural stocks through sea ranching.

References

Abramowitz, R. K., & Abramowitz, A. A. (1940). Molting and growth after eye-stalk removal in *Uca pugilator*. *Biological Bulletin, 78*, 179–188.

Accioly, I. V., Lima-Filho, P., Santos, T. L., Barbosa, A. C. A., Santos Campos, L. B., Souza, J. V., Araujo, W. C., & Molina, W. F. (2013). Sexual dimorphism in *Litopenaeus vannamei* (Decapoda) identified by geometric morphometrics. *Pan-American Journal of Aquatic Sciences, 8*(4), 276–281.

Aiken, D. E. (1980). Molting and growth. In J. S. Cobb & B. F. Phillips (Eds.), *The biology and management of lobsters* (Vol. 1, pp. 91–147). New York: Academic.

Aiken, D. E., & Waddy, S. L. (1975). Induction and inhibition of molting in *Homarus*. In: J.S. Cobb (Ed.), *Recent advances in lobster aquaculture* (Sea Grant Lobster Aquaculture Workshop, p. 13). Kingston: University of Rhode Island.

Aiken, D. E., & Waddy, S. L. (1980). Reproductive biology. In J. S. Cobb & B. F. Phillips (Eds.), *The biology and Management of Lobsters* (Vol. 1, pp. 215–276). New York: Academic.

Aiken, D. E., & Waddy, S. L. (1982). Cement gland development, ovary maturation and reproduction cycles in the American lobster *Homarus americanus*. *Journal of Crustacean Biology, 2*(3), 315–327.

Al-Marzouqi, A., Al-Nahdi, A., Jayabalan, N., & Groeneveld, J. C. (2007). An assessment of the spiny lobster *Panulirus homarus* fishery in Oman – Another decline in the Western Indian Ocean? Western Indian Ocean. *Journal of Marine Science, 6*(2), 159–174.

Almog-Shtayer, G. 1988. Behavioural-ecological aspects of Mediterranean slipper lobsters in the past and of the slipper lobster *Scyllarides latus* in the present. MA thesis, University of Haifa, Israel, p. 165.

Bauchau, A. G. (1948). Phenomenes de croissance et glande sinusaire chez Ericheir sinensis H. M-Edw. *Annales de la Société royale zoologique de Belgique, 79*, 125.

Bell, R. S., Channels, P. W., & MacFarlane, J. W. (1987). Movements and breeding of the ornate rock lobster, *Panulirus ornatus*, in Torres Strait and on the north-east coast of Queensland. *Australian Journal of Marine & Freshwater Research, 38*, 197–210.

Berry, P. F. (1970). Mating behaviour, oviposition and fertilization in the spiny lobster *Panulirus homarus* (Linnaeus). *South African Oceanography Research Institute of Investigation Report, 24*, 1–16.

Berry, P. F. (1971). The biology of the spiny lobster *Panulirus homarus* (Linnaeus) off the east coast of Southern Africa. *South African Oceanography Research Institute of Investigation Report, 28*, 1–75.

Berry, P. F., & Heydorn, A. E. F. (1970). A comparison of the spermatophoric masses and mechanisms of fertilization in southern African spiny lobsters (Palinuridae). *South African Oceanography Research Institute of Investigation Report, 25*, 1–8.

Bertelsen, R., & Horn, L. (2000). Night-time courting behaviour of *Panulirus argus* in South Florida. *The Lobster Newsletter, 13*(1), 13.

Branford, J. R. (1980). Notes on the scyllarid lobster *Thenus orientalis* (Lund, 1793) off the Tokar Delta (Red Sea). *Crustaceana, 38*(2), 221–224.

Braun, M. (1875). Reproductive activities of decapod crustacean. *The American Naturalist, 81*, 392–398.

Briones-Fourzan, P., & Lozano-Alvarez, E. (1992). Aspects of the reproduction of *Panulirus inflatus* (Bouvier) and *P. gracilis* streets (Decapoda: Palinuridae) from the Pacific coast of Mexico. *Journal of Crustacean Biology, 12*, 41–50.

Burton, T. E., & Davie, P. J. F. (2007). A revision of the shovel-nosed lobsters of the genus *Thenus* (Crustacea: Decapoda: Scyllaridae), with descriptions of three new species. *Zootaxa, 1429*, 1–38.

Butler IV, M. J., Heisig-Mitchell, J. S., MacDiarmid, A. B., & Swanson, R. J. (2011). The effect of male size and spermatophore characteristics on reproduction in the Caribbean spiny lobster, *Panulirus argus*. In: Akira Asakura et al. (Eds.), *New Frontiers in Crustacean biology* (pp. 69–84).

CMFRI. (2010). *Annual report 2009–10* (p. 62). Cochin: Central Marine Fisheries Research Institute.

Campbell, A., & Robinson, D. G. (1983). Reproductive potential of three lobster (*Homarus americanus*) stocks in the Canadian Maritimes. *Canadian Journal of Fisheries and Aquatic Sciences, 40*, 1958–1967.

Cano, H. (1891). *Aspects of reproduction of the female shore crab Carcinus maenas (L) and some related decapods*. PhD thesis, University of Glasgow, UK.

Cau, A., Davini, M. A., & Deiana, A. M. (1988). Gonadal structure and gametogenesis in *Scyllarus arctus* (L.) (Crustacea, Decapoda). *Bulletin of Zoology, 55*, 299–306.

Chang, Y.-J., Sun, C.-L., Chen, Y., Yeh, S.-Z., & Chiang, W.-C. (2007). Reproductive biology of the spiny lobster, *Panulirus penicillatus*, in the southeastern coastal waters off Taiwan. *Marine Biology, 151*, 553–564. https://doi.org/10.1007/s00227-006-0488-9.

Chhapgar, B. F., & Deshmukh, S. K. (1964). Further records of lobsters from Bombay. *Journal of the Bombay Natural History Society, 61*(1), 203–207.

Chittleborough, R. G. (1974). Western rock lobster reared to maturity. *Australian Journal of Marine & Freshwater Research, 25*, 221–227.

Chittleborough, R. G. (1976). Breeding of *Panulirus longipes cygnus* George under natural and controlled conditions. *Australian Journal of Marine & Freshwater Research, 27*, 499–516.

Chubb, C. F. (1991). Measurement of spawning stock levels for the western rock lobster *Panulirus cygnus*. *Revista Investigaciones Marina, 12*, 223–233.

Chubb, C. F. (1994). Reproductive biology: Issues for management. In B. F. Phillips, J. S. Cobb, & J. Kittaka (Eds.), *Spiny lobster management* (pp. 181–212). Oxford: Blackwell Scientific Publications.

Chubb, C. F. (2000). Reproductive biology: Issues for management. In B. F. Phillips & J. Kittaka (Eds.), *Spiny lobsters: Fisheries and culture* (pp. 245–275). Oxford: Blackwell Scientific Publications.

Claverie, T., & Smith, I. P. (2009). Morphological maturity and allometric growth in the squat lobster *Munida rugosa*. *Journal of the Marine Biological Association of the United Kingdom, 89*, 1189–1194.

Comeau, M., & Savoie, F. (2002). Maturity and reproductive cycle of the female American lobster *Homarus americanus* in the Southern Gulf of St. Lawrence, Canada. *Journal of Crustacean Biology, 22*(4), 762–774.

Creaser, E. P. (1950). Repetition of egg laying and number of eggs of the Bermuda spiny lobster. *Proceedings of the Gulf and Caribbean Fisheries Institute, 2*, 30–31.

Cruz, R., & Bertelsen, R. D. (2008). The spiny lobster (*Panulirus argus*) in the wider Caribbean: A review of life cycle dynamics and implications for responsible fisheries management. In *Proceedings of the 61st Gulf and Caribbean Fishery Institute*, November 10–14, 2008, Gosier, Guadeloupe, French West Indies.

Dall, W. (1977). Review of physiology of growth and moulting in rock lobsters. *Circ.- CSIRO Division of Fisheries Oceanography(Aust.), 7*, 75–81.

De Bruin, G. H. P. (1962). Spiny lobsters of Ceylon. *Bulletin Fisheries Research Station Department of Fisheries Ceylon, 14*, 1–28.

de Lima, A. V. P., & Gesteira, T. C. V. (2008). Morphology of the male gonads of the spiny lobster *Panulirus laevicauda* (Latreille, 1817). *Brazilian Archives of Biology and Technology, 51*(4), 501–509.

De Martini, E. E., & Williams, H. A. (2001). Fecundity and egg size of *Scyllarides squammosus* (Decapoda: Scyllaridae) at Maro reef, Northwestern Hawaiian islands. *Journal of Crustacean Biology, 21*, 891–896.

Deguchi, Y. (1988). II. Copulation and spawning. 4. Spiny lobster. In R. Hirano (Ed.), *Seed production of decapod crustaceans* (pp. 64–76). Tokyo: KoscishaKoscikaku. (in Japanese).

Deguchi, Y., Sugita, H., & Kamemoto, F. (1991). Spawning control of Japanese spiny lobster. *Memoirs of the Queensland Museum, 31*, 449.

DeMartini, E. E., McCracken, M. L., Moffitt, R. B., & Wetherall, J. A. (2005). Relative pleopod length as an indicator of size at sexual maturity in the slipper (*Scyllarides squammosus*) and spiny Hawaiian (*Panulirus marginatus*) lobsters. *Fishery Bulletin, 103*, 23–33.

Dennis, D. M., Pitcher, C. R., Skewes, T. D., & Prescott, J. H. (1992). Severe mortality of breeding tropical rock lobsters, *Panulirus ornatus*, near Yule Island, Papua New Guinea. *Journal of Experimental Marine Biology and Ecology, 162*, 143–158.

Donahue, J. K. (1951). What makes a lobster shed? *Maine Department of Sea Shore Fisheries Fish Circular, 7*.

Evans, C. R., Lockwood, A. P. M., Evans, A. J. & Free, E. (1995). Field studies of the reproductive biology of the spiny lobsters *Panulirus argus* (Latreille) and *P. guttatus* (Latreille) at Bermuda. *Journal of Shellfish Research, 14*, 371–381.

Farmer, A. S. D. (1974). Reproduction in *Nephrops norvegicus* (Decapoda: Nephropidae). *Journal of Zoology London, 174*, 161–183.

Fernandez, R. (2002). Neuroendocrine control of vitellogenesis in the spiny lobster *Panulirus homarus* (Linnaeus, 1758). PhD thesis, Central Institute of Fisheries Education, Mumbai, India. p. 189.

Fielder, D. R. (1964). The spiny lobster *Jasus lalandii* (H. Milne-Edwards) in South Australia. II. Reproduction. *Australian Journal of Freshwater Reseach, 15*(1), 133–144.

Flint, R. W. (1972). The effect of eyestalk removal and ecdysterone infusion on molting in *Homarus americanus*. *Journal of the Fisheries Research Board of Canada, 29*, 1229–1233.

Freitas, R., Medina, A., Correa, S., & Castro, M. (2007). Reproductive biology of spiny lobster *Panulirus regius* from the northwestern Cape Verde Islands. *African Journal of Marine Science, 29*(2), 201–208.

Frisch, A. J. (2007). Growth and reproduction of the painted spiny lobster (*Panulirus versicolor*) on the Great Barrier Reef (Australia). *Fisheries Research, 85*(1–2), 61–67.

Gendron, L. (2003). Determination of sexual maturity of female American lobster (*Homarus americanus*) in the Magdalene Islands (Quebec) based on cement gland development. In: M. Comeau (Ed.), Workshop on lobster (*Homarus americanus*) and *H. gammarus*). Reference point for fishery management held in Tracadie-Sheila, New Brunswick, 8–10 September 2003: Abstracts and proceedings. *Canadian Technical Report of Fisheries and Aquatic Sciences, 2506*, 30–32.

George, R. W. (1958). *The biology of the Western Australian crayfish, Panulirus longipes*. PhD thesis, University of Western Australia.

George, R. W. (1963). Report to the government of Aden on the crawfish resources of the eastern Aden protectorate. In *Expanded Programme of Technical Assistance* (Report, 1696, p. 23). Rome: FAO.

George, M. J. (1965). Observations on the biology and fishery of the spiny lobster *Panulirus homarus* (Lin.). *Proceedings of the Symposium on Crustacea MBAI, 4*, 1308–1316.

George, R. W. (2005). Comparative morphology and evolution of the reproductive structures in spiny lobsters, Panulirus. *New Zealand Journal of Marine and Freshwater Research, 39*, 493–501.

George, R. W., & Morgan, G. R. (1979). Linear growth stages in the rock lobster (*Panulirus versicolor*) as a method for determining size at first physical maturity. *Rapports des Proces-Verbaux des Reunions. Conseil International pour l'exploration de la Mer, 175*, 182–185.

Gomez, E. D., Juinio, M. A. R., & Bermas, N. A. (1994). Reproduction of *Panulirus longipes longipes* in Calatagan, Batangas, Philippines. *Crustaceana, 67*, 110–120.

Gopal, C., Gopikrishna, G., Krishna, G., Jahageerdar, S. S., Rye, M. C., Hayes, B. J., Paulpandi, S., Kiran, R. P., Pillai, S. M., Ravichandran, P., Ponniah, A. G., & Kumar, D. (2010). Weight and time of onset of female-superior sexual dimorphism in pond reared *Penaeus monodon*. *Aquaculture, 300*, 237–239.

Gregory, D. R., Labisky, R. F., & Combs, C. L. (1982). Reproductive dynamics of the spiny lobster *Panulirus argus* in South Florida. *Transactions of the American Fisheries Society, 111*, 575–584.

Grey, K. A. (1979). Estimates of the size at sexual maturity of the western rock lobster, Panulirus cygnus, using secondary sexual characteristics. *Australian Journal of Marine & Freshwater Research, 30*, 785–791.

Haddy, J. A., Courtney, A. J., & Roy, D. P. (2005). Aspects of the reproductive biology and growth of Balmain bugs (*Ibacus* spp.) (Scyllaridae). *Journal of Crustacean Biology, 25*, 263–273.

Hardwick, C. W., & Cline, G. B. (1990). Reproductive status, sex ratios and morphometrics of the slipper lobster *Scyllarides nodifer* (Stimpson) (Decapoda: Scyllaridae) in Northeastern Gulf of Mexico. *Northeast Gulf Science, 11*(2), 131–136.

Hearn, A., & Toral-Granda, M. V. (2007). Reproductive biology of the red spiny lobster, *Panulirus penicillatus* and the Galapagos lobster *Scyllarides astori* in the Galapagos Islands. *Crustaceana, 80*(3), 297–312.

Herrick, F. H. (1894). The reproduction of the lobster. *Zoologischer Anzeiger, 17*, 289–292.

Herrnkind, W. F., & Lipcius, R. N. (1989). Habitat use and population biology of Bahamian spiny lobster. *Proceedings of the Gulf and Caribbean Fisheries Institute, 39*, 265–278.

Heydorn, A. E. F. (1965). The rock lobster of the south African west coast *Jasus lalandii* (H. Milne-Edwards). 1. Notes on the reproductive biology and the determination of minimum size limits for commercial catches. *South African Division of Sea Fish Investigation Report, 53*, 1–32.

Heydorn, A. E. F. (1969). The rock lobster of the south African west coast *Jasus lalandii* (H. Milne-Edwards). 2. Population studies, behaviour, reproduction, moulting, growth and migration. *South African Division of Sea Fish Investigation Report, 71*, 1–52.

Hogarth, P. J., & Barratt, L. A. (1996). Size distribution, maturity and fecundity of the spiny lobster *Panulirus penicillatus* (Olivier 1791) in the Red Sea. *Tropical Zoology, 9*, 399–408.

Hossain, M. A. (1975). On the Squat Lobster, *Thenus orientalis* (Lund) off Visakhapatnam (Bay of Bengal). *Current Science, 44*(5), 161.

Hossain, M. A. (1976). Studies on some aspects of biology of the sand lobster *Thenus orientalis* (Lund) (Decapoda: Scyllaridae) from Andhra coast. PhD thesis, Andhra University, p. 257.

Hossain, M. A. (1978a). Few words about the sand lobster *Thenus orientalis* (Lund) (Decapoda: Scyllaridae) from the Andhra coast. *Seafood Export Journal, 10*(1), 43–46.

Hossain, M. A. (1978b). Appearance and development of sexual characters of sand lobster *Thenus orientalis* (Lund) (Decapoda: Scyllaridae) from the Bay of Bengal. *Bangladesh Journal of Zoology, 6*, 31–42.

Hossain, M. A. (1978c). Telson setae and sexual dimorphism of the sand lobster *Thenus orientalis* (Lund). *Current Science, 47*, 644–645.

Hossain, M. A. (1979). On the fecundity of the sand lobster, *Thenus orientalis* from Bay of Bengal. *Bangladesh Journal of Scientific Research, 2*(A), 25–32.

Hussain, M., & Amjad, S. (1980). A study of breeding and fecundity of the spiny lobster *Panulirus polyphagus* Herbst (Decapoda: Palinuridae) occurring along the Pakistan coast. *Pakistan Journal of Agricultural Research, 1*(1), 9–13.

Ino, S. (1950). Observations on spawning cycle of Ise-ebi (*Panulirus japonicus* (v. Siebold)). *Bulletin of the Fishery Society Japan, 15*, 725–727. (in Japanese).

Jawahar, P., Sundaramoorthy, B., & Chidambaram, P. (2014). Studies on breeding biology of *Panulirus homarus* (Linnaeus, 1758) Thoothukudi, southeastern coast of India. *Journal of Experimental Zoology, India, 17*(1), 175–181.

Jayakody, D. S. (1989). Size at onset of sexual maturity and onset of spawning in female *Panulirus homarus* (Crustacea: Decapoda: Palinuridae) in Sri Lanka. *Marine Ecology Progress Series, 57*, 83–87.

Jayakody, D. S. (1994). Reproductive biology of the spiny lobster *Panulirus homarus* on the south coast of Sri Lanka. *Proceedings of theThird Asian Fisheries Forum*, 454–456.

Jeena, N. S., Gopalakrishnan, A., Kizhakudan, J. K., Radhakrishnan, E V., Kumar, R., & Asokan, P. K. (2015). Population genetic structure of the shovel-nosed lobster *Thenus unimaculatus* (Decapoda, Scyllaridae) in Indian waters based on RAPD and mitochondrial gene sequences. *Hydrobiologia*, 1–12. doi:https://doi.org/10.1007/s10750-015-2458-z.

Johnson, W. D., & Al-Abdulsalaam, T. Z. (1991). The scalloped spiny lobster (*Panulirus homarus*) fishery in the Sultanate of Oman. *The Lobster Newsletter, 4*, 1–4.

Jones, C. M. (1988). The biology and behaviour of bay lobsters, *Thenus* spp. (Decapoda: Scyllaridae) in Northern Queensland, Australia. PhD thesis, University of Queensland, Brisbane.

Juinio, M. A. R. (1987). Some aspects of the reproduction of *Panulirus penicillatus* (Decapoda: Palinuridae). *Bulletin of Marine Science, 41*, 241–252.

Kagwade, P. V. (1988a). Reproduction in the spiny lobster *Panulirus polyphagus* (Herbst). *Journal of the Marine Biological Association of India, 30*(1 & 2), 37–46.

Kagwade, P. V. (1988b). Fecundity in the spiny lobster *Panulirus polyphagus* (Herbst). *Journal of the Marine Biological Association of India, 30*(1 & 2), 114–120.

Kagwade, P. V., & Kabli, L. M. (1996). Reproductive biology of the sand lobster *Thenus orientalis* (Lund) from Bombay waters. *Indian Journal Fisheries, 43*(1), 13–25.

Kanciruk, P. (1980). Ecology of juvenile and adult Palinuridae (spiny lobsters). In J. S. Cobb & B. F. Phillips (Eds.), *The biology and management of lobsters, Vol. 2. Ecology and management* (pp. 59–96). Oxford: Blackwell Scientific Publications.

Kanciruk, P., & Herrnkind, W. F. (1976). Autumnal reproduction in *P. argus* at Bimini, Bahamas. *Bulletin of Marine Science, 26,* 417–432.

Kensler, C. B. (1967). The distribution of spiny lobsters in New Zealand waters (Crustacea; Decapoda; Palinuridae). *New Zealand Journal of Marine and Freshwater Research, 1*(4), 404–420.

Kittaka, J. (1987). Ecological survey of rock lobster *Jasus* in the southern hemisphere. Ecology and distribution of *Jasus* along the coast of Australia and New Zealand (Report to the Ministry of Education, Culture and Science, Vol. 232).

Kittaka, J., & MacDiarmid, A. (1994). Breeding. In: B. F. Phillips, J. S. Cobb & J. Kittaka (Eds.), *Spiny lobster management* (pp. 384–400). Oxford: Blackwell Scientific Publications.

Kizhakudan, J. K. (2005). Culture potential of the sand lobster *Thenus orientalis* (Lund). International symposium on improved sustainability of fish production by appropriate technologies for utilization. *Sustainable Fishery, 2005,* 16–18.

Kizhakudan, J. K. (2007). Reproductive biology, ecophysiology and growth in the Mudspiny lobster, *Panulirus polyphagus* (Herbst, 1793) and the sand lobster, *Thenus orientalis* (Lund, 1793). PhD thesis, Bhavnagar University, Bhavnagar, Gujarat. p. 169.

Kizhakudan, J. K. (2013). Effect of eyestalk ablation on moulting and growth in the mud spiny lobster *Panulirus polyphagus* (Herbst, 1793) held in captivity. *Indian Journal of Fisheries, 60*(1), 77–81.

Kizhakudan, J. K. (2014). Reproductive biology of the female shovel-nosed lobster *Thenus unimaculatus* (Burton and Davie, 2007) from north-west coast of India. *Indian Journal of Geo-Marine Sciences, 43*(6), 927–935.

Kizhakudan, J. K., & Patel, S. K. (2010). Size at maturity in the mud spiny lobster Panulirus polyphagus (Herbst, 1793). *Journal of the Marine Biological Association of India, 52*(2), 170–179.

Kizhakudan, J. K., Thirumilu, P., Rajapackiam, S., & Manibal, C. (2004). Captive breeding and seed production of scyllarid lobsters – Opening new vistas in crustacean aquaculture. *Marine Fisheries Information Service, Technical and Extension Series, 181,* 1–4.

Kneipp, I. J. (1974). A preliminary study of reproduction and development in *Thenus orientalis* (Crustacea: Decapoda: Scyllaridae). Honours Thesis, James Cook University, Townsville.

Kooda-Cisco, M. J., & Talbot, P. (1982). A structural analysis of the freshly extruded spermatophore from the lobster, *Homarus americanus*. *Journal of Morphology, 172,* 193–207.

Kulmiye, A. J. (2004). Growth and Reproduction of the Spiny Lobster *Panulirus homarus homarus* (Linnaeus, 1758) in Kenya. PhD thesis, University of Nairobi.

Kulmiye, A. J., Mavuti, K. M., & Groeneveld, J. C. (2006). Size at onset of maturity in spiny lobsters *Panulirus homarus* from Mambrui, Kenya. *African Journal of Marine Science, 28,* 51–55.

Lindberg, R. G. (1955). Growth, population dynamics and field behaviour in the spiny lobster *Panulirus interruptus* (Randall). *University of California Publications in Zoology, 59,* 157–248.

Lipcius, R. N. (1985). Size dependent reproduction and molting in spiny lobsters and other long-lived decapods. In: A. M. Wenner (Ed.), *Crustacean Issues Vol. 3. Factors in adult growth* (pp. 129–148). Rotterdam: Balkema.

Lipcius, R. N., & Herrnkind, W. F. (1987). Control and coordination of reproduction in the spiny lobster *Panulirus argus*. *Marine Biology, 96,* 207–214.

Lipcius, R. N., Edwards, M. L., Herrnkind, W. F., & Waterman, S. A. (1983). *In situ* mating behaviour of the spiny lobster *Panulirus argus*. *Journal of Crustacean Biology, 3,* 217–222.

Lizarraga-Cubedo, H. A., Tuck, I., Bailey, N., Pierce, G. J., & Kinnear, J. A. M. (2003). Comparisons of size at maturity and fecundity of two Scottish populations of the European lobster, *Homarus gammarus*. *Fisheries Research, 65,* 137–152.

Lyons, W. G. (1970). Memoirs of the hourglass cruises: Scyllarid lobsters (Crustacea, Decapoda). *Florida Department of Natural Resources Marine Research Laboratory*, (IV), 1–74.

Lyons, W. G., Barber, D. G., Foster, S. M., Kennedy, F. S., & Milano, G. R. (1981). The spiny lobster of the middle and upper Florida keys: Population structure, seasonal dynamics and reproduction. *Florida Marine Research Publications, 38,* 1–45.

MacDiarmid, A. B. (1987). The ecology of *Jasus edwardsii* (Hutton) (Crustacea: Palinuridae). PhD thesis, University of Auckland, New Zealand.

MacDiarmid, A. B. (1989a). Moulting and reproduction of the spiny lobster *Jasus edwardsii* (Decapoda: Palinuridae) in Northern New Zealand. *Marine Biology, 103,* 303–310.

MacDiarmid, A. B. (1989b). Size at onset of maturity and size dependent reproductive output of female and male spiny lobsters *Jasus edwardsii* (Hutton) (Decapoda, Panuliradae) in Northern New Zealand. *Journal of Experimental Marine Biology and Ecology, 127,* 229–243.

MacDiarmid, A. B., & Butler, M. J., IV. (1999). Sperm economy and limitation in spiny lobsters. *Behavioral Ecology and Sociobiology, 146,* 14–24.

MacDiarmid, A. B., & Sainte-Marie, B. (2006). Reproduction. In: B. F. Phillips (Ed.), *Lobsters: Biology, management, aquaculture and fisheries* (pp. 45–77). Oxford: Blackwell Publishing.

MacDiarmid, A. B., Hickey, B., & Maller, R. A. (1991). Daily movement patterns of the spiny lobster *Jasus edwardsii* (Hutton) on a shallow reef in northern New Zealand. *Journal of Experimental Marine Biology and Ecology, 147,* 185–205.

MacDonald, C. D. (1982). Catch composition and reproduction of the spiny lobster *Panulirus versicolor* at Palau. *Transactions of the American Fisheries Society, 111,* 694–699. https://doi.org/10.1577/1548-8659(1982)111<694:CCAROT>2.0.CO;2

MacFarlane, J. W., & Moore, R. (1986). Reproduction of the ornate rock lobster, *Panulirus ornatus* (Fabricius), in Papua New Guinea. *Australian Journal of Marine & Freshwater Research, 37,* 55–65.

Martins, H. R. (1985). Biological studies of the exploited stock of the Mediterranean locust lobster *Scyllarides latus* (Latreille, 1803) (Decapoda: Scyllaridae) in the Azores. *Journal of Crustacean Biology, 5*(2), 294–305.

Mathews, D. C. (1951). The origin, development and nature of the spermatophoric mass of the spiny lobster, *Panulirus penicillatus* (Oliver). *Pacific Science, 5,* 359–371.

Mauviot, J. C., & Castell, J. D. (1976). Molt-enhancing and growth-enhancing effects of bilateral eyestalk ablation on juvenile and adult American lobsters (*Homarus americanus*). *Journal of the Fisheries Research Board of Canada, 33,* 1922–1929.

McGinnis, F. (1972). *Management investigation of two species of spiny lobster, Panulirus japonicus and Panulirus penicillatus* (p. 51). Honolulu: Report of Hawaii Division of Fish and Game.

Melville-Smith, R., & de Lestang, S. (2006). Spatial and temporal variation in the size at maturity of the western rock lobster *Panulirus cygnus* George. *Marine Biology, 150,* 183–195.

Melville-Smith, R., de Lestang, S., Beale, N.E., Groth, D., & Thompson, A. (2009). *Investigating reproductive biology issues relevant to managing the Western rock lobster Broodstock.* Final FRDC Report – Project 2003/005.Fisheries Research Report No. 193. Department of Fisheries, Western Australia. 122p.

Miller, R. J., & Watson, F. L. (1991). Change in lobster size at maturity among years and locations. *Journal of Shellfish Research, 10*(1), 286–287.

Minagawa, M., & Higuchi, S. (1997). Analysis of size, gonadal maturation and functional maturity in the spiny lobster *Panulirus japonicus* (Decapoda: Palinuridae). *Journal of Crustacean Biology, 17,* 70–80.

Minagawa, M., & Sano, M. (1997). Oogenesis and ovarian development cycle of the spiny lobster *Panulirus japonicus* (Decapoda: Palinuridae). *Marine and Freshwater Research, 48,* 875–887.

Mohan, R. (1997). Size structure and reproductive variation of the spiny lobster *Panulirus homarus* over a relatively small geographic range along the Dhofar coast in the Sultanate of Oman. *Marine and Freshwater Research, 48,* 1085–1091.

Moore, R., & MacFarlane, J. W. (1984). Migration of the ornate rock lobster, *Panulirus ornatus* (Fabricius), in Papua New Guinea. *Australian Journal of Marine & Freshwater Research, 35*(2), 197–212.

Morgan, G. R. (1972). Fecundity in the western rock lobster *Panulirus longipes Cygnus* (George) (Crustacea: Decapoda: Palinuridae). *Australian Journal of Marine and Freshwater Research, 23,* 133–141.

Mota-Alves, M. I., & Tomé, G. S. (1965). On the histological structure of gonads of Panulirus argus (Latr.). *Arquivos Estatísticos de Biologia Marinha – Universidade Federal do Ceará, 5*(1), 15–26.

Mota-Alves, M. I., & Tomé, G. S. (1966). Estudo sobre as gônadas da lagosta *Panulirus laevicauda* (Latr.). *Arquivos Estatísticos de Biologia Marinha- Universidade Federal do Ceará, 6*(1), 1–9.

Musbir, M., Sudirman, Mallawa, A., & Bohari, R. (2018). Egg quantity of wild breeders of spiny lobster (*Panulirus ornatus*) caught from southern coastal waters of Bulukumba, South Sulawesi, Indonesia. *AACL Bioflux, 11*(1), 295–300. http://www.bioflux.com.ro/aacl.

Nakamura, K. (1990). Maturation of the spiny lobster *Panulirus japonicus*. *Memoirs of the Faculty of Fisheries Kagoshima University, 39*, 129–135.

Nelson, K., Hedgecock, D., & Borgeson, W. (1988a). Factors influencing egg extrusion in the American lobster (*Homarus americanus*). *Canadian Journal of Fisheries and Aquatic Sciences, 45*(5), 797–804.

Nelson, K., Hedgecock, D., & Borgeson, W. (1988b). Effects of reproduction upon molting and growth in female American lobsters (*Homarus americanus*). *Canadian Journal of Fisheries and Aquatic Sciences, 45*(5), 805–821.

Oliveira, G., Freire, A. S., & Bertuol, P. R. K. (2008). Reproductive biology of the slipper lobster *Scyllarides deceptor* (Decapoda: Scyllaridae) along the southern Brazilian coast. *Journal of the Marine Biological Association of the United Kingdom, 88*(7), 1433–1440. https://doi.org/10.1017/S0025315408001963.

Passano, L. M. (1953). Neurosecretory control of molting in crabs by the X-organ sinus gland complex. *Physiologia Comparata et Oecologia, 3*, 155–189.

Passano, L. M. (1960). Molting and its control. In T. H. Waterman (Ed.), *The physiology of Crustacea* (Vol. 1, pp. 473–536). New York: Academic.

Paterson, N. F. (1969). Fertilization in the Cape rock lobster, *Jasus lalandii* (H. Milne Edwards). *South African Journal of Science, 65*, 163.

Pérez-González, R., Puga-López, D., & Castro-Longoria, R. (2009). Ovarian development and size at sexual maturity of the Mexican spiny lobster *Panulirus inflatus*. *New Zealand Journal of Marine and Freshwater Research, 43*(1), 163–172. https://doi.org/10.1080/00288330909509990.

Pillai, S. L. (2007). Reproductive biology of the male lobster *Panulirus homarus*. PhD thesis, Department of Zoology, University of Calicut, Kerala, India. p. 119.

Pillai, S. L., Nasser, M., & Sanil, N. K. (2014). Histology and ultrastructure of male reproductive system of the Indian spiny lobster *Panulirus homarus* (Decapoda: Palinuridae). *International Journal of Tropical Biology and Conservation / Revista de Biologia Tropical, 62*(2), 533–541.

Pinheiro, A. P., & Lins-Oliveira, J. E. (2006). Reproductive biology of *Panulirus echinatus* (Crustacea: Palinuridae) from São Pedro and São Paulo Archipelago, Brazil. *Nauplius, 14*(2), 89–97.

Pitcher, C. R., Dennis, D. M., Skewes, T. D., & Prescott, J. H. (1991). Catastrophic mortality of breeding tropical rock lobsters. *Proceedings of the 1990 International Crustacean Conference; Memoirs of the Queensland Museum, 31*, 397.

Plaut, I. (1993). Sexual maturity, reproductive season and fecundity of the spiny lobster *Panulirus penicillatus* from the Gulf of Eilat (Aqaba), Red Sea. *Australian Journal of Marine and Freshwater Research, 44*, 527–535.

Pollock, D. E. (1982). The fishery for and population dynamics of west coast rock lobster related to the environment in the Lambert's Bay and Port Nolloth areas. *Investigational Report on Sea Fisheries Institute of South Africa, 124*, 57.

Pollock, D. E. (1997). Egg production and life-history strategies in some clawed and spiny lobster populations. *Bulletin of Marine Science, 61*, 97–109.

Pollock, D. E., & Augustyn, C. J. (1982). Biology of the rock lobster *Palinurus gilchristi* with notes on the South African fishery. *Fisheries of Bulletin of South Africa, 16*, 57–73.

Pollock, D. E., & Goosen, P. C. (1991). Reproductive dynamics of two *Jasus* species in the South Atlantic region. *South African Journal of Marine Science, 10*, 141–147.

Prescott, J. (1988). Tropical spiny lobsters: An overview of their biology, the fisheries and the economics with particular reference to the double-spined rock lobster *P. penicillatus* (SPC workshop on Pacific inshore fisheries research), New Caledonia, WP 12, 36 pp.

Prescott, J. H., & Pitcher, C. R. (1991). Deep water survey for *Panulirus ornatus* in Papua New Guinea and Australia. *The Lobster Newsletter, 4*(2), 8–9.

Quackenbush, L. S., & Herrnkind, W. F. (1981). Regulation of molt and gonadal development in the spiny lobster, *Panulirus argus* (Crustacea: Palinuridae). Effect of eyestalk ablation. *Comparative Biochemistry and Physiology, 69*, 523–527.

Quackenbush, L. S., & Herrnkind, W. F. (1983). Partial characterization of eyestalk hormones controlling molt and gonad development in the spiny lobster, *Panulirus argus*. *Journal of Crustacean Biology, 3*, 34–44.

Radha, T., & Subramoniam, T. (1985). Origin and nature of spermatophoric mass of the spiny lobster *Panulirus homarus*. *Marine Biology, 86*, 13–19.

Radhakrishnan, E. V. 1977. Breeding of laboratory reared spiny lobster *Panulirus homarus* (Linnaeus) under controlled conditions. Indian Journal of Fisheries, 24 (1&2): 269–270.

Radhakrishnan, E. V. & Manisseri, M. K. (2003) In: M. Mohan Joseph & A. A. Jayaprakash (Eds.), *Status of exploited marine fishery resources of India*. Central Marine Fisheries Research Institute, Kochi – 682014, India.

Radhakrishnan, E. V., & Vijayakumaran, M. (1982). Unprecedented growth induced in spiny lobsters. *The Marine Fisheries Information Service: Technical and Extension Series, 43*, 6–8.

Radhakrishnan, E. V., & Vijayakumaran, M. (1984). Effect of eyestalk ablation in the spiny lobster *Panulirus homarus* (Linnaeus) 3. On gonadal maturity. *Indian Journal of Fisheries, 31*, 209–216.

Radhakrishnan, E. V., Thangaraja, R., & Vijayakumaran, M. (2015). Ontogenetic changes in morphometry of the spiny lobster, *Panulirus homarus homarus* (Linnaeus, 1758) from southern Indian coast. *Journal of the Marine Biological Association of India, 57*(1), 5–13.

Rahman, M. K. (1989). Reproductive biology of female sand lobster *Thenus orientalis* (Lund) (Decapoda: Scyllaridae). PhD thesis, University of Madras.

Rahman, M. K., & Subramoniam, T. (1989). Molting and its control in the female sand lobster *Thenus orientalis* (Lund). *Journal of Experimental Marine Biology and Ecology, 128*(2), 105–115.

Rahman, M. K., Prakash, E. B., & Subramoniam, T. (1987). Studies on the egg development stages of sand lobster *Thenus orientalis* (Lund). In: P. Natarajan, H. Suryanarayana, & P. K. A. Aziz (Eds.), *Advances in aquatic biology and fisheries. Prof N. Balakrishnan Nair Felicitation Volume*. Department of Aquatic Biology and Fisheries, University of Kerala, Trivandrum, pp. 327–335.

Rao, K. R., Fingerman, S. W., & Fingerman, M. (1973). Effects of exogenous ecdysones on the molt cycles of fourth and fifth stage American lobsters, *Homarus americanus*. *Comparative Biochemistry and Physiology Series A, 44*, 1105–1120.

Reeves, A., Choi, J., & Tremblay, J. (2011). Lobster size at maturity estimates in http://www.dfo-mpo.gc.ca/Csas-sccs/publications/resdocs-docrech/2011/2011_079-eng.pdf. Eastern Cape Breton, Nova Scotia. DFO Can.Sci.Advis.Sec.Res.Doc. 2011/079: vi+18 p.

Robertson, D. N., & Butler, M. J., IV. (2003). Growth and size at onset of maturity in the spotted spiny lobster *Panulirus guttatus*. *Journal of Crustacean Biology, 23*, 265–272.

Schade, M. S., & Shivers, R. R. (1980). Structural modulation of the surface of the cytoplasm of oocytes during vitellogenesis in the lobster *Homarus americanus*: An electron microscope-protein tracer study. *Journal of Morphology, 163*, 13–26.

Scudamore, H. H. (1948). Factors influencing molting and the sexual cycles in the crayfish. *The Biological Bulletin, 95*, 229–237.

Sharp, W. C., Hunt, J. H., & Teehan, W. H. (2007). Observations on the ecology of *Scyllarides aequinoctialis*, *Scyllarides nodifer*, and *Parribacus antarcticus* and a description of the Florida scyllarid lobster fishery. In: K. L. Lavalli, & E. Spanier (Eds.), *The biology and fisheries of the Slipper Lobster* (pp. 231–242). New York: CRC Press/Taylor & Francis Group.

Silberbauer, B. E. (1971). The biology of the South African rock lobster *Jasus lalandii* (H. Milne-Edwards). 2. The reproductive organs, mating and fertilisation. *Investigational Report. iv. Sea Fishery of South Afrrica, 93*, 1–49.

Silva, J. R. F., & Landim, C. C. (2006). Macroscopic aspects and scanning electron microscopy of ovaries of the spiny lobsters *Panulirus* (Crustacea: Decapoda). *Brazilian Journal of Morphological Sciences, 23*(3–4), 479–486.

Silva, M. A., Tremblay, M. J., & Pezzack, D. S. (2012). Recent studies of lobster *Homarus americanus* size at onset of sexual maturity in Nova Scotia 2008–2011: *Canso, Tangier, Port Mouton and Lobster Bay -A progress report* (Working Paper Draft, 35p).

Smith, R. I. (1940). Studies on the effect of eyestalk removal upon young crayfish (*Cambarus clarkii* Girard). *Biological Bulletin, 79*, 145–152.

Soares, C. N. C., & Peret, A. C. (1998a). Tamanho médio de primeira maturação da lagosta Panulirus laevicauda (Latreille), no litoral do Estado do Ceará, Brasil. *Arquivos de Ciência do Mar, 31*, 12–27.

Soares, C. N. C., & Peret, A. C. (1998b). Tamanho médio de primeira maturação da lagosta Panulirus argus (Latreille), no litoral do Estado do Ceará, Brasil. *Arquivos de Ciência do Mar, 31*, 5–16.

Soares, C. N. C., Fonteles-Filho, A. A., & Gesteira, T. C. V. (1998). Reproductive dynamics of the Spiny lobster *Panulirus argus* (Latreille, 1804) (Decapoda, Palinuridae), from Northeastern Brazil. *Revista Brasileira de Biologia, 58*, 181–191.

Sochasky, J. B., Aiken, D. E., & McLeese, D. W. (1973). Does eyestalk ablation accelerates molting in the lobster *Homarus americanus*? *Journal of the Fisheries Research Board of Canada, 30*, 1600–1603.

Spanier, E., & Lavalli, K. L. (1998). Natural history of *Scyllarides latus* (Crustacea: Decapoda): A review of the contemporary biological knowledge of the Mediterranean slipper lobster. *Journal of Natural History, 32*, 1769–1786.

Spanier, E., & Lavalli, K. L. (2006). *Scyllarides* spp. In B. F. Phillips (Ed.), *Lobsters: Biology, management, aquaculture and fisheries* (pp. 462–496). Oxford/London: Blackwell Publishing.

Stephens, G. C. (1952). The control of cement gland development in the crayfish, Cambarus. *The Biological Bulletin, 103*, 242–258.

Stewart, J., & Kennelly, S. J. (1997). Fecundity and egg size of the Balmain Bug *Ibacus peronii* leach (Decapoda: Scyllaridae) off the east coast of Australia. *Crustaceana, 70*(2), 191–197.

Stewart, J., Kennelly, S. J., & Hoegh-Guldberg, O. (1997). Size at sexual maturity and the reproductive biology of two species of scyllarid lobster from New South Wales and Victoria, Australia. *Crustaceana, 70*(3), 344–367.

Street, R. J. (1969). The New Zealand crayfish *Jasus edwardsii* (Hutton, 1875). *New Zealand Marine Department Fisheries Technical Report, 30*, 53 pp.

Sutcliffe, W. H. (1953). Further observations on the breeding and migration of the Bermuda spiny lobster, *Panulirus argus*. *Sears Foundation for Journal of Marine Research, 12*, 173–183.

Thangaraja, R. (2011). Ecology, reproductive biology and hormonal control of reproduction in the female spiny lobster *Panulirus homarus* (Linnaeus, 1758). PhD thesis, Mangalore University, India, pp 172.

Thangaraja, R., & Radhakrishnan, E. V. (2012). Fishery and ecology of the spiny lobster *Panulirus homarus* (Linnaeus, 1758) at Kadiyapattanam in the south-west coast of India. *Journal of the Marine Biological Association of India, 54*(2), 69–79.

Thangaraja, R., & Radhakrishnan, E. V. (2017). Reproductive biology and size at onset of sexual maturity of the spiny lobster *Panulirus homarus homarus* (Linnaeus, 1758) in Kadiyapattanam, southwest coast of India. *Journal of the Marine Biological Association of India, 59*(2), 19–28.

Tidu, C., Sarda, R., Pinna, M., Cannas, A., Meloni, M. F., Lecca, E., & Savarino, R. (2004). Morphometric relationships of the European spiny lobster *Palinurus elephas* from northwestern Sardinia. *Fisheries Research, 69*, 371–379.

Travis, D. F. (1951). The control of the sinus gland over certain aspects of calcium metabolism in *Panulirus argus* Latreille. *The Anatomical Record, 111*, 503. (abstr.).

Travis, D. F. (1954). The molting cycle of the spiny lobster, *Panulirus argus* Latreille. I. Molting and growth in laboratory-maintained individuals. *The Biological bulletin (Woods Hole, MA), 107*, 433–450.

Trendall, J. T., & Prescott, J. H. (1989). Severe physiological stress associated with the annual breeding emigration of *Panulirus ornatus* in the Torres Strait. *Marine Ecology Progress Series, 58*, 29–39.

Tuck, I. D., Atkinson, R. J., & Chapman, C. J. (2000). Population biology of the Norway lobster, *Nephrops norvegicus* (L.) in the Firth of Clyde, Scotland. 2. Fecundity and size at onset of sexual maturity. *ICES Journal of Marine Sciences, 57*, 1227–1239.

Tully, O. Roantree, V., & Robinson, M. (2001). Maturity, fecundity and reproductive potential of the European lobster (*Homarus gammarus*) in Ireland. *Journal of the Marine Biological Association of the United Kingdom, 81*, 61–68.

Turner, K., Gardner, C., & Swain, R. (2002). Onset of maturity in male southern rock lobsters *Jasus edwardsii* in Tasmania, Australia. *Invertebrate Reproduction & Development, 42*, 129–135.

Ugarte, R. A. (1994). Temperature and the distribution of mature females (*Homarus americanus* Milne Edwards) off Canso, N.S. PhD thesis dissertation, Dalhousie University, Halifax, Nova Scotia. Canada.

Velazquez, A. V. (2003). Reproductive strategies of the spiny lobster *Panulirus interruptus* related to the marine environmental variability off Central Baja California, Mexico: Management implications. *Fisheries Research, 65*, 123–135.

Vijayakumaran, M., Senthilmurugan, T., Remany, M. C., Mary Leema, T., Dilip Kumar, J., Santhanakumar, T., Venkatesan, R., & Ravindran, M. (2005). Captive breeding of the spiny lobster, *Panulirus homarus*. *New Zealand Journal of Marine and Freshwater Research, 39*, 325–334.

Vijayakumaran, M., Maharajan, A., Rajalakshmi, S., Jayagopal, P., Subramanian, M. S., & Remani, M. C. (2012). Fecundity and viability of eggs in wild breeders of spiny lobsters, *Panulirus homarus* (Linnaeus, 1758), *Panulirus versicolor* (Latrielle, 1804) and *Panulirus ornatus* (Fabricius, 1798). *Journal of the Marine Biological Association of India, 54*(2), 18–22.

Waddy, S., & Aiken, D. E. (1992). Seasonal variation in spawning by preovigerous American lobster *Homarus americanus,* in response to temperature and photoperiod condition. *Canadian Journal of Fisheries and Aquatic Sciences, 49*, 1114–1117.

Waddy, S. L., & Aiken, D. E. (2005). Impact of invalid biological assumptions and misapplication of maturity criteria on size at maturity estimates for American lobster. *Transactions of the American Fisheries Society, 134*(5), 1075–1090.

Watson, F. L., Miller, R. J., & Stewart, S. A. (2013). Spatial and temporal variation in size at maturity for female American lobster in Nova Scotia. *Canadian Journal of Fisheries and Aquatic Sciences, 70*(8), 1240–1251. https://doi.org/10.1139/cjfas-2012-0480.

Yonge, C. M. (1932). On the nature and permeability of chitin. I. The chitin lining the foregut of decapod crustacean, and the function of the tegumental glands. *Proceedings of the Zoological Society of London, Series B, 3*, 298–329.

Yonge, C. M. (1937). The nature and significance of the membrane surrounding the developing eggs of *Homarus vulgaris* and other decapods. *Proceedings of the Zoological Society of London, Series A, 107*, 499–517.

Zakaria, M. Z., & Kassim, A. (1999). Size at maturity stages of lobster *Panulirus ornatus* (Fabricius). *Journal of the Marine Biological Association of India, 41*(1&2), 125–129.

Breeding, Hatchery Production and Mariculture

10

E. V. Radhakrishnan, Joe K. Kizhakudan, Vijayakumaran M, Vijayagopal P, Koya M, and Jeena N. S

Abstract

Marine lobsters are a diverse group of large crustaceans distributed almost throughout the world oceans that support commercial fisheries in many countries with significant economic benefits. While rapid progress and phenomenal success were achieved in shrimp aquaculture throughout the world, progress in commercial-scale aquaculture of lobsters was relatively slow due to various biological and technical problems, especially the hatchery production of seeds. Although a marginal increase in total global landings was evident in recent years, production from world capture fisheries has generally been almost stagnant for many years. Lobster aquaculture research began almost 115 years ago, and post-larval production of several species of spiny and slipper lobsters, both temperate and tropical, has been achieved, though in limited quantities. Several diets for each developing stage of the larva were evaluated, and the physiological and nutritional requirements determined in order to successfully rear the phyllosoma larvae through the prolonged larval phase. Different models of rearing tanks including the shape, volume and flow rate of water in the larviculture systems were tested, as maintenance of the fragile phyllosoma larvae in a pathogen-free environment is the one of the most difficult aspects of larval rearing. Significant advances have been made in captive breeding and mass-scale seed production of the tropical fast-growing spiny lobster, *Panulirus ornatus*, and the slipper lobster, *Thenus* spp. Two generations of *P. ornatus* have been produced in captivity by the Australian scientists, which may pave way for commercialisation of hatchery production and genetic improvement of the species in future. The breakthrough has been achieved due to focused research on two key areas, the nutrition and health management of the larval culture systems.

E. V. Radhakrishnan (✉) · J. K. Kizhakudan · Vijayakumaran M · Vijayagopal P
Koya M · Jeena N. S
ICAR-Central Marine Fisheries Research Institute, Cochin, Kerala, India
e-mail: evrkrishnan@gmail.com

Lobster aquaculture research in India began in 1975 with the establishment of a Field Laboratory by the ICAR-Central Marine Fisheries Research Institute (CMFRI) at Kovalam, south of Chennai, Tamil Nadu, on the southeast coast of India. Later, a few other Government research institutions and universities have also been associated with lobster aquaculture research. Different types of puerulus collectors were developed and tested to study the seasonal pattern of settlement. The spiny lobster *P. homarus homarus* was reared to sexual maturity and successfully induced to breed under controlled conditions. The phyllosoma larvae of the slipper lobster *T. unimaculatus* has been successfully cultured through to the juvenile stage. Studies on food and environmental requirements and growth of three species of spiny lobsters, *P. homarus*, *P. ornatus* and *P. polyphagus*, and the scyllarid lobster, *T. unimaculatus*, were conducted in indoor growout systems, and the economics worked out. The feasibility of capture-based growout of pueruli and juveniles in marine floating cages was tested along the southern and northwestern coastal regions of India, and the technology was transferred to local lobster fishermen. The chapter reviews the historical development of lobster culture across the world with recent advances in hatchery production and farming of lobsters in different aquaculture systems and the opportunities and challenges ahead to make it a commercial reality.

Keywords

Breeding · Larval culture · Nutrition · *Panulirus ornatus* · *Thenus* sp. · Cage culture · Economics

10.1 Introduction

Lobster aquaculture has been receiving a lot of serious attention in recent years as the outlook for increasing production from wild fisheries is not very promising. Approximately 66% of world lobster landing is contributed by nephropids, specifically *Homarus*, *Nephrops* and *Metanephrops*, and 32% by palinurids, with scyllarids contributing just 2% (FAO 2012). Although some new lobster fisheries have been discovered, they may not contribute significantly towards the global catch (Webber and Booth 1995). Existing fisheries, except the American clawed lobster *Homarus americanus*, have almost reached the maximum sustainable yield (MSY), and in most countries, lobster fisheries are either poorly managed or overexploited (Lipcius and Eggleston 2000). Environmental degradation and habitat loss are also emerging as key factors in reducing the productivity of lobster fisheries (Jeffs 2010). Apart from anthropogenic intervention, a number of diseases in wild populations have been threatening the productivity of some fisheries. Long-term climate change and large interannual fluctuations in recruitment processes leading to poor settlement of puerulus are also likely to adversely influence fishery production (Lipcius and Eggleston 2000; Wahle et al. 2009; Caputi et al. 2010).

Kittaka and Booth (2000), Phillips and Liddy (2003), Phillips and Matsuda (2011) and Francis et al. (2014) provide excellent reviews of global research developments and recent advances on spiny lobster aquaculture. Jeffs (2010) discussed the major challenges ahead for developing lobster aquaculture into a commercially viable industry. Research carried out on slipper lobster breeding and aquaculture worldwide has also been reviewed (Mikami 2007; Vijayakumaran and Radhakrishnan 2011).

The primary research issues in development of a hatchery technology for palinurid lobsters have been reviewed by Hall et al. (2013). The amenability to culture conditions, relatively less susceptibility to pathogens, low dietary protein requirement and good food conversion ratio are positive traits for considering spiny lobsters as an ideal species for aquaculture. However, these favourable attributes are offset by their complex and prolonged larval phase, the longest among the crustaceans (Thorson 1950). Larval phase of spiny lobsters compared to homarid lobsters is much prolonged (4–12 months), with the exception of *Palinurus elephas*. Among the tropical palinurids, in which the larvae have been reared to settlement, *Panulirus ornatus*, *P. argus* and *P. homarus homarus* had the shortest larval phase (Goldstein et al. 2008; Smith et al. 2009a, b; Murakami and Sekine 2011). The prolonged larval phase and complexity of the life cycle of spiny lobsters pose the biggest challenge for the lobster scientists to develop a robust and reliable commercial hatchery production technology (Kittaka and Booth 2000; Kittaka 2000; Francis et al. 2014). Unlike the palinurids, larval phase of scyllarids is much shorter and, therefore, slipper lobster is fast emerging as a potential species for aquaculture.

Japan has pioneered research on lobster larval culture for more than 116 years (Hattori and Oishi 1899). Australia, New Zealand, India and the USA also have been focusing research on developing a viable aquaculture system for spiny lobsters over 40 years, especially to culture lobsters through their larval stages. Research on the feasibility of culturing spiny lobsters in captivity was initiated way back in the late 1970s by the Central Marine Fisheries Research Institute, Cochin, India, at their Field Laboratory in Kovalam, Tamil Nadu (Radhakrishnan and Vijayakumaran 1982). Partial success in phyllosoma culture was achieved by many researchers (Hattori and Oishi 1899; Nonaka et al. 1958a, b; Saisho 1962a, b; Ong 1967; Mitchell 1971; Dexter 1972; Inoue 1978; Nishimura 1983; Nishimura and Kawai 1984; Nishimura and Kamiya 1985, Nishimura and Kamiya 1986; Radhakrishnan and Vijayakumaran 1995). However, the complete culture from phyllosoma larva to puerulus of a spiny lobster was first accomplished successfully by Kittaka et al. (1988). The phyllosoma larvae obtained from the broodstock of the South African spiny lobster *Jasus lalandii* transferred from South Africa to Japan were reared to settlement. Following this success, larvae of several spiny lobster species, including tropical species, were reared to settlement (Kittaka and Ikegami 1988; Kittaka et al. 1988; Kittaka and Kimura 1989; Yamakawa et al. 1989; Kittaka et al. 1997, 2005; Calverley 2006; Matsuda et al. 2006; Ritar and Smith 2005; Goldstein et al. 2008; Smith et al. 2009b; Murakami and Sekine 2011). Remarkable progress has been made in Australia in seed production of the tropical species, *Panulirus ornatus*, not only in laboratories but also on a semicommercial scale, and hatchery-produced juveniles sufficient enough to stock a farm are likely to be available in the near future (Rogers et al. 2010; Barnard et al. 2011a, b, c; Anon 2015). The first successful rearing of a slipper lobster (*Scyllarus americanus*) was accomplished by Robertson (1968). Later, phyllosoma larvae of several slipper lobster species were also reared to settlement (Takahashi and Saisho 1978; Marinovic et al. 1994; Mikami and Greenwood 1997; Pessani et al. 1999; Matsuda and Yamakawa 2000; Kizhakudan et al. 2004; Wakabayashi et al. 2012, 2016; Wakabayashi and Phillips 2016).

Problems associated with culturing spiny lobster larvae, especially the prolonged larval phase, prompted aquaculturists to find an alternate method of aquaculture, the capture-based growout of puerulus, juveniles and subadults. Lobster fishery in India is constituted mainly by three species of palinurids, *P. homarus*, *P. polyphagus* and *P. ornatus*, and the slipper lobster *T. unimaculatus*. Experimental growout of spiny lobsters indicated that the prospects of growing lobsters in indoor culture system is promising, provided the infrastructure and energy costs could be reduced (Radhakrishnan 1995, 2015). Development of aquaculture of the commercially important species of spiny lobsters, *P. homarus*, *P. polyphagus* and *P. ornatus*, sourced from wild fisheries in cages, indoor tanks and intertidal pits has been actively pursued since 1975 in India (Radhakrishnan and Vijayakumaran 1982; Silas 1982; Srikrishnadhas et al. 1983; Radhakrishnan and Devarajan 1986; Tholasilingam and Rengarajan 1986; Sarvaiya 1987; Radhakrishnan 1994; Vijayakumaran et al. 2009; Mojjada et al. 2012; Radhakrishnan 2015). This method of culture using wild source of juveniles has been termed as 'fattening' and later as 'growout' or 'on-growing'. Similar aquaculture production practices were also known from the Philippines with an average annual production of about 27 tonnes between 1994 and 2010 (Francis et al. 2014). Seacage culture production by Vietnam was 1900 t in 2006 valued at US$ 90 million (FAO 2016). Jeffs (2003) made an assessment of aquaculture potential of *P. argus* in the Caribbean. In the Caribbean, *Panulirus argus* aquaculture has been reported from Belize and Cuba with an annual production of just 8 tonnes.

Global production from aquaculture reached 1610 metric tonnes in 2010, which represents only 0.5% of total annual wild fishery production of 2,89,567 metric tonnes for all marine lobsters and 2% of the 77,532 metric tonnes of palinurid lobsters (Sibeni and Calderini 2012). According to FAO, 73% of the total world aquaculture production is by Vietnam (Table 10.1). Although there are several advantages of using wild-caught juveniles in aquaculture, caution must be exercised to ensure harvesting juveniles and subadults from wild fisheries will not negatively impact wild fisheries (Hair et al. 2002).

Seafood is a valuable source of nutrients and is of fundamental importance in terms of diversity and healthy diets. Its nutritional quality is related to its high-quality protein, essential elements, antioxidants and vitamins. The intake of these nutrients has decreased in industrialized societies and, therefore, seafood is

Table 10.1 Aquaculture production of lobsters (FAO 2016) in metric tonnes

Country	2000	2001	2002	2003	2004	2005	2006	2007	2008	2009	2010	2011	2012	2013	2014
Philippines	27	17	10	18	19	19	23	64	72	64	89	68	38	13	10
Singapore	6	10	6	11	14	5	8	2	4	4	9	13	9	52	43
Vietnam	600	1000	1127	1200	1120	1200	1900	1400	720	1003	631	742	803	705	693
Indonesia	6	10	10	10	20	60	100	100	292	338	311	225	488	914	202
India								3	3	6					
Total	**639**	**1037**	**1153**	**1239**	**1173**	**1284**	**2031**	**1569**	**1091**	**1415**	**1040**	**1048**	**1338**	**1684**	**948**

recommended as a healthy diet, and the USDA dietary guidelines recommend that 20% of total protein intake come from seafood (about 8–12 oz., i.e. 224–336 g) per week. Danish investigators in the 1970s found that the Eskimos of Greenland had a low mortality rate from coronary heart disease (CHD), despite a high intake of fat (about 40 % of their total caloric intake) from their fish diet (Middaugh 1990). Subsequently many researchers endorsed this view of the relationship between seafood and low incidence of coronary artery disease (CAD) leading to increased consumption of seafood (Dyerberg et al. 1978; Bang and Dyerberg 1985).

Crustaceans are among the most highly priced seafood and are regarded as a valuable source of high-quality proteins and minerals (Elvevoll et al. 2008; Mathew et al. 1999). Among the crustaceans, lobsters are the least available ones caught in a very limited quantity, and only a small quantity is produced in aquaculture systems, owing to the special attributes of its biology such as protracted larval life of spiny lobsters and the requirement of individual compartments to prevent cannibalism, which is highly prevalent among the clawed lobsters. It is considered as a gourmet food of the resourceful upper middle class, who of late are very conscious of the nutritional quality of the food they consume. They seek low-fat, low-cholesterol food with high n3 and n6 PUFA and HUFA in the diet, and lobsters fit into this category.

Information on nutritional quality of shellfish, lobsters in particular, is generally scattered in the literature and is often limited to general proximate composition of finished products (Venugopal and Gopakumar 2017). A large volume of lobsters are currently marketed as live, and the key market qualities influencing the price include shell colouration, size and body shape of the lobster with least priority given to the variability in meat quality. The most favourable attributes seem to satisfy the consumer are lobsters of high tail meat yield, with edible tissues rich in protein, omega-3 (n3) lipids and good taste and texture of the meat (Simon et al. 2016). Species, sex, habitats, harvesting season, feed, moult stage and the reproductive condition during capture can influence the tail meat quality and weight of lobster. While tail meat is a rich source of protein and glycogen, hepatopancreas is the major storehouse of lipids. Several analytical techniques are used for biochemical analysis of shellfishes. Rapid, noninvasive monitoring of nutritional condition of meat has considerable application in the live market and spiny lobster aquaculture. Simon et al. (2016) successfully used near-infrared spectroscopy (NIRS) tool to assess the nutritional quality of spiny lobster. Providing safe seafood to the consumer is a great responsibility of the fishing and aquaculture industry.

This chapter reviews research information gathered over the past years on lobster breeding, larval culture and the recent advances and success achieved in seed production technology. It also summarizes our current understanding on various fields leading to aquaculture of lobsters, including the culture systems, on-growing attempts in the laboratory and in the open sea, food and nutrition, economics, constraints and the future challenges for commercial lobster culture in the world.

10.2 Breeding and Hatchery Production of Spiny Lobsters

10.2.1 Broodstock Development and Management

Since culture of spiny lobsters from hatchery-produced seeds is yet to be commercialized, the broodstock source for larval production is only from the wild fisheries. Though there are restrictions for capture of berried lobsters in most of the lobster-fishing countries, permission to capture them for experimental purposes can be obtained from fishery authorities. Berried lobsters from trap, pot or gillnet fisheries are mostly taken rather than from commercial trawlers. Broodstock lobsters in their first year of maturity are generally chosen, as there are no advantages in selecting larger sizes, which need more resources to maintain in captivity (Jones 2009b). For the breeding programme, breeders may be procured from either of the three sources: (1) egg-bearing lobsters captured from the wild; (2) females with newly deposited spermatophoric mass, which may oviposit within 2–7 days; or (3) juvenile lobsters held in the laboratory and reared to maturity.

10.2.1.1 Breeders from the Wild

Tropical species of spiny lobsters such as *P. homarus*, *P. inflatus*, *P. gracilis*, *P. versicolor*, *P. polyphagus* and *P. penicillatus* breed more or less continuously throughout the year (Macdonald 1982; Juinio 1987; Briones-Fourzan and Lozano-Alvarez 1992; Kagwade 1988). *P. homarus homarus* along the southern India coast attains first sexual maturity at a carapace length (CL) of 55 mm (Radhakrishnan 2005), and breeders used for breeding experiments were mostly above 60-mm CL. *P. ornatus* females from northeast Australia mature at around 90-mm CL or 800 g, but lobsters above 1 kg are used for breeding (MacFarlane and Moore 1986; Sachlikidis et al. 2005).

The lobster breeding programme at the Field Laboratory of the Central Marine Fisheries Research Institute at Kovalam near Chennai, India, began during 1976. Initially, breeding experiments were conducted with berried *Panulirus ornatus* procured from fishermen (Anon 1978). The laboratory conducted lobster breeding experiments mostly with *P. homarus*, and owing to the proximity of the laboratory to the artisanal lobster-fishing village, egg-bearing lobsters were brought to the laboratory without any stress. In 1977, juveniles of *P. homarus* collected from the wild and maintained in indoor tanks attained maturity and spawned (Radhakrishnan 1977; Anon 1978). Radhakrishnan (2005) suggested locating the lobster hatchery nearer to the landing centre to avoid long-distance transportation and bacterial infection of the breeders.

Radhakrishnan et al. (2009) cautioned that breeders from holding centres are likely to be of poor quality due to undesirable water quality in the holding tanks and probable pathogenic infections. Such breeders shed the eggs within 2–3 days after bringing to the laboratory. Berried lobsters kept outside water for a prolonged period also may shed the eggs due to stress.

Although *P. homarus homarus* breeds throughout the year, peak breeding period is from December to March along the Chennai coast (Radhakrishnan et al. 2009).

The newly deposited egg mass may not have any fouling; however, berried lobsters with advanced-stage eggs may have external fouling organisms such as filamentous bacteria, fungi or ciliates (see the chapter on pathogens and health management of lobsters). Kizhakudan (2016) suggested a 10 ppm of formalin bath for the breeders, before releasing into the broodstock holding tanks (BHT) to prevent entry of any disease-causing pathogens into the hatchery system, whereas exposure to 50 ppm formalin for 30 min has also been recommended (Radhakrishnan et al. 2009). The incubation period depends upon the developmental stage of egg during procurement and the water temperature. The colour of newly deposited eggs turns from orange to brick red and then to brown before hatching.

10.2.1.2 Oviposition, Fertilization and Egg Development

Another source of breeders is the lobsters procured from the wild with a fresh spermatophoric mass. *P. homarus* oviposit fertilized eggs within 2–7 days after bringing to the laboratory (Radhakrishnan, personal observation). Oviposition normally takes place without any external stimulation. In bilaterally eyestalk-ablated *P. homarus*, failure to oviposit eggs results in resorption of the ovary, turning the haemolymph in pink colour (Radhakrishnan and Vijayakumaran 1984b).

10.2.1.3 Maturation, Mating and Spawning

The third source of breeders is the captive broodstock developed in the laboratory from juveniles. Many species of palinurids have been reared from juveniles or subadults to maturity under captive conditions. Laboratory-held pueruli of the western rock lobster *Panulirus cygnus* have been reared to maturity (Chittleborough 1974a). Berry (1970) studied the mating behaviour and breeding of the spiny lobster *Panulirus homarus* in captivity. In all the 17 matings observed, the exoskeleton of both male and female lobsters was in a hard condition. Radhakrishnan (2005) report year-round breeding activity in captive *P. homarus* maintained either in the flow-through or in the recirculation systems under low light intensity (500 Lux) and fed on green mussel *Perna viridis*. In *P. ornatus*, year-round matings were obtained with manipulation of water temperature and photoperiod (Sachilikidis et al. 2005).

10.2.1.3.1 Management of Captive Broodstock

Radhakrishnan (1977) reared juvenile lobsters of *Panulirus homarus* procured from fishermen to sexual maturity in indoor tanks and conducted breeding experiments under laboratory conditions (Fig. 10.1). The broodstock was raised from juveniles collected from wild (35mm CL), maintained in 10 t circular cement tanks and fed daily with mussels (*Perna viridis*), clams (*Katelysia* sp. and *Meretrix casta*) and fishes (*Thryssocles* sp.). Two females (carapace length 57.8 mm and 57.9 mm) mated and spawned in captivity releasing viable eggs for the first time. Regular spawning has been obtained from the captive broodstock throughout the year without any specific environmental manipulation, though the ambient temperature of the broodstock tank water ranged from 26.1 to 29.8 °C.

Fig. 10.1 The spiny lobster *Panulirus homarus homarus* reared, matured and spawned in the laboratory. (Photo by: E. V. Radhakrishnan, CMFRI)

The lobster released healthy phyllosoma larvae after an estimated incubation period of 15 days. Though the exact incubation period from oviposition to hatching could not be estimated, it is likely to be more. Chittleborough (1974a) reared pueruli of the western rock lobster *P. cygnus* collected from the wild to maturity under controlled conditions. Vijayakumaran et al. (2005) report 20–30 days as incubation period for *P. homarus* kept at temperatures of 25.6–31.5°C. Berry (1971) observed an incubation period of 59 to 29 days for captive *P. homarus rubellus* maintained at temperatures of 20.2 to 25.9°C, showing an inverse relationship between temperature and incubation period. The incubation period of *P. ornatus* maintained at water temperatures of 24.9 to 31.2°C ranges from 23 to 27 days (Murugan et al. 2005).

10.2.1.3.2 Brood Sizes and Spawning

Vijayakumaran et al. (2005) conducted breeding experiments with 14 numbers of *P. homarus* (carapace length 52.7–90.0 mm) collected from the wild during 2000–2004 and observed breeding activity throughout the year with peaks in February, May and August. Repetitive spawning was obtained with an average of four spawnings in a year and a maximum of seven spawnings by an individual. The number of broods produced may be influenced by various environmental factors. While only 12% of wild breeders spawned twice per year, 77% of captive breeders of *P. cygnus* provided with surplus food spawned twice annually in the laboratory. An average of six spawnings per year was obtained in *P. cygnus* at elevated temperatures and on feeding abundant food (Chittleborough 1976b). However, considerable reduction in brood size with repetitive spawning within the same season has been reported for several species of temperate and tropical species including *P. homarus* (Creaser 1950; Briones-Fourzán and Lozano-Alvarez 1992; MacFarlane and Moore 1986; Juinio 1987; Vijayakumaran et al. 2005). On an average, *P. homarus* took 19.4 days to mate after ecdysis, 2.2 days to oviposit after mating, 26 days for egg incubation and 28.5 days to moult again after hatching out the eggs, completing one breeding cycle in 76 days (Vijayakumaran et al. 2005). In captive *P. ornatus*, the interval between moulting and mating was 63 days for the first and 25 days for the second spawning. However, moulting did not precede the third mating (Murugan et al. 2005). Berry (1971) observed an interval of 22 to 33 days between ecdysis and mating and 3 to 44 days between mating and oviposition (mostly between 3 and 9 days) in laboratory-held *P. homarus*. Occasionally, the female may mate again for a second time, after hatching out the first batch of eggs (Vijayakumaran et al. 2005). Repetitive spawning of viable eggs from a single mating has also been observed in *P. homarus*. In such cases, the spawning took place within 2–3 days after hatching out the first batch of eggs.

10.2.1.3.3 Fecundity

The number of eggs in a single brood in wild population of spiny lobsters ranged from 1,20,544 to 4,49,585 in *P. homarus*, 5.18,181 to 1,979,522 in *P. ornatus* and 1,70,212 to 7,33,752 in *P. versicolor* (Vijayakumaran et al. 2012). The phyllosoma released as percentage of fecundity was 85.7% in *P. homarus*, 49.7% in *P. ornatus* and 74% in *P. versicolor*. The hatch percentage depends upon various extrinsic and intrinsic factors. The egg size, percentage of yolk in the ova and the level of fertilization are certain intrinsic factors that may influence hatch percentage. The extrinsic factors are the stress due to long-term exposure of eggs outside water, infestation of egg mass by fungi, nemertean egg predators, other parasites or bacterial infection and other environmental stressors. In a captive *P. homarus* lobster (mean carapace length 70.2 ± 18.3 mm and weight 398.7 ± 132 g), the average number of eggs per spawning was an estimated 2, 59,226 ± 1, 08,637, with a hatch percentage of 53.90 ± 31.49. Though the average number of eggs produced by a wild breeder of almost the same size (72.96 ± 13.4-mm CL) lobster was marginally higher

Table 10.2 Number of eggs and percentage hatch of berried lobsters collected from the wild (*Panulirus homarus, Panulirus versicolor* and *Panulirus ornatus*) and those spawned (*Panulirus homarus*) in captivity (Vijayakumaran et al. 2012)

Species	Mean CL (mm)	Mean weight (g)	No. of eggs ± SD	% hatch
Panulirus homarus (wild)	72.96 ± 13.4	437.4 ± 201.6	278919 ± 84586	23.38 ± 20.93
Panulirus versicolor (wild)	75.2 ± 5.5	523.8 ± 162.0	406006 ± 174121	31.9 ± 22.98
Panulirus ornatus (wild)	120.8 ± 13.22	1649 ± 511.2	1132894 ± 516819	34.2 ± 26.07
Panulirus homarus (captive)	70.20 ± 18.32	398.70 ± 132.04	259226.6 ± 108637	53.90 ± 31.49

(2, 78,919 ± 84,586), the hatch percentage was much lower (23.38 ± 20.9). The hatching percentage for wild breeders of *P. versicolor* and *P. ornatus* was 31.9 ± 22.98 and 34.2 ± 26.07, respectively (Table 10.2) (Vijayakumaran et al. 2012). The percentage hatching in lobsters procured from secondary holding centre is always lower than that procured directly from the landing centre, as these breeders are exposed to outside water for a prolonged period.

10.3 Hatchery Production

The closed cycle production of palinurid lobsters on a commercial scale is a highly challenging task facing modern-day aquaculture (Francis et al. 2014). The prolonged larval phase (4–12 months) and scanty information on the natural food of the larvae exacerbate the problems, posing major hurdles to their successful aquaculture. The complex morphology, the diverse food requirements and feeding practices of the phyllosoma larvae were the key factors that delayed the successful completion of the larval phase.

Japanese researchers were the first to attempt rearing of lobster larvae under controlled conditions. Culture of phyllosoma larvae has been carried out in Japan since the end of the nineteenth century (Hattori and Oishi 1899). Experimental rearing was continued through the twentieth century by many researchers across the globe with majority studies by Japanese workers (Oshima 1936; Nonaka et al. 1958b; Saisho 1962b; Inoue and Nonaka 1963; Inoue 1978, 1981; Dexter 1972). Several temperate and tropical species such as *Jasus lalandii*, *Jasus edwardsii*, *Sagmariasus verreauxi*, *Palinurus elephas*, *Panulirus japonicus*, *P. argus*, *P. longipes bispinosus*, *P. penicillatus*, *P. homarus homarus* and *P. ornatus* have been bred in captivity, and pueruli produced, though with low survival (Kittaka and Ikegami 1988; Kittaka and Kimura 1989; Kittaka et al. 1988, 1997, 2001, 2005; Matsuda and Yamakawa 2000; Calverley 2006; Matsuda et al. 2005, 2006; Goldstein et al. 2008; Smith et al. 2009a, b; Murakami and Sekine 2011). Among the six species of shallow-water spiny lobsters distributed along the Indian coast, the scalloped spiny lobster *Panulirus homarus homarus*, the ornate lobster *P. ornatus* and the mud spiny lobster *P. polyphagus* stand out as candidate species for aquaculture because of the high market demand.

Panulirus ornatus has a limited distribution and biomass along the Indian coast, and therefore, capture-based lobster growout has to depend upon the two species, *P. homarus homarus* and *P. polyphagus*. However, *P. ornatus* being the most valuable of all the tropical species with a relatively shorter larval phase of 'only' five months (Barnard et al. 2011c), hatchery production of seed lobsters is probably the only solution for successful commercial farming of this species in India.

In India, descriptions of phyllosoma larval stage of *P. homarus* and *P. ornatus* were initially made from plankton collections (Prasad and Tampi 1959). The first published record of successful larval culture was of phyllosoma of *P. homarus*, which moulted to Stage VI when fed with *Artemia* nauplii (Radhakrishnan and Vijayakumaran 1995). In a later experiment, the larvae reared in a microalgal culture system attained Stage VIII in 42 days when fed with enriched *Artemia* juveniles and plankton (Radhakrishnan et al. 2009). Larval culture experiments with other species such as *P. ornatus*, *P. versicolor* and *P. polyphagus* were also partially successful (Radhakrishnan and Vijayakumaran 1995; Vijayakumaran et al. 2014).

10.3.1 Larval Biology

The planktonic phyllosoma larva of the spiny lobster has a delicate, flattened, transparent body with several long appendages. The Stage I phyllosoma has an unstalked eye and three pereiopods with only pereiopods one and two having exopods. The temperate spiny lobster, *Jasus lalandii*, has a 'prephyllosoma' stage known as the naupliosoma (Gilchrist 1916; Von Bonde 1936). Von Bonde (1936) working on *Jasus lalandii* showed that a prenaupliosoma stage existed, which lasted for 8 h, before moulting into the naupliosoma. There are few reports of palinurid and scyllarid lobster larvae hatching out as prenaupliosoma and naupliosoma (Deshmukh 1968; Robertson 1979; Mohammed et al. 1971). Phillips and Melville-Smith (2006a) are of the opinion that these intermediate stages in the genus *Panulirus* are likely to be due to physical or chemical factors, such as damages to egg membranes or adverse salinities. In *P. homarus*, prolonged exposure of eggs outside water has led to either premature egg shedding or hatching the larvae as naupliosoma (Radhakrishnan, personal observation). Robertson (1979) was of the view that a prenaupliosoma stage exists in scyllarid lobsters, although naupliosoma was considered as the first valid stage. Radhakrishnan (1977) and Radhakrishnan and Vijayakumaran (1995) reported that the larvae of *P. homarus* hatch out as first-stage phyllosoma. Vijayakumaran et al. (2012) observed eggs of one of the breeders of the spiny lobster *P. homarus* collected from the wild hatching out as naupliosoma with wrinkled pereiopods and with little swimming capability. The naupliosoma after 10–12 h was found to stretch out their pereiopods assuming the phyllosoma form. However, none of the phyllosoma larvae survived for more than 24 h. Healthy phyllosoma larvae pass through several morphologically distinct stages before finally metamorphosing to the post-larva (puerulus). The larva after each moult is considered to be an 'instar', and each 'stage' may have one or two instars depending upon their scale of development. The phyllosoma larvae drift to offshore regions during their long pelagic life; after several moults, they metamorphose finally into the puerulus and return to inshore areas for settlement.

Laboratory culture of phyllosoma larvae was first attempted by the Japanese as early as 1899 (Hattori and Oishi 1899). Early experiments to rear phyllosoma larvae through the entire cycle were not successful due to difficulties in providing the required environmental conditions and suitable food to later stages (Saisho 1962b; Ong 1967; Dexter 1972). Inoue (1978) was the first to succeed and culture the phyllosoma of *P. japonicus* to the final gilled-stage larva of 29.6-mm body length, by feeding early larvae with *Artemia* nauplii. Advanced stages of larvae require larger food items such as chopped mussel, arrow worms or fish larvae (Mitchell 1971; Dexter 1972).

10.4 Larval Culture Attempts

The first mass culture trials of phyllosoma larvae with local (*P. japonicus*) and exotic species of spiny lobsters (*J. lalandii, J. edwardsii, S. verreauxii, P. elephas*) were commenced in 1980 at Sanriku, Northeast Japan, adopting the 'ecosystem' larval culture method (Hudinaga and Kittaka 1967) developed for penaeid shrimp larvae. These species were successfully cultured from egg to puerulus (Kittaka et al. 1988; Kittaka and Ikegami 1988; Kittaka and Kimura 1989; Yamakawa et al. 1989; Kittaka et al. 1997). The larval duration varied from 132 to 417 days depending upon the species (Table 10.3).

The number of instars for *J. verreauxi* (*Sagmariasus verreauxi*), *J. lalandii* and *J. edwardsii* were 17 during the phyllosoma development. Survival from the first instar to puerulus for *S. verreauxii* was on an average of 11.1% from two rearing trials. The total number of instars for *P. japonicus* was an estimated 25 in a total of 11 stages (Kittaka 1994b). Sekine et al. (2000) produced 325 pueruli of *P. japonicus* from larval culture experiments conducted during 1989 to 1997. The duration of phyllosoma stage ranged from 231 to 417 days (mean, 319.4 days). The number of instars varied from 20 to 31. A total of 136 pueruli moulted to juvenile stage, and first moulting to juvenile stage took 9–26 days. Survival up to 80% from first instar to puerulus was achieved at the Minamiizu Station, Fisheries Research Agency,

Table 10.3 Successful culture of palinurid lobster phyllosoma larvae

Species	Larval duration (days)	Number of instars	Authors
Jasus lalandii	306	15	Kittaka et al. (1988)
Jasus edwardsii	303–319	15–23	Kittaka et al. (1997), Ritar and Smith (2005), Kittaka et al. (2005)
Sagmariasus verreauxi	189–359	16–17	Kittaka et al. (1997), Moss et al. (2000), Ritar et al. (2006)
Palinurus elephas	132	6–9	Kittaka and Ikegami (1988), Kittaka et al. (2001)
Panulirus japonicus	231–417	20–31	Kittaka and Kimura (1989), Yamakawa et al. (1989), Matsuda and Yamakawa (2000), Sekine et al. (2000), Matsuda and Takenouchi (2005)
Panulirus penicillatus	256–294	20	Matsuda et al. (2006)
Panulirus longipes bispinosus	281–294	17	Matsuda and Yamakawa (2000)
Panulirus homarus homarus	166–235 103–192		Murakami and Sekine (2011)
Panulirus ornatus	120–150	23–24	Calverley (2006) and Smith et al. (2009a, b)
Panulirus argus	140–198	18–21	Goldstein et al. (2008)

Japan (Murakami and Sekine 2011). Phyllosoma of *P. elephas* metamorphosed into pueruli after nine moults in 132–148 days (average 138 days). By optimizing the environmental, health and nutritional conditions, the larval duration was shortened from 132 days (Kittaka and Ikegami 1988; Kittaka 1997a) to 65 days (Kittaka et al. 2001) showing that phyllosoma development duration can be shortened by environmental and feed manipulation.

Several tropical species such as *P. ornatus*, *P. homarus homarus*, *P. argus*, *P. penicillatus* and *P. longipes* were also cultured from egg to puerulus in other parts of the world (Calverley 2006; Matsuda and Yamakawa 2000; Matsuda et al. 2006; Smith et al. 2009a, b; Goldstein et al. 2008; Murakami and Sekine 2011) (Table 10.3). Murakami and Sekine (2011) succeeded in completing the larval cycle of *Panulirus homarus homarus* in 1998 and by applying advance larval culture technology were able to shorten the larval phase from an average of 200.7 days to a mean 126.5 days with higher survival (30%).

10.4.1 Larval Culture Systems

The important considerations for larval culture are minimal contact between animals while ensuring accessibility to feeds, optimum hygiene and high-quality water (Barnard et al. 2011c). Larvae were cultured in static systems, flow-through system, recirculation systems and water inoculated with microalgae with varying results. Culture tanks differed in various dimensions and shapes, from flat-bottom (Radhakrishnan and Vijayakumaran 1995), to cylindro-conical (Radhakrishnan et al. 2009), to concave (Kittaka 1994b), to elliptical-shaped tanks (Matsuda and Takenouchi 2005). In New Zealand, phyllosoma larvae of *J. edwardsii* and *S. verreauxi* were cultured in an upwelling system (Illingworth et al. 1997). The need for keeping the larvae in suspension was addressed in all the systems by providing water jets from the bottom of the tanks to create upwelling (Kittaka 1994b) or circular motion (Ritar 2001).

10.4.2 Environmental Requirements

Optimal environmental quality is of paramount importance for culture of phyllosoma larvae. The environmental requirements for rearing the phyllosoma larvae were recently reviewed by Phillips and Matsuda (2011).

10.4.2.1 Temperature

The phyllosoma larvae are very sensitive to temperature of culture water. *Panulirus homarus homarus* larvae maintained in culture water with temperatures varying between 26.0 and 30.0 °C attained Stage VIII in 42 days (Radhakrishnan et al. 2009). The ambient temperature in larval culture tanks for rearing early phyllosoma larvae of *P. homarus*, *P. versicolor* and *P. ornatus* ranged from 25 to 31°C (Vijayakumaran et al. 2014). The intermoult period was increased, and the moult increment decreased with decreasing temperature in *P. japonicus* phyllosoma larvae (Matsuda and Yamakawa 1997).

10.4.2.2 Salinity

Cultured phyllosoma larvae appear to show tolerance to a wide range of salinity but may be susceptible to sudden and abrupt fluctuations (Phillips and Matsuda 2011). The salinity tolerance by early-, mid- and late-stage larvae may be different (Matsuda et al. 2005). Larvae of *P. homarus* exposed to salinities of 31 psu moulted to Stage V in 55–60 days with 90% survival, whereas only 60% of larvae kept at 25 psu survived after 5 days (Jha et al. 2010).

10.4.2.3 Light Intensity

Light is an important factor that affects the physiology and behaviour of larval decapods (Rimmer and Phillips 1979). Phyllosoma larvae of *J. edwardsii* exposed to lower light intensities (0.001 µmols) showed no effect on growth and survival, but higher light intensities (0.1 or 10 µmols^{-1} m^{-2}) had a negative effect on growth (Moss et al. 1999). Newly hatched larvae are positively phototactic, and the photopositive reaction disappears with development. Behavioural separation of larvae and the *Artemia* nauplii may lead to reduction in food intake, slower growth and poor survival (Mikami and Greenwood 1997; Moss et al. 1999). Bermudes and Ritar (2008a) found no difference in feeding, oxygen consumption and nitrogen excretion in phyllosoma larvae exposed to low and higher light intensities. Therefore, rearing larvae under low light intensity may be more feasible as it helps in maintaining even distribution of larvae and feed in the culture systems. Phyllosoma larvae of *P. homarus homarus* never reached beyond Stage IV when reared in black-coloured culture tanks (Radhakrishnan, personal observation).

10.4.2.4 Photoperiod

Photoperiod may influence feeding and growth of phyllosoma larvae, and this may differ with species and development stage of the larvae. Phyllosoma larvae of *P. homarus homarus* exposed to natural day/night (12D:12L) periodicity completed fifth moult in 31.2 days compared to 37 days in 24 hour darkness and 30.5 days in 24 hour light regimes (Radhakrishnan and Vijayakumaran 1986). The faster growth of the larvae under natural day-night periodicity was due to higher food intake by the larvae. However, *P. japonicus* larvae exposed to continuous darkness had the largest mean body length after four moults, although there was no difference in survival (Matsuda et al. 2012).

10.4.2.5 Ammonia

Among the environmental parameters, ammonia may be the most critical parameter to be continuously monitored and controlled in phyllosoma culture water. Ammonia accumulates in the culture system from decomposed food and catabolism in larvae (Phillips and Matsuda 2011). Phyllosoma larvae of *P. homarus homarus* reared in microalgal culture water had lower ammonia concentration (0.12–0.35 mg/L) compared to non-algal culture tanks (0.14–0.68 mg/L), and larvae attained Stage VIII in

42 days in the former compared to Stage V in non-algal tanks (Radhakrishnan et al. 2009). Improved water quality may be one of the reasons for faster growth of the larvae and higher survival. The median lethal concentration (96-hLC_{90}) for total ammonia for *J. edwardsii* phyllosoma larvae (and corresponding NH_3-N) were 31.6 (0.97 mg l^{-1}), 45.7 (1.40 mgl^{-1}), 52.1 (1.59 mgl^{-1}) and 35.5 (1.01 mgl^{-1}) at Stage I, II, III and IV, respectively (Bermudes and Ritar 2008b).

10.5 Water Treatment

Maintaining high-quality seawater and minimizing bacterial loading are the keys to successful control of water quality in larval rearing of spiny lobsters. Ozonation was successfully used in lobster larval culture systems to control water quality and prevent bacterial loading. Ozone (O_3) is an unstable, water-soluble gas and a powerful oxidizing agent, which degrades organic compounds and inactivates microorganisms (Scolding et al. 2012). Ritar et al. (2006) report significantly higher larval survival when culture water was ozone treated during rearing of *J. edwardsii* from egg to juvenile. Ozonation at lower levels (330 mV) with filtration through coral sand and charcoal was found to be effective. Ozonation of seawater (400 mV ozone, 20 ppb TRO) for a period of 3 months reduced *Vibrio* sp. load in culture water and improved survival of *H. gammarus* larvae (Scolding et al. 2012). However, significant reduction in length and weight of the larvae, but not the post-larvae, was observed. Mortality of lobster larvae in aquaculture systems is reported to be due to *Vibrio* sp. loading in culture water. Ozonation may reduce *Vibrio* sp. load of brine shrimp nauplii offered as food to the larvae (Wietz et al. 2009).

An alternative method is UV treatment of microfiltered water (Gemende et al. 2008; Castaing et al. 2011), which is less expensive and less complex than ozonation. However, the effect of UV in water sterilization will depend upon the light intensity of the bulb, the average distance of water from the bulb and the flow rate. It is advisable to 'oversize' the unit, selecting one with a greater stated capacity (expressed as flow rate) than what the system recommends.

10.6 Culture Tank Designs

Larval culture tank designs are crucial for successful culture of phyllosoma larvae. The shape, depth and volume of the tank are critical in deciding the flow characteristics, food contact and survival of the larvae during the course of their development (Goldstein and Nelson 2011). Several tank designs have been tested from the Greve plankton kreisel to the more recent editions of tanks (gelatinous plankton kreisels) used at public aquariums, particularly for rearing and exhibiting jellyfish and ctenophores (Greve 1975; Hamner 1990; Raskoff et al. 2003). The modified plankton kreisel (Massachusetts Institute of Technology) used in culturing *H. americanus* larvae provided mixed results (Hughes et al. 1974; Illingworth et al. 1997; Kittaka 1997a) in rearing spiny lobster larvae. The hydrodynamics of the system minimizes physical stress to the larvae and prevents larval congregation and entanglement (Calado et al. 2003). The tanks typically used for culturing gelatinous zooplankton provide a suitable environment for rearing phyllosoma, and co-culturing jellyfish and phyllosoma in the same tank may benefit both these organisms. For example, phyllosoma may obtain nutrition by feeding on the jellyfish and/or use jellyfish for transportation and thus energy conservation (Wakabayashi et al. 2012; O'Rorke et al. 2015).

Early-stage phyllosoma larvae are highly phototactic and, in the absence of water movement, tend to congregate against the light source. Late-stage larvae are negatively phototactic and aggregate in some parts of the tank. The aggregation often results in entanglement and loss of appendages resulting in decreased feeding, slow growth and even mortality (Matsuda and Takenouchi 2005). For successful culture, larvae should be prevented from aggregating by adjusting the light, water current and tank design. The 30-L hemispherical tank designed by Inoue (1981) with many tiny holes in the water inlet pipe through which seawater circulates on the bottom was able to keep the larvae in suspension and prevent aggregation. Sekine et al. (2000) fabricated similar-shaped tanks without holes at the bottom of the inlet pipe, and the water flow was directed through air tubes fitted to air stones. Illingworth et al. (1997) developed a larval culture system consisting of four square tanks that allowed phyllosoma of *J. edwardsii* to be transferred without handling. Using this system, the larvae have been grown to Stage VIII in 90 days with a survival of over 60%. Radhakrishnan et al. (2009) used cylindro-conical and semitransparent fibre-reinforced plastic (FRP) tanks connected to a recirculation system for larval rearing of *P. homarus homarus* (Fig. 10.2). Though these tanks were successful to a great extent in rearing early- and mid-stage phyllosoma larvae, Matsuda and Takenouchi (2005) found tendency for the larvae to aggregate in intermediate-type cylindrical tanks. Therefore, it is unsuitable for rearing the late-stage larvae.

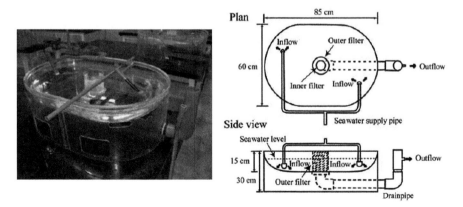

Fig. 10.2 Photo and schematic diagram of a 40-L elliptical tank for larval culture of *Panulirus japonicus*. (Reproduced from Matsuda & Takenouchi, 2005, NZJ of Mar. & Freshwater Res., 39 with permission form Taylor & Francis Group, Royal Society of New Zealand)

10.6.1 Modified Plankton Kreisel

Takushi Horita of the Toba Aquarium introduced a modified upwelling system (called the 'plankto kreisel') originally developed by Greve (1968) to rear the phyllosoma larvae and successfully produced several pueruli (Matsuda and Takenouchi 2007). The circular current created in the culture system keeps the larvae in suspension without entangling each other. Murakami (2004) further improved the system and designed a rotatory tank of 70 L capacity, which rotates itself to provide a more moderate circular current. A survival of about 28.3% was obtained in *P. japonicus* larvae from hatching to juvenile stage. Kittaka (1994a, b) designed an upwelling system with 'U'-shaped glass tanks to culture phyllosoma larvae of several cold-water species of spiny lobsters. Upwelling tanks (16, 30 and 100 L capacity), similar to those used by Hughes et al. (1974) for the culture of *Homarus americanus* larvae, were used to culture phyllosoma. The 16-L tanks with concave bottom and 'U' shape are made of glass, whereas the 30-L and 100-L tanks with flat bottoms are made of transparent plastic to get a better view of the larvae. The diatom culture water introduced into the bottom compartment was forced up through several holes drilled into the bottom container, creating an upwelling motion. The velocity of current in the tank was about 5 cm/s on the bottom and about 3 cm/s on the surface. The excess water passes through a screen covering the central drain pipe (ranging from 100 to 25 per 2.5 cm depending upon the size of the larvae) back into the reservoir preventing the larvae and *Artemia* from escaping. A double-bottom upwelling container was also tested. However, the water movement was insufficient to keep the late-stage larvae in motion even when the flow rate was more than ten times higher than the classical Hughes-type container. By upscaling the capacity of the culture tanks from 30 L to 100 L, mortality of puerulus stage was considerably reduced (Kittaka 1997a). The first successful culture of phyllosoma larvae was achieved for *J. lalandii* in 1986–1987 (Kittaka et al. 1988) followed by *J. edwardsii*, *S. verreauxi*, *Palinurus elephas* and *Panulirus japonicus* (Kittaka and Ikegami 1988; Kittaka and Kimura 1989; Kittaka et al. 1997, 2005). Pueruli produced in the laboratory had poor survival percentage to the juvenile stage: 17% for *S. verreauxi* and 42% for *P. japonicus* (Sekine et al. 2000). Goldstein and Nelson (2011) tested the feasibility of culturing phyllosoma larvae of *P. argus* in large (180 L) modified acrylic plankton kreisels used by aquarium to culture gelatinous zooplankton. No significant differences in growth (Stages V and VI mean body length, 7.5 mm and 10.2 mm, respectively) and survival of larvae, reared through Stage I to Stage VI, were observed at stocking densities of 28 larvae/L (60.7%) and 14 larvae/L (54.5%).

10.6.2 Elliptical Tank

Matsuda and Takenouchi (2005) designed larval culture tanks made from transparent acrylic resin with a capacity of 40 L. The elliptical-shaped tank with concave bottom has smoothly curved corners designed to prevent aggregation of phyllosoma larvae and a flat area of 1500 cm^2 at the centre (Fig. 10.2). Seawater enters the tank through two 12-mm diameter PVC pipes, to which faucet cups were fitted at each pipe outlet and the water flow rate was adjusted to 60–70 L/h. The faucet cups assist in creating moderate water current in the tanks. The drainage pipe was mounted at the centre of the tank fitted with double filters. The filters were constructed of PVC frames around which nylon screens were fixed (the outer filter with 3 mm mesh screen and the inner one with 0.2 mm mesh screen); the outer filter prevents the larvae from escaping and the inner one prevents the outflow of *Artemia*. The phyllosoma larvae of *P. japonicus* were successfully cultured in the system with high survival rates of 37–54% from the middle phyllosoma stage to the puerulus stage (mean BL, 11.5 mm). The larvae approaching final metamorphosis to puerulus were transferred to 5 L circular tanks, one for each larval culture tank, until the morning after metamorphosis.

The advantages of this system over previous designs were the following: i) the shallow depth allows phyllosoma to be conveniently observed and transferred by hand to clean tanks, ii) excessive aggregation of larvae was avoided because of the concave bottom and central flat area and iii) uneaten food (*Artemia* and mussel gonads) can be easily removed through a drain pipe or by a siphon. The survival of *P. japonicus* larvae from the first instar to puerulus in this system was 79–86%.

Matsuda and Takenouchi (2005) compared the advantages and disadvantages of the three types of phyllosoma larval culture systems, horizontal, vertical and intermediate, used by various researchers. The horizontal type has the advantages of easy operation and convenient observation. However, the major disadvantage is that the larvae are always in contact with the bottom of the tank, increasing the probability of diseases and mortality. Further, these types of tanks need more space and are difficult to scale up. The vertical type has the advantage that the larvae could be maintained always in suspension with moderate water current and are easier to scale up. However, the operation and management of this type of system is complicated and needs more attention. Intermediate-type cylindrical tanks have been used in many studies to examine the optimal conditions for flow-through culture of palinurid phyllosoma (Inoue 1981; Kittaka 2000). This type of tank has the disadvantage of larvae accumulating at the small bottom, irrespective of moderately strong current. An ideal system for mass culture of phyllosoma larvae with good survival is yet to be designed and put in operation. Such a system will be required for establishing an economically profitable hatchery-based commercial lobster farming industry.

10.6.3 Japanese Ecosystem Larval Culture Method

Several studies (Kittaka et al. 1988; Kittaka and Ikegami 1988; Kittaka and Kimura 1989) have demonstrated the effectiveness of microalgae in controlling water quality in phyllosoma larval culture recirculation systems and successfully rearing several cold-water species of spiny lobsters to settlement. About 5–10 L (60–70 million cells per ml) of a *Nannochloropsis* sp. culture was inoculated into newly prepared phyllosoma culture water (capacity, 450 L; total initial water volume, about 250 L) at about 5–10% of the total quantity (Kittaka 1994b). The microalgae were recirculated in the culture system, during which the microalgal cells multiplied from an initial 1–2 million cells/ml to 5–20 million cells/ml (Kittaka 1997b). The diatoms controlled the ammonia-N in larval culture water below the lethal limits (median lethal limit for a 72-hour period was an estimated 8 mg/L). Addition of *Nannochloropsis* sp. may increase the COD levels in the culture water, and the upper limit was set at 1.2 mg/L. However, *Nannochloropsis* sp. had a regulatory effect on bacterial load in the culture system, especially the swarming bacteria. Although the bacterial number initially increased rapidly from 101 CFU ml^{-1} to 103–104 CFU ml^{-1} for 1–3 days, it was then decreased to 103 CFU/ml. Higher survival of the phyllosoma was attributed to the domination of *Pseudomonas* sp. in culture water. Though *Tetraselmis* sp. and diatoms such as *Phaeodactylum* sp. and *Nitzschia* sp. were also tested as an alternative to *Nannochloropsi*s sp., none of them were as successful as *Nannochloropsis* sp. in controlling water quality in phyllosoma larval rearing systems (Kittaka 1994b).

10.6.4 Gelatinous Zooplankton Tank

The tanks are modified acrylic plankton kreisels originally used by aquariums to culture and exhibit gelatinous zooplankton. Goldstein and Nelson (2011) used the tanks to mass culture of phyllosoma larvae of the tropical spiny lobster *P. argus*. The working volume of the tank is 180 L (122.1 cm L × 32 cm W × 122 cm H X 96 cm dia). The speciality of the tank is that it has a downward flow, curved sides and bottom and the placement of fine mesh screens that separate the outflow of the tank and large fine-meshed screen partitions that separate the drainage out of the tank. Kreisel tanks were linked to a common FRP seawater reservoir containing a bio-wheel filter apparatus and mechanical filtration up to 2µm, protein foam fractionation and UV sterilization. Heating elements were mounted on to the reservoir tank to keep the temperature stable. Overhead lighting was provided to maintain a 12:12 hour light–dark cycle. A daily water turnover of 75% of the total volume of the tank is maintained. Larvae were reared from hatch to Stage VI. No significant difference in growth and survival (60.7%) of larvae was observed at initial stocking densities of 5000 and 2500 numbers. The application of such culture tank designs with high larval survival contributes to future designs for mass production systems.

10.7 Larval Culture Systems in India

Phyllosoma culture was first attempted at the Field Laboratory, Kovalam, Chennai, of the Central Marine Fisheries Research Institute in 1978. Phyllosoma larvae of *P. homarus homarus* were maintained in 150-ml individual plastic containers with microfiltered seawater. For mass rearing, 150 larvae were maintained in 5-L capacity glass troughs with filtered seawater. The larvae were transferred to fresh culture water daily. The larvae moulted nine times and attained Stage VI in 52–64 days on a diet of *Artemia* nauplii (Radhakrishnan & Vijayakumaran, 1995). Parasitic infestation with *Vorticella*-like protozoans resulted in mortality. Later an upwelling tank system (Kittaka 1994b) was adopted with addition of *Nannochloropsis* sp. However, fast multiplication of diatom cells resulted in high levels of COD and mortality.

10.7.1 Flow-Through Recirculation System

A larval culture experimental system for larval rearing of the spiny lobster *P. homarus homarus* was developed at CMFRI, Cochin (Radhakrishnan, personal communication). Black, circular flat-bottomed Perspex tanks of 100-L capacity were serially connected to a reservoir system with filters and a magneto pump. Each vessel had a T-shaped PVC pipe (3/4" diameter) (1.9 cm) at the centre, with holes on the bottom side walls, to create movement of seawater in the tank. A tap controlled the rate of flow in each vessel which was adjusted to 500 ml min^{-1}. Water exited via a drain attached to 5 cm below the tank edge through a screen. Tanks were cleaned daily and the larvae fed with *Artemia* nauplii. Larvae did not survive beyond Stage IV in this system.

A series of cylindro-conical tanks made of translucent FRP connected to a reservoir with filters and UV sterilizer were also developed at CMFRI for rearing phyllosoma larvae of the spiny lobster *P. homarus homarus* (Fig. 10.3). The larvae were

Fig. 10.3 Clear water recirculation system for culturing phyllosoma larvae of *Panulirus homarus homarus* at CMFRI, Kochi, India. (Photo by: E. V. Radhakrishnan, CMFRI)

reared up to sixth stage on a diet of *Artemia* nauplii initially and subadults later (Radhakrishnan, pers. comm.). The larger larvae could not feed on the fast-moving subadult *Artemia* sp.

10.7.2 Static Ecosystem Culture Method

Larval culture trials were conducted at the Central Marine Fisheries Research Institute, Cochin, using static water systems to rear the phyllosoma larvae of *P. homarus* (Radhakrishnan et al., 2009). Soon after hatching, the female was removed from the spawning tank, and the newly hatched phyllosoma larvae were transferred (10 larvae/L) to two cylindro-conical larval rearing tanks containing 200 L of seawater each. The larval rearing tank was previously chlorinated, sundried and then filled with fresh microfiltered and chlorinated seawater. The seawater used for larval rearing was regularly treated with chlorine (990 ml, 100 g L^{-1} stock) for 24 h followed by neutralization with sodium thiosulphate (1 M, 300 ml) and was aerated vigorously. The larval rearing tank received 30% water exchange every 24 h with fresh seawater.

The microalga *Nannochloropsis salina* (30–40 million cells/ml) was added to one tank, and the other was maintained without microalga. The microalga was added daily to maintain the initial concentration. Phyllosoma larvae were fed with freshly hatched live *Artemia salina* nauplii (0.32 mm) for an initial period of 10 days, followed by juvenile *Artemia* (1.45–1.54 mm) enriched with microalgae for the subsequent 15 days. *Artemia* enriched with commercial polyunsaturated fatty acid concentrate (DHA-Selco) was fed to the larvae for the rest of the culture period. Larvae maintained in microalga-inoculated tank attained Stage VIII in 42 days compared to Stage V in the non-algal tank. Larvae reared in microalgal tank and fed with enriched *Artemia* took a shorter time to reach Stage VIII (42 days) compared to 60 days to reach Stage VI in an earlier experiment (Radhakrishnan et al. 2009).

The rearing tanks inoculated with *Nannochloropsis salina* exhibited a sharp decrease in total heterotrophic bacteria count of 1000 CFU ml^{-1} and total *Vibrio* count to 20 CFU ml^{-1} after Stage IV, whereas non-algal culture tank had high bacterial load (total heterotrophic marine bacterial count, $6.4*10^4$ CFU ml^{-1} and vibrio count, $2.2*10^3$ CFU ml^{-1}) (Radhakrishnan et al. 2009). Igarashi et al. (1990) attributed low bacterial count in microalgal larval culture system to the antibacterial polysaccharides present in the cell wall of the microalgae, which probably stimulate the non-specific immune system in early larval stages and also appear to affect the feeding activity of the phyllosoma larvae. Microalgal larval culture system was efficiently used for successful phyllosoma culture of four temperate species of spiny lobsters (Kittaka 1997b).

10.8 Hatchery Production of *P. ornatus*

Hatchery production of *P. ornatus* was initially attempted in Vietnam during the 1990s, without reported success (Tuan and Mao 2004). Australia initiated research in this area in 1999 (Benzie and Yoshimura 1999) with Duggan and McKinnon (2003) documenting the production of early-stage larvae at the Australian Institute of Marine Science (AIMS), Townville, in 2003. *Panulirus ornatus* pueruli were first produced by M.G. Kailis Group in Australia in 2006 from captive breeding at a commercial prawn hatchery in Exmouth, Western Australia (Calverley 2006). More numbers of pueruli were produced in subsequent years by Lobster Harvest Ltd., the company established for the purpose of commercializing lobster propagation (Barnard et al. 2011a, b, c). In 2007–2008, five consecutive batches of pueruli were produced at the Australian Institute of Marine Science (AIMS), facilitating complete morphological description and staging of the phyllosoma larvae (Smith et al. 2009a, b). Phyllosoma larvae obtained from laboratory-held breeders were stocked at 10,000 larvae/500 L in rearing tanks and fed initially on newly hatched *Artemia* nauplii. Juvenile *Artemia* and blue mussel gonad were fed to late-stage larvae. The larvae metamorphosed to puerulus in 120–150 days, passing through 11 stages involving 20 moults. Smith et al. (2009a, b) believe that by optimizing the rearing conditions and standardizing the feeds, it is possible to reduce the hatchery phase from 4 months to even 3 months, as was achieved in *P. elephas* (Kittaka et al., 2001) and *P. homarus homarus* (Murakami & Sekine, 2011), thus opening the way for an aquaculture industry based on hatchery-produced seeds.

10.9 Final Metamorphosis to Puerulus

Phillips and Matsuda (2011) have reviewed the metamorphosis of final-stage phyllosoma larvae to puerulus. McWilliam and Phillips (1997, 2007) believe that there may not be any external stimuli triggering the final metamorphosis. However, a high-energy diet and nutritional levels in the final or even the penultimate phyllosoma stage may be implicated for successful metamorphosis. Longer photoperiod is presumed to improve the chances of successful metamorphosis in larval *P. japonicus*. Murakami et al. (2007) divided the process of final metamorphosis of phyllosoma larvae to puerulus into five stages. Water current and flow rate may influence the physical process of metamorphosis from phyllosoma to the puerulus stage. The success rate of metamorphosis of *P. japonicus* phyllosoma to puerulus was 38% in a circular 5-L tank with a water flow rate of 0.5 l/min, whereas the success rate of metamorphosis was 85% in 100-L capacity plankton kreisel with rotating current and a water flow rate of 1.5 l/min (Phillips and Matsuda 2011).

10.10 Larval Feed Development

Commercial production of lobster seeds is possible only with the development of formulated diets that can deliver the correct balance of essential nutrients, which is provided in a manner acceptable to phyllosoma (Cox and Johnston 2003). The feed should have the essential nutrients and physical characteristics such as size, shape and texture and should be maintained in suspension. The nutritional requirements of phyllosoma larvae may change depending on the developmental stage, especially the final stage, where the nutritional demand of the larvae is high, not only to meet the energy demanding final metamorphosis but also for the storage of energy reserves in hepatopancreas required by the non-feeding puerulus.

10.10.1 Live and Fresh Feeds

Identification of suitable feeds for different larval stages is a critical factor in successful rearing of the phyllosoma larvae. Scientific information on natural food and nutritional requirements of phyllosoma larva is limited (Phillips and Sastry 1980; Kittaka 1994b; Tong et al. 1997). Recent studies have identified several pelagic zooplankton species such as jellyfish, salps, chaetognaths, siphonophores, polychaetes and crustaceans as the prey for lobster larvae (Wang and Jeffs 2014). Several foods such as *Artemia* nauplii, mussel gonad, fish larvae, hydromedusae, chaetognaths, crab zoea, trochophore veliger larvae and polychaete and ascidian larvae were used for feeding phyllosoma larvae (Lebour 1925; Oshima 1936; Dexter 1972). These food items, except *Artemia* nauplii, are found abundantly in nature, especially in the nearshore regions, water mass boundaries and areas of upwelling, where primary productivity is high (Phillips et al. 2006b). Appropriate feeds are also available in offshore regions, where late-stage phyllosoma larvae are found. Salps have been suggested as a possible food of phyllosoma larvae because of their high tissue density, nitrogen content and inability to escape capture (Heron et al. 1988). Kittaka (1994a, 2005) suggested that the jellyfishes *Aurelia aurita* and *Dactylomela pacifica* are usable as feed for the culture of early stages of palinurid phyllosoma larvae. Wakabayashi et al. (2012, 2016) and Wakabayashi and Phillips (2016) reared the phyllosoma larvae of three species of scyllarid lobsters (*Ibacus novemdentatus*, *I. ciliatus* and *Thenus australiensis*) on an exclusive diet of jellyfish. Laboratory studies offering a wide range of wild plankton showed that *P. interruptus* displayed a preference for medusa, ctenophores and chaetognaths (Mitchell 1971). Identifying the food from digestive tract of wild-caught phyllosoma larvae was not successful, as the food is completely masticated resulting in a homogenized amorphous mass, making microscopic identification impossible (Smith et al. 2009a, b). Biochemical analysis of body tissues of larvae also could not give any lead to the natural diets of the phyllosoma larvae (Phleger et al. 2001; Wells et al. 2001; Jeffs et al. 2004; Hall et al. 2013).

While live organisms are the current source of feed for larval lobsters, inadvertent introduction of bacterial pathogens to the rearing systems through live feed is a possibility (Høj et al. 2009). *Artemia* used as feed harbours bacteria mostly in their gut, and enrichment with lipid concentrates and microalgae increases the bacterial load. Studies show that *vibrio*-dominated bacterial community poses the potential danger to the cultured larvae. The use of disinfected *Artemia* in combination with probiotics is probably the practical solution to prevent introduction of pathogenic bacteria into the rearing system.

A recent approach is the analysis of DNA remains of the food found in the gut using nucleotide sequence analysis and PCR (Suzuki et al. 2006, 2007, 2008; Chow et al. 2011; O'Rorke et al. 2013). Food items identified by molecular approaches in palinurid phyllosoma larvae were fish larvae, chaetognaths, gastropods and gelatinous zooplankton, particularly siphonophora and ctenophora (Hall et al. 2013; O'Rorke et al. 2013).

In an attempt to find the natural prey of phyllosoma larvae, the abundance of phyllosoma in oceanic waters was correlated with their prey. The abundance of mid-stage phyllosoma (Stages V–VIII) was correlated with shrimp biomass, while the abundance of late-stage larvae (Stages IX–XI) was strongly associated with amphipod biomass. The movement of late-stage larvae of *S. verreauxi* during the day (20–50 m) along with the dominant potential prey category, the gelatinous zooplankton, possibly shows a shift in dietary preference of phyllosoma larvae (Jeffs 2007).

The phyllosoma larvae of palinurid lobsters are opportunistic predators that feed on a variety of prey (Jeffs 2007). Early-stage larvae can be described as active and rapacious feeders (Abrunhosa and Kittaka 1997; Cox and Bruce 2003). However, late-stage larvae of *P. homarus homarus* are passive feeders, and their capacity to capture prey depends upon the size and their proximity to the food. The feeding activity of the phyllosoma larvae may depend upon the nature and nutritional quality of the prey with the late-stage larvae taking on a more active predatory role, pursuing more active prey such as larger crustaceans with higher nutritive returns (Jeffs 2007). The ability of the larvae to capture prey also depends upon the size, body texture and activity of the prey. Majority of researchers used *Artemia* nauplii for feeding early larvae as they were found to accept the feed readily and moulted successfully to the next stage. *Artemia* nauplii are slow moving with a soft and fleshy body, which can be easily struck by the sharp dactylus on the first and second pereiopods. However, the 2-day-old *Artemia* with no yolk is of lower nutritional quality and fast moving resulting in larvae struggling to capture the *Artemia*. Larvae fed with 2-day-old *Artemia* consumed more nauplii (19 *Artemia* nauplii/day) but took longer time to reach Stage IV (34 days) compared to 31.2 days by the larvae fed with newly hatched *Artemia* nauplii (Radhakrishnan and Vijayakumaran 1986). As the larvae grow in size, they had difficulty in catching newly hatched smaller *Artemia* nauplii, indicating preference for larger prey size with increase in size of the larvae (Cox and Johnston 2003).

Mussel gonad, especially from *Mytilus* sp. and *Artemia* nauplii have been the most widely used feeds in phyllosoma culture, as these feeds seem to satisfy the nutritional requirement of the phyllosoma larvae to a great extent (Kittaka 1994a, b). While *Artemia* nauplii (0.2–0.3 mm) are the most suitable food for early-stage larvae, *Mytilus* gonad (1–2 mm) is fed to advanced stages as the larvae would have by then possessing well-developed pereiopods. Juvenile and adult *Artemia* represent a poor source of protein and fatty acids, which are required for optimum growth of phyllosoma larvae and therefore need enrichment with protein and marine oils to provide fortified nutrition (Liddy et al. 2003; Ritar et al. 2003; Nelson et al. 2004; Matsuda et al. 2009; Chakraborty et al. 2010). In comparison to *Artemia*, *Mytilus* gonad is characterized by high protein (30–70% on dry weight) and high concentrations of long-chain omega-3 fatty acids (1.5–3.9% dry weight) (Takeuchi and Murakami 2007). However, preparation of the feed daily is cumbersome, especially for feeding the larvae in mass culture systems.

The quantity of food decides the survival and moult frequency of phyllosoma larvae (Lavens et al. 1989; Vijayakumaran and Radhakrishnan 1986). Prey density trials have shown a direct relationship between food quantity and food consumption rates in laboratory-held phyllosoma larvae of *P. homarus homarus* up to a threshold level, with no further increase in food consumption with increase in food density (Vijayakumaran and Radhakrishnan 1986). Further, feeding early-stage phyllosoma larvae with *Artemia* at high densities reduces intermoult duration resulting in higher growth rate (Moss et al. 1999). High density of *Artemia* in an alginate/*Artemia* pizza has shown promise as the large surface area probably reduces the energy involved in capture and manipulation (Tong et al. 1997). Delayed or accelerated moults may be related to nutritional status of the larvae (Smith et al. 2009a, b). Reduced duration in intermoult period has been observed in well-fed laboratory-reared phyllosoma larvae, indicating the importance of nutrition in larval development of palinurid lobsters (Kittaka 1997a; Matsuda and Takenouchi 2007). Perhaps the best example is that of the larval development of *P. elephas*, in which the larval duration was reduced by 50% (132 days to 65 days) (Kittaka 1997a; Kittaka et al. 2001).

10.10.2 Starvation and Recovery Ability of Phyllosoma Larvae

Starvation of *P. ornatus* and *P. homarus homarus* larvae resulted in a PNR_{50} for both species of 5.9 days, with no significant reduction in the ability to moult when starved for 4–5 days (Smith et al. 2010). Moulting commenced after 5–6 days of feeding. The PRS_{50} (the point at which larval reserves have been stored to enable them to complete the next scheduled moulting independent of additional feeding) was 3.8 and 4.4 days for *P. ornatus* and *P. homarus homarus* larvae, respectively. Stage I phyllosoma larvae of *P. ornatus* and *P. homarus* were able to hold on for a few days without any substantial reduction in weight (20%) in the absence of food, if they are able to continuously feed for 7 days after hatch (Smith et al. 2010). This may be a strategy to sustain them through periods of variable temporal and geographical prey abundance in nature as was also noted in other larval crustaceans (Dawirs and Dietrich 1986). The greater PNR_{50} for phyllosoma larvae of these two species of lobsters maybe an adaptation to endure variable concentrations of primary and secondary productivity in the tropical environment, when compared to temperate waters (Kirk 1994). Mikami et al. (1994) reported PNR_{50} at 3.4 days for first phyllosomas of *P. japonicus* with 6.18 days as the intermoult period for the first instar. The average interval between the first day of feeding and the first day of moulting was 12 and 10 days, respectively, for phyllosoma larvae of *S. verreauxi* and *J. edwardsii*. The starvation tolerance period (50% survival) for *S. verreauxi* and *J. edwardsii* was 8 and 4 days, respectively, at temperatures of 16–17°C (Abrunhosa and Kittaka 1997).

10.11 Mouthpart Morphology and Feeding Behaviour

Studies on feeding behaviour, mouthpart morphology and accessory feeding appendages in phyllosoma larvae are useful in understanding their feeding preferences (Agrawal 1965; Nishida et al. 1990; Mikami and Takashima 1993; Johnston and Ritar 2001; Nelson et al. 2002; Cox and Johnston 2003; Arndt et al. 2005; Johnston et al. 2005; Konishi 2007). The morphology of mouthparts may provide indications of the size and texture of natural diets that can be consumed by the larvae. The mouthpart structure and feeding behaviour of early-stage phyllosoma of palinurid lobsters are basically the same with minor differences. The preferred prey size is likely to markedly increase with age (Cox and Johnston 2003). The barbed setation on maxillae and molar parts of early-stage larvae is suited to capture and masticate softer food items such as *Artemia* nauplii, whereas large oral field, robust setae on first maxillae and increased spinose projections on maxillae of late instars suggest more efficient handling of larger, softer and muscular prey (Cox and Johnston 2003; Chakraborty et al. 2011). Feeding studies in phyllosoma larvae of *P. ornatus* show that the larvae capture food using the spines and dactyl on the second and third pereiopods, which is then brought to the oral cavity assisted by the second and third maxillipeds. Optimal capture was obtained when diets are in a gelatinous-muscular consistency and when hard particles are embedded in a muscular carrier. A range of food textures, from gelatinous to hard, were masticated in the oral cavity by Stage V phyllosoma of *P. ornatus* (Smith et al. 2009a, b). The largest *Artemia* that the instars 1 and 2 phyllosoma of *S. verreauxi* were able to successfully grasp and manipulate was around 1 mm, while the instar 13 were able to grasp and manipulate 5-mm size *Artemia* (Cox and Johnston 2003). The efficiency of the larvae to capture more active and larger prey also seems to increase with increase in size of the oral field. Studies on the ultrastructure of the mouthpart morphology of phyllosoma larvae and ontogenetic changes in feeding behaviour may help in identifying appropriate diet characteristics and dietary preferences for phyllosoma as they age (Cox and Johnston 2003; Chakraborty et al. 2011).

10.12 Artificial Larval Diets

Formulated diets have several advantages over natural fresh foods: they are cheaper; can be modified to suit the feeding requirements of the species, stored and fed easily; and can also reduce the risk of bacterial introduction in culture systems.

A micro-encapsulated diet made primarily of plant protein, lipid and carbohydrate was readily accepted by phyllosoma larvae of *S. verreauxi*, and the mid-stage larvae survived for 119 days and underwent three moults (Otawa and Kittaka unpublished; cited in Kittaka and Booth 2000). Studies on the chemical components and texture of gelatinous organisms in the ocean may be useful in formulating artificial foods, as they represent a significant proportion of wild phyllosoma diet (Kittaka 1997b). An effective formulated diet for the phyllosoma larvae of any species is yet to be developed, though active research is progressing in Australia. The addition of chemoattractants (L-glutamic acid, inosine-5-monophosphoric acid, adenosine-5-monophosphoric acid, trimethylamine or mussel extract (Kurmaly et al. 1990)) may increase feeding rates and attractability of formulated feeds (Cox and Johnston 2003). Hall et al. (2013) taking the cue from crustacean nutrition, especially the penaeid prawns, concluded that the protein requirement of phyllosoma larvae may fall in the range of 23–57% dry weight and lipid from 5 to 10%. Based on the feeding studies in phyllosoma larvae of *P. ornatus*, Smith et al. (2009a, b) concluded that phyllosoma require a gelatinous-muscular food texture to facilitate capture, although mastication and digestion capability also extends to harder dietary components. They also suggested that the development of formulated diets may include a range of food textures encapsulated within a soft matrix.

10.13 Larval Diseases

Diseases in phyllosoma larvae are generally caused by bacteria, fungi, protozoans and metazoans (Shields 2011). Disease-causing agents enter into the larval culture systems mainly through infected broodstock, fresh larval feeds and incoming water. Larvae are mostly infected by opportunistic pathogens when the stressors weaken the larvae. The initial symptoms of disease are reduced activity of the larvae, interrupted feeding behaviour and change in colour (Hall et al. 2013).

Viral diseases are yet to be a major problem for phyllosoma larvae, though vertical transmission from mother to the larvae is a possibility through the infected ovary and the oocyte (Quintana et al. 2011). On the other hand, bacteria may form complex communities and exist as bacterioplankton in the water column (Payne et al. 2006), on tank surfaces as biofilm (Bourne et al. 2006; Wietz et al. 2009), on the feed (Hoj et al. 2009) and on the phyllosoma larvae themselves (Payne et al. 2007). The bacterial community on the wild and laboratory-reared larvae is entirely different (Payne et al. 2008), and this may be due to the high larval biomass and nutrient load in culture water.

Vibriosis is a major disease problem in rearing the phyllosoma larva of palinurid lobsters (Webster et al. 2006). Many pathogenic species have been isolated (Diggles et al. 2000; Payne et al. 2007; Yoshizawa et al. 2012). *Vibrio* species isolated from early phyllosoma of palinurid lobsters include *V. alginolyticus, V. campbellii, V. harveyi, V. jasicida, V. natriegens, V. owensii, V. parahaemolyticus, V. proteolyticus,*

V. rotiferianus and *V. tubiashii* (Diggles et al. 2000; Bourne et al. 2004; Payne et al. 2007; Cano-Gómez et al. 2010; Yoshizawa et al. 2012). Mass mortality in phyllosoma larvae of *S. verreauxi* was attributed to a new species of *Harveyi* clade, *V. jasicida* (Yoshizawa et al. 2012). A new species *V. owensii* of highly virulent *Vibrio* belonging to *Harvey*i clade was responsible for mass mortality of 80–90% of *P. ornatus* phyllosoma larvae within 72 h (Cano-Gómez et al. 2010; Goulden et al. 2013).

Gram-positive bacteria, which are pathogenic to palinurid lobsters, constitute only 5% of marine bacteria (Hall et al. 2013). *Aerococcus viridans* is a highly virulent strain known to infect adult homarid as well as spiny lobsters, but not reported in phyllosoma. Other pathogenic strains found in phyllosoma hatcheries are *Bacillus* (Bourne et al. 2004) and *Clostridium* (Payne et al. 2007). Filamentous bacteria fouling the phyllosoma larvae often interfere with feeding and thereby deny proper nutrition to the larvae leading to longer intermoult period. Though *Leucothrix mucor* was often identified as the causative agent, recent work shows that *Thiothrix* sp. is a major threat for survival of *P. ornatus* phyllosoma larvae (Payne et al. 2007).

The common pathogenic oomycetes infecting eggs and phyllosoma larvae is *Lagenidium* sp. (Nilson et al. 1975). Though a number of chemotherapeutics such as malachite green, formalin and copper sulphate were applied to control the infection, only trifluralin was found to be effective (Hall et al. 2013).

The ciliate protozoans, *Vorticella* sp. and *Zoothamnium* sp., though not pathogenic, interfere with locomotion and feeding in phyllosoma (Kittaka 1997a; Vijayakumaran and Radhakrishnan 2003) (for more information, see Chap. 13). Chemotherapy has limited success in the treatment of ciliate infections (Boghen 1982). Infected larvae exposed to a short bath treatment of 10 ppm malachite green were found to be relieved of the parasite. Longer exposure with lesser concentration of the chemical is suggested from the health point of view of the larvae. A dip treatment of *Artemia* nauplii in 100 ppm malachite green before feeding the larvae was found to ward off the ciliate infection in phyllosoma larvae of *P. homarus homarus* (Vijayakumaran and Radhakrishnan 2003). The ciliates mostly enter the culture system through the *Artemia* given as food to the larvae. Disinfection of the *Artemia* cysts with sodium hypochlorite is effective in killing the microbes attached to the egg shell. Helminths, such as polychaetes, nematodes and turbellarians, may not be a problem in larval rearing of lobsters, due to their size (Shields et al. 2006; Shields 2011).

Vibrios create maximum problem in larval rearing of lobsters, and successful larval culture requires management of disease-causing agents in the hatchery. Indiscriminate use of antibiotics for disease treatment has resulted in proliferation of antibiotic-resistant bacteria. Several alternative control options have emerged recently, which include the use of bacteriophages (Nakai and Park 2002), probiotics (Verschuere et al. 2000) and quorum sensing disruption (Defoirdt et al. 2004). However, practical application of quorum sensing disruption techniques is yet to be available to control luminous bacterial infestations in larval culture.

10.14 Breeding and Seed Production of Scyllarid Lobsters

10.14.1 Introduction

Slipper lobsters contribute only less than 2% of the total global lobster landings (Vijayakumaran and Radhakrishnan 2011). Seed production of scyllarid lobsters was not a priority among the lobster researchers, as majority of the species was small and not economically important. Among the 92 species, only less than 30 are commercially important. Although the complete larval culture of the scyllarid lobster *Scyllarus americanus* was achieved in 1968 (Robertson 1968), there were no serious attempts in seed production of scyllarids, probably due to the lower demand for the group in the international market. The renewed interest in culturing the lobsters in this group came up during 1990s, as the markets started picking up and had increased demand for them. Further, the larval cycle of scyllarids was much shorter compared to palinurids and therefore easier to produce seeds. Several species were reared to settlement by the Japanese and Australian scientists, though survival was low, especially from nisto to juvenile. Vijayakumaran and Radhakrishnan (2011) have reviewed the global status of slipper lobster farming and growout and discussed the problems and potential for developing commercial-scale slipper lobster culture. The recent advances made in breeding and culture of some commercially valuable species are also discussed.

Complete larval rearing has been successfully achieved in scyllarid lobsters in different parts of the world: *I. ciliatus* and *I. novemdentatus* (Takahashi & Saisho 1978; Mikami & Takashima 1993; Wakabayashi et al. 2012; Wakabayashi et al. 2016), *I. peronii* (Marinovic et al.1994), *S. demani* (Ito & Lucas 1990) and *S. americanus* (Robertson 1968), *Thenus orientalis* and *T. indicus* (Mikami & Greenwood 1997), *T. unimaculatus* (Kizhakudan & Krishnamoorthi 2014), *Petrarctus rugosus* (Kizhakudan et al. 2004) and *T. australiensis* (Wakabayashi & Phillips, 2016). The length of larval phase varies from species to species (Table 10.4).

Larval development, distribution and partial rearing successes were recorded in *S. kitanoviriosus* (Higa & Saisho, 1983; Wada et al., 1985) and *T. orientalis* (Barnett et al., 1986). Robertson (1968) has described the larval life span of scyllarid lobsters to last from 30 days to nine months, depending on the species and environmental factors.

Compared to spiny lobsters, captive breeding of scyllarid lobsters has been much less studied. Studies on the effects of environmental conditions and nutrition on larval development have been mostly restricted to spiny lobsters. Mikami and Greenwood (1997) attributed variability in larval instar duration to the feed used for rearing.

In India, the Central Marine Fisheries Research Institute (CMFRI) has been conducting pioneering research on mariculture development of lobsters since 1976. In January 2004, captive breeding and complete culture of phyllosoma larvae to nisto of *T. unimaculatus* and another scyllarid lobster, *P. rugosus*, was successfully achieved (Kizhakudan et al. 2004; Kizhakudan and Krishnamoorthi 2014).

Table 10.4 Successful culture of scyllarid lobster phyllosoma larvae

Species	No. of stages	Length (days) of larval life	Feed	Authors
Arctidinae				
Ibacus ciliatus (von Siebold, 1824)	7–8 + nisto (complete)	54–76	Fed initially on *Artemia* sp. nauplii; later stages supplemented with the meat of the short-necked clam *Tapes philippinarum*	Takahashi and Saisho (1978)
Ibacus ciliatus (von Siebold, 1824)	7–8 + nisto (complete)	17	*Artemia* sp. nauplii; later chopped mussel *Mytilus edulis* meat	Mikami and Takashima (1993)
	7–8 + nisto (complete)		Fish larvae, mussel, clam, jellyfish	Matsuda et al. (1987, 1988) and Matsuda and Yamakawa et al. (1989)
	7 + nisto (complete) 7 + nistos + juvenile	66.8 85	Exclusively on jellyfishes, *Aurelia aurita* and *Chrysaora pacifica*	Wakabayashi et al. (2016)
Ibacus novemdentatus (Gibbes, 1850)	7 + nisto (complete) 7 + nisto + juvenile (complete)	65 41–44 47.7–48.2 (average)	Initially fed on *Artemia* nauplii; later stages supplemented with the meat of the short-necked clam *Tapes philippinarum* Exclusively on jellyfishes, *Aurelia aurita* and *Chrysaora pacifica*	Takahashi and Saisho (1978) and Wakabayashi et al. (2012)
Ibacus peronii (Leach, 1815)	6 + nisto + juveniles (complete)	79	Initially *Artemia* sp. nauplii; after the third instar fed with ovaries of the mussel *Mytilus edulis*	Marinovic et al. (1994)
Scyllarinae				
Chelarctus cultrifer (Ortmann, 1897)	Stage X + nisto (complete)	159	Initially *Artemia* nauplii; after 30 days supplemented with chopped gonads of the mussel *Mytilus galloprovincialis*	Matsuda and Mikami (unpublished)
Crenarctus bicuspidatus	Stage XI + nisto + juveniles (complete)	51–62	*Artemia* nauplii with chopped *Mytilus edulis*	Matsuda (unpublished)

(continued)

Table 10.4 (continued)

Species	No. of stages	Length (days) of larval life	Feed	Authors
Scyllarus americanus (SI) (Smith, 1869)	6–7 post-larvae (complete)	32–40	*Artemia* nauplii	Robertson (1968)
Scyllarus arctus (Linnaeus, 1758)	16 post-larvae (complete)	192 at 20 ± 1 °c	*Artemia salina* for ~80 days, then whisked fish and beef	Pessani et al. (1999)
Petrarctus demani (Holthuis, 1946)	8 nisto + juvenile lobster instar 1 (complete)	46 to nisto	*Artemia* sp. nauplii supplemented after the fifth instar with chopped *Gafrarium* sp.	Ito and Lucas (1990)
Petrarctus rugosus (H. Milne Edwards, 1837)	8 + nisto (complete)	32	*Artemia* nauplii during the first three instars and chopped meat of the clam *Meretrix casta* from phyllosoma IV onwards	Kizhakudan et al. (2004)
Theninae				
Thenus orientalis (Lund, 1793)	4 + nisto + juveniles (complete)	28	Fresh flesh of *Donax brazieri*	Mikami and Greenwood (1997)
Thenus indicus (Leach, 1816)	4 + nisto + juveniles (complete)	28	Fresh flesh of *Donax brazieri*	Mikami and Greenwood (1997)
Thenus unimaculatus (Burton and Davie, 2007)	4 + nisto + juvenile (complete)	26	Chopped meat of the clam *Meretrix casta* and live ctenophores	Kizhakudan and Krishnamoorthi (2014)
Thenus australiensis (Burton and Davie, 2007)	4 + nisto + juvenile (complete)	32–41 days	Exclusively on jellyfish *Aurelia aurita*	Wakabayashi and Phillips (2016)

Environmental and nutritive parameters, such as water quality, photoperiod and type of feed have been shown to influence breeding of sand lobsters in captivity (Kizhakudan et al. 2004). Rahman et al. (1987) described egg development in *T. orientalis*. Mass-scale larval production has also been accomplished, but the major bottleneck was mortality, particularly in late phyllosoma stages due to *Vibrio* infection. Apart from reviewing the literature on breeding and hatchery production of scyllarid lobsters, attempts to grow *Thenus* sp. in indoor growout system are also highlighted.

10.15 Hatchery Production of Seeds

The major challenge to successful phyllosoma larval rearing is the final metamorphosis from the last planktonic larval phyllosoma stage (with thoracic propulsion) to the planktonic post-larval (abdominal propulsion) and pre-juvenile larval (decapodid) phase, termed 'nisto' in scyllarid lobsters. This decapodid phase is a non-feeding planktonic stage that ends with a moult into the juvenile phase. The metamorphosis from larva to post-larva is the most profound transformation at a single moult known among Decapoda. The *nisto* stage is characterized by lack of calcium in the carapace, and once the post-larvae moults to juvenile, the carapace becomes calcified.

While CMFRI has been able to complete nisto production in captivity, the survival rate remains a major bottleneck in developing the rearing technology into a commercially viable package. It is necessary to evaluate the biological performance of the animals during various stages of their life cycle in different rearing environments in order to identify the optimum physico-chemical environmental parameters and nutrition standards that will result in the attainment of marketable sizes at economically viable growth rates.

First stage phyllosoma stocked in raceways at stocking densities of 5–7 larvae/L^{-1} progresses to nisto in 29–32 days of culture with restricted water exchange and using clam tissue as larval feed. The feed was distributed four times daily in equal proportion. The physico-chemical parameters of the rearing water was maintained with temperature at 28°C, pH 8.1 and salinity 35 ppt.

10.15.1 Breeding in Captivity

Maturation and breeding in captivity are major challenges in the evolution of husbandry packages for lobsters. Establishing culture conditions that ensure continuous maturation and breeding cycles for maintaining a year-round supply of larvae holds the key to successful lobster aquaculture. Compared to spiny lobsters, captive breeding of scyllarid lobsters is a less explored area. Kizhakudan et al. (2004) observed high incidence of maturation and breeding in scyllarids held in captivity. Water quality and photoperiod were found to play a major role, and the animals were reared in larger tanks with increased water depth.

Broodstock maintenance and development in *T. unimaculatus* can be accomplished in a closed recirculatory system with fluidized bed filter and minimum light exposure (LD 1:23). Food is a major factor determining the performance of the animals in captivity. Booth and Kittaka (2000) mention the preference for shellfish, particularly mussels, over finfish by juvenile spiny lobsters. This was found to be true in the case of *T. unimaculatus* also, as the animals show good reception to fresh clam meat. Juvenile (<30-mm CL) and subadult (30–40-mm CL) lobsters collected from the wild and reared in recirculatory systems were developed into mature adult

Fig. 10.4 Broodstock of *Thenus unimaculatus* developed in laboratory. (Photo by Joe K Kizhakudan, CMFRI)

lobsters (65–70-mm CL) over a period of about 6–8 months (Fig. 10.4). Regulation of light exposure and feeding at 5% of body weight in two divided doses daily give good results.

Males are generally smaller, that is, they mature at a smaller size (55–65-mm CL) and their life span gets reduced after about four to five successive mating cycles. Berried female lobsters collected from the wild showed amenability to captive conditions. Phyllosoma that hatched from the eggs of laboratory-developed broodstock are more viable than the ones that hatch from berried females collected from the wild.

The fecundity in these animals has been reported to range from 22,050 to 53,280 (Kagwade and Kabli 1996), 15,000 to 55,000 (Kizhakudan 2007) and 14,750 to 33,250 (Radhakrishnan et al. 2013). The incubation period lasts for about 35 to 37 days, and hatching occurs over an extended duration of 30–36 h. The rate of egg pruning by the brooder and the length of the incubation period are dependent on the quality of the water in which the animals are held.

Sand lobsters of >65-mm CL with clear external secondary sexual traits and orange pigmentation in the pleopod tips and ventral sinus gland area are sorted out and held in a 3-m long maturation tanks, with inner dark-coloured side walls, 1-m deep water column and closed in situ bed filter recirculatory system. Females along with active males, grown separately, were pooled at the time of mating in the sex ratio of 1M:2–4F. These lobsters were found to mate and spawn in captivity at 26–27°C water temperature and 35–36 ppt salinity with short photoperiodicity. The breeding is synchronized with a sudden dip in the ambient temperature by 2°C.

Juveniles of *P. rugosus*, raised in soft sand substrates in reduced photoperiodicity and kept at 26–28°C temperature, mature in captivity when fed on the Wedge clam (*Donax cuneatus*) meat. Adult females have been induced to mate and spawn two to three times in a single moult phase with the help of environmental manipulation, particularly water temperature and photoperiodicity.

10.15.2 Broodstock Tank Design and Water Quality

Marine lobsters can be broadly classified based on their substrate preference, shelter requirement and behavioural responses. Community living and solitary residency are a character shift seen in the life cycle as the lobster turns from juvenile to adult. In general, all lobsters are nocturnal with a feed preference for molluscs, particularly bivalves and gastropods. Holding system design is a critical hurdle in the success of lobster domestication at every stage of its life cycle. The design of broodstock holding system may require modification to suit behavioural characteristics of the broodstock, such as aggression at moulting and mating, autotomy and regeneration and cannibalism and feeding behaviour.

Black-coloured tanks are preferred for lobster holding since it may improve feeding and reduce shelter stress. Tanks with curved sides reduce incidences of tail injuries. Hanging screens and suspended mazes provide protection during moulting. Cement pipes, slates, tiles and compartments also provide crevices and shelters for the lobsters to hide.

Fine sand substratum of 2–3″ depth enables slipper lobsters to remain buried during daytime hours, while long tanks give them more scope for fast movement during night hours. In situ fluidized bed filters with activated charcoal and shell grit ensures best control of ammonia and nitrate levels in the water particularly when the lobsters are fed on raw bivalve meat (shell-on). Airlift pumps aerating the in situ filters help in oxygenation and removal of toxic gases.

Stocking density is a major factor to be considered while designing lobster holding systems. Scyllarids (*Thenus* spp.) can be stocked at 40–50/m^2 (1–3 kg). Water column depth of 0.7–1 m is ideal for coastal species, while 1–1.2 m is preferred for deep-sea scyllarids. The evolution of specific protocols for holding systems to cater to the requirements of the lobsters at each stage of the life cycle will hold the key to overcoming the hurdles that restrict the perfection of a domestication technology package for lobster aquaculture at present.

Water quality is a major factor that affects success of scyllarid lobster larval rearing. Phyllosoma larvae stocked at 5 larvae/L of seawater in clear water systems with minimum light exposure and 100% water exchange daily, are reared on a combination of live zooplankton and clam meat (*M. casta*). The optimal water conditions suitable for scyllarid larval rearing and growout in India include maintaining rearing water at a temperature range of 25–27°C, salinity, 37–39 ppt, pH, 8–8.2 and lightto exposure ratio of 1:3.r Water exchange of 200% during the larval phase and a shift to closed recirculatory systems during nursery rearing and growout are other essential requirements. Nutrient levels are to be maintained at <10 µg for nitrate/nitrite and <50 µg for ammonia. Care should be taken to ensure that there is no build-up of hydrogen sulphide in the rearing system.

10.15.3 Holding System Specifications for *T. unimaculatus*

10.15.3.1 Broodstock Holding Tanks (BHT)

Kizhakudan et al. (2004) suggest, 1X1X3 m black-coloured rectangular FRP tank with recirculatory system, including an *in situ* bed filter with fine sea sand as substrate for holding the broodstock of *T. unimaculatus*. The stocking density of adult lobsters was 10–15 lobsters per sq.m (Figs. 10.5 and 10.6).

10.15.3.2 Growout Tanks

For growout purposes, 1X1.5X1 m black-coloured rectangular FRP tanks with in situ gradient biofilter and river sand as substrate with minimal water exchange were used. Tanks with holding capacity of 40–50 lobsters per sq.m (floor space) are ideal. An estimated filter throughput at 10–12 times per day keeps the water quality optimal. Cement tanks (3X3X1.2 m) with *in situ* biofilters of sand and charcoal beds separated by screens can be used for high-density fattening (3 to 5 months) with a stocking rate of 40–100 juvenile lobsters (25–35 mm CL/m^2).

Fig. 10.5 *Thenus unimaculatus* broodstock holding facility at Field Laboratory, CMFRI, Kovalam, Chennai, India. (Photo by: Joe K KIzhakudan, CMFRI)

Fig. 10.6 Recirculating seawater system for broodstock maturation and rematuration of *Thenus unimaculatus* at Field Laboratory CMFRI, Kovalam, Chennai, India (Photo by: Joe K Kizhakudan, CMFRI)

10.15.3.3 Spawner Tanks

For holding spawners, 1X1X1 m black-coloured rectangular FRP tanks with gradient bed filters at the bottom and sterile sand substrate t (stocking desnity, 10 per sq.m) are suggested. The tanks were kept in a dark room with necessary lighting, and larval scoops and collection chamber for collecting the hatched larvae were used.

10.15.3.4 Larval Rearing Tank

Larval rearing tanks (LRT) of different dimensions and water flow were used at the Kovalam Field Laboratory for larval rearing of scyllarid lobsters. FRP tanks, with dimension 8X0.4X0.4 m with a perforated lid, a central drain pipe with larval separation screens and peripheral water delivery ports/nozzles directed in one direction, were tested. A raceway larval rearing system has also been made functional. The raceway chamber measures 20X6 m and houses four raceway LRTs, two of 5-m length and two of 8-m length. The nursery system (8X5 m) houses four nursery raceways of 5-m length (Fig. 10.7). The hatchery chamber is temperature and light controlled. Dynamics of the raceway systems has been established at various flow rates, heights, temperatures and drain speed. Stocking densities of 350–450 slipper lobster larvae per 100 L is ideal in running water raceways. A flow rate with

Fig. 10.7 Prototype of raceway system for larval culture of *Thenus unimaculatus* at Field Laboratory of CMFRI at Kovalam, Chennai, India

replacement of a minimum of 24 times of the volume of tank water per day gave better larval suspension and metabolite exchange. Larval survival was found to be high up to final-stage phyllosoma.

10.15.3.5 Nursery Tank
For nursery culture of juveniles, the tank requirements are 5X 0.4X 0.4 m black-coloured FRP tank with perforated lid and suspended bottom platform.

10.15.4 Rearing System Specifications for *Petrarctus rugosus*

10.15.4.1 Breeder Tank
Black FRP holding tanks with the dimension 1X1.5X1 m with in situ bed filters with internal airlift and a holding capacity of 200 per sq.m were used to hold broodstock.

10.15.4.2 Larval Rearing Tank
For larval rearing, FRP cylindro-conical tanks with central drainage with unidirectional water flow were used.

10.15.4.3 Nursery Tank
To hold the nisto and juveniles, 400-L rectangular tank with vertical gradient sand filter and recirculatory system, shallow-water column, external airlift and 4″ (10.16 cm) sand bed has been used.

10.15.5 Larval Development

Complete larval development of 13 species of scyllarid lobsters has so far been achieved (Table 10.4). In Scyllarids, the naupliosoma stage is embryonized unlike the palinurid *S. verreauxi* and some species under the genera *Jasus*, *Scyllarides* and *Ibacus* (Phillips et al. 2006a). A short larval phase is characteristic of most scyllarids, unlike the spiny lobsters (Fig. 10.6). In *S. americanus* (Smith, 1869), the larval life consists of six or seven phyllosoma stages and is of relatively short duration, post-larvae being obtained at 25 °C in a minimum of 32 and a maximum of 40 days after hatching (Robertson 1968). Takahashi and Saisho (1978) reared the phyllosoma larvae of *I. ciliatus* to settlement in 54–76 days. Matsuda et al. (1987, 1988) and Matsuda and Yamakawa et al. (1989) also completed the larval phase in captivity but had problems in rearing the nisto further. Complete culture from hatching to juvenile stage of the same species was achieved in an average of 68.8 days on feeding an exclusive diet of jellyfish (Wakabayashi et al. 2016) (Table 10.4). Wakabayashi et al. (2012) have reported the shortest phyllosomal duration in *I. novemdentatus* (Gibbes, 1850) to be 44 and 41 days (average 47.7 and 48.2 days) in closed recirculatory or static conditions, respectively, with six phyllosoma and one nisto stage. Larval phase of *Ibacus* sp. is relatively longer compared to other scyllarids (for more information, see Chap. 11).

The phyllosoma of slipper lobsters are characteristically flattened, leaflike and transparent (Fig. 10.8). The cephalic shield is much broader than the thorax. The abdomen is very short and narrow. The pereiopods arise from the thoracic region. phyllosoma of *Thenus* sp. is hardy compared to other palinurid or scyllarid phyllosoma but are still vulnerable at the time of moulting (Mikami and Kuballa 2007). Maximum mortality occurs at moulting from one phyllosomal stage to the next.

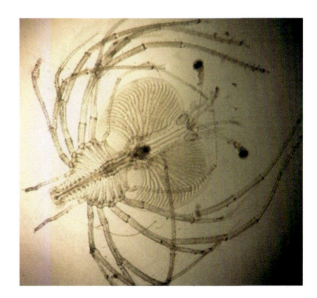

Fig. 10.8 Stage III phyllosoma larvae of *Thenus unimaculatus* reared in laboratory. (Photo by: Joe K Kizhakudan, CMFRI)

The nisto is a non-feeding post-larval stage resembling the adult lobster but with a transparent exoskeleton. Metamorphosis to nisto marks the end of the planktonic larval phase, and the nisto buries into the sandy substratum. It is a dormant non-feeding stage with no active movement unless disturbed. The nisto moults into the first seed or juvenile stage in 2–3 days. Unlike the nisto, the first juvenile stage has a hard exoskeleton and begins feeding on fresh clam meat.

10.15.6 Larval Rearing of *Thenus* spp.

Until now, phyllosoma larvae of four species of *Thenus* were reared from hatching to settlement, and to juvenile stage. The larval phase of *Thenus* spp. is relatively shorter compared to other scyllarids (Fig. 10.9). Larval development in *Thenus* spp. progresses through four phyllosomal stages. This was first suggested by Barnett et al. (1984) and Ito (1988) and confirmed by Mikami and Greenwood (1997), when they completed the larval cycle of *T. orientalis* and *Thenus* sp. in laboratory trials. Larval development of *T. unimaculatus* has also been found to progress through four phyllosomal stages (Table 10.5) but with two instars in the first stage (Kizhakudan et al. 2004; Kizhakudan and Krishnamoorthi 2014). Moulting occurs late in the evening or at night. Metamorphosis into the post-larval nisto stage takes place within 28 days in

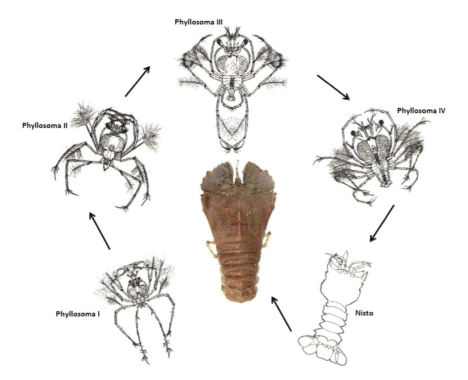

Fig. 10.9 Schematic representation of larval cycle of *Thenus* sp.

Table 10.5 The average duration of the intermoult period (days) (hours for first instar) for each stage of larval *Thenus unimaculatus*

Larval stage	Intermoult duration (days)
Phyllosoma I (first instar)	24 h
Phyllosoma I (second instar)	6
Phyllosoma II	5
Phyllosoma III	7
Phyllosoma IV	7
Nisto	4

Table 10.6 Larval stage and intermoult duration of laboratory-reared phyllosoma larvae of *Petrarctus rugosus*

Larval stage	Intermoult duration (days)
Phyllosoma I	2
Phyllosoma II	4
Phyllosoma III	4
Phyllosoma IV	5
Phyllosoma V	3
Phyllosoma VI	4
Phyllosoma VII	4
Phyllosoma VIII	4
Nisto	5

T. orientalis and *Thenus* sp. (Mikami & Greenwood, 1997) and within 26–30 days in *T. unimaculatus* (Kizhakudan & Krishnamoorthi, 2014). Wakabayashi and Phillips (2016) completed the larval phase of *T. australiensis* under laboratory conditions. The larvae metamorphosed to nisto and to juvenile in 32–41 days.

10.15.7 Larval Rearing of *P. rugosus*

The phyllosoma of *P. rugosus* advance through eight stages before the larvae settle as the post-larval nisto stage (Kizhakudan et al. 2004). The nisto moults into the juvenile stage on the fifth day after settlement. The larval phase lasts up to 32 days after hatching (Table 10.6). Kumar et al. (2009) obtained the final phyllosoma stage in 51 days.

10.15.8 Larval Feeds

Identification of suitable food for different stages of phyllosoma larvae was found to be the most difficult task in rearing the larvae under captivity. Several varieties of natural food were tested, and some were found to be most acceptable to the larvae. *Artemia* nauplii were considered to be the most convenient food for feeding earlier larvae of both palinurid and scyllarid lobsters, though this is not their natural food. Phyllosoma of scyllarids were fed with mussel and clam meat, fish larvae and whisked fish and beef during rearing (Table 10.6). Kizhakudan et al. (2004) fed phyllosoma of

P. rugosus with *Artemia* nauplii initially for the first three stages and then with chopped clam meat until metamorphosis to nisto and clam meat and live ctenophores to *T. unimaculatus* (Kizhakudan & Krishnamoorthi, 2014). Wakabayashi et al. (2012, 2016) exclusively fed the larvae of *I. ciliatus* and *I. novemdentatus* with jellyfish (*Aurelia aurita* and *Chrysaora pacifica*) throughout the culture period and *A. aurita* alone for feeding *T. australiensis* larvae (Wakabayashi & Phillips, 2016). Jellyfish seems to be a viable diet for feeding phyllosoma larvae of scyllarid lobsters in culture (Wakabayashi and Phillips 2016). Techniques for mass culture of moon jellyfishes and sea nettles have already been developed (Purcell et al. 2013), and their quality can be improved by feeding with enriched *Artemia* (Fukuda & Naganuma, 2001).

10.16 Spiny Lobster Growout

10.16.1 Seed Source

Spiny lobsters have a complex and lengthy larval phase (5–12 months) at the end of which they metamorphose into the post-larva (puerulus) and swim from offshore regions to the coastal areas for final settlement. The puerulus moults several times to become a benthic dwelling juvenile. Survival through the first year of life in the wild is estimated to be between 3% and 25% for temperate species and as low as 3% for tropical species (Gardner et al. 2001). First-year survival of puerulus removed from the wild into culture is 93–99% (Crear 1998). Studies on puerulus settlement along the Indian coast is very limited, except for the collection attempts made at Kovalam, south of Chennai, on the southeast coast of India (Vijayakumaran and Radhakrishnan 1988). Considering the prolonged larval phase and low larval survival under captive conditions for most of the species, it is unlikely that hatchery-produced seeds will be available for commercial-scale aquaculture in the near future. However, researchers in Japan, New Zealand and Australia have achieved success in rearing the newly hatched lobster larvae to the puerulus (Kittaka et al. 1988, Kittaka and Booth 1994; Tong et al. 1997, Matsuda et al. 2006; Smith et al. 2009a, b; Goldstein et al. 2008; Murakami and Sekine 2011) in the laboratory. Recent advances in seed production of the most sought-after tropical spiny lobster *P. ornatus* in Australia show that lobster culture using hatchery-produced seeds may be a reality (Barnard et al. 2011c). In India, larval culture attempts were only partially successful (Radhakrishnan and Vijayakumaran 1995; Radhakrishnan et al. 2009; Vijayakumaran et al. 2014). Therefore, culture attempts have been purely based on the juveniles and subadults incidentally caught in gillnets during fishing operations.

10.16.2 Collection of Pueruli

A wide variety of artificial habitat collectors has been used for catching pueruli of *P. argus* for research (Phillips and Booth 1994). Although these artificial habitats have taken many forms, they are usually constructed of low-cost materials and designed to provide an abundance of structural complexity on a fine scale in which the pueruli can hide (Williams 2009). These include fibrous material (folded airconditioning filter fabric made from pigs' hair) in the Witham and Hunt collectors (Phillips and Booth 1994; Phillips et al. 2005), masses of synthetic fibre tassels (frayed polypropylene rope) in the GuSi collector (Gutiérrez-Carbonell et al. 1992) and artificial seaweed or polypropylene tassels in the Phillips and Sandwich collectors (Phillips et al. 2005; Cruz et al. 2006). Radhakrishnan and Vijayakumaran (unpublished) used local materials such as corrugated roofing tiles (30x20 cm), tyres and country tiles (semi-circular) wound with coir ropes for pueruli settlement studies at Kovalam near Chennai.

There is good potential for collection of large number of pueruli from the wild for aquaculture purposes in some countries. Ecological studies indicate that a good proportion of pueruli is lost to the fishery mainly through predation (Butler and Herrnkind 1988; Bannerot et al. 1988; Ryther et al. 1988; Butler and Herrnkind 1989; Forcucci et al. 1994). A large number of pueruli and juveniles have been caught for research purposes at various locations using a variety of simple and lowcost puerulus collectors (Witham et al. 1964; Field and Butler 1994; Phillips and Booth 1994). The number of pueruli collected often varies, depending upon the location, season and lunar phase (Bannerot et al. 1987, Monterrosa 1987; BrionesFourzan and Gutierrez-Carbonell 1988; Butler and Herrnkind 1988; Aguilar et al. 1992; Quinn et al. 1992). Lellis (1991) estimated that a single floating pueruli collector could trap an average of 300–400 pueruli of *P. argus* a year in Florida and Antigua. Nearly 30,000 pueruli of *P. argus* were captured around Antigua on collectors in one year (Lellis 1990). The successful lobster culture operations in Vietnam and Indonesia are purely dependent upon pueruli collected from wild (Chap. 12). Pueruli of three species, *P. homarus homarus*, *P. polyphagus* and *P. ornatus*, were collected from Kovalam, near Chennai, southeast coast of India, and roofing tiles wound with coir ropes were found to be the most successful collecting device in catching the pueruli. All the species had maximum settlement on tiles suspended at 4-m depth compared to 1- and 2-m depth.

Several studies have examined the timing of pueruli arrivals in relation to the lunar cycle. The greatest catches of pueruli appear to be around the new moon period, although the precise timing varies with different studies or locations (Acosta et al. 1997; Eggleston et al. 1998). At Kovalam, maximum settlement of *P. homarus homarus* and *P. ornatus* pueruli was recorded in the first quarter approaching full moon, whereas *P. polyphagus* pueruli settlement was highest in the last quarter approaching the new moon, showing influence of lunar cycle in puerulus settlement. Settlement *of P. homarus homarus* pueruli was observed throughout the year with a peak in February. However, maximum settlement of *P. polyphagus* (Fig. 10.10) and *P. ornatus* pueruli was in March. Since only small quantities of pueruli could be collected, the study could not conclude the quantum of pueruli that could be harvested to start a commercial lobster farm. Vijayakumaran et al. (2009) observed settlement of fairly good numbers of *P. homarus homarus* pueruli on sea cages suspended for lobster growout experiments at Tharuvaikulam, Tuticorin, southeast coast of India. Pueruli settled inside the cages, a month after seeding with the

Fig. 10.10 Pueruli of *Panulirus polyphagus* collected from wild. (Photo by E. V. Radhakrishnan, CMFRI)

macroalga *Kappaphycus alvarezii*. In New Zealand, permission to private companies for harvesting commercial quantities of wild pueruli was granted in 1996.

In the Caribbean, *P. argus* pueruli were collected using modified 'Witham' and 'GUSI' collectors suspended from a floating raft. GUSI collector was originated in Mexico and is a surface collector suspended at about 2-m deep (Gutiérrez-Carbonell et al. 1992). The tassel fibres resembling artificial seaweed are the most important part of the collector as they actually formed the post-larval shelter. The tassels were 35 cm in length and black in colour as earlier studies revealed a preference for dark-coloured habitats. They were made by shredding nylon rope 0.94 cm in diameter and placing this shredded rope into bundles 0.5 cm in diameter. In all, one hundred and ten tassels were tied to the rope encircling each of the buckets of each collector (Meggs et al. 2011). The Witham collector (Witham et al. 1968) is a surface collector constructed of a polyurethane float, 30 cm square and 2.5-cm thick, with leaves of 3M (Minnesota Mining and Manufacturing Co., Minnesota, USA) nylon webbing, 30-cm wide and 15-cm long attached to one side. While pueruli was observed to reach the inshore regions throughout the year, specific seasonal fluxes were observed. Numerous studies on puerulus collection suggest that there is potential for commercial collection of pueruli at some locations within the natural range of the species.

10.16.3 Potential Impacts to Fisheries from Harvesting Juvenile Lobsters

Scientific studies have indicated that, in the wild, only a very small proportion of pueruli of *P. argus* survive to 1 year after settlement (Marx 1986; Herrnkind and Butler 1994). One study estimated that between 0.6% and 4.1% of all settling pueruli survived to 1 year after settlement (Herrnkind and Butler 1994). Survival of

P. cygnus pueruli one year after settlement was an estimated 2–20% (Phillips et al. 2003). Overall, these data suggest that there is very significant loss of within 1 year after settlement, and as a consequence, the harvesting of pueruli for aquaculture could provide a means of greatly increasing production of lobsters from a wild fishery (Williams 2009). In Australia, removal of 20 million pueruli of the western rock lobster *P. cygnus*, which has a total settlement of 600 million pueruli in an year, represents a decrease of only 0.62% of the wild capture fishery for harvest-sized lobsters (Phillips et al. 2003). In some countries, where wild seed lobster harvesting has been permitted, the potential for reduction in the recruitment of lobsters to wild populations has been offset, either through reduced commercial harvest of adult lobsters or through the release of larger cultured juveniles which are thought to have higher survival due to their size to the wild (Phillips et al. 2003). Therefore, harvesting pueruli from dense settlement areas may help the remaining population survive better due to less competition for space and food. However, harvesting larger juveniles and subadults, which have lower mortality rates, has the potential to impact on the recruitment to the fishery and long-term implications on the sustainability of the natural resource (Diaz-Iglesias et al. 2010; Radhakrishnan 2015). Improving efficiency in large-scale puerulus collection will rely upon our knowledge of the behaviour and ecology of pueruli and of coastal ocean currents aiding highly variable spatial and temporal movement of the pueruli from offshore areas to coastal areas (Jeffs and Davis 2009).

10.17 Juvenile Lobster Collection

Juveniles and subadults have been used as the seed source for lobster growout in many countries. The two promising species that could be considered for growout in India are *P. homarus homarus* and *P. polyphagus*. On the southern part of the Indian coast, gillnets, trammel nets and traps are employed for lobster fishing. Juveniles and subadult lobsters are incidentally caught in gillnets and trammel nets. Nearly 50% of the catch in trammel nets consists of juveniles and subadults (23–48 mm CL), which are not suitable for export (Radhakrishnan 2015) (Fig. 10.11).

An estimated 25 t of juvenile *P. homarus homarus* has been reported to be landed annually along the Tamil Nadu coast (Radhakrishnan 2015). In Maharashtra, during 2000–2008, an estimated average 184 t of *P. polyphagus* was landed by bottom-set gillnets, of which more than 70% were undersized (below 150 g) (Fig. 10.12). In Gujarat, the lobster catch in a trap net called bandhan (*Wada net*) operated in intertidal areas mostly consists of juvenile lobsters (Fig. 10.13).

At Veraval, Gujarat, juvenile lobsters of *P. polyphagus* procured from fishermen and used for stocking in open sea floating cages ranged from 31 to 121 g (Mojjada et al. 2012). At Kanyakumari, southwest coast of India, juveniles and subadults of *P. homarus homarus* collected from the wild for stocking in a floating experimental cage were in the range of 45–90 g (Lipton et al. 2010). At Vizhinjam, on the southwest coast of India, juveniles and subadults of *P. homarus homarus* collected from

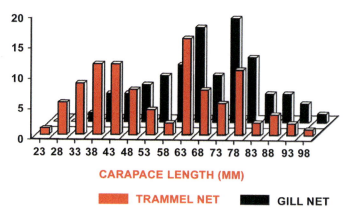

Fig. 10.11 Length composition of *Panulirus homarus homarus* in gillnet and trammel net operated at Kovalam, Chennai, India

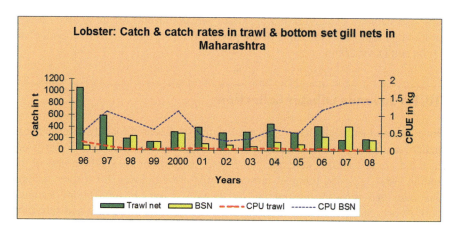

Fig. 10.12 Annual landing and catch rates of *Panulirus polyphagus* by trawl and bottom-set-gillnets in Maharashtra, India

gillnet fishermen of Kanyakumari and Vizhinjam for stocking in a floating cage and in land-based tanks (1100 nos.) were in the size range of 32.5 to 42 mm CL (72.3–114.8 g) (Rao et al. 2010).

10.18 Transport and Holding of Seed Lobsters

Seed stock health is the primary requirement for successful growout of spiny lobsters. Very often, juveniles and sub-adult lobsters get very low priority from fishermen in India as they fetch low price (Radhakrishnan 2015). The lobster fishermen

Fig. 10.13 Wada net, a local gear used for catching *Panulirus polyphagus* juveniles from intertidal mudflat at Veraval, Gujarat, India. (Photo by Mohammed Koya, CMFRI)

bring the gillnet and trammel net to the shore to remove the lobsters from the net. Compared to gillnets, removing lobsters without damage from a trammel net is a tedious job. The lobster fishermen keep the juvenile and undersized lobsters removed from the net in a pit dug on the sandy beach until the traders collect the lobsters. Lobsters are kept alive for 4–5 h in the pit, provided the pits have sufficient moisture and are not directly exposed to the sun. Some traders keep the lobsters in holding tanks maintained either near the landing centre or in seawater recirculating systems. The health and survival of lobsters depend upon their overall condition, quality of seawater and the stocking density. In coastal areas of Tamil Nadu, traders keep the juvenile lobsters at very high densities in tanks with limited water exchange facility. Therefore, these lobsters are either weak or are infected with bacterial diseases, making them unsuitable for growout operations. The seed lobsters are transported either in wet packs or in seawater tanks with aeration.

Holding systems should provide the optimal environmental conditions for lobsters. Critical environmental parameters include the concentrations of dissolved oxygen, ammonia, nitrite, nitrate, pH, salinity, carbon dioxide and alkalinity levels (Losordo et al. 1998). In flow-through systems, the main limiting water quality parameter is dissolved oxygen (Crear and Allen 2002). In recirculation systems, unionized ammonia (NH_3) is the second important limiting factor after dissolved oxygen. The pH, ammonia and oxygen levels in the holding tanks should be monitored periodically to keep the lobsters in a healthy condition. For a detailed protocol on holding procedures for lobsters, refer to the *Guide for Rock Lobster Industry No. 2* (Crear et al. 2003). In India, no specific protocols are followed for holding and transportation of seed lobsters from the landing centre to the growout facility, and this often results in poor quality of seeds and mortality during growout operations (Radhakrishnan 2015). The incidence of Gaffkemia-like bacterial disease reported in lobsters grown in land-based growout system is often due to poor health of seed lobsters and water quality in holding tanks (Vijayan et al. 2010a, b). High lobster densities, which negatively affect water quality, may result in the prevalence of bacterial shell disease, since there is a periphytic association of bacteria on particulate matter (Rayns 1991).

10.19 Quarantine

Currently, lobster growout in India is solely an open-life-cycle production, where the seed source is juvenile lobsters collected from the wild. The growout facility should take care of quarantine and acclimatization procedures for new seed stock arrivals. The entry of disease in the facility will normally be through the new stock brought to the hatchery. Therefore, the seed should be tracked for the presence of bacteria, fungi and viruses. Each lobster should be examined for injury in the abdominal region, which may happen when removing the entangled animal from nets. The injured animals may be kept in separate tanks and treated with antibiotics. The lobsters should not be fed during the quarantine period. After three days, both the injured and the uninjured lobsters may be given a formalin bath of 100 ppm concentration on alternate days for a period of four days and fed with mussels. The growout facility should be stocked only with healthy lobsters. Lobsters with wide difference in weight should not be stocked together to avoid dominance by larger individuals during feeding.

10.20 Growout of Pueruli in Laboratory Conditions

In India, studies on growth of spiny lobsters starting from puerulus are limited. Tholasilingam and Rangarajan (1986) expect *P. polyphagus* pueruli to reach 200 g in 12–15 months. However, Radhakrishnan and Devarajan (1986) found that *P. polyphagus* pueruli took about 27 months to reach 300 g at 21–30°C on an exclusive diet of the clam *M. casta*. Vijayakumaran (unpublished data) indicated growth of 73 g in 310 days for the same species. The observed growth of pueruli for other species is 80 g in 380 days for *P. homarus homarus* and 80 g in 250 days for *P. ornatus* (Radhakrishnan & Vijayakumaran, 1990). Michel (1979) reared five post-larval *P. ornatus* to 300 g within 10 months. It is estimated that *P. cygnus* pueruli take 2.1 years to reach the commercial size of 76-mm CL at about the elevated water temperature of 25°C (Chittleborough 1974b; Phillips 1985; Phillips et al. 1977, 1983). *Artemia* have been used to feed on early juveniles of *P. argus*. However, mortalities were high on feeding with both live and frozen *Artemia* (Pardee, 1992a, b; Pardee & Foster, 1999). For information on stocking densities, shelter, survival and influence of temperature and feeding frequencies on growth of *P. cygnus* fed with natural and formulated diets, refer to Johnston et al. (2006, 2007, 2008).

10.21 Growout of Pueruli in Cages

The post-puerulus phase is characterized by its short duration of 2–4 weeks and high mortality. There is no particular husbandry applied to this stage. In Vietnam, post-pueruli are housed in small cages, generally 3.5 post-pueruli/m^3, stocked at up to 100/cage (28/m^3) and fed with finely chopped trash fish. Juveniles are transferred to larger cages, up to 9 m^3, at densities of up to 20/m^3 and fed with trash fish. In turn, the larger juveniles at around 100 g are transferred to larger cages for growth through to market size at densities of up to 10/m^3 and fed with finely chopped trash fish. Mortality during the post-puerulus phase is very high, 50–60%, and is thought to be primarily attributable to cannibalism. Survival of lobsters through the juvenile phase is likely to be 60–90%. It takes about 6 months for a 2-cm juvenile to reach market size of 200 g for *P. homarus* and 8 months for *P. ornatus* to reach a marketable 350 g (Priyambodo and Sarifin 2009). Vijayakumaran et al. (2009) obtained a lower growth rate for pueruli and post-pueruli of *P. homarus* (mean initial weight 1.58 g) maintained in a sea cage at Tharuvaikulam, Tuticorin, on the southeast coast of India. Pueruli attained 123.1 ± 0.62 g in 266 days (8.8 months) registering a growth rate (weight increase/day) of 0.46 g. By combining growth data obtained from different size groups, they estimated a growth rate of 200 g in 365 days and 350 g in 490 to 520 days for the species. Jones et al. (2001) demonstrated that good growth rates of *P. ornatus* at densities as high as 5 kg/m^2 could be sustained and enable a 1-kg harvest size to be achieved in 18 months from a 3-g stocking size. This concurs with the experience of the Vietnam farmers who generally harvest >1 kg lobsters after 18–20 months of growout at densities of up to 80 lobsters per cage (4X 4 m).

10.22 Land-Based Indoor Systems for Juvenile Growout

The indoor lobster culture technology involves raising lobsters in cement tanks or raceways either with a water recirculating system or flow-through system (Radhakrishnan 2015). A semi-flow-through growout system with self-cleaning facility has been established at the Calicut Research Centre of Central Marine Fisheries Research Institute, India (Fig. 10.14). The tanks and raceways can be of varying sizes and shapes: circular, rectangular, square or oval. Circular tanks may take more space, but with a central drain, the tank will be most efficient in removing solid wastes. Rectangular and square tanks can save space and construction cost by interconnecting the tanks with common walls. However, the waste is likely to accumulate in the tank corners. Oval-shaped raceways with the drain at one end and slope towards the drain are also effective in waste removal. Square tanks with corners tapered are the most advantageous with characteristics of both circular and the square tanks. Each tank has a two-way waste removal facility. The dual drain system has a central stand pipe, which opens outside at the base of the tank, through which the wastes in suspension like the faecal matter will be automatically flushed out. The second drain pipe, which is parallel to the first pipe, has a stand pipe outside the

Fig. 10.14 A self-cleaning partial flow-through lobster growout system at Calicut Research Centre of CMFRI, India. (Reprinted from Radhakrishnan E.V. 2015, Advances in Marine and Brackishwater culture, P. Santhanam, A. R. Thirunavukkarasu, P. Perumal (eds), Springer with permission)

tank. The solid wastes settling at the bottom such as unfed feed will pass through the bottom drain and be flushed out through the stand pipe. The entire tank water can be drained by removing this stand pipe. The inlet water entering into the tank through a vertical pipe with holes on one side creates a vortex, which concentrates both the suspended and solid wastes towards the central drainage. The system can be either flow-through or partial flow-through with 60–75% water replacement with fresh seawater in the evening and in the morning. The advantage of a self-cleaning tank is least disturbance to the lobsters while the wastes are automatically flushed out from the system.

10.23 Growout of Juveniles in Laboratory

Nair et al. (1981) reported average growth increment per moult of 2.3 to 3.4-mm CL for males (size ranging from 45.0- to 56.0-mm CL) and 2.3–3.0 mm for females (size ranging from 33.0- to 57.0-mm CL) for *P. homarus* under aquarium conditions. On the other hand, Kulmiye and Mavuti (2005) obtained an average growth increment of 1.72-mm CL/moult for the subspecies *P. homarus homarus* from Kenya (size class 46–55- to 76–85-mm CL). In *P. homarus rubellus* from South Africa, increase in carapace length per moult varied from 0.7 mm (size class 7–9-mm CL) to 3.7 mm (size class 45–49-mm CL) (Berry 1971). While average carapace length per moult decreased with increase in length in both males and females, intermoult duration increased with increase in size (36 to 68 days for lobsters measuring 7 to 59 mm in CL). Berry (1971) observed the most rapid rate of growth between carapace length of 15 mm and 50 mm in *P. homarus rubellus*, with growth rate slowing down beyond this size, which corresponds to the size at first sexual maturity. Growth rate was similar in both males and females up to 60-mm CL, beyond which males show faster growth. Mass-rearing trials of *P. homarus* under controlled conditions conducted at Tuticorin showed higher mean-specific daily food consumption (26.9%) and faster growth rate in smaller size groups (<15 g), compared to lower consumption (10.95%) and growth in larger size groups (51–100 g) (Rahman et al. 1997). Rodriguez-Viera and Perera (2012) report high specific growth rate (SGR) in juvenile *P. argus* fed on fresh squid meat compared to prepared pellets. Lobsters attained four times more final weight than those fed on dry pellets. In general, specific growth rate is higher for tropical species compared to temperate lobsters (Rodrguez-Viera and Perera 2012). *P. polyphagus* juveniles fed on the gastropod *Turbo* sp. exhibited maximum daily weight increments of 0.47 g and 0.33 g in males and females, respectively. The lobsters showed highest preference for gastropods and clams compared to crab, squid and fish (Kizhakudan and Patel 2011). In *P. polyphagus*, mean increment in CL per moult in males was found to increase from about 2 mm (size range of 16–20-mm CL) to about 5–7 mm (size range of 51–80-mm CL) and then decrease to about 1.5 mm (size range of 106–110-mm CL). In females, the mean increment in CL per moult was reported to increase from about 3.8 mm (size range of 31–35-mm CL) to a peak of about 4 mm (size range of 41–65-mm CL) and decrease to about 1 mm (size range of 86–90-mm CL) (Kizhakudan et al. 2013). Radhakrishnan (2015) estimated that *P. homarus homarus* juveniles weighing 80 g (Fig. 10.15) can be grown to 400 g in 12 months (Fig. 10.16) on a mixed diet of mussel, clam (Fig. 10.17) and fish. Chittleborough (1974b) successfully reared *P. cygnus* juveniles (average 45 g) to the minimum legal size (MLS) of 387 g (76-mm CL) in 17 months under near-optimal conditions in the laboratory. Lellis (1990) was able to grow similar-sized juveniles of *P. argus* to 454 g in 14 months on a mixed diet of ground fish, clam and squid. An

Fig. 10.15 *Panulirus homarus homarus* juveniles collected from wild for stocking in an indoor growout system (Photo by: Joe K Kizhakudan, CMFRI)

Fig. 10.16 Harvest of *Panulirus homarus homarus* from an indoor growout system, India (Reprinted from Radhakrishnan E. V 2015, Advances in Marine and Brackishwater culture, P. Santhanam, A. R. Thirunavukkarasu, P. Perumal (eds), Springer with permission)

Fig. 10.17 Live clams for feeding lobsters in indoor growout systems (Photo by: Joe K Kizhakudan, CMFRI)

average weight gain of 30–50 g in 220 days with a survival of 50% was obtained for *P. homarus homarus* under high-density stocking (3–5 kg/m^2) growout experiment in the laboratory (Kizhakudan et al. 2007).

The studies on the effect of density on growout of *P. ornatus* juveniles (3.24 g, 13.8 ± 0.13-mm CL) showed survival and size unaffected by the densities tested (14, 29 and 43/m^2), and this density is equivalent to the highest suggested as suitable for other palinurids also (Jones et al. 2001). After 272 days, the lobsters measured an average of 225.3+4.68 g (61.8 ± 4.7-mm CL) at harvest in all the treatments. A specific growth rate of 1.56% per day indicates that the lobster is capable of achieving 1 kg in 18 months.

10.23.1 Environmental Requirements

The lobsters are subjected to changes in environmental conditions such as temperature, salinity, pH, oxygen and ammonia even in a culture environment. The animal counteracts the external environmental variations through a physiological process called homeostasis. If the variations are extreme, cell damage occurs and the state of health of the animal is compromised, and it becomes susceptible to disease due to immunoincompetence. Therefore, it is necessary to provide near-optimal conditions for maximizing growth and survival of the cultured organisms. The ability to tolerate changes in environmental conditions varies with moult stage, age and size of the organisms, and this also depends upon whether the changes are sudden or gradual. Knowledge of the immune response of the lobsters to water quality and stress may provide insight into developing proper management strategies (Verghese et al. 2007).

10.23.1.1 Temperature

Spiny lobsters can tolerate a wide range of temperatures without any adverse effects. However, it is better to avoid exposing them to extreme temperatures (chilling is commonly used to prepare lobsters for aerial transport) (Crear and Allen 2002). Jones (2009a) suggested 27°C as the optimal temperature for growth of *P. ornatus*, as, at this temperature, the moult increment has been the maximum and intermoult period the shortest. A laboratory trial to examine the effect of temperature on growth showed that *P. ornatus* juveniles grew 31% faster at 30°C (0·88-mm CL week 1) compared with growth at 26°C (0·67-mm CL week 1) as a result of shorter moult intervals in large juveniles (Dennis et al. 1997). Crear et al. (2000) observed significantly faster growth ($P < 0.05$) of *J. edwardsii* juveniles at 18°C (SGR = 1.32% BW day^{-1}) than at ambient temperature (13–18°C) (1.21% BW day^{-1}), obviously due to significantly higher ($P < 0.05$) apparent feed intake. (Lellis and Russell 1990) reported 29–30°C as the optimal temperature at which intermoult period is shorter with no significant effect on moult increment in the Caribbean spiny lobster *Panulirus argus*. Witham (1973), however, found the most rapid growth among juveniles of the species at lower temperatures of 25–27°C, and that survival was affected when exposed to temperature below 15.6°C or above 32.2°C. Some males reach 450 g in 12 months and 1.4 kg in 2 years at 29°C (Lellis 1991). Radhakrishnan (2015) suggested 25–30°C as the temperature range for growout of the cultivable species of lobsters in Indian waters (Table 10.6).

10.23.1.2 Salinity

Though the spiny lobsters *P. ornatus* and *P. homarus homarus* have the ability to tolerate relatively wide range of salinities, they have clear optima for maximum growth and survival. Jones (2009a) reported significantly low growth rate for *P. ornatus* (daily growth coefficient 0.26) reared for 91 days at 20 ppt salinity compared to 1.29 for those reared at 35 ppt, and these lobsters were 284% larger than those grown at 20 ppt. However, survival was lowest in lobsters grown at 35 ppt, which is considered to be the most ideal salinity regime for on-growing *P. ornatus*.

Low survival at the optimum salinity was attributed to three reasons: relatively high ammonia content in culture water, frequent moulting leading to high cannibalism of newly moulted lobsters and increased potential of diseases. Success on-growing for over 3 months at salinities of 25 and 30 ppt suggests that *P. ornatus* is capable of hyperosmotic regulation, to the extent that it may be considered marginally euryhaline (Rockel and Watson III 1996). Studies on salinity tolerance of *P. homarus* also indicate that salinity variations up to 25 ppt are tolerable by the species below which growth and survival are drastically affected (Vidya and Joseph 2012). The lower growth rate at lower salinities may be due to high energy spent by the lobster to maintain the osmotic balance (Parry and Potts 1965). Decreasing salinity has also been shown to affect the metabolic rate in spiny lobsters (Schmitt and Uglow 1997; Villareal et al. 2003).

The Mud spiny lobster *P. polyphagus*, unlike *P. homarus homarus* and *P. ornatus*, can withstand wider fluctuations in salinity. Laboratory studies show that *P. polyphagus* is hyperosmotic and therefore euryhaline and can tolerate salinities ranging from 10 to 55 ppt; therefore, this may be a suitable species for culture in coastal areas and shrimp ponds (Kasim 1986).

Sudden exposure to extreme environmental conditions can lead to reduced disease resistance, growth impairment, poor reproductive performance and lower survival in stenohaline marine organisms (Verghese et al. 2007). In a land-based growout system, sudden exposure to seawater salinity of 20 ppt and consequent *Gaffkemia* infection resulted in heavy mortality of *P. homarus homarus* (Samraj et al. 2010). Rain and consequent land run-off may lead to sudden reduction in salinity in coastal areas, and therefore, the salinity and pH should be checked prior to addition of water into the culture system. Jayagopal and Vijayakumaran (2010) reported incidence of *Gaffkemia* in secondary lobster holding systems having poor water quality. Therefore, management of water quality and maintenance of optimal environmental conditions are crucial in disease prevention and successful aquaculture.

Verghese et al. (2007) reported significant reduction in total haemocyte count (THC) and decreased PO activity in *P. homarus homarus* exposed to lower (20 and 25 ppt) and higher (45 ppt) salinities, which indicates that these lobsters are under stress and that their susceptibility to opportunistic pathogens is high. Prolonged reduction in THC in cultured lobsters may lead to reduced immunocompetence. *P. homarus homarus* maintained in 25% and 50% daily water exchange regimes had lower THC compared to those maintained in 75% and 90% water exchange regime (Mary Leema et al. 2010).

10.23.1.3 Dissolved Oxygen

Spiny lobsters are oxygen regulators (Winget 1969), but they are oxygen independent only down to some critical oxygen tension (Booth and Kittaka 1994). Oxygen consumption and the lethal oxygen level depend on body weight and moult state of the lobster and on environmental parameters such as temperature, salinity and partial pressure of oxygen in the water. At non-lethal salinities, *P. polyphagus* was observed to survive in oxygen partial pressures as low as 5 mm Hg, whereas in lethal salinities mortality occurs even at higher oxygen partial pressures owing to added osmotic stress (Kasim 1986). The recommended optimal dissolved oxygen level in the lobster culture system is >80% saturation but should not be below 70% saturation (Crear and Allen 2002). Supersaturation by air can cause gas bubble disease in spiny lobsters (Cobb 1976; Brisson 1985). In a growout system, the most critical period for a fully fed lobster and those in the process of moulting is during early morning (between 3 and 5 am) when oxygen level is likely to fall below the lethal level due to high oxygen demand by the fully fed lobster and the unfed food. Mortalities normally happen during this period. The lethal low oxygen level is estimated to vary between 0.5 and 3.0 mg/L, depending upon the species (Booth and Kittaka 1994).

10.23.1.4 Ammonia

In the recirculating culture system for *J. edwardsii* and for other species, the recommended levels of total ammonia is below 0.5 mg/l and NH_3-N (unionized ammonia) not to exceed 0.1 mg/L, nitrite levels below 1 mg NO_2-N per/L and nitrates below 100 mg/L (Forteath 1990). Crear and Allen (2002) have set the tolerance limits of ammonia to 2 mg/L and nitrite to 5 mg/L for *J. edwardsii* and *P. cygnus* stocked in holding system prior to live packing. Biological filtration is the most widely used method for removing ammonia from culture water.

10.23.2 Stocking Density, Growth and Survival

The communal behaviour of spiny lobsters makes them ideal candidate species for aquaculture. However, at higher densities, growth and survival can be adversely affected. For profitable aquaculture, information on the optimum stocking density for maximum growth and survival is important. Economics of production require high-density stocking in the culture system that gives maximum growth and survival (Jones et al. 2001).

10.23.2.1 Effect of Stocking Density on Survival

The highest growth rate obtained in indoor culture of *P. homarus* juveniles (45 g wet weight) was 0.75 g per day at a stocking density of 7 lobsters/m^2 (Radhakrishnan and Vijayakumaran 1990). Jones et al. (2001) obtained an average of 52.5% survival for juvenile *P. ornatus* (mean initial weight 3.24 + 0.9 g) grown for 272 days in a flow-through system and fed with a prawn pellet. Growth in length and weight in *J. edwardsii* was slower when stocking densities increased from 50 to 200 m^{-2}. However, stocking density has no bearing on overall mortality, which was generally lower in all densities tested (James et al. 2002). They suggested stocking of 50–100 lobsters/m^2 with adequate shelter for good growth and survival. In *P. argus*, increases in stocking densities, equivalent to 266, 533 and 666 juvenile lobsters/m^2 of tank floor, resulted in lower survival due to cannibalism (Diaz-Iglesias et al. 1991). Mortalities of 22–33% over nine weeks were reported at very low stocking densities of 17.4 and 34.8 juvenile lobsters/m^2 suggesting feed quality as a prerequisite for high-density growout systems (Pardee 1992b; Pardee and Foster 1992). Jones et al. (2001) found survival and growth of *P. ornatus* maintained in a flow-through holding system unaffected by stocking densities (14, 29 and 43/m^2). Probably, the densities tested may not be high enough to inhibit growth, particularly for 1- and 2-year-old lobsters (Johnston et al. 2006). Growth and survival were significantly lower in post-pueruli at higher densities (150/m^2) compared with lowest density (30/m^2) in *P. cygnus* (Moyle 2005). Therefore, it was suggested that a stocking density of 100/m^2 may be optimal for this species to get good growth.

Vijayakumaran et al. (2009) stocked *P. homarus* of varying sizes at densities ranging from 21 to 80 lobsters per m^2 in floating FRP sea cages along the southeast coast of India to study growth and survival. Survival was not affected by increase in stocking densities. For pueruli and post-pueruli, 70% survival was obtained after a growth period of 266 days, whereas in juveniles and subadults, percent survival ranged from 72 ± 6% to 100%. In Vietnam, lobster farmers obtained survival rates of 70–95% in *P. ornatus* stocked at 15 individuals/m^2 in fixed sea cages (Tuan and Mao 2004). Survival of *P. homarus* stocked at 24 lobsters/m^3 in marine cages in Indonesia was around 70%. Jones and Shanks (2008) investigated growth and survival of *P. ornatus* weighing a mean 750.2 + 6 g stocked at a stocking density of 6.2 lobsters/m^2 in four net cages maintained in a shrimp pond. The mean survival was 77.5% after 126 days of culture.

10.23.2.2 Effect of Stocking Density on Growth

Johnston et al. (2006) observed that growth in *P. cygnus* was unaffected by stocking density and opined that the pueruli and juveniles of the species collected from the wild are suited for aquaculture. A stocking density of 50 pueruli/m^2 and between 20 and 25/m^2 for 1- and 2-year-old juveniles was recommended to maximize survival and production. Jones et al. (2001) also recorded similar observation in *P. ornatus* where growth was not affected by stocking density. Though survival was an average of 52%, rapid growth and high specific growth rate of 1.56% per day is sufficient enough to assume that the lobster can be grown from 3 g to 1 kg in 18 months.

Pueruli and post-pueruli of *P. homarus homarus* (mean weight 1.58 g) stocked at a density of 60 individuals/m^2 in a marine cage reached 123 g in 266 days (Vijayakumaran et al. 2009).

10.23.2.3 Effect of Shelter on Growth and Survival

In growout tanks, shelter improved survival of *P. ornatus* juveniles and reduced cannibalism of newly moulted lobsters, whereas it had no significant effect on growth (Chau et al. 2009). No significant difference in growth but low survival was reported in *P. homarus homarus* provided with shelter (Vijayakumaran et al. 2010). However, 75% of the lobsters were found to occupy the shelter in day time. In contrast, increased feeding and faster growth were reported in *P. cygnus* provided with communal shelter (Chittleborough 1976a). Radhakrishnan (personal observation) observed crowding of *P. homarus* lobsters inside PVC tubes provided as shelter, leading to loss of appendages and injury. Such crowding may interfere with oxygen consumption, metabolism and slow growth. The amenability of *P. ornatus* and *P.*

homarus homarus to high-density stocking and production in sea cage culture systems indicates that shelter is not a prerequisite for optimum growth and survival in these two species (Priyambodo and Jaya 2009; Jones 2010; Vijayakumaran et al. 2010). However, for species with intense aggressive behaviour and territoriality, shelter may be required, and such species may not be suitable for high-density farming.

10.24 Food Consumption and Conversion

Although spiny lobsters accept a wide range of fresh food, they are selective feeders preferring shellfish to finfish. Among the shellfish, they have preference for mussels than clams. The use of fresh feeds comprising molluscs and fish has provided excellent growth rates in *P. homarus homarus*. The mean gross food conversion ratio (on a wet weight basis) was 5.0 for mussel-fed lobsters, 5.8 for clam fed, 6.6 for fish fed and 4.8 for a mixed diet comprising all the three food items in equal proportions (Radhakrishnan 1995). Lobster farmers in Vietnam depended almost entirely on trash fish to feed lobsters in cages, generating high organic load in the culture site, resulting in prevalence of disease and mortality. The feed conversion ratio (FCR) has been 20:1 and in some cases even 50:1 in using the traditional trash fish feeding (Jones et al. 2010). For successful culture of lobsters, transition from trash fish to manufactured diet is essential for optimizing their aquaculture practices. The most significant advantage from pellet feeding will be minimal environment impact, good growth and sustained production. Providing a nutritionally balanced cost-effective food will be one of the major challenges in environment-friendly aquaculture. A substantial body of information is already available on nutritional requirements of *P. ornatus* (Williams 2004), and developing an artificial diet for growout of pueruli and juveniles of all species is a very high priority (Smith et al. 2005). Cox and Davis (2006) studied the effect of feeding frequency on growth of juvenile spiny lobster *P. argus*. Irwin and Williams (2007) examined the apparent digestibility of selected marine and terrestrial feed ingredients for *P. ornatus*. Evaluation of seven potential artificial diets for the culture of post-pueruli of *P. argus* showed fastest growth rate over a 28-day period (Cox and Davis 2009).

10.25 Lobster Farming in Shrimp Pond

Spiny lobster farming in land-based systems such as tanks and raceways with flow-through or recirculating water supply has been proven to be technically and economically feasible (Radhakrishnan 2005; Jones and Shanks 2009). Growth and survival in such systems are promising and suggests commercial production could be done, although the production cost is likely to be relatively high (Jones et al. 2010). A production trial of *P. ornatus* in the intake channel of a shrimp farm in northern Queensland was carried out to study the biological feasibility of farming lobsters in ponds, where the environmental factors such as salinity and turbidity are likely to fluctuate. Laboratory studies have already indicated that *P. ornatus* is marginally euryhaline and can withstand salinity ranges of 20–35 ppt, although growth is faster in 30–35 ppt range (Jones 2009a). The trial system consisted of four rectangular cages, 1.8 mX 1.8 mX 0.9 m deep, suspended from the surface using a floating frame. Cages were each stocked with 20 lobsters (6.17 m^2) of mean size 750.2 ± 6.0 g. A control group of 13 lobsters (mean size 717.8 ± 9.4 g) was stocked in a fibreglass tank and managed equivalently. Although growth was monitored from 750 g to 1 kg only, growth rate was good (8 g/week), survival 78% and colour and vigour at harvest excellent, showing that *P. ornatus* can be cultured in ponds with good survival and growth (Jones and Shanks 2009). *P. polyphagus* also may be a suitable species for pond culture as their natural habitat is muddy bottom and has the ability to tolerate wide fluctuations in salinity (Kasim 1986). In pond culture, trash fish cannot be used as feed as there is no mechanism to remove the unfed food. Pellet feed, which provides good nutrition and minimal waste, is compulsory. Jones and Shanks (2008, 2009) used a laboratory-manufactured moist pellet as feed, which was readily accepted by the cultured lobsters. Radhakrishnan and Vijayakumaran (unpublished) made an attempt to grow the more salinity-tolerant *P. polyphagus* in a pond at Muttukadu, south of Chennai, on the southeast coast of India. However, unprecedented rain and consequent sudden reduction in salinity below their tolerance level resulted in mortality.

10.26 Lobster Culture in Intertidal Ponds

In Bhavnagar district of Gujarat state, where tidal amplitude is high, culture of wild-caught juveniles of *P. polyphagus* was attempted in small ponds or 'pits' dug in intertidal limestone (Sarvaiya 1987; Philipose 1994). Juveniles of lobsters caught in 'bandhan', a net operated in coastal areas of Gujarat, are stocked in the pit. The pits are covered by nets to prevent escape of lobsters. The pits are flushed two times a day during high tide. Lobsters weighing an average of 30–50 g stocked at a density of 20 individuals/m^2 and fed with trash fish, gastropods and crabs attained 100–125 g in 90 days. Seeds purchased at ₹ 20/kg from fishermen were sold at Rs. 250/kg after fattening (Radhakrishnan et al. 2009).

10.27 Growth Enhancement

Radhakrishnan and Vijayakumaran (1984a, b) investigated the effect of bilateral eyestalk ablation on moulting and weight gain in different size groups of the spiny lobster *P. homarus homarus*. Ablation resulted in increased frequency of moulting and higher weight gain in all size groups irrespective of the reproductive status. In 25–40-mm CL size groups, ablated lobsters gained an average weight of 1.2 g/day, whereas normal lobsters gained an average weight of 0.28 g/day. In 41–66-mm CL groups, the average weight gain per day for ablated lobsters was 2.4 g compared to 0.35 g in normal lobsters. While survival was 70% in ablated *P. homarus*, 95% of control group survived under experimental conditions. In *P. polyphagus*, ablated juveniles showed a maximum weight gain of 370% in 114 days, while in the control group, an increase of 174% in weight gain was observed (Kizhakudan 2013). Eyestalk-ablated juveniles of *P. ornatus* attained an increase of 550 g in 19 weeks compared to 65.4 g increase in a control group (Silas et al. 1984). Radhakrishnan (1995) reported that an eyestalk-ablated *P. ornatus* weighing 100 g could reach 1500 g in 8 months. Juinio-Menez and Ruinata (1996) studied the effect of unilateral and bilateral eyestalk ablation on *P. ornatus* from the Philippines. At the end of their 4-month study period, the average CL of bilaterally ablated lobsters measuring 44.8–45.4-mm CL increased to 60.3–68.8-mm CL and weight from 90.3–92.8 g to 210.0–265.0 g. Despite the significantly higher growth rates of individuals, the total yield in biomass of bilaterally ablated lobsters (0.8 kg) was very much less than the unilaterally ablated (8.0 kg) and control group at the end of the 4-month experiment because of very low survivorship of bilaterally ablated lobsters. Under low oxygen conditions, ablated lobster is likely to succumb faster because of their higher metabolic rate and oxygen demand compared to normal lobsters, although the lethal oxygen levels for ablated *P. homarus* is not known (Radhakrishnan 1989).

10.28 Growout of Juveniles and Subadults in Cages

The technical feasibility of sea cage culture of lobsters collected from the wild was tested in several countries. In Vietnam, the spiny lobsters *P. ornatus* and *P. homarus* are cultured in fixed and floating cages since 1992. Until 2006, 1900 t of lobsters valued at US$90 million was produced (Thuy and Ngoc 2004: Hung and Tuan 2009). Experimental cage farming using wild harvested lobster juveniles has also been conducted at Tuticorin and Vizhinjam on the southern Indian coast and at Veraval on the northwest coast (Srikrishnadhas et al. 1983; Vijayakumaran et al. 2009; Rao et al. 2010; Mojjada et al. 2012).

10.28.1 Open Sea Growout of Lobsters in Sea Cages in India

Experimental growout of spiny lobsters in floating open sea cages was attempted along the northwest and southwest coast of India by the Central Marine Fisheries Research Institute, Cochin. Though there is a minimum legal size for export of spiny lobsters in India, there is no legal ban on fishing. On the southwest coast, fishermen operate gillnets and trammel nets for lobster fishing in inshore rocky areas, and nearly 50% of the catch consists of undersize lobsters, which have no market value and cannot be legally exported (Radhakrishnan 2015). On the northwest, fishermen catch young lobsters from inshore areas using gillnets and a traditional stake net called *Wada net*. *Panulirus homarus homarus* on the southwest and southeast coast and *P. polyphagus* on the northwest coast are the two promising spiny lobster species that may be considered for on-growing. *T. unimaculatus*, the slipper lobster, is another candidate species suitable for culture.

10.28.1.1 Cage Growout of *Panulirus homarus homarus* at Tuticorin, Southeast Coast of India

The first trial on cage growout of spiny lobster was reported by Srikrishnadhas et al. (1983) at Tuticorin, southeast coast of India. Forty numbers of wild-caught *P. homarus* juveniles (mean weight 20 g) were stocked (stocking density 10 nos./m^2) in a floating galvanised iron (GI) cage of 2.5X1.6X4 m dimension. The lobsters attained an average weight of 165 g after 8 months, with a survival of 57.5%.

Pilot-scale sea farming of lobsters was also carried out by the National Institute of Ocean Technology (NIOT) at Tuticorin. While the lobster cages in Vietnam were bigger net cages, NIOT has designed small cages for distribution to more beneficiaries under the socioeconomic programme of livelihood security for coastal fishers. Another important factor was to make the cage design simple and small, so that it can be moored and operated at a low depth by the fishers, including women. Thus, low cost, durability, ability to withstand unfavourable sea conditions and ease of operation by coastal fishers were the main factors that influenced cage design and development. Growth and survival of lobsters were similar in both mild steel and fibre-reinforced plastic cages, but the latter, even though more expensive, have performed better owing to higher durability and higher stability in unfavourable sea conditions.

Initially, mild steel cage frames with pulleys to lift the inner compartments and capable of accommodating four 1-m^3 compartments were designed. As the growout operations progressed, it became cumbersome to lift the inner cages as even mild fouling made them difficult to operate. Hence, fibre-reinforced plastic (FRP) material was used to design cages with floatation devices, and four such 1-m^3 cages were tied together and moored with a single anchor. As no metal part was exposed, it was corrosion resistant and was easier to operate and clean. Algae, barnacles, bryozoans and bivalves such as edible oysters were the main fouling organisms in both mild steel and fibre-reinforced plastic cages.

Lobster fattening trials were conducted in mild steel cages at stocking densities ranging from 21 to 38 individuals/m^2 at Tharuvaikulam, Thoothukudi district (Tuticorin), Tamil Nadu. Initial lobster weights ranged from 60 to 160 g in all trials. Auto-stocking of pueruli and early juveniles was noticed in all cages, and one experiment was conducted with auto-stocked pueruli and post-pueruli (1.58 ± 0.62 g) with a stocking density of 60 individuals/m^2 at this site. FRP cages were also used for lobster-fattening trials at Kulasekharapatnam in the same district and at Erwadi (Ramanathapuram district) and Parangipettai (Chidambaram district). At Erwadi, the stocking density of juveniles and subadults (30–150 g) in these trials was 50/m^2 and 80/m^2. The duration of the growout trials varied from 62 to 266 days (Vijayakumaran et al. 2009).

Panulirus homarus homarus was the main species of lobster stocked in all the sites, though a few *P. ornatus* were also reared alongside when available in wild collections. Marine clams, *Donax* spp., collected from the intertidal zone near the cage sites, were the main food. Other foods, such as gastropod meat, *Xancus pyrum* (at Tharuvaikulam), marine crab (*Charybdis* sp.), mantis shrimp (*Squilla* sp.), fish (clupeids and *Leiognathus* sp.) and squid heads (*Loligo* sp.), collected by fishers during their daily fishing voyages, were fed once in the evening daily. Green mussel, *Perna viridis*, was cultured at Tharuvaikulam and also used to feed lobsters.

For pueruli and post-pueruli, the survival was 72 ± 6%, mainly due to escape from the cages. Increase in stocking density from 21 to 80 individuals/m^2 did not affect survival of juvenile and subadult lobsters, which varied from 81 to 100% at different culture sites.

10.28.1.1.1 Growth of Pueruli, Post-Pueruli and Juveniles

Pueruli and post-pueruli ($n = 33$) reached 123.10 ± 0.62 g in 266 days, and subadults (123.61 ± 29.26 g, $n = 12$) attained 341.25 ± 46.2 g in 225 days. The growth rate (weight increase per day) increased from 0.46 ± 0.10 g in pueruli and post-pueruli to 0.97 ± 0.20 g in subadults. However, the specific growth rate (SGR) was highest (1.64) for post-pueruli, which was reduced as the lobster's weight increased. It was lowest (0.43) for lobsters with an initial average weight of 138.10 g. SGRs of juveniles at the open sea sites of Erwadi (0.82) and Kulasekharapatnam (0.96) were higher compared with those lobsters at the semi-enclosed site at Tharuvaikulam (0.43 to 0.67) indicating that growth rate would be better at the sites with better seawater flushing rates, and at such sites, the stocking density can be increased without affecting survival.

In *P. ornatus* juveniles, the daily growth rate increased from 0.94 ± 0.33 g in 71.0 ± 21.8 g of lobsters to 1.18 ± 0.11 g in 165.0 ± 9.6 g of lobsters, whereas the SGR decreased from 0.70 to 0.43. Survival varied between 81 and 100%. The survival rates obtained at all open sea sites (72–100%) at higher stocking densities (21–80 individuals/m^2) in this study are comparable to the results obtained in larger cages in Vietnam for *P. ornatus* (Tuan and Mao 2004).

The growth of animals tends to decrease when they attain maturity, and this has been one of the problems in aquaculture of many fish species. In all trials in small cages, males and females were reared together, and spawning was recorded only once, in a female *P. homarus* weighing 220 g, at Tharuvaikulam, which had low flushing rate. At higher stocking rates and higher flushing rates in the open sea sites, no mating was recorded, though the lobsters attained sizes at which they usually reproduce.

The highest growth rate obtained in indoor culture of *P. homarus* juveniles (45 g of wet weight) was 0.75 g per day at a stocking density of 7 individuals/m^2 (Radhakrishnan and Vijayakumaran 1990). For subadults of this species (90–120 g of wet weight), 0.80 g growth per day was recorded at a lower stocking density of 3 individuals/m^2 (NIOT unpubl. data). The highest growth rate of 0.97 g per day for juveniles and subadults of *P. homarus* (30–170 g of wet weight) in cages at open sea sites in this study exceeded these growth rates. This growth rate almost equals the highest growth rate (0.99 g per day) of spiny lobster juveniles in India, obtained with *P. polyphagus* grown in intertidal pits at Saurashtra (Sarvaiya 1987).

Growth rates of 200 g in 365 days and 350 g in 490–520 days for *P. homarus homarus* pueruli in sea cages indicate that lobster farming in India is potentially economically viable. These growth rates were considerably higher than those obtained in onshore studies of pueruli of the same species (80 g in 380 days,

Radhakrishnan and Vijayakumaran 1990). The observed growth of pueruli of other species are 73 g in 310 days for *P. polyphagus* (Vijayakumaran, unpubl. data), 80 g in 250 days for *P. ornatus* (Radhakrishnan & Vijayakumaran, 1990), 80 g in 455 days for *P. cygnus* (estimated from Chittleborough 1974b), 16 g in 432 days for *Palinustus waguensis* (Vijayakumaran, unpubl. data) and 36.9 g in 365 days in sea cages for *J. edwardsii* (Jeffs & James, 2001). However, Tamm (1980) raised pueruli of *P. ornatus*, the fastest-growing tropical spiny lobster, to 300 g in 365 days, and Vietnamese lobster farmers are growing post-pueruli of *P. ornatus* to approximately 1000 g in 540–600 days (Anon 2004; see Williams 2007).

10.28.1.2 Cage Culture at Vizhinjam, Southwest Coast of India

At Vizhinjam, on the southwest coast, the trial experiments involved both sea-based and land-based on-growing. The objective was to compare the growth in sea cages as well as in an onshore facility (Rao et al. 2010).

The experiment used lobster seeds from a natural source. The holding centres purchase lobsters from fishermen and maintain them until the exporters lift the stock. The holding centres may or may not be near the seacoast. Holding centres, which are away from seacoast, use recirculation systems to maintain the lobsters. However, due to inadequacy of the filter or water, the seawater in such systems has very high ammonia content. Under such conditions, the lobsters either become weak or are infected with disease due to physiological stress. The bacterial disease Gaffkemia has been reported from some holding centres. These lobsters on stocking gradually become weak and end up with heavy mortality. There is no protocol for holding and transport of juvenile lobsters that could be utilized for on-growing (Radhakrishnan 2015).

In onshore FRP tanks of 10-t capacity, 100 lobsters were stocked. In a cylindrical floating cage, 1100 juveniles and subadults of *P. homarus homarus* ranging in size from <100 g to >150 g (mean body weight 114.8 ± 25.67 g) were also stocked. The cage of 7-m diameter and 4-m depth was moored at a depth of 10 m and about 75 m away from the shore (Fig. 10.18). The experiment used live brown mussel *Perna*

Fig. 10.18 Circular floating net cages for spiny lobster growout (Photo, CMFRI)

10 Breeding, Hatchery Production and Mariculture

Fig. 10.19 Harvest of cage grown *Panulirus homarus homarus* by CMFRI at Vizhinjam, south west coast of India (Photo by E. V. Radhakrishnan)

indica as feed, and it was fed daily at 16–20% of body weight. The lobsters stocked in the indoor tanks and fed on the clam *Meretrix* sp. attained 137 g from an initial average weight of 77.88 g. The lobsters in the cage attained a mean weight of 223 ± 42.86 g (size range of 220–350 g) in 135 days (Fig. 10.19). Survival rates of 75% and 71% were obtained in the floating cage and in the land-based system, respectively. The mean weight increase in cage-grown lobsters was an average of 0.82 g/day compared to 0.48 g/day in tank-grown lobsters.

No economic comparison of the farming operations was available. The results of this study show that lobsters had a significantly higher growth rate in sea cage than in the land-based system. It is obvious that the cost of production from onshore tanks will be much higher than from a sea cage operation due to the high infrastructure cost and operational cost of maintaining the stock in onshore tanks. The commercial feasibility of on-growing lobsters in a land-based facility will therefore depend upon reducing these costs (Jeffs and Hooker 2000). The major hurdles in sea cage growout are cleaning the cages from fouling and removing the feed remains (mussel shells) after feeding the lobsters. The other area which needs improvement is to increase the durability of the cage for at least two to three cycles for economic viability. The growout operation cannot entirely depend upon natural feeds, and development of cost-effective suitable artificial feeds appears to be a priority for improving the economic outlook for culturing spiny lobsters.

10.28.1.3 Cage Growout at Kanyakumari, Southwest Coast of India

Lipton et al. (2010) reported on cage growout of *P. homarus homarus* at Kanyakumari, Tamil Nadu, on the southwest coast of India. Seeds obtained from local sources are stocked at 85 nos./m^2 (2400 numbers) in a floating net cage of 6-m diameter moored at a depth of 4.5 m. The stocked lobsters measured an average of 68.5 g (45–90 g). Food consists of locally available brown mussel, and chopped trash fish fed daily once. The low survival (40.8%) may be due to high stocking density, crowding and possible cannibalism of newly moulted lobsters. The lobsters gained an average weight of 172 g after 94 days of rearing. Growth and survival rates in this study suggest that information on stocking density in relation to stocking size is essential for obtaining optimum growth and improving survival.

10.28.1.4 Cage Growout at Veraval, Northwest coast of India

Experimental cage growout of lobsters was reported from Veraval, on the northwest coast of India (Mohammed et al. 2010). Seed lobsters of the Mud spiny lobster *P. polyphagus* procured from local sources were stocked in a 6-m diameter floating net cage. An average weight gain of 145 g was obtained in 120 days. Mojjada et al. (2012) also reported the growout of *P. polyphagus* from Veraval. Juveniles form a substantial component in all the operating gears in many places in Gujarat especially along the Saurashtra and Kachch coasts at varying rates over the seasons. Lobsters that weigh less than 150 g fetch at least three times less value compared to those that weigh above 150 g, and such juveniles/subadults form considerable part of the landing (Mohammed Koya, personal communication).

Two growout trials of *P. polyphagus* were conducted in floating net cages. The cage frame made of 1.5" (3.6 cm) galvanised iron (GI) pipe reduced the construction cost, but without compromising the strength of cages to withstand the sea conditions (Fig. 10.20). A set of two nets, viz., the inner net (culture net) and outer net (protection net), made, respectively, of twisted HDPE twines of 0.75-/26-mm webbings (4.6X 4.6X 2 m) and braided HDPE twines of 3/80 mm webbings (6-m diameter, 2.7-m depth), were used for the construction of each cage. All the inner nets were lined with velon screen tied at the bottom of the cage to provide artificial substratum in order to facilitate crawling by the animal. Sufficient numbers of 2–3" PVC pipe pieces were tied over the velon screen as hideouts for the lobsters to take shelter. Trapezoid-shaped reinforced cement concrete (RCC) blocks having a weight of 2.5 t each were used as anchors for mooring the cages.

Fig. 10.20 Cages for lobster growout at Veraval, Gujarat, India (Photo by: Mohammed Koya, CMFRI)

Fig. 10.21 Stocking of juvenile *Panulirus polyphagus* in intertidal pits for selling to lobster farmers at Veraval, Gujarat, India. (Photo by: Mohammed Koya, CMFRI)

10.28.1.4.1 Collection and Stocking

Seeds from two major sources were used for stocking the cages: one from the stake net (Wada net) fishery in Mahua and adjacent talukas of Bhavnagar and the other from the trawl bycatch of Veraval and adjacent harbours. In case of seed collection from Mahua areas, the juvenile lobsters were caught by local artisanal fishermen by using stake nets and kept in limestone rock pits (Fig. 10.21). On demand, the fishermen hand-pick the required numbers of lobsters from the rock pools during low tide (when the pits are visible) by draining out water from the pits. The lobsters collected from this system are least stressed and perform better in the cages compared to that from the trawl and gillnet bycatch.

The lobsters form a bycatch in the bottom trawls, and the fishermen generally maintain the lobster catch in live condition by keeping it in plastic bins or vessels with seawater. Due to the limited space and facilities, the lobsters are maintained on board in stressful conditions including crowded tanks, shallow water and increased temperature, though fresh seawater is poured into the bin occasionally. As the fishermen intend to keep the lobsters alive only till the fishing harbour to fetch better price, least care is given to reduce the stress. The lobsters caught by the single-day trawlers are less stressed compared to multiday trawlers as they land the catch by evening of the same day. Therefore, live and healthy lobsters intact with all appendages landed by single-day trawlers were collected from the fishing harbours.

10.28.1.4.2 Seed Transportation

The lobster seeds were generally transported in water-free conditions to reduce the transportation cost as the farm is nearly 200 km away from the seed collection site, involving road transport of over 5 h. Healthy lobster seeds with all appendages

Fig. 10.22 Packing of *Panulirus polyphagus* juveniles collected from wild in moist sand for transportation to culture farm (Photo by: Mohammed Koya, CMFRI)

intact were collected and arranged neatly with the abdomen flexed inward and the antennae tucked outward in shallow plastic crates lined with moistened sea sand (Fig. 10.22). The plastic crates were then covered with seawater-moistened gunny bags and loaded on to a well-covered vehicle for transport. Additional seawater was carried in plastic cans to intermittently moisten the gunny bags covering the crates.

10.28.1.4.3 Stocking

Each of the 6-m-diameter cages (total bottom area of net cage 103.6 m^2) were stocked with 1000 (Cage 1) or 1500 numbers of juveniles (Cage 2), respectively (Mojjada et al. 2012). The initial body weight of lobsters stocked in Cage 1 was on average 100 g (range 87–121 g), and the final average weight of a lobster on harvest was 204.6 g (range 188–260 g). Cage 2 was stocked with juveniles weighing on average 46.4 g (range 31–61 g), and the final weight averaged 180 g (range 138–225 g) after a growth period of 90 days. The lobsters were fed daily with trash fish (*Saurida tumbil*, *Decapterus* spp., squid waste, shrimp waste and *Turbo* sp.) at 8% of their body weight. Survival in Cages 1 and 2 was 94.5% and 92.9%, respectively. Lobsters in Cage 1 showed a body weight increase of 1.17 g d^{-1}, whereas those in Cage 2 registered a body weight increase of 1.49 g d^{-1}. The specific growth rate (% body weight d^{-1}) of lobsters in Cages 1 and 2 was 0.80 and 1.51, respectively. The estimated harvest from Cage 1 was 194 kg (stocking weight 100 kg) with a survival of 94.5% and 251 kg from Cage 2 (stocking weight 69 kg) with a survival of 92.9%.

The floating sea cage technology was later transferred to Siddi tribes of African origin in Veraval, who migrated from Somalia centuries ago. CMFRI provided 20 floating cages free of cost to the cooperative society, Bharat Adin Jyot Matsya Udyog Mandali, formed by them. The farming was conducted with financial support from the Government of India. During the first operation, 2.5-t *Panulirus polyphagus* was harvested generating around Rs. 26 lakhs (US$40628) (Figs. 10.23 and 10.24).

Fig. 10.23 Harvesting cage grown *Panulirus polyphagus* from a floating cage, Veraval, Gujarat, India (Photo by: Mohammed Koya, CMFRI)

10.28.2 Pathogens and Diseases

Tail fan damage makes lobsters 'unsightly and unmarketable' (Lorkin et al. 1999; Bryars and Geddes 2005). Infection was minimal in *P. homarus homarus* and *P. ornatus* even at a high stocking density of 80 individuals/m^2 (Vijayakumaran et al. 2009). However, tail fan damage, ulceration and mass mortality occurred in one of the cages in Kulasekharapatnam, when the culturist stocked more than 500 juveniles in a single cage (1 m^3). A sudden drop in salinity due to overnight rain and subsequent drainage into the intertidal area at Tharuvaikulam resulted in mass mortality of more than 1500 lobsters in cages moored at 1.5-m depth (for more information, refer chapter on Pathogens and Health Management). Abraham et al. (1996) report bacterial infection in the haemolymph and exoskeleton lesions in *Panulirus homarus* lobsters reared under controlled conditions. The causative pathogen was identified as *Vibrio alginolyticus* and *V. harveyi*.

10.29 Growout Feeds

Finding suitable food in adequate quantities is critical to the successful aquaculture of any organism. Lobsters prefer live and fresh rather than frozen food. Commercial aquaculture cannot entirely depend upon natural food due to difficulties in collecting and storing food on a daily basis, diminishing supply of bycatch from trawlers and increase in cost, together with environmental pollution from uneaten food (Williams 2007). In Vietnam, lobsters are exclusively fed on bycatch obtained from

Fig. 10.24 Siddis with harvest of cage grown *Panulirus polyphagus* at Veraval, Gujarat, India (Photo by: Mohammed Koya, CMFRI)

trawlers. Apart from low conversion efficiency and high cost, feeding with fresh fish, crustaceans and molluscs has led to environmental degradation. The immediate solution is to develop practical feeds for different growout stages.

10.29.1 Natural Growout Feed

Lobsters are opportunistic carnivores of benthic invertebrates. Molluscs ranging from bivalves, to gastropods, to chitons; crustaceans such as barnacles, crabs and decapods; polychaete worms; echinoderms and occasionally macroalgae are reported to be present in the foregut (Joll and Phillips 1986; Jernakoff et al. 1993; Booth and Kittaka 1994; Barkai et al. 1996; Cox et al. 1997; Mayfield et al. 2000; Griffiths et al. 2000; Goni et al. 2001). Apparently, they have evolved in a protein-rich environment, and glycogen is the source of carbohydrates from molluscs and crustaceans.

Molluscan meat is the most preferable feed for lobsters. It remains attractive to the lobster even after 10 h of immersion in water (Williams et al. 2005). Using marine and brackish water clams along with trash fish, it was found that feeding of molluscan meat resulted in better growth in *P. homarus homarus* (Radhakrishnan and Vijayakumaran 1984a, b). However, mussel meat is nutritionally incomplete for lobsters. Smith et al. (2005) reported progressive decline of growth and paleness in natural colour suggesting nutritional inadequacies leading to subnormal pigmentation. Freezing and thawing of mussel meat is another factor suspected to affect nutrition (James and Tong 1997). Díaz-Iglesias et al. (2002) have demonstrated that gastropod, pelecypods and crustaceans provide the most efficient use of dietary proteins for growout of *P. argus*. However, none of these items are available in the amounts needed for sustaining large-scale growout. Mohammed et al. (2010) fed *P. polyphagus* stocked in floating sea cages off Gujarat, northwest coast of India, a mixture of fish and molluscan meat in a ratio of 1:1 at 10% of the biomass. In another growout operation off Vizhinjam, on the southwest coast of India, *P. homarus homarus* stocked in sea cages were fed on live brown mussel *Perna indica* at 16–20% of body weight daily (Rao et al. 2010). Cox and Davis (2006) report significantly better growth in juvenile *P. argus* in Florida, fed on frozen clams, shrimp, squid and oysters at 100% of their body weight once at dusk, than in those fed 50% of their body weight twice daily. Rope culture of green mussel inside the lobster cages or adjacent to the lobster culture cages is probably a practical approach to reap double benefits: proximity to live food nearer to the cages and improvement in environmental quality at the culture site. Du et al. (2004) carried out combined culture of *P. ornatus* and the green mussel *Perna viridis* in Vietnam and found reduction in organic matter in water and sediment (83% and 63%, respectively, compared to 65% and 45% in culture sites without mussels). Significant reduction in faecal coliforms (94%) and *Vibrio* load (76%) was also observed compared to culture site of lobsters alone.

10.29.2 Development of Artificial Feeds

Almost 38 years of research all over the world attempted development of compounded feed mainly for 'on-growing' lobsters. Essentially known as fattening in Asia, the process involves capturing young ones and growing them to a marketable size in captivity. This circumvents two bottlenecks in lobster mariculture: (1) hatchery production of seeds and (2) high natural mortality (Conklin 1980; Booth and Kittaka 1994; Brown et al. 1995).

None will dispute that the availability of a nutritionally complete formulated diet holds the key to success of lobster mariculture. Application of shrimp feed formulation principles and practices in lobster feed formulation, production and evaluation has not been successful. Researchers then went on to learn several new lessons which can be summarized as follows:

1. Inadequacies in the amino acid profiles when fishmeal is used as the major source of protein in the diets (Fordyce 2004).
2. Deficiencies in the certain lipid classes such as cholesterol (Williams 2007).
3. Lobsters fed with considerable amounts of marine protein require no cholesterol supplementation, but a high level of vegetable proteins requires dietary cholesterol supplementation at 4 g/kg^{-1} (Simon et al. 2010).
4. Incorrect physical form of the pellet hampers adequate consumption (Geddes et al. 2001; Sheppard et al. 2002).
5. Poor water stability and leaching of chemoattractants reduce consumption (Williams et al. 2005).
6. Poor digestibility and utilization of carbohydrates (Johnston et al. 2003; Ward et al. 2003).

7. Absence of growth stimulants from certain ingredients such as krill, squid and mussels (Cruz-Ricque et al. 1987; Williams 2007).
8. Inadequacies in feeding frequency (Smith et al. 2005; Williams 2007; Johnston et al. 2008).
9. Reduction in feed intake due to poor palatability and attractability (Glencross et al. 2001; Williams et al. 2005; Nelson et al. 2006; Johnston et al. 2007).
10. Intensified intracellular digestive activity in lobsters fed with formulated diets, leading to a marked decline in digestive capacity of the animal over a period of time, suggesting difficulties in digestive processing of formulated feeds (Simon 2009).
11. Digestible carbohydrate in diets has a positive effect on dry matter intake over a period of 5 h and does not affect appetite, despite prolonged hyperglycaemia in lobsters.
12. Glucose-derived carbohydrates appear to be stored as glycogen for utilization during short-term food deprivation (Simon and Jeffs 2013).

Consumption of formulated diets at high levels cannot be achieved by lobsters because foregut capacity is small. Foregut filling time is reported to be 1–2 h after which feed expands, and its clearance takes place in another 10 h. With a gut throughput time of 34–42 h, appetite revival is reported after 18 h. With fresh mussel meat, even though the dry matter intake and foregut evacuation rates are same, the appetite revival is reported to be within 10 h. It is also reported that formulated diets enter midgut directly, which prolongs the gut evacuation time and appetite revival (Simon and Jeffs 2008).

10.29.2.1 Macronutrient Requirements

Species-specific requirements for macronutrients have been reported in lobsters. Australian workers reported growth of lobsters as curvilinear with increments in dietary protein (Glencross et al. 2001; Smith et al. 2003b; Ward et al. 2003) and depended upon the lipid levels and species. With *P. ornatus*, the optimum growth was reported at 47–53% protein with 6–10% lipids (Smith et al. 2003b). Glencross et al. (2001) reported 55% crude protein (CP) optimum for *P. cygnus*. *J. edwardsii* showed no significant effect on growth with dietary lipid levels, and the optimum protein content required in feeds was reported to be 42–47% on a dry matter basis (Ward et al. 2003). What is interesting in all these studies is that the control feeds used were molluscan meat which resulted in twice the growth reported with experimental feeds. It is also noteworthy that growth in the wild is also similar to the growth obtained with molluscan meat, which points towards the deficit of nutrient/s in formulated feeds. Later on, Smith et al. (2003b, 2005) demonstrated that inclusion of krill in the feeds resulted in growth better than with molluscan meat feeding. The formulated feed used by Smith et al. (2005) had more than 60% protein and with 12% internal fat from fishmeal, krill meal and krill hydrolysate. Whether the improvement in growth in *P. ornatus* was due to the high level of protein or the appropriate P/E ratio is yet to be verified. Therefore, macronutrient interactions in lobster nutrition have to be studied more comprehensively, species-wise.

Capuzzo and Lancaster (1979) reported protein-sparing action of carbohydrates in post-larval *P. homarus*. Johnston (2003) reported a carbohydrate-to-lipid ratio of 2:1 as appropriate, and lipid levels as high as 18.7 supported good growth, contrary to the generalized level of not more than 10% lipid crustacean feeds in general, beyond which growth depression is reported in majority of the crustaceans, especially shrimp. The American lobster, *Homarus americanus*, is reported to use carbohydrates effectively in another study reported by Brown (2006).

There is a total absence of information regarding the essential fatty acid requirement in lobsters. In such a situation, the requirements elucidated from comprehensive studies in shrimp (Glencross et al. 2002a, b) can be only resorted to at present. A total lipid content of 7.5%, 0.9, 1.5, 0.3 and 0.3% of linoleic acid, linolenic acid, eicosapentaenoic acid (EPA) and docosahexaenoic acid (DHA), respectively, is considered ideal.

10.29.2.2 Micronutrient Requirements

No comprehensive information exists on the micronutrient requirements in lobsters. A requirement of 0.4% cholesterol has been reported when the feed does not contain marine protein (Irvin et al. 2010). Castell et al. (1975) reported that lobster feeds providing 0.5% cholesterol promoted growth and survival in *H. americanus*. Levels above 2% and below 0.2% depressed growth. Equivocal findings on the requirements of crustaceans in general with respect to total lipids, phospholipids and cholesterol can be seen in the case of lobsters also (Teshima 1997; Smith et al. 2003a; Sheen 2000).

Supplementation of lobster feed with carotenoids is common, especially when prolonged rearing in captivity leads to loss of natural colour and, thereby, market value. The inability of crustaceans, in general, to synthesize them de novo is known. Astaxanthin is the predominant carotenoid in fish and crustaceans with putative roles in sexual maturation and reproduction other than pigmentation. D'Abramo et al. (1983) found that β-carotene supplementation was not as effective as astaxanthin supplementation in improving pigmentation. A dietary carotenoid level of 115 mg/kg was advocated by Crear et al. (2002).Since a dose-response could not be demonstrated in lobsters by carotenoid supplementation in feeds, for maintenance of normal pigmentation in captivity, Williams (2007) recommends a supplementation of 70 mg/kg astaxanthin for spiny lobsters.

Vitamin and mineral requirements are practically unknown in lobsters. Isolated studies indicate no absolute requirement of ascorbic acid in lobsters (Kean et al. 1985), and an absolute dietary requirement demonstrated in shrimp (Conklin 1997) seems incorrect. Therefore, until information on vitamin requirements of lobsters are available, requirements accepted as standard in shrimp feeds can be adopted. Lobsters are stenohaline (Lucu et al. 2000), and the richness of seawater in minerals can be only accepted as a solution for lobsters to fulfil their mineral requirement as of now or till the demonstration of dietary requirement of minerals in lobster feeds.

With the information available in the public domain, it can be concluded by summarizing that, traditionally, lobsters are fattened with snails, green mussel, clams, crabs, lizardfish, red bigeye or trash fish. Pelleted feeds improve survival by providing balanced nutrition and retain the natural colour when carotenoids are incorporated in the feed. Fattening of lobsters is popular in Asian nations, and Vietnam appears to be the major producer. Until hatchery technology and commercial-scale seed production is perfected, wild collection of seeds and their culture will continue (Radhakrishnan 2015) to be a lucrative option. Only one feed company says they have lobster feeds (Lucky Star, Taiwan. http://www.luckystarfeed.net/ls-product/shrimpgrowout/lobster). Research on lobster nutrition has to be promoted internationally both in the areas of start nutrition or larviculture and growout nutrition.

10.30 Economics

Countries such as New Zealand and Australia permit harvesting of pueruli from the wild in large quantities, and the seeds have been used in experimental growout studies. Based on 20 years of research data from growout studies of the spiny lobster *Jasus edwardsii* in land-based experimental systems, Jeffs and Hooker (2000) designed a hypothetical spiny lobster farm to conduct financial analysis for 10 years of farm operation. The puerulus is assumed to grow to the commercial size of 300 g in 4 years at an ambient water temperature of 15–23°C. The hypothetical farm was expected to produce 5 metric tonnes of 300 g of lobsters (16,600 lobsters) in the fourth year and each year thereafter. For the land-based culture systems, the major capital input is the infrastructure development for seawater supply, storage and purification system, buildings to hold the growout tanks and construction costs of quarantine and growout tanks. The analysis suggests significant reduction in infrastructure and operating costs, especially the feed and labour costs, to run the farm profitably. Increasing productivity through faster growth rate and higher survival will only have a marginal effect on profitability, unless the infrastructure costs can be reduced substantially. The high expenditure on infrastructure could be minimized by developing spiny lobster culture in conjunction with existing land-based farming operations, such as abalone farming, or by utilizing suitable rental premises for the land-based farm. The major operating costs in decreasing order of importance are feed, freight and packing, labour, electricity and seed lobsters (puerulus). A total of 6% per year was estimated to be the cumulative loss of earnings on capital, cash input and inflation.

Commercial lobster farming cannot entirely depend upon natural foods, such as mussels or trash fish, due to uncertainty in their availability throughout the culture period and the increased labour costs associated with daily feeding and cleaning of the tanks, to avoid bacterial contamination. Development of a cost-effective artificial diet is the only solution to increase growth rate and reduce labour input, and a great deal of research is required before a complete diet can be made available. Another major operating cost identified by the analysis is packaging and airfreight costs, which could be circumvented by developing premium markets closer to the farm rather than sending live lobsters to distant markets such as China or Japan to get the premium price. High energy costs for pumping large volumes of seawater to maintain lobsters in healthy condition is one area which needs serious consideration to improve profitability. Significant reduction in consumption of electricity through refinement of holding conditions and possibly through the partial recirculation systems will have significant impact on the production cost. Improvement in the cost of puerulus collection and maintenance was not considered to have a major economic gain, when compared to the high cost of infrastructure development and other input costs such as feed.

The results of this analysis suggest that culture of *J. edwardsii* is biologically feasible but will be a commercial success only if significant reductions in input costs could be achieved. The analysis may hold good for other temperate species of lobsters in another part of the world. A major disadvantage for land-based lobster culture in temperate regions is the slow growth of cold-water species of spiny lobster such as *J. edwardsii*, though they fetch higher price in export markets. The cost of increasing the water temperature to increase growth may not offset the price advantage.

10.30.1 Economics of Tropical Species of Lobster Culture in Land-Based Culture Systems

The infrastructure and operational procedures for farming tropical lobsters are not different from culturing cold-water species. The three potential species to be considered for culture in the Southeast Asian region are *P. homarus homarus*, *P. polyphagus* and *P. ornatus*. Culture operations starting with pueruli or juveniles are the only option in the absence of a hatchery technology. Studies on commercial growout of these tropical species from puerulus to commercial size in indoor systems are not available, though there are several reports of laboratory-scale rearing (Tholasilingam and Rangarajan 1986; Radhakrishnan and Devarajan 1986; Balkhair et al. 2012). Small-scale growout of the tropical *P. argus* has been accomplished in a variety of systems including flow-through, semi-recirculation and full recirculation systems (Travis 1954; Witham 1973; Warner et al. 1977; Quackenbush and Herrnkind 1981; Lipcius and Herrnkind 1982; Ryther et al. 1988, Lellis and Russel 1990; Pardee 1992a, b; Pardee and Foster 1992; Field and Butler 1994; Sjoken 1999; Sharp et al. 2000). In spite of a large number of studies on small-scale juvenile growout of tropical species, economic feasibility studies of commercial-scale farming of lobsters in land-based systems are not available, except the production cost analysis of a hypothetical lobster growout facility with an estimated annual production of 10 t (Radhakrishnan 1994). For a commercial growout operation (presumably *P. ornatus*), beginning from an initial stocking size of 150 g to a harvestable size of 500 g in a period of 8 months, the feed and seed cost were estimated to be 71% and 13%, respectively, of the total production cost. The growout starting from early juveniles or subadults (100–150 g) looks more promising than rearing the pueruli to commercial size, as the entire culture operation may take a longer duration with lesser profitability. The estimated profitability has been worked out to be 84% of the operational cost over a period of 8 months.

10.30.2 Economics of Sea Cage Culture System

Sea cage culture or sea ranching of lobsters is suggested as a better option for lobster farming to circumvent the high infrastructure costs associated with land-based farming operations (Lozano-Alvarez 1996).

In India, capture-based, open sea cage growout of spiny lobsters was attempted by the Central Marine Fisheries Research Institute at Veraval on the northwest coast and Vizhinjam and Kanyakumari on the southwest coast. The National Institute of Ocean Technology, Chennai, initiated a programme on experimental sea cage farming of lobsters at Tuticorin along the southeast coast of India. However, economic feasibility analysis of the growout operations was not available. At Veraval along the Gujarat coast, subadults of the spiny lobster *P. polyphagus* incidentally caught in gillnets, stocked at two different stocking densities (1000 and 1500 numbers) and initial sizes (80–120 g and <80 g) in two 6-m-diameter cages, on harvesting after 90 days of growout, recorded a specific growth rate of 0.8 and 1.51, respectively. However, economic analysis of culture operations is not available. At Kanyakumari, 2400 numbers of subadult *P. homarus homarus* (mean initial weight 68.5 g) stocked in a 6-m-diameter sea cage attained an average of 173 g after 94 days of growout. The operational expenses formed 23% of total investment, which include cost of cage and other input costs such as seed, feed and labour. The estimated production cost was estimated at Rs. 687/kg, and at a selling price of Rs. 1200/kg, the net income was Rs. 89,725 per harvest, which excludes the depreciation of the cage and mooring cost. At Vizhinjam, 1100 numbers of *P. homarus homarus* subadults weighing an average of 114.8 ± 25.7 g stocked in a 7-m-diameter floating net cage, on harvesting after 135 days, reached an average of 226 ± 42.9 g. Vijayakumaran et al. (2009) reported faster growth of pueruli, post-pueruli and subadults of the two tropical species, *P. homarus homarus* and *P. ornatus*, in floating sea cages maintained along the southeast coast of India. However, economic feasibility of these culture operations is not available.

A complete bioeconomic analysis of lobster growout operations in land-based as well as cage culture systems is required to determine the net profitability. The profitability of lobster growout using subadults as seed lobsters is likely to be significantly impacted by the price of seed lobsters and the fresh feed (mussel). However, the short duration required to attain the commercial size of 200 g (*P. homarus*) makes it a little more attractive.

Harvesting subadults from a low-key fishery for growout operations may have significant impact on the wild fishery, which needs further investigation. Lobster farming starting with puerulus may be a better option as studies on the western rock lobster *P. cygnus* have shown 80–98% mortality of puerulus during the first year after settlement (Phillips et al. 2003). Information on level of puerulus settlement in lobster-fishing grounds along the Indian coast is limited except some studies conducted at Kovalam bay, south of Chennai (Radhakrishnan & Vijayakumaran, personal communication).

10.31 Slipper Lobster GrowOut

10.31.1 Introduction

Slipper lobsters (family: Scyllaridae) contribute to about 8% of the world's lobster production. Lobsters of the genus *Thenus* (subfamily: Theninae) acquire significance in the Indo-West Pacific region (from the east coast of Africa to the Red Sea and India, up to Japan and the northern coast of Australia). *Thenus* spp. contributes to the demersal trawl fisheries which operate along the tropical coasts of the Indian Ocean and the Western Pacific Region (Ben-Tuvia 1968; Prasad and Tampi 1968; Shindo 1973; Jones 1984; Courtney 2002). Production from lobster fisheries has declined in recent years due to the combined effects of overfishing, climate change, disease, habitat destruction and coastal pollution. Non-judicious fishery of the shovel-nosed lobster in India has, over the years, resulted in a drastic decline in catch and export, indicating the paucity of the resource in the fishing grounds.

Earlier identified as *T. orientalis*, descriptions by Burton and Davie (2007) have indicated the occurrence of five species, including *T. orientalis*, *T. indicus*, *T. unimaculatus*, *T. australiensis* and *T. parindicus*. Using COI barcodes, Jeena et al. (2015) have reported that the species distributed and caught widely along the Indian coast is *T. unimaculatus*, which can be identified by the presence of purple stripes on the pereopods (Radhakrishnan et al. 2013; Kizhakudan and Krishnamoorthi 2014). The presence of a less-abundant species, *T. indicus*, along the east coast, has also been confirmed. The Australian species have now been identified as *T. australiensis* and *T. parindicus*.

The desirable characteristics of a candidate lobster species for commercially viable aquaculture ideally include short larval duration, high fecundity, fast growth rate, high market value and lack of cannibalism, all of which will impact directly on bottom-line profits in a commercial aquaculture enterprise (Rogers et al. 2010). The candidature of slipper lobsters for aquaculture came into focus in the late 1990s, with Australia making strides in larval development and seed production of *T. orientalis* and *Thenus* sp. (Mikami and Greenwood 1997) and in the development from egg to juvenile of *T. australiensis* in 2004 (Rogers et al. 2010). Prospects for aquaculture of the slipper lobster in India were opened with CMFRI completing larval development and seed production of *T. unimaculatus* in laboratory conditions (Kizhakudan and Krishnamoorthy 2014).

One of the major bottlenecks in developing seed production techniques for spiny lobsters of the genus *Panulirus* is the prolonged larval phase with a number of stages. The delicate and fragile nature of the larvae makes them particularly vulnerable to stress stimuli originating from extrinsic factors such as rearing environment, nutrition and pathogens. Slipper lobster larvae, while as fragile as the spiny lobster larvae, have fewer stages to develop through before they enter into the post-larval nisto stage, which is also not a very prolonged stage. The relative duration of larval rearing is thus much shorter in slipper lobsters, and this encouraged research on prospects of captive rearing of these lobsters, particularly in Australia and India.

The time taken for eggs of the slipper lobster to develop to market size of about 150 g is less than 10 months, while the hatchery phase is completed in less than 30 days. Nursery rearing of this species is easier, compared to the palinurid lobsters. Besides, growout of slipper lobsters can be carried out at high densities since they do not exhibit cannibalistic behaviour and have high feed conversion ratio (FCR). These animals can be transported live for up to 6 h on a wet surface/moist cloth or for 12–16 h in water at slightly lowered temperatures. Slipper lobsters are a prized export commodity which has a high market value in countries such as Australia, Japan, China and Hong Kong. All these factors make them a suitable candidate for culture.

10.31.1.1 Seeds

Availability of seeds for stocking is the primary requirement for commercial farming of any aquaculture species. Slipper lobster farming was not an important priority for lobster aquaculturists, until mass production of large number of seeds is achieved. The hatchery production of *Thenus* spp. with a fairly good survival rate (80%) was the first major breakthrough that has generated interest in farming of slipper lobsters (Mikami 2007; Mikami and Kuballa 2007). Preliminary growout trials with wild juveniles showed that the species is amenable to culture conditions and can attain commercial size in a reasonable period of time even at high stocking densities. Since they live in muddy/sand substrate in the ocean, it may be even possible to grow them in shrimp ponds. The preferred diet of scyllarids is molluscan meat, and they can crush soft-shelled molluscs easily with their mandibles, but not hard-shelled bivalves like palinurids.

10.31.1.2 High-Density Growout of *T. unimaculatus* in Indoor System

There is not much information on juvenile habitat and ecology of *Thenus* sp. However, good numbers of *T. unimaculatus* juveniles could be collected from artisanal fishermen operating gillnets near the Chennai coast (Fig. 10.25). Juveniles of *T. unimaculatus* collected from the wild showed good growth and survival in indoor recirculating systems (Anon 2008) at the Kovalam Field Laboratory of the Central Marine Fisheries Research Institute (CMFRI), Chennai, India. Juveniles measuring 20 mm CL and weighing 5 g were stocked in two cement tanks with floor area of 12.5 m^2 and water depth of 0.5 m. The tank floor with a sand bed and trickling filter was stocked at a stocking density of 30–35 lobsters/m^2. The environmental parameters in growout culture water were as follows: temperature, 27–29°C; salinity, 36–38 ppt; and pH, 7.8–8.2. The lobsters were fed *ad libitum* with clam meat (*Donax cuneatus*) daily. The lobsters attained 140–175 g in 250 days with an overall survival of 90%.

Growout of *T. unimaculatus* was also conducted at the Field Laboratory of NIOT, Chennai. Lobsters weighing an average of 10 g stocked (7 lobsters/m^2) in FRP tanks with partially covered sand bed and fed with mussel (*Perna viridis*) and clam meat (*D. cuneatus*) at 5% body weight per day attained 150 g in 350 days (Vijayakumaran

10 Breeding, Hatchery Production and Mariculture

Fig. 10.25 *Thenus unimaculatus* juveniles collected from wild and stocked in FRP tanks for high density growout at Field Laboratory of CMFRI, Kovalam, Chennai, India. (Photo by: Joe K Kizhakudan, CMFRI)

and Radhakrishnan 2011). The intermoult duration varied from 25 to 30 days initially and from 111 to 114 days in later stages. The environmental parameters were as follows; temperature, 26–32°C; salinity, 30–33 ppt; pH, 7.8–8.2; total ammonia, <1 ppm; and dissolved oxygen, >4 mg/l.

In another experimental trial (Vijayanand et al. 2004), *T. unimaculatus* weighing an average of 5 g and stocked (20 numbers/m^2 in 3 tanks) in 200-L FRP tanks attained 95–100 g in 6 months. The lobsters in each of the tanks were fed with mussel, clam and oyster meat at 10% of body weight. The experiments show that ongrowing *T. unimaculatus* in indoor tanks is feasible.

10.31.1.3 Culture in Raceways

A detailed description of farming of *Thenus* sp. in raceway systems in Australia has been given by Vijayakumaran and Radhakrishnan (2011). Pilot-scale production trials in shallow raceways were carried out by the Australian Fresh Research and Development Corporation with 40,000 seeds produced in each batch (Anon 2008). The production time from egg to a commercial-size lobster (250 g) with 80% survival was 12 months. The presence of bacteria in seawater was the main factor for initial mortalities in hatchery, and with modification to larval production system and optimization of feeds, the problems were overcome (Mikami 2007). The project is designed to produce two products: hard-shelled live lobsters of an average 218 g and soft-shelled lobsters (frozen or fresh chilled) of an average 45 g (Mikami 2007). The annual production target is 1000 tonnes in Stage I and 3000 tonnes in Stages II and III.

10.32 Nutritional Quality of Lobsters

Nutritional compositions of many important food fishes are well documented, but only few references are available on nutrition facts on lobsters. While the data on the two most important clawed lobsters *H. americanus* and *H. gammarus* are available, the same is not the case with individual species of the spiny lobsters. There are few references on the biochemical composition of an Indian spiny lobster, *Panulirus homarus*, which is the most important lobster for live export from India (Radhakrishnan 1989; Vijayakumaran 1990; Anil Kumar 2002). The chemical composition of lobsters shows variation in relation to species, size, sex, season and food as in many other crustaceans (Radhakrishnan 1989; Vijayakumaran 1990; Chang 1995; Anil Kumar 2002; Vijayakumaran and Radhakrishnan 2002).

The main body part consumed is the tail muscle in all lobsters. While tail muscle is the only part consumed in spiny and slipper lobsters, claw, knuckle and leg muscle also form valuable edible portion of the clawed lobster. Hepatopancreas and ripe ovary of *P. homarus homarus* form 3.0–5.3% of the wet weight, respectively (Vijayakumaran 1990), and the egg (roe) of tropical spiny lobsters may weigh 7.54–9.83% body weight (Vijayakumaran et al. 2012). There is no appreciable change in water content of the tail muscle during different moult stages (72.4 to 76.6 %) in the spiny lobster *P. homarus homarus*, though total energy content exhibits variation from 18.6 to 22.6 kJ/g (Radhakrishnan 1989). As these tissues are also rich in protein and total fat content, these are also being considered as possible edible portions of spiny lobster. Lobster with freshly deposited egg (roe) is in demand in few Southeast and East Asian countries. However, egg-bearing lobsters are protected and cannot be caught in most lobster-fishing countries. Edible tissues of *H. americanus* are available for human consumption in a wide variety of forms such as processed hepatopancreas, presented as a green-coloured meat (named tomalley), and roe in addition to muscle, which is the main body part consumed (Holmyard and Franz 2006).

Though hepatopancreas, ovary and roe are rich in nutritional quality, only the composition of muscle, which forms the main edible portion of lobsters, is evaluated here. Protein content shows the general crustacean pattern and varies from 17.1 to 20.6%, whereas the total fat content ranges from an average of 0.3% in *H. gammarus* to 2.25 in *P. homarus* (Table 10.7). The highest value of 22.1% protein and 3.4% for unspecified boiled lobster (Paul et al. 1980) appears to be an exception from all other data. In general, spiny lobsters have higher protein and fat in the tail meat compared to the clawed lobsters, and the fat content is very much below 5%, a value generally considered to characterize a low-fat food (Sikorski et al. 2004). The proximate composition of the Indian *P. homarus* is similar to the data for spiny lobsters from other places in the world. Fish and shellfish are especially important for providing omega-3 (n3) fatty acids, which are found mainly in animal foods. Although the n3 fatty acid content in lobster is low compared to fishes such as sardines and salmon, it should be considered as a source to complement its high protein content. Omega-3 fatty acids have also been shown to decrease aggression, impulsivity and depression in adults. This association is even stronger for kids with mood disorders and disorderly conduct issues, such as attention deficit hyperactivity disorder (ADHD) (Megan Ware 2016).

Tail meat of both the clawed and the spiny lobsters has a similar pattern of fatty acids dominated by PUFA, followed by MUFA and SFA. The main saturated fatty acid in the muscle of the all lobsters was palmitic acid. Among MUFA, oleic acid (18:1n-9) was the dominating fatty acid mostly in the muscle of all lobsters.

The main n-3 PUFAs in the muscle of all lobsters are eicosapentaenoic acid (EPA, 20:5n-3) and docosahexaenoic acid (DHA, 22:6n-3). The n-3 PUFA ranges from 34.5 to 36.5 in the clawed lobsters and is much higher (38.1%) in the spiny lobsters (Table 10.8). The major n-6 PUFA is arachidonic acid (AA, 20:4n-6) in the muscle of all lobsters. The ratios of n-3/n-6 are 4.2:6.2 in the clawed lobsters and lower (2.28) in the spiny lobster as it contained n-6 also in high quantities. Incidentally, the spiny lobster has a higher PUFA to SFA ratio (2.49), compared to the clawed lobsters (1.9:2.0). Glutamate and aspartate dominated the amino acid profile of tail meat of *H. gammarus*, *H. americanus* and spiny lobster followed by glycine in the clawed lobsters and phenylalanine plus tyrosine in the spiny lobster, among the non-essential amino acids (NEAA) (Barrento et al. 2009; skipthe pie.org 2016). The concentration patterns of essential amino acids (EAA) in the muscle of all lobsters are similarly dominated by arginine, leucine and lysine. The amino acid content in muscle is very homogeneous between sexes and species in the clawed lobsters (Barrento et al. 2009). In hepatopancreas, *H. gammarus* had more tyrosine and taurine than *H. americanus* but less histidine.

The ratio of EAA to NEAA is significantly higher in the spiny lobster muscle (1.03) compared to *H. gammarus* (0.8) and *H. americanus* (0.78). Protein digestibility-corrected amino acid score (PDCAAS) is a method of evaluating the protein quality based on both the amino acid requirements of humans and their ability to digest it. The PDCAAS rating was adopted by the US Food and Drug Administration (FDA) and the Food and Agricultural Organization of the United Nations/World Health Organization (FAO/WHO) in 1993 as 'the preferred 'best" method to determine protein quality (Boutrif 1991). The highest amino acid scores were obtained for threonine, phenylalanine plus tyrosine and histidine in the clawed lobsters (Barrento 2010) and phenylalanine plus tyrosine, tryptophan and lysine in the spiny lobster (Skypthe pie.org 2016). The limiting amino acids of the spiny lobster tail meat are valine, leucine and histidine, though more than 50% of the requirements of these EAA are met in 100-g serving of spiny lobster tail muscle.

Cholesterol level is one of the major concerns about food quality and nutrition in most of the countries. Cholesterol is synthesized in the liver of humans and is not a dietary requirement and may enhance the chance of coronary artery disease (CAD) caused by atherosclerosis, which is the build-up of plaque inside the artery walls. This build-up causes the inside of the arteries to become narrower and slows down the flow of blood. Lobster meat does contain cholesterol, 31–43 mg/100 g in the clawed lobsters and 70 mg/100 g in the spiny lobsters, which is lesser than in other decapod species such as shrimps and squid (140 mg/100 g) (Barrento et al. 2009; Skypthe pie.org 2016; Megan Ware 2016). However, recent studies have suggested that the cholesterol content in foods does not necessarily increase harmful

Table 10.7 Nutrition facts (proximate composition) of different groups of lobsters per 100 g of wet weight (whole)

Country/lobster Species	Body part/tissue	Moisture (%)	Protein (g)	Fat (g)	Carbohydrate (g)	Ash (%)	Energy (cal/100 g)	References
Imported to Europe American lobster *Homarus americanus*	Wild Tail meat	79.2	17.1	0.7		1.3	82	1
Europe European lobster *Homarus gammarus* (female)	Wild Tail meat	78.1	18.3	0.3		1.8	87	1
England European lobster *Homarus gammarus* (female)	Wild Tail meat boiled	75.7	17.5	1.2		1.9		2
Lobster Species unspecified (spiny)	Boiled	72.4	22.1	3.4	0	1.4	119	3
USA Spiny lobster Mixed species (spiny)	Raw	74.1	20.6	1.44	2.4	1.4	112	4
South Pacific Island *Panulirus* spp.	Raw	74.0	18	1.0	0.5		84	5
Chennai, India Spiny lobster *Panulirus homarus* (female)	Wild Tail meat	75.0	20.4	2.25	0.4	1.9	103	6
Slipper lobster Species unspecified	Wild Tail meat		20.1	1.3	2.7		107	7

1. Barrento, S., Marques, A, Teixeira, B, Vaz-Pires, P & Nunes, M.L. 2009. Nutritional quality of the edible tissues of European lobster *Homarus gammarus* and American Lobster *Homarus americanus*, *Journal of Agricultural and Food Chemistry*. 57, 3645–3652
2. Stroud, G. D., & Dalgarno, E. J. 1982. Wild and farmed lobsters (*Homarus gammarus*). A comparison of yield, proximate chemical composition and sensory properties. *Aquaculture*. 29, 147-154
3. Paul, A.A., Southgate, D.A.T & Russell, J. 1980: First supplement to Mc Cance and Widdowson's. The composition of foods: amino acid composition (mg per 100-g food) and fatty acid composition (g per 100-g food). Her Majesty's Stationery Office, London and Elsevier/North-Holland Biomedical Press, Amsterdam/New York/Oxford
4. http://skipthepie.org/finfish-and-shellfish-products/crustaceans-spiny-lobster-mixed-species-raw/, Nutritional Data for Crustaceans, spiny lobster, mixed species, raw. Skip the Pie. Org., The Nutrition search Engine
5. Dignan, C., Burlingame, B, Kumar, S & Aalbersberg, W. 2004. The Pacific sea food composition table, FAO
6. Vijayakumaran, M. 1990. "Energetics of a few marine crustaceans", Ph D Thesis, Cochin University of Science and Technology, Kochi, India
7. http://www.Fatsecret.com/calories-nutrition/colesons-catch/slipper-lobster-tail

Table 10.8 Lipid composition: total fats (%), cholesterol content (mg/100 g of wet weight), energy content (kcal/g of tissue on a wet-weight basis), fatty acid profile (%) and amino acids content (g/100 g of wet weight) in the muscle of *Homarus gammarus* and *Homarus americanus* and spiny lobsters

Parameters	European lobster[1]	American lobster[1]	Spiny lobster[2]
	Homarus gammarus	*Homarus americanus*	Spiny lobster (mixed species)
Total fat	0.3	0.7	1.44
Cholesterol (%)	36.6	43.2	70.0
Energy	87	82.3	112
SFA	22.7	21.6	23.7
MUFA	28.8	31.5	27.5
PUFA	43.6	42	59
PUFA/SFA	1.9	2	2.49
n3	34.3	36.5	38.1
n6	8.2	5.8	16.7
(n3/n6)	4.2	6.2	2.28
(n6/n3)	0.2	0.2	0.44
TAA	17.04	16.93	20.38
EAA	7.58	7.42	10.45
NEAA	9.46 g	9.51	10.17
EAA/NEAA	0.8	0.78	1.12

cholesterol in the body and that saturated fat intake is more directly related to an increase in harmful cholesterol levels. Lobster is not a significant source of saturated fat (Megan Ware 2016).

Elemental composition of seafood is getting growing attention owing to its nutritional benefits and toxicological concerns related to bioaccumulation in contaminated waters. Impaired enzyme-mediated metabolic functions are attributed to deficiency of essential elements (e.g. Na, Mg, Cl, P, S, K, Ca, Mn, Fe, Cu, Zn, Se) leading to organ malfunctions, chronic diseases and ultimately death (Simopoulos 1997; FAO/WHO 2002). Therefore, regular intake of these elements via food ingestion is vital. Heavy metal (e.g. As, Cd, Hg and Pb) contamination is another concern in aquatic animal food since they cause severe human health disorders (Carvalho et al. 2005). Lobsters and other crustaceans have proved to be good sources of most of the wanted minerals and trace elements, though few toxic elements are also occasionally implicated in the tissues of these animals.

Sodium and potassium were the most abundant elements in all tissues of clawed and spiny lobsters (Vijayakumaran 1990; Vijayakumaran and Radhakrishnan 2002; Barrento 2010). The lobster muscle has significantly higher concentrations of Na, Mg, Ca and Se in the clawed lobsters and Na, K, Ca, Mg, Zn and Cu in *P. homarus* and other spiny lobsters (Table 10.9). In contrast, Fe is in very low concentration in this tissue and Cd is close to below detectable level (BDL) in clawed lobsters, whereas Fe is significantly high in the muscle of *P. homarus*. Most of these minerals are at very much higher level in the hepatopancreas and ovary, especially in mature *P. homarus* (Vijayakumaran 1990).

Table 10.9 Mineral (mg/100 g) and trace element (μg/100 g) (wet weight) composition of clawed and spiny lobster

Mineral/trace element	RDA/DRI	Homarus americanus Barrento 2010		Homarus gammarus Barrento 2010		Spiny lobster, mixed species Skypthepie.org		Spiny lobster Panulirus sp. Dignan et. al. 2005		Spiny lobster P. homarus (India) Vijayakumaran 1990	
		Value	%RDA	Value	%RDA	Value	%RDA	Value	%RDA	Value	%RDA
Na (mg)	2400	314	13.08	213	8.8	117	4.88	97	4.04	314	12.92
K (mg)	3500	263	7.51	240	6.86	180	5.14	209	5.97	285	8.14
Ca (mg)	1300	44	3.38	44	3.38	49	3.77	50	3.85	86	6.53
Mg (mg)	420	24	5.71	22	5.24	40	9.52	40	9.52	35	8.33
Fe (mg)	18	0.63	3.5	0.42	2.33	1.22	6.78	0	0	4.6	25.56
Zn (mg)	11	2.9	26.36	2.9	26.36	5.67	51.55	5.6	50.91	4.56	41.45
P (mg)	1250					238	19.04			496	39.68
Mn (mg)	2.3	0.09	3.91	0.06	2.61	0.015	0.65			0.65	28.26
Cu (μg)	900	2706	300.67	301	33.44	381	42.33			736	81.78
Se (μg)	55	103	187.27	187	340	138	250.91				
Cr (μg)	35				0					33.1	94.57
Co (μg)										8.4	
Cd (μg)		156		119						9.3	
Pb (μg)		10		8.8						≤0.001	

The mineral values indicate that all lobsters are good sources of Na, Cu, Mg and Zn. Zinc is an essential component of several enzymes, participating in the synthesis and degradation of carbohydrates, lipids, proteins and nucleic acids (FAO/WHO 2002). Lobsters are a very good source for Zn, which plays a role in wound healing and blood clotting and also supports immunity. Cu concentrations were above the daily recommended intake (DRI) in clawed lobsters, while it satisfies 80% DRI in the muscle and much above the DRI in hepatopancreas, ovary and egg of the spiny lobster *P. homarus* (Vijayakumaran 1990). Cadmium is considered as a contaminant in sea food, but absorption and utilization of Cd during maturation and egg development in *P. homarus* and the penaeid shrimp *Fenneropenaeus indicus* suggest that it plays a prominent role during maturation and development (Vijayakumaran 1990). The nutritional value of the spiny lobsters seems to be better than that of the clawed lobsters in terms of minerals and trace elements.

The nutrient quality of lobsters has been evaluated with minimum data available for different species and groups. Considering the vide variation in the nutrient composition of lobsters in relation to age, sex, maturation, tissue/organ, species and geographic location, it is important that more analyses should be undertaken to consolidate the findings on the nutritional value of lobsters.

10.33 Future Prospects

Spiny lobster aquaculture development has been rather slow in spite of lobster being a high-value product. The reason is obviously biological, though they are reasonably good candidate species for culture, especially the tropical species. Intensive culture in indoor systems or extensive farming in growout ponds is a practical possibility. Temperate species of lobsters tend to be slow growing and require longer time to grow to commercial size. Unlike the homarid lobsters (*H. americanus* and *H. gammarus*), which are solitary, aggressive and cannibalistic, spiny lobsters are gregarious and generally prefer communal living. They can tolerate a wide range of environmental conditions, are fast growing and accept a variety of natural feeds (Booth and Kittaka 2000; Jeffs and Davis 2003). Major hurdle for the development of lobster aquaculture has been the prolonged and complex larval phase, which takes many months and varies among different species. The duration of larval period of tropical species is relatively shorter (4.0–9.5 months) compared to the cold-water species (6.1 to 13.4 months), except the cold-water *P. elephas* (2.0–4.2 months) (Phillips and Matsuda 2011). Among the tropical species, *P. ornatus* and *P. argus* have the shortest larval period (4–5 months) (Calverley 2006; Goldstein et al. 2008; Smith et al. 2009a, b; Phillips and Matsuda 2011). A few other subtropical species, *P. japonicus* (Kittaka & Kimura, 1989) and *P. longipes bispinosus* (Matsuda & Yamakawa, 2000), and tropical species such as *P. penicillatus* (Matsuda et al. 2006) and *P. homarus homarus* (Murakami & Sekine, 2011) were also successfully cultured from egg to puerulus.

Though larval culture of spiny lobsters is technically feasible, developing a mass production system for spiny lobster seeds has been a major challenge. The critical areas pertaining to larval development had been the focus of research for many years, and some encouraging results were obtained for the tropical species, *P. ornatus*, the fast-growing tropical spiny lobster, which fetches nearly US$ 80 per kilogram during certain seasons for live lobster that weigh more than 1 kg. *P. ornatus* is a hardy species and very adaptable to growout conditions. A private company in Australia 'Lobster Harvest' has ventured into R&D since 2007 to develop commercial lobster aquaculture technologies and has reached 'commercial-ready status' for the production of *Thenus* spp. Lobster Harvest produced viable pueruli of *P. ornatus* in several batches, and the F_1 juveniles thus produced in captivity from 2006 to 2008 were grown to sexual maturity leading to production of the F_2 generation. This advancement has paved way for selective breeding and genetic improvement of *P. ornatus* in the near future (Barnard et al. 2011a, b). Several thousands of slipper lobster seeds produced in the hatchery are also under growout trials in Malaysia.

The future of lobster culture based on hatchery-produced seeds seems to be highly promising with these recent developments. *P. ornatus*, though not a large fishery in India, forms 5–8% of the total lobster landing in Tamil Nadu. Since Australia and Vietnam have demonstrated the potential of *P. ornatus* as the most promising aquaculture species, breeding and hatchery production trials initiated in the Field Laboratory of CMFRI, Kovalam, in 1980s must continue to focus on this species and develop the hatchery technology if India proposes to venture into commercial lobster farming. On 3 November 2016, the Institute for Marine and Antarctic Studies (IMAS) in Tasmania, Australia, reported that they were on their way to making sustainable lobster farming a reality. They can annually produce thousands of lobster juveniles in hatchery suitable for stocking commercial growout facilities (Anon 2015).

Technology for growout of spiny lobsters in land-based indoor tanks has been developed by CMFRI. Stocking density, feeding and growth of four species of spiny lobsters, *P. homarus*, *P. ornatus*, *P. polyphagus* and *P. versicolor*, were studied (Radhakrishnan 2015). The major input costs are infrastructure development (growout tanks, seawater pumping, filtration and storage) and energy cost for pumping, aeration and water recirculation. The advantages are the growth and health of lobsters could be monitored and food wastes minimized during the growout operation. Harvesting operation is much easier unlike the open sea cage growout system. The cost of production in land-based systems is relatively higher due to the infrastructure development and energy cost. Farming systems using solar energy for operation of pumps and aeration may offset the cost of high energy cost. Polyculture systems or multitrophic aquaculture systems using spiny lobsters as the top carnivore may also be considered (Tamm 1980; Sumbing et al. 2016).

While development of hatchery technology is the highest priority for commercial aquaculture of spiny lobsters, lobster farming using wild captured pueruli and juveniles has been practiced since the early 1980s in India and other Southeast Asian countries. This practice will continue to flourish because of the large-scale availability of seed lobsters, minimal capital cost for the seeds and simple growout technology. The sustainability will largely depend on environmental health of the farming ecosystem, continued market demand and if the farming activity continues as a village-based enterprise. It can provide significant economic benefit to the coastal fishing community and the ancillary small-scale enterprises supporting the aquaculture industry. The aquaculture production of lobsters in Vietnam rapidly increased from a small beginning in 1992, with a few hundred cages, to a large export-oriented, village-based industry in 2006, producing 1900 tonnes from 49,000 sea cages valued at US$ 65 million (Anon. 2009). The success in sea cage farming of lobsters in Vietnam has aroused interest in Australia and Indonesia. Vietnam has recently faced some serious setbacks in farming due to environmental degradation and disease outbreaks (Anon. 2009).

On the southeast coast of India, nearly 50% of the catch from trammel nets consists of juveniles and subadults which are not preferred for live export due to their small size and vulnerability during live transport (Radhakrishnan 2015). Further, it is illegal to export lobsters below the minimum legal size (MLS) prescribed by the Government of India (Radhakrishnan et al. 2005). These juveniles and undersized lobsters, if carefully collected and transported to the growout facility, can be grown to commercial size. Growout of juvenile *P. homarus homarus*, incidentally captured during gillnet operation by fishermen, in indoor tank systems and sea cages, was demonstrated by the Madras and Calicut Research Centres and Vizhinjam Research Centre of CMFRI, respectively (Radhakrishnan 1994; Radhakrishnan 2015; Rao et al. 2010). The NIOT, Chennai, conducted cage growout of lobsters involving local fishermen with fairly good success at Tuticorin (Vijayakumaran et al. 2009) and by CMFRI at selected locations on the east and west coasts of India. On the northwest coast, in Maharashtra alone, nearly 180 tonnes of juveniles of *P. polyphagus* are landed annually as a bycatch in artisanal gears operating in nearshore areas. Open sea cage farming using this seed source was carried out at Veraval along the Gujarat coast by CMFRI (Mohammed et al. 2010; Mojjada et al. 2012). The CMFRI demonstrated cage farming of lobsters involving the Siddi community in Gujarat. A major bottleneck is sourcing the required quantity of seeds, as the juveniles are landed in several lobster-fishing villages along the coast, and there is no organized agency to collect seeds from different lobster-landing villages and supply them to the farmers. Feeding the lobsters was mainly with locally available mussels, gastropods and low-value fishes landed as bycatch by trawlers.

A key issue in all aquaculture activities is food requiring a balanced nutrition for the cultured species. Other important factors are cost-effectiveness of the feed, environmental impacts of the feeding practices and protein and lipid conversion ratios (Jones et al. 2010). Feeding of a large stock requires a huge quantity of trash fish to be procured and delivered to the farming site daily. Though reasonably good growth rate has been obtained, a nutritionally complete prepared feed is required for profitable aquaculture (Irvin and Williams 2009). Trash fish of 20–50 kg was fed to produce 1 kg of lobster in Vietnam (Jones et al. 2010). Formulated feed development programmes are progressing to remove dependency on natural fresh feeds, which in turn will reduce the production costs and increase profitability. Formulated feeds will have the advantage of lower environmental impact compared to feeding with trash fish, a majority of which will go as waste, creating water quality problems in the culture site (Hoang et al. 2009). A large volume of information on nutritional requirements, feed formulation and feeding of *P. ornatus* is available (Williams 2004).

Lobster mariculture has proven to be a commercially profitable village-based enterprise in many parts of Southeast Asia, because of the ease of capturing large quantities of pueruli from inshore sea, minimal capital cost of cage culture and good market demand and price in international markets, especially in China and Hong Kong. In India, since lobster farming is entirely based on large juveniles and sub-adults captured from the wild, the impact of harvesting lobsters at a higher level of their life cycle needs careful monitoring to ensure that the fishery is not negatively impacted. Since it offers a high return in a short time, more entrepreneurs may enter into this field resulting in deliberate fishing of juveniles for aquaculture. The scope is limited, and lobster farming at the current level should be promoted only as a small-scale village-based enterprise.

10.34 Summary

The opportunities and challenges facing lobster breeding, larval rearing and aquaculture have been discussed in this chapter. Recent developments and technological advances in the management of larval culture systems and research information on food, nutrition, feeding behaviour, environmental requirements and disease management have provided the basis for the successful production of lobster seeds on a limited scale. However, the greatest challenge is mass production of seeds on a consistent scale. Recent advances in breeding and larval culture of *P. ornatus* and *Thenus* sp. in Australia demonstrated promising results. Success in mass production systems relies on filling the technological gaps in our understanding on the hydrodynamics in the rearing containers in relation to the swimming and food capture abilities of the larvae. Large-scale production systems cannot entirely depend upon the natural feeds, and therefore, suitable, nutritionally complete, practical formulated diets have to be developed. Lack of proper larval nutrition and increased susceptibility to opportunistic pathogens may lead to delayed moulting and poor health of the larvae. A major reason for puerulus/nisto mortality is poor nutritional reserves in the hepatopancreas, which is the major energy source during the non-feeding period and during the energy-demanding moulting process to the post-puerulus stage.

Capture-based growout of spiny and slipper lobsters on an experimental and semicommercial scale in indoor tank systems or ponds in flow-through or static systems with partial water replacement was found to be technically feasible. However, the cost-effectiveness of indoor aquaculture will depend upon reducing the energy costs for pumping seawater and provision of aeration. Sea cage culture of lobsters is comparatively more economically advantageous being less capital intensive. However, a lot more information on growout systems, culture cage/tank designs, stocking densities, optimum feed levels and shelter requirements in high-density culture is to be generated. Studies have shown that fresh feed is likely to make up 25–71% of the production cost. A number of commercial feeds for lobsters are available in the market, though the growth performance of these feeds is below optimum. Nutrition research in New Zealand and Australia is focusing on developing cost-effective artificial growout feeds for lobsters, and it is likely that practical low-cost feeds will be available in the near future to make lobster farming commercially successful. In India, juveniles and subadult lobsters incidentally caught in trawl or indigenous gears are currently utilized for growout. Deliberate catching of juveniles and subadults for aquaculture may have serious implications on natural fishery and should not be encouraged. In Vietnam, lobster culture using naturally settling pueruli, post-pueruli and juveniles is carried out, and research elsewhere has indicated harvesting pueruli from heavy settlement areas may not have a negative impact on the wild fishery, as majority of the pueruli perish in their early years of settlement. Further, heavy settlement of pueruli along the Vietnam coast favours harvesting it for growout operation. There is no organized supply chain of good-quality lobster seeds in India unlike Vietnam and Indonesia, as juveniles supplied by lobster-holding centres are of low quality due to poor handling and high-density stocking in limited tank space. Lobster growout in India may be promoted as a small-scale village-level activity involving the local fishermen.

References

Abraham, T. J., Rahman, K. M., & Mary Leema, T. J. (1996). Bacterial disease in cultured spiny lobster, *Panulirus homarus* (Linnaeus). *Journal of Aquaculture in the Tropics, 11*, 187–192.

Abrunhosa, F. A. & Long, J. (1997). Effect of starvation on the first larvae of *Homarus americanus* (Decapoda: Nephropidae) and phyllosomas of *Jasus verreauxi* and *J. edwardsii* (Decapoda: Palinuridae). *Bulletin of Marine Science, 61*(1), 73–80.

Acosta, C. A., Matthews, T. R., & Butler, M. J., IV. (1997). Temporal patterns and transport processes in recruitment of spiny lobster (*Panulirus argus*) postlarvae to South Florida. *Marine Biology, 129*(1), 79–85.

Agrawal, V. P. (1965). Feeding appendages and the digestive system of *Gammarus pulex*. *Acta Zoologica, 46*, 67–81.

Aguilar, O. D., Salas, M. S., & Cabrera, M. A. V. (1992). Characterization of settlement areas of postlarval lobster, *Panulirus argus*, on the northeast coast of Yucatan. *Proceedings of the Gulf and Caribbean Fisheries Institute, 45*, 743–758.

Anil Kumar, P. K. (2002). Biochemical studies and energetics of the spiny lobster *Panulirus homarus* (Linnaeus. 1758). PhD thesis, Central Institute of Fisheries Education, Versova, Mumbai, India, 148pp.

Anon. (1978). CMFRI Annual Report, 1978.

Anon. (2004). Marine life information network: biology and sensitivity key information subprogramme (online) Plymouth: *Marine Biological Association of the United Kingdom* (cited 08/02/2006), http://www.marlin.ac.uk

Anon. (2008). CMFRI Annual Report, 2007–08. pp. 62–63.

Anon. (2015). Lobster farming a step closer thanks to Tasmanian research team. *ABC News, Australia*.

Anon. (2009). *Sustainable tropical spiny lobster spiny lobster aquaculture in Vietnam and Australia, FIS/2001/058*. Canberra: Australian Centre for International Agricultural Research.

Arndt, C. E., Berge, J., & Brandt, A. (2005). Mouthpart-atlas of arctic sympagic amphipods-trophic niche separation based on mouthpart morphology and feeding ecology. *Journal of Crustacean Biology, 25*, 401–412.

Balkhair, M., Al-Mashiki, A., & Chesalin, M. (2012). Experimental rearing of spiny lobster, *Panulirus homarus* (Palinuridae) in land-based tanks at Mirbat Station (Sultanate of Oman) in 2009-2010. *Journal of Agricultural and Marine Sciences, 17,* 33–43.

Bang, H. O., & Dyerberg, J. (1985). Fish consumption and mortality from coronary heart disease. *New England Journal of Medicine, 313,* 822–823.

Bannerot, S., Fox, W. W., Jr., & Powers, J. E. (1987). Reproductive strategies and the management of snappers and groupers in the Gulf of Mexico and Caribbean. In J. J. Polovina & S. Ralston (Eds.), *Tropical snappers and groupers: Biology and fisheries management* (pp. 561–603). Boulder: Westview Press.

Bannerot, S. P., Ryther, J. H., & Clark, M. (1988). Large-scale assessment of recruitment of postlarval spiny lobsters, *Panulirus argus*, to Antigua, West Indies. In *Proceedings of the Gulf and Caribbean fisheries institute* (Vol. 41, pp. 471–486).

Barkai, A., Davis, C. L., & Tugwell, S. (1996). Prey selection by the South African cape rock lobster Jasus lalandii: Ecological and physiological approaches. *Bulletin of Marine Science, 58,* 1–8.

Barnard, R. M., Johnston, M. D., & Phillips, B. F. (2011a). Exciting developments: Generation F_2 of the tropical *Panulirus ornatus*. *Aquaculture Asia Pacific Magazine, 7*(1), 37–38.

Barnard, R. M., Johnston, M. D., & Phillips, B. F. (2011b, March). *Spiny lobster aquaculture in Australia produces second generation (F2) of the tropical lobster, Panulirus ornatus*. Presentation at World Aquaculture Society Meeting in New Orleans, USA.

Barnard, R. M., Johnston, M. D., Phillips, B. F., & Ritar, A. J. (2011c). Hatchery production of spiny lobsters: Meeting growing demand for premium product. *Global Aquaculture Advocate*, September/October, 92–95.

Barnett, B. M., Hartwick, R. F., & Milward, N. E. (1984). Phyllosoma and nisto stage of the Moreton Bay bug, *Thenus orientalis* (Lund) (Crustacea: Decapoda: Scyllaridae), from shelf waters of the great barrier reef. *Marine and Freshwater Research, 35*(2), 143–152.

Barnett, B. M., Hartwick, R. F., & Milward, N. W. (1986). Descriptions of the nisto stage of Scyllarus demani Holthuis, two unidentified Scyllarus species, and the juvenile of *Scyllarus martensii* Pfeffer (Crustacea: Decapoda: Scyllaridae), reared in the laboratory; and behavioural observation of the nistos of *S. demani, S. martensii*, and *Thenus orientalis*. *Australian Journal of Marine & Freshwater Research, 37,* 595–608.

Barrento, S. (2010). *Nutritional quality and physiological responses to transport and storage of live crustaceans traded in Portugal*. PhD thesis, Universidade do Porto. Porto, 261pp.

Barrento, S., Marques, A. N., Teixeira, B. R., Vaz-Pires, P., & Nunes, M. L. (2009). Nutritional quality of the edible tissues of European lobster Homarus gammarus and American lobster Homarus americanus. *Journal of Agricultural and Food Chemistry, 2009*(57), 3645–3652.

Ben-Tuvia, A. (1968). Report on the fisheries investigations of the Israel South Red Sea expedition, 1962. *Bulletin of the Sea Fisheries Research Station, Haifa, 52,* 21–55.

Benzie, J. A. H. & Yoshimura, T. (1999). Rock lobster aquaculture workshop. *Australia-Japan Joint Science and Technology Consultative Committee,* Perth, 26–27 January, Australian Department of Industry, Science and Resources, Canberra, p. 16.

Bermudes, M., & Ritar, A. J. (2008a). Response of early stage spiny lobster *Jasus edwardsii* phyllosoma larvae to changes in temperature and photoperiod. *Aquaculture, 281*(1), 63–69.

Bermudes, M., & Ritar, A. J. (2008b). Tolerance for ammonia in early stage spiny lobster (*Jasus edwardsii*) phyllosoma larvae. *Journal of Crustacean Biology, 28*(4), 695–699.

Berry, P. F. (1970). Mating behaviour, oviposition and fertilization in the spiny lobster: *Panulirus homarus* (Linnaeus). *Oceanographic Research Institute (Durban) Investigational Report, 24,* 1–16.

Berry, P. F. (1971). The biology of the spiny lobster *Panulirus homarus* (Linnaeus) off the east coast of Southern Africa. *Oceanographic Research Institute, (Durban) Investigational Report, 28,* 1–75.

Boghen, A. D. (1982). Effects of Wescodyne and malachite green on parasitic ciliates of juvenile American lobsters. *Progressive Fish-Culturist, 44,* 97–99.

Booth, J. D., & Kittaka, J. (2000). Spiny lobster growout. In B. F. Phillips & J. Kittaka (Eds.), *Spiny lobsters: Fisheries and culture* (2nd ed., pp. 556–585). Oxford: Fishing News Books.

Booth, J. D., & Kittaka, J. (1994). Growout of juvenile spiny lobster. In B. F. Phillips, L. S. Cobb, & J. Kittaka (Eds.), *Spiny lobster management* (pp. 424–445). London: Fishing News Books, Blackwell Scientific Publication.

Booth, J. D., & Phillips, B. F. (1994). Early life history of spiny lobster. *Crustaceana, 66*(3), 271–294.

Bourne, D. G., Young, N., Webster, N., Payne, M., Salmon, M., Demel, S., & Hall, M. (2004). Microbial community dynamics in a larval aquaculture system of the tropical rock lobster, *Panulirus ornatus*. *Aquaculture, 242*(1), 31–51.

Bourne, D. G., Høj, L., Webster, N. S., Swan, J., & Hall, M. R. (2006). Biofilm development within a larval rearing tank of the tropical rock lobster, *Panulirus ornatus*. *Aquaculture, 260*(1), 27–38.

Boutrif, E. (1991). Recent developments in protein quality evaluation. *Food Nutrition Agriculture Iss*, 2/3, FAO, Rome.

Briones-Fourzan, P., & Gutierrez-Carbonell, D. (1988). Postlarval recruitment of the spiny lobster, *Panulirus argus* (Latrielle 1804) in Bahia de la Ascension, QR. *Proceedings of the Annual Gulf and Caribbean Fisheries Institute, 41*, 492–507.

Briones-Fourzán, P. & Lozano-Alvarez, E. 1992. Aspects of the reproduction of *Panulirus inflatus* (Bouvier) and P. gracilis streets (Decapoda: Palinuridae) from the Pacific coast of MexicoJournal of Crustacean Biology, 12(1): 41–50.

Brisson, S. (1985). Gas-bubble disease observed in pink shrimps, *Penaeus brasiliensis* and *Penaeus paulensis*. *Aquaculture, 47*(1), 97–99.

Brown, A. C. (2006). Effect of natural and laboratory diet on O: N ratio in juvenile lobsters (*Homarus americanus*). *Comparative Biochemistry and Physiology, 144A*, 93–97.

Brown, P. B., Leader, R., Jones, S., & Key, W. (1995). Preliminary evaluations of a new water-stable feed for culture and trapping of spiny lobsters (Panulirus argus) and fish in the Bahamas. *Journal of Aquaculture in the Tropics, 10*, 177–183.

Bryars, S. R., & Geddes, M. C. (2005). Effects of diet on the growth, survival, and condition of sea-caged adult southern rock lobster, *Jasus edwardsii*. *New Zealand Journal of Marine and Freshwater Research, 39*(2), 251–262.

Burton, T. E., & Davie, P. J. F. (2007). A revision of the shovel-nosed lobsters of the genus Thenus (Crustacea: Decapoda: Scyllaridae), with descriptions of three new species. *Zootaxa, 1429*(1), 1–38.

Butler, M. J., & Herrnkind, W. F. (1988). Large-scale assessment of recruitment of Spiny lobster recruitment in South Florida; quantitative experiments and management implications. *Proceedings of the Annual Gulf and Caribbean Fisheries Institute, 41*, 508–515.

Butler, M. J. & Herrnkind, W. F. (1989). Are artificial "Witham" surface collectors adequate indicators of Caribbean spiny lobster, *Panulirus argus*, recruitment?. *Biological Sciences Faculty Publications. Paper, 84*. pp. 135-136

Calado, R., Narciso, J., Morais, S., Rhyne, A. L., & Lin, J. (2003). A rearing system for the culture of ornamental decapod crustacean larvae. *Aquaculture, 218*, 329–339.

Calverley, A. (2006). Kailis rears tropical rock lobster eggs to juveniles. *Aust-Asia Aquaculture, 20*(5), 63.

Cano-Gómez, A., Goulden, E. F., Owens, L., & Høj, L. (2010). *Vibrio owensii* sp. nov., isolated from cultured crustaceans in Australia. *FEMS Microbiology Letters, 302*(2), 175–181.

Caputi, R., Melville-Smith, R., de Lestang, S., Pearce, S. A., & Feng, M. (2010). The effect of climate change on the western rock lobster (*Panulirus cygnus*) fishery of Western Australia. *Canadian Journal of Fisheries and Aquatic Sciences, 57*(1), 85–96.

Capuzzo, J. M., & Lancaster, B. A. (1979). The effects of dietary carbohydrate levels on protein utilization in the American lobster. *Proceedings of the World Mariculture Society, 10*, 689–700.

Carvalho, M. L., Santiago, S., & Nunes, M. L. (2005). Assessment of the essential element and heavy metal content of edible fish muscle. *Analytical and Bioanalytical Chemistry., 382*(2), 426–432.

Castaing, J. B., Masse, A., Sechet, V., Sabiri, N. E., Pontie, M., Haure, J., & Jaouen, P. (2011). Immersed hollow fibres microfiltration (MF) for removing undesirable micro-algae and protecting semi-closed aquaculture basins. *Desalination, 276*(1–3), 386–396.

Castell, J. D., Mason, E. G., & Covey, J. F. (1975). Cholesterol requirements in the juvenile lobster *Homarus americanus*. *Journal of the Fisheries Research Board of Canada, 32*, 1431–1435.

Chakraborty, K., Chakraborty, R. D., Radhakrishnan, E. V., & Vijayan, K. K. (2010). Fatty acid profiles of spiny lobster (*Panulirus homarus*) phyllosoma fed enriched Artemia. *Aquaculture Research, 41*(10), 393–403.

Chakraborty, R. D., Radhakrishnan, E. V., Sanil, N. K., Thangaraja, R., & Unnikrishnan, C. (2011). Mouthpart morphology of phyllosoma of the tropical spiny lobster *Panulirus homarus* (Linnaeus, 1758). *Indian Journal of Fisheries, 58*(1), 1–7.

Chang, E. S. (1995). Physiological and biochemical changes during the molt cycle in decapod crustaceans: An overview. *Journal of Experimental Marine Biology and Ecology, 193*, 1–14.

Chau, N. M., Ngoc, N. T. B., & Williams, K. C. (2009). Effect of different types of shelter on growth and survival of Panulirus ornatus juveniles. In: *ACIAR proceedings series, 132* (pp. 85–88). Australian Centre for International Agricultural Research (ACIAR).

Chittleborough, R. G. (1974a) Western rock lobsters reared to maturity. *Marine and Freshwater Research, 25*(2), 221–225.

Chittleborough, R. G. (1974b). Review of prospects for rearing rock lobsters. *Australian Fisheries, 33*(4), 1–5.

Chittleborough, R. G. (1976a). Growth of juvenile Panulirus longipes cygnus George on coastal reefs compared with those reared under optimal environmental conditions. *Australian Journal of Marine and Freshwater Research, 27*(2), 279–295.

Chittleborough, R. G. (1976b). Breeding of Panulirus longipes cygnus under natural and controlled conditions. *Australian Journal of Marine & Freshwater Research., 27*, 499–516.

Chow, S., Suzuki, S., Matsunaga, T., Lavery, S., Jeffs, A., & Takeyama, H. (2011). Investigation on natural diets of larval marine animals using peptide nucleic acid-directed polymerase chain reaction clamping. *Marine Biotechnology, 13*(2), 305–313.

Cobb, J. S. (1976). The American lobster: The biology of *Homarus americanus*. *The University of Rhode Island Marine Technical Report, 48*, 1–32.

Conklin, D. E. (1980). Nutrition. In J. S. Cobb & B. F. Phillips (Eds.), *The biology and management of lobsters* (pp. 277–300). New York: Academic.

Conklin, D. E. (1997). Vitamins. In L. R. D'Abramo, D. E. Conklin, & D. M. Akiyama (Eds.), *Crustacean nutrition* (*Advances in world aquaculture*) (Vol. 6, pp. 123–149). Rouge: Baton World Aquaculture Society.

Courtney, A. J. (2002). The status of Queensland's Moreton Bay bug (*Thenus* spp.) and Balmain bug (*Ibacus* spp.) stocks: Department of Primary Industries, Queensland Government, Brisbane, Australia.

Cox, S. L., & Bruce, M. P. (2003). Feeding behaviour and associated sensory mechanisms of stage I–III phyllosoma of *Jasus edwardsii* and *Jasus verreauxi*. *Journal of the Marine Biological Association of the United Kingdom, 83*(3), 465–468.

Cox, S. L., & Davis, M. (2006). The effect of feeding frequency on growth of juvenile spiny lobster, *Panulirus argus* (Palinuridae). *Journal of Application in Aquaculture, 18*(4), 33–43.

Cox, S. L., & Davis, M. (2009). An evaluation of potential diets for the culture of postpueruli spiny lobster *Panulirus argus* (Palinuridae). *Aquaculture Nutrition, 15*, 152–159.

Cox, S. L., & Johnston, D. J. (2003). Feeding biology of spiny lobster larvae and implications for culture. *Reviews in Fisheries Science, 11*(2), 89–106.

Cox, C., Hunt, J. H., Lyons, W. G., & Davis, G. E. (1997). Nocturnal foraging of the Caribbean spiny lobster (*Panulirus argus*) on offshore reefs of Florida, USA. *Marine and Freshwater Research, 48*, 671–679.

Crear, B. J. (1998). A physiological investigation into methods of improving the post-capture survival of both the southern rock lobster, *Jasus edwardsii*, and the western rock lobster, *Panulirus cygnus*, p. 219. PhD thesis, School of Aquaculture, University of Tasmania.

Crear, B. J. & Allen, G. W. (2002). Guide for the rock lobster industry no. 1. Optimising water quality, rock lobster post-harvest subprogram. *Fisheries Research Development Corporation, Contract Report.*

Crear, B. J., Thomas, C. W., Hart, P. R., & Carter, C. G. (2000). Growth of juvenile southern rock lobsters, *Jasus edwardsii*, is influenced by diet and temperature, whilst survival is influenced by diet and tank environment. *Aquaculture, 190*(1), 169–182.

Crear, B., Hart, P., Thomas, C., & Barclay, M. (2002). Evaluation of commercial shrimp grow-out pellets as diets for juvenile southern rock lobster, *Jasus edwardsii*: Influence on growth, survival, color, and biochemical composition. *Journal of Applied Aquaculture, 12*, 43–57.

Crear, B., Cobcroft, J., & Battaglene, S. (2003). Guide for the rock lobster industry no. 2. Recirculating systems NH_3 (p. 29). Tasmanian Aquaculture & Fisheries Institute, University of Tasmania.

Creaser, E. P. (1950). Repetition of egg laying and number of eggs of the Bermuda spiny lobster. In: *Proceedings of the Gulf and Caribbean Fisheries Institute, 2*, 30–31. Institute of Marine Science, University of Miami.

Cruz, R., Lalana, R., Perera, E., Báez-Hidalgo, M., & Adriano, R. (2006). Large scale assessment of recruitment for the spiny lobster, *Panulirus argus*, aquaculture industry. *Crustaceana, 79*(9), 1071–1096.

Cruz-Ricque, L. E., Guillaume, J., & Cuzon, G. (1987). Squid protein effect on growth of four penaeid shrimp. *Journal of the World Aquaculture Society, 18*, 209–217.

D'Abramo, L. R., Baum, N. A., Bordner, C. E., & Conklin, D. E. (1983). Carotenoids as a source of pigmentation in juvenile lobsters fed a purified diet. *Canadian Journal of Fisheries and Aquatic Sciences, 40*, 699–704.

Dawirs, R. R. & Dietrich, A. (1986). Temperature and laboratory feeding rates in *Carcinus maenas* L. (Decapoda: Portunidae) larvae from hatching through metamorphosis. *Journal of Experimental Marine Biology and Ecology, 99*(2), 133–147.

Defoirdt, T., Boon, N., Bossier, P., & Verstraete, W. (2004). Disruption of bacterial quorum sensing: An unexplored strategy to fight infections in aquaculture. *Aquaculture, 240*(1), 69–88.

Dennis, D. M., Skewes, T. D., & Pitcher, C. R. (1997). Habitat use and growth of juvenile ornate rock lobsters, *Panulirus ornatus* (Fabricius, 1798), in Torres Strait, Australia. *Marine and Freshwater Research, 48*(8), 663–670.

Deshmukh, S. (1968). On the first phyllosomae of the Bombay spiny lobsters (*Panulirus*) with a note on the unidentified first *Panulirus* phyllosoma from India (Palinuridae). *Crustaceana (Supplement), 2*, 47–58.

Dexter, D. M. (1972). Molting and growth in laboratory reared phyllosomas of the Californian spiny lobster, *Panulirus interruptus. California Fish and Game, 58*, 107–115.

Diaz-Iglesias, E., Brito, R., & Baez-Hidalgo, M. (1991). Cria de postlarvas de langosta *Panulirus argus* en condiciones de laboratorio. *Revista Investigaciones Marinas, Cuba, 12*(1–3), 323–331.

Díaz-Iglesias, E., Ba'ez-Hidalgo, M., Perera, E., & Fraga, I. (2002). Respuesta metabo'lica de la alimentacio'n naturaly artificialen juveniles de la langosta Espinosa *Panulirus argus* (Latreille, 1804). *Hidrobiolo'gica, 12*(2), 101–112.

Diaz-Iglesias, E., Baez-Hidalgo, M., & Murillo-Valenzuela, L. A. (2010). Capture and rearing of pueruli of the red spiny lobster *Panulirus interruptus* from northern Pacific coast of Mexico. *Journal of the Marine Biological Association of India, 52*(2), 286–291.

Diggles, B. K., Moss, G. A., Carson, J., & Anderson, C. D. (2000). Luminous vibriosis in rock lobster *Jasus verreauxi* (Decapoda: Palinuridae) phyllosoma larvae associated with infection by *Vibrio harveyi. Diseases of Aquatic Organisms, 43*(2), 127–137.

Du, P. T., Hoang, D. H., Du, H. T. & Thi, V. T. H. (2004). Combined culture of mussel: A tool for providing live feed and improving environmental quality for lobster aquaculture in Vietnam. In: K.C. Williams (Ed.), *Spiny lobster ecology and exploitation in the South China sea region. ACIAR Proceedings, No. 120*, CSIRO, Australia, pp. 57–58.

Duggan, S., & McKinnon, A. D. (2003). The early larval developmental stages of the spiny lobster *Panulirus ornatus* (Fabricius, 1798) cultured under laboratory conditions. *Crustaceana, 76*(3), 313–332.

Dyerberg, J., Bang, H. O., Stofferson, E., Moncada, S., & Vane, J. R. (1978). Eicosapentaenoic acid and prevention of thrombosis and atherosclerosis? *Lancet, 2*, 117–119.

Eggleston, D. B., Lipcius, R. N., Marshall, L. S., Jr., & Ratchford, S. G. (1998). Spatiotemporal variation in postlarval recruitment of the Caribbean spiny lobster in the Central Bahamas: Lunar and seasonal periodicity, spatial coherence, and wind forcing. *Marine Ecology Progress Series, 174*, 33–49.

Elvevoll, E. O., Eilertsen, K., Brox, J., Dragnes, B. T., Falkenberg, P., Olsen, J. O., Kirkhus, B., Lamglait, A., & Østerud, B. (2008). Seafood diets: Hypolipidemic and antiatherogenic effects of taurine and n-3 fatty acids. *Atherosclerosis, 200*, 396–402.

FAO- Fishstat Database Plus. (2012). Fisheries and Aquaculture Department, Food and Agriculture Organization of the United Nations, Rome, www.fao.org/fishery/statistics/software/fishstat

FAO/WHO. (2002). *Human vitamin and mineral requirements* (Report of a joint Food and Agriculture Organization of the United Nations). Bangkok: World Health Organization Expert Consultation.

FAO. (2016). *The state of world fisheries and aquaculture 2016* (200 pp). Rome: Contributing to Food Sand Nutrition for all.

Field, J. M., & Butler, M. J., IV. (1994). The influence of temperature, salinity, and postlarval transport on the distribution of juvenile spiny lobsters, *Panulirus argus* Latreille, 1804, in Florida bay. *Crustaceana, 67*(1), 26–44.

Forcucci, D., Butler, I. V., Mark, J., & Hunt, J. H. (1994). Population dynamics of juvenile Caribbean spiny lobster, *Panulirus argus*, in Florida bay, Florida. *Bulletin of Marine Science, 54*(3), 805–818.

Fordyce, M. S. (2004). Potential protein sources for use in an artificial diet for juvenile southern spiny lobsters, *Jasus edwardsii*. Auckland: MSc thesis, University of Auckland.

Forteath, N. (1990). *A handbook on recirculating systems for aquatic organisms*. Fishing Industry Training Board of Tasmania.

Francis, D. S., Salmon, M. L., Kenway, M. J., & Hall, M. R. (2014). Palinurid lobster aquaculture: Nutritional progress and considerations for successful larval rearing. *Reviews in Aquaculture, 6*(3), 180–203.

Fukuda, Y., & Naganuma, T. (2001). Potential dietary effects on the fatty acid composition of the common jelly fish *Aurelia aurita*. *Marine Biology, 138*, 1029–1035.

Gardner, C., Frusher, S. D., Kennedy, R. B., & Cawthorn, A. (2001). Relationship between settlement of southern rock lobster pueruli, *Jasus edwardsii*, and recruitment to the fishery in Tasmania, Australia. *Marine and Freshwater Research, 52*(8), 1271–1275.

Geddes, M. C., Bryars, S. R., Jones, C. M. et al. (2001). *Final report of project 98/305 to fisheries*. Research and Development Corporation (Australia).

Gemende, B., Gerbeth, A., Pausch, N., & von Bresinsky, A. (2008). Tests for the application of membrane technology in a new method for intensive aquaculture. *Desalination, 224*, 57 pp.

Gilchrist, J. D. F. (1916). Larval and postlarval stages of Jasus lalandii (Milne Edw.) Ortmann. *Zoological Journal of the Linnean Society, 33*, 1050125.

Glencross, B., Smith, M., Curnow, J., Smith, D., & Williams, K. (2001). The dietary protein and lipid requirements of post-puerulus western rock lobster, *Panulirus cygnus*. *Aquaculture, 199*, 119–129.

Glencross, B. D., Smith, D. M., Thomas, M. R., & Williams, K. C. (2002a). Optimising the essential fatty acids in the diet for weight gain of the prawn, *Penaeus monodon*. *Aquaculture, 204*(1–2), 85–99.

Glencross, B. D., Smith, D. M., Thomas, M. R., & Williams, K. C. (2002b). The effect of dietary n-3 and n-6 fatty acid balance on the growth of the prawn *Penaeus monodon*. *Aquaculture Nutrition*, 43–51.

Goldstein, J. S., & Nelson, B. (2011). Application of a gelatinous zooplankton tank for the mass production of larval Caribbean spiny lobster, *Panulirus argus*. *Aquatic Living Resources, 24*, 45–51.

Goldstein, J. S., Matsuda, H., Takenouchi, T., & Butler, M. J., IV. (2008). The complete development of larval Caribbean spiny lobster *Panulirus argus* (Latreille, 1804) in culture. *Journal of Crustacean Biology, 28*(2), 306–327.

Goni, R., Quetglas, A., & Renones, O. (2001). Diet of the spiny lobster *Palinurus elephas* (Decapoda: Palinuridea) from the Columbretes Islands marine reserve (North-Western Mediterranean). *Journal of the Marine Biological Association of the United Kingdom, 81*, 347–348.

Goulden, E. F., Høj, L., & Hall, M. R. (2013). Management for bacterial pathogen control in invertebrate aquaculture hatcheries. In A. G. Burnell (Ed.), *Advances in aquaculture hatchery technology* (Series in food science, technology and nutrition no. 242, pp. 246–286). Cambridge: Woodhead Publishing Limited.

Greve, W. (1968). The 'planktokreisel', a new device for culturing zooplankton. *Marine Biology, 1*(3), 201–203.

Greve, W. (1975). The "plankton kreisel", a new device for culturing zooplankton. *Marine Biology, 1*, 201–203.

Griffiths, M. K. P., Mayfield, S., & Branch, G. M. (2000). A note on a possible influence of traps on assessment of the diet of *Jasus lalandii* (H.Milne-Edwards). *Journal of Experimental Marine Biology and Ecology, 247*, 223–232.

Gutiérrez-Carbonell, D., Simonín-Díaz, J., & Briones-Fourzán, P. (1992). A simple collector for postlarvae of the spiny lobster *Panulirus argus*. *Proceedings of the Gulf and Caribbean Fisheries Institute, 41*, 516–527.

Hair, C., Bell, J., & Dohorty, P. (2002). The use of wild-caught juveniles in coastal aquaculture and its application to coral reef fishes. In R. R. Stickney & J. P. McVey (Eds.), *Responsible marine aquaculture* (pp. 327–354). Oxon: CABI Publishing.

Hall, M. R., Kenway, M., Salmon, M., Francis, D., Goulden, E. F., & Høj, L. (2013). Palinurid lobster larval rearing for closed-cycle hatchery production. In: G. Allan, & G. Burnell (Eds.), *Advances in aquaculture hatchery technology* (Woodhead Publishing Series in Food Science, Technology and Nutrition. 242, pp. 289–328).

Hamner, W. M. (1990). Design developments in the planktonkreisel, a plankton aquarium for ships at sea. *Journal of Plankton Research, 12*, 397–402.

Hattori, T., & Oishi, Y. (1899). Hatching experiment on Ise lobster. *Report of the Imperial Fisheries Institute, 1*, 76–132.

Heron, A. C., McWilliam, P. S., & Dal Pont, G. (1988). Length-weight relation in the salp *Thalia democratica* and potential of salps as a food source. *Marine Ecology Progress Series, 42*, 125–132.

Herrnkind, W. F., & Butler, M. J. (1994). Settlement of spiny lobster, *Panulirus argus* (Latreille, 1804), in Florida: Pattern without predictability? *Crustaceana, 67*(1), 46–64.

Higa, T., & Saisho, T. (1983). Metamorphosis and growth of the late-stage phyllosoma of *Scyllarus kitanoviriosus* Harada (Decapoda, Scyllaridae). *Memoirs of the Kagoshima University Research Center for the South Pacific, 3*, 86–98.

Hoang, D. H., Sang, H. M., Kien, N. T., & Bich, N. T. K. (2009). Culture of *Panulirus ornatus* lobster fed fish by-catch or cocultured Perna viridis mussel in sea cages in Vietnam. In K. C. Williams (Ed.), *Proceedings of a workshop on spiny lobster ecology and exploitation in the South China Sea Region. ACIAR Proceedings No. 120* (pp. 118–125). Canberra: Australian Centre for International Agricultural Research.

Høj, L., Bourne, D. G., & Hall, M. R. (2009). Localization, abundance and community structure of bacteria associated with Artemia: Effects of nauplii enrichment and antimicrobial treatment. *Aquaculture, 293*(3), 278–285.

Holmyard, N., & Franz, N. (2006). *Lobster markets*. FAO, GLOBEFISH, Fishery Industries Division Viale delle Terme di Caracalla, 00153 Rome.

Hudinaga, M., & Kittaka, J. (1967). Large scale production of the young Kuruma prawn *Penaeus japonicus*. *Information Bulletin of Plankton*. Japan, no. 13.

Hughes, J. T., Shleser, R. A., & Tchobanoglous, G. (1974). A rearing tank for lobster larvae and other aquatic species. *Progressive Fish-Culturist, 36*(3), 129–132.

Hung, L. V., & Tuan, L. A. (2009). Lobster sea-cage culture in Vietnam. In K. C. Williams (Ed.), *Spiny lobster aquaculture in the Asia-Pacific region* (Vol. 132, pp. 9–17). Canberra: Australian Centre for International Agricultural Research.

Igarashi, M. A., Kittaka, J., & Kawahara, E. (1990). Phyllosoma culture with inoculation of marine bacteria. *Nippon Suisan Gakkaishi, 56*, 1781–1786.

Illingworth, J. L., Tong, J., Moss, G. A., & Pickering, T. D. (1997). Upwelling tank for culturing rock lobster (*Jasus edwardsii*) phyllosomas. *Marine and Freshwater Research, 48*(8), 911–914.

Inoue, M. (1978). Studies on the cultured phyllosoma larvae of the Japanese spiny lobster *Panulirus japonicus*. I. Morphology of the phyllosoma. *Bulletin of the Japanese Society for the Science of Fish, 44*(5), 457–475.

Inoue, M. (1981). Studies on the cultured phyllosoma larvae of the Japanese spiny lobster, *Panulirus japonicus* (V. Siebold). *Special Report Kanagawa Prefecture Fish Exp Stn, 1*, 1–91.

Inoue, M., & Nonaka, M. (1963). Notes on the cultured larvae of the Japanese spiny lobster, *Panulirus japonicus* (V. Siebold). *Bulletin of the Japanese Society for the Science of Fish, 29*(3), 211–218.

Irvin, S. J., & Williams, K. C. (2009). *Panulirus ornatus* lobster feed development: From trash fish to formulated feeds. In K. C. Williams (Ed.), *Proceedings of an international symposium on spiny lobster aquaculture in the Asia-Pacific region* (pp. 147–156). Canberra: Australian Centre for International Agricultural Research.

Irvin, S. J., Williams, K. C., Barclay, M. C. & Tabrett, S. J. (2010). Do formulated feeds for juvenile *Panulirus ornatus* lobsters require dietary cholesterol supplementation? *Aquaculture, 307*, 341–347.

Ito, M. (1988). Mariculture-related laboratory studies on the early life histories of the scyllarid lobster (Crustacea: Decapoda: Scyllaridae): Two forms of *Thenus* leach; and *Scyllarus demani* Holthuis. Doctoral dissertation, MSc thesis, James Cook University of North Queensland, Queensland, Australia.

Ito, M., & Lucas, J. S. (1990). The complete larval development of the scyllarid lobster, Scyllarus demani Holthuis, 1946 (Decapoda, Scyllaridae), in the laboratory. *Crustaceana, 58*(2), 144–167.

James, P. J., & Tong, I. J. (1997). Differeces in growth and moult frequency among post-pueruli of *Jasus edwardsii* fed fresh, aged or frozen mussels. *Marine and Freshwater Research, 48*, 931–934.

James, P. J., Tong, L., & Paewai, M. (2002). Effect of stocking density and shelter on the growth and mortality of rock lobsters Jasus edwardsii held in captivity. *Marine and Freshwater Research, 52*(8), 1415–1417.

Jayagopal, P. & Vijayakumaran, M. (2010). Studies on stress during live transport of spiny lobsters *Panulirus* sp. In: *International conference on recent advances in lobster biology, aquaculture and management*, 5–8 January 2010, Chennai, India. Abstract, pp. 103–104.

Jeena, N. S., Gopalakrishnan, A., Radhakrishnan, E. V., Kizhakudan, J. K., Basheer, V. S., Asokan, P. K., & Jena, J. K. (2015). Molecular phylogeny of commercially important lobster species from Indian coast inferred from mitochondrial and nuclear DNA sequences. *Mitochondrial DNA*, 1–10. doi:https://doi.org/10.3109/19401736.2015.1046160.

Jeffs, A. (2007). Revealing the natural diet of the phyllosoma larvae of spiny lobster. *Bulletin of Fisheries Research Agency Japan, 20*, 9.

Jeffs, A. G. (2010). Status and challenges for advancing lobster aquaculture. *J. Mar. Biol. Ass. India, 52*(2), 320–326.

Jeffs, A. G., & Davis, M. (2003). An assessment of the aquaculture potential of the Caribbean spiny lobster, *Panulirus argus*. *Proceedings of the Gulf and Caribbean Fisheries Institute, 54*, 413–426.

Jeffs, A., & Davis, M. (2009). The potential for harvesting seed of *Panulirus argus* (Caribbean spiny lobster*). Spiny lobster aquaculture in the Asia–Pacific region*, 46. Canberra: Australian Centre for International Agricultural Research.

Jeffs, A., & Hooker, S. (2000). Economic feasibility of aquaculture of spiny lobsters Jasus edwardsii in temperate waters. *Journal of the World Aquaculture Society, 31*(1), 30–41.

Jeffs, A. G., & James, P. (2001). Sea-cage culture of spiny lobster *Jasus edwardsii*. *New Zealand Marine and Freshwater Research, 52*, 1419–1424.

Jeffs, A. G., Nichols, P., Mooney, B., Phillips, K., & Phleger, C. (2004). Identifying potential prey of the pelagic larvae of the spiny lobster *Jasus edwardsii* using signature lipids. *Comparative Biochemistry and Physiology – Part B, 137*, 487–507.

Jernakoff, P., Phillips, B. F., & Fitzpatrick, J. J. (1993). The diet of postpuerulus western rock lobster, *Panulirus cygnus* George, at seven Mile Beach, Western Australia. *Marine and Freshwater Research, 44*, 649–655.

Jha, D. K., Vijaykumaran, M., Murugan, T. S., Santhanakumar, J., Kumar, T. S., Vinithkumar, N. V., & Kirubagaran, R. (2010). Survival and growth of early phyllosoma stages of *Panulirus homarus* under different salinity regimes. *Journal of the Marine Biological Association of India, 52*, 215–218.

Johnston, D. J. (2003). Ontogenetic changes in digestive enzyme activity of the spiny lobster, *Jasus edwardsii* (Decapoda: Palinuridae). *Marine Biology, 143*, 1071–1082.

Johnston, D. J., Calvert, K. A., Crear, B. J., & Carter, C. G. (2003). Dietary carbohydrate/lipid ratios and nutritional condition in juvenile southern rock lobster, *Jasus edwardsii. Aquaculture, 220*(1–4), 667–682.

Johnston, D. J., & Ritar, A. J. (2001). Mouthpart and foregut ontogeny in phyllosoma larvae of the spiny lobster *Jasus edwardsii* (Decapoda: Palinuridae). *Marine and Freshwater Research, 52*(8), 1375–1386.

Johnston, M. D., Johnston, D. J., Knott, B., & Jones, C. M. (2005). Mouthpart and foregut ontogeny in phyllosomata of *Panulirus ornatus* and their implications for development of a formulated larval diet. In: C. I. Henry, G. Van Stappen, M. Willie, & P. Sorgeloos (Eds.), *Proceedings of Larvi'05 – Fish & Shellfish larviculture symposium, European Aquaculture Society Special Publication* No. 36, pp. 223–226.

Johnston, D. J., Melville-Smith, R., Hendriks, B., Maguire, G. B., & Phillips, B. F. (2006). Stocking density and shelter type for the optimal growth and survival of western rock lobster *Panulirus cygnus* (George). *Aquaculture, 260*(1), 114–127.

Johnston, D. J., Melville-Smith, R., & Hendriks, B. (2007). Survival and growth of western rock lobster *Panulirus cygnus* (George) fed formulated diets with and without fresh mussel supplement. *Aquaculture, 273*(1), 108–117.

Johnston, D., Melville-Smith, R., Hendriks, B., & Phillips, B. F. (2008). Growth rates and survival of western rock lobster (*Panulirus cygnus*) at two temperatures (ambient and 23 °C) and two feeding frequencies. *Aquaculture, 279*(1), 77–84.

Joll, L. M., & Phillips, B. F. (1986). Foregut contents of the ornate rock lobster, Panulirus ornatus. In A. K. Haines, G. C. Williams, & D. Coates (Eds.), *Torres Strait fisheries seminar, Port Moresby* (pp. 212–217). Canberra: Australian Government Publishing Service.

Jones, C. M. (1984). Development of the bay lobster fishery in Queensland. *Australian Fisheries, 43*, 19–21.

Jones, C. M. (2009a). Temperature and salinity tolerances of the tropical spiny lobster, *Panulirus ornatus*. *Journal of World Aquaculture Society, 40*(6), 744–752.

Jones, C. M. (2009b). Advances in the culture of lobsters. In: G. Burnell & G. Allan (Eds.), *New technologies in aquaculture: Improving production, efficiency, quality and environmental management* (pp. 822–844). Cambridge: Woodhead Publishing Limited.

Jones, C. M. (2010). Tropical spiny lobster aquaculture development in Vietnam, Indonesia and Australia. *Journal of the Marine Biological Association of India, 52*(2), 304–315.

Jones, C. M., & Shanks, S. (2008). Grow-out of tropical rock lobster in shrimp pond conditions in Australia. *The Lobster Newsletter, 21*(2), 18–21.

Jones, C. M., & Shanks, S. (2009). Requirements for the aquaculture of *Panulirus ornatus* in Australia. *Spiny Lobster Aquaculture in the Asia–Pacific Region, 98*.

Jones, C. M., Linton, L., Horton, D., & Bowman, W. (2001). Effect of density on growth and survival of ornate rock lobster, *Panulirus ornatus* (Fabricius, 1798), in a flow-through raceway system. *Marine and Freshwater Research, 52*(8), 1425–1329.

Jones, C. M., Long, N. V., Hoc, D. T., & Priyambodo, B. (2010). Exploitation of puerulus settlement for the development of tropical spiny lobster aquaculture in the Indo-West Pacific. *Journal of the Marine Biological Association of India, 52*(2), 292–303.
Juinio, M. A. R. (1987). Some aspects of the reproduction of *Panulirus penicillatus* (Decapoda: Palinuridae). *Bulletin of Marine Science, 41*, 242–252.
Juinio, M. A. R., & Ruinata, J. (1996). Survival, growth and food conversion efficiency of *Panulirus ornatus* following eyestalk ablation. *Aquaculture, 146*(3–4), 225–236.
Kagwade, P. V. (1988). Reproduction in the spiny lobster *Panulirus polyphagus* (Herbst). *Journal of the Marine Biological Association of India, 30*(1&2), 37–46.
Kagwade, P. V., & Kabli, L. M. (1996). Reproductive biology of the sand lobster *Thenus orientalis* (Lund) from Bombay waters. *Indian Journal of Fisheries, 43*, 13–25.
Kamiya, N., Yamakawa, T., & Tsujigado, A. (1985). Studies on larval rearing of Ise lobster. *A Rep Mie Prefecture Fish Technology Centre, 36-37*, 1985.
Kasim, H. M. (1986). Effect of salinity, temperature and oxygen partial pressure on the respiratory metabolism of *Panulirus polyphagus* (Herbst). *Indian Journal of Fisheries, 33*(1), 66–75.
Kean, J. C., Castell, J. D., Bogen, A. G., D'Abramo, L. R., & Conklin, D. E. (1985). A re-evaluation of the lecithin and cholesterol requirements of juvenile lobster (*Homarus americanus*) using crab protein-based diets. *Aquaculture, 47*, 143–149.
Kirk, J. T. O. (1994). *Light and photosynthesis in aquatic ecosystems* (3rd ed.). Cambridge: Cambridge University Press.
Kittaka, J. (1988). Culture of the palinurid Jasus lalandii from egg stage to puerulus. *Nippon Suisan Gakkaishi, 54*, 87–93.
Kittaka, J. (1994a). Culture of phyllosomas of spiny lobster and its application to studies of larval recruitment and aquaculture. *Crustaceana, 66*(3), 258–270.
Kittaka, J. (1994b). Larval rearing. In B. F. Phillips, J. S. Cobb, & J. Kittaka (Eds.), *Spiny lobster management* (pp. 402–423). Oxford: Fishing News Books.
Kittaka, J. (1997a). Culture of larval spiny lobsters: A review of work done in northern Japan. *Marine and Freshwater Research, 48*(8), 923–930.
Kittaka, J. (1997b). Application of ecosystem culture method for complete development of phyllosomas of spiny lobster. *Aquaculture, 155*(1), 319–331.
Kittaka, J. (2000). Culture of larval spiny lobsters. In B. F. Phillips & J. Kittaka (Eds.), *Spiny lobster management* (2nd ed., pp. 508–532). Oxford: Fishing News Books.
Kittaka, J. (2005). Jellyfish as food organisms to culture phyllosoma larvae. *Bulletin of the Plankton Society of Japan, 52*(2), 91–99.
Kittaka, J., & Booth, J. D. (1994). Prospectus for aquaculture. In B. F. Phillips, J. S. Cobb, & J. Kittaka (Eds.), Ž. *Spiny Lobster Management* (pp. 365–373). Oxford: Fishing News Books.
Kittaka, J., & Booth, J. D. (2000). Prospects for aquaculture. In *Phillips, B.F & Kittaka, J (eds.). Spiny Lobster Fisheries and Culture* (2nd ed., pp. 465–473). Oxford: Fishing News Books.
Kittaka, J., & Ikegami, E. (1988). Culture of the palinurid *Palinurus elephas* from egg stage to puerulus. *Nippon Suisan Gakkaishi, 54*(7), 1149–1154.
Kittaka, J., & Kimura, K. (1989). Culture of the Japanese spiny lobster *Panulirus japonicus* from egg to juvenile stage. *Nippon Suisan Gakkaishi, 55*, 963–970.
Kittaka, J., Iwai, M., & Yoshimura, M. (1988). Culture of a hybrid of spiny lobster genus *Jasus* from egg to puerulus stage. *Nippon Suisan Gakkaishi, 54*(3), 413–417.
Kittaka, J., Ono, K., & Booth, J. D. (1997). Complete development of the green rock lobster *Jasus verreauxi* from egg to juvenile. *Bulletin of Marine Science, 61*(1), 57–71.
Kittaka, J., Kudo, R., Onoda, S., Kanemaru, K., & Mercer, J. P. (2001). Larval culture of the European spiny lobster *Palinurus elephas*. *Marine and Freshwater Research, 52*(8), 1439–1444.
Kittaka, J., Ono, K., Booth, J. D., & Webber, W. R. (2005). Development of the red rock lobster, *Jasus edwardsii*, from egg to juvenile. *New Zealand Journal of Marine and Freshwater Research, 39*(2), 263–277.
Kizhakudan, J. K. (2007). Reproductive biology, ecophysiology and growth in the mud spiny lobster, *Panulirus polyphagus* (Herbst, 1793) and the sand lobster, Thenus orientalis (Lund, 1793). PhD thesis. Bhavnagar University, Bhavnagar, Gujarat, 169 pp.

Kizhakudan, J. K. (2013). Effect of eyestalk ablation on moulting and growth in the mud spiny lobster *Panulirus polyphagus* (Herbst, 1793) held in captivity. *Indian Journal of Fisheries, 60*(1), 77–81.

Kizhakudan, J. K. (2016, January 5–25). Hatchery technology and seed production of lobsters. In I. Joseph & B. Ignatius (Eds.), *Course manual, Winter school on technological advances in mariculture for production enhancement and sustainability* (pp. 106–113). Kochi: CMFRI.

Kizhakudan, J. K., & Krishnamoorthi, S. (2014). Complete larval development of Thenus unimaculatus Burton & Davie, 2007 (Decapoda, Scyllaridae). *Crustaceana, 87*(5), 570–584.

Kizhakudan, J. K., & Patel, S. K. (2011). Effect of diet on growth of the mud spiny lobster Panulirus polyphagus (Herbst, 1793) and the sand lobster *Thenus orientalis* (Lund, 1793) held in captivity. *Journal of the Marine Biological Association of India, 53*(2), 167–171.

Kizhakudan, J. K., Radhakrishnan, E. V., George, R. M., Thirumilu, P., Rajapackiam, S., Manibal, C., & Xavier, J. (2004). Phyllosoma larvae of *Thenus orientalis* and *Scyllarus rugosus* reared to settlement. *The Lobster Newsletter, 17*(1).

Kizhakudan, J. K., Margaret, A. M. R., & Kandasami, D. (2007). High density growout techniques in tropical spiny lobsters. *8th international conference and workshop on lobster biology and management*, Charlottetown, Canada, p. 39.

Kizhakudan, J. K., Kizhakudan, S. J., & Patel, S. K. (2013). Growth and moulting in the mud spiny lobster, *Panulirus polyphagus* (Herbst, 1793). *Indian Journal of Fisheries, 60*(2), 79–85.

Konishi, K. (2007). Morphological notes on the mouthparts of decapod crustacean larvae, with emphasis on palinurid phyllosomas. *Bulletin of Fisheries Research Agency, Japan, 20*, 73.

Kulmiye, A. J., & Mavuti, K. M. (2005). Growth and moulting of captive Panulirus homarus homarus in Kenya, western Indian Ocean. *New Zealand Journal of Marine and Freshwater Research, 39*(3), 539–549.

Kumar, T.S., Vijayakumaran, M., Senthil Murugan, T., Jha, D.K., Sreeraj, G., & Muthukumar, S. (2009). Captive breeding and larval development of the scyllarine lobster *Petrarctus rugosus*. *New Zealand Journal of Marine and Freshwater Research, 43*(1), 101–112.

Kurmaly, K., Jones, D. A., & Yule, A. B. (1990). Acceptability and digestion of diets fed to larval stages of Homarus gammarus and the role of dietary conditioning behaviour. *Marine Biology, 106*(2), 181–190.

Lavens, P., Leger, P., & Sorgeloos, P. (1989). Manipulation of the fatty acid profile in *Artemia* offspring produced in intensive culture systems. In N. De Pauw, E. Jaspers, H. Ackefors, & N. Wilkins (Eds.), *Aquaculture – A biotechnology in progress* (pp. 731–739). Belgium: European Aquaculture Society.

Lebour, M. V. (1925). Young anglers in captivity and some of their enemies. A study in a plunger jar. *Journal of the Marine Biological Association of the United Kingdom, 13*, 721–734.

Lellis, W. (1990). Early studies on spiny lobster mariculture. *The Crustacean Nutrition Newsletter, 6*(1), 70.

Lellis, W. (1991). Spiny lobster, a mariculture candidate for the Caribbean. *World Aquaculture, 22*(1), 60–63.

Lellis, W. A., & Russell, J. A. (1990). Effect of temperature on survival, growth and feed intake of postlarval spiny lobsters, *Panulirus argus. Aquaculture, 90*(1), 1–9.

Liddy, G. C., Phillips, B. F., & Maguire, G. B. (2003). Survival and growth of instar 1 phyllosoma of the western rock lobster, *Panulirus cygnus*, starved before or after periods of feeding. *Aquaculture International, 11*(1–2), 53–67.

Lipcius, R. N., & Eggleston, D. B. (2000). Ecology and fishery biology of spiny lobsters. In B. F. Phillips & J. Kittaka (Eds.), *Spiny lobsters: Fisheries and culture* (pp. 1–41). Oxford: Fishing News Books.

Lipcius, R. N., & Herrnkind, W. F. (1982). Molt cycle alterations in behavior, feeding and diet rhythms of a decapod crustacean, the spiny lobster *Panulirus argus. Marine Biology, 68*(3), 241–252.

Lipton, A. P., Rao, G. S., Kingsly, H. J., Imelda, J., Mojjada, S. K., Rao, G., & Rajendran, P. (2010). Open Sea floating cage farming of lobsters. Successful demonstration by CMFRI off Kanyakumari coast. *Fishing Chimes, 30*(2), 11–13.

Lorkin, M., Geddes, M., Bryars, S., Leech, M., Musgrove, R., Reuter, R., & Clark, S. (1999). *Sea-based live holding of the southern rock lobster, Jasus edwardsii: A pilot study on long term holding feeding* (SARDI research report series no, 46) (p. 22).

Losordo, T. M., Masser, M. P., & Rakocy, J. (1998). Recirculating aquaculture tank production systems. *An overview of critical considerations* (SPAC Publications, no. 451).

Lozano-Alvarez, E. (1996). Ongrowing of juvenile spiny lobsters, *Panulirus argus* (Latreille, 1804) (Decapoda, Palinuridae), in portable sea enclosures. *Crustaceana, 69*(8), 958–973.

Lucu, C., Devescovi, M., Skaramuca, B., & Kozul, V. (2000). Gill Na, K-ATPase in the spiny lobster *Palinurus elephas* and other marine osmoconformers—Adaptiveness of enzymes from osmoconformity to hyperregulation. *Journal of Experimental Marine Biology and Ecology, 246*, 163–178.

Macdonald, C. D. (1982). Catch composition and reproduction of the spiny lobster *Panulirus versicolor* at Palau. *Transactions of the American Fisheries Society, 111*(6), 694–699.

MacFarlane, J. W., & Moore, R. (1986). Reproduction of the ornate rock lobster, *Panulirus ornatus* (Fabricius), in Papua New Guinea. *Marine and Freshwater Research, 37*(1), 55–65.

Marinovic, B., Lemmens, J. W., & Knott, B. (1994). Larval development of *Ibacus peronii* leach (Decapoda: Scyllaridae) under laboratory conditions. *Journal of Crustacean Biology, 14*(1), 80–96.

Marx, J. M. (1986). Settlement of spiny lobster, *Panulirus argus* pueruli in South Florida: An evaluation from two perspectives. *Canadian Journal of Fisheries and Aquatic Sciences, 43*(11), 2221–2227.

Mary Leema, J. T., Kirubagaran, R., Vijayakumaran, M., Remany, M. C., Kumar, T. S., & Babu, T. D. (2010). Effects of intrinsic and extrinsic factors on the haemocyte profile of the spiny lobster, *Panulirus homarus* (Linnaeus, 1758) under controlled conditions. *Journal of the Marine Biological Association of India, 52*(2), 219–228.

Mathew, S., Ammu, K., Viswanathan Nair, P. G., & Devadasan, K. (1999). Cholesterol content of Indian fish and shellfish. *Food Chemistry, 66*, 455–461.

Matsuda, H., & Takenouchi, T. (2005). New tank design for larval culture of Japanese spiny lobster, *Panulirus japonicus*. *New Zealand Journal of Marine and Freshwater Research, 39*(2), 279–285.

Matsuda, H., & Takenouchi, T. (2007). Development of technology for larval culture in Japan: A review. *Bulletin of Fisheries Research Agency, 20*, 77–84.

Matsuda, H., & Yamakawa, T. (1997). Effects of temperature on growth of the Japanese spiny lobster, *Panulirus japonicus* (V. Siebold) phyllosomas under laboratory conditions. *Marine and Freshwater Research, 48*(8), 791–796.

Matsuda, H., & Yamakawa, T. (2000). The complete development and morphological changes of larval *Panulirus longipes* (Decapoda, Palinuridae) under laboratory conditions. *Fisheries Science, 66*(2), 278–293.

Matsuda, H., Yamakawa, T., & Tsujigado, A. (1987). Seedling production of Japanese fan lobster, *Ibacus ciliatus*-I. *Bulletin of the Fisheries Research Division Mie Prefectural Science and Technology Promotion Center, 2*, 78–82. (in Japanese).

Matsuda, H., Yamakawa, T., & Tsujigado, A. (1988). Seedling production of Japanese fan lobster, *Ibacus ciliatus*-II. *Bulletin of the Fisheries Research Division Mie Prefectural Science and Technology Promotion Center, 3*, 70–73. (in Japanese).

Matsuda, H., Takenouchi, T., & Yamakawa, T. (2005). New tank design for larval culture of Japanese spiny lobster, *Panulirus japonicus*. *N.Z. Journal of Marine and Freshwater Research, 39*, 279–285.

Matsuda, H., Takenouchi, T., & Goldstein, J. S. (2006). The complete larval development of the pronghorn spiny lobster *Panulirus penicillatus* (Decapoda: Palinuridae) in culture. *Journal of Crustacean Biology, 26*(4), 579–600.

Matsuda, H., Takenouchi, T., Tanaka, S., & Watanabe, S. (2009). Relative contribution of Artemia and mussel as food for cultured middle-stage *Panulirus japonicus* phyllosomata as determined by stable nitrogen isotope analysis. *N.Z.J. Marine and Freshwater Research, 43*, 217–224.

Matsuda, H., Abe, F., & Tanaka, S. (2012). Effect of photoperiod on metamorphosis from phyllosoma larvae to puerulus postlarvae in the Japanese spiny lobster *Panulirus japonicus*. *Aquaculture, 326-329*, 136–140.

Mayfield, S., Atkinson, S., Branch, G. M., & Cockcroft, A. C. (2000). Diet of the west coast rock lobster *Jasus lalandii:* Influence of lobster size, sex, capture depth, latitude and moult stage. *South African Journal of Marine Science, 22*, 57–69.

McWilliam, P. S., & Phillips, B. F. (1997). Metamorphosis of the final phyllosoma and secondary lecithotrophy in the puerulus of *Panulirus cygnus* George: A review. *Marine and Freshwater Research, 48*(8), 783–790.

McWilliam, P. S., & Phillips, B. F. (2007). Spiny lobster development: Mechanisms inducing metamorphosis to the puerulus: A review. *Reviews in Fish Biology and Fisheries, 17*(4), 615–632.

Megan Ware, R. D. N. L. D. (2016). Lobster: Nutritional Information, health benefits. RDN LD Knowledge Centre. http://www.medicalnewstoday.com/articles/303332.php. *Crustaceans, spiny lobster, mixed species, cooked, moist heat. Self nutrition data: know what you eat* http://nutritiondata.self.com/facts/finfish-and-shellfish-products/4250/2

Meggs, L. G. C., Steele, R. D., & Aiken, K. A. (2011). Settlement patterns of spiny lobster (*Panulirus argus*) postlarvae on collectors in Jamaican waters and culture of juveniles. *Proceedings of the Gulf and Caribbean Fisheries Institute, 63*, 472–481.

Michel, A. (1979). Personal Communication to Tamm, G.R. 1980. Spiny lobster culture: An alternative to natural stock assessment. *Fisheries, 5*(4), 59–62.

Middaugh, J. P. (1990). Cardiovascular deaths among Alaskan natives, 1980-86. *American Journal of Public Health, 80*(3), 282–285.

Mikami, S. (2007). Prospects of aquaculture of bug lobsters (Thenus spp.). *Bulletin of Fisheries Research Agency, 20*, 45–50.

Mikami, S., & Greenwood, J. G. (1997). Complete development and comparative morphology of larval *Thenus orientalis* and *Thenus* sp. (Decapoda: Scyllaridae) reared in the laboratory. *Journal of Crustacean Biology, 17*, 289–308.

Mikami, S., & Kuballa, A. V. (2007). Factors important in larval and postlarval molting, growth and rearing. In K. L. Lavalli & E. Spanier (Eds.), *The biology and fisheries of slipper lobster* (pp. 91–111). Boca Raton: CRC Press.

Mikami, S., & Takashima, F. (1993). Development of the proventriculus in larvae of the slipper lobster, *Ibacus ciliatus* (Decapoda: Scyllaridae). *Aquaculture, 116*(2–3), 199–217.

Mikami, S., Greenwood, J. G., & Gillespie, N. C. (1994). The effect of starvation and regimes on survival, growth and the moulting interval of cultured *Panulirus japonicus* and *Thenus* sp. phyllosomas (Decapoda, Palinuridae & Scyllaridae). *Crustaceana, 68*, 160–169.

Mitchell, J. R. (1971). Food preferences, feeding mechanisms, and related behaviour in phyllosoma larvae of the California spiny lobster, *Panulirus interruptus* (Randall). Master's thesis, San Diego State University, San Diego, CA.

Mohammed, K. H., Vedavyasa Rao, P., & Suseelan, C. (1971). The first phyllosoma stage of the Indian deep sea spiny lobster *Puerulus sewelli* Ramadan. *Proceedings of the Indian Academy of Sciences Part A, 74*, 208–215.

Mohammed, G., Rao, G. S., & Ghosh, S. (2010). Aquaculture of spiny lobsters in sea cages in Gujarat, India. *Journal of the Marine Biological Association of India, 52*(2), 316–319.

Mojjada, S. K., Imelda, J., Koya, M., Sreenath, K. R., Dash, G., Dash, S. S., Fofandi, M., Anbarasu, M., Bhint, H. M., Pradeep, S., & Shiju, P. (2012). Capture based aquaculture of mud spiny lobster, *Panulirus polyphagus* (Herbst, 1793) in open sea floating net cages off Veraval, north-west coast of India. *Indian Journal of Fisheries, 59*(4), 29–33.

Monterrosa, O. E. (1987). Postlarval recruitment of the spiny lobster *Panulirus argus* (Latreille), in southwestern Puerto Rico. *Proceedings of the Gulf and Caribbean Fisheries Institute, 40*, 434–451.

Moss, G. A., James, P. J., Allen, S. E., & Bruce, M. P. (2004). Temperature effects on the embryo development and hatching of the spiny lobster *Sagmariasus verreauxi*. *New Zealand Journal of Marine and Freshwater Research, 38*(5), 795–801.

Moss, G. A., Tong, L. J., & Illingworth, J. (1999). Effects of light intensity and food density on the growth and survival of early-stage phyllosoma larvae of the rock lobster *Jasus edwardsii*. *Marine and Freshwater Research, 50*(2), 129–134.
Moss, G., James, P., & Tong, L. (2000). *Jasus verreauxi* phyllosomas cultured. *The Lobster Newsletter, 13*(1), 9–10.
Moyle, K. (2005). *Effect of stocking density on the growth, survival and behaviour of Panulirus cygnus post puerulus.* Honours thesis. University of Western Australia.
Murakami, K. (2004). Culturing technology for phyllosoma of the Japanese spiny lobster *Panulirus japonicus*. *Youshoku, 6*, 31–33.
Murakami, K. & Sekine, S. (2011). Application of advanced rearing technology of *Panulirus japonicus* phyllosoma to *P. homarus homarus* larvae. Abstract. *The 9th International Conference and Workshop on lobster biology and management*, Bergen 19–24 June, 2011, p. 21.
Murakami, K., Jinbo, T., & Hamasaki, K. (2007). Aspects of the technology of phyllosoma rearing and metamorphosis from phyllosoma to puerulus in the Japanese spiny lobster *Panulirus japonicus* reared in the laboratory. *Bulletin of Fisheries Research Agency, 20*, 59–67.
Murugan, T. S., Remany, M. C., Mary Leema, T., Jha, D. K., Santhanakumar, J., Vijayakumaran, M., Venkatesan, R., & Ravindran, M. (2005). Growth, repetitive breeding, and aquaculture potential of the spiny lobster, *Panulirus ornatus*. *New Zealand Journal of Marine and Freshwater Research, 39*, 311–316.
Nair, R. V., Soundarajan, R., & Nandakumar, G. (1981). Observations on growth and moulting of spiny lobsters *Panulirus homarus* (Linnaeus), *P. ornatus* (Fabricius) and *P. penicillatus* (Oliver) in captivity. *Indian Journal of Fisheries, 28*, 25–35.
Nakai, T., & Park, S. C. (2002). Bacteriophage therapy of infectious diseases in aquaculture. *Research in Microbiology, 153*(1), 13–18.
Nelson, M. M., Cox, S. L., & Ritz, D. A. (2002). Function of mouthparts in feeding behavior of phyllosoma larvae of the packhorse lobster, *Jasus verreauxi* (Decapoda: Palinuridae). *Journal of Crustacean Biology, 22*(3), 595–600.
Nelson, M. M., Crear, B. J., Nichols, P. D., & Ritz, D. A. (2004). Growth and lipid composition of phyllosomata of the southern rock lobster, *Jasus edwardsii*, fed enriched Artemia. *Aquaculture Nutrition, 10*(4), 237–246.
Nelson, M. M., Bruce, M. P., Nichols, P. D., Jeffs, A. G., & Phleger, C. F. (2006). Nutrition of wild and cultured lobsters. In *Lobsters: Biology, management, aquaculture and fisheries* (pp. 205–230). Oxford: Blackwell Publishing.
Nilson, E. H., Fisher, W. S., & Shleser, R. A. (1975). Filamentous infestations observed on eggs and larvae of cultured crustaceans. *Proceedings of the Annual Meeting-World Mariculture Society, 6*, 367–375.
Nishida, S., Quigley, B. D., Booth, J. D., Nemoto, T., & Kittaka, J. (1990). Comparative morphology of the mouthparts and foregut of the final-stage phyllosoma, puerulus, and postpuerulus of the rock lobster *Jasus edwardsii* (Decapoda: Palinuridae). *Journal of Crustacean Biology, 10*(2), 293–305.
Nishimura, M. (1983). Studies on larval production of Ise lobster-I. *Annual Report of Mie Prefecture Hamajima Fisheries Experimental Station, 1981*, 6–69. (in Japanese).
Nishimura, M. & Kamiya, N. (1985). Studies on larval production of Ise lobster- III. Results of phyllosoma culture in 1983. *Annual Report of Mie Prefecture Fisheries Technological Center, 1983*, 1–6. (in Japanese).
Nishimura, M., & Kamiya, N. (1986). Studies on the larval production of Ise lobster – IV. *Annual Report of Mie Prefecture Fisheries Technological Center*, 38–39. (in Japanese).
Nishimura, M., & Kawai, H. (1984). Studies on larval production of Ise lobster – II. Food value of Artemia cultured with yeast enriched with fat on phyllosomas. *Annual Report of Mie Prefecture Hamajima Fisheries Experimental Station, 1982*, 1–6. (in Japanese).
Nonaka, M., Oshima, Y., & Hirano, R. (1958a). Rearing of phyllosoma of Ise lobster and moulting. *Suisan Zoshoku (Aquiculture), 5*, 13–15. (in Japanese).
Nonaka, M., Oshima, Y., & Hirano, R. (1958b). Cultured and ecdysis of spiny lobster at phyllosoma stage. *Aquaculture Science, 5*(3), 13–15. (in Japanese).

O'Rorke, R., Lavery, S. D., Wang, M., Nodder, S. D., & Jeffs, A. G. (2013). Determining the diet of larvae of the red rock lobster (*Jasus edwardsii*) using high throughput DNA sequencing techniques. *Marine Biology*. https://doi.org/10.1007/s00227-013-2357-7.

O'Rorke, R., Lavery, S. D., Wang, M., Gallego, M., Waite, M., Beckley, L. E., Thompson, P. A., & Jeffs, A. G. (2015). Phyllosomata associated with large gelatinous zooplankton: Hitching rides and stealing bites. *ICES Journal of Marine Science*. https://doi.org/10.1093/icesjms/fsu163.

Ong, K. S. (1967). A preliminary study of the early larval development of the spiny lobster Panulirus polyphagus (Herbst). *Malaysian Agricultural Journal, 46*, 183–190.

Oshima, Y. (1936). Feeding habit of Ise lobster. *Suisan Gakkai Ho, 7*, 16–21. (in Japanese).

Pardee, M. G. (1992a). *Culture of young spiny lobster (Panulirus argus): Effects of density and feed type on growth and survivorship*. Master's thesis, Florida Institute of Technology, USA.

Pardee, M. G. (1992b). Culture of puerulus through juvenile spiny lobster (Panulirus argus): Evaluation of live and supplemental feeds on growth and survivorship. *Aquaculture, 92*, 21–25.

Pardee, M. G., & Foster, S. M. (1992). Culture of young spiny lobster (*Panulirus argus*): Effects of density and feed type on growth and survivorship. In *Proceeding of the Gulf and Caribbean fisheries institute* (Vol. 45, pp. 778–789).

Pardee, M. G., & Foster, S. M. (1999). Culture of young lobster (*Panulirus argus*): Effects of density and feed type on growth and survivorship. In: M. H. Goodwin, & G. T. Waugh (Eds.). *Proceedings of the forty-fifth annual Gulf and Caribbean Fisheries Institute, Mexico*, November 1992, pp. 778–789. Charleston, South Carolina.

Parry, G., & Potts, W. T. (1965). Sodium balance in the freshwater prawn, *Palaemonetes antennarius*. *The Journal of Experimental Biology, 42*, 415–421.

Paul, A.A., Southgate, D.A.T., & Russell, J. 1980. First supplement to Mc Cance and Widdowson's. *The composition of foods: Amino acid composition (mg per 100 g food) and fatty acid composition (g per 100 g food)*. London/Amsterdam/New York/Oxford: Her Majesty's Stationery Office/Elsevier/North-Holland Biomedical Press.

Payne, M. S., Hall, M. R., Bannister, R., Sly, L., & Bourne, D. G. (2006). Microbial diversity within the water column of a larval rearing system for the ornate rock lobster (*Panulirus ornatus*). *Aquaculture, 258*(1), 80–90.

Payne, M. S., Hall, M. R., Sly, L., & Bourne, D. G. (2007). Microbial diversity within early-stage cultured *Panulirus ornatus* phyllosomas. *Applied and Environmental Microbiology, 73*(6), 1940–1951.

Payne, M. S., Høj, L., Wietz, M., Hall, M. R., Sly, L., & Bourne, D. G. (2008). Microbial diversity of mid-stage Palinurid phyllosoma from Great Barrier Reef waters. *Journal of Applied Microbiology, 105*(2), 340–350.

Pessani, D., Pisa, G., & Gattelli, R. (1999). The complete larval development of *Scyllarus arctus* (Decapoda, Scyllaridae) in the laboratory. In *7th colloquium Crustacea Decapoda Mediterranea*, Lisbon, 143–144.

Phillips, B. F., Campbell, N. A., & Rea, W. A. (1977). Laboratory growth of early juveniles of the western rock lobster Panulirus longipes cygnus. *Marine Biology, 39*, 31–39.

Phillips, B. F., Joll, L. M., Sandland, R. L., & Wright, D. W. (1983). Longevity, reproductive condition and growth of the western rock lobster *Panulirus cygnus* reared in aquaria. *Marine and Freshwater Research, 34*, 419–429.

Phillips, B. F. (1985). Aquacuture potential for rock lobsters in Australia. *Australian Fisheries, 44*(6), 2–7.

Phillips, B. F., & Booth, J. (1994). Design, use, and effectiveness of collectors for catching the puerulus stage of spiny lobsters. *Reviews in Fisheries Science, 2*(3), 255–289.

Phillips, B. F. & Liddy, G. C. (2003). Recent developments in spiny lobster aquaculture. In: B.F. Phillips, B. Magrey, & Y. Zhou (Eds.), *Proceedings of third world fisheries congress*. American Fisheries Society.

Phillips, B. F., & Matsuda, H. (2011). A global review of spiny lobster aquaculture. In R. K. Fotedar & B. F. Phillips (Eds.), *Recent advances and new species in aquaculture* (pp. 22–84). Sussex: Blackwell Publishing Ltd.

Phillips, B. F., & Sastry, A. N. (1980). Larval ecology. In J. S. Cobb & B. F. Phillips (Eds.), *The biology and management of lobsters* (pp. 11–57). London: Academic.

Phillips, B. F., Campbell, N. A., & Rea, W. A. (1977). Laboratory growth of early juveniles of the western rock lobster *Panulirus longipes cygnus*. *Marine Biology, 39*, 31–39.

Phillips, B. F., Melville-Smith, R., & Cheng, Y. W. (2003). Estimating the effects of removing *Panulirus cygnus* pueruli on the fishery stock. *Fisheries Research, 65*(1–3), 89–101.

Phillips, B. F., Cheng Y. W., Cox C., Hunt J., Jue, N. K., & Melville-Smith, R. (2005). Comparison of catches on two types of collector of recently settled stages of the spiny lobster (*Panulirus argus*), Florida, United States. *New Zealand Journal of Marine and Freshwater Research, 39*(3), 715–722.

Phillips, B. F., & Melville-Smith, R. (2006a). Panulirus species. In B. F. Phillips (Ed.), *Lobsters: Biology, management, aquaculture and fisheries* (pp. 359–384). Oxford, UK: Blackwell Publishing Ltd.

Phillips, B. F., Jeffs, A. G., Melville-Smith, R., Chubb, C. F., Nelson, M. M., & Nichols, P. D. (2006b). Changes in lipid and fatty acid composition of late larval and puerulus stages of the spiny lobster (*Panulirus cygnus*) across the continental shelf of Western Australia. *Comparative Biochemistry and Physiology – Part B Biochemistry and Molecular Biology, 143*(2), 219–228.

Phleger, C. F., Nelson, M. M., Mooney, B. D., Nichols, P. D., Ritar, A. J., Smith, G. G., Hart, P. R., & Jeffs, A. G. (2001). Lipids and nutrition of the southern rock lobster, *Jasus edwardsii*, from hatch to puerulus. *Marine and Freshwater Research, 52*(8), 1475–1486.

Prasad, R. R., & Tampi, P. R. S. (1959). A note on the first phyllosoma of *Panulirus burgeri* (de Haan). *Proceedings: Plant Sciences, 49*(6), 397–401.

Prasad, R. R., & Tampi, P. R. S. (1968). On the distribution of palinurid and scyllarid lobsters in the Indian Ocean. *Journal of the Marine Biological Association of India, 10*(1), 78–87.

Priyambodo, B., & Jaya, I. B. M. (2009). Lobster aquaculture in eastern Indonesia. Part I. Methods evolve for fledgling industry. *Global Aquaculture Advocate, 12*, 36–39.

Priyambodo, B. & Sarifin. (2009). Lobster aquaculture industry in eastern Indonesia: Present status and prospects. In: K. C. Williams (Ed.), *Spiny lobster aquaculture in the Asia-Pacific Region. Proceedings of the International Symposium,* Nha Trang, Vietnam, 9–10 December 2008. *ACIAR Proceedings 132*. Australian Centre for International Agricultural Research, Canberra. pp. 36–45.

Purcell, J. E., Baxter, E. J., & Fuentes, V. L. (2013). Jellyfish as products and problems of aquaculture. In G. Allan & G. Burnell (Eds.), *Advances in hatchery technology* (pp. 404–430). Cambridge: Woodhead Publishing.

Quackenbush, L. S., & Herrnkind, W. F. (1981). Regulation of molt and gonadal development in the spiny lobster, *Panulirus argus* (Crustacea: Palinuridae): Effect of eyestalk ablation. *Comparative Biochemistry and Physiology Part A: Physiology, 69*(3), 523–527.

Quinn, N. J., Kojis, B. L., & Chapman, G. (1992). Spiny lobster (*Panulirus argus*) recruitment to artificial habitats in waters off St. Thomas, U.S. Virgin Islands. *Proceedings of the Gulf and Caribbean Fisheries Institute, 45*, 759–777.

Quintana, Y. C., Canul, R. R., & Martínez, V. M. V. (2011). First evidence of *Panulirus argus* virus 1 (PaV1) in spiny lobster from Cuba and clinical estimation of its prevalence. *Diseases of Aquatic Organisms, 93*(2), 141–147.

Radhakrishnan, E. V. (1977). Breeding of laboratory reared spiny lobster *Panulirus homarus* (Linnaeus) under controlled conditions. *Indian Journal of Fisheries, 24*(1&2), 269–270.

Radhakrishnan, E.V. (1989). Physiological and biochemical studies in the spiny lobster *Panulirus homarus*. PhD thesis, University of Madras, pp.

Radhakrishnan, E. V. (1994). Commercial prospects for farming spiny lobsters. In: K. P. P. Nambiar & T. Singh (Eds.), *Aquaculture towards the 21st century: Proceedings of the INFOFISH-AQUATECH '94 Conference,* Colombo, Sri Lanka, 29-31 August 1994. pp. 96–102.

Radhakrishnan, E. V. (1995). Lobster fisheries of India. *The Lobster Newsletter, 8*(2), 1.

Radhakrishnan, E. V. (2005). Breeding and hatchery technology development of spiny lobsters and crabs- a review. In *Proceedings International Conference and Exposition on Marine Living*

resources of India for Food & Medicine, 27–29 February, 2004, aquaculture Foundation of India (pp. 265–272).
Radhakrishnan, E. V. (2009). Overview of lobster farming. In: K. Madhu, & R. Madhu (Eds.), *Winter school on recent advances in breeding and larviculture of marine finfish and shellfish.* 30.12.2008 to 19.1.2009, CMFRI, Kochi, pp. 129–138.
Radhakrishnan, E. V. (2015). Review of prospects for lobster farming. In S. Perumal, A. R. Thirunavukkarasu, & P. Perumal (Eds.), *Advances in marine and brackish water aquaculture* (pp. 173–186). New Delhi: Springer.
Radhakrishnan, E. V. & Devarajan, K. (1986). Growth of the spiny lobster *Panulirus polyphagus* (Herbst) reared in the laboratory. In: *Proceedings of the Symposium on Coastal Aquaculture*, part 4; 12–18 January 1980, Cochin.
Radhakrishnan, E. V., & Vijayakumaran, M. (1982). Unprecedented growth induced in spiny lobsters. *The Marine Fisheries Information Service: Technical and Extension Series, 43*, 6–8.
Radhakrishnan, E. V., & Vijayakumaran, M. (1984a). Effect of eyestalk ablation in the spiny lobster *Panulirus homarus* (Linnaeus) 1. On moulting and growth. *Indian Journal of Fisheries, 31*, 130–147.
Radhakrishnan, E. V., & Vijayakumaran, M. (1984b). Effect of eyestalk ablation in the spiny lobster *Panulirus homarus* (Linnaeus) 3. On gonadal maturity. *Indian Journal of Fisheries, 31*, 209–216.
Radhakrishnan, E. V. & Vijayakumaran, M. (1986). Observations on the feeding and moulting of laboratory reared phyllosoma larvae of the spiny lobster *Panulirus homarus* (Linnaeus) under different light regimes. In: *Proceedings of the symposium on coastal aquaculture*, part 4, 12–18 January 1980, Cochin.
Radhakrishnan, E. V. & Vijayakumaran, M. (1990). An assessment of the potential of spiny lobster culture in India. In: *CMFRI Bulletin-National symposium on research and development in marine fisheries*, CMFRI, Kochi, 44(2), 416–426.
Radhakrishnan, E. V., & Vijayakumaran, M. (1995). Early larval development of spiny lobster, *Panulirus homarus* (Linnaeus, 1758) reared in the laboratory. *Crustaceana, 68*(2), 151–159.
Radhakrishnan, E. V., Deshmukh, V. D., Manisseri, M. K., Rajamani, M., Kizhakudan, J. K., & Thangaraja, R. (2005). Status of the major lobster fisheries in India. *New Zealand Journal of Marine and Freshwater Research, 39*(3), 723–732.
Radhakrishnan, E. V., Chakraborty, R. D., Thangaraja, R., & Unnikrishnan, C. (2009). Effect of Nannochloropsis Salina on the survival and growth of phyllosoma of the tropical spiny lobster, *Panulirus homarus* L. under laboratory conditions. *Journal of the Marine Biological Association of India, 51*(1), 52–60.
Radhakrishnan, E. V., Chakraborty, R. D., Baby, P. K., & Radhakrishnan, M. (2013). Fishery and population dynamics of the sand lobster *Thenus unimaculatus* (Burton & Davie, 2007) landed by trawlers at Sakthikulangara fishing harbour in the south-west coast of India. *Indian Journal of Fisheries, 60*(2), 7–12.
Rahman, M. K., Prakash, E. B., & Subramoniam, T. (1987). Studies on the egg development stages of the sand lobster *Thenus orientalis* (Lund). In: P. Natarajan, H. Suryanarayana, & P. K. A Aziz (Eds.), *Advances in aquatic biology and fisheries,* Prof. N. Balakrishnan Nair Felicitation Volume, Department of Aquatic Biology and Fisheries, University of Kerala, Trivandrum, pp. 327–335.
Rahman, M. K., Joseph, M. T. L., & Srikrishnadhas, B. (1997). Growth performance of spiny lobster *Panulirus homarus* (Linnaeus) under mass rearing. *Journal of Aquaculture in the Tropics, 12*(4), 243–253.
Rao, G. S., George, R. M., Anil, M. K., Saleela, K. N., Jasmine, S., Jose Kingsly, H., & Rao, G. H. (2010). Cage culture of the spiny lobster *Panulirus homarus* (Linnaeus) at Vizhinjam, Trivandrum along the southwest coast of India. *Indian Journal of Fisheries, 57*(1), 23–29.
Raskoff, K. A., Sommer, F. A., Hamner, W. M., & Cross, K. M. (2003). Collection and culture techniques for gelatinous zooplankton. *The Biological Bulletin, 204*, 68–80.
Rayns, N. D. (1991). *The growth and survival of juvenile rock lobster Jasus edwardsii held in captivity.* PhD thesis, University of Otago.

Rimmer, D. W., & Phillips, B. F. (1979). Diurnal migration and vertical distribution of phyllosoma larvae of the western rock lobster *Panulirus cygnus*. *Marine Biology, 54*, 109–114.

Ritar, A. J. (2001). The experimental culture of phyllosoma larvae of southern rock lobster (*Jasus edwardsii*) in a flow-through system. *Aquacultural Engineering, 24*(2), 149–156.

Ritar, A. J. & Smith, G. (2005). Hatchery production of southern rock lobster in Tasmania. *Austasia Aquaculture*, February/March, 42–43.

Ritar, A. J., Dunstan, G. A., Crear, B. J., & Brown, M. R. (2003). Biochemical composition during growth and starvation of early larval stages of cultured spiny lobster (*Jasus edwardsii*) phyllosoma. *Comparative Biochemistry and Physiology – Part A: Molecular & Integrative Physiology, 136*(2), 353–370.

Ritar, A. J., Smith, G. G., & Thomas, C. W. (2006). Ozonation of seawater improves the survival of larval southern rock lobster, *Jasus edwardsii*, in culture from egg to juvenile. *Aquaculture, 261*(3), 1014–1025.

Robertson, P. B. (1968). The complete larval development of the sand lobster, *Scyllarus americanus* (Smith)(Decapoda, Scyllaridae) in the laboratory, with notes on larvae from the plankton. *Bulletin of Marine Science, 18*(2), 294–342.

Robertson, P. B. (1979). Larval development of the scyllarid lobster *Scyllarus planorbis* Holthuis reared in the laboratory. *Bulletin of Marine Science, 29*(3), 320–328.

Rockel, C. M., & Watson, W. H., III. (1996). A comparison of the osmoregulatory capabilities of coastal and estuarine lobsters. *Journal of Shellfish Research, 15*, 495.

Rodriguez-Viera, L., & Perera, E. (2012). *Panulirus argus* postlarva performance fed with fresh squid. *Revista de Investigaciones Marinas, 32*(1), 9–15.

Rogers, P. P., Barnard, R., & Johnston, M. (2010). Lobster aquaculture a commercial reality: A review. *Journal of the Marine Biological Association of India, 52*(2), 327–335.

Ryther, J. H., Lellis, W. A., Bannerot, S. P., Chaiton, J. A. (1988). *Spiny lobster mariculture* (Final Report, US Aid Grant No. 538–0140.03(1)). Harbour Branch Oceanographic Institute, Ft. Pierce, FL, p. 42.

Sachlikidis, N. G., Jones, C. M., & Seymour, J. E. (2005). Reproductive cues in *Panulirus ornatus*. *New Zealand Journal of Marine and Freshwater Research, 39*(2), 305–310.

Saisho, T. (1962a). Notes on the early development of a scyllarid lobster, *Parribacus antarcticus* (Lund). *Memorial of Faculty of Fisheries, Kagoshima University, 11*(2), 174–178.

Saisho, T. (1962b). Notes on the early development of phyllosoma of *Panulirus japonicus*. *Memorial Faculty of Fisheries, Kagoshima University, 11*(1), 18–23.

Samraj, Y. C., Jayagopal, P., & Kamalraj, K. (2010). Farming of spiny lobsters in onshore facility. Abstract. *International conference on recent advances in lobster biology, aquaculture and management, RALBAM, 2010*, NIOT, Chennai, p. 69.

Sarvaiya, R. T. (1987). Successful spiny lobster culture In Gujarat. *Fishing Chimes, 7*(7), 18–23.

Schmitt, A. S. C., & Uglow, R. F. (1997). Effects of ambient ammonia levels on blood ammonia, ammonia excretion and heart and scaphognathite rates of *Nephrops norvegicus*. *Marine Biology, 127*, 411–418.

Scolding, J. W. S., Powell, A., Boothroyd, O. P., & Shields, R. J. (2012). The effect of ozonation on the survival, growth and microbiology of the European lobster *Homarus gammarus*. *Aquaculture, 364-365*, 217–223.

Sekine, S., Shima, Y., Fushimi, H., & Nonaka, M. (2000). Larval period and molting in the Japanese spiny lobster *Panulirus japonicus* under laboratory conditions. *Fisheries Science, 66*(1), 19–24.

Sharp, W. C., Lellis, W. A., Butler, M. J., Herrnkind, W. F., Hunt, J. H., Pardee-Woodring, M., & Matthews, T. R. (2000). The use of coded microwire tags in mark-recapture studies of juvenile Caribbean spiny lobster, *Panulirus argus*. *Journal of Crustacean Biology, 20*, 510–521.

Sheen, S.-S. (2000). Dietary cholesterol requirement of juvenile mud crab *Scylla serrata*. *Aquaculture, 189*, 277–285.

Sheppard, J. K., Bruce, M. P., & Jeffs, A. G. (2002). Optimal feed pellet size for culturing juvenile spiny lobster *Jasus edwardsii* (Hutton, 1875) in New Zealand. *Aquaculture Research, 33*, 913–916.

Shields, J. D. (2011). Diseases of spiny lobsters: A review. *Journal of Invertebrate Pathology, 106*(1), 79–91.

Shields, J. D., Stephens, F. J., & Jones, B. (2006). Pathogens, parasites and other symbionts. In: *Lobsters: biology, management, aquaculture and fisheries* (pp. 146–204). Oxford: Blackwell.

Shindo, S. (1973). *General review of the trawl fishery and the demersal fish stocks of the South China Sea* (F.A.O. Fisheries technical paper. no. 120). Rome: FAO.

Sibeni, F., & Calderini, F. (2012). *FishStatJ: A tool for fishery statistics analysis*. Version 2.0. Food and Agriculture Organization of the United Nations, Rome, Italy.

Sikorski, Z. E., Kolakowska, A., & Pan, B. S. (2004). The nutritive composition of the major groups of marine food organisms. In Z. E. Sikorski (Ed.), *Seafood: Resources nutritional composition and preservation* (pp. 30–47). Boca Raton: CRC Press Inc.

Silas, E. G. (1982). Major breakthrough in spiny lobster culture. *The Marine Fisheries Information Service: Technical and Extension Series, 43*, 1–5.

Silas, E. G., Radhakrishnan, E. V., & Vijayakumaran, M. (1984). Eyestalk ablation. A new idea boosts the growth of the Indian spiny lobster. *Fish Farming International, 2*(7), 10–11.

Simon, C. J. (2009). Digestive enzyme response to natural and formulated diets in cultured juvenile spiny lobster, *Jasus edwardsii*. *Aquaculture, 294*(3–4), 271–281.

Simon, C. J., & Jeffs, A. G. (2008). Feeding and gut evacuation of cultured juvenile spiny lobsters, *Jasus edwardsii*. *Aquaculture, 280*, 211–219.

Simon, C. J., & Jeffs, A. G. (2013). The effect of dietary carbohydrate on the appetite revival and glucose metabolism of juveniles of the spiny lobster, *Jasus edwardsii*. *Aquaculture, 384–387*, 111–118.

Simon, J. I., Williams, K. C., Barclay, M. C., & Tabrett, S. J. (2010). Do formulated feeds for juvenile *Panulirus ornatus* lobsters require dietary cholesterol supplementation? *Aquaculture, 307*, 241–246.

Simon, C. J., Rodemann, T., & Carter, C. G. (2016). Near-infrared spectroscopy as a novel invasive tool to assess spiny lobster nutritional condition. *PLoS One, 11*(7), e0159671. https://doi.org/10.1371/journal.pone.0159671.

Simopoulos, A. P. (1997). ω-3 fatty acids in the prevention management of cardiovascular disease. *Canadian journal of physiology and pharmacology. NRC Research Press, 75*(3), 234-239.

Sjoken, R. 1999. *The formulation and evaluation of an artificial feed for the Growout of spiny lobsters*. MS thesis, graduate School of the Florida Institute of technology, Florida, USA.

Smith, G., Ritar, A. J., Carter, C. G., Dunstan, G. A., & Brown, M. R. (2003a). Photothermal manipulation of reproduction in broodstock and larval characteristics in newly hatched phyllosoma of the spiny lobster, *Jasus edwardsii*. *Aquaculture, 220*, 299–311.

Smith, D. M., Williams, K. C., Irvin, S., Barclay, M., & Tabrett, S. (2003b). Development of a pelleted feed for juvenile tropical spiny lobster (*Panulirus ornatus*): Response to dietary protein and lipid. *Aquaculture Nutrition, 9*, 231–237.

Smith, D. M., Williams, K. C., & Irvin, S. J. (2005). Response of the tropical spiny lobster *Panulirus ornatus* to protein content of pelleted feed and to a diet of mussel flesh. *Aquaculture Nutrition, 11*(3), 209–217.

Smith, D. M., Irvin, S. J., & Mann, D. (2009a). *Optimising the physical form and dimensions of feed pellets for tropical spiny lobsters*. Spiny lobster aquaculture in the Asia–Pacific region, 157.

Smith, G., Salmon, M., Kenway, M., & Hall, M. (2009b). Description of the larval morphology of captive reared *Panulirus ornatus* spiny lobsters, benchmarked against wild-caught specimens. *Aquaculture, 295*(1), 76–88.

Smith, G., Kenway, M., & Hall, M. (2010). Starvation and recovery ability of phyllosoma of the tropical spiny lobsters *Panulirus ornatus* and *P. homarus* in captivity. *Journal of the Marine Biological Association of India, 52*(2), 249–256.

Srikrishnadhas, B., Sunderraj, V., & Kuthalingam, M. D. K. 1983. Cage culture of spiny lobster *Panulirus homarus*. In *Proceedings of the National Seminar on Cage and Pen culture* (pp. 103–106). Tuticorin: Fisheries College and Research Institute.

Sumbing, M. V., Al-Azad, S., Estim, A., & Mustafa, S. (2016). Growth performance of spiny lobster *Panulirus ornatus* in land based integrated multitrophic aquaculture (IMTA) system. *Translation on Science and Technology, 3*(1–2), 143–149.

Suzuki, N., Murakami, K., Takeyama, H., & Chow, S. (2006). Molecular attempt to identify prey organisms of lobster phyllosoma larvae. *Fisheries Science, 72*, 342–349.

Suzuki, N., Murakami, K., Takeyama, H., & Chow, S. (2007). Eukaryotes from the hepatopancreas of lobster phyllosoma larvae. *Bulletin of Fisheries Research Agency (Japan), 20*, 1–7.

Suzuki, N., Hoshino, K., Murakami, K., Takeyama, H., & Chow, S. (2008). Molecular diet analysis of phyllosoma larvae of the Japanese spiny lobster *Panulirus japonicus* (Decapoda: Crustacea). *Marine Biotechnology, 10*, 49–55.

Takahashi, M., & Saisho, T. (1978). The complete larval development of the scyllarid lobster *Ibacus ciliatus* (Von Siebold) and *Ibacus novemdentatus* Gibbes in the laboratory. *Memoirs of the Faculty of Fisheries Kagoshima University, 27*, 305–353.

Takeuchi, T., & Murakami, K. (2007). Crustacean nutrition and larval feed, with emphasis on Japanese spiny lobster, *Panulirus japonicus*. *Bulletin of Fisheries Research Agency, 20*, 15–23.

Tamm, G. R. (1980). Spiny lobster culture: An alternative to natural stock assessment. *Fisheries, 5*, 4.

Teshima, S.-I. (1997). Phospholipids and sterols. In L. R. D'Abramo, D. E. Conklin, & D. M. Akiyama (Eds.), *Crustacean nutrition, advances in world aquaculture* (Vol. 6, pp. 85–107). Baton Rouge: World Aquaculture Society.

Tholasilingam, T. & Rengarajan, K.1986. Prospects on spiny lobster Panulirus spp. culture in the east coast of India. In: *Proceedings of the Symposium on Coastal Aquaculture*, part 4, 12-18 January 1980, Cochin.

Thorson, G. (1950). Reproductive and larval ecology of marine bottom invertebrates. *Biological Reviews, 25*, 1–45.

Thuy, N. T. B & Ngoc, N. B. (2004). Current status and exploitation of wild spiny lobsters in Vietnamese waters. In: K. C. Williams (Ed.), *Spiny lobster ecology and exploitation in the South China Sea Region,* Australian Centre for International Agricultural Research, Canberra, 120: pp. 13–16.

Tong, L. J., Moss, G. A., Paewai, M. M., & Pickering, T. D. (1997). Effect of brine-shrimp numbers on growth and survival of early-stage phyllosoma larvae of the rock lobster *Jasus edwardsii*. *Marine and Freshwater Research, 48*(8), 935–940.

Travis, D. F. (1954). The molting cycle of the spiny lobster, *Panulirus argus* Latreille. I. Molting and growth in laboratory-maintained individuals. *The Biological Bulletin, 107*(3), 433–450.

Tuan, L.A., & Mao, N. D. (2004). Present status of lobster cage culture in Vietnam. In: K. C. Williams (Ed.). *Spiny lobster ecology and exploitation in the South China Sea Region. ACIAR Proceedings, 120* (pp. 21–25). Canberra: Australian Centre for International Agriculture Research.

Venugopal, V., & Gopakumar, K. (2017). Shellfish: Nutritive value, health benefits and consumer safety. *Comprehensive Reviews in Food Science and Food Safety, 16*, 1219–1242.

Verghese, B., Radhakrishnan, E. V., & Padhi, A. (2007). Effect of environmental parameters on immune response of the Indian spiny lobster, *Panulirus homarus* (Linnaeus, 1758). *Fish & Shellfish Immunology, 23*, 928–936.

Verschuere, L., Rombaut, G., Sorgeloos, P., & Verstraete, W. (2000). Probiotic bacteria as biological control agents in aquaculture. *Microbiology and Molecular Biology Reviews, 64*(4), 655–671.

Vidya, K., & Joseph, S. (2012). Effect of salinity on growth and survival of juvenile Indian spiny lobster, *Panulirus homarus* (Linnaeus). *Indian Journal of Fisheries, 59*(1), 113–118.

Vijayakumaran, M. (1990). *Energetics of a few marine crustaceans*. PhD thesis, Cochin University of Science & Technology, Cochin, India, 138 pp.

Vijayakumaran, M. & Radhakrishnan, E. V. (1986). Effects of food density on feeding and moulting of phyllosoma larvae of the spiny lobster *Panulirus homarus* (Linnaeus). In: *Proceedings of the symposium on coastal aquaculture*, Part 4, MBAI, 12–18 January 1980, Cochin.

Vijayakumaran, M. & Radhakrishnan, E.V. (1988). Ecological and behavioural studies of the pueruli of the palinurid lobsters along the Madras coast. Abstract. Symposuim on Tropical Marine Living Resources, Cochin.

Vijayakumaran, M., & Radhakrishnan, E. V. (2002). Changes in biochemical and mineral composition during ovarian maturation in the spiny lobster, *Panulirus homarus* (Linnaeus). *Journal of the Marine Biological Association of India, 44*(1&2), 85–96.

Vijayakumaran, M. & Radhakrishnan, E. V. (2003). Control of epibionts with chemical disinfectants in the phyllosoma larvae of the spiny lobster *Panulirus homarus* (Linnaeus). In: *Aquaculture Medicine*. CUSAT, Kochi, pp. 69–72.

Vijayakumaran, M., & Radhakrishnan, E. V. (2011). Slipper lobsters. In R. K. Fotedar & B. F. Phillips (Eds.), *Recent advances and new species in aquaculture* (pp. 85–114). Blackwell Publishing Ltd..

Vijayakumaran, M., Murugan, T. S., Remany, M. C., Mary Leema, T., Jha, D. K., Santhanakumar, J., Venkatesan, R., & Ravindran, R. (2005). Captive breeding of the spiny lobster *Panulirus homarus* (Linnaeus, 1758). *New Zealand Journal of Marine and Freshwater Research, 39*, 325–334.

Vijayakumaran, M., Venkatesan, R., Murugan, T. S., Kumar, T. S., Jha, D. K., Remany, M. C., Mary Leema, T., Syed Jahan, S., Dharani, G., Selvan, K., & Kathiroli, S. (2009). Farming of spiny lobsters in sea cages in India. *New Zealand Journal of Marine and Freshwater Research, 43*, 623–634.

Vijayakumaran, M., Anbarasu, M., & Kumar, T. S. (2010). Moulting and growth in communal and individual rearing of the spiny lobster, *Panulirus homarus*. *Journal of the Marine Biological Association of India, 52*(2), 274–281.

Vijayakumaran, M., Maharajan, A., Rajalakshmi, S., Jayagopal, P., Subramanian, M. S., & Remani, M. C. (2012). Fecundity and viability of eggs in wild breeders of spiny lobsters, *Panulirus homarus* (Linnaeus, 1758), *Panulirus versicolor* (Latreille, 1804) and *Panulirus ornatus* (Fabricius, 1798). *Journal of the Marine Biological Association of India, 54*(2), 5–9.

Vijayakumaran, M., Maharajan, A., Rajalakshmi, S., Jayagopal, P., & Remani, M. C. (2014). Early larval stages of the spiny lobsters, *Panulirus homarus*, *Panulirus versicolor* and *Panulirus ornatus* cultured under laboratory conditions. *International Journal of Development Research, 4*(2), 377–383.

Vijayan, K. K., Sharma, K., & Kizhakudan, J. K. et al. (2010a). Gaffkemia (red tail disease) – An emerging disease problem in lobster holding facilities in southern India. *International conference in recent advances in lobster biology, aquaculture and management*, 4–8 January, 2010, Chennai, Abstract, pp. 61–62.

Vijayan, K. K., Sanil, N. K., & Krupesh, S. (2010b). Health management concepts in lobster mariculture. In: M. Vijayakumaran, & E.V. Radhakrishnan, (Eds.), *Lobster research in India. International Conference in Recent advances in Lobster Biology, Aquaculture and Management*, 4–8 January, 2010, Chennai. Lobster Research in India, pp. 41–45.

Vijayanand, P., Murugan, A., Saravanakumat, K. et al. (2004). Experimental farming of sand lobster *Thenus orientalis* (Lund): Abstract. In: *Ocean Life Food & Medicine Expo 2004*. International Conference & Exposition on marine living resources of India for Food & Medicine. Aquaculture Foundation of India.

Villareal, H., Hernandez-Llamas, A., & Hewitt, R. (2003). Effects of salinity on growth, survival, and oxygen consumption of 118 juvenile brown shrimp *Farfantepenaeus californiensis* (Holmes). *Aquaculture Research, 34*, 187–193.

Von Bonde, C. (1936). The reproduction, embryology and metamorphosis of the cape crawfish: *Jasus lalandii* (Milne Edwards) Ortmann. *Department of Commerce and Industries, South Africa, Investigational Report*, no. 6, 25 pp.

Wada, Y., Kawahara, A., Munekiyo, M., & Sobajima, N. (1985). Distribution and larval stages of the phyllosoma larvae of a scyllarid lobster, *Scyllarus kitanoviriosus*, in the western Wakasa Bay. *Bulletin of the Kyoto Institute of Oceanic and Fishery Science, 9*, 51–57. (in Japanese).

Wahle, R. A., Gibson, M., & Fogarty, M. (2009). Distinguishing disease impacts from larval supply effects in a lobster fishery collapse. *Marine Ecology Progress Series, 376*, 185–192.

Wakabayashi, K., & Phillips, B. F. (2016). Morphological descriptions of laboratory reared larvae and post-larvae of Australian shovel-nosed lobster, *Thenus australiensis* Burton & Davie, 2007 (Decapoda, Scyllaridae). *Crustaceana, 89*(1), 97–117.

Wakabayashi, K., Sato, R., Ishii, H., Akiba, T., Nogata, Y., & Tanaka, Y. (2012). Culture of phyllosomas of *Ibacus novemdentatus* (Decapoda: Scyllaridae) in a closed recirculating system using jellyfish as food. *Aquaculture, 330*, 162–166.

Wakabayashi, K., Nagai, S., & Tanaka, Y. (2016). The complete larval development of *Ibacus ciliatus* from hatching to the nisto and juvenile stages using jellyfish as the sole diet. *Aquaculture, 450*, 102–107.

Wang, M., & Jeffs, A. G. (2014). Nutritional composition of potential zooplankton prey of spiny lobster larvae: A review. *Reviews in Aquaculture, 6*(4), 270–299.

Ward, L. R., Carter, C. G., Crear, B. J., & Smith, D. M. (2003). Optimal dietary protein level for juvenile southern rock lobster, *Jasus edwardsii*, at two lipid levels. *Aquaculture, 217*, 483–500.

Warner, R. E., Combs, C., & Gregory, D. R. (1977). Biological studies of the spiny lobster, *Panulirus argus* (Decapoda: Palinuridae), South Florida. *Proceedings of the Gulf and Caribbean Fisheries Institute, 29*, 166–183.

Webber, W. R., & Booth, J. D. (1995). A new species of *Jasus* (Crustacea: Decapoda: Palinuridae) from the eastern South Pacific Ocean. *New Zealand Journal of Marine and Freshwater Research, 29*, 613–622.

Webster, N. S., Bourne, D. G., & Hall, M. (2006). Vibrionaceae infection in phyllosomas of the tropical rock lobster *Panulirus ornatus* as detected by fluorescence in situ hybridisation. *Aquaculture, 255*(1), 173–178.

Wells, R. M. G., Lu, J., Hickey, A. J. R., & Jeffs, A. G. (2001). Ontogenetic changes in enzyme activities associated with energy production in the spiny lobster, *Jasus edwardsii*. *Comparative Biochemistry and Physiology B: Biochemistry & Molecular Biology, 130*(3), 339–347.

Wietz, M., Hall, M. R., & Høj, L. (2009). Effects of seawater ozonation on biofilm development in aquaculture tanks. *Systematic and Applied Microbiology, 32*(4), 266–277.

Williams, K. C. (2004). In: (Ed./Eds.), *Proceedings of a Workshop on Spiny Lobster Ecology and Exploitation in the South China Sea Region*. ACIAR Proceedings no. 120. Australian Government, Australian Centre for International Agricultural Research, Canberra.

Williams, K. C. (2007). Nutritional requirements and feeds development for post-larval spiny lobster: A review. *Aquaculture, 263*, 1–14.

Williams, K. C. (2009). Spiny lobster aquaculture in the Asia-Pacific region. Proceedings of an international symposium held at Nha Trang, Vietnam, 9–10, December, 2008. *ACIAR proceedings no. 132*. Australian Center for International Agricultural Research, Canberra.

Williams, K. C., Smith, D. M., Irvin, S. J., Barclay, M. C., & Tabrett, S. J. (2005). Water immersion time reduces the preference of juvenile tropical spiny lobster *Panulirus ornatus* for pelleted dry feeds and fresh mussel. *Aquaculture Nutrition, 11*, 415–426.

Winget, R. R. (1969). Oxygen consumption and respiratory energetics in the spiny lobster, *Panulirus interruptus* (Randall). *The Biological Bulletin, 136*(2), 301–312.

Witham, R. (1973). Preliminary thermal studies on young *Panulirus argus*. *Florida Scientist, 36*, 154–158.

Witham, R., Ingle, R. M., & Sims, H. W., Jr. (1964). Notes on postlarvae of *Panulirus argus*. *Quarterly Journal of the Florida Academy of Sciences, 27*, 289–297.

Witham, R., Ingle, R., & Joyce, E. A. (1968). Physiological and ecological studies of *Panulirus argus* from the St. Lucie estuary. *Florida Board Conservation Technical Series, 53*, 31 pp.

Yamakawa, T., Nishimura, M., Matsuda, H., Tsujigadou, A., & Kamiya, N. (1989). Complete larval rearing of the Japanese spiny lobster *Panulirus japonicus*. *Nippon Suisan Gakkaishi, 55*, 745.

Yoshizawa, S., Tsuruya, Y., Fukui, Y., Sawabe, T., Yokota, A., Kogure, K., Higgins, M., & Thompson, F. L. (2012). *Vibrio jasicida* sp. nov., a member of the Harveyi clade, isolated from marine animals (packhorse lobster, abalone and Atlantic salmon*)*. *International Journal of Systematic and Evolutionary Microbiology, 62*(8), 1864–1870.

11

Culture of Slipper Lobster Larvae (Decapoda: Achelata: Scyllaridae) Fed Jellyfish as Food

Kaori Wakabayashi, Yuji Tanaka, and Bruce F. Phillips

Abstract

Planktonic larvae of slipper and spiny lobsters, so-called phyllosoma, are known to be associated with various kinds of gelatinous zooplankton such as jellyfish and salps in the wild. Phyllosoma larvae likely utilise the gelatinous zooplankton for food, transport, and protection. Based on knowledge of the natural association behaviour of phyllosoma larvae with gelatinous zooplankton, a seed production technique for lobsters using gelatinous zooplankton as food can be established. In tank conditions, the complete larval development from newly hatched phyllosoma to juvenile stage has been achieved with cnidarian jellyfish as the sole food in three slipper lobster species, *Ibacus novemdentatus*, *I. ciliatus*, and *Thenus australiensis*. Understanding of the biophysical and biochemical compositions of jellyfish and their effect on growth and survival of phyllosoma larvae may result in new knowledge and techniques for successful achievement of mass seed production, as well as development of a sustainable jelly-like artificial diet for phyllosoma larvae.

K. Wakabayashi (✉)
Graduate School of Integrated Sciences for Life, Hiroshima University, Higashi-Hiroshima, Hiroshima, Japan
e-mail: kaoriw@hiroshima-u.ac.jp

Y. Tanaka
Graduate School of Marine Science and Technology, Tokyo University of Marine Science and Technology, Minato, Tokyo, Japan
e-mail: ytanaka@kaiyodai.ac.jp

B. F. Phillips
School of Molecular and Life Sciences, Curtin University, Perth, WA, Australia
e-mail: B.Phillips@curtin.edu.au

Keywords

Phyllosoma larvae · Gelatinous zooplankton · Jellyfish riding behaviour · Larviculture

11.1 Introduction

The infraorder Achelata consists of two families: Scyllaridae (known as slipper lobsters) and Palinuridae (known as spiny lobsters). They are characterised by lacking the chelae on adult pereiopods and passing through a larval form of dorsoventrally flattened zoea, phyllosoma (Palero et al. 2014a; see also Chap. 1). Nineteen genera are currently recognised in Scyllaridae, and they belong to one of four subfamilies (Chan 2010; see also Chap. 2). All these slipper lobsters are benthic during the adult phase, whereas they are planktonic during the larval phase.

The life history of the slipper lobsters is similar to that of the spiny lobsters. In general, the larva hatches as phyllosoma from an egg attached externally to the female's abdomen. It develops through a series of moultings via several different stages; subsequently, the larva at final stage of phyllosoma metamorphoses into a nisto, which is a comparable stage of puerulus for the spiny lobsters. The nisto moults into the first juvenile stage, and then it grows through successive juvenile and young stages to become an adult lobster (Lavalii and Spanier 2007; see also Chap. 1).

A phyllosoma larva of slipper lobsters often associates with various kinds of gelatinous zooplankton, such as pelagic cnidarians and ctenophores; therefore, it is also known as a 'jellyfish rider' (Fig. 11.1). Many scientists have recorded the occurrences of slipper lobster larvae clinging onto the gelatinous zooplankton (Table 11.1). Moreover, it is a trend for divers to photograph a slipper lobster larva accompanied with a gelatinous zooplankton underwater (e.g. Yoshino 2015; Wakabayashi et al. 2017a; Ianniello and Mears 2018). The larvae undoubtedly rely on the gelatinous zooplankton for their nutrients as explained in this chapter. Morphological and histological analyses have indicated that the larvae of both slipper and spiny lobsters possess mouthparts and foregut suitable for feeding and digesting soft materials (Nishida et al. 1990; Mikami and Takashima 1993; Mikami et al. 1994), such as gelatinous zooplankton.

Despite knowledge of the natural diet of phyllosoma larvae, brine shrimp, clam flesh and mussel gonads have been traditionally used as the principal food materials for research on seed production of slipper and spiny lobsters (reviewed in Kittaka 2000; Mikami and Kuballa 2007). Attempts were made to culture at least 21 species of slipper lobsters larvae using those food materials, and 12 out of the 21 species successfully completed metamorphosing into the nisto stage (Higa et al. 2005; Mikami and Kuballa 2007; Kumar et al. 2009; Kizhakudan and Krishnamoorthi 2014); therefore, those food materials have been considered a key contributor to the

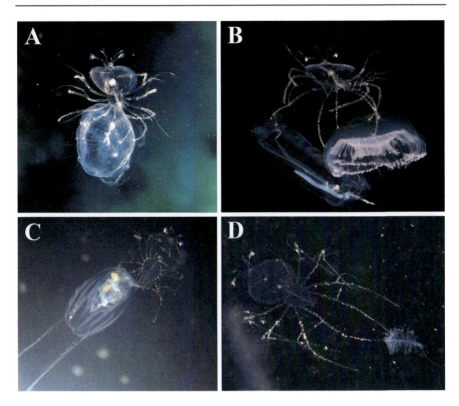

Fig. 11.1 Direct observations of the association behaviour of scyllarid phyllosoma larvae with gelatinous zooplankton. (**a**) *Ibacus ciliatus* with a schyphozoan, *Aurelia coerulea*, photographed by Kaori Wakabayashi; (**b**) *Ibacus ciliatus* with a hydrozoan, *Aequorea victoria*, and a salp, *Salpa fusiformis*, photographed by Hideki Abe (in Wakabayashi et al. 2017a, reproduced with a permission of Bun-ichi); (**c**) *Eduarctus martensii* with a ctenophore, *Hormiphora* sp., photographed by Miki Sasagawa (in Wakabayashi et al. 2017b, reproduced from the *Journal of Crustacean Biology* with a permission of Oxford University Press); (**d**) *Galearctus kitanoviriosus* with a pelagic annelid, *Tomopteris* sp., photographed by Yusuke Yoshino

complete rearing of slipper lobster larvae (Mikami and Kuballa 2007). On the other hand, there are disadvantages to using those materials. Brine shrimp is not ideal in size and nutritional quality for late-stage larvae (Mikami and Kuballa 2004), and it can be a probable vector of pathogenic bacteria (Fisher et al. 1978; Goulden et al. 2012). Clams and mussels cause a labour to remove the shells. Having other types of food materials or even developing a formulated food would make slipper and spiny lobster larviculture sustainable.

This chapter reviews the knowledge of the association of slipper lobster larvae with gelatinous zooplankton, both in the wild and in the laboratory. It also outlines the attempt at culturing the slipper lobster larvae using gelatinous zooplankton as food. Finally, we discuss the potential of gelatinous zooplankton to be utilised as a nutritional resource for larviculture of commercially important slipper lobsters.

Table 11.1 Records of scyllarid phyllosomas associating with gelatinous zooplankton in the wild

Species	Gelatinous zooplankton	Locality	References
Scyllaridae			
Ibachinae			
Ibacus ciliatus	*Aurelia coerulea* (Scyphozoa: Cnidaria)	Nagasaki, Japan	Shojima (1963, 1973)
		Yamaguchi, Japan	Wakabayashi et al. (2017a)
	Chrysaora pacifica (Scyphozoa: Cnidaria)	Nagasaki, Japan	Shojima (1963, 1973)
		Yamaguchi, Japan	Wakabayashi et al. (2017a)
	Liriope tetraphylla (Hydrozoa: Cnidaria)	Nagasaki, Japan	Shojima (1973)
	Aequorea coerulescens (Hydrozoa: Cnidaria)	Yamaguchi, Japan	Wakabayashi et al. (2017a)
	Unidentified species of Laodiceidae (Hydrozoa: Cnidaria)	Nagasaki, Japan	Shojima (1963)
	Bolinopsis mikado (Tentaculata: Ctenophora)	Nagasaki, Japan	Shojima (1973)
	Beroe cucumis (Nuda: Ctenophora)	Nagasaki, Japan	Shojima (1973)
	Salpa fusiformis (Thaliacea: Chordata)	Yamaguchi, Japan	Wakabayashi et al. (2017a)
Ibacus novemdentatus	*Aequorea macrodactyla* (Hydrozoa: Cnidaria)	Nagasaki, Japan	Shojima (1973)
	Unidentified species of Pelagiidae (Scyphozoa: Cnidaria)	Nagasaki, Japan	Shojima (1973)
	Beroe cucumis (Nuda: Ctenophora)	Nagasaki, Japan	Shojima (1973)
Unidentified species of *Ibacus*	*Pelagia panopyra* (Scyphozoa: Cnidaria)	Sydney, Australia	Thomas (1963)
	Unidentified species of Semaeostomeae (Scyphozoa: Cnidaria)	Off Ballina, Australia	Thomas (1963)
	Catostylus mosaicus (Scyphozoa: Cnidaria)	Hawkesbury River, Australia	Thomas (1963)
Scyllarinae			
Eduarctus martensii	Jellyfish or other gelatinous zooplanktons	Off Townsville, Australia	Barnett et al. (1986)
	Clytia languida (Hydrozoa: Cnidaria)	Yamaguchi, Japan	Wakabayashi et al. (2017a)
	Unidentified species of *Hormiphora* (Tentaculata: Ctenophora)	Yamaguchi, Japan	Wakabayashi et al. (2017b)

(continued)

Table 11.1 (continued)

Species	Gelatinous zooplankton	Locality	References
E. martensii[a]	Liriope tetraphylla (Hydrozoa: Cnidaria)	Nagasaki, Japan	Shojima (1963)
Galearctus kitanoviriosus	Unidentified species of Tomopteris (Phyllodocida: Annelida)	Yamaguchi, Japan	Figure 11.1d of this chapter
G. kitanoviriosus[b]	Liriope tetraphylla (Hydrozoa: Cnidaria)	Nagasaki, Japan	Shojima (1963)
Petrarctus demani	Jellyfish or other gelatinous zooplanktons	off Townsville, Australia	Barnett et al. (1986)
Scyllarus chacei	Unidentified species of Muggiaea (Hydrozoa: Cnidaria)	North of the Gulf of Mexico, USA	Greer et al. (2017)
	Unidentified species of Liriope (Hydrozoa: Cnidaria)	North of the Gulf of Mexico, USA	Greer et al. (2017)
Unidentified species of Scyllarus	Aurelia aurita (Scyphozoa; Cnidaria)	Bimini, Bahamas	Herrnkind et al. (1976)
	Unidentified species of Prayidae (Hydrozoa: Cnidaria)	Gran Canaria, Spain	Ates et al. (2007)
Unidentified species of Scyllaridae	Jellyfish or other gelatinous zooplanktons	Off Townsville, Australia	Barnett et al. (1986)
	Liriope tetraphylla (Hydrozoa: Cnidaria)	Nagasaki, Japan	Shojima (1973)
	Unidentified species of Abylidae (Hydrozoa: Cnidaria)	Yamaguchi, Japan	Wakabayashi et al. (2017a)
Theninae			
Thenus orientalis[c]	Jellyfish or other gelatinous zooplanktons	Off Townsville, Australia	Barnett et al. (1986)

[a]The phyllosoma was originally identified as *Scyllarides* sp., but the authors correctly identified it as *E. martensii* from the image
[b]The authors identified the species from the image in the reference
[c]*Thenus orientalis* is not distributed in the Australian waters (Burton and Davie 2007). It should be either *T. australiensis* or *T. parindicus*

11.2 Phyllosoma Larvae of the Slipper Lobsters Associated with Gelatinous Zooplankton

11.2.1 In the Wild

The association behaviour of phyllosoma larvae with gelatinous zooplankton was scientifically described for the first time in Australia (Thomas 1963) and in Japan (Shojima 1963), respectively. There is evidence suggesting that this phenomenon is not an accidental event: approximately 20% of *Aurelia aurita* were associated with the late-stage larvae of *Scyllarus* spp. (probably *S. americanus* or *S. chacei*) in the

Florida Strait, western shore of Bimini, Bahamas (Herrnkind et al. 1976); 49 out of 51 slipper lobster larvae were found in association with pelagic cnidarians or other gelatinous zooplankton off Townsville, Australia (Barnett et al. 1986); and 30% of 347 larvae of *S. chacei*, recorded using an in situ imaging system, were associated with gelatinous zooplankton in the northern Gulf of Mexico (Greer et al. 2017). The *S. chacei* larvae are likely to be abundant at greater depths and higher salinity along the shelf (Greer et al. 2017).

The adaptive advantages of association for phyllosoma larvae with gelatinous zooplankton are likely to be for food, protection, transport, or a combination of these (Thomas 1963; Booth et al. 2005; Sekiguchi et al. 2007). A giant phyllosoma larva caught in the north of Bermuda (characteristics of this larva overall correspond to the subfinal larval stage of *Parribacus* sp. described by Palero et al. 2014b) was noted to be extruding faecal matter consisting entirely of undigested nematocysts (Sims and Brown 1968). Digestive organs of wild-caught larvae of *Scyllarus* spp. were demonstrated to contain tissues assigned to pelagic cnidarians, larvaceans or thaliaceans by DNA barcoding (Suzuki et al. 2007; Connell et al. 2014). This evidence reveals that the gelatinous zooplankton are likely to be the major prey organisms for scyllarid larvae in the wild. Although phyllosoma larvae have been observed to carry a damaged gelatinous zooplankton likely bitten by the larvae (e.g. Greer et al. 2017), the direct observations of larval feeding behaviour on gelatinous zooplankton have not been made in situ until now.

Gelatinous zooplankton is probably used as a predator avoidance by phyllosoma larvae (Thomas 1963; Booth et al. 2005), as is known for some fish juveniles (e.g. Masuda et al. 2008). The larvae riding on pelagic cnidarians often direct their own 'vehicle' towards divers when they go closer to the larvae (H. Abe, pers. comm.). Nevertheless, phyllosoma larvae are known to be consumed by a wide variety of predatory fish (Sekiguchi et al. 2007). For example, *Ibacus alticrenatus* larvae were found in the stomachs of sunfish (*Mola mola*), rudderfish (*Centrolophus niger*), albacore (*Thunnus alalunga*), and pilot fish (*Naucrates ductor*) (Bailey and Habib 1982). Sunfish and rudderfish are listed as typical predators of pelagic cnidarians (Arai 1988; Ates 1988), and albacore is also a possible consumer of gelatinous zooplankton (Cardona et al. 2012). Phyllosoma larvae may use gelatinous zooplankton to protect themselves from predators, but the protection does not seem to be perfect against the gelatinous zooplankton consumers.

Phyllosoma larvae may be able to save energy from swimming and to reach a site for settlement by means of associating with gelatinous zooplankton in the water column; however, there are no available scientific data from investigations in the wild.

11.2.2 In the Laboratory

Phyllosoma larvae usually capture any type of gelatinous zooplankton by accidental contact. Tactile sense seems to be necessary for the larvae to initiate a capture of any gelatinous zooplankton around (Wakabayashi et al. 2012a), in the same manner as observed for the spiny lobster larvae capturing brine shrimp as food (Cox and Bruce 2003). Chemical sense can be another trigger for hunting gelatinous zooplankton. Kamio et al. (2015) demonstrated that the larvae of *Ibacus novemdentatus* showed their responses to a solution of glycine which is the principal free amino acid of moon jellyfish (see Sect. 12.4). Phyllosoma larvae may use a combination of tactile and chemical senses when associating with gelatinous zooplankton.

The principal purpose for phyllosoma larvae to associate with gelatinous zooplankton in tank is likely to be feeding. They are capable of eating various kinds of gelatinous zooplankton. Eleven different species of pelagic cnidarians have been demonstrated to be a food for the larvae of *I. novemdentatus* (Wakabayashi et al. 2012a). The following species of gelatinous zooplankton have also been consumed entirely or partly by the larvae of the genus *Ibacus*: *Eutonia indicans* (Ishii et al. 2016), *Phacellophora chatschatica*, *Cassiopea ornata*, *Catostylus mosaicus* (Wakabayashi, unpublished) (cnidarians), *Bolinopsis mikado*, *Ocyropsis fusca*, *Beroe cucumis* (ctenophores) (R. Sugimoto & Wakabayashi, unpublished), *Pterotrachea coronata* (pterotracheoids), and *Sagitta* sp. (chaetognaths) (Wakabayashi, unpublished).

The association behaviour varies depending on the form and size of the gelatinous zooplankton being captured. When in contact with a pelagic cnidarian, the phyllosoma larva moves invariably to the edge of the exumbrella of the cnidarian and pierces itself on the exumbrella with dactylus of pereiopods, regardless of the part where it clings first. The larva then hauls up a tentacle or contacts its mouthpart to the edge of the cnidarian umbrella to prey on it (Fig. 11.2). A phyllosoma larva immediately preys on food after capture or hangs the food on its dactylus of third to

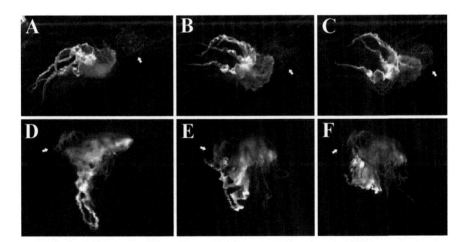

Fig. 11.2 Time-lapse video images showing the process of association between phyllosoma larva (*Ibacus novemdentatus*) and Japanese sea nettle (*Chrysaora pacifica*). (**a**) before contact; (**b**) contact; (**c**) clinging; (**d**) positioning; (**e**) hauling up the oral arms and tentacles; (**f**) preying on a tentacle. Arrows indicate the larva. The images were captured from the original video in Wakabayashi et al. (2012a) of The Biological Bulletin by Marine Biological Laboratory with a permission of Lancaster Press

fifth pereiopods for eating later when happens to contact a smaller gelatinous zooplankton than the larva itself. In any case, phyllosoma larvae usually eat the whole body of gelatinous zooplankton eventually (Wakabayashi et al. 2012a). The association behaviour of the larvae has been analysed in tanks using a video recording; however, those recordings were made in a small planktonkreisel (8 L) for a short period (less than 1 h), providing limited information of the association. Video recordings for longer periods under different conditions may allow us to better understand the ecological implication of the association between phyllosoma larvae and gelatinous zooplankton.

Phyllosoma larvae use their pereiopod exopods as swimming legs. It is expected that the exopod motion can be reduced when the larvae rely on gelatinous zooplankton for swimming in their association. Shojima (1973) observed that swimming propulsion of phyllosoma larvae was greater than that of pelagic cnidarians when the cnidarians are smaller than the larvae. However, quantitative studies on the larval exopod motion have not been carried out until now.

We just set about examining the appendage motion of the slipper lobster *Ibacus ciliatus*; preliminary observations on the Stages III and IV larvae of *I. ciliatus* suggested the following:

1. Of the five pairs of pereiopods, the exopods of the anterior three or four pairs are effectively moving to swim (Tanaka, W. Cheng, Wakabayashi, unpublished). The most posterior pair does not seem to be so effective for swimming, but more useful for clinging onto something like a jellyfish bell.
2. The pereiopod exopods show sculling oscillations to push water. This motion seems to generate propulsion, and also to steer the jellyfish on which the larva is clinging. By doing so, the larva seems to be manoeuvring a vehicle, i.e. the jellyfish producing jet propulsion by each pulsation.
3. The sculling of pereiopod exopods is of constant frequency, being likely to depend on the stages of the phyllosoma larvae. Our unpublished observations suggest that the pereiopod exopods of Stages III and IV phyllosoma larvae of *I. ciliatus* beat with a frequency of 4.2–4.9 Hz and 3.2–3.8 Hz, respectively (Tanaka, W. Cheng, Wakabayashi, unpublished). It looks natural that the pereiopod exopods of larger (later stage) larvae move with lower frequency.

As it would be of further interest for us to investigate the appendage motion from a hydromechanical point of view, we are going to investigate to see if the way of motion varies depending on whether the larva is riding on the jellyfish or not. The stage-dependent difference on the manner of the exopod motion would also be examined. In addition, environmental effects such as temperature dependence and day-night difference on the motion frequency and amplitude should be studied. These observations would produce quantitative information about the larval motion to allow calculations on the energy consumption.

Carrying such behavioural observations forwards, together with the morphological and nutritional studies, we may get a clue to find the significance of the way of life of phyllosoma in clinging to the jellyfish.

11.3 Larviculture of Slipper Lobsters Fed Jellyfish as Food

Shojima (1973) kept a wild-caught larva of *Ibacus ciliatus* in a small tank together with the medusae of Japanese sea nettle (*Chrysaora pacifica*) and moon jellyfish (*Aurelia coerulea*), and observed the larva preying on the oral arms and the edge of the medusa umbrella. The larva showed its digestive organs filled with the tissues in the same colour of the medusae (Shojima 1973). Feeding behaviour of *I. ciliatus* larvae on pelagic cnidarians was also observed by Matsuda (2010), who suggested the possibility of using gelatinous zooplankton as food in rearing the slipper lobster larvae. To date, three commercial species of slipper lobsters have successfully been cultured from hatching to metamorphosis with pelagic cnidarians as the sole food (Tables 11.2 and 11.3). However, some risks from pathogens and non-standardised nutritional quality and quantity of jellyfish are pointed out (Francis et al. 2014).

Table 11.2 Slipper lobsters attempted to rear phyllosomas using wild-caught gelatinous zooplankton as food

Species	Developmental stages	Gelatinous zooplankton	References
Ibachinae			
Ibacus ciliates	N/A (used wild caught larvae)	*Aurelia coerulea*, *Chrysaora pacifica*	Shojima (1973)
	N/A	*Carybdea brevipedalia*	Matsuda (2010)
	Newly hatched phyllosoma to metamorphosis	*Aurelia coerulea*, *Chrysaora pacifica*	Wakabayashi et al. (2016a)
Ibacus novemdentatus	Newly hatched phyllosoma to metamorphosis	*Aurelia coerulea* wild-caught and cultured individuals, *Chrysaora pacifica*	Wakabayashi et al. (2012b)
	Newly hatched phyllosoma to metamorphosis	*Aurelia coerulea*	Wakabayashi et al. (2016b)
	Newly hatched phyllosoma to Stage VI (gilled) phyllosoma	*Chrysaora pacifica*	Wakabayashi et al. (2016b)
Scyllarinae			
Petrarctus demani	Newly hatched phyllosoma to Stage IX (gilled) phyllosoma	*Aurelia coerulea*	Wakabayashi and Phillips (unpublished)
Theninae			
Thenus australiensis	Newly hatched phyllosoma to metamorphosis	*Aurelia coerulea*	Wakabayashi and Phillips (2016)
Thenus unimaculatus	Stage II, III, and IV (gilled) phyllosomas	Ctenophores	Kizhakudan and Krishnamoorthi (2014)

Table 11.3 Survival and growth of slipper lobster phyllosomas fed jellyfish as food

No. of moulting	Slipper lobsters										
	Ibacus novemdentatus[a]			*Ibacus novemdentatus*[b]			*Ibacus ciliatus*[c]			*Thenus australiensis*[c]	
	Stage	SR (%)	DAH	Stage	SR (%)	DAH	Stage	SR (%)	DAH	Stage	SR (%)	DAH
1	I→II	100	6.0	I→II	77.5	6.1	I→II	100	6.4	I→II	90.0	7.9
2	II→III	88.9	11.4	II→III	75.0	10.1	II→III	100	13.0	II→III	70.0	16.7
3	III→IV	83.3	17.1	III→IV	57.5	16.2	III→IV	95.5	21.0	III→V	20.0	25.4
4	IV→V	66.7	22.9	IV→V	37.5	22.7	IV→V	90.9	29.9	IV→N	10.0	40.0
5	V→VI	44.4	30.4	V→VI	27.5	31.0	V→VI	86.4	37.2	—		
6	VI→N	36.1	48.2	VI→N	12.5	44.7	VI→VII	90.0	46.0	—		
7	—			—			VII→N	65.0	65.6	—		

[a]Individual culture with static water (Wakabayashi et al. 2012b)
[b]Group culture in planktonkreisel with recirculating water (Wakabayashi et al. 2012b)
[c]Individual culture in mesh container with recirculating water (Wakabayashi et al. 2016a; Wakabayashi and Phillips 2016)

11.3.1 Design of Tank System

Hydrodynamics is one of the key issues for rearing the phyllosoma larvae (Mikami and Kuballa 2004). Among the different types of tanks designed so far (e.g. Takahashi and Saisho 1978; Ritar 2001; Kittaka 2000; Matsuda and Takenouchi 2005; Murakami et al. 2007; Horita 2007; Goldstein and Nelson 2011; Wakabayashi et al. 2012b), using a modified planktonkreisel appears to provide good success for rearing larvae (Phillips and Matsuda 2011). The planktonkreisel was originally designed for keeping fragile zooplankton suspended with the aid of providing an adequate water movement without using air bubbles (Greve 1968, 1970). The modified planktonkreisel tanks allow creation of a continuous gentle vertical water current, keeping larvae suspended in the water column (Horita 2007; Goldstein and Nelson 2011; Phillips and Matsuda 2011; Wakabayashi et al. 2012b). An advanced planktonkreisel with a vertically revolving tank is able to regulate the rotating water current by revolving the tank itself (Murakami et al. 2007). On the other hand, raceways were also designed and tested for mass culture of phyllosoma larvae of *Thenus* spp. (Mikami and Kuballa 2007). The raceways are the only system currently able to produce a large number of slipper lobster juveniles (Mikami and Kuballa 2004; Vijayakumaran and Radhakrishnan 2011).

In terms of using gelatinous zooplankton as food, the modified planktonkreisel is likely more advantageous for large-scale rearing to recreate the natural behaviour of association between phyllosoma larvae and gelatinous zooplankton. Indeed, a modified planktonkreisel was used for rearing larvae fed pelagic cnidarians as food, with good results for both survival and growth of the larvae (Wakabayashi and Phillips 2016; Wakabayashi et al. 2012b, 2016a). The accessibility of the larvae and debris can be solved by using a smaller planktonkreisel. Even a rectangular tank with vertical water movement appears to be useful at least for rearing the larvae of the genus *Ibacus* and *Thenus* (Wakabayashi and Phillips 2016; Wakabayashi et al. 2012b). With an ordinary rectangular tank, rearing systems can be easily reassembled to fit the space and function of facilities wherever used.

11.3.2 Survival Rate of Phyllosoma Larvae

The larvae of slipper lobsters are capable of developing by feeding them pelagic cnidarians as their sole food source. At least three species of slipper lobsters (*I. novemdentatus*, *I. ciliatus*, and *Thenus australiensis*) are able to complete their planktonic stages by feeding only on pelagic cnidarians (Wakabayashi and Phillips 2016; Wakabayashi et al. 2012b, 2016a) (Table 11.3).

Larvae of *I. ciliatus* were kept with an approximately 60% survival rate, from hatching to metamorphosis, when fed medusae of moon jellyfish and Japanese sea nettle as the sole food in a tank with individual care (Wakabayashi et al. 2016a). More than 80% of larvae survived from the first to the final phyllosoma stages, and 72.2% of the larvae at the final phyllosoma stage metamorphosed into the nisto stage. However, only 38.5% of the nistos moulted into the first juvenile stage. In other words, 22.7% of the newly hatched larvae survived to the juvenile stage. One nisto of *T. australiensis* successfully moulted into the first juvenile stage; however, the juvenile showed an abnormal form (Wakabayashi and Phillips 2016). The mortality at the nisto stage may implicate the shortage of energy reserves that should be accumulated during the final phyllosoma stage. The shortage of required nutrients may adversely affect the nisto developing into the juvenile. Lipids are the primary form of energy storage for the non-feeding nisto stage (Jeffs et al. 1999, 2001; Mikami and Kuballa 2007). Nutritionally optimised pelagic cnidarians may be required, particularly for the larvae at the final phyllosoma stage, to reach the nisto and juvenile stages in a healthy condition (see Sect. 11.4).

A modified planktonkreisel with vertical water movement was used to rear phyllosoma larvae of *I. novemdentatus* fed jellyfish as food (Wakabayashi et al. 2012b). The period of each developmental stage and body dimensions of the larvae reared in the modified planktonkreisel were not significantly different from those of larvae reared in static water. Approximately 10% of larvae successfully metamorphosed into the nisto stage in the modified planktonkreisel: the survival rate was lower than that of the larvae reared individually in the static water (Table 11.3). The mortality of later stage larvae was likely caused by cannibalism (Wakabayashi et al. 2012b). It may be possible to prevent the larvae from attacking each other by controlling the amount of food introduced. Other rearing environments, such as flow rate of vertical water movement and light conditions, may also require improvement (see also Chap. 10).

11.3.3 Required Amount of Gelatinous Zooplankton

It is important to understand the adequate amount of food necessary for phyllosoma larvae to maintain the tank environment with a smaller amount of excess food, to prevent cannibalism, and to preserve the food resources. The required amount of dietary moon jellyfish needed for the larvae of *I. novemdentatus* to be able to develop from the newly hatched stage to metamorphosis was examined (Wakabayashi et al. 2013). Consumption rates of the larvae increased from approximately 0.1 g d^{-1} at Stage I to 7.5 g d^{-1} at Stage VI (final stage) as they grow (Table 11.4), and no remarkable difference in consumption rates was observed between light and dark conditions.

Table 11.4 Daily consumption on jellyfish by *Ibacus novemdentatus* phyllosomas

Stage	Consumption (g ind^{-1} d^{-1})	
	Light	Dark
I	0.07 ± 0.03	0.08 ± 0.04
II	0.19 ± 0.09	0.16 ± 0.08
III	0.40 ± 0.20	0.39 ± 0.20
IV	1.09 ± 0.54	1.10 ± 0.55
V	2.10 ± 1.05	2.35 ± 1.18
VI	7.13 ± 3.57	7.49 ± 3.74

11.4 Nutritional Advantages and Disadvantages of Gelatinous Zooplankton as Food for Phyllosomas

Gelatinous zooplanktons are composed of diverse taxa showing different physical, chemical, and biological characteristics. Although phyllosoma larvae are likely capable of preying on any kinds of gelatinous zooplankton with non-selective manner, there are advantages and disadvantages of gelatinous zooplankton as food materials for the larvae.

11.4.1 Water Content

Water content of pelagic cnidarians and ctenophores is typically more than 95%, whereas that of non-gelatinous zooplankton, such as crustaceans, is generally 60–85% (Sullivan and Kremer 2011; Wang and Jeffs 2014). The high water content permits a gelatinous zooplankton to grow into a large body in a short period. The large body allows symbiotic crustaceans, including phyllosoma larvae, to stay on it (Madin and Harbison 2001). The density of body components is close to that of seawater, resulting in neutral buoyancy (Madin and Harbison 2001). The high water content reduces the food value of their tissue but makes them easily digested on the other hand (Arai et al. 2003). Phyllosoma larvae might be able to utilise the scarce nutrients by feeding continuously (see Sect. 11.2.2).

11.4.2 Proteins and Amino Acids

The ratio of carbon to other compositions is variable depending on species and food availability, but organic tissues, such as muscle and gonad, usually show a higher ratio than the other body parts in gelatinous zooplankton. Some species have been reported to contain 20% or more of carbon as dry mass; however, most of the species contain 1–8% of carbon as dry mass (Sullivan and Kremer 2011). The optimal protein concentration of diet for crustacean larvae is between 30 and 60% of dry mass (Guillaume 1997; Cahu 1999). The larvae use proteins to construct muscle tissue as well as to produce energy particularly during the early stages. On the other hand, the larvae have an ability to spare proteins when a diet containing a high ratio of digestible carbohydrates is available (Cahu 1999).

It is notable that pelagic cnidarians (and salps) contain all ten of the essential amino acids for crustacean larvae (Kittaka 2005; Wakabayashi et al. 2016b). Taurine, known as one of the vital nutrients for fish larvae (Takeuchi 2014), is also abundant in pelagic cnidarians (Kittaka 2005). The principal free amino acid in pelagic cnidarians is glycine, the main element of collagen, followed by glutamine, and alanine or proline (Kittaka 2005; Wakabayashi et al. 2016b). Glycine likely stimulates the appetite of phyllosoma larvae (Kamio et al. 2015), and it is also known to play a bacteriostatic role (Wake et al. 1974).

11.4.3 Lipids and Fatty Acids

The phyllosoma larvae are known to accumulate lipids in their body to use as an energy during metamorphosis (Jeffs et al. 1999, 2001; Phleger et al. 2001; Phillips et al. 2006). Polar lipids and polyunsaturated fatty acids (PUFA) are particularly essential (Phleger et al. 2001). However, the elemental ratio of carbon to nitrogen by weight in gelatinous zooplankton is typically about 4, reflecting the high protein content and presence of scarce lipids (Sullivan and Kremer 2011). The lipid content reported in gelatinous zooplankton was less than 6% of dry mass (Wang and Jeffs 2014).

Low lipid content in gelatinous zooplankton can be a cause of mortality of phyllosoma larvaewhen they grow by feeding on gelatinous zooplankton (see Sect. 11.3.2).

The principal lipid class of pelagic cnidarians and ctenophores is polar lipids and is usually followed by wax esters (Nelson et al. 2000). The dominant free fatty acids contained are C16:0, C18:0, C20:5 and C20:4 in scyphozoans (Wakabayashi et al. 2016b). These dominant lipids and fatty acids are required by phyllosoma larvae, implying that gelatinous zooplanktons contain appropriate kinds of lipids, although the concentrations are low.

Missing lipid components, such as PUFA, can be encapsulated into brine shrimp and then transferred to the target pelagic cnidarians by feeding on enriched brine shrimp. It has been demonstrated to take approximately 10 days for moon jellyfish to reflect their dietary fatty acid compositions (Fukuda and Naganuma 2001). Optimisation techniques for nutritional compositions of gelatinous zooplankton are desired.

11.4.4 Minor Nutrient Elements

Ash content of pelagic cnidarians and ctenophores is relatively high compared to that of non-gelatinous zooplankton (Sullivan and Kremer 2011; Wang and Jeffs 2014). Boron, selenium, and zinc are the dominant trace minerals detected from scyphozoans (Wakabayashi et al. 2016b). Minor elements, such as trace minerals and vitamins, may play an important role in improving the survival and growth of phyllosoma larvae. The critical key factors affecting survival remain unclear.

11.5 Defence of Phyllosomas to the Venomous Stinging by Pelagic Cnidarians

Cnidarians are characterised by the possession of unique tubular structures, cnidae, which are contained within cellular capsules that aid in prey capture, defence, locomotion and attachment (Brusca et al. 2016). Cnidae can be sorted into three basic types: nematocysts, spirocysts, and ptychocysts. Nematocysts penetrate a tubule armed with spines into the target organism and inject a toxic mixture; other types of cnidae have a volvent or glutinant function rather than a penetrative function (Jouiaei et al. 2015; Brusca et al. 2016). The tubules are likely to be discharged in response to mechanical and chemical stimuli. The protein and polypeptide toxins contained exhibit diverse functions of cytolytic, enzymatic, neurotoxic, and other activities. Although nematocysts are mostly located on the tentacles, they also exist on the exumbrella, as well as on the oral arms in some species (Jouiaei et al. 2015). Nematocysts occur in members of all cnidarian classes except within the Myxozoa (Brusca et al. 2016).

Small pelagic crustaceans, including copepods, amphipods, and decapod larvae, can be dominant prey organisms of pelagic cnidarians (Sullivan and Kremer 2011). These crustaceans are usually paralysed or killed by the nematocyst venoms immediately after contact with pelagic cnidarians. Nevertheless, several groups of crustaceans, including slipper lobster larvae, are capable of associating with such venomous animals without being envenomated. It remains to be elucidated whether there is a key mechanism for prevention of stinging on their external chitinous shells.

The foregut and hindgut of phyllosoma are known to be cuticle-lined (Mikami and Takashima 1993), and are likely to be mechanically protected from stinging. In contrast, the midgut is not cuticle-lined (Mikami et al. 1994). The midgut gland allows epithelial cells to function in digestion and absorption. The filter-press located at the proventriculus probably permits only fluids and fine particles (less than 1 μm in diameter) to enter the midgut gland, and selectively eliminates the larger particles (Simon et al. 2012), such as nematocysts with diameters of 5 μm or larger (Brusca et al. 2016). The midgut trunk is also mechanically protected by a peritrophic membrane, which functions to separate ingested materials from the gut epithelium (Forster 1953). Phyllosoma larvae of *I. novemdentatus* are demonstrated to encase the cnidae within this membrane. No single tubular of a nematocyst resulting from a diet of moon jellyfish penetrates the peritrophic membrane, and the venomous cnidae, as well as other undigested materials, are evacuated while enclosed in the peritrophic membrane (Kamio et al. 2016).

A physiological test showed that the phyllosoma larvae did not resist the nematocyst venoms injected into the abdominal muscle, suggesting that the defence of the larvae against the nematocyst venoms is mechanical protection, chemical digestion, or a combination of both, rather than a physiological resistance (Kamio et al. 2016).

11.6 Conclusions and Perspectives

The dorsoventrally flattened body of phyllosoma is considered an adaptation to buoyancy in water (Sverdrup et al. 1947). On the other hand, this unique form may also be advantageous for association with gelatinous zooplankton. By clinging with spreading pereiopods, the flat body may allow phyllosoma larvae to stay continuously on the external surface of gelatinous zooplankton without being shaken off by a hydraulic drag, even when attached to those making strong swimming pulses.

Furthermore, the mouth of the larva is located in the centre of the ventral body surface, allowing it to consume the gelatinous zooplankton while clinging. A phyllosoma larva may save its energy consumption by relying on the buoyancy and propulsion of the prey.

A phyllosoma larva is an opportunistic predator, but its ideal foods are likely to consist of soft materials, such as gelatinous zooplankton. Various recent studies have supported this hypothesis as mentioned above. Using gelatinous zooplankton as a nutritional resource can be a breakthrough in large-scale commercial seed production of slipper lobsters. However, problems still remain to be overcome: (1) variable nutritional compositions of wild gelatinous zooplankton, (2) difficulties of handling the fluid body of live gelatinous zooplankton for long periods and (3) the existence of cannibalism by phyllosoma larvae on each other. Furthermore, viable grow-out of juveniles resulting from larvae fed on gelatinous zooplankton as food has not been examined yet.

Considering the survival rate, developmental duration of each stage, and consumption rate, the minimum required amount of moon jellyfish for large-scale culture of *I. novemdentatus* phyllosoma larvae can be estimated. When obtaining 100 individuals in the nisto stage, at least 33 kg of moon jellyfish would be required in total from hatching to metamorphosis (Fig. 11.3). This means at least 330 tonnes of moon jellyfish would be required to achieve the production of a million juveniles. It is not practical to produce this amount of moon jellyfish in tanks; however, 'jellyfish blooms' in the ocean may make it realistic. In Uwakai Sea, Japan, approximately 9400 metric tonnes (corresponding to 583 million individuals) of moon jellyfish appeared in an area of 2.34 km^2 in a day (Uye et al. 2003). Although this highly dense aggregation was reported as an unusual event (Uye et al. 2003), massive blooms of gelatinous zooplankton in the coastal waters have recently increased and caused serious collapses of coastal activities around the world (Purcell et al. 2007; Purcell 2012). The blooming individuals of gelatinous zooplankton are mostly unutilised. Lobster seed production may take advantage of the jellyfish blooms as phyllosoma larvae are capable of growing with feeding on various species of gelatinous zooplankton. Alternatively, a jelly-like formulated diet using gelatinous zooplankton as materials may be developed.

It had been previously unknown whether the spiny lobster larvae also possess an association behaviour with gelatinous zooplankton under natural conditions (Phillips and Sastry 1980; Sekiguchi et al. 2007; Wakabayashi and Tanaka 2012). Feeding trials (Thomas 1963 as a pers. comm. with Dr. D.I. Williamson; Mitchell 1971; Kittaka 1997, 2000, 2005), field surveys of plankton (Saunders et al. 2012), and DNA analyses of gut contents of wild-caught larvae (Suzuki et al. 2006, 2007; O'Rorke et al. 2012) indirectly suggested the same association behaviour of palinurid larvae. Finally, direct observation of the association between the phyllosoma of *Palinurellus wieneckii* and a siphonophore was recently made underwater (Mańko et al. 2017; Ianniello and Mears 2018). This evidence suggests that gelatinous zooplankton can be a feasible nutrient resource for spiny lobster larviculture as well.

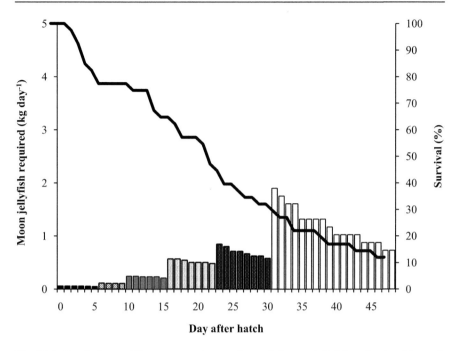

Fig. 11.3 Survival rate of *Ibacus novemdentatus* larvae fed moon jellyfish as food and the estimated amount of daily required moon jellyfish to obtain hundred individuals in the nisto stage. The data of survival and consumption rates were obtained from the original researches by Wakabayashi et al. (2012b) and Wakabayashi et al. (2013), respectively (see also Tables 11.3 and 11.4). Different patterns of bar show the different developmental stages of phyllosoma larvae: black (Stage I), oblique lines (Stage II), grey (Stage III), white with black dots (Stage IV), black with white dots (Stage V), and white (Stage IV)

Acknowledgements The authors express their gratitude to the underwater photographers Mr. Hideki Abe, Mr. Yusuke Yoshino, Mr. Tsutomu Sasagawa, Ms. Miki Sasagawa, and Ms. Kumiko Maki who always provided us their observations with wonderful photographs. The gratitude also goes to Dr. Wanting Cheng and Mr. Ryuichi Sugimoto who agreed with providing unpublished data in this chapter. A special acknowledgement to the late Dr. Jiro Kittaka who was encouraging the authors to seek a possibility of gelatinous zooplankton as food for slipper and spiny lobster phyllosomas.

References

Arai, M. N. (1988). Interactions of fish and pelagic coelenterates. *Canadian Journal of Zoology, 66*, 1913–1927.

Arai, M. N., Welch, D. W., Dunsmuir, A. L., Jacobs, M. C., & Ladouceur, A. R. (2003). Digestion of pelagic Ctenophora and Cnidaria by fish. *Canadian Journal of Fisheries and Aquatic Sciences, 60*, 825–829.

Ates, R. M. L. (1988). Medusivorous fishes, a review. *Zoologische Mededelingen Leiden, 62*, 29–42.

Ates, R., Lindsay, D. J., & Sekiguchi, H. (2007). First record of an association between a phyllosoma larva and a prayid siphonophore. *Plankton & Benthos Research, 2*, 67–69.

Bailey, K. N., & Habib, G. (1982). Food of incidental fish species taken in the purse-seine skipjack fishery, 1976–1981. *Fisheries Research Divison Occasional Publication Data Series, 6*, 1–24.

Barnett, B. M., Hartwick, R. F., & Milward, N. E. (1986). Descriptions of the nisto stage of *Scyllarus demani* Holthuis, two unidentified *Scyllarus* species, and the juvenile of *Scyllarus martensii* Pfeffer (Crustacea: Decapoda: Scyllaridae), reared in the laboratory; and behavioural observations of the nistos of *S. demani*, *S. martensii* and *Thenus orientalis* (Lund). *Australian Journal of Marine & Freshwater Research, 37*, 595–608.

Booth, J. D., Webber, W. R., Sekiguchi, H., & Coutures, E. (2005). Diverse larval recruitment strategies within the Scyllaridae. *New Zealand Journal of Marine & Freshwater Research., 39*, 581–592.

Brusca, R. C., Moore, W., & Shuster, S. M. (2016). *Invertebrates* (3rd ed., p. 1104). Sunderland: Sinauer Associates.

Burton, T. E., & Davie, P. J. F. (2007). A revision of the short-nosed lobsters of the genus *Thenus* (Crustacea: Decapoda: Scyllaridae), with descriptions of three new species. *Zootaxa, 1429*, 1–38.

Cahu, C. (1999). Nutrition and feeding of penaeid shrimp larvae. In J. Guillaume, S. Kaushik, P. Bergot, & R. Métailler (Eds.), *Nutrition and feeding of fish and crustaceans* (pp. 253–263). Chichester: Paris Publishing.

Cardona, L., Álvarez de Quevedo, I., Borrell, A., & Aguilar, A. (2012). Massive consumption of gelatinous plankton by mediterranean apex predators. *PLoS One, 7*, e31329. https://doi.org/10.1371/journal.pone.0031329.

Chan, T. Y. (2010). Annotated checklist of the world's marine lobsters (Crustacea: Decapoda: Astacidea, Glypheidea, Achelata, Polychelida). *The Raffles Bulletin of Zoology, 23*, 153–181.

Connell, S. C., O'Rorke, R., Jeffs, A. G., & Lavery, S. D. (2014). DNA identification of the phyllosoma diet of *Jasus edwardsii* and *Scyllarus* sp. Z. *New Zealand Journal of Marine and Freshwater Research, 48*, 416–429.

Cox, S. L., & Bruce, M. P. (2003). Feeding behaviour and associated sensory mechanisms of stage I–III phyllosoma of *Jasus edwardsii* and *Jasus verreauxi*. *Journal of the Marine Biological Association of the United Kingdom, 83*, 465–468.

Fisher, W. S., Nilson, E. H., Steenbergen, J. F., & Lightner, D. V. (1978). Microbial diseases of cultured lobsters: A review. *Aquaculture, 14*, 115–141.

Forster, G. R. (1953). Peritrophic membranes in the Caridea (Crustacea Decapoda). *Journal of the Marine Biological Association of the United Kingdom, 32*, 315–318.

Francis, D. S., Salmon, M. L., Kenway, M. J., & Hall, M. R. (2014). Palinurid lobster aquaculture: Nutritional progress and considerations for successful larval rearing. *Reviews in Aquaculture, 6*, 180–203.

Fukuda, Y., & Naganuma, T. (2001). Potential dietary effects on the fatty acid composition of the common jellyfish *Aurelia aurita*. *Marine Biology, 138*, 1029–1035.

Goldstein, J. S., & Nelson, B. (2011). Application of a gelatinous zooplankton tank for the mass production of larval Caribbean spiny lobster, Panulirus argus. *Aquatic Living Resources, 24*, 45–51.

Goulden, E. F., Hall, M. R., Bourne, D. G., Pereg, L. L., & Høj, L. (2012). Pathogenicity and infection cycle of *Vibrio owensii* in larviculture of the ornate spiny lobster (*Panulirus ornatus*). *Applied and Environmental Microbiology, 78*, 2841–2849.

Greer, A. T., Briseno-Avena, C., Deary, A. L., Cowen, R. K., Hernandez, F. J., & Graham, W. M. (2017). Associations between lobster phyllosoma and gelatinous zooplankton in relation to oceanographic properties in the northern Gulf of Mexico. *Fisheries Oceanography, 26*, 693–704.

Greve, W. (1968). The "planktonkreisel", a new device for culturing zooplankton. *Marine Biology, 1*, 201–203.

Greve, W. (1970). Cultivation experiments on North Sea ctenophores. *Helgoländer Wissenschaftliche Meeresuntersuchungen, 20*, 304–317.

Guillaume, J. (1997). Protein and amino acids. In L. R. D'Abramo, D. E. Conklin, & D. M. Akiyama (Eds.), *Crustacean nutrition* (pp. 26–50). Baton Rouge: The World Aquaculture Society.

Herrnkind, W., Halusky, J., & Kanciruk, P. (1976). A further note on phyllosoma larvae associated with medusae. *Bulletin of Marine Science, 26*, 110–112.

Higa, T., Fujita, Y., & Shokita, S. (2005). Complete larval development of a scyllarine lobster, Galearctus kitanoviriosus (Harada, 1962) (Decapoda: Scyllaridae: Scyllarinae), reared under laboratory conditions. *Crustacean Research, 34*, 1–26.

Horita, T. (2007). A challenge toward rearing, exhibiting spiny lobster larvae. In G. Nishi & T. Saruwatari (Eds.), *Work at aquariums* (pp. 84–98). Hatano: Tokai University Press.

Ianniello, L., & Mears, S. (2018). *Blackwater creatures. 170 pp. Ianniello & Mears*. USA.

Ishii, H., Morishita, A., & Yamaguchi, Y. (2016). Productive ecology and utilization as food of hydrozoan jellyfish *Eutonia indicans*. In *Abstract book of the 5th international jellyfish bloom symposium* (p. 117). Barcelona: L'Aquàrium de Barcelona.

Jeffs, A. (2007). Revealing the natural diet of the phyllosoma larvae of spiny lobster. *Bulletin of Fisheries Research Agency, 20*, 9–13.

Jeffs, A. G., Willmott, M. E., & Wells, R. M. G. (1999). The use of energy stores in the puerulus of the spiny lobster *Jasus edwardsii* across the continental shelf of New Zealand. *Comparative Biochemistry and Physiology. A, Comparative Physiology, 123*, 351–357.

Jeffs, A. G., Nichols, P. D., & Bruce, M. P. (2001). Lipid reserves used by pueruli of the spiny lobster *Jasus edwardsii* in crossing the continental shelf of New Zealand. *Comparative Biochemistry and Physiology. A, Comparative Physiology, 129*, 305–311.

Jouiaei, M., Yanagihara, A. A., Madio, B., Nevalainen, T. J., Alewood, P. F., & Fry, B. G. (2015). Ancient venom systems: A review on cnidaria toxins. *Toxins, 7*, 2251–2271.

Kamio, M., Furukawa, D., Wakabayashi, K., Hiei, K., Yano, H., Sato, H., Yoshie-Stark, Y., Akiba, T., & Tanaka, Y. (2015). Grooming behavior by elongated third maxillipeds of phyllosoma larvae of the smooth fan lobster riding on jellyfishes. *Journal of Experimental Marine Biology and Ecology, 463*, 115–124.

Kamio, M., Wakabayashi, K., Nagai, H., & Tanaka, Y. (2016). Phyllosomas of smooth fan lobsters (*Ibacus novemdentatus*) encase jellyfish cnidae in peritrophic membranes in their feces. *Plankton & Benthos Research, 11*, 100–104.

Kittaka, J. (1997). Application of ecosystem culture method for complete development of phyllosomas of spiny lobster. *Aquaculture, 115*, 319–331.

Kittaka, J. (2000). Culture of larval spiny lobsters. In B. F. Phillips & J. Kittaka (Eds.), *Spiny lobsters: Fisheries and culture* (pp. 508–532). Oxford: Fishing News Books.

Kittaka, J. (2005). Jellyfish as food organisms to culture phyllosoma larva. *Bulletin of the Plankton Society of Japan, 52*, 91–99.

Kizhakudan, J. K., & Krishnamoorthi, S. (2014). Complete larval development of *Thenus unimaculatus* Burton & Davie, 2007 (Decapoda, Scyllaridae). *Crustaceana, 87*, 570–584.

Kumar, T. S., Vijayakumaran, M., Murugan, T. S., Jha, D. K., Sreeraj, G., & Muthukumar, S. (2009). Captive breeding and larval development of the scyllarine lobster *Petrarctus rugosus*. *New Zealand Journal of Marine and Freshwater Research, 43*, 101–112.

Lavalii, K. L., & Spanier, E. (2007). Introduction to the biology and fisheries of slipper lobsters. In K. L. Lavalli & E. Spanier (Eds.), *The biology and fisheries of the slipper lobster* (pp. 3–21). Boca Raton: CRC Press.

Madin, L. P., & Harbison, G. R. (2001). Gelatinous zooplankton. In S. A. Thorpe & K. K. Turekian (Eds.), *Encyclopedia of ocean sciences* (Vol. 2, pp. 1120–1130). Amsterdam: Elsevier.

Mańko, M. K., Słomska, A. W., & Jażdżewski, K. (2017). Siphonophora of the Gulf of Aqaba (red sea) and their associations with crustaceans. *Marine Biology Research, 13*, 480–485.

Masuda, R., Yamashita, Y., & Matsuyama, M. (2008). Jack mackerel *Trachurus japonicus* juveniles use jellyfish for predator avoidance and as a prey collector. *Fisheries Science, 74*, 276–284.

Matsuda, H. (2010). *Ise-ebi wo tsukutu* (178 pp). Seizan, Tokyo. (In Japanese).

Matsuda, H., & Takenouchi, T. (2005). New tank design for larval culture of Japanese spiny lobster, *Panulirus japonicus*. *New Zealand Journal of Marine and Freshwater Research, 39*, 279–285.

Mikami, S., & Kuballa, A. V. (2004). Overview of lobster aquaculture research. In S. Kolkovski, J. Heine, & S. Clarke (Eds.), *Proceedings of the second hatchery feeds and technology workshop* (pp. 127–130). Sydney: Novotel Century Sydney.

Mikami, S., & Kuballa, A. V. (2007). Factors important in larval and postlarval molting, growth, and rearing. In K. L. Lavalli & E. Spanier (Ed.), *The biology and fisheries of the slipper lobster* (pp. 91–110). CRC Press.

Mikami, S., & Takashima, F. (1993). Development of the proventriculus in larvae of the slipper lobster, *Ibacus ciliatus* (Decapoda: Scyllaridae). *Aquaculture, 116*, 199–217.

Mikami, S., Greenwood, J. G., & Takashima, F. (1994). Functional morphology and cytology of the phyllosomal digestive system of *Ibacus ciliatus* and *Panulirus japonicus* (Decapoda, Scyllaridae and Palinuridae). *Crustaceana, 67*, 212–225.

Mitchell, J. R. (1971). Food preference, feeding mechanisms, and related behavior in phyllosoma larvae of the California spiny lobster, *Panulirus interruptus* (Randall) (110 pp). Master Thesis in San Diego State College.

Murakami, K., Jinbo, T., & Hamasaki, K. (2007). Aspects of the technology of phyllosoma rearing and metamorphosis from phyllosoma to puerulus in the Japanese spiny lobster *Panulirus japonicus* reared in the laboratory. *Bulletin of Fisheries Research Agency, 20*, 59–67.

Nelson, M. M., Phleger, C. F., Mooney, B. D., & Nichols, P. D. (2000). Lipids of gelatinous Antarctic zooplankton: Cnidaria and Ctenophora. *Lipids, 35*, 551–559.

Nishida, S., Quigley, B. D., Booth, J. D., Nemoto, T., & Kittaka, J. (1990). Comparative morphology of the mouthparts and foregut of the final-stage phyllosoma, puerulus, and postpuerulus of the rock lobster *Jasus edwardsii* (Decapoda: Palinuridae). *Journal of Crustacean Biology, 10*, 293–305.

O'Rorke, R., Lavery, S., Chow, S., Takeyama, H., Tsai, P., Beckley, L. E., Thompson, P. A., Waite, A. M., & Jeffs, A. G. (2012). Determining the diet of larvae of western rock lobster (*Panulirus cygnus*) using high-throughput DNA sequencing techniques. *PLoS One, 7*, e42757. https://doi.org/10.1371/JOURNAL.PONE.0042757.

Palero, F., Clark, P. F., & Guerao, G. (2014a). Achelata. In J. W. Martin, J. Olesen, & J. T. Høeg (Eds.), *Atlas of Crustacean Larvae* (pp. 272–278). Maryland: Johns Hopkins University Press.

Palero, F., Guerao, G., Hall, M., Chan, T. Y., & Clark, P. F. (2014b). The 'giant phyllosoma' are larval stages of *Parribacus antarcticus* (Decapoda: Scyllaridae). *Invertebrate Systematics, 28*, 258–276.

Phillips, B. F., & Matsuda, H. (2011). A global review of spiny lobster aquaculture. In R. K. Fotedar & B. F. Phillips (Eds.), *Recent advances and new species in aquaculture* (pp. 22–84). West Sussex: Blackwell.

Phillips, B. F., & Sastry, A. N. (1980). Larval ecology. In J. S. Cobb & B. F. Phillips (Eds.), *The biology and management of lobsters. II: Ecology and management* (pp. 11–57). New York: Academic Press.

Phillips, B. F., Jeffs, A. G., Melville-Smith, R., Chubb, C. F., Nelson, M. M., & Nichols, P. D. (2006). Changes in lipid and fatty acid composition of late larval and puerulus stages of the spiny lobster (*Panulirus cygnus*) across the continental shelf of Western Australia. *Comparative Biochemistry and Physiology. B, 143*, 219–228.

Phleger, C. F., Nelson, M. M., Mooney, B. D., Nichols, P. D., Ritar, A. J., Smith, G. G., Hart, P. R., & Jeffs, A. G. (2001). Lipids and nutrition of the southern rock lobster, *Jasus edwardsii*, from hatching to puerulus. *Marine and Freshwater Research, 52*, 1475–1486.

Purcell, J. E. (2012). Jellyfish and ctenophore blooms coincide with human proliferations and environmental perturbations. *Annual Review of Marine Science, 4*, 209–235.

Purcell, J. E., Uye, S., & Lo, W. T. (2007). Anthropogenic causes of jellyfish blooms and their direct consequences for humans: A review. *Marine Ecology Progress Series, 350*, 153–174.

Ritar, A. J. (2001). The experimental culture of phyllosoma larvae of southern rock lobster (*Jasus edwardsii*) in a flow-through system. *Aquacultural Engineering, 24*, 149–156.

Saunders, M. I., Thompson, P. A., Jeffs, A. G., Säwström, C., Sachlikidis, N., Beckley, L. E., & Waite, A. M. (2012). Fussy feeders: Phyllosoma larvae of the western rocklobster (*Panulirus cygnus*) demonstrate prey preference. *PLoS One, 7*, e36580. https://doi.org/10.1371/journal. pone.0036580.

Sekiguchi, H., Booth, J. D., & Webber, W. R. (2007). Early life histories of slipper lobsters. In K. L. Lavalli & E. Spanier (Eds.), *The biology and fisheries of the slipper lobster* (pp. 69–90). Boca Raton: CRC Press.

Shojima, Y. (1963). Scyllarid phyllosomas' habit of accompanying the jelly-fish (preliminary report). *Bullettin of the Japanese Society of Scientific Fisheries, 29*, 349–353.

Shojima, Y. (1973). The phyllosoma larvae of Palinura in the East China Sea and adjacent waters—I. *Bulletin of Seikai Regulation Fisheries Research Laboratory, 43*, 105–115. (In Japanese).

Simon, C. J., Carter, C. G., & Battaglene, S. C. (2012). Development and function of the filter-press in spiny lobster, *Sagmariasus verreauxi*, phyllosoma. *Aquaculture, 370–371*, 68–75.

Sims, H. W., Jr., & Brown, C. L., Jr. (1968). A giant scyllarid phyllosoma larva taken north of Bermuda (Palinuridea). *Crustaceana (Supplement), 2*, 80–82.

Sullivan, L. J., & Kremer, P. (2011). Gelatinous zooplankton and their trophic roles. In E. Wolanski & D. McLusky (Ed.), *Treatise on estuarine and coastal science, Vol. 6, Trophic relationships of coastal and estuarine ecosystems* (pp. 127–171). London: Academic Press.

Suzuki, N., Murakami, K., Takeyama, H., & Chow, S. (2006). Molecular attempt to identify prey organisms of lobster phyllosoma larvae. *Fisheries Science, 72*, 342–349.

Suzuki, N., Murakami, K., Takeyama, H., & Chow, S. (2007). Eukaryotes from the hepatopancreas of lobster phyllosoma larvae. *Bulletin of Fisheries Research Agency, 20*, 1–7.

Sverdrup, H.U., Johnson, M.W. & Fleming, R.H. 1947. The oceans: Their physics, chemistry, and general biology. 1087 pp. Prentice-Hall, Englewood Cliffs.

Takahashi, M., & Saisho, T. (1978). The complete larval development of the scyllarid lobster, *Ibacus ciliatus* (von Siebold) and *Ibacus novemdentatus* Gibbes in the laboratory. *Members Faculty of Fisheries Kagoshima University, 27*, 305–353.

Takeuchi, T. (2014). Progress on larval and juvenile nutrition to improve the quality and health of seawater fish: A review. *Fisheries Science, 80*, 389–403.

Thomas, L. R. (1963). Phyllosoma larvae associated with medusae. *Nature, 198*, 208.

Uye, S., Fujii, N., & Takeoka, H. (2003). Unusual aggregations of the scyphomedusa *Aurelia aurita* in coastal waters along western Shikoku, Japan. *Plankton Biology and Ecology, 50*, 17–21.

Vijayakumaran, M., & Radhakrishnan, E. V. (2011). Slipper lobsters. In R. K. Fotedar & B. F. Phillips (Eds.), *Recent advances and new species in aquaculture* (pp. 85–114). West Sussex: Blackwell.

Wakabayashi, K., & Phillips, B. F. (2016). Morphological descriptions of laboratory reared larvae and post-larvae of the Australian shovel-nosed lobster *Thenus australiensis* Burton and Davie, 2007 (Decapoda, Scyllaridae). *Crustaceana, 89*, 97–117.

Wakabayashi, K., & Tanaka, Y. (2012). The jellyfish-rider: Phyllosoma larvae of spiny and slipper lobsters associated with jellyfish. *TAXA, Proceedings of Japanese Society Systematic Zoology, 33*, 5–12. (In Japanese).

Wakabayashi, K., Sato, R., Hirai, A., Ishii, H., Akiba, T., & Tanaka, Y. (2012a). Predation by the phyllosoma larva of *Ibacus novemdentatus* on various kinds of venomous jellyfish. *The Biological Bulletin, 222*, 1–5.

Wakabayashi, K., Sato, R., Ishii, H., Akiba, T., Nogata, Y., & Tanaka, Y. (2012b). Culture of phyllosomas of *Ibacus novemdentatus* (Decapoda: Scyllaridae) in a closed recirculating system using jellyfish as food. *Aquaculture, 330–333*, 162–166.

Wakabayashi, K., Matsumura, K., & Tanaka, Y. (2013). Consumption rates of jellyfish by phyllosoma larvae of the smooth fan lobster *Ibacus novemdentatus*. In *Abstracts of aquaculture conference: To the next 40 years of sustainable global aquaculture*. Palacio de Congresos de Canarias Convention Centre, Las Palmas de Gran Canaria. O6.5.

Wakabayashi, K., Nagai, S., & Tanaka, Y. (2016a). The complete larval development of *Ibacus ciliatus* from hatching to the nisto and juvenile stages using jellyfish as the sole diet. *Aquaculture, 450*, 102–107.

Wakabayashi, K., Sato, H., Yoshie-Stark, Y., Ogushi, M., & Tanaka, Y. (2016b). Differences in the biochemical compositions of two dietary jellyfish species and their effects on the growth and survival of *Ibacus novemdentatus* phyllosomas. *Aquaculture Nutrition, 22*, 25–33.

Wakabayashi, K., Tanaka, Y., & Abe, H. (2017a). *Field guide to marine plankton* (Vol. 180). Tokyo: Bun-ichi. (In Japanese).

Wakabayashi, K., Yang, C. H., Shy, J. Y., He, C. H., & Chan, T. Y. (2017b). Correct identification and redescription of the larval stages and early juveniles of the slipper lobster *Eduarctus martensii* (Pfeffer, 1881) (Decapoda: Scyllaridae). *Journal of Crustacean Biology, 37*, 204–219.

Wake, F., Izumimoto, M., Mikami, M., & Miura, H. (1974). On the bacteriostatic action by glycine. *Research Bullettin of Obihiro University Series I, 9*, 159–163.

Wang, M., & Jeffs, A. G. (2014). Nutritional composition of potential zooplankton prey of spiny lobster larvae: A review. *Reviews in Aquaculture, 6*, 270–299.

Yoshino, Y. (2015). Sekai de ichi-ban utsukushii umi no ikimono zukan (Most beautiful creatures in the ocean). 231 pp. Sogensha, Osaka. (In Japanese).

Lobster Aquaculture Development in Vietnam and Indonesia

12

Clive M. Jones, Tuan Le Anh, and Bayu Priyambodo

Abstract

Development of spiny (rock) lobster aquaculture is of special interest because market demand continues to increase while capture fisheries production remains static and with little likelihood of any increase. This chapter provides a synopsis of information about the history, development, status and future of tropical spiny lobster aquaculture with a particular focus on Vietnam and Indonesia, where considerable development has already occurred. Vietnam is the only country in the world where farming of lobsters is fully developed and commercially successful. The Vietnamese industry is based on a natural supply of seed lobsters – the puerulus stage, as hatchery supply is not yet available due to the difficult technical demands of rearing spiny lobster larvae in captivity. Vietnam currently produces around 1600 tonnes of premium grade lobsters, primarily of the species *Panulirus ornatus*, that are exported to China where the price is higher. The industry is valued at over $US120 million. This success led to significant interest in Indonesia where a fishery for seed lobsters has become well developed, with a catch 10–20 times greater than that of Vietnam. However, growout of lobster in Indonesia remains insignificant due to adverse government policy and lack of farmer knowledge and skills. The seed lobsters available in Indonesia are primarily *Panulirus homarus*, a species with excellent production characteristics like *P. ornatus*, although with lesser value. Extraordinarily high abundance of naturally settling seed lobsters is appar-

---

C. M. Jones (✉)
Centre for Sustainable Tropical Fisheries and Aquaculture, James Cook University, Cairns, QLD, Australia
e-mail: clive.jones@jcu.edu.au

T. Le Anh
Nha Trang University, Nha Trang, Vietnam

B. Priyambodo
Marine Fisheries and Aquaculture Development Centre, Lombok, Indonesia

© Springer Nature Singapore Pte Ltd. 2019
E. V. Radhakrishnan et al. (eds.), *Lobsters: Biology, Fisheries and Aquaculture*,
https://doi.org/10.1007/978-981-32-9094-5_12

ent in selected areas due to a confluence of suitable conditions that create a high concentration of late-stage larvae near the coast. These areas have been termed hotspots, as the availability of settling seed is much higher than other areas. Such hotspots are now recognised in the central northern coast of Vietnam – supplying their growout industry – and the central southern coast of Indonesia. Natural mortality of the seed lobsters in these areas is correspondingly high due to insufficient settlement habitat and fish predation. Consequently, responsible fishing of these seed is sustainable, providing a valuable resource that can be on-grown for benefit of impoverished coastal communities. Innovative and inexpensive techniques have been developed to effectively catch the seed as they swim towards the coast seeking suitable habitat. In Vietnam, the seed are typically sold by fishers to dealers, who aggregate supplies and then on-sell to nursery farmers. Nursing consists of rearing the seed lobsters in small suspended or submerged cages, with a diet of fresh seafood – crabs, mollusc and fish. Advanced juvenile lobsters are produced that are in turn on-sold to growout farmers who stock them to larger floating cages, suspended from simple floating frames.

The economics involve relatively low capital and operating costs and production of high-value product that provides significant economic and social benefit to the communities involved. Although several health and disease issues have impacted spiny lobster farming, they can be effectively managed through good nutrition and husbandry. Market demand for spiny lobster from China is strong and growing, far exceeding supply. There appears to be great scope for much larger farm production of spiny lobsters with little impact on price. The future for tropical spiny lobster aquaculture appears to be very positive, particularly for developing countries in the Asian region, where seed are available, suitable growout locations are present and where costs of production are relatively low. It is expected that lobster aquaculture will continue to develop in the region, expanding beyond Vietnam and Indonesia.

Keywords
Rock lobster · Aquaculture · *Panulirus* · Vietnam · Indonesia

12.1 Introduction

Interest in farming lobster to meet growing demand has been evident for several decades, but has developed commercially to only a small extent due to biological constraints (Jones 2009). Supply of lobsters from fishery sources is largely static or in decline for all commercial species across the world (Phillips 2013). For spiny or rock lobsters (Palinuridae), there have been several cases of well-managed fisheries, declining sharply with uncertain attributions to the causes. The western rock lobster *Panulirus cygnus* (George 1962) fished commercially along the southwestern coast of Australia, for example, experienced a significant decline in catch in recent years, with revised quota now less than half what it once was, and despite a very well-credentialed management regime (Plagányi et al. 2017). Meanwhile, the demand from markets for spiny lobster continues to grow – particularly from China (Jones 2015a). It is clearly evident that the only option for increasing spiny lobster supply is aquaculture.

To date the only significant aquaculture production of spiny lobsters has come from Vietnam (Jones et al. 2010). The on-growing of the ornate lobster, *Panulirus ornatus* (Fabricius, 1798), has been a successful village-based industry along the central south coast of Vietnam since 1995, based on an abundance of naturally settling lobster seed and the establishment of up to 49,000 lobster sea cages (Hung and Tuan 2009). According to advice from Nha Trang University in Vietnam (Dr. Le Anh Tuan, September 2017, pers. comm), production of cultured lobsters in 2016 was estimated to be about 1600 tonnes, worth more than $US120 M.

The success of the Vietnam industry has led to a similar interest in Indonesia, where there is potential for an equivalent, if not larger, lobster farming industry. To date there has been little development elsewhere within the natural distribution of the preferred species, although this may change, particularly if hatchery technology is commercialised. There is strong evidence of significant positive impact of lobster farming on impoverished coastal communities (Bell 2004), and thus great interest in extending such benefit to other poor areas (Petersen and Phuong 2010). Although the Vietnam lobster farming industry is successful, it has experienced significant challenges from disease and environmental degradation (Anh and Jones 2015a; Jones 2015b). Ongoing research (Jones 2015c) is working to resolve problems and support extension of sustainable practices to new lobster farming areas. This chapter describes the development of tropical spiny lobster aquaculture in Vietnam and Indonesia.

12.2 History

Development of lobster aquaculture has been keenly sought for many decades (Jones 2009; Wickins and Lee 2002), and several commercially important fishery species have been intensively researched to assess and develop their commercial aquaculture. The clawed lobster *Homarus americanus* (H. Milne Edwards, 1837) from the north-west Atlantic, which supports the world's largest lobster fishery, has been comprehensively researched to develop aquaculture technology. Due to biological constraints, primarily aggressive behaviour, and unattractive economics, efforts to establish clawed lobster aquaculture have been largely abandoned (Addison and Bannister 1994; Aiken 1988; Aiken and Waddy 1995; Tlusty 2004; Wahle et al. 2013). Nevertheless, a number of *Homarus* hatcheries in North America and Northern Europe (*Homarus gammarus* Linnaeus, 1758) continue to produce juveniles for restocking purposes.

Interest in developing aquaculture of spiny lobsters (Palinuridae) has grown strongly in recent decades (Groeneveld et al. 2013; Jeffs and Hooker 2000; Jeffs and Davis 2003; Jones 2009; Kittaka and Booth 1994, 2000), although advances have been constrained by the protracted larval phase of all species and the difficulty of managing this phase in aquaculture systems. Commercial rock lobster hatchery technology has reached proof of concept, but is yet to become a commercial reality. Based on recent progress, commercial production of hatchery-produced pueruli may be realised in the next few years (Smith 2017), but its economic viability is yet to be tested.

The only established and significant lobster aquaculture industry in the world is that of Vietnam, based on the growout of wild-caught juveniles (Jones 2015c). Its development dates from the late 1970s in response to growing demand for lobsters from China (Jones et al. 2010) (Table 12.1). In its earliest iteration, smaller sub-adult lobsters were held in cages and fattened over a period of weeks or months to achieve larger and higher priced sizes. This rapidly expanded to full growout from the post-larval phase to 1 kg + lobsters with more advanced husbandry and feeding. The primary species representing more than 90% of lobsters farmed in Vietnam is *P. ornatus*, which is the most readily available along the Vietnam coastline as a settling puerulus. This is fortunate, as *P. ornatus* is the most valuable of the various tropical species in the China market. For Vietnam, *Panulirus homarus* (Linnaeus,

Table 12.1 History of rock lobster farming development in Vietnam (after Anh and Jones (2015a))

Period	Main factors and statistics
1975–1985	Annual fishery catch of marketable rock lobsters <100 tonnes
	P. ornatus represents moderate part of the supply
	Local demand and price for rock lobster is low
1986–1991	Chinese demand for rock lobsters grew rapidly
	Of the various tropical species, *P. ornatus* > 1 kg is the premium
1992–1995	Vietnam lobster fishers started fattening small lobsters to reach the 1 kg premium size
	Production mostly in fixed cages in shallow water
	Farmed lobster production <100 tonnes
1996–1999	Vietnam lobster fishers started collecting swimming pueruli, in addition to juveniles and sub-adults
	Farmed lobster production <300 tonnes
2000–2006	Most lobster farming now conducted in floating cages in deeper water
	Significant resource of pueruli identified and exploited
	Farmed lobster production of 500–2000 tonnes
	Indications of overload – disease and environmental degradation
2007–2009	Significant losses on farm due to disease and health problems
	Farmed lobster production ~860 tonnes (2008/2009)
	Government-led industry planning implemented
2010–2014	Nearly all lobster farming based on wild puerulus supply
	Lobster growout recovers from disease
	Farmed lobster production for 2010–2014, about 1500 tonnes per year
2015–2017	Vietnam wild puerulus catch steady at 2–4 million per year
	Puerulus supply supplemented with imported wild puerulus from Indonesia
	Farmed lobster production 1600 tonnes per year
	Constraints on further growout production due to limited sea cage sites and conflicting use (tourism)

1758) is a secondary species, representing the remaining 10% of farm production, but also attracting high price.

By 2004, over 30,000 net cages had been established along the south central coastline of Vietnam, producing more than 2000 tonnes of farmed lobsters (Jones et al. 2010; Thuy and Ngoc 2004), and this subsequently expanded to 49,000 cages, although production has moderated to around 1500 tonnes per annum over recent years (Anh and Jones 2015a).

Despite the success of lobster aquaculture in Vietnam, equivalent development has not occurred elsewhere within the Asian region. In Taiwan and the Philippines, there has been some fattening of smaller lobsters (Arcenal 2004; Juinio-Menez and Gotanco 2004), but in both countries, the supply of such lobsters is small and has not expanded through an identified puerulus resource. In India, Vijayakumaran et al. (2009) reported on a small-scale lobster aquaculture industry based on capture of wild seed, but this was destroyed by the tsunami of 2004 and has never recovered. Open sea cage farming of *P. homarus* along the southern Indian coast and *P. polyphagus* on the north-west coast using juveniles and sub-adults captured from the wild has also been carried out by the Central Marine Fisheries Research Institute, India (Lipton, Rani Mary, Mojjada). However, the economic viability and sustainability of the capture-based farming operation are yet to be proven. Indonesia appears to be the only other country in the South East Asian region to initiate lobster aquaculture using a wild puerulus supply (Jones et al. 2010; Priyambodo and Jaya 2009; Priyambodo and Sarifin 2009). As of 2018, a substantial puerulus resource for *P. ornatus* has been identified along the eastern coast of the Philippines, although to date there is very limited growout.

On the central Indonesian island of Lombok in the early 2000s, pueruli (primarily *P. homarus*) were observed settling naturally on seaweed farms and the floating cages used for grouper culture. The seaweed and fish farmers recognised that these small lobsters presented an opportunity, given the high market value of lobsters. They began to seek out the pueruli and develop specific methods to fish for them and to stock them to dedicated lobster growout cages. By the mid-2000s, puerulus fishing had developed using methods similar to Vietnam, and more than 600,000 were caught in 2008/2009 (Jones et al. 2010). Over subsequent years, the fishing methods for pueruli were refined, and although components of the Vietnam approach were applied, e.g. use of lights to attract the swimming pueruli, the fishing methods of Lombok developed their own unique character (Priyambodo et al. 2015, 2017). By 2015, fishing for pueruli with the best practice methods developed in Lombok had expanded to the entire southern coast of Java and Sumbawa, representing some 1500 km of coastline. The puerulus catch in 2016 was estimated to be 100 million pieces (B. Priyambodo, Indonesian Ministry of Marine Affairs and Fisheries, unpublished data).

12.3 Species

In the absence of commercially viable hatchery technology, spiny lobster aquaculture is reliant on a supply of naturally settling pueruli. Vietnam has seven species of rock lobster: *Panulirus ornatus, Panulirus homarus, Panulirus versicolor* (Latreille, 1804), *Panulirus penicillatus* (Oliver, 1791), *Panulirus stimpsoni* Holthuis, 1963, *Panulirus longipes* (A. Milne Edwards, 1868) and *Panulirus polyphagus* (Herbst, 1793), but it is only *P. ornatus* and *P. homarus* that are abundant as pueruli (Long and Hoc 2009) and consequently these are the only species used for aquaculture. The preponderance of *P. ornatus* pueruli in Vietnam is fortuitous, as this is the most valuable species in the Chinese market.

Indonesia's native rock lobster species include *Panulirus ornatus, Panulirus homarus, Panulirus versicolor, Panulirus penicillatus, Panulirus longipes* and *Panulirus polyphagus*, all of which are fished commercially for consumption-sized lobsters. However, the most abundant species available as settling pueruli are *P. homarus* and *P. ornatus*. The fishery for pueruli was first established in the southeast of Lombok Island where their natural abundance is sufficient to support a fishery. Adult lobster stocks in Lombok are scarce as there are few reef areas suited to spiny lobster habitation. In Lombok, more than 99% of the pueruli caught are either *P. homarus* or *P. ornatus*, and the relative proportion of these two species changes throughout the year and between years (Bahrawi et al. 2015b). Overall, pueruli of *P. homarus* are the most abundant (between 65 and 85%), and the proportion of *P. ornatus* varies between 15% and 35%. In recent years, the fishery for lobster pueruli expanded to the southern coasts of Java and Sumbawa, where the species composition was consistent with that of Lombok.

Although there has been no scientific assessment of the various tropical rock lobster species in regard to their suitability for aquaculture, anecdotal evidence, primarily from marketing information, suggests that the most robust of the species are *P. ornatus* and *P. homarus* – the two species that are currently farmed. Given that abundance of pueruli of the other native species in Vietnam and Indonesia appears to be quite low and therefore insufficient to support their aquaculture, there is no justification to assess their aquaculture credentials. This however may change in the future when hatchery technology is commercialised and when further consideration may be given to farming the most robust and aquaculture-suitable species.

In summary, there are two lobster species currently farmed in Vietnam and Indonesia, *P. ornatus* and *P. homarus* (Photo 12.1), based on their abundance as naturally settling pueruli and their suitability for captive on-growing. Although both are available and farmed in both countries, in Vietnam, the predominant species is *P. ornatus* and *P. homarus* in Indonesia.

Photo 12.1 Photo of *P. ornatus* and *P. homarus* adults

12.4 Seed Supply

Aquaculture based on natural seed supply is not unique to spiny lobster. Substantial aquaculture operations have been established for other crustaceans and fish (Lucas and Southgate 2012; Naylor et al. 2000), using wild seed supply, although in most cases their long-term sustainability can only be assured if a hatchery supply is established. In the case of tropical spiny lobster farming, Vietnam has demonstrated that aquaculture based on a wild seed supply can be successful and sustained. For Indonesia, a significant seed resource has been identified upon which a significant growout industry could be established.

12.4.1 Lobster Seed Fishing

Methods for the fishing of lobster pueruli (seed) were first developed in Vietnam in the mid-1990s (Jones et al. 2010). Entrepreneurial Vietnamese fishers recognised that small lobsters could be fattened to a more valuable product, as the Chinese demand and price per kilogram was greatest for lobsters larger than one kilogram. In the earliest years, methods were developed for catching small juveniles, typically by creating habitat in which juvenile lobsters would settle. Small-diameter holes were drilled into coral rocks and timber posts, and these materials were placed in shallow waters along the coastline. Fishers would periodically dive on to these habitats and manually remove settled juveniles. This method was progressively replaced with fishing for the puerulus stage, using various nets to capture the swimming pueruli as they actively move through inshore waters seeking suitable habitat. Puerulus fishing quickly proved to be more effective than juvenile fishing, as the abundance of pueruli was often much higher. Methods evolved and catch rates increased as the canny fishers came to understand the oceanographic conditions that matched the highest abundance. These conditions were characterised by inshore areas protected from larger swells, inside embayments often with fringing islands, moderate current, against which the puerulus would swim, and often in proximity to river mouths where turbidity was elevated. Nets set across the current would effectively intercept the pueruli as they swam through the hours of darkness (Jones et al. 2010). Today, the most common and effective method for fishing of pueruli is a set seine, deployed in a V-shape with opening facing the prevailing current, and using lights positioned near the net apex to attract the pueruli (Anh and Jones 2015b) (Fig. 12.1). These nets are set in the hours around dusk and retrieved twice each night, around midnight and again at dawn, with pueruli hand collected from the net as it is hauled aboard.

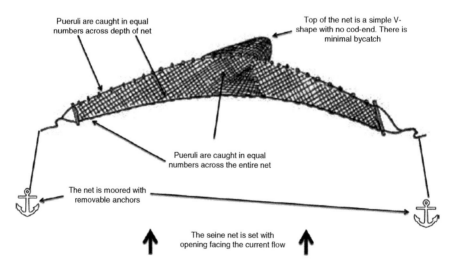

Fig. 12.1 Illustration of seine net used for capture of lobster pueruli in Vietnam

Interest in farming lobsters in Indonesia began in the mid-2000s on the island of Lombok. Although the Vietnam rock lobster aquaculture industry was well established by this time, the coastal communities of Lombok were unaware of the methods the Vietnamese employed. Consequently, the Indonesians developed their own unique methods for catching seed and on-growing them. It was only in 2013, following a study tour of Vietnam by a group of Indonesian farmers (Priyambodo et al. 2015), that some cross fertilisation of methods occurred, that led to the most effective puerulus fishing methods now employed. The Indonesian method shares with Vietnam the practice of intercepting pueruli as they swim; however, rather than using a seine net, artificial habitats are deployed through the water column, into which the pueruli settle. The development of the Indonesian methods is well described by Bahrawi et al. (2015a, c), and the effectiveness of various habitat materials and their positioning to maximise catch is detailed in the reports of Priyambodo et al. (2015, 2017). At present, the most common and effective method employed in Indonesia consists of an array of small fabricated habitats that are suspended in the water column from a floating frame. The frame is typically rectangular, ranging in size from 2.5 m × 2.5 m up to 12 m × 12 m, constructed with bamboo supported by styrofoam floats. The frame is moored semi-permanently in locations known to have high abundance of seed lobsters, and these locations share the same characteristics as described for Vietnam, namely protected bays with distinct current and elevated turbidity from terrestrial outflow (Photo 12.2).

The array of habitats are attached to a rectangular panel of netting approximately 1.5 m across and 2 m in depth, its top edge held rigid with a timber frame and its bottom edge weighted. Across the panel, up to 25 individual habitats are attached, each consisting of a folded piece of plastic-lined paper or rice bag material. The paper is from used cement bags that have been cut open to form a rectangle, and

Photo 12.2 Aerial photo of floating lobster seed frames in Lombok

Photo 12.3 Typical bowtie collector from Indonesia and floating frame from which they are suspended

similarly the rice bag material (woven polyurethane) from disused rice bags. The material is folded across its width in a concertina fashion to create a series of folds and the folded material is then tied at its centre, so the resulting item resembles a bowtie (Photo 12.3). Such fabrication results in a series of sharply creased crevices that are attractive to pueruli as a suitable habitat in which to settle. These bowtie habitats have much in common with the crevice traps widely used for resource management assessment of commercial lobster fisheries settlement throughout the world (Booth and Tarring 1986; Phillips and Booth 1994), but in Indonesia, they are used for fishing rather than management.

Research (Priyambodo et al. 2015) and commercial trial and error have determined that catch is greatest near the sea floor, and consequently, the bowtie habitat arrays are positioned just off the sea floor. Priyambodo et al. (2015, 2017) have also determined that cement bag paper is superior to other materials (including rice bag), as it is quite robust through immersion and supports an optimal amount of biofouling that appears to contribute to its attractiveness to pueruli. The combination of dual plastic and paper layers provides dual benefits. Cement bag paper is ideal for the Indonesian farmers as it is in ready supply and inexpensive.

Unlike Vietnam where the nets are actively deployed and removed each day, the seed fishing frames in Indonesia are moored in place and only shifted or removed if rough conditions develop. One of the most significant methods borrowed from the Vietnamese was the application of lights to the seed fishing devices, first applied in 2013 (Priyambodo et al. 2015). A single fluorescent or incandescent light mounted above the frame results in significantly increased catch rates, presumably due to positive photo-taxis of the swimming pueruli. Indonesian seed fishers visit their fishing frames early each morning to lift each panel of habitats and manually remove the settled seed hiding in the habitat crevices.

In both Vietnam and Indonesia, the captured seed are immediately placed into containers with fresh seawater, sometimes aerated. These containers are returned to shore and often sold to seed dealers (middlemen) who then move the seed to holding facilities where they can be packed for transport to farmers. Holding facilities are typically tanks at the dealers' household, with rudimentary filtration. Seed are most often graded by species and quality, and housed in plastic colanders floating in the tank. Once sold, they are counted and placed into plastic bags of around 4–5 L capacity, nearly always with aeration and sometimes with oxygen injected. The plastic bags are then placed in styrofoam boxes for transport by road to their destination (Photo 12.4). In Vietnam, the transport may be up to 1000 km from the more

Photo 12.4 Bag of lobster seed ready for transport

northerly seed fishing grounds to the central and southern coast where the bulk of farming occurs. In Indonesia, seed may be transported short distance to adjacent farms, or more commonly to the airport for export to Vietnam. Details of seed handling and transport are provided in the reports of Jones et al. (2010), Bahrawi et al. (2015c) and Anh and Jones (2015b).

12.4.2 Lobster Seed Fishery Statistics

Fishing for lobster seed in Vietnam is not well regulated, and consequently there are no formal government statistics on the fishery. However, through a collaborative research programme aimed at development and sustainability of lobster aquaculture in Vietnam from 2005 to 2014 (Jones 2015c; Williams 2009), detailed data were collected that provided quantitative description of the resource and its exploitation. Much of this detail is documented in Dao and Jones (2015) and Long and Hoc (2009). For Indonesia, where there is a similar lack of formal government statistics, a seed census equivalent to that of Vietnam was implemented over several years from 2007 to 2014, to collect detailed data on the seed fishery in Lombok (Bahrawi et al. 2015a, b; Idris and Bahrawi 2015; Priyambodo and Bahrawi 2015). This was subsequently enhanced with a survey of lobster seed resources beyond Lombok.

In Vietnam, the seed supply has been relatively stable at around 3–4 million seed lobsters caught each year (Anh and Jones 2015b; Dao and Jones 2015) and, despite inter-annual variation, the catch has stayed at this level for the past 10 years (Fig. 12.2). From this seed supply, Vietnam generates around 1500 tonnes of annual production of on-grown lobsters (Anh and Jones 2015a), representing an approximate overall survival of 40–50% of lobsters from seed to 1 kg saleable product.

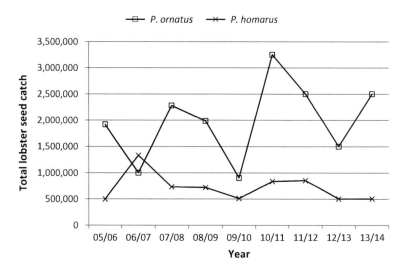

Fig. 12.2 Seed catch from 2005 to 2014 in Vietnam. After Dao and Jones (2015)

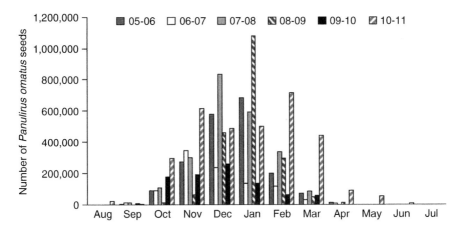

Fig. 12.3 Monthly seed catch in Vietnam for *P. ornatus*

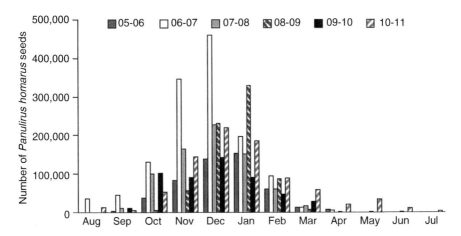

Fig. 12.4 Monthly seed catch in Vietnam for *P. homarus*

There is distinct seasonality to the lobster seed catch of Vietnam, with most seed caught between October and March (Figs. 12.3 and 12.4). The price paid to the fisher per seed during the period 2005–2011 is presented in Figs. 12.5 and 12.6 for *P. ornatus* and *P. homarus*. Since 2011 the price has continued to trend upwards, with some variation through each season in accordance with supply and demand. Most recently, in 2017, price per seed for *P. ornatus* has exceeded $US15.00 and $US5.00 for *P. homarus*. The increasing supply of imported seed from Indonesia has moderated prices a little, although a premium is paid for the local Vietnam-caught seed as they tend to be of superior quality, presumably because of the shorter distance and duration of transport. There are no data on catch per unit effort.

Indonesia's lobster seed fishery originated on the island of Lombok where annual seed catch was estimated to have peaked at around three million pieces in 2013/2014 (Bahrawi et al. 2015b). In 2015, the Indonesian government introduced a new

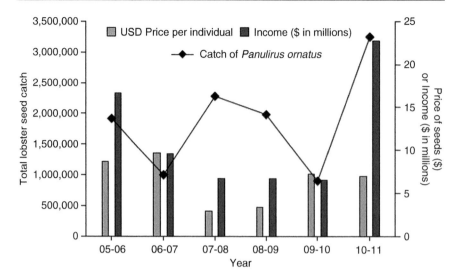

Fig. 12.5 Seed price stats in Vietnam for *P. ornatus*

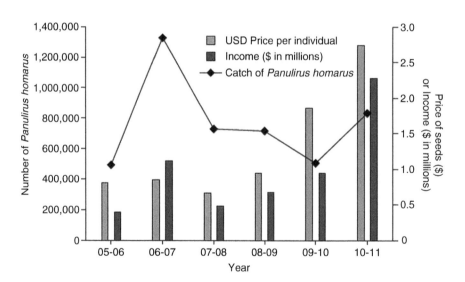

Fig. 12.6 Seed price stats in Vietnam for *P. homarus*

fishery regulation specifying a minimum legal size for rock lobster of 200 grams (all species). Although this was aimed at management of the adult resources, it effectively made fishing of seed illegal. At this time, there was a burgeoning lobster seed export industry developing, as Indonesian seed dealers became aware of the high demand for seed in Vietnam. From 2015 to 2017, enforcement of the regulation increased, as did frequency of seizure of shipments and the magnitude of penalties to the offenders. Consequently, there was a decrease in seed fishing activity and of

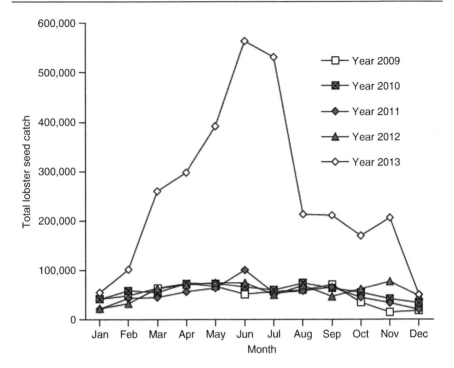

Fig. 12.7 Monthly total seed catch in Indonesia, 2009–2013

catch and current catch data are uncertain. Data collected in Lombok from 2007 to 2014 revealed some seasonality to catch, although not as marked as that of Vietnam. Figure 12.7 shows the peak catch is between March and November, although fishing continues throughout the year. Monthly catch data for each of *P. homarus* and *P. ornatus* in 2013 are presented in Fig. 12.8, showing the relative proportions of the two species. Price data for each species over the period 2009–2013 are shown in Figs. 12.9 and 12.10. The marked increase in price from 2012 to 2013 is attributed to the increased demand as export from Indonesia to Vietnam began. The price continued to increase into 2015 with greater disparity between the species. Higher demand for *P. ornatus* saw price increase to more than $US5.00 per piece, while that for *P. homarus* reached a high of $US2.00.

In 2015 and 2016, semi-quantitative lobster seed catch surveys were performed beyond Lombok (B. Priyambodo, Indonesian Ministry of Marine Affairs and Fisheries, pers. comm), along the southern coastline of Java and Sumbawa. They showed that lobster seed fisheries developed more widely, driven by the demand from Vietnam, and the total annual catch along the southern coastline of Java, Bali, Lombok and Sumbawa, over a distance of 1500 km, is greater than 100 million. As the seed capture methods are not exhaustive, the recorded catch is likely to be a small proportion of the entire resource which may number in the order of many hundreds of million settling each year. Although the species composition between the two countries is different, the seed supply of Indonesia is far greater than that of Vietnam and represents a significant opportunity for lobster aquaculture.

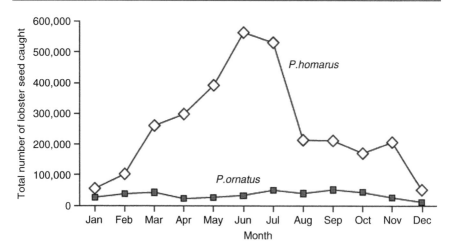

Fig. 12.8 Monthly seed catch in Indonesia for *P. homarus* and *P. ornatus* for 2013

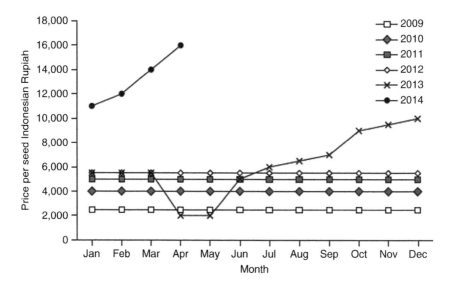

Fig. 12.9 Monthly lobster seed price in Indonesia from 2010 to 2014 for *P. homarus*

12.4.3 Sustainability of Lobster Seed Fishing

The sustainability of fishing the puerulus stage is an important consideration in the management of lobster resources. For Vietnam, there has been no scientific effort to quantify lobster resources and their recruitment. Only anecdotal data are available and they suggest that the annual influx of settling pueruli, as detailed above, is disconnected from the adult stocks in Vietnam waters which are at historically low abundance. Dao et al. (2015) have surmised that the source of pueruli settling in

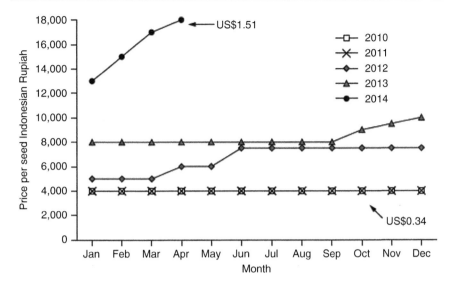

Fig. 12.10 Monthly lobster seed price in Indonesia from 2010 to 2014 for *P. ornatus*

Vietnam each year is likely to be far to the north and east, and the protracted larval period and oceanic life habit dictate that the larval stages travel significant distances from point of hatch to location of settlement.

Depending on the location of the seed captured, the effect of removing them from nature may have no effect on adult populations and the fishery. Research of the Caribbean rock lobster *Panulirus argus* (Lipcius et al. 1997) showed that in some locations, adult abundance is low and post-larval (puerulus) abundance is high. In such locations, the pueruli are concentrated at the location by the nature of regional and local oceanic currents. These are termed 'sink' populations, because most of those pueruli will not survive nor contribute to adult populations. The puerulus supply is effectively disconnected or decoupled from adult abundance. Occurrence of such sink populations of lobster pueruli is greatly enhanced by the biology of the species and particularly the long duration (4–6 months for tropical species) of the larval stages (phyllosoma) that are released into oceanic currents and physically transported very long distances from where they were spawned.

The *P. homarus* and *P. ornatus* puerulus resource along the central southern coast of Vietnam is likely to represent a sink population, and similarly that of southern central Indonesia along the southern coastline of Java, Bali, Lombok and Sumbawa. In Indonesia, there is a confluence of geographic and oceanographic conditions that would likely lead to a concentration of late-stage phyllosoma larvae, particularly of *P. homarus* and *P. ornatus* in the Java Sea, generated from the pull of the Indonesian Through flow – a powerful current running south, through the strait between Bali and Lombok. As this current enters the Timor Sea to the south, it slows and eddies west to Java and east to Lombok and Sumbawa, and the larval lobsters complete their development, transform to puerulus and then settle along the coastline in this region. Their concentration along this coastline appears to be far higher than elsewhere, based upon semi-quantitative surveys of other Indonesian provinces (Bahrawi et al. 2015a). In addition, there is very limited suitable natural habitat in the settlement locations to support juveniles and adults. The southern coastline of Java, Bali, Lombok and Sumbawa is characterised by steep topography, with a narrow strip of fringing reef adjacent to the coast before immediate drop off to depths unsuitable for lobster habitation. Consequently, the bulk of the lobster seed settling in these locations are likely to die from natural attrition in the absence of suitable habitat. This lobster seed population might accurately be described as a sink population, as it is disconnected from the reproductive stock that it arose from (Dao et al. 2015). On this basis, the resource can be exploited to supply seed for farming, with no significant impact on adult lobster populations. Although the study of Milton et al. (2014) suggested a high degree of local recruitment in the south-central coast of Java, this is not inconsistent with the hypothesis of a sink, as their modelling also suggested more than 25% of recruitment to their study location was sourced remotely. It is noteworthy that Milton et al. (2014) recorded very few *P. ornatus* within the commercial catch, and yet *P. ornatus* pueruli are a significant component of the seed settlement in southern Java (B. Priyambodo, Indonesian Ministry of Marine Affairs and Fisheries, pers. comm). Presumably the *P. ornatus* pueruli have originated elsewhere and are transported to this location through the Indonesian Through flow model as suggested.

The puerulus resource (of *P. homarus* and *P. ornatus*) in south central Indonesia is particularly abundant, significantly greater than other coastal areas of Indonesia and this can be explained by a concentration of pueruli generated by the power and volume of the Indonesian Through flow. There is very limited suitable adult habitat in the region where this puerulus population occurs, and on balance there is a high probability that the bulk of these pueruli perish. Exploitation of them for the purposes of aquaculture is likely to be sustainable.

12.5 Growout

On-growing of small tropical rock lobsters to a size suitable for marketing was first developed to a substantial commercial level in Vietnam. The history of lobster growout in Vietnam is summarised in Table 12.1, after that reported by Anh and Jones (2015a). Beyond Vietnam, there has been interest throughout the Southeast Asian region in developing rock lobster aquaculture, but no significant production. It would appear that the primary constraint is the supply of seed lobsters. Although a formal assessment of lobster seed resources in most countries has not been made, anecdotal evidence suggests that abundant seed resources, suitable for routine fishing, are not widespread. In the case of Indonesia, however, recent development of seed fishing and scientific assessment have confirmed a high abundance of seed in parts of Indonesia that can support significant growout. Although the opportunity is very strong, Indonesia has not yet established a growout sector, and the small growout industry that began to develop around 2010–2014 has now been largely abandoned, primarily because of new fisheries regulations that prohibit the catch of lobsters less than 200 g.

12.5.1 Growout Phases

Growout of lobster seed to market size is performed in two distinct phases, an initial nursery phase from puerulus to around 3 grams or larger, and then a growout phase through to a size suitable for marketing to consumers. In Vietnam this phased approach is quite distinct, and, in many cases, there are nursery farmers who only perform the nursery phase, on-selling the juveniles to growout farmers who perform the on-growing to market size. In Indonesia, where currently there is minimal farming to market size, the distinction between the nursery and growout phases has been less evident. As the Indonesian lobster farming industry develops and matures, it is expected that the same phased approach will emerge, reflecting the distinct systems and methods used for each to achieve high survival and growth rate.

12.5.2 Nursery Systems

Because the seed lobsters are very small, around 12 mm in body length and < 5 mm in body diameter, the cages used to house and nurture them are relatively small with small mesh size. In Vietnam there has been considerable development of cage type and management regime to achieve the greatest survival and growth rate of juveniles. Currently, there are two methods that represent the bulk of juvenile production in Vietnam. The first is a submerged cage, typically rectangular that is positioned on the sea floor in relatively shallow water less than 5 m. The cage is box shaped, with a frame of 10–15 mm diameter steel rod covered in shade-cloth-type mesh with a mesh size of less than 3 mm (Photo 12.5). Cage size varies from 1 m × 1 m × 1 m high to 3 m × 2 m × 2 m high. The box cage is equipped with a PVC pipe feeding tube, 100 mm in diameter and long enough to extend to the

Photo 12.5 Square submerged nursery cage in Vietnam

surface (Photo 12.5). Seed lobsters are placed in the cage at a density of between 50 and 100 per m^2, and the cage is left on the sea floor for the duration of the nursing phase, typically 6–16 weeks. Fresh food is provided daily. At harvest, the cage is lifted to the sea surface to a boat and the juvenile lobsters removed through a drawstring opening in the mesh. The second and more recently developed method consists of cages suspended in deeper water, typically greater than 10 m depth of water with the cage suspended at around 5 m below the surface. These cages are typically round with 1 m diameter and 0.8 m height, their shape maintained via a circle of 10–15 mm steel rod at top and bottom (Photo 12.6). There is a central opening in the centre of the top surface that is operated with a draw string arrangement. These cages are typically stocked at 100 pueruli per m^2, and are lifted to the surface daily to check the juveniles, clean uneaten food and waste and provide fresh food. This method appears to support better survival and growth of juveniles than the submerged system. Food for both systems consists of fresh, finely chopped fresh fish, crustacean and mollusc, with each nursery farmer following a preferred combination. It is apparent that crustacean and mollusc food supports superior production to fish, but this must be balanced against the higher cost of these materials. The fresh food materials tend to be purchased daily from local fish markets.

Survival and growth of juveniles through the nursery phase is highly variable, and the best operators can achieve greater than 90% survival and a harvest size of around 3–5 g after 6 weeks, 10–30 g after 12 weeks and 30–50 g after 16 weeks.

Photo 12.6 Round suspended nursery cage in Vietnam

12.5.3 Growout Systems

The initial production systems of Vietnam consisted of very simple fixed cages in shallow water, less than 3 m depth. Rectangular box nets were supported within an outer frame made of rough-hewn timber or bamboo, 10–15 cm diameter and 4–5 m in length, fixed to vertical posts embedded into the sediment. Each cage typically had a mesh cover to provide some shading, and its base was either resting on or suspended above the sea floor. Cages placed on the seabed had a layer of sand across the floor, while those fixed off the bottom had a gap of about 0.5 m from the seabed (Photo 12.7).

Issues with fouling and build-up of uneaten food and waste forced farmers to move their cages to deeper water where they could be held above the seafloor. Such sites were more exposed to wave and wind action, and materials for the cage frames and cages themselves became more robust including milled timbers, steel fixings and stronger netting. The staked cages soon gave way to floating pontoon structures suitable for even deeper water, anchored to the seafloor with moorings of steel or concrete. The profitability of the industry allowed farmers to invest in bigger cages made from more durable materials so that by the mid-2000s, floating farms were often as sophisticated as any in the world (Photo 12.8) (Hung and Tuan 2009).

Currently the typical Vietnamese lobster growout farm consists of 30–60 floating cages, suspended from a floating frame in an area protected from prevailing wind and wave action, adjacent to land (often an island) in water depth of 10–30 m. Cage size varies from farm to farm and depending on the size of lobsters stocked, but on average the cages are around 3 m × 3 m × 4 m deep. More details on farm specification including cage number and type are presented in Petersen and Phuong (2010).

Photo 12.7 Early iteration of fixed growout cage in Vietnam

Photo 12.8 Typical Vietnamese floating lobster farm

Although manufactured diets have been formulated for lobsters, there has been little uptake to date, and food for the lobsters consists of fresh sea food including shrimp, crab, fish and molluscs. Smaller lobsters tend to be fed twice per day while larger lobsters are fed once per day. There is daily cleaning of the cages of excessive bio-fouling, uneaten food and moulted lobster shells. Periodically, the cage net is removed and replaced by a clean net, and the used net is cleaned with a high-pressure hose. Grading of lobsters is applied to minimise the size variation within a cage, and as such growout is often characterised by 3 phases between gradings, from stocking at 50 g to 200 g, 200 g to 700 g and 700 g to 1 kg.

12.5.4 Economics

An economic analysis of lobster farming in Vietnam was performed by Petersen and Phuong (2010), providing a detailed description of costs and benefits and indicating a highly profitable industry. In the intervening years since that analysis, some costs have increased but these have been outpaced by increases in demand and price paid for farmed lobsters, and consequently, profitability has increased. In contrast, an equivalent economic analysis of lobster aquaculture in Indonesia was performed in 2013 indicating at that time a marginally viable business. The contrasting profitability may be attributed to the differences in experience and knowledge, with Vietnam having a significantly longer history and Indonesian farmers with relatively limited knowledge of best practice farming methods.

Indonesia's developing lobster aquaculture industry is advantaged by having access to the Vietnam experiences, and through Australian-based research projects (Jones 2015c; Williams 2009), the transfer and adaptation of Vietnamese lobster farming technology has been implemented. Nevertheless, a number of factors have negatively impacted Indonesia's nascent lobster farming industry. In the 5 years to 2009, no more than 50 tonnes of lobsters was produced each year due to limited knowledge of effective farming practices and nutritionally deficient feed. Due to greater abundance, the primary species farmed was *P. homarus*, and it was typically on-grown for less than 12 months to a mean size of 100–200 g, fetching a farm-gate price of around 350,000 Indonesian rupiah per kg, equivalent to less than $US30. Although there was a concerted effort to educate the farmers involved, their personal financial circumstances and risk-averse nature resulted in inefficient (harvest size too small) and ineffective (use of cheaper, nutritionally deficient feeds) practices (Petersen et al. 2013, 2015). Coincidentally, the demand from Vietnam for Indonesian lobster seed began to develop, and prices began to increase. The Indonesian lobster farmers, who were often also the fishers for the seed, quickly surmised that catching and selling the seed was a better alternative to on-growing. The risk was low, cash flow was immediate and the capital outlay was very small. Consequently, the emerging lobster growout industry, which was effectively restricted to Lombok alone, diminished as more and more of those involved focussed on seed catch only. By 2013 there were very few farmers raising lobsters, while the seed fishing sector had expanded considerably. From 2013 to 2015, lobster seed fishing increased in efficiency (with application of lights at night to attract swimming pueruli) (Bahrawi et al. 2015a) and production, as it spread from village to village, westwards to Bali and Java and east to Sumbawa. In late 2015 a new national fishery regulation was introduced, applying a minimum legal size for lobster (all species) of 200 g. This regulation effectively prohibited the taking of seed and marked the final demise of lobster growout in Indonesia.

12.5.5 Health and Disease

Health issues and diseases of tropical lobsters are primarily the result of opportunistic infection and physiological degradation rather than from primary pathogens. Those most often encountered in Vietnam lobster farm environments are all preventable and, with best practice husbandry and nutrition, can be minimised or avoided. As such, best practice farming should comprise provision of an environment and nutrition that maintains lobsters in optimal condition such that their susceptibility is as low as possible.

The health conditions and diseases of lobsters recorded in Vietnam are summarised in Table 12.2.

The diseases and health issues of farmed lobsters have collectively caused substantial losses to lobster production each year. In Vietnam significant additional loss occurred specifically from milky disease in 2007 and in 2008 when production losses of 31–71% were experienced across all provinces. The average across the entire industry was a 50% reduction in production due to milky disease in 2007–2008, valued at US$50 million loss and more than 5000 households affected. Prior to 2002, survival rate through the growout phase was 70%, and by 2008, this was less than 50% due to disease. By 2011, overall survival was back to 70% as a result of disease prevention measures. However, in 2012 a further milky disease outbreak had occurred, although with a much lower impact than 2007–2008. In recent years, milky disease has continued to be the most important in lobster farming, although its impact has been relatively small as farmers have improved their surveillance and management of lobster health.

Lobster disease in the Indonesian lobster farming industry has been a significant issue even though overall production has been quite small. Milky disease was confirmed in lobsters in the village of Telong Elong in eastern Lombok in 2012, with the same Rickettsia-like bacteria involved. This suggests that the milky disease

Table 12.2 Health issues and diseases impacting farmed lobsters in Vietnam and Indonesia

Name	Symptoms	Cause
Red body disease	Red colouration initially to the carapace and/or abdomen, which ultimately affects the whole body, and necrotic tissue in the hepatopancreas	Bacteria *Vibrio alginolyticus*
Black gill disease	Gradual darkening in colour of the gill filaments from brown to black, followed by progressive tissue breakdown and shedding of gill filaments	Fungus *Fusarium* spp.
Milky haemolymph disease	Abdominal tissue colour turns from translucent to opaque white, as the cells become engorged with the pathogen	Rickettsia-like bacterium
Big head syndrome	Abnormally big carapace relative to the abdomen, and affected lobsters appear to have their growth retarded, and experience difficulty in moulting	Nutrient deficiency
Separate head syndrome	Separation of the carapace from the abdomen, caused by excess body fluid which appears under the epidermis at the junction of the head and tail	Exposure to low salinity (<25 ppt)

agent, the Rickettsia-like bacteria, are endemic, and that milky disease is an ongoing threat to lobsters in compromised condition due to poor nutrition and environmental stress. Although milky disease-infected lobsters respond positively to antibiotic treatments, prevention is key to the long-term sustainability of the industry.

There are a number of preventative measures that can be taken to reduce disease susceptibility and, when integrated, are likely to reduce the risk of losses from disease and health issues to a negligible level. Firstly, site selection is important, with a recommended depth of greater than 2.5 m from cage bottom to sea floor and a location with good water movement from tide and currents that enables constant flushing of the cages with clean water. Density of farming (i.e. the distance between farms and spacing of cages within each farm) is important to minimise the ratio of lobster biomass to volume of water. Cage design and mesh size are important to maximise the flow of water through the cage.

It is likely that the use of fresh seafood (fish, crustacean and mollusc) as lobster food is a major contributing factor to poor condition of lobsters and increased susceptibility to disease. Pelleted feeds will provide a much cleaner and nutritionally complete diet, and increase disease resistance.

Effective maintenance of cages is beneficial to health management, including daily cleaning of waste and uneaten food from the cages, and periodic exchange of cage nets, including sun drying. Farm workers should be well trained in disease symptom awareness and observation, with corresponding skills on how best to respond.

12.6 Markets for Farmed Lobster

As spiny lobsters are a premium seafood, their aquaculture is not driven by food security or provision of protein, as much of global aquaculture is. Although Vietnam and Indonesia have food security issues, their interest in lobster aquaculture is principally about wealth creation, particularly for impoverished communities (Hambrey et al. 2001; Pahlevi 2009). The attraction of lobster farming is that it involves simple technology and moderate capital input and produces a very high-value product (Petersen and Phuong 2010). Markets for lobster around the world are generally characterised by unmet demand (Hart 2009; Jones 2015a), and the Chinese market for both *P. ornatus* and *P. homarus* is strong and price per kilogram paid by wholesalers is very high. For both Vietnam and Indonesia, these two lobster species represent the bulk of the naturally settling puerulus, so marketable lobsters produced from on-growing these seed can sustain viable businesses. Nevertheless, the farmed product must meet market specifications. Unpublished market intelligence suggests the farmed *P. homarus* product from Lombok attracted a lower price than wild-caught product of the same species. This is attributed to pale shell colouration of the farmed lobster and poor vigour, which makes live transport difficult and the product unattractive to the customer. Fortunately both problems can be rectified in aquaculture through improved nutrition and husbandry, and ultimately farmed product may attract a premium because of consistency in supply and quality (Hart 2009).

12.7 Summary

In the absence of a hatchery supply of seed lobsters, rock lobster aquaculture is reliant on naturally settling pueruli as a source of seed. In Vietnam, and more recently in Indonesia, the natural supply of such seed is sufficiently abundant to support industrial-scale aquaculture. Despite valid concerns that using a wild supply may not be sustainable, particularly in regard to possible impacts on existing adult lobster populations, the source of lobster seed in both countries appears to be sink populations, whose abundance is unrelated to adult populations. The lobster seeds, although very small and delicate, are sufficiently robust to be transported from catching locations to areas suitable for growout and easily adapt to high-density culture in sea cage systems. As naturally social species, both *P. ornatus* and *P. homarus* – the two prominent species widely available as seed, have proven to be excellent candidates for farming, and their high value in various Asian markets provides strong economic basis for commercial production. Relatively poor coastal communities, who have engaged in lobster farming in both Vietnam and Indonesia, have gained great economic and social benefit from this enterprise. While the Vietnam lobster aquaculture industry appears to have matured and is now relatively stable, that of Indonesia is at an emerging stage, and has the capacity to grow to a very large industry with potential production likely to exceed the fishery supply. As with all intensive aquaculture production, lobster farming faces several challenges, particularly in improving survival of the early stages and with health and disease impacting the growout stage. These challenges will be significantly mitigated with the uptake of manufactured diets that will improve survival, growth rates and disease resistance, and provide benefits to the environmental impact of farming through reduced nutrient leaching. Vietnam and Indonesia are likely to remain the major producers of farmed lobster over the next decade due to their established seed supply. Nevertheless, other countries in the Asian region may also engage in lobster farming if additional, local seed resources are identified. It seems likely that there will be other locations in Asia where the confluence of conditions supports heavy settlement of lobster pueruli, sufficient to enable a sustainable supply for farming. Other countries will benefit from the farming technology developed and perfected in Vietnam and Indonesia.

For Indonesia, there will need to be policy revision that supports lobster aquaculture using the natural supply of seed. There is a growing body of knowledge about the seed resource in Indonesia that will enable informed decisions on policy adjustments that allow reasonable levels of seed fishing to support lobster farming. At present, the minimum legal size (200 g) for all species of rock lobster remains active in Indonesia. The fishing for lobster seed is therefore illegal, although significant fishing continues, driven by the high price paid for the seed by exporters sending them to Vietnam. Although there are regular seizures of seed shipments and apprehension of the dealers who are subjected to heavy penalties (fines and incarceration), the poor coastal villagers involved in the fishing continue to fish in the face of no alternative livelihood. Based on the contentions outlined above, that the seed resource of southern Java, Lombok and Sumbawa represents a sink population, a revision of the fisheries policy to support limited lobster seed fishing is justified. This policy should be accompanied with an export ban on the seed captured and a requirement that they be used for on-growing within Indonesia to expand the social and economic benefits domestically. To maximise both the sustainability of lobster resources and the socio-economic benefit to the coastal communities with seed fishing opportunity, seasonal limitations could be applied to Indonesia that would allow a proportion of the entire seed resource (of 100 million) to be captured. If this were to represent, for example, ten million seed captured per year, it would be sufficient, based on the productivity of Vietnam, to produce in excess of 2000 tonnes of on-grown, marketable product, with a value likely to exceed $US150 million. Given the unusually high abundance of lobster seed in this part of Indonesia, managed exploitation of the resource is unlikely to have any impact on adult populations.

Acknowledgements The authors acknowledge the support of colleagues and associates involved in a series of research projects funded by the Australian Government through the Australian Centre for International Agricultural Research (ACIAR) in providing data, insights and advice.

References

Addison, J. T., & Bannister, R. C. A. (1994). Re-stocking and enhancement of Clawed Lobster n stocks: A review. *Crustaceana, 67*, 2131–2155.

Aiken, D. E. (1988). Lobster farming: Fantasy or opportunity. In Anon (Ed.), *Proceedings of the aquaculture international congress and exposition, September 6–9, 1988, Vancouver, Canada* (pp. 575–582).

Aiken, D. E., & Waddy, S. L. (1995). Aquaculture. In J. Factor (Ed.), *Biology of the Lobster Homarus americanus* (pp. 153–175). New York: Academic Press.

Anh, T. L., & Jones, C. (2015a). Status report of Vietnam lobster grow-out. Chapter 4.2. In C. M. Jones (Ed.), *Spiny lobster aquaculture development in Indonesia, Vietnam and Australia. Proceedings of the International Lobster Aquaculture Symposium held in Lombok, Indonesia, 22–25 April 2014* (pp. 82–86). Australian Centre for International Agricultural Research, Canberra.

Anh, T. L., & Jones, C. (2015b). Lobster seed fishing, handling and transport in Vietnam. Chapter 2.4. In C. M. Jones (Ed.), *Spiny lobster aquaculture development in Indonesia, Vietnam and Australia. Proceedings of the International Lobster Aquaculture Symposium held in Lombok, Indonesia, 22–25 April 2014* (pp. 31–35). Australian Centre for International Agricultural Research, Canberra.

Arcenal, J. M. M. (2004). Sustainable farming of spiny lobster in Western Mindanao, Philippines. In K. C. Williams (Ed.), *Spiny lobster ecology and exploitation in the South China sea region. Proceedings of a workshop held at the Institute of Oceanography, Nha Trang, Vietnam, July 2004. ACIAR Proceedings No. 120* (pp. 19–20). Australian Centre for International Agricultural Research, Canberra.

Bahrawi, S., Priyambodo, B., & Jones, C. (2015a). Assessment and development of the lobster seed fishery of Indonesia. Chapter 2.3. In C. M. Jones (Ed.), *Spiny lobster aquaculture development in Indonesia, Vietnam and Australia. Proceedings of the international lobster aquaculture symposium held in Lombok, Indonesia, 22–25 April 2014* (pp. 27–30). Canberra: Australian Centre for International Agricultural Research.

Bahrawi, S., Priyambodo, B., & Jones, C. (2015b). Census of the lobster seed fishery of Lombok. Chapter 2.1. In C. M. Jones (Ed.), *Spiny lobster aquaculture development in Indonesia, Vietnam and Australia. Proceedings of the international lobster aquaculture symposium held in Lombok, Indonesia, 22–25 April 2014* (pp. 12–19). Canberra: Australian Centre for International Agricultural Research.

Bahrawi, S., Priyambodo, B., & Jones, C. (2015c). Lobster seed fishing, handling and transport in Indonesia. Chapter 2.5. In C. M. Jones (Ed.), *Spiny lobster aquaculture development in Indonesia, Vietnam and Australia. Proceedings of the international lobster aquaculture symposium held in Lombok, Indonesia, 22–25 April 2014* (pp. 36–38). Canberra: Australian Centre for International Agricultural Research.

Bell, J. D. (2004). Key issues for sustaining aquaculture production of the spiny lobster, *Panulirus ornatus* in Vietnam. In K. C. Williams (Ed.), *Spiny lobster ecology and exploitation in the South China sea region. Proceedings of a workshop held at the Institute of Oceanography, Nha Trang, Vietnam, July 2004. ACIAR Proceedings No. 120* (pp. 59–62). Australian Centre for International Agricultural Research, Canberra.

Booth, J. D., & Tarring, S. C. (1986). Settlement of the red rock lobster, *Jasus edwardsii*, near Gisborne, New Zealand. *New Zealand Journal of Marine and Freshwater Research., 20*, 291–297.

Dao, H. T., & Jones, C. (2015). Census of the lobster seed fishery of Vietnam. Chapter 2.2. In C. M. Jones (Ed.), *Spiny lobster aquaculture development in Indonesia, Vietnam and Australia. Proceedings of the International Lobster Aquaculture Symposium held in Lombok, Indonesia, 22–25 April 2014. ACIAR Proceedings 145* (pp. 21–26). Canberra: Australian Centre for International Agricultural Research.

Dao, H. T., Smith-Keune, C., Wolanski, E., Jones, C. M., & Jerry, D. R. (2015). Oceanographic currents and local ecological knowledge indicate, and genetics does not refute, a contemporary pattern of larval dispersal for the ornate spiny lobster, *Panulirus ornatus* in the South-East Asian archipelago. *PLoS One, 10*, e0124568.

Groeneveld, J. C., Goni, R., & Diaz, D. (2013). Chapter 11. *Panulirus* species. In B. F. Phillips (Ed.), *Lobsters biology, management, aquaculture and fisheries* (pp. 326–356). Oxford: Wiley.

Hambrey, J., Tuan, L., & Thuong, T. (2001). Aquaculture and poverty alleviation II. Cage culture in coastal waters of Vietnam. *World Aquaculture*, 34–38.

Hart, G. (2009). *Assessing the South-East Asian tropical lobster supply and major market demands. ACIAR Final Report (FR-2009-06)*. Canberra: Australian Centre for International Agricultural Research.

Hung, L. V., & Tuan, L. A. (2009). Lobster seacage culture in Vietnam. In K. C. Williams (Ed.), *Spiny lobster aquaculture in the Asia-Pacific region. Proceedings of an international symposium held at Nha Trang, Vietnam, 9–10 December, 2008. ACIAR Proceedings 132* (pp. 10–17). Canberra: Australian Centre for International Agricultural Research.

Idris, M., & Bahrawi, S. (2015). Assessment of tropical spiny lobster aquaculture development in South Sulawesi, Indonesia. Chapter 5.11. In C. M. Jones (Ed.), *Spiny lobster aquaculture development in Indonesia, Vietnam and Australia. . Proceedings of the international lobster aquaculture symposium held in Lombok, Indonesia, 22–25 April 2014* (pp. 148–149). Canberra: Australian Centre for International Agricultural Research.

Jeffs, A., & Davis, M. (2003). An assessment of the aquaculture potential of the Caribbean spiny lobster, *Panulirus argus*. *Proceedings of the Gulf of Carribean Fisheries Institute, 54*, 413–426.

Jeffs, A., & Hooker, S. (2000). Economic feasibility of aquaculture of spiny lobsters *Jasus edwardsii* in temperate waters. *Journal of the World Aquaculture Society., 31*, 30–41.

Jones, C. M. (2009). Advances in the culture of lobsters. In G. Burnell & G. L. Allan (Eds.), *New technologies in aquaculture: Improving production efficiency, quality and environmental management* (pp. 822–844). Cambridge: Woodhead Publishing Ltd/CRC Press.

Jones, C. M. (2010). Tropical rock lobster aquaculture development in Vietnam, Indonesia and Australia. *Journal of the Marine Biological Association of India, 52*, 304–315.

Jones, C. (2015a). Market perspective on farmed tropical spiny lobster. Chapter 5.9. In C. M. Jones (Ed.), *Spiny lobster aquaculture development in Indonesia, Vietnam and Australia. Proceedings of the international lobster aquaculture symposium held in Lombok, Indonesia, 22–25 April 2014* (pp. 142–144). Canberra: Australian Centre for International Agricultural Research.

Jones, C. (2015b). Summary of disease status affecting tropical spiny lobster aquaculture in Vietnam and Indonesia. Chapter 5.4. In C. M. Jones (Ed.), *Spiny lobster aquaculture development in Indonesia, Vietnam and Australia. Proceedings of the international lobster aquaculture symposium held in Lombok, Indonesia, 22–25 April 2014* (pp. 111–113). Canberra: Australian Centre for International Agricultural Research.

Jones, C. M. (2015c). *Spiny lobster aquaculture development in Indonesia, Vietnam and Australia. Proceedings of the international lobster aquaculture symposium held in Lombok, Indonesia, 22–25 April 2014*. Australian Centre for International Agricultural Research (ACIAR), Canberra, 163 plus appendices pp.

Jones, C. M., Long, N. V., Hoc, D. T., & Priyambodo, B. (2010). Exploitation of puerulus settlement for the development of tropical rock lobster aquaculture in the Indo-West Pacific. *Journal of the Marine Biological Association of India, 52*, 292–303.

Juinio-Menez, M. A., & Gotanco, R. R. (2004). Status of spiny lobster resources of the Philippines. In K. C. Williams (Ed.), *Spiny lobster ecology and exploitation in the South China sea region. Proceedings of a workshop held at the Institute of Oceanography, Nha Trang, Vietnam, July 2004. ACIAR Proceedings No. 120* (pp. 3–6). Canberra: Australian Centre for International Agricultural Research.

Kittaka, J., & Booth, J. D. (1994). Prospectus for aquaculture. In B. F. Phillips, J. S. Cobb, & J. Kittaka (Eds.), *Spiny lobster management* (pp. 365–373). Oxford: Blackwell Scientific Publications.

Kittaka, J., & Booth, J. D. (2000). Prospectus for aquaculture. In B. Phillips & J. Kittaka (Eds.), *Spiny lobster: Fisheries and culture* (pp. 465–473). Oxford: Blackwell Science Ltd.

Lipcius, R. N., Stockhausen, W. T., Eggleston, D. B., Marshall, L. S., Jr., & Hickey, B. (1997). Hydrodynamic decoupling of recruitment, habitat quality and adult abundance in the Caribbean spiny lobster: Source-sink dynamics? *Marine & Freshwater Research, 48*, 807–815.

Long, N. V., & Hoc, D. T. (2009). Census of lobster seed captured from the central coastal waters of Vietnam for aquaculture grow-out, 2005-2008. In K. C. Williams (Ed.), *Spiny lobster aquaculture in the Asia-Pacific region. Proceedings of an international symposium held at Nha Trang, Vietnam, 9–10 December, 2008. ACIAR Proceedings 132* (pp. 52–58). Canberra: Australian Centre for International Agricultural Research.

Lucas, J. S., & Southgate, P. C. (2012). *Aquaculture: Farming aquatic animals and plants* (2nd ed., p. 648). Oxford: Wiley Blackwell.

Milton, D. A., Satria, F., Proctor, C. H., Prasetyo, A. P., Utama, A. A., & Fauzi, M. (2014). Environmental factors influencing the recruitment and catch of tropical Panulirus lobsters in southern Java, Indonesia. *Continental Shelf Research, 91*, 247–255.

Naylor, R. L., Goldburg, R. J., Primavera, J. H., Kautsky, N., Beveridge, M. C. M., Clay, J., Folke, C., Lubchenco, J., Mooney, H., & Troell, M. (2000). Effect of aquaculture on world fish supplies. *Nature, 405*, 1017.

Pahlevi, R. S. (2009). Potential for co-management of lobster seacage culture: A case study in Lombok, Indonesia. In K. C. Williams (Ed.), *Spiny lobster aquaculture in the Asia-Pacific region. Proceedings of an international symposium held at Nha Trang, Vietnam, 9–10 December, 2008. ACIAR Proceedings 132* (p. 26). Canberra: Australian Centre for International Agricultural Research.

Petersen, E. H., & Phuong, T. H. (2010). Tropical spiny lobster (*Panulirus ornatus*) farming in Vietnam – bioeconomics and perceived constraints to development. *Aquaculture Research, 41*, 634–642.

Petersen, E. H., Jones, C., & Priyambodo, B. (2013). Bioeconomics of spiny lobster farming in Indonesia. *Asian Journal of Agriculture and Development, 10*, 25–39.

Petersen, L., Jones, C., & Priyambodo, B. (2015). Bio-economics of spiny lobster farming in Indonesia. Chapter 4.5. In C. M. Jones (Ed.), *Spiny lobster aquaculture development in Indonesia, Vietnam and Australia. Proceedings of the international lobster aquaculture symposium held in Lombok, Indonesia, 22–25 April 2014* (pp. 92–96). Canberra: Australian Centre for International Agricultural Research.

Phillips, B. F. (2013). *Lobsters: Biology, management, aquaculture and fisheries* (2nd ed., p. 474). Oxford: Wiley- Blackwell.

Phillips, B. F., & Booth, J. D. (1994). Design, use, and effectiveness of collectors for catching the puerulus stage of spiny lobsters. *Reviews in Fisheries Science, 2*, 255–289.

Plagányi, É. E., McGarvey, R., Gardner, C., Caputi, N., Dennis, D., de Lestang, S., Hartmann, K., Liggins, G., Linnane, A., Ingrid, E., Arlidge, B., Green, B., & Villanueva, C. (2017). Overview, opportunities and outlook for Australian spiny lobster fisheries. *Reviews in Fish Biology and Fisheries, 28*(1), 57–87.

Priyambodo, B. (2015). Study tour of Indonesian farmers to Vietnam lobster aquaculture industry in 2013. Chapter 5.8. In C. M. Jones (Ed.), *Spiny lobster aquaculture development in Indonesia, Vietnam and Australia. . Proceedings of the International Lobster Aquaculture Symposium held in Lombok, Indonesia, 22–25 April 2014* (pp. 136–141). Canberra: Australian Centre for International Agricultural Research.

Priyambodo, B., & Bahrawi, S. (2015). *Puerulus assessment and survey in Banyuwangi East Java. Trip report 30 July to 2 August 2015., ACIAR FIS/2014/059 Expanding spiny lobster farming in Indonesia*. Lombok: Marine Aquaculture Development Centre.

Priyambodo, B., & Jaya, S. (2009). Lobster aquaculture in Eastern Indonesia. Part 1. Methods evolve for fledgling industry, *Global Aquaculture Advocate* (pp. 36–40). St Louis: Global Aquaculture Alliance.

Priyambodo, B., & Sarifin. (2009). Lobster aquaculture industry in eastern Indonesia: Present status and prospects. In K. C. Williams (Ed.), *Spiny lobster aquaculture in the Asia-Pacific region. Proceedings of an international symposium held at Nha Trang, Vietnam, 9–10 December, 2008. Proceedings No. 132* (pp. 36–45). Canberra: Australian Centre for International Agricultural Research.

Priyambodo, B., Jones, C., & Sammut, J. (2015). The effect of trap type and water depth on puerulus settlement in the spiny lobster aquaculture industry in Indonesia. *Aquaculture, 442*, 132–137.

Priyambodo, B., Jones, C. M., & Sammut, J. (2017). Improved collector design for the capture of tropical spiny lobster, *Panulirus homarus* and *P. ornatus* (Decapoda: Palinuridae), pueruli in Lombok, Indonesia. *Aquaculture, 479*, 321–332.

Smith, G. (2017). *A dream soon to become a reality? Sustainable farming of lobsters, international aquafeed* (pp. 34–36). Cheltenham: Perendale Publishers Ltd.

Thuy, N. T. B., & Ngoc, N. T. B. (2004). Current status and exploitation of wild spiny lobsters in Vietnamese waters. In K. C. Williams (Ed.), *Spiny lobster ecology and exploitation in the South China Sea region. Proceedings of a workshop held at the Institute of Oceanography, Nha Trang, Vietnam, July 2004. ACIAR Proceedings No. 120* (pp. 13–16). Canberra: Australian Centre for International Agricultural Research.

Tlusty, M. (2004). Refocusing the American lobster (*Homarus americanus*) stock enhancement program. *Journal of Shellfish Research, 23*, 313–314.

Vijayakumaran, M., Venkatesan, R., Murugan, T. S., Kumar, T. S., Jha, D. K., Remany, M. C., Thilakam, J. M. L., Jahan, S. S., Dharani, G., Kathiroli, S., & Selvan, K. (2009). Farming of spiny lobsters in sea cages in India. *New Zealand Journal of Marine and Freshwater Research., 43*, 623–634.

Wahle, R. A., Castro, K. M., Tully, O., & Cobb, J. S. (2013). Chapter 8 *Homarus*. In B. F. Phillips (Ed.), *Lobsters: Biology, management, aquaculture and fisheries* (2nd ed., pp. 221–258). Chichester: Wiley.

Wickins, J. F., & Lee, D. O. C. (2002). *Crustacean farming. Ranching and culture* (2nd ed.). Oxford: Blackwell Scientific Publications.

Williams, K. C. (2009). *Spiny lobster aquaculture in the Asia-Pacific region. Proceedings of an international symposium held at Nha Trang, Vietnam, 9–10 December, 2008*. Canberra: Australian Centre for International Agricultural Research. 162 pp.

Health Management in Lobster Aquaculture

13

E. V. Radhakrishnan and Joe K. Kizhakudan

Abstract

This chapter discusses different diseases and pathogens that lobsters are susceptible to, in the wild and in rearing systems. Although reports of disease outbreaks in lobsters are scarce, there are several known organisms that cause pathogenicity in lobsters, particularly under stressful conditions. Lobsters held in captive conditions are more prone to attack by pathogens and parasites, with known susceptibility in the larval phase. Among the known diseases in lobsters are viral diseases like *Panulirus argus* virus 1 (PaV1) and White Spot Syndrome Virus (WSSV), bacterial diseases like Gaffkaemia, shell disease, Vibriosis, red-body disease, tail necrosis and Milky White Disease Syndrome and fungal infections like Oomycetes, Burnspot disease and Lagenidium disease. Also described in this chapter are dinoflagellate blood disease, paramoebiasis, infections caused by microsporidians and several other invertebrate parasites like copepods. Egg-bearing lobsters are also prone to predation by Carcinonemertean worms which feed on the eggs. Epibiont fouling and ciliate diseases are also a major concern in lobster holding systems. This chapter highlights some of the remedial and prophylactic measures to prevent lobster diseases and infestation that are commonly reported in holding systems.

Keywords

Pathogens · Gaffkaemia · Vibriosis · Prophylactic measures

E. V. Radhakrishnan (✉) · J. K. Kizhakudan
ICAR-Central Marine Fisheries Research Institute, Cochin, Kerala, India
e-mail: evrkrishnan@gmail.com

© Springer Nature Singapore Pte Ltd. 2019
E. V. Radhakrishnan et al. (eds.), *Lobsters: Biology, Fisheries and Aquaculture*,
https://doi.org/10.1007/978-981-32-9094-5_13

13.1 Introduction

Lobsters like other decapods are also susceptible to diseases when they are under stress. Though the reported pathogens and parasites in lobsters are surprisingly few, they do have adverse fauna composed of pathogenic virus, several bacteria, protozoans, helminths and even symbiotic crustaceans (Shields 2011). Lobsters held in aquaculture systems are more prone to bacterial and fungal infections, especially in poor environmental conditions; however, only one pathogen, *Panulirus argus* virus 1 (PaV1), is thought to have damaged a fishery for a spiny lobster in the wild (Shields and Behringer 2004). Several reviews and syntheses of crustacean diseases are available (Couch 1983; Johnson 1983; Overstreet 1983; Brock and Lightner 1990; Sindermann 1990; Shields and Overstreet 2007), but those exclusively dealing with lobster diseases are by Stewart (1980) and Shields (2011). The Aquatic Science Research Unit of the Curtin University of Technology, Australia has published a Guide for the benefit of the lucrative lobster industry, to deal with health and disease management in rock lobster (Stephens et al. 2003).

An animal is presumed to be 'healthy' when the physiological processes controlling growth, maintenance, defence against disease and reproduction are functioning normally (Stephens et al. 2003). In a 'diseased' condition, the body functions abnormally and the animal responds to the infection by eliciting the immune system to counteract the threat. Disease-causing agents are mostly opportunistic pathogens, whereas non-infectious diseases are caused by nutritional deficiencies, genetic disorders, immune diseases and exposure to toxins. Lobsters fight diseases through a number of processes, by changing the levels of circulating haemocytes, aggregation of haemocytes around the foreign body, clotting of blood, phagocytosis and antibacterial activity in the blood. The hard exoskeleton of the lobster prevents entry of the infectious organisms to a great extent. However, the pathogen can enter the body through an injury on the soft lower abdomen.

13.2 Homeostasis

Lobsters are exposed to fluctuating environmental conditions such as temperature, pressure, light, oxygen levels and food intake. The internal environment of the lobster responds to these changes by initiating various physiological processes called homeostasis. If the external changes are extreme, the physiological parameter initiates a stress reaction counteracting the internal environmental damage, which may lead to cell damage. If the cell damage is too severe, this may lead to organ failure and death of the lobster. If the cell damage is minimal, the damaged cells cease to function at an optimal level and the lobster slips into poor health and becomes susceptible to diseases and finally death, if it is exposed to further stress.

No disease outbreaks have been reported from wild lobster fisheries in India. However, following injury or stress, many opportunistic pathogens such as bacteria and fungi present in the natural environment of lobster can invade them and cause disease. Further, when lobsters are not carefully handled during postharvest, it is likely that they get injured and contract disease. Disease is encountered in lobsters kept in crowded condition in holding tanks with poor quality water. Holding tank water with high ammonia and low dissolved oxygen contains pathogenic bacteria, which may infect already stressed lobsters. For this reason, health management in lobsters should always consider the factors that pre-dispose lobsters to developing disease.

13.3 Viral Diseases

13.3.1 *Panulirus argus* Virus 1 (PaV1)

PaV1 is the first naturally occurring pathogenic virus found to infect a lobster and is specific to the spiny lobster *P. argus*, as the name implies. This virus has been reported from lobsters caught from the Caribbean Sea including the Florida Keys, US Virgin Islands, Mexico and Belize. Lobsters heavily infected with PaV1 are often lethargic and their haemolymph is milky in colour and does not clot (Shields and Behringer Jr. 2004). PaV1 initially infects the fixed phagocytes in the hepatopancreas. The tubules of the hepatopancreas atrophy in late stages of infection. Overtly diseased lobsters had significantly lower haemolymph protein as a general indicator of health.

13.3.2 White Spot Syndrome Virus (WSSV)

Although spiny lobsters are not naturally infected with the White spot syndrome virus, there is a possibility of this virus infecting spiny lobsters co-inhabiting with infected shrimp stocks. Indeed, at least 42 crustaceans, including three species of spiny lobsters of the genus *Panulirus*, are known to serve as experimental reservoir hosts for WSSV (Supamattaya et al. 1998; Rajendran et al. 1999; Musthaq et al. 2006). The virus has been detected in several species of spiny lobsters (*P. versicolor*, *P. penicillatus*, *P. ornatus* and *P. longipes longipes*) fed with infected shrimp, but none developed the disease. *Panulirus homarus* inoculated with WSSV shrimp tissues died, presumably from the infection (Rajendran et al. 1999; Musthaq et al. 2006). In the latter study, all the lobsters injected with the virus died after 160 days.

It is surprising that no other viruses are known from lobsters. Probably, the paucity of information on viruses in lobsters may be that they are considered for aquaculture only now, and that the focus on lobsters has been mostly on the fisheries (Shields 2011).

13.4 Bacterial diseases

13.4.1 Gaffkaemia

Gaffkaemia or red-tail is a serious disease of clawed lobsters, primarily *Homarus americanus* and *H. gammarus*, caused by a tetrad-forming gram-positive bacteria *Aerococcus viridans*. The variant pathogenic to lobsters is known as *A. viridans homari*. It is not common in the wild population and is often found under holding conditions (Stewart et al. 1966; Stewart 1980; Keith et al. 1992; Lavallee et al. 2001). The infected animals are lethargic, have a reddish exoskeleton, watery pink haemolymph with reduced haemocyte count and delayed clotting. The animals cease to feed and exhibit the *spread eagle syndrome*, spreading the legs and antennae and lying flat at the tank bottom.

The transmission route is through the abdominal wound into the haemolymph of the animal and not through ingestion. The disease is highly infectious and spreads fast in other animals with some injury. The bacterium can remain in the mud for a long time where it can act as a reservoir of infection (Stewart 1984). The presence of a similar bacterium was reported in *Panulirus homarus, P. ornatus* and *P. polyphagus* kept in indoor growout tanks in Chennai, (southeast coast of India), resulting in high mortality (Fig. 13.1). Infected juvenile lobsters collected from holding centres, when subjected to sudden exposure to environmental stress factors such as salinity, were mostly affected (Samraj et al. 2010; Vijayan et al. 2010a). *Panulirus homarus* exposed to sudden reduction in salinity (20 ppt) has significantly lower THC in haemolymph, making them highly susceptible to opportunistic pathogens (Verghese et al. 2007).

Gaffkaemia can be experimentally induced in spiny lobsters but the lethal dose required is much higher than that of *H. americanus*, probably indicating a high degree of resistance to experimental infections in spiny lobsters (Schapiro et al. 1974). Temporary reduction in mortality may be obtained by keeping the lobsters at a lower temperature, which may reduce the proliferation of the bacterium (Vijayan et al. 2010a). Vancomycin when given at high concentrations (25 mg/kg lobster bodyweight) to *H. americanus* during early stages of the infection gave complete protection against gaffkaemia for 15 days (Stewart and Arie 1974). Oral administration of 10 mg terramycin in solution twice daily for 4 days to *H. americanus* showed absence of the bacteria in the haemolymph (Bayer et al. 1983). Further clinical studies on the efficacy of treatment with antibiotics are required for recommending terramycin or other antibiotics as a chemotherapeutic or chemoprophylactic agent for the treatment of gaffkaemia in infected lobsters. Vaccines prepared from virulent

Fig. 13.1 Juvenile *Panulirus homarus* infected by gaffkaemia disease. (Photo by K.K. Vijayan, CIBA, Chennai)

strains of the pathogen were found to confer protection as well as an increase in the phagocytic index against the bacterium (Stewart 1980).

In India, juvenile lobsters for aquaculture are mostly procured from middlemen, holding lobsters in indoor tanks. Juvenile spiny lobsters caught by gillnets are likely to be injured below the soft abdomen while removing the lobster from the net. These lobsters may be given an antibiotic or formalin bath and kept in a less crowded condition in good quality and well-aerated seawater for at least 48 h before they are stocked in cages or in indoor lobster culture tanks. Lobsters kept in overcrowded condition in poor quality seawater are likely to be infected by the opportunistic pathogens present in the holding tanks.

13.4.2 Shell Disease

Shell disease in crustaceans is a progressive chitinolysis and necrosis (bioerosion) condition of the exoskeleton (Rosen 1970). The syndrome first appears as a small pit in the cuticle and progresses into large lesions (Getchell 1989; Noga et al. 1994). For comprehensive reviews on shell disease, refer to Shields et al. (2006) and Shields and Overstreet (2007).

The shell disease is caused by chitinoclastic gram-negative bacteria, including species from the genera *Pseudomonas*, *Vibrio* (*V. vulnificus*, *V. alginolyticus* and *V. parahaemolyticus*), *Beneckea* and *Flavobacteriaceae*. Shell disease is quite uncommon in natural and cultured spiny lobsters in India. However, it is advised to maintain good water quality and reduce crowding of lobsters in culture systems.

13.4.3 Vibriosis

Vibriosis can be caused by any species of *Vibrio*. The pathogenic species of *Vibrio* implicated in disease in spiny lobsters are *V. alginolyticus*, *V. harveyi*, *V. parahaemolyticus* and *V. anguillarum* (Browser et al. 1981; Jawahar et al. 1996). *Vibrio* can cause serious infection in spiny lobsters but only few instances are reported. These bacteria have been encountered in phyllosoma larval cultures causing significant mortalities (Kittaka 1997; Diggles et al. 2000; Radhakrishnan et al. 2009). Infected larvae were opaque and their hepatopancreas atrophied. The source of *Vibrio* infection in phyllosoma larvae is presumed to be *Artemia nauplii* given as feed.

Sand lobster larvae reared in CMFRI laboratory at Kovalam near Chennai (southeast coast of India) showed signs of luminescent bacterial invasions (Kizhakudan, personal observation) (Fig. 13.2). Detailed analysis of isolates collected from the berry and infected animals and larvae using the basic microbiological methods in developing cultures and 16 s RNA techniques indicated that the pathogens were *Vibrio harveyi* from dead larvae and *Micrococcus luteus, Thioclava pacifica, Bacillus niacin, Novosphingobium aromaticivorans, Shimwellia blattae, Shewanella indica* and *Enterobacter aerogenes* from moribund animals. The treated water showed a reduced total count, but the total plate counts increased as the larvae progressed to moulting stages in the rearing medium (Kizhakudan, personal observation).

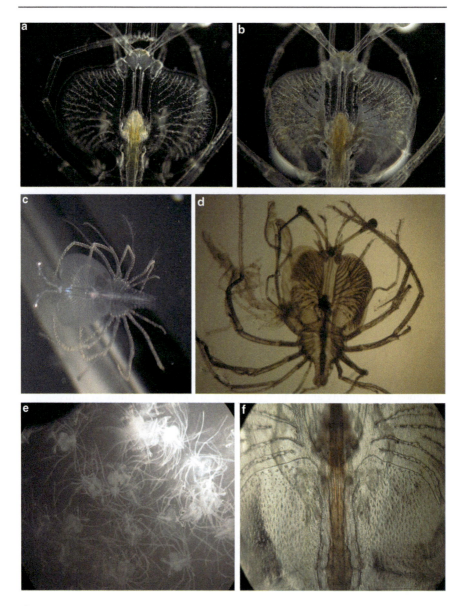

Fig. 13.2 Healthy and *Vibrio*-infected phyllosoma larvae of *Thenus unimaculatus* reared in the laboratory. (**a**). Healthy sand lobster larva (Pi); (**b**). Live Pi phyllosoma *Vibrio* invasion begins; (**c**). Healthy sand lobster larva (Piii); (**d**). Piii larva choked by *Vibrio* sp.; (**e**). Mass mortality – fresh dead larvae; (**f**). Midgut reddening syndrome with starvation. (Photos by Joe K. Kizhakudan, CMFRI)

Vibrio sp. exists in the water used for shrimp culture facilities (Lavilla-Pitogo et al. 1990) and the biofilm, which is formed on different water contact structures of hatcheries and farms. Bacteria enter crustaceans via wounds or cracks in the cuticle and are also ingested with food (Paynter 1989; Lavilla-Pitogo et al. 1990). The primary source of *V. harveyi* in hatcheries appears to be the midgut contents of female broodstock, which are shed during spawning (Lavilla-Pitogo et al. 1992). There is recent evidence to suggest that *V. harveyi* can survive in pond sediment even after chlorination or treatment with lime (Karunasagar et al. 1996).

In different larval rearing experiments at CMFRI laboratory at Kovalam near Chennai (southeast coast of India) (Kizhakudan 2009–2015, personal observation), broodstock of *Thenus unimaculatus* sourced from the wild was usually found to test positive for the *Vibrio* bacteria, and samples collected from the lower abdomen swabs and faecal matter showed high plate counts. The studies indicated that if the late-stage spawners (holding eggs in the eyed stage, heartbeat and beyond) are detected positive for the bacteria, the larval batches suffer severely at the start itself; if the brooder infection rates are mild, then the larvae progresses up to the fifth day without much damage. If larval development progresses beyond 7 days, with low attrition rates, larval mortality is usually less than 50%. Treatments of brooders with 40–50 ppm formalin (chronic exposure) did not control the proliferation at the rearing stages. Chlorination of rearing water has some control initially but affects larval efficiency. Maintaining the brooders in preheated (70°C) and cooled water has been found to suppress the infection rates and colonies in the breeders. But this is not practical in larval culture systems. The breeders maintained in UV-treated and ozonated water produced relatively healthy larvae. Use of pre-ozonated water for larval rearing gave good results but post-moult deformities and Moult Death Syndrome (MDS) were almost 90% at the P-II Stage. *Vibrio*-related mortality was significantly higher just after feeding in seemingly healthier larvae. Mortality of the larvae is preceded by sudden rupture of the hepatic tubules and whitening of the carapace and alimentary system. Microscopic examination reveals swarming motile bacteria flowing out through the ruptured hepatic tubules (Fig. 13.3).

Fig. 13.3 Rupture of hepatic tubules with swarm of motile bacteria. (Photo by Joe K. Kizhakudan, CMFRI)

The combination of two drugs sulphadimidine (1/5) and trimethoprim at a concentration of 120mg/l^{-1} (20 mg/l^{-1} trimethoprim) were found to improve the survival of lobster larvae during luminescent vibriosis outbreaks in larval culture systems (Diggles et al. 2000).

The prophylactic use of antibiotics can result in multiple resistances in several pathogens (Teo et al. 2000, 2002). Rampant misuse of antibiotics has led researchers to look for alternative strategies in disease management. The strategy shall be not killing the bacteria but inhibiting their growth. *Artemia* fed to the larvae is the main carrier of pathogens to the larva. Short-chain fatty acids (SCFAs) are known to inhibit the growth of pathogenic bacteria, but delivering it to the particle feeder *Artemia* is a big challenge. The beneficial effects of poly-β-hydroxybutyrate (PHB) for aquaculture have been shown in several studies (Thai et al. 2014). Defoirdt et al. (2007) tested PHB as an elegant method to deliver SCFAs to *Artemia*, since the effective concentration of SCFAs sufficient enough to protect them from luminescent vibrio is very high. Addition of 1000 mg/l commercial PHB particles (average diameter 30 μm) to culture water was found to offer a complete protection from vibriosis. Other treatment methods such as use of bacteriophage therapy, quorum sensing disruption, probiotics and green water treatment (Defoirdt et al. 2007) have their own advantages and disadvantages, and further research is needed to provide a complete protection protocol from pathogenic attacks to the larvae of spiny lobsters.

13.4.4 Red-Body Disease

Red-body disease has been reported in both juveniles and adults of four species of cultured lobsters in Vietnam: *P. homarus*, *P. ornatus*, *P. longipes* and *P. polyphagus*. The symptom is a distinct red colouration initially to the carapace and/or abdomen, which affects the whole body later. *Vibrio alginolyticus* is the causative pathogen, although other viral pathogens and stress factors also may be involved (Jones 2015). More than 90% mortality is reported.

13.4.5 Tail necrosis

Tail necrosis associated with muscle erosion is found mostly in lobsters held in captivity. The lesions are the result of physical damage, inflammation or an infection. Bacteria, including several species of *Vibrio*, are found in the lesions, which produce enzymes that dissolve chitin. In laboratory-held *P. homarus*, the luminous bacterium, *V. harveyi*, was found to cause tail fan necrosis (Leslie et al. 2013). *V. alginolyticus* was implicated predominantly in tail necrosis of *J. edwardsii* held in holding facilities in Australia (Reuter et al. 1999). Inflammation involves the accumulation of large numbers of haemocytes at the site of injury. The blood produces melanin, a black pigment which is thought to help limit infections (Stephens et al. 2003). Large amounts of melanin deposited on the infected tissue cause blackening of the wound.

13 Health Management in Lobster Aquaculture

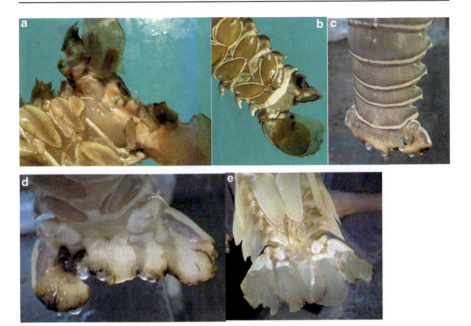

Fig. 13.4 Tail necrosis in lobsters. (**a**). *Thenus unimaculatus*. (**b**). *Panulirus homarus homarus*. (**c**). *Panulirus polyphagus*. (**d**). *Panulirus ornatus*. (**e**). *Panulirus versicolor.* (Photo by Joe K. Kizhakudan, CMFRI)

Tail fan necrosis has been found to affect both spiny and sand lobsters held in captivity (Kizhakudan, personal observation) (Fig. 13.4). Adult male sand lobsters have been found to be particularly susceptible to tail necrosis following intense courtship behaviour during mating, while adult female sand lobsters have been found to develop tail necrosis in senile stages, particularly when held in rearing systems with high organic load.

13.4.6 Milky White Disease Syndrome

Milky haemolymph disease of spiny lobster, also known as milky haemolymph syndrome (MHS) is caused by a *Rickettsia*-like bacterium. The disease affects juveniles and adults. Horizontal transmission occurs by direct contact with infected lobsters (Lightner et al. 2008). The disease has been experimentally transmitted among lobsters by cohabitation and by injection of unfiltered haemolymph from infected lobsters into healthy lobsters. There are reports of incidence of the disease from lobster culture cages and pens in Vietnam. It is suspected that fresh foods such as trash fish, molluscs and decapod crustaceans are the sources of these bacteria.

The symptoms include lethargy, anorexia, cessation of feeding and mortality. Gross pathological signs are milky haemolymph exuding from wounds or visible under swollen abdominal pleura of exoskeleton. Injection of the antibiotic oxytetracycline has been found to be effective in treating and preventing the disease.

In Vietnam, aquaculture production loss due to milky white disease was to the tune of US$50 million in 2007–2008, and livelihood of more than 5000 households has been reported to be affected (Jones 2015).

13.5 Dinoflagellate Blood Disease

The causative organism is Haematodinium-like dinoflagellates in the haemolymph. The parasites invade most organs and haemolymph leading to death.

13.6 Microsporidians

The Microsporidia is a phylum of small intracellular and intranuclear parasites. They were once considered as a phylum within Protozoa, but they are now closely related to the true fungi and are rarely found in lobsters. The parasite invades the muscle tissue and forms large number of microscopic spores within the muscle bundles. The infected muscle looks 'milky' or 'cooked' and is termed as 'white tail'. Microsporidiosis was reported in the western rock lobster *P. cygnus* (Stephens et al. 2003) and in *P. argus* from Florida (Bach and Beardsley 1976). In Australia, one species of *Ameson* sp. is known to infect spiny lobsters, *P. cygnus* and *P. ornatus* (Dennis and Munday 1994; Owens and Glazebrook 1988). Since live spiny lobsters are shipped to markets in different overseas destinations, the potential transport of microspores in infected hosts should be investigated as a risk factor, as many microsporidians show high host specificity (Shields 2011).

13.7 Paramoebiasis

The disease is caused by parasites of the genus *Paramoeba*. The infection can lead to mortality under stressful environmental conditions (Vijayan et al. 2010b).

13.8 Parasitic copepods

Nicothoid copepods are a family of highly specialised symbionts that live on other crustaceans (Shields 2011). They are considered as egg mimics (Bowman and Kornicker 1967). *Choniomyzon panuliri* is a nicothoid copepod that infects the egg clutches of *Panulirus* spp. (Pillai 1962). Three species of parasitic copepods (Figs. 13.5 and 13.6) were found on the spiny lobster *P. homarus homarus*. The harpacticoid copepod *Paramphiascopsis* sp. is symbiotic on the gills of *J. edwardsii* and *S. verreauxi* (Shields 2011). Some parasitic copepods attach to the gills of lobsters and feed on the haemolymph (Vijayan et al. 2010b).

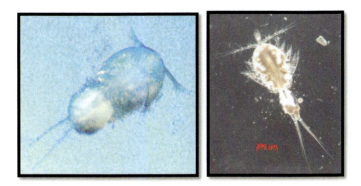

Fig. 13.5 Copepod parasites attached to the gills of the spiny lobster *Panulirus homarus homarus*. (Photo by Joe K. Kizhakudan, CMFRI)

Fig. 13.6 Copepod ectoparasite seen on the spiny lobster *Panulirus homarus homarus*. (Photo by Joe K. Kizhakudan, CMFRI)

13.9 Helminths

Few platyhelminthes use lobsters as hosts. There are no reports of turbellarians on spiny lobsters. However, Shields (2011) observed them on the mouthparts and gills of several species of *Panulirus* from the Great Barrier Reef.

13.9.1 Digenetic trematode infections

Lobsters are secondary intermediate hosts to digenetic trematodes. The microphallid *Thaulakiotrema genitale* was found encysted as metacercariae in the gonads of *P. cygnus* from Western Australia (Deblock et al. 1991). Nematode worms and trematodes (flukes) are usually noticed on adult scyllarid lobsters from Chennai on the southeast coast of India, which are in constant contact with soft sediments (Figs. 13.7 and 13.8).

Fig. 13.7 Nematode worms seen in (**a**). spiny lobsters and (**b**). sand lobsters. (Photo by Joe K. Kizhakudan, CMFRI)

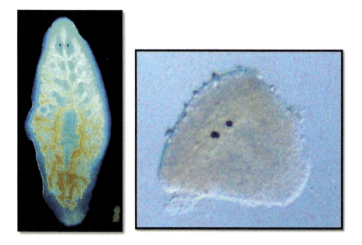

Fig. 13.8 Digenetic trematodes found on adult *Thenus unimaculatus*. (Photo by Joe K. Kizhakudan, CMFRI)

13.9.2 Cestoda

Cestodes may use lobsters as an intermediate host, bearing the metacestode, which is normally encysted in the connective tissues (Shields 2011). Shields (unpublished data) observed metacestodes of a tetraphyllidean cestode in the foreguts of *Panulirus* spp. and *Scyllarides* sp. from the Great Barrier Reef.

13.9.3 Ticks and Mites

Ticks and mites (Subclass Acari) are parasitic members of the Class Arachnida and family Halacaridae (Bartsch 1979). They are mostly attached to the gills and feed on blood (haemolymph) (Figs. 13.9 and 13.10). *Copidognathus matthewsi* is parasitic in the gill chambers of the slipper lobster *Parribacus antarcticus* from Hawaiian

Fig. 13.9 Ticks seen on the gills of *Thenus unimaculatus*. (Photo by Joe K. Kizhakudan, CMFRI)

Fig. 13.10 Mites found in the gill chamber of *Thenus unimaculatus*. (Photo by Joe K. Kizhakudan, CMFRI)

Islands (Newell 1956). Although there is no evidence of detrimental effects on infested invertebrates, severe infestations may lead to mortality.

13.10 Nemertea

Egg predators in lobsters are particularly detrimental to developing eggs on the setose pleopodal tips of the incubating females. Two species of *Carcinonemertes* are known from spiny lobsters. They are egg predators; at least on crabs, they can occur in epidemic scale where they affect reproduction on a population level

(Wickham 1979, 1986; Kuris et al. 1991). Their presence has to be assessed when lobsters carry eggs. *P. cygnus* from Western Australia is infested by *C. australiensis* (Campbell et al. 1989) and *P. interruptus* by *C. wickhami* (Shields and Kuris 1990). Simpson et al. (2017) described a new species of nemertean worm, *C. conanobrieni*, from the egg mass of the Caribbean spiny lobster, *P. argus* from the Florida Keys. *Carcinonemertes* spp. are considered to be voracious egg predators and have been associated with the collapse of various crustacean fisheries.

Kizhakudan et al. (2014) reported that several egg-bearing females of the scyllarid lobsters *Thenus unimaculatus* and *Petrarctus rugosus* and the spiny lobsters *Panulirus homarus*, *P. ornatus* and *P. versicolor* held in the lobster breeding facility at Kovalam Field laboratory of CMFRI, Chennai, were found to discard the complete lot of developing eggs. Sampling of eggs and the clutch revealed infestation by a *Carcinonemertes* sp. worm (Fig. 13.11). The female lobsters mostly discard the entire lot of eggs irrespective of the stage of embryonic development. Symptoms of nemertean egg predation include white eggs (Fig. 13.12), empty egg shells, worms

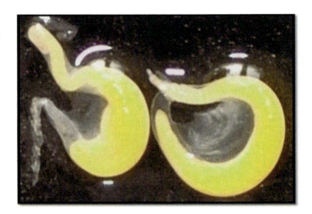

Fig. 13.11 Carcinonemertean egg predator in *T. unimaculatus*. (Photo by Joe K. Kizhakudan, CMFRI)

Fig. 13.12 *Thenus unimaculatus* eggs infected with *Carcinonemertes* sp. (Photo by Joe K. Kizhakudan, CMFRI)

inside the egg shells and frequent shedding of eggs. Often the colour of the live worm is the same as that of the host eggs. Nearly 20–30 adult worms are seen on each brooder egg clutch. They lay their eggs in mucus tubes. The juveniles are found as small larvae on the body and in pilidium larvae associated with ascidian epifauna on the carapace. The intensity of loss of eggs caused by *Carcinonemertes* sp., is severe with each adult worm producing several hundred eggs (Figs. 13.11 and 13.12). These worms consume lobster eggs during the entire incubation period and larvae of the worms were found to survive in the gill chamber of the adult lobster.

The substrate used for scyllarid lobster rearing needs to be treated periodically with 50 ppm formalin to reduce the load of trematodes, nematodes, polychaetes and amphipods. It is possible to remove the bulk of the epifauna on spiny lobsters and sand lobsters by exposing them to a fresh water bath once in a month. A formalin dip of 10–15 ppm dosage for 20 min removes juvenile nemerteans. However, adult nemerteans and eggs inside the lobster egg walls are not removed in this process and require repeated treatments (Kizhakudan et al. 2014).

13.11 Fungi

13.11.1 Oomycetes

Oomycetes typically infect eggs and larvae, but a few infect adult lobsters (Alderman 1973). *Atkinsiella panulirata* is an oomycete reported from the phyllosoma of *P. japonicus* (Kitancharoen and Hatoi 1995). *Haliphthoros mildfordensis* has been implicated in mortalities in aquaculture facilities for spiny and clawed lobsters (Fisher et al. 1975, 1978; Nilson et al. 1975). In New Zealand, *Haliphthoros* sp. caused heavy mortalities in pueruli and juveniles of *J. edwardsii* (Diggles 2001). Melanisation of infected tissues is a common defence against *Haliphthoros* and other fungal infections in crustaceans (Shields 2011). Antibiotic treatment using malachite green and formalin was effective in containing the spread of this disease.

Maintaining optimum water quality is the preferred method to control water mould outbreaks in cultures (Shields 2011). Ozonation and chlorination are recommended and water changes to reduce ammonia and organic loads in the culture system may be strictly followed.

13.11.2 Burnspot Disease

The *Burnspot* disease of lobsters is caused by a fungus *Fusarium* sp. resulting in black spots on exoskeleton and abdomen, uropods, telson, pereiopods and brownish discolouration of the gills. Incidence of this disease is due to poor water quality in culture tanks. The fungus penetrates through the cuticle, which causes the formation of lesions (Lightner and Fontaine 1975; Fisher et al. 1978). High organic load in the rearing system often delays moulting activity, and the stressed lobsters are easily prone to attack by fungal spores which cause gill rot or darkening of the gills with necrosis of the underlying tissue. The affected lobsters become dull and reduce feed intake (Kizhakudan, personal observation). The western rock lobster *P. cygnus* in the wild has been reported to be infected with *F. solani*. The telson and uropods are severely damaged, and the black pigment melanin, produced by the lobster's host defence mechanism, is visible in the affected areas (Stephens et al. 2003).

13.11.3 *Lagenidium* disease

The disease is caused by the fungus *Lagenidium* sp. The fungus penetrates through the cuticle and fills larvae with mycelia giving a white, opaque appearance and is usually lethal. Maintenance of optimum water quality is the solution to prevent the outbreak of the disease in larval culture systems.

13.12 Fouling Organisms

13.12.1 Epibiont Fouling

Fouling of shell and gills is due to growth of other living organisms or inorganic debris (Stephens et al. 2003). Fouling indicates that the lobster is weak and lethargic and has not moulted for long. Fouling under captive conditions may be due to continued exposure to poor water quality, which may pre-dispose the lobster to infection with opportunistic pathogens. Healthy lobsters and the egg-bearing lobsters normally clean the external body and the egg mass free of debris and parasites using their fifth pereiopods. High organic load in holding water attracts pathogens and the lobster becomes the target of attack when they become physically weak and with poor immunity. While some fouling agents are visible to the naked eye, the presence of others can only be identified through microscopic examination of the affected parts.

Filamentous bacteria are ubiquitous on marine crustaceans including lobsters (Johnson et al. 1971; Bland and Brock 1973). *Thiothrix* sp. and *Leucothrix mucor* are common filamentous species observed in larval lobsters (Kittaka 1997; Bourne et al. 2007; Payne et al. 2007). Bourne et al. (2007), using a *Thiothrix*-specific rRNA probe, detected the presence of filamentous bacteria on the phyllosoma larva of *P. ornatus*, and, surprisingly, majority belonged to the *Thiothrix* group. *L. mucor* is a saprophyte of dead algae and is a common fouling organism in larval lobsters. They are difficult to control as they proliferate in organic-rich culture environments. Fouling with *Leucothrix*-like bacteria was common within static culture vessels and was associated with low oxygen levels and high ammonia resulting from poor water flow patterns within holding tanks. Sand lobster larvae reared in controlled conditions at CMFRI laboratory near Chennai (southeast coast of India) were found to host filamentous bacteria when moulting between stages was delayed (Figs. 13.13a

Fig. 13.13a Filamentous bacterial infection in sand lobster larvae. (Photo by Joe K. Kizhakudan, CMFRI)

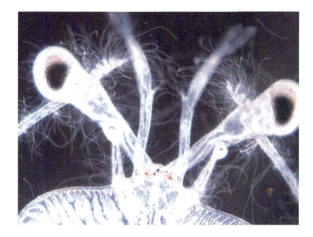

Fig. 13.13b Eggs of *Thenus unimaculatus* infected with filamentous bacteria. (Photo by Joe K. Kizhakudan, CMFRI)

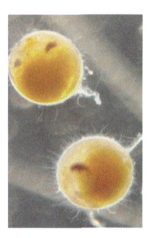

and 13.13b) (Kizhakudan, personal observation). Heavy infections often lead to mortality. *Leucothrix* has been controlled by antibiotics (Fisher et al. 1976a, b; Sadusky and Bullis 1994). It has been suggested that the filamentous bacteria outbreaks in larval culture systems can be controlled by strict attention to nutrition and maintenance of good water quality (Shields 2011).

Phyllosoma larvae of *P. homarus homarus* cultured in static systems in the laboratory were found to be infected with a Cyanobacteria, *Lyngbya* sp. (Fig. 13.13c). The filamentous blue green alga grows on eyes and pereiopods and interferes with feeding and swimming and ultimately during moulting (Radhakrishnan et al. 2009).

Many wild specimens of the rock lobsters *P. homarus* and to an extent, *P. polyphagus* collected from shallow reef areas along the Tamil Nadu (southeast coast of India) have shown small encrusted algal shoots in tiny patches projecting on the carapace in varying levels of density (Kizhakudan, personal observation). This is sometimes observed to liken the colour of the shells to the adjacent reef surfaces

Fig. 13.13c Phyllosoma larva of *Panulirus homarus homarus* infected with *Lyngbya* sp., a filamentous blue green alga. (Photo by E.V. Radhakrishnan, CMFRI)

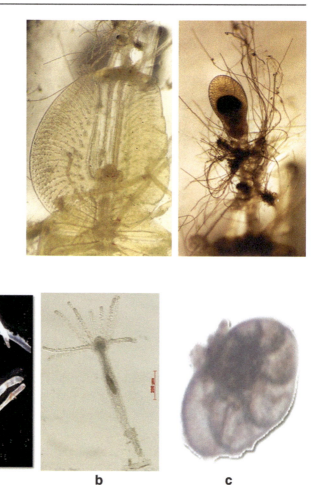

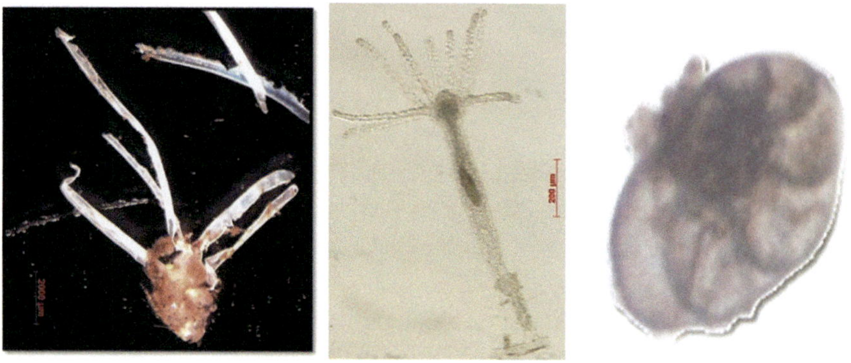

Fig. 13.14 Epibionts on wild specimens of spiny lobsters. (Photo by Joe K. Kizhakudan, CMFRI). (**a**) Encrusted algae (**b**) Hydrozoan (**c**) Foraminifera

(Fig. 13.14a). Some hydroid attachments (Fig. 13.14b), foraminiferans (Fig. 13.14c) and free-moving protozoans and ciliates are also found rarely on lobsters. These protozoan forms are seen in greater numbers on lobsters from loose sediment habitats and captive system facilities.

Older specimens of *P. ornatus* caught from deeper waters are occasionally seen with attached rock barnacles (Fig. 13.15). Specimens of the sand lobster *T. unimaculatus* moving from deeper waters to the coastal areas during breeding and the painted spiny lobster *P. versicolor* have been often observed with the goose barnacle *Lepas* (Fig. 13.16). The attached goose barnacles incidentally have the same colour as that of the developing ovary or released eggs of the female sand lobster, possibly indicating some correlation with the reproductive cycles and vitellogenin synthesis

Fig. 13.15 Barnacles on exoskeleton following delayed moulting. (Photo by Joe K. Kizhakudan, CMFRI)

Fig. 13.16 Goose barnacles *Lepas* sp. (Photo by Joe K. Kizhakudan, CMFRI)

or parasitic extraction (Fig. 13.17). Fatihah et al. (2014) reported the occurrence of five parasites and ecto-symbionts (pedunculate barnacle genus *Octolasmis*, nematode, copepod, nemertean worm and cestode larvae) on the mud spiny lobster *P. polyphagus* from Peninsular Malaysia. The pedunculate barnacle constituted 69%, followed by nemertean worms (21%), nematodes (5.2%), cestode larvae (3.4%) and copepods (1.7%), respectively. The occurrence of barnacles on lobsters has been documented in several studies (Bowers 1968; Newman 1960; Jeffries et al. 1982).

Polychaete worms (Fig. 13.18) cased in calcareous tubes attached to the lobster carapace are common in larger *P. homarus* and *P. ornatus* spiny lobster specimens

Fig. 13.17 *Lepas* showing colour of lobster egg. (Photo by Joe K. Kizhakudan, CMFRI)

Fig. 13.18 Tubiculous polychaete worm. (Photo by Joe K. Kizhakudan, CMFRI)

possibly because of the longer intermoult durations. Through a simple technique of repeated washing of infested lobster pleopods, egg clutches and gills, followed by vacuum filtration, it was possible to isolate and identify several epibionts living on lobsters collected for captive rearing at CMFRI laboratory at Kovalam near Chennai (southeast coast of India). Non-tubiculous forms (Fig. 13.19) were also observed in lesser numbers on berried lobsters during egg washing (Kizhakudan, personal observation). In general, scyllarid lobsters have a larger diversity of egg foulers like nematodes, trematodes, ticks and mites, parasitic copepods, filamentous bacteria, fungi and other invertebrate larvae. Bivalve and gastropod spat (Fig. 13.20) are common epibionts on lobsters.

Fig. 13.19 Non-tubiculous polychaete worm. (Photo by Joe K. Kizhakudan, CMFRI)

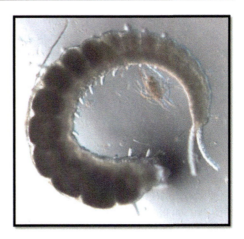

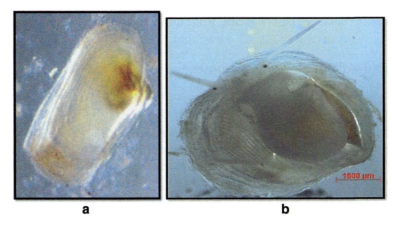

Fig. 13.20 Ectobionts on *Thenus unimaculatus*. (**a**). Bivalve spat. (**b**). Gastropod larva. (Photo by Joe K. Kizhakudan, CMFRI)

13.12.2 Ciliate Disease

Peritrich and suctorian ciliates are often found as commensals on the external surfaces of embryos and larvae of crustaceans in culture, often due to poor water quality (Shields 2011). They are not usually pathogenic but heavy infestations may interfere with respiration (Schuwerack et al. 2001). Peritrich ciliate in the genera *Zoothamnium* sp. was found to foul the appendages of phyllosoma larvae of *P. homarus homarus* (Vijayakumaran and Radhakrishnan 2003). *Vorticella* sp. and *Navicula* sp. fouled the phyllosoma larvae of *J. edwardsii* in culture (Kittaka 1997). Heavy infestation interferes with swimming and feeding and ultimately leads to mortality. The holotrich ciliate *Anophryoides* sp., which attaches to and destroys haemocytes, causes mortality in lobster larvae (Vijayan et al. 2010b). Baths of formalin, oxytetracycline, erythromycin and streptomycin have been used to control fouling organisms in phyllosoma larvae of *P. ornatus* (Bourne et al. 2004). Malachite green, potassium permanganate and formalin at different concentrations have been used to control *Zoothamnium* sp. in phyllosoma larvae of *P. homarus* (Vijayakumaran and Radhakrishnan 2003). A dip treatment with 100 ppm or a short bath treatment for 10 min in 10 ppm concentration of malachite green was found to control *Zoothamnium* infection in phyllosoma larvae of *P. homarus*. Long-term exposure (5 hrs) at a lower concentration of formalin (25 ppm) seems to be effective in killing the ciliates, and at this concentration the larvae are found to be safe (Vijayakumaran and Radhakrishnan 2003).

Ciliate infection in phyllosoma larvae is through feeding of non-sterilised *Artemia* nauplii. *Artemia* cysts are carriers of bacterial, fungal and protozoan pathogens. The cysts have to be hydrated for about 45 min and treated with sodium hypochlorite solution (50–100 g/l) containing 200 ppm of active chlorine for 20 min and washed thoroughly to remove chlorine before keeping it for hatching. The nauplii also should be washed before feeding to the larvae to avoid any contamination.

13.13 Health-Related Syndromes

13.13.1 Bighead Syndrome

The 'Bighead syndrome' is more of a health issue than a disease and has been reported in adults of *P. homarus*, *P. ornatus*, *P. longipes*, *P. penicillatus* and *P. versicolor*. The head becomes abnormally large relative to the abdomen and the affected lobsters appear to have a retarded growth. Such lobsters have a poor market value and the condition is believed to be due to nutrient deficiency. Improved nutrition through supplementary feeding with pelleted feed is found to reverse the condition.

13.13.2 Separate Head Syndrome

Separate head syndrome affects all life stages with separation of the carapace from the abdomen, caused by excess body fluid under the epidermis at the junction of the cephalothorax and tail. It affects the marketing quality and is believed to be caused by exposure to low salinity (<25 ppt). Exposure to sea water of salinity 25 ppt and above reverses the condition.

13.13.3 Moult Death Syndrome (MDS)

Moult death syndrome during or just after ecdysis is a commonly reported phenomenon in spiny lobsters and have symptoms consistent with MDS in homarid lobsters. The lobster gets entangled in the old exoskeleton during moulting and death is often related to nutritional deficiencies and increased stress due to some physiological anomalies. The mortality also may be related to the protein source or vitamins and minerals (Castell et al. 1991; Booth and Kittaka 2000). MDS also has been related to calcinosis, a metabolic disorder in which calcium deposited in between the old and the new exoskeleton prevents the lobster to exuviate (Bowser and Rosemark 1981).

13.13.4 Pink Lobster Syndrome

In the western rock lobster *P. cygnus* and in *P. ornatus* held under captive conditions, pink flesh or haemolymph that is associated with an unpleasant taste have been reported. The affected lobsters become weak and may not withstand live transport. Development of black or grey discolouration of flesh in cooked lobsters has also been noticed, which is caused by melanin, a product of prophenoloxidase system. It may also be due to mobilisation of carotinolipoprotein from the hepatopancreas to ovary through haemolymph during secondary vitellogenesis in female lobsters (Fernandez 2002; Dove et al. 2004; Thangaraja 2011). Investigation into the biochemical and physiological processes responsible for the discolouration of the flesh and haemolymph is needed.

13.13.5 Turgid Lobster Syndrome

Turgid lobster syndrome is an unusual condition reported in *P. ornatus* from Australia (Evans et al. 2000) and *P. homarus* from India (Radhakrishnan, personal observation), in which the thin arthrodial membrane between the carapace and the first abdominal tergite swell from fluid pressure, causing the animals to become stiff and lifeless (Shields 2011). The condition occurs mostly in indoor lobster growout facilities and is not very common. The occurrence of this syndrome has been found to coincide with sudden salinity fall (<25 ppt) or drastic pH variations in the rearing medium (Kizhakudan, personal observation). Though the aetiology of the syndrome is unknown, histology revealed an inflammation of the myocardium as well as infiltration of haemocytes into the affected cardiac muscles (Wada et al. 1994). Sucking the fluid using a syringe was found to relieve fluid pressure, and such lobsters normally survive (Shields 2011).

13.14 Defence mechanisms in lobsters

Invertebrates, including crustaceans, do not possess an acquired immunity; instead they have an innate immune system, which includes melanisation by activation of the prophenoloxidase-activating system (proPO system), a clotting process, phagocytosis, encapsulation of foreign material, antimicrobial action and cell agglutination (Soderhall 1999). Circulating haemocytes, fixed phagocytes and fibrocytes play a pivotal role in cellular defence mechanisms. Hence, total haemocyte count (THC) is considered as an indicator to monitor stress/health condition in crustaceans (Lorenzon et al. 2001).

Three types of haemocytes are generally reported from crustaceans: hyalinocytes, semi-granulocytes and granulocytes (Hearing and Vernick 1967; Hose et al. 1990; Johansson et al. 2000; Verghese 2003). Environmental parameters such as temperature, salinity, oxygen, ammonia content and pH have significant impacts on immune system of crustaceans (Cheng and Chen 2000; Le Moullac and Haffner 2000). Significant reduction in THC was reported in *P. homarus* exposed to lower (20 ppt) and higher (45 ppt) salinities and pH (pH 5 and pH 9.5). The significant reduction in THC and PO activity at both lower and higher salinity and pH indicates that the animals are under stress, and susceptibility to opportunistic pathogens is probably high at such extreme salinity levels (Verghese et al. 2007). The THC and PO activity significantly decreased when the DO was reduced to 1 mg L^{-1}. At 1.5 mg L^{-1} and 3 mg L^{-1} ammonia-N concentrations, the THC in *P. homarus* showed a decline of 58.9% and 51.63%, respectively, over the control group (Verghese et al. 2007). Such reduction in THC in lobsters exposed to high levels of ammonia is likely to decrease the phagocytic potential and other antimicrobial abilities of the organism. Haemocyte counts below 4×10^6 cells/ml have been reported as an indication of poor health in spiny lobsters (Jussila et al. 1997).

Starvation for a period of 3 weeks results in significant reduction in THC in *P. homarus* indicating a direct relationship between haemocyte numbers and nutritional status of the lobster (Jussila et al. 1997; Verghese et al. 2008). Bilateral eyestalk ablation results in increase in THC after 2 hrs. in *P. homarus*. The increase may be due to the release of haemocytes into circulation from storage sites such as haematopoietic tissue and could be a defensive reaction to repair the damage caused by injury. Long distance live transportation causes significant reduction in THC in *P. homarus*, which is an indicator of reduced immune status.

13.15 Conclusion

Adult spiny lobsters have few reported parasites, diseases and symbionts (Shields 2011). Diseases are more common and serious in larval and juvenile lobsters. They are not fraught with many epizootic diseases in the wild, but there are several pathogens which have exacted high mortalities (Behringer et al. 2012). It is likely that emerging diseases will become more pronounced in nascent aquaculture industries for lobsters. Several microbial pathogens and protozoan parasites are problematic in larval culture of spiny lobsters, all of which are difficult to control in closed systems (Shields 2011).

Emergence of disease outbreaks is an indication that the population is under stress and it is difficult to precisely identify the underlying stress factors. Experimental induction of stress and the physiological and immune response of lobsters to such stress factors have been studied in spiny lobsters (Jussila et al. 1997; Verghese et al. 2007). These studies showed that lobsters cannot tolerate combinations of severe salinity, hypoxic, poor water quality and temperature stress and they become highly susceptible to opportunistic pathogens (Dove et al. 2004, 2005; Robohm et al. 2005; Verghese et al. 2007).

Though diseases in lobsters in the wild are rather few compared to many other crustaceans, outbreaks of diseases have been reported from lobster holding centres and culture systems from India, southeast Asian countries and Australia (Handlinger et al. 1999). With increased transportation of live lobsters to international markets, there is an increased threat of accidental introduction of diseases to new regions with potential for consequences in other species (Shields et al. 2006). Understanding pathogenicity and transmission pathways is critical to minimising their spread, and this should not be taken lightly as introduced viruses have played havoc in shrimp farming industry worldwide (Flegel 1997; Lightner and Redman 1998).

13.15.1 Managing Health Problems in Lobsters

Identifying the causative factors is the first step in controlling health problems in lobsters (Stephens et al. 2003). There may be many predisposing factors that have stressed the lobsters, such as water quality problems, handling injury or overcrowding and these factors have to be identified and corrective measures taken.

Preventing disease infection rather than treatment is economically advantageous, as no fool-proof treatment protocol with high survival of the infected population has been developed yet. Maintenance of optimum water quality, quarantine of population brought from unknown destinations and preventive treatments may be followed while practicing capture-based aquaculture of lobsters. As most of the pathogens are opportunistic, they affect only those lobsters which are already in a weakened condition. Therefore, exposure to extreme stress factors has to be minimised or avoided to prevent poor immunity and mortality. Disease outbreaks which are serious in nature may require treatment, and use of antibiotics or exposure to any chemical agent is to be practiced only on the recommendation of an expert so as to avoid potential toxicity to lobsters, residues in lobster meat, harm to the environment and harm to humans applying the treatment (Stephens et al. 2003). Though several probiotics are available in the market, the efficacy of such products on the species under culture has to be tested before its use in mass culture systems. Lobster culture is heavily dependent upon natural feeds, which bring in lot of diseases. Therefore, suitable artificial diets have to be developed for profitable and sustainable lobster culture.

References

Alderman, D. J. (1973). Fungal infection of crawfish (*Palinurus elephas*) exoskeleton. *Transactions of the British Mycological Society, 61*, 595–597.

Bach, S. D., & Beardsley, G. L. (1976). A disease of the Florida spiny lobster. *Sea Frontiers, 22*, 52–53.

Bartsch, I. (1979). Halacaridae (Acari) von der Atlantikküste Nordamerikas: Beschreibung der Arten. *Mikrofauna Meeresbod, 79*, 1–62.

Bayer, R. C., Reno, P. W., & Lunt, M. W. (1983). Terramycin as a chemotherapeutic or chemoprophylactic agent against gaffkemia in the American lobster. *Progressive Fish Culturist, 45*, 167–169.

Behringer, D. C., Butler, M. J., IV, & Stentiford, G. D. (2012). Disease effects on lobster fisheries, ecology, and culture: Overview of DAO Special 6. *Diseases of Aquatic Organisms, 100*, 89–93.

Bland, J. A., & Brock, T. D. (1973). The marine bacterium *Leucothrix mucor* as an algal epiphyte. *Marine Biology, 23*, 283–292.

Booth, J. D., & Kittaka, J. (2000). Spiny lobster grow out. In B. F. Phillips & J. Kittaka (Eds.), *Spiny lobsters: Fisheries and culture* (pp. 556–585). Oxford: Blackwell Science.

Bourne, D. G., Young, N., Webster, N., Payne, M., Salmon, M., Demel, S., & Hall, M. (2004). Microbial community dynamics in a larval aquaculture system of the tropical rock lobster, *Panulirus ornatus. Aquaculture, 242*, 31–51.

Bourne, D. G., Hoj, L., Webster, N., Payne, M., Skindersoe, M., Givskov, M., & Hall, M. (2007). Microbiological aspects of phyllosoma rearing of the ornate rock lobster *Panulirus ornatus. Aquaculture, 268*, 274–287.

Bowers, R. L. (1968). Observations on the orientation and feeding behaviour of barnacles associated with lobsters. *Journal of Experimental Marine Biology and Ecology, 2*, 105–112.

Bowman, T. E., & Kornicker, L. S. (1967). Two new crustaceans: The parasitic copepod *Sphaeronellopsis monothrix* (Choniostomatidae) and its myodocopid ostracod host *Parasterope pollex* (Cylindroleberidae) from the southern New England coast. *Proceedings of the United States National Museum, 123*, 1–28.

Bowser, P. R., & Rosemark, R. (1981). Mortalities of cultured lobsters, *Homarus*, associated with a moult death syndrome. *Aquaculture, 23*(1–4), 11–18.

Brock, J. A., & Lightner, D. V. (1990). Diseases caused by microorganisms. In O. Kinne (Ed.), *Diseases of marine animals, diseases of Crustacea* (Biologische Anstalt Helgoland, Vol. III, pp. 245–349). Hamburg.

Browser, P. R., Rosemark, R., & Reiner, C. R. (1981). A preliminary report of vibriosis in cultured American lobsters, *Homarus americanus. Journal of Invertebrate Pathology, 37*, 80–85.

Campbell, A., Gibson, R., & Evans, L. H. (1989). A new species of *Carcinomertes* (Nemertea: Carcinonemertidae) ectohabitant on *Panulirus cygnus* (Crustacea: Palinuridae) from Western Australia. *Zoological Journal of the Linnean Society, 95*, 257–268.

Castell, J. D., Boston, L. D., Conklin, D. E., & Baum, N. (1991). Nutritionally induced molt death syndrome in aquatic crustaceans: ii. The effect of B vitamin and manganese deficiencies in lobster (*Homarus americanus*). *Crustaceans Nutrition Newsletter, 7*, 108–114.

Cheng, W., & Chen, J. C. (2000). Effects of pH, temperature and salinity on immune parameters of the freshwater prawn, Macrobrachium rosenbergii. *Fish & Shellfish Immunology, 10*, 387–391.

Couch, J. A. (1983). Diseases caused by protozoa. In A. J. Provenzano Jr. (Ed.), *The biology of the crustacea, pathobiology* (Vol. 6, pp. 79–111). New York: Academic Press.

Deblock, S., Williams, A., & Evans, L. H. (1991). Contribution a l'etude des Microphallidae Travassos 1920 (Trematoda). Description de *Thulakiotrema genitale* n. gen., n sp., metacercaire parasite de langoustes australiennes. *Bulletin du Museum National d'histoire naturelle Paris, 12*, 563–576.

Defoirdt, T., Boon, N., Sorgeloos, P., Verstraete, W., & Bossier, P. (2007). Alternatives to antibiotics to control bacterial infections: Luminescent vibriosis in aquaculture as an example. *Trends in Biotechnology, 25*(10), 472–479.

Dennis, D. M., & Munday, B. L. (1994). Microsporidiosis of palinurid lobsters from Australian waters. *Bulletin of the European Association of Fish Pathologists, 14*, 16–18.

Diggles, B. K. (2001). A mycosis of juvenile spiny rock lobster *Jasus edwardsii* (Hutton, 1875) caused by *Haliphthoros* sp., and possible methods of chemical control. *Journal of Fish Diseases, 24*, 99–110.

Diggles, B. K., Moss, G. A., Carson, J., & Anderson, C. (2000). Luminous vibriosis in rock lobster *Jasus verreauxi* (Decapoda: Palinuridae) phyllosoma larvae caused by infection with *Vibrio harveyi*. *Diseases of Aquatic Organisms, 43*, 127–137.

Dove, A. D. M., LoBue, C., Bowser, P., & Powell, M. (2004). Excretory calcinosis: A new fatal disease of wild American lobsters *Homarus americanus*. *Diseases of Aquatic Organisms, 58*, 215–221.

Dove, A., Allam, B., Powers, J. J., & Sokolowski, M. S. (2005). A prolonged thermal stress experiment on the American lobster, *Homarus americanus*. *Journal of Shellfish Research, 24*, 761–765.

Evans, L. H., Jones, J. B., & Brock, J. A. (2000). Diseases of spiny lobsters. In B. F. Phillips & J. Kittaka (Eds.), *Spiny lobsters- fisheries and culture* (pp. 586–600). Oxford: Blackwell Science.

Fatihah, M., Hassan, M., Ihwan, M. Z., Wahab, W., & Ikhwanuddin, M. (2014). Parasites and ecto-symbiont of mud spiny lobster, *Panulirus polyphagus* from Peninsular Malaysia. In *Proceedings of International Fisheries Symposium*, Surabaya, Indonesia (Vol. 4).

Fernandez, R. (2002). *Neuroendocrine control of vitellogenesis in the spiny lobster Panulirus homarus (Linnaeus, 1758)*. PhD thesis, Central Institute of Fisheries Education, Mumbai, India, p. 189.

Fisher, W. S., Nilson, E. H., & Shelser, R. A. (1975). Effect of fungus *Haliphthoros milfordensis* on the juvenile stages of the American lobster *Homarus americanus*. *Journal of Invertebrate Pathology, 26*, 41–45.

Fisher, W. S., Nilson, E. H., Follett, L. F., & Shelser, R. A. (1976a). Hatching and rearing lobster larvae (*Homarus americanus*) in a disease situation. *Aquaculture, 7*, 75–80.

Fisher, W. S., Rosemark, R., & Nilson, E. H. (1976b). The susceptibility of cultured American lobsters to a chitinolytic bacterium. *Proceedings of the World Mariculture Society, 7*, 511–520.

Fisher, W. S., Nilson, E. H., Steenbergen, J. F., & Lightner, D. V. (1978). Microbial diseases of cultured lobsters: A review. *Aquaculture, 14*, 115–140.

Flegel, T. W. (1997). Major viral diseases of the black tiger prawn (*Penaeus monodon*) in Thailand. *World Journal of Microbiology and Biotechnology, 13*, 433–442.

Getchell, R. G. (1989). Bacterial shell disease in crustaceans: A review. *Journal of Shellfish Research, 8*, 1–6.

Handlinger, J., Carson, J., Ritar, A. J., Crear, B. J., Taylor, D. P., & Johnston, D. (1999). Disease conditions of cultured phyllosoma larvae and juveniles of the southern rock lobster (*Jasus edwardsii*, Decapoda; Palinuridae). In L. H. Evans & J. B. Jones (Eds.), *Proceedings, international symposium on lobster health management*, 19–21 September, 1999, Adelaide, Curtin University of Technology (pp. 75–87).

Hearing, V., & Vernick, S. H. (1967). Fine structure of the blood cells of the lobster *Homarus americanus*. *Chesapeake Science, 8*, 170–186.

Hose, J. E., Martin, G. G., & Gerard, A. S. (1990). A decapod hemocyte classification scheme integrating morphology, cytochemistry and function. *The Biological Bulletin, 178*, 33–45.

Jawahar, A., Kaleemur, R., & Leema, J. (1996). Bacterial disease in cultured spiny lobster, *Panulirus homarus* (Linnaeus). *Journal of Aquaculture in the Tropics, 11*, 187–192.

Jeffries, W. B., Voris, H. K., & Yang, C. M. (1982). Diversity and distribution of the pedunculate barnacle *Octolasmis* in the seas adjacent to Singapore. *Journal of Crustacean Biology, 2*, 562–569.

Johansson, M., Keyser, W., Sritunyalucksana, P., & Soderhall, K. (2000). Crustacean hemocytes and haematopoiesis. *Aquaculture, 191*, 45–52.

Johnson, P. T. (1983). Diseases caused by viruses, rickettsiae, bacteria, and fungi. In A. J. Provenzano Jr. (Ed.), *The biology of crustacea, pathology* (Vol. 6, pp. 1–78). New York: Academic Press.

Johnson, P. W., Sieburth, J. M., Sastry, A., Arnold, C. R., & Doty, M. S. (1971). *Leucothrix mucor* infestation of benthic crustacean, fish eggs, and tropical algae. *Limnology and Oceanography, 16*, 962–969.

Jones, C. M. (2015). Summary of disease status affecting tropical spiny lobster aquaculture in Vietnam and Indonesia. In C. M. Jones (Ed.), *Spiny lobster aquaculture development in Indonesia, Vietnam and Australia. Proceedings of the International lobster aquaculture symposium held in Lombok, Indonesia, 22–25 April, 2014, ACIAR Proceedings, No. 145* (pp. 111–113). Canberra: Australian Centre for International Agricultural Research.

Jussila, J., Jago, J., Tsvetnenko, E., Dunstan, B., & Evans, L. H. (1997). Total and differential haemocyte counts in western rock lobsters (*Panulirus cygnus* George) under post-harvest stress. *Marine and Freshwater Research, 48*, 863–867.

Karunasagar, I., Otta, S. K., & Karunasagar, I. (1996). *Effect of chlorination on shrimp pathogenic Vibrio harveyi*. World Aquaculture '96, book of abstracts (p. 193). Baton Rouge: The World Aquaculture Society.

Keith, I. R., Paterson, W. D., Airdrie, D., & Boston, L. D. (1992). Defense mechanisms of the American lobster (*Homarus americanus*): Vaccination provided protection against gaffkemia infections in laboratory and field trials. *Fish & Shellfish Immunology, 2*, 109–119.

Kitancharoen, N., & Hatoi, K. (1995). A marine oomycete *Atkinsiella panulirata* sp. nov. from phyllosoma of spiny lobster, *Panulirus japonicus*. *Myoscience, 36*, 97–104.

Kittaka, J. (1997). Culture of larval spiny lobsters: A review of work done in northern Japan. *Marine and Freshwater Research, 48*, 923–930.

Kizhakudan, J. K., Krishnamoorthi, S., Jasper, B., Xavier, V. J., Sundar, R., & Manibal, C. (2014). Incidence of egg predators and epibionts in lobsters. In P. U. Zacharia, P. Kaladharan, M. Varghese, N. K. Sanil, J. Rekha J. Nair and N. Aswathy (Eds.), *Marine Ecosystems Challenges and Opportunities (MECOS 2), book of abstracts. Marine Biological Association of India, December 2–5, 2014, Kochi* (pp. 97–98).

Kuris, A. M., Blau, S. F., Paul, A. J., Shields, J. D., & Wickham, D. E. (1991). Infestation by brood symbionts and their impact on egg mortality in the red king crab, *Paralithoides camtschatica*, in Alaska: A geographic and temporal variation. *Canadian Journal of Fisheries and Aquatic Sciences, 48*, 559–568.

Lavallee, D., Hammell, K. L., Spangler, E. S., & Cawthorn, R. J. (2001). Estimated prevalence of *Aerococcus viridans* and *Anophryoides haemophilia* in American lobsters *Homarus americanus* freshly captured in the waters of Prince Edward Island Canada. *Diseases of Aquatic Organisms, 46*, 231–136.

Lavilla-Pitogo, C. R., Baticados, C. L., Cruz-Lacierda, E. R., & de la Pena, L. (1990). Occurrence of luminous bacteria disease of *Penaeus monodon* larvae in the Philippines. *Aquaculture, 91*, 1–13.

Lavilla-Pitogo, C. R., Albright, L. J., Paner, M. G., & Sunaz, N. A. (1992). Studies on the sources of luminescent *Vibrio harveyi* in *Penaeus monodon* hatcheries. In M. Shariff, R. P. Subasinghe, & J. R. Authur (Eds.), *Diseases in Asian Aquaculture 1* (pp. 157–164). Manila: Fish Health Section, Asian Fisheries Society.

Le Moullac, G., & Haffner, P. (2000). Environmental factors affecting immune responses in crustaceans. *Aquaculture, 191*, 121–131.

Leslie, V. A., Margaret, M. R., & Balasingh, A. (2013). Rapid identification of *Vibrio harveyi* isolates from *Panulirus homarus*. *International Journal of Current Microbiology and Applied Sciences, 2*(3), 6–10.

Lightner, D. V., & Fontaine, C. T. (1975). A mycosis of the American lobster *Homarus americanus*, caused by *Fusarium* sp. *J. Invetebr. Pathol., 25*, 239–245.

Lightner, D. V., & Redman, R. M. (1998). Strategies for the control of viral diseases of shrimp in the Americas. *Fish Pathologists, 33*, 165–180.

Lightner, D. V., Pantoja, C. P., Redman, R. M., Poulos, B. T., Nguyen, H. D., Do, T. H., & Nguyen, T. C. (2008). Collaboration on milky disease of net-pen-reared spiny lobsters in Vietnam. *OIE Bulletin, 2*, 46–47.

Lorenzon, S., Francese, M., Smith, V. J., & Ferrero, E. A. (2001). Heavy metals affect the circulating haemocyte number in the shrimp *Palaemon elegans*. *Fish & Shellfish Immunology, 11*, 459–472.

Musthaq, S. S., Sudhakaran, R., Balasubramanian, G., & Shahul Hameed, S. (2006). Experimental transmission and tissue tropism of white spot syndrome virus (*WSSV*) in two species of lobsters, *Panulirus homarus* and *Panulirus ornatus*. *Journal of Invertebrate Pathology, 93*, 75–80.

Newell, I. M. (1956). A parasitic species of *Copidognathus* (Acari: Halacaridae), P. Hawaii. *Entomological Society, 16*, 122–125.

Newman, W. A. (1960). Octolasmis californiana, spec. nov., a pedunculate barnacle from the gills of the California spiny lobster. *Veliger, 3*, 9–11.

Nilson, E. H., Fisher, W. S., & Shelser, R. A. (1975). Filamentous infestations on eggs and larvae of cultured crustaceans. *Proceedings of the World Mariculture Society, 6*, 367–375.

Noga, E. J., Engel, D. P., Arroll, T. W., McKenna, S., & Davidian, M. (1994). Low serum antibacterial activity coincides with increased prevalence of shell disease in blue crabs *Callinectes sapidus*. *Diseases of Aquatic Organisms, 19*, 121–128.

Overstreet, R. M. (1983). Metazoan symbionts of crustaceans. In A. J. Provenzano Jr. (Ed.), *The biology of the crustacea, pathobiology* (Vol. 6, pp. 156–250). New York: Academic Press.

Owens, L., & Glazebrook, J. S. (1988). Microsporidiosis in prawns from northern Australia. *Australian Journal of Marine & Freshwater Research, 39*, 301–305.

Payne, M. S., Hall, M. R., Sly, L., & Mourne, D. G. (2007). Microbial diversity within early-stage cultured *Panulirus ornatus* phyllosomas. *Applied and Environmental Microbiology, 73*, 1940–1951.

Paynter, J. L. (1989). *Invertebrates in aquaculture. Refresher course for veterinarians, Proceedings 117*. The University of Queensland, Australia.

Pillai, N. K. (1962). *Choniomyzon* gen. nov. (Copepoda: Choniostomatidae) associated with *Panulirus*. *Journal of the Marine Biological Association of India, 4*, 95–99.

Radhakrishnan, E. V., Chakraborty, R. D., Thangaraja, R., & Unnikrishnan, C. (2009). Effect of *Nannochloropsis salina* on the survival and growth of phyllosoma of the tropical spiny lobster, *Panulirus homarus* L. under laboratory conditions. *Journal of the Marine Biological Association of India, 51*(1), 52–60.

Rajendran, K. V., Vijayan, K. K., Santiago, T. C., & Krol, R. M. (1999). Experimental host range and histopathology of white spot syndrome virus (*WSSV*) infection on shrimp, prawns, crabs, and lobsters from India. *Journal of Fish Diseases, 22*, 183–191.

Reuter, R. E., Geddes, M. C., Evans, L. H., & Bryars, S. R. (1999). Tail disease in southern rock lobsters (*Jasus edwardsii*). In L. H. Evans & J. B. Jones (Eds.), *Proceedings, international symposium on lobster health management, 19–21 September, 1999* (pp. 88–91). Adelaide: Curtin University of Technology.

Robohm, R. A., Draxler, A. F. J., Wieczorek, D., Diane Kapareiko, D., & Pitchford, S. (2005). Effects of environmental stressors on disease susceptibility in American lobsters: A controlled laboratory study. *Journal of Shellfish Research, 24*, 773–880.

Rosen, B. (1970). Shell disease of aquatic crustaceans. In S. F. Snieszko (Ed.), *A symposium on diseases of fishes and shellfishes. American Fisheries Society Special Publication, 5* (pp. 409–415).

Sadusky, T. J., & Bullis, R. A. (1994). Experimental disinfection of lobster eggs infected with *Leucothrix mucor*. *The Biological Bulletin, 187*, 254–255.

Samraj, Y. C. T., Jayagopal, P., & Kamalraj, K. (2010). Farming of spiny lobsters in onshore facility. Abstracts. In *International conference on Recent Advances in Lobster Biology, Aquaculture and Management (RALBAM 2010), 5–8 January, 2010, Chennai, India* (p. 69).

Schapiro, H. C., Mathewson, J. H., Steenbergen, J. F., Kellog, S., Ingram, C., Nierengarten, G., & Rabin, H. (1974). Gaffkemia in the California spiny lobster, *Panulirus interruptus*: Infection and immunization. *Aquaculture, 3*, 403–408.

Schuwerack, P. M., Lewis, J. W., & Jones, P. W. (2001). Pathological and physiological changes in the south African freshwater crab *Potamonautes warreni* Calman induced by microbial gill infections. *Journal of Invertebrate Pathology, 77*, 269–279.

Shields, J. D. (2011). Diseases of spiny lobsters: A review. *Journal of Invertebrate Pathology, 106*, 79–91.

Shields, J. D., & Behringer, D. C., Jr. (2004). *A new pathogenic virus in the Caribbean* spiny lobster *Panulirus argus* from the Florida keys. *Diseases of Aquatic Organisms, 59*, 109–118.

Shields, J. D., & Kuris, A. M. (1990). *Carcinonemertes wickhami* n. sp. (Nemertea), an egg predator on the California lobster, *Panulirus interruptus*. *Fishery Bulletin, 88*, 279–287.

Shields, J. D., & Overstreet, R. M. (2007). Parasites, symbionts, and diseases. In V. Kennedy (Ed.), *The biology and management of the blue crab* (pp. 299–417). University of Maryland Sea Grant Press.

Shields, J. D., Stephens, F. J., & Jones, J. B. (2006). Chapter 5: Pathogens, parasites and other symbionts. In B. F. Phillips (Ed.), *Lobsters: Biology, management, aquaculture and fisheries* (pp. 146–204). Chichester: Blackwell Scientific.

Simpson, L. A., Ambrosio, L. J., & Baeza, J. A. (2017). A new species of *Carcinonemertes*, *Carcinonemertes conanobrieni* sp. nov. (Nemertea: Carcinonemertidae), an egg predator of the Caribbean spiny lobster, Panulirus argus. *PLOS ONE, 12*(5), e0177021. https://doi.org/10.1371/journal.pone.0177021.

Sindermann, C. J. (1990). Responses of shellfish to pathogens. In C. J. Sindermann (Ed.), *Principal diseases of marine fish and shellfish, 2. Diseases of marine shellfish* (2nd ed., pp. 247–300). San Diego: Academic Press.

Soderhall, K. (1999). Editorial. Invertebrate immunity. *Developmental and Comparative Immunology, 23*, 263–266.

Stephens, F., Fotedar, S., & Evans, L. (2003). *Rock lobster health and diseases: A guide for the lobster industry* (p. 28). Perth: Curtin University of Technology.

Stewart, J. E. (1980). Diseases. In J. S. Cobb & B. F. Phillips (Eds.), *The biology and management of lobsters* (pp. 301–342). New York: Academic Press.

Stewart, J. E. (1984). Lobster diseases. *Helgolander Meeresuntersuchungen, 37*, 243–254.

Stewart, J. E., & Arie, B. (1974). Effectiveness of vancomycin against gaffkemia, the bacterial disease of lobsters (genus *Homarus*). *Journal of the Fisheries Research Board of Canada, 31*, 1873–1879.

Stewart, J. E., Cornick, J. W., & Spears, D. I. (1966). Incidence of *Gaffkya homari* in natural lobster (*Homarus americanus*) populations of the Atlantic region of Canada. *Journal of the Fisheries Research Board of Canada, 23*, 1325–1330.

Supamattaya, K., Hoffmann, R. W., Boonyaratpalin, S., & Kanchanaphum, P. (1998). Experimental transmissions of white spot syndrome virus (WSSV) from black tiger shrimp *Penaeus monodon* to the sand crab *Portunus pelagicus*, mud crab *Scylla serrata* and krill *Acetes* sp. *Diseases of Aquatic Organisms, 32*, 79–85.

Teo, J. W. P., Suwanto, A., & Poh, C. L. (2000). Novel β-lactamase genes from two environmental isolates of *Vibrio harveyi*. *Antimicrobial Agents and Chemotherapy, 44*, 1309–1314. https://doi.org/10.1128/AAC.44.5.1309-1314.2000.

Teo, J. W. P., Tan, T. M. C., & Poh, C. L. (2002). Genetic determinants of tetracycline resistance in *Vibrio harveyi*. *Antimicrobial Agents and Chemotherapy, 46*, 1038–1045. https://doi.org/10.1128/aac.46.4.1038-1045.2002.

Thai, T. Q., Wille, M., Garcia-Gonzalez, L., Sorgeloos, P., Bossier, P., & Schryve, P. D. (2014). Poly-ß-hydroxybutyrate content and dose of the bacterial carrier for *Artemia* enrichment determine the performance of giant freshwater prawn larvae. *Applied Microbiology and Biotechnology, 98*(11), 5205–5215.

Thangaraja, R. (2011). *Ecology, reproductive biology and hormonal control of reproduction in the female spiny lobster Panulirus homarus (Linnaeus, 1758)*. Ph D Thesis, Mangalore Biosciences University, India, pp. 172.

Verghese, B. (2003). *Some immunobiological aspects of the spiny lobster* Panulirus homarus *(Linnaeus, 1758)*. PhD thesis, Central Institute of Fisheries Education, Mumbai, India, pp. 118.

Verghese, B., Radhakrishnan, E. V., & Padhi, A. (2007). Effect of environmental parameters on immune response of the Indian spiny lobster, *Panulirus homarus* (Linnaeus, 1758). *Fish & Shellfish Immunology, 23*, 928–936.

Verghese, B., Radhakrishnan, E. V., & Padhi, A. (2008). Effect of moulting, eyestalk ablation, starvation and transportation on the immune response of the Indian spiny lobster, Panulirus homarus. *Aquaculture Research, 39*, 1009–1013.

Vijayakumaran, M., & Radhakrishnan, E. V. (2003). Control of epibionts with chemical disinfectants in the phyllosoma larvae of the spiny lobster, *Panulirus homarus* (Linnaeus). In I. S. Bright Singh, S. Pai, R. Philip, & A. Mohandas (Eds.), *Aquaculture medicine, Centre for Fish Disease Diagnosis and Management, CUSAT, Kochi, India* (pp. 69–72).

Vijayan, K. K., Sharma, K., Kizhakudan, J. K., Sanil, N. K., Saleela, K. N., Alvandi, S. V., Margaret, M. R., & Radhakrishnan, E. V. (2010a). Gaffkemia (*Red tail disease*)- an emerging disease problem in lobster holding facilities in southern India. In *Abstracts. International Conference on Recent Advances in Lobster Biology, Aquaculture and Management (RALBAM 2010), 5–8 January, 2010, Chennai, India* (p. 61).

Vijayan, K. K., Sanil, N. K., & Sharma, K. (2010b). Health management concepts in lobster mariculture. Abstracts. In *International Conference on Recent Advances in Lobster Biology, Aquaculture and Management (RALBAM 2010), 5–8 January, 2010, Chennai, India* (p. 41–45)

Wada, S., Takayama, A., Hatai, K., Shima, Y., & Fushima, H. (1994). A pathological study on cardiac disease found in lobsters. *Fisheries Science, 60*, 129–131.

Wickham, D. E. (1979). Predation by *Carcinonemertes errans* on eggs of the Dungeness Crab, *Cancer magister. Marine Biology, 55*, 45–53.

Wickham, D. E. (1986). Epizootic infestations by nemertean brood parasites on commercially important crustaceans. *Canadian Journal of Fisheries and Aquatic Sciences, 43*, 2295–2302.

14. Post-harvest Processing, Value Addition and Marketing of Lobsters

Vijayakumaran M, E. V. Radhakrishnan, G. Maheswarudu, T. K. Srinivasa Gopal, and Lakshmi Pillai S

Abstract

Seafood is the most commonly traded commodity in the world and the export value of world trade in fish was $136 billion in 2013. Lobster remains a highly prized delicacy the world over. World lobster catch has been steadily increasing over the years and 3,08,926 tonnes were harvested from capture fisheries in 2015. The four main commercial lobster species contributing to 80% of the total world catch are the American lobster (*Homarus americanus*), the European lobster, *H. gammarus*, the Norway lobster *Nephrops norvegicus* and the Spiny lobster *Panulirus argus*. World trade in lobster has grown steadily with both exports and imports showing an increasing trend. During the past 13 years, world trade in lobster grew substantially, from 1,10, 000 tonnes in 2001 to over 1,70, 000 tonnes in 2014. Total lobster trade in 2014 was valued at US$3.3 billion, almost double that of 13 years earlier. The USA was the largest importer of lobster products worth US$1.29 billion in 2014, (36.2%), followed by China with US$576.7 million (17.4%) and Canada with 334.5 million (10.2%). Lobsters are marketed live, processed and frozen form, and the demand for these products is guided by the consumer preference and price. Chinese and Japanese consumers prefer live lobsters, as seafood in live condition ensures freshness and good quality of the meat. The processed lobster market has also developed considerably in recent years, with companies seeking to make lobster products more easily accessible and attractive to the consumer.

Annual average capture fisheries production of lobsters in India is 1696 tonnes (2000–2016), mainly constituted by the spiny and slipper lobsters. The annual average lobster export from India between 1997 and 2014 was 1417 tonnes val-

Vijayakumaran M (✉) · E. V. Radhakrishnan · G. Maheswarudu · Lakshmi Pillai S
ICAR-Central Marine Fisheries Research Institute, Cochin, Kerala, India
e-mail: manambrakatv@gmail.com

T. K. Srinivasa Gopal
ICAR-Central Institute of Fisheries Technology, Cochin, Kerala, India

ued at US$ 20.1 million. India exports lobsters to 22 countries in Europe and Asia and China followed by UAE are the largest importers from India. Nearly 25 different lobster products including live lobsters are exported from India. Fresh-chilled whole spiny lobster followed by frozen whole spiny lobster form the major export. Recently IQF lobsters (spiny and slipper) have found good export market, especially in Europe. The volume of plate-frozen whole cooked lobster in Japan has drastically come down from 1000 tonnes in 2000 to 361 tonnes in 2014, probably due to increased demand for IQF whole cooked lobster. Online marketing of sea food is creating waves especially in the western countries and India too should take advantage of this to market the lobster products to boost export. The volume of live lobster export have to be substantially increased to fetch the premium price offered to this product in the international market.

Keywords
Seafood · Export · Marketing · Live lobsters

14.1 Introduction

Fish including finfish, shellfish and other aquatic products contribute about 16.6% of the total animal protein requirements of the world, and the demand for it is in the ascending order as it also provides other health nutrients such as essential amino acids (EAA), polyunsaturated fatty acids (PUFA), highly unsaturated fatty acids (HUFA), vitamins and minerals and trace elements (FAO 2014). Beneficial effects of fish consumption have been evidenced in relation to coronary heart disease, stroke, age-related macular degeneration and mental health. It has also been correlated with growth and development especially for women and children during gestation and infancy for optimal brain development of children (FAO 2014).

Seafood is the most commonly traded commodity in the world. The export value of world trade in fish was $136 billion in 2013, which is more than the combined value of net exports of rice, coffee, sugar and tea (United Nations Food and Agriculture Organization) (Plaganyi et al. 2017). The fisheries sector, marine as well as freshwater, expanded exponentially till it reached the maximum level of exploitation of around 90 million metric tonnes per annum, by the turn of the twentieth century. As both freshwater and marine fish are exploited at the maximum level, the growing demand for fish for the expanding population is being met by aquaculture production which has been growing at an average annual growth rate of 4.7% in the period 1990–2010. In 2010, aquaculture contributed about 47% of the fishery output for human consumption. World per capita fish consumption increased from an average of 9.9 kg in the 1960s to 11.5 kg in the 1970s, 12.6 kg in the 1980s, 14.4 kg in the 1990s, 17.0 kg in the 2000s and reached 18.4 kg in 2009 (FAO 2014). World per capita fish food consumption is projected to reach 20.6 kg in 2022, up from nearly 19 kg in 2010–2012. Of the 18.4 kg of fish per capita available for consumption in 2009, about 74% came from finfish. Shellfish supplied 26% (or about 4.5 kg per capita, subdivided into 1.7 kg of crustaceans, 0.5 kg of cephalopods and 2.3 kg of other molluscs) (FAO 2014). Shrimps, crabs, lobsters and crayfish constitute the crustacean fishery products (10.36 million tonnes in 2010; capture 4.66 and aquaculture 5.7 million tonnes) with shrimp contributing about 70% of the catch and lobsters just about 6%.

Lobster remains a highly prized delicacy the world over, though it used to be a poor man's food in the seventeenth century, as the colonists in Massachusetts considered lobster shells in a home to be a sign of poverty and only fed lobster to their servants (Megan Ware 2016). World lobster catch has been steadily increasing over the years and 3,08,926 tonnes (less landing of ghost shrimps not considered as lobster) were harvested from capture fisheries in 2015 (FAO 2016a, FAO 2016b). The major portion of the world production of lobsters is shared by three countries, Canada (34%), the USA (29%) and Australia (11%) (Annie and McCarron 2006). The four main commercial lobster species contributing to 80% of the total world catch are the American lobster (*Homarus americanus*), the European lobster, *H. gammarus*, the Norway lobster *Nephrops norvegicus* and the Spiny lobster *Panulirus argus*. In India, the lobster catch has been fluctuating between 1245 and 2976 tonnes during 2000–2016 (CMFRI 2017), and the average production was 1591 tonnes, mainly constituted by the spiny and slipper lobsters.

As lobsters are consumed globally, there is a rapidly growing export market for lobsters. World trade in lobster has grown steadily with both exports and imports showing an increasing trend. Although Asian consumers prefer live lobsters, with the complexity in packing and transport and higher risk involved, the export of live lobsters is limited. On the contrary, processed lobsters are easier to store, handle and transport and more than half of the lobsters landed in lobster producing countries are processed into various products (Barker and Rossbach 2013; Ilangumaran 2014). The average unit value of lobster is US$20 per kg, compared to around US$10 per kg for shrimp and below US$5 per kg of finfish. During the past 13 years, world trade in lobster grew substantially, from 1,10, 000 tonnes in 2001 to over 1,70, 000 tonnes in 2014. Total lobster trade in 2014 was valued at US$3.3 billion, almost double that of 13 years earlier (FAO 2017). The USA was the largest importer of lobster products worth US$1.29 billion in 2014 (36.2%), followed by China with US$576.7 million (17.4%), Canada with 334.5 million (10.2%), France with US$181.1 million (5.1%) and Japan with US$94.2 million (2.6%) (FAO 2017). In terms of volume, Canada stands second whereas value-wise, China is in the second position.

India exports lobsters mainly to the Southeast Asian countries, China and Japan, followed by the Arabian Gulf and the Indian Ocean island nations, Maldives, Mauritius and Seychelles. Processed and preserved lobsters are also being exported to Europe and Australia showing an expansion in the sphere of export.

This chapter presents the various processing methods for the production of lobster products, global export markets and the lobster exports from India.

14.2 Lobster Products

Food service accounts for about 80% of lobster consumption in the USA (Tselikis and McCarron 2009), which is the largest lobster exporter, importer and consumer of the world. About 55% of lobsters exported from the USA go to Canada, and most of it comes back to the USA, as Canada is the largest exporter of lobsters to the USA (Tselikis and McCarron 2009). The situation is much similar in European countries such as France and Italy, which are the major consumers of lobster in Europe. Japan and China are the major importers of spiny/rock lobsters. In Japan, about 48% of lobsters consumed are through food service, 35% is sold by retail and the rest by institutional sales (http://www.tropicseafood.com; Guilbault 2015). Most of the live lobsters imported into China are sold through more than 340 wholesale seafood markets, which are then resold to high-end Chinese restaurants that demand premium. Hong Kong, which is part of the People's Republic of China with seven million people, is an avid consumer of seafood and prefers it live and fresh. Hong Kong imports large quantity of live spiny/rock lobsters and most of it is served in the high-end restaurants and the rest is exported to other parts of China. The Chinese and Japanese proudly serve lobster as a statement delicacy at wedding receptions and banquets, signifying luxury and prestige (http://www.tropicseafood.com).

14.3 Marketing

Lobsters are marketed live or processed and frozen, and the demand for these products is guided by the consumer preference and price. The main lobster products sold through wholesale/retail segments are:

14.3.1 Live Lobster

The market for live crustaceans is complex and the product can be sold by only those with a proper image and perception of the market demands (Kriz 1995). Eating a rock lobster in a high class restaurant is far more important for an elite Japanese consumer than physically eating the product. The live product always assures freshness, quality and superiority in look and taste. The demand for live crustaceans is ever increasing and the burgeoning demand for live product in Asia suggests an extremely positive outlook for live crustaceans in future. Live lobster is such a valuable export commodity that considerable time and effort have been devoted to developing its storage methods and transport systems in order to ensure that it reaches the consumer in prime condition (FAO 2017).

Lobsters are generally caught alive in traps and are active and healthy when landed. In addition, lobsters are capable of surviving outside water for long periods, if proper aerial environment with high humidity and optimum temperature are provided. This aspect of lobster biology has been fully utilized for its live transport within a country and across the continents. Live lobster is in maximum demand in

North America, Europe and Asia and is the most cost-effective way to sell lobster, where no weight loss is involved. In addition, live lobster is the most expensive product and gives good value for the fisher and the trader.

The clawed lobsters are transported from the harbour in tanks, wooden boxes or plastic buckets inside refrigerated and insulated trucks or through airplane to holding facilities and kept alive in earthen impoundments or in well aerated and clean seawater in indoor tanks, which need not necessarily be along the coast as the seawater requirement is low and can be met by transportation from the sea coast. Live packing for air transport and shipping would be done few hours before the departure of the flight/ship/insulated truck. At the receiving centre, the live lobsters are released to tanks with clean seawater to recover from the transport stress and retained there till it is further moved to wholesale/retail markets/ restaurants, which will have seawater aquaria to keep the lobsters alive. To avoid cannibalism and physical damage following capture, the chelae of clawed lobsters are immobilized with elastic bandages during transport as well as holding (Barrento 2010).

Similar protocol is used for live transport of spiny/rock lobsters in Australia, New Zealand and many other countries. In India, spiny and slipper lobsters are caught in trawl net in the north-west coast and are mostly brought dead to the harbour. In the southwest and southeast coasts also, slipper lobsters and deep sea lobsters are caught in trawls and are not fit for live transport. But the palinurid lobsters along the southern coast are largely caught using bottom-set-gillnet and are active and healthy when removed from the gill net carefully, so that the appendages are not damaged. Unlike in other major lobster fishing countries where lobsters are fished by a small number of fishers and landed in few harbours, lobsters are caught by a large number of artisanal fishers spread all along the coast in India and are landed in many landing centres in small quantities. The live lobsters are kept in small aquaria often in high density at the landing centre up to 24 h before it is transported to the lobster holding facilities. Recently, in few landing centres, the lobsters are kept alive buried in intertidal sand in indoor tanks and covered with jute material. Sea water is sprinkled occasionally over it to keep the sand moist. Vijayakumaran and Radhakrishnan (1997) report that *P. homarus* can be kept alive for more than 48 h using moist intertidal sand and jute material. Few holding facilities are far away from the airport necessitating transport of lobsters to a secondary holding facility near the airport from where it is packed and exported. Mortality of lobsters at these holding facilities and during transport is dependent on the size, health condition, method of transport and species. Mortality was almost none in *P. ornatus* and among *P. homarus*, it was more in the juveniles (Vijayakumaran and Radhakrishnan 1997).

14.3.1.1 Live Lobster Packing

Packing of lobsters for air transport is essentially similar all over the world for cold water clawed (*H. americanus* and *H. gammarus*) and spiny lobsters (*Panulirus* and *Jasus* spp.) and warm water spiny/rock lobsters (*Panulirus* spp.), with minor differences owing to the requirements of the species. The main difference in treatment is the temperature in the holding tank and the transport pack. The shipping temperature will depend upon the species and the ambient temperature of the harvest area, but in general, cold water species can be chilled to 4 °C and shipped at temperatures ranging from 1 to 7 °C. Tropical species may be rendered dormant at temperatures as high as 14 °C (Holmyard and Franz 2006). Warm water lobsters will not tolerate temperature below 10 °C, and the shipping temperature has to be adjusted 13–14 °C for 24 h air transport. The average holding tank water temperature during most of the months in India would be 26–29 °C, and when it goes beyond 29 °C, the water is cooled by placing ice in floating stainless steel containers. One of the main exporters of live lobster in India introduced a chiller in his holding facility to reduce the water temperature to 25 °C. Sudden mass mortalities were observed in lobsters kept in tanks (Maharajan and Vijayakumaran 2004) with chilled water apparently with no external disease symptom. The problem was identified as copper toxicity due to leaching of copper from the cupro-nickel cooling element used in the chiller. In India, temperature used to be gradually brought down to 13–14 °C in 2–3 h before packing. This was then changed to a sudden 1–2-min dip in 13 °C cold water just before packing for export. Incidentally, it also gave an anaesthetic effect to the lobster being packed. In the same holding facility where copper toxicity was detected, more than 10% mortality was observed during transport to Hong Kong from Chennai in 20–24 h. The problem was traced to very high level of ammonia in the cold water used to give cold shock to the lobsters before packing, as the same water was used repeatedly. This problem was solved by changing the water used for cold shock after dipping 100 kg of lobster.

Fibreboard boxes lined with expanded polystyrene is used for shipment. An absorbent is placed in the bottom of the box and the lobsters are kept over it. Moist paper, chilled wet sponges or gel packs are kept in between so that the lobsters do not touch each other (Holmyard and Franz 2006). Vertical packing in individual compartments is the latest technology introduced for clawed lobsters in Northern America. It is a 'Cloud Pack' system, approved by IATA with a bottom section that holds the live lobster body and folded tail, a centre section that holds the claws and a cover that contains gel ice and a moisture medium. This packaging is expensive but can be reused, and one point in its favour is that it reduces mortality (Holmyard and Franz 2006).

In India, polystyrene boxes (60 × 37 × 23 cm) are used for air transport. About 8.7 kg of lobsters are packed in one box and the total weight of the packed box would be around 11 kg. An average loss of 200 g was recorded at the destination port after about 24 h of transport. Jute material or bloating paper with seawater sprinkled to keep it moist is spread at the bottom of the box and a layer of lobsters with flexed abdomen and individually wrapped in newsprint is kept on it. Frozen seawater in a 750 ml polypropylene sealed bottle wrapped with news print is placed at the middle of the box in between the two rows of lobsters. Depending on the size of lobsters, one more layer of lobster is placed above the bottom layer with another bottle of frozen seawater at the centre. When large lobsters (*P. ornatus*) are packed, about 400 g sawdust is often spread at the bottom. After closing the lid, the box is strongly fastened with wide adhesive tape and labelled appropriately. When packed for export, the temperature inside the box was 13–14 °C and a humidity of about

70% (Vijayakumaran and Radhakrishnan 1997). However, at the destination port after air transport, the temperature inside the pack increased to 16–18 °C, depending on the duration of transport. Frozen seawater was replaced by gel ice for keeping the box cool at the desired temperature during air transport. Incidentally, gel ice was not found to be as effective as frozen seawater which is again used at present for packing.

14.3.1.2 Methods to Reduce Metabolism

14.3.1.2.1 Use of Anaesthetics for Immobilization

Live lobster market is very sensitive, as the price is 2–3 times that of frozen lobsters and the Chinese importers would not tolerate a mortality of more than 4%. The primary cause of mortality during air transport is the health of lobsters prior to packing. Many physical and chemical methods are used to check the activity of lobster for live transport. The physical methods used to analyse stress in crustaceans are: eyestalk response and movement of claws/antennae and legs. Anaesthetics are not generally used in crustaceans as they are amenable to handling, but use of it during holding and transport to reduce cannibalism in few species such as clawed lobsters are getting attention nowadays. The FDA-approved anaesthetic for fish, MS-222, is not effective on many crustaceans (Coyle et al. 2004). Aqui-S® is a relatively new anaesthetic for fish developed by the Seafood Research Laboratory in New Zealand. This compound has been proven to be effective for short duration transport (<2 h) of salmon smolts to ranch at sea, or longer transports (> 40 h) of live lobster, eel or cod to distant international markets (Barrento 2010). Foley et al. (1966) compared isobutyl alcohol and pentynol for anaesthetizing *H. americanus* and found isobutyl alcohol as superior in that toxic effects were not evident even at high concentration. He suggested for the use of isobutyl alcohol concentration of 1.5–7.0 ml/l for shorter duration transport. Jayagopal (2004) also found that isobutyl alcohol is an effective anaesthetic for spiny lobsters during short duration (22 h) live transport. Four species of palinurids, *P. homarus, P. ornatus, P. versicolor* and *P. polyphagus*, consistently recorded lower haemolymph ammonia when transported live after anaesthetizing with isobutyl alcohol (Jayagopal 2004). However, more studies are required to confirm its utility and this compound requires clearance for use in animals for human consumption from appropriate authorities in countries dealing with lobster.

14.3.1.2.2 Cooling

Cooling is an effective way to immobilize crustaceans, but one must be careful since excessive cooling can also kill the animals (Coyle et al. 2004). Tropical spiny lobsters may not tolerate cooling below 10 °C (Jayagopal 2004). A temperature shock at 13 °C has proven to be a good method to inactivate *P. homarus* and *P. ornatus* during live transport in India. However, it tends to increase mortality during transport in the other two species, *P. polyphagus* and *P. versicolor* (Jayagopal 2004; Vijayakumaran and Radhakrishnan 1997).

14.3.1.3 Stress Evaluation in Live Lobster Transport

14.3.1.3.1 Chemical Methods

Among the chemical methods to determine stress, glucose concentration in the haemolymph is a good tool, as hypoxia tends to induce hyperglycaemia and reverts to normal when normal oxygen supply is restored (Barrento 2010). Lactate is the main end product of anaerobic metabolism in crustaceans (Zebe 1982). Lactate increases in the tissue such as hepatopancreas, heart and muscle during anaerobic metabolism, which can then be released into the haemolymph (Frederich and Pörtner 2000). During recovery in well-aerated water after air transport, lactate is converted back into other metabolic intermediates and either oxidized further to carbon dioxide or used to re-synthesize storage sugars such as glycogen (Gäde et al. 1986, Barrento 2010). Anaerobic metabolism reduces the pH in crustacean haemolymph due to acidosis by lactate and bicarbonate-carbonic acid (Spanoghe and Bourne 1997). Therefore, pH in haemolymph is a good tool to determine stress in decapod crustaceans.

Studies on palinurid lobsters in India have indicated that high ammonia concentration in the cold water used to give a cold dip to acclimatize the lobster to the packing temperature might increase the mortality during air transport (Vijayakumaran, personal observation). Subsequent studies on warm water palinurids revealed that haemolymph ammonia increases during air transport and can be a good indicator of recovery after transport (Jayagopal 2004).

14.3.1.4 Physiological Changes during Live Transport

Distinctive changes were observed in haemolymph ammonia and lactic acid during simulated live transport in four commercially important species of palinurid lobsters – *P. homarus, P. ornatus, P. versicolor* and *P. polyphagus* – and Δ L-lactate in the clawed lobsters, *H. gammarus* and *N. norvegicus* (Table 14.1).

Acclimatizing temperature, cooling temperature, cooling duration and their relationship with haemolymph ammonia and lactic acid vary among the four species of palinurid lobsters. Duration of live transport was directly proportional to the haemolymph ammonia in all the species. In the case of haemolymph lactic acid, in all the species except *P. versicolor*, a direct relationship was observed with the duration of transport. Humidity of the pack was also positively related to haemolymph ammonia of all the species and haemolymph lactic acid in all the species except *P. versicolor*. In the case of *Jasus edwardsii*, haemolymph values for glucose, lactate and ammonia during 24-h air exposure and following a 12-h period of re-immersion reach highest during exposure, but return towards initial values following re-immersion (Speed et al. 2001). In the clawed lobster *H. gammarus* also, lactate in haemolymph increases during air transport and is restored to normal values within a short duration during recovery (Barrento 2010). Although, high haemolymph ammonia and lactic acid were observed in *P. homarus* and *P. ornatus*, high mortality was observed among *P. versicolor* and *P. polyphagus* compared to the other two species. This indicates the inability of *P. versicolor* and *P. polyphagus* to survive under stressful condition compared to that of *P. homarus* and *P. ornatus*. Rahman and Srikrishnadhas (1994) had also reported higher mortality during live transport of *P. polyphagus*.

Table 14.1 Haemolymph ammonia and lactic acid levels in different species of spiny lobsters after live transport (mean values ± SD) *Δ L-lactate in mM

Species	Cooling temp/ minutes °C	Packing temperature °C	Humidity inside the pack %	Duration of transport (h)	Initial Ammonia (mg/10 ml)	Final Ammonia (mg/10 ml)	Initial Lactic acid (mg/ml)	Final lactic acid (mg/ml)	References
P. homarus	10/5	20	70%	28	0.01	3.6	0.01	3.28	Jayagopal (2004)
P. ornatus	18/15	20	70	25		2.12		1.54	Jayagopal (2004)
P. versicolor	12/7.5	13	62%	20		1.98		0.9	Jayagopal (2004)
P. polyphagus	15/15	27.5	58%	19		1.32		2.16	Jayagopal (2004)
H. gammarus		15		48				4.7*	Lorenzon et al. (2007)
N. norvegicus		10		3				6.0*	Spicer et al. (1990)

Crustaceans have a poor system for lactate metabolism (Ellington 1983). Hence, it takes longer period for recovery. Bridges and Brand (1980) suggested that species that encounter aerial exposures in their natural environments are better adapted for removing accumulated lactate when the aerobic conditions return. Spiny lobster, being a sub-tidal non-burrowing species, probably removes lactate slowly even though normal concentrations were restored within 24 h (Vermeer 1987).

Schmitt and Uglow (1997) have found that most of the ammonia accumulated in the haemolymph of *N. norvegicus* during exposure to increased ambient ammonia disappeared within 10 min of recovery. When oxygen packing was used, haemolymph ammonia and lactic acid levels were lower in the lobster pack of all the four species of palinurids (Jayagopal 2004) (Table 14.2). This may be suggestive of the ability of spiny lobster to adapt to bi-modal breathing to a limited extent as described by many authors in intertidal crustaceans (Defur 1988; Morris and Bridges 1994; Greenaway et al. 1996).

Use of cold shock for immobilization during air transport varied in the four species of palinurids. A short dip in 13 °C seawater is enough to immobilize the lobsters, and it did not have any deleterious effect during air transport in *P. homarus* and *P. ornatus*. *P. polyphagus* can tolerate only a very short duration dip in cool water with lower temperature, whereas *P. versicolor* is sensitive to cooling temperature. Hence, *P. versicolor* should not be cooled prior to packing and an alternate method of immobilization should be developed for this species. Therefore, it is not possible to standardize a common cooling protocol for all species for immobilization during live transport.

14.3.2 Frozen and Canned Secondary Processed Lobster Products

The processed lobster market has developed considerably in recent years, with companies seeking to make lobster products more easily accessible and attractive to the consumer. Simple processes, ready-to-eat products and meat selection packs in attractive packaging have contributed to an increase in demand. Processors are able to maximize yield per lobster by producing a wide variety of frozen and sterilized shelf stable products. These include the liver of clawed lobsters, which is processed as a green coloured paste/spread known as tomalley; lobster roe, which is also called red caviar; lobster concentrates/extracts and lobster meat paste.

Hot pack canned lobster are sterilized or retorted and are therefore shelf stable. Concentrates, pastes and extracts are sought after by the food service industry as they provide a standardized, consistent product that can be used to make sauces, pasta fillings, soups, bisques, pates, terrines and mousses, pie fillings and seafood noodles (Holmyard and Franz 2006).

14.3.2.1 High Hydrostatic Pressure Processing of Lobster

High hydrostatic pressure processing (HPP), which got its start over a century ago, is finally reaching the point where it can be commercially applied on a large

Table 14.2 Ammonia and lactic acid levels in haemolymph during recovery of spiny lobsters in lobster holding facility (from Jayagopal 2004) (mean values ± SD)

Duration in hours	P. homarus Ammonia (mg/10 ml)	Lactic acid (mg/ml)	P. ornatus Ammonia (mg/10 ml)	Lactic acid (mg/ml)	P. versicolor Ammonia (mg/10 ml)	Lactic acid (mg/ml)	P. polyphagus Ammonia (mg/10 ml)	Lactic acid (mg/ml)
0	0.55 ± 0.06	0.43 ± 0.03	0.52 ± 0.06	0.47 ± 0.03	0.43 ± 0.036	0.36 ± 0.043	0.4 ± 0.036	0.4 ± 0.03
0.5	0.17 ± 0.02	1.5 ± 0.09	0.4 ± 0.05	1.73 ± 0.13	0.2 ± 0.03	1.8 ± 0.111	0.18 ± 0.01	1.6 ± 0.132
1.5	0.34 ± 0.04	0.88 ± 0.05	0.34 ± 0.03	0.95 ± 0.06	0.17 ± 0.02	0.73 ± 0.055	0.24 ± 0.03	0.8 ± 0.09
9	0.36 ± 0.03	0.21 ± 0.03	0.28 ± 0.02	0.33 ± 0.04	0.3 ± 0.05	0.3 ± 0.03	0.34 ± 0.03	0.25 ± 0.04
27	0.4 ± 0.04	0.16 ± 0.03	0.3 ± 0.05	0.19 ± 0.03	0.36 ± 0.04	0.085 ± 0.005	0.34 ± 0.04	0.11 ± 0.36
48	0.06 ± 0.01		0.04 ± 0.01		0.02 ± 0.01		0.09 ± 0.01	
60	BDL		BDL		BDL		BDL	

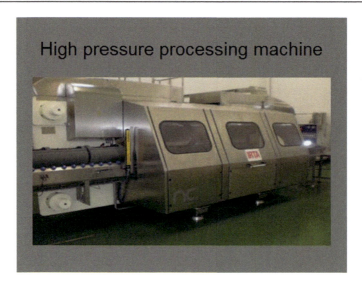

Fig. 14.1 High-pressure processing machine. (Photo by Srinivasa Gopal, CIFT)

scale. The HPP technology is gaining prominence in the seafood industry as a non-thermal technology for controlling microorganisms and improving the processing of shellfish and crustaceans (Fig. 14.1). The technology, particularly in the lobster marketing, provides the opportunity to market fresh, shucked lobster meat without the use of heat. Additionally, this process increases meat recovery by as much as 50% over traditional cooking methods, improves product weight by as much as 10% from the natural hydration of proteins and improves product quality. The technology subjects foods, which can be liquid or solid, packaged or unpackaged, to pressures between 40,000 and 80,000 pounds/square inch (PSI), usually for 5 min or less, which significantly increases product throughput of processing plants (Raghubeer 2007). The high pressure does not destroy the food, because it is applied evenly from all sides. The typical HPP system consists of a high-pressure vessel, a means to close the vessel off, a system for pressure generation, a system for temperature and pressure control, and a material handling system. HPP lobster meat is delicate to cook, has moisture loss and expensive at present (Sackton 2007).

Live lobsters are loaded into a high-pressure processing machine, which uses pressure at or above 40,000 psi (pounds per inch) to kill the lobsters almost instantaneously. This process also separates the meat from the shell inside the lobster, making it exceedingly easy to shuck while still raw. Because the pressure is equal from all sides and no heat is involved, the taste and texture of the lobster's meat are not harmed. After this first trip to the HPP machine, the claw, tail and knuckle meat is expertly removed from the shell by hand (except for HPP processed shell-on lobsters) and then effectively 'cold pasteurized' by a second trip to the HPP machine.

14.3.2.2 Modified Atmospheric Packaging

Modified atmospheric packaging is a commonly used processing method for the shelf life extension of seafood, and the data obtained from fish and other shellfish shows a general shelf life increase of 30–60% (Fig. 14.2). In MAP the air inside the packaging is replaced by a specific gas or a mixture of gases that differ from the air composition. The proportions of each gas are established, the mixture is introduced into the packaging and no further control is carried out during storage. The success of the MAP depends on various factors such as: good initial quality of the product, good hygiene practices during fishing, selection of the right packaging material, a safe packing equipment, good maintenance and control of temperature, a proper gas mixture for the product and the gas/product ratio. The ideal CO_2 concentration depends on the fish species, initial microbial population, gas/fish ratio and on the packing method. Norway lobster *Nephrops norvegicus* packed in flexible film with good barrier properties and packed in modified atmosphere packaging containing 80% carbon dioxide, 10% oxygen and 10% nitrogen had a shelf life of 13 days compared to 3–4 days in control air-packed samples at 1 °C.

14.3.3 Products Range of Cold Water Clawed and Spiny Lobsters

Processing can add value to the raw product and it increases the shelf life and shippability. Processed products are acceptable for food service, wholesale and retail markets. It satisfies consumer preferences as they are convenient for easy preparation of food. Consumers, especially in western countries seek sustainable sources (ecolabels) and are interested in the traceability of the food. They are health conscious and are interested in seafood with low fat, low cholesterol, high n 3 and n 6 fatty acids. The demand for wild, organic, fresh and chemical-free fish is on the rise, and people are willing to pay more for such products.

Fig. 14.2 Modified atmospheric packaging machine. (Photo by Srinivasa Gopal, CIFT)

Coldwater-clawed lobsters are often marketed as raw-frozen or whole-cooked frozen with a size range of 8 oz. (225 g) to 6 lb. (2.75 kg). Frozen whole lobsters often include cooked Cull (one claw and one arm missing) and Pistol (claws and arms missing) and also whole cooked split. Whole (raw) and split (raw) lobster tails of size 2/3 oz. to 20 oz. or more.

Lobster Tails

Other shell-on frozen products are claw and arm, double scored 3/9, 5/9. 9/11 counts per lb. (Tselikis and McCarron 2009).

Shucked products are also in the market as lobster meat of tail, claw, knuckle, legs, body and their combinations. Trade name for these products are minced (body and leg), salad (body and leg), broken (tail, claw, knuckle, body and legs).

Of the processed lobster, 78% is in-shell product with a break up of raw tail in shell 31%, cooked in shell 42%, in shell raw 2% and 3 knuckle claw in shell. Meat forms 15% followed by blocks 5% and Knuckle claw tail meat 2% (Tselikis and McCarron 2009).

The product range of spiny/rock lobsters, both cold water and warm water are similar in America, Europe, Australia and Asia. The main products are whole tail (raw) and split tail (raw) 5–20 oz. (140–560 g) size. Whole (raw), whole (cooked) and whole split (raw or cooked) of size ranging from 12 oz. to 2 lb. (336 g to 900 g) are also important products (Tselikis and McCarron 2009).

The product ranges of lobsters exported from India are *spiny lobster live, whole (raw-frozen/chilled/glazed IQF/IWP), whole (cooked)* and *slipper lobster whole (raw, frozen and cooked, IQF/IWP)*.

14.3.3.1 Whole Lobster Cooking and Freezing

Live or fresh and chilled lobsters are commonly used for the preparation of whole cooked lobsters. The Mud spiny lobster, *P. polyphagus* is the most preferred species for whole cooking. This lobster turns to deep reddish colour after cooking and the white transverse bands on the abdomen in the red background makes the product very attractive.

The lobsters are washed and cleaned before keeping in a freezer for 15 min to kill them. They are then cooked in 3–5% brine for 15–25 min and immediately chilled. Fresh-chilled lobsters are also used for whole cooking. The cooked lobsters are packed individually in polythene film, frozen at −35 °C to −40 °C, and stored. The Popsicle pack contains a small whole cooked lobster packaged in a vacuum sealed cello bag with saltwater brine, which is then blast frozen to preserve the lobster's fresh caught taste. Frozen popsicles are generally available in 300–450 g sizes. Lobsters may be packed several to a carton, individually held in net bags, or separated with waxed paper or plastic strips. Individual skin packs can be used with raw, partly cooked and cooked lobster and present particularly well to the consumer (Holmyard and Franz 2006).

The ICAR-Central Institute of Fisheries Technology, Cochin has standardized the method for whole cooked lobster marketing in India. The lobster is cooked as described above and the freezing is carried out in minimum possible time. Frozen material is packed in 10 or 20 lbs. lots in corrugated fibreboard containers lined inside with moisture vapour-proof paper and can be stored in frozen storage room at −20 °C.

Frozen whole lobster is available raw, blanched or steam cooked, in a number of different presentation packs and may be natural, glazed or brined. Processing frozen lobster products utilizes the latest technological advances in blast freezing that ensures the highest quality and flavour of frozen whole lobsters and lobster meat. Under ideal frozen temperatures of −26 to −30 °C (−15 to −20 °F) or below, frozen lobster can be stored with no quality loss for up to 9 months. Lobsters may also be frozen using liquid nitrogen. Cold pack frozen lobster meat is packed in cans and requires frozen storage as it is not sterilized or retorted (Holmyard and Franz 2006).

Frozen whole blanched lobster is first cooked for 2 min in a vacuum skin pack and then vacuum packed and frozen immediately. Cooking is completed by the end user, for a fresh-boiled taste. The specialized technology allows an extended shelf life of 24 months.

14.3.3.2 Frozen Lobster Head

Interestingly few consignments of frozen spiny lobster head were exported to the United Arab Emirates in 2015. The end use of this product is uncertain. It could have been used as a feed supplement/ingredient.

The general range of American lobsters is Chix at 11.20 lb., Quarters at 1.20–1.45 lb., Halves at 1.45–1.75 lb., Selects at 1.75–3 lb. and Jumbos at 3–6+ lb. At the small end of the scale are Canners, which weigh approximately 1/2–1 lb. American consumers are the most partial to Jumbos, while the highest world demand is for Chix and Quarters, although the very largest specimens are permanently on request by upmarket restaurants in, for example, Japan and Hong Kong, to enhance seafood displays.

Rock and Spiny lobsters are generally sold whole or as tails weighing from 1 to 5 pounds (Holmyard and Franz 2006).

14.3.4 Value-Added Products

Sustainability labels through the use of international schemes, such as the Marine Stewardship Council's (MSC) sustainability label, which shows consumers that a product has been produced/fished in a responsible manner, is one of the means for value addition for all lobster products. The Australian Western Rock Lobster was the first product to be certified by the MSC. There is no sufficient data for certification through other such agencies though many such agencies are in operation (Holmyard and Franz 2006).

Value addition can also be realized by producing new secondary processed products, ready-to-eat meals catering to the retail consumers as well as chefs in high-end restaurants. Few of the emerging value-added products are listed below.

14.3.4.1 Lobster Pizza

It has a thin and crispy crust with a white Newberg sauce loaded with lobster and delicious mozzarella cheese.

14.3.4.2 Premium Lobster Salad

It is a preparation with premium North Atlantic lobster meat along with mayonnaise, lemon juice and special seasoning (Tselikis and McCarron 2009).

Frozen medallions for retails and foodservice customers, providing a sashimi grade meat for Japan, and bottling lobster oil, which is proving immensely popular with consumers and chefs. Carefully targeted market research and focus group research provided much of the inspiration for these new developments by an Australian company.

14.3.5 Processed Lobster Pricing

Coldwater frozen tails are used primarily in foodservice and are preferred by chefs in North America. Cold water whole lobster command higher price whereas warm water spiny lobster tails are often priced higher. The market price of spiny lobster tails in 2009 was N. American (5–6 oz) $14.85, Brazil tails $13.53 and Australia tails $23.20.

In India, the prices realized by the lobster fisher at the landing centre in September 2016 are variable with species and size: Live *P. homarus* (90–250 g, Rs. 750), *P. ornatus* (up to 900 g, Rs. 2400; 900 g–2 kg, Rs. 3400), *P. versicolor* (100–300 g, Rs. 300; 300–400 g, Rs. 700; 400–500 g, Rs. 800; and above 500 g, Rs. 1400) and sand lobster *T. unimaculatus* (40 counts per kg, Rs. 500; 15 counts per kg, Rs. 900–1000; and 6 counts per kg, Rs. 1300) (US$ = Rs. 66.6). Sand lobsters are caught in trawlers and are brought dead to the landing centre. Though there are size limit for export of lobsters from India, many under-sized lobsters are caught and traded (Radhakrishnan 1989; Vijayakumaran, personal observation).

Almost the entire volume of wild harvested lobsters is exported from India and there is no internal market except a few high-end restaurants. Though there is no retail market in India for lobsters, few online traders are marketing lobster products. The price of whole raw-frozen lobster varied from Rs. 1800–2290 per piece in individual packing with a size range of 800–1000 g. For IQF lobster meat, the price quoted is Rs, 550 per ½ kg. The declared price realization during export gives a fairly good idea of the price range of lobsters in India. The maximum price per kg (Rs. 1455–3693) is gained by live lobsters of different sizes. Whole raw-frozen products are priced at Rs. 870–2230 with the maximum price for those exported to Maldives. Lobster tails earn Rs. 842–1281 and whole cooked lobster Rs. 703–1241. Chilled spiny lobster price is Rs. 600–1269, while that of sand lobster is Rs. 226–824. The meat of only sand lobster is exported from India at a price of Rs. 626–957. Compared to all these products, the price of deep sea lobster exported as whole raw frozen is very low ranging from Rs. 269 to 550/kg.

14.4 Global and Domestic Market for Lobsters

Canada and the USA are the top producers of the clawed lobsters where as Australia are the biggest producers of spiny/rock lobsters. The USA is also the largest consumer of lobsters and is topping the import chart by more than 2 times in value than that of China, the second largest importer. Canada, the European Union countries and Japan are the other countries importing large quantities of lobsters. Canada tops in export of lobsters, mainly to the US markets and earns two times the value of the next big exporter, Australia.

14.4.1 World's Largest Buyer of Live Spiny/Rock Lobsters

The value of lobster export is governed by the item of export. Live lobsters earn more than twice that of any other forms of lobsters exported. The total share of Indian lobster export to China (14%) and Hong Kong (7%) is 21%, but in terms of value it is 36% (China 25% and Hong Kong 11%). Incidentally, almost 95% of lobsters exported to China and Hong Kong are live lobsters and that explains the value. The spiny or rock lobster is much more popular in China than the clawed lobsters, due in part to the higher meat yield per lobster. The Chinese proudly serve lobster as a statement delicacy at wedding receptions and banquets, signifying luxury and prestige. When cooked, the lobster shell turns red, which is a lucky colour in China.

Prior to 2009, there was not much interest by the Chinese in live clawed lobster imports. The combined dollar value of all US lobster (*H. americanus*), sold to China totalled just $74,651 in 2008. However, by 2010, there was exponential growth of 400% to $1,308,401 and then to almost $three million in 2011. Currently, China buys 80% of the spiny lobster harvested in Australia – a cash value of well over US$300 million annually. Neighbouring New Zealand also is part of an active Chinese lobster trade, earning close to US$200 by shipping 90% of its live rock lobster to China.

Most of the live lobster imported into China is sold to one of 300+ wholesale seafood markets, which are then resold to high-end Chinese restaurants that demand premium, fresh seafood. Live seafood imports are one of the most highly regulated industries in China, and importers must meet stringent requirements for sales process, food safety, labelling and proper documentation. Chinese importers are highly critical of mortality rates and it is very expensive for a shipper to export live lobster only to have the shipment denied or refused. No more than 3–5% mortality is considered acceptable to the buyer, and on the shipper's end anything less is not financially feasible. Therefore, operators must take every precaution to reduce shipping and wait times so that the product arrive fresh and alive, ensuring a successful transaction.

14.5 Lobster Export from India

The first export consignment of frozen lobster tails was sent to the USA in 1956; 0.02 million pounds of lobster tails valued at 0.009 million USD was exported (Smith 1958). Since then, India was one of the prominent lobster producing and exporting countries. The main lobster export products from India during the 1960s and 1970s were frozen lobster tails and lobster meat. Between 1962 and 1966, 329 tonnes of lobster tails valued at US$ 97,720 was exported (George 1973). Lobster meat had good demand and export of this product beginning in 1962, and between 1962 and 1970, 185 tonnes valued at US$ 130293 was exported. Exports steadily increased, and between 1997 and 2006, an average 1280 tonnes of lobsters worth US$ 15.1 million was exported annually from India. Live lobster export began in 1991 and since then spiny lobsters were exported in live condition to Singapore and Hong Kong from Chennai and Thiruvanathapuram airports. Presently, live lobsters are mainly exported to China and Hong Kong. Other countries such as Singapore, Vietnam, Taiwan and Malaysia also import small quantities of live lobsters from India.

The diversification in lobster processed products was started during 1980s. Frozen whole cooked and frozen lobster has the biggest share in Indian export until 2003. The frozen whole lobster captured the export market from 2004 onwards. A major share of chilled lobsters, mainly sand/slipper lobster, *T. unimaculatus*, is being exported primarily to the Middle East. The deep sea lobster, *Puerulus sewelli*, is also exported in small quantities as whole (raw) frozen mainly to Italy, and it is the least valued product. The IQF whole lobster was a new product introduced in 2002, and since then several value-added IQF products such as IQF head on, IQF whole cooked, IQF lobster meat, IQF lobster tail and whole round lobsters had gained export demand. During 1997, only four lobster products were exported from India, and by 2014, 25 different products are traded in the export market.

The annual average lobster export from India between 1997 and 2014 was 1417 tonnes valued at US$ 20.1 million (Fig. 14.3). Lobsters are caught throughout the year in India with a peak from October to January which is reflected in the export also.

14.5.1 Live Lobsters

Live export of spiny lobster from India began in 1990s. The demand is maximum from Southeast Asian countries, especially from mainland China and Hong Kong. Live lobsters fetch almost twice the price of a dead lobster (US$37 for live and $18 for frozen whole or whole cooked) in the export market. An average 154 tonnes was annually exported during 1997–2014, valued at US$3.9 million (Fig. 14.4). The volume exceeded 200 tonnes only in 2014, and during 2015, 286 tonnes was exported (MPEDA 2016).

In a country like India, more than 75% of the lobsters are landed either dead or moribund and are unfit for live transport. Of the 1417 tonnes of lobster exported annually between 1997 and 2014 from India, only about 11% or 154 tonnes included live lobsters.

Fig. 14.3 Lobster export by volume (t) and value (US$ million) from India from 1997 to 2014 (MPEDA Data 2016)

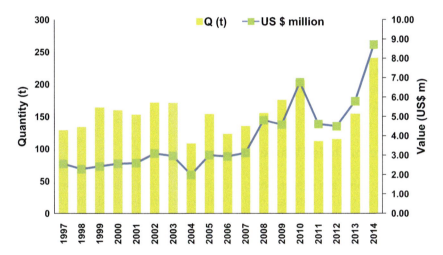

Fig. 14.4 Export of live lobsters in quantity (t) and value (US $ million) during 1997–2014

14.5.2 Whole-Cooked Frozen Lobster

Fresh, whole cooked lobster is a natural, ready-to-eat product, which can also be steamed or boiled if required hot. Generally sold in a vacuum packed pouch, a fresh whole cooked lobster will have a 7–10 day shelf life from the time of production (Holmyard and Franz 2006). An average 475 tonnes whole-cooked plate-frozen lobsters valued at US$ 6.13 million was exported annually from India between 1997 and 2014, which forms about 34% of the total lobster export (MPEDA)

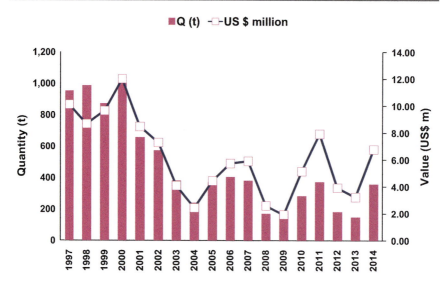

Fig. 14.5 Annual export of whole cooked and frozen lobsters from India (1997–2014)

(Fig. 14.5). The bulk of the product is exported to Japan. The volume of export declined after 2000, and between 2007 and 2014, an average 262 tonnes was exported. With application of advanced freezing technology, lobsters are exported more as whole raw frozen nowadays, as it also fetches the same or more price than the whole cooked lobster.

14.5.3 Fresh, Chilled Whole Lobster

Fresh whole spiny and slipper lobster preserved in ice is a product much in demand in the Arabian Gulf countries. Local as well as imported chilled lobsters are marketed in supermarkets as well as in other fish retail markets in these countries. During 2015, 55 tonnes of chilled whole lobster with a value of US$ 0.8 million was exported to Middle East countries.

14.5.4 Frozen Whole Lobster

The export of frozen whole rock lobster from India began during 2000, and an annual average of 392 tonnes valued at US$ 6.34 million was exported from 1999 to 2014, constituting 27% of the total lobster export (Fig. 14.6). Frozen whole rock lobster had consistent demand throughout. The unit value of the product almost doubled in a span of 10 years.

The frozen whole slipper lobster (*T. unimaculatus*) also had export demand, and an average 92 tonnes worth US$ 0.7 million was exported from 2002 to 2014 (Fig. 14.7).

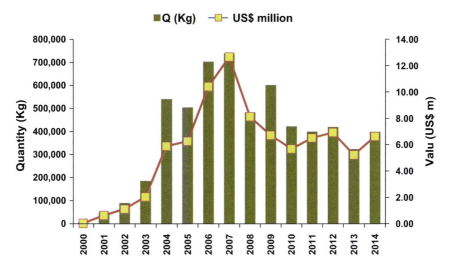

Fig. 14.6 Export figures of frozen whole rock lobster from India

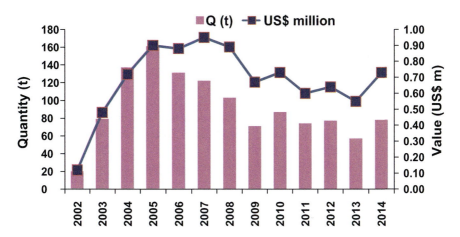

Fig. 14.7 Export figures of frozen whole slipper lobster (*T. unimaculatus*) from India

14.5.5 Frozen Rock and Slipper Lobster Tails

During late 90s and early years of this century, frozen rock lobster and slipper lobster tails was the second most exported item in quantity and value. During 1997–2014, an average 52 tonnes of frozen rock lobster tails per year valued at US$ 0.64 million was exported from India. However, during 1997–2002, an annual average 132 tonnes of rock lobster tails valued at US$ 1.4 million was exported (Fig. 14.8). During 2014, the annual export of frozen rock lobster tails was just 4 tonnes valued at US$ 0.06 million. The slipper lobster tails fetched a lesser price, and an average

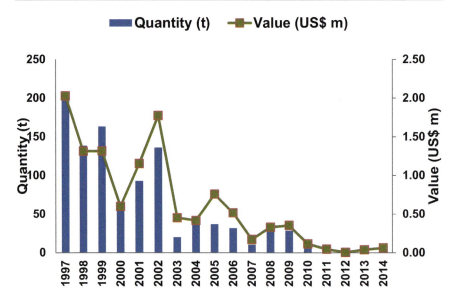

Fig. 14.8 Annual export figures of rock lobster tail from India

34 tonnes valued at US$ 0.3 million was exported during the same period (1997–2014) (Fig. 14.9). The export of slipper lobster tails during 1997–2002 was an average 80 tonnes valued at US$ 0.65 million.

The drop in export of frozen lobster tails may be due to lesser demand for this item in the international market and increased demand for the frozen whole lobster.

Spiny lobster tails are sometimes glazed (up to 20% of tail weight), which can lead to short net weights, whereas cold water lobster tails are generally sold dry, but protected with plastic to prevent drying and freezer burn.

14.5.6 Frozen Lobster Meat

An average 17 tonnes of frozen lobster meat valued at US$ 0.14 million was exported during 1999–2014 (Fig. 14.10). The market shows high fluctuation in demand for the product. In a period of 10 years, the international price for the product has almost doubled. Whilst a tonne of the product fetched US$ 4286 in 2003–2004, the price was US$ 8750/tonnes during 2013–2014.

Lobster meat is available in a wide variety of forms, including block frozen, blast frozen and flash frozen forms, prepared with or without brine and with or without glazing. It may be raw, boiled or steam cooked and then packed in vacuum packs, pouches, skin-wrap packs, clear plastic trays, or cans, depending on the product and the intended market.

Fresh lobster meat is popular in caterers as it offers a simple way to increase menu options with a high-value product. Lobster meat is generally handpicked and

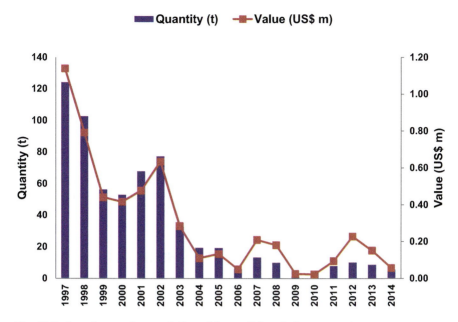

Fig. 14.9 Annual export figures of slipper lobster tail from India

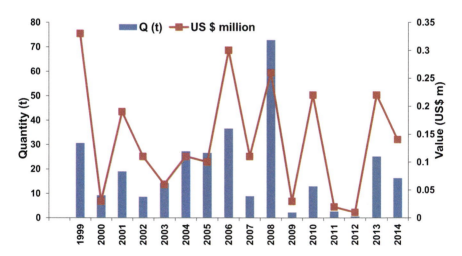

Fig. 14.10 Export of frozen rock lobster meat during 1999–2014

sold in tamper-proof containers or vacuum packs, and may contain a combination of tail, claw and knuckle meat ready for use. Tail meat is also sold on its own as a higher value product (Holmyard and Franz 2006). High-end restaurants in North America are slowly shifting their preference from live lobsters to this ready-to-prepare lobster product.

14.5.7 IQF Products

The export of IQF lobster products from India began in 2002 with a meagre consignment of 5 tonnes valued at US$ 0.02 million. The IQF products have great demand in the international market. The volume of exports gradually increased from 2002, and in 2014, 586 tonnes of IQF products (31.4% of total exports) valued at US$ 6.21 million (19.2%) was exported (Fig. 14.11). Among the IQF products, the IQF whole lobster was in maximum demand.

Presentation and packaging is playing an increasingly significant role in the marketing of lobster and lobster products, especially where retail packs are concerned. Effective use is made of attractive window sleeve packs, through which the skin or vacuum packed product can be clearly seen, but on which is printed the product and cooking information (Holmyard and Franz 2006).

14.5.8 Month-Wise Lobster Export from India in 2015

The latest month-wise lobster export in 2015 varied from 50.02 tonnes in August to a maximum of 217.46 tonnes in October (Fig. 14.12). Lobster export is maximum between September and December which is the peak season for lobster production in India.

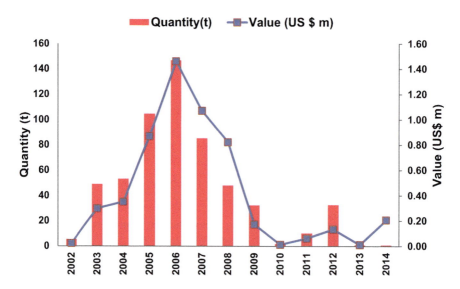

Fig. 14.11 Export figures of IQF lobster products from India

14 Post-harvest Processing, Value Addition and Marketing of Lobsters

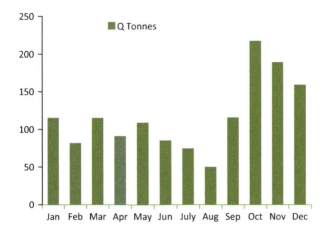

Fig. 14.12 Monthly export of lobsters from India

14.5.9 Country-Wise Export of Lobsters from India

India exports lobsters to 22 countries in Europe and Asia. Recently, the export of lobsters to Japan and Europe has reduced owing more lucrative markets in the Indian Ocean islands Maldives and Mauritius and the increase of the export of chilled and frozen lobsters to the Middle East. Another important change in the trend is the direct export of live lobsters to China. China is the most important lobster importing country for Indian lobsters with 14% share in export by quantity. The share of export of Indian lobster to main importing countries by volume and Value are given in Figs. 14.13 and 14.14.

The maximum return for the export of frozen lobsters from India is from Maldives which yields 8% by value though by quantity it is only 7%. Minimum value realized is from the chilled lobsters exported to the middle yeast, and the export to this segment is predominantly of sand/slipper lobster, *Thenus unimaculatus*.

14.5.10 Port-Wise Export of Lobsters from India

The port-wise export data of lobsters by quantity as well as value in 2015 are given in Figs. 14.15 and 14.16. The north-west ports, Nhava Sheva sea port, Pipavav sea port and Mundra sea port, and Mumbai air cargo together share a quantity share of 68%, whereas by value the share is only 57%. In contrast, the southern air cargo from Thiruvananthapuram and Chennai earn a value of 34%, though the quantity exported is only 18% as live lobsters are exported from these ports.

There is a change in the export of lobster products as lobster tails were the only product exported in the beginning, and in 1980s, India started the export of whole cooked lobsters to Japan which yielded almost double the price of frozen tails. Live

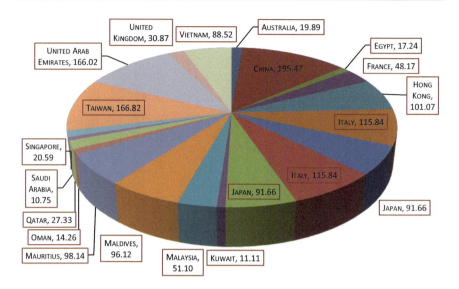

Fig. 14.13 Lobster export by quantity (tonnes) to different countries from India in 2015

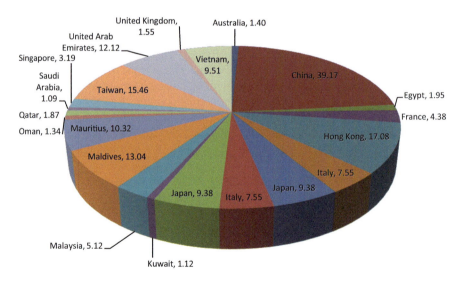

Fig. 14.14 Lobster export by value (lakh rupees) to different countries from India in 2015

lobster export was initiated in early 1990s and is ever on the increase since then. The major change in lobster export in the early twenty-first century is the increase in export of whole raw lobsters using latest technologies in processing. The quality of the frozen products has been enhanced and the raw-frozen lobster is priced more than that of whole cooked lobster now.

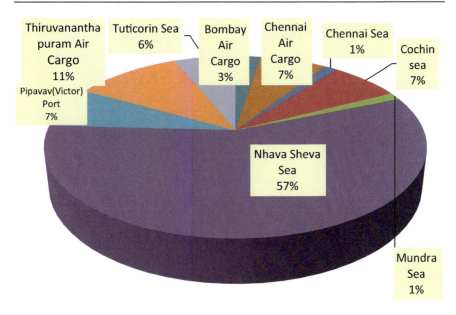

Fig. 14.15 Export of lobsters by quantity (tonnes) by air cargo and sea from India in 2015

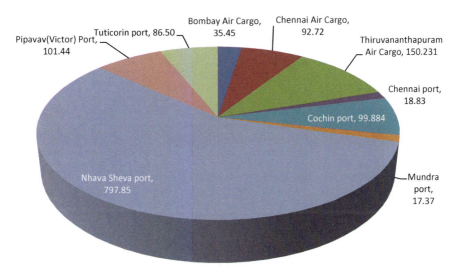

Fig. 14.16 Lobster export by value through Indian sea and airports (in lakh rupees) in 2015

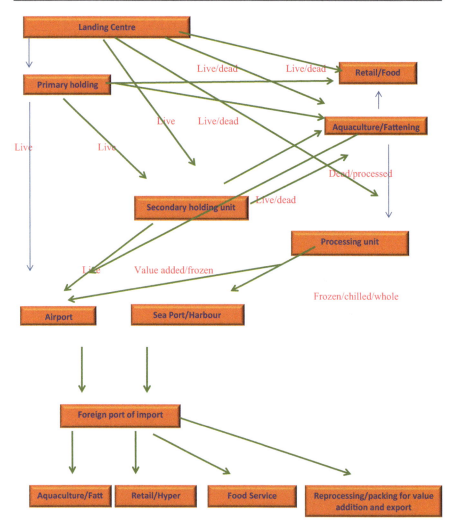

Fig. 14.17 Flow chart of lobster supply chain in India

14.6 The Lobster Supply Chain in India

The supply chain of lobsters from landing centre to the export house is represented in the flow-through chart (Fig. 14.17). The live lobsters are either held in a primary holding centre from where they are packed and sent to airport or are held in a secondary holding centre and packed and sent to the airport. Sometimes live lobsters are directly sent to secondary holding centres. Juveniles and subadults procured and held in primary holding centres are at times sent to aquaculture farms for ongrowing and value addition. Live lobsters from either the landing centre or the primary holding centre are also procured and served by five star hotels. *P. polyphagus* landed by trawlers are generally dead and are sent to processing centres for export as whole (raw) frozen, frozen whole cooked and chilled. The slipper lobster *T. unimaculatus* normally is landed dead and spiny lobsters which are also landed as dead are immediately sent to processing centres to export as chilled whole, frozen whole or tails in plate frozen or IQF form. The processing centres also make value-added products such as individually frozen lobster meat or IQF meat or IQF whole and market to niche markets where they get higher price.

14.7 Conclusion

The world market for lobsters is ever increasing, as average prices have gone up in recent years, especially for imports into China, where the unit value exceeds now US$30/kg. The relatively strong euro led to a decline in French lobster prices (FAO 2017). The export of lobsters from India has expanded to many countries in Asia and Europe. However, efforts should be made to export more lobsters live as there is great market for it in China and East Asian countries. Another important initiative to increase earnings from lobsters should be production and marketing of value-added products directly catering to the food service and retail market as more such products are being introduced. American lobster market is facing a serious challenge as the demand of its lobster product is decreasing in China and other countries (FAO) and still more threatening is the ban of live American lobster export to European Union countries. The European Union considered banning the import of live American lobster as there was a complaint from Swedish scientists that the American lobster is becoming invasive as live American or hybrid lobsters and few with eggs were caught in Swedish Waters (Stebbing et al. 2012; Anderson 2016). In a great relief to the American lobster market, the European Union has rejected the invasive theory, and no ban was imposed on lobster import from America. There is no threat to spiny lobster live export at present and India should take advantage of this factor and try to expand the live lobster export to more countries. Online marketing of sea food is creating waves especially in the western countries and India too should take advantage of this to market the lobster products to boost export.

References

Anderson, J. C. (2016, March 18). *Sweden seeks to stop imports of live Maine lobster into Europe*. Business.

Annie, T., & McCarron, P. (2006). *Lobster market overview*. Maine Lobstermen's Association.

Barker, E., & Rossbach, M. (2013). Western rock lobster fishery—2013/2014 season. *Comm Fish Product Bulletin, 48*, 1–8. Department of Fisheries, Government of Western Australia.

Barrento, S. (2010). *Nutritional quality and physiological responses to transport and storage of live crustaceans traded in Portugal*. Ph. D. thesis, Universidade do Porto. Porto. 261 pp.

Bridges, C. F., & Brand, A. R. (1980). Oxygen consumption and oxygen independence in Marine crustaceans. *Marine Ecology Progress Series, 2*, 133–141.

CMFRI, K. (2017). *CMFRI Ann. Rep. 2016–2017. Technical report* (p. 284). Kochi: CMFRI.
Coyle, S. D., Durborow, R. M., & Tidwell, J. M. (2004). *Anesthetics in aquaculture*. Stoneville: Southern Regional Aquaculture Center, Publication. No. 3900.
Defur, P. L. (1988). Systemic respiratory adaptations to air exposure in intertidal decapod crustaceans. *American Zoologist, 28*, 115–124.
Ellington, W. R. (1983). The recovery from anaerobic metabolism in invertebrates. *The Journal of Experimental Zoology, 228*, 431–444.
FAO. (2014). *FAO year book 2012: Fish aquaculture statistcs*. Rome: FAO.
FAO. (2016a). *Strong lobster supplies but weakening demand in China and Europe*. http://www.fao.org/in-action/globefish/market-reports/resource-detail/en/c/429238/
FAO. (2016b). FAO year book 2015: Fish aquaculture statistics. FAO, Rome 2016.
FAO. (2017). *Globefish Research Programme, The World Lobster Market*, Vol. 123. Food Export Association of the Midwest USA 309 W Washington Street, Suite 600 Chicago, IL 60606 USA http://www.tropicseafood.com
Foley, D. M., Stewart, J. E., & Holley, R. A. (1966). Iso-butyl alcohol and methyl pentynol as general anaesthetics for the lobster, *Homarus americanus* Milne-Edwards. *Canadian Journal of Zoology, 44*, 141–143.
Frederich, M., & Pörtner, H. O. (2000). Oxygen limitation of thermal tolerance defined by cardiac and ventilatory performance in spider crab, *Maja squinado*. *American Journal of Physiology—Regulatory, Integrative and Comparative Physiology, 279*, 1531–1538.
Gäde, G., Graham, R. A., & Ellington, W. R. (1986). Metabolic disposition of lactate in the horseshoe crab *Limulus polyphemus* and the stone crab *Menippe mercenaria*. *Marine Biology, 91*, 473–479.
George, M. J. (1973). The lobster fishery resources of India. In *Proceedings symposium on living resources of the seas around India. Cochin, December 1968* (pp. 570–580). Special Publication, Central marine Fisheries research Institute.
Greenaway, P., Morris, S., McMahon, B. R., Farrelly, C. C., & Gallagher, K. L. (1996). Air breathing by the purple shore crab *Hemigrapsus nudus* (Dana). 1. Morphology, behaviour and respiratory gas exchange. *Physiological Zoology, 69*, 785–805.
Guilbault, K. (2015). Global analysis report. *Agriculture and Agri-Food in Canada*. Global Analysis Division 1341 Baseline Road, Tower 5, 4th floor Ottawa, ON Canada K1A 0C5.
Holmyard, N., & Franz, N. (2006). Lobster Markets. *FAO, GLOBEFISH*, Fishery Industries Division Viale delle Terme di Caracalla, 00153 Rome.
Ilangumaran, G. (2014). *Microbial degradation of lobster shells to extract chitin derivatives for plant disease management*. Master of Science, Dalhousie, Halifax, Nova Scotia.
Jayagopal, P. (2004). Studies on stress during live transport of spiny lobsters-*Panulirus* spp., Ph.D. Thesis. University of Madras, Chepauk, Chennai – 600005, India.
Kriz, A. (1995, August 29–31). Live crustaceans: Seeking a competitive advantage. Aquaculture towards the 21st century. K. P. P. Nambiar, & T. Singh (Eds.), Proceedings of the INFOFISH-AQUATECH '94 conference (pp. 249–253). Colombo: INFOFISH.
Lorenzon, S., Giulianini, P. G., Martinis, M., & Ferrero, E. A. (2007). Stress effect of different temperatures and air exposure during transport on physiological profiles in the American lobster *Homarus americanus*. *Comparative Biochemistry and Physiology – Part A: Molecular Integrative Physiology, 147*, 94–102.
Maharajan, A., & Vijayakumaran, M. (2004). Copper toxicity in the spiny Lobster, *Panulirus homarus* (Linnaeus). *Proc. MBR 2004 Natl. Sem. New Front. Mar. Biosci. Res.* (pp. 205–212).
Megan Ware. (2016). *Lobster: Nutritional information, health benefits. RDN LD Knowledge Centre*. http://www.medicalnewstoday.com/articles/303332.php.
Morris, S., & Bridges, C. R. (1994). Properties of respiratory pigments in bimodal breathing animals: Air and water breathing by fish and crustaceans. *American Zoologist, 34*, 216–228.
MPEDA. (2016). *Annual Report*. Kochi: Marine Products Export Development Authority.
Plaganyi, E. E., McGarvey, R., Gardner, C., Caputi, N., Dennis, D., de Lestang, S., Hartmann, K., Liggins, G., Linnane, A., Ingrid, E., Arlidge, B., Green, B., & Villanuev, C. (2017). Overview,

opportunities and outlook for Australian spiny lobster fisheries. *Reviews in Fish Biology and Fisheries, 28*(1), 57–87. ISSN 0960-3166 (2017) [Refereed Article].

Radhakrishnan, E. V. (1989). Physiological and biochemical studies on the spiny lobster *Panulirus homarus*. Ph. D Thesis, University of Madras, Madras, pp. 168.

Raghubeer, E. V. (2007, September). *High hydrostatic pressure processing of seafood*. Avure Technologies. https://www.researchgate.net/topics?ev=nav_discussions

Rahman, K. M., & Srikrishnadhas, B. (1994). Packing of live lobsters-the Indian experience. *INFOFISH International, 6/94*, 47–49.

Sackton, J. (2007). *Global supply, demand and markets for lobster: What are the opportunities for N. American Lobster. Seafood.com*. Halifax.

Schmitt, A. S. C., & Uglow, R. F. (1997). Haemolymph constituent levels and ammonia efflux rates of *Nephrops norvegicus* during emersion. *Marine Biology, 127*, 403–410.

Smith, W.F.G. 1958. *The spiny lobster fishery in Florida*. Univ Miami Educat Ser No. 11 p. 18.

Spanoghe, P. T., & Bourne, P. K. (1997). Relative influence of environmental factors and processing techniques on *Panulirus cygnus* morbidity and mortality during simulated live shipments. *Marine and Freshwater Research, 48*(8), 839–844.

Speed, S. R., Baldwin, J., Wong, R. J., & Wells, R. M. G. (2001). Metabolic characteristics of muscles in the spiny lobster, *Jasus edwardsii* and responses to emersion during simulated live transport. *Comparative Biochemistry and Physiology, 128 B*, 435–444.

Spicer, J. I., Hilll, A. D., Taylor, A. C., & Strang, R. H. C. (1990). Effect of aerial exposure on concentrations of selected metabolites in the blood of the Norwegian lobster *Nephrops norvegicus* (Crustacea: Nephropidae). *Marine Biology, 105*, 129–135.

Stebbing, P., Johnson, P., Delahunty, A., Clark, P. F., McCollin, T., Hale, C., & Clark, S. (2012). Reports of American lobsters, *Homarus americanus* (H. Milne Edwards, 1837), in British waters. *BioInvasions Records, 1*(1), 17–23.

Tselikis, A., & McCarron, P. (2009). *Lobster market overview. Trade adjustment assistance for farmers*. Kennebunk: Maine Lobstermen's Association.

Vermeer, G. K. (1987). Effects of air exposure on desiccation rate, haemolymph chemistry, and escape behaviour of the spiny lobster, *Panulirus argus*. *Fishery Bulletin, 85*(1), 45–51.

Vijayakumaran, M., & Radhakrishnan, E. V. (1997). Live transport and marketing of spiny lobsters in India. *Marine and Freshwater Research, 48*, 823–828.

Zebe, E. (1982). Anaerobic metabolism in *Upogebia pugettensis* and *Callianassa californiensis* (Crustacea, Thalassinidea). *Comparative Biochemistry and Physiology, 72*(B), 613–618.

Perspectives and Future Directions for Research

15

Gopalakrishnan A, E. V. Radhakrishnan, and Bruce F. Phillips

Abstract

Lobster fishery, considered as a highly valued and most profitable marine resource, encounters problem of overexploitation and collapse in regions where fishery is not well managed. This chapter begins with the fishery database and stock assessment of lobster resources along the Indian coast and suggestions in this aspect. The current management measures and regulations in lobster fishery along with the limitations for implementation and scope for improvement in various sectors are discussed. Integration of genetic stock structure data into lobster fisheries governance, utility of molecular tools for evaluating response of lobsters to climate change, need for the development of traceability markers and microarray chips in lobsters, scope for nutrigenomics in larviculture, need for understanding molecular mechanisms of metamorphosis of larvae, quantitative genetic aspects and requirement of DNA-based assays for pathogen diagnosis are explained afterwards. The chapter stresses the importance of cross-border collaborations in research for the globally distributed lobsters. The necessity of database on carrying capacity, water currents as well as climatic and hydrographic conditions in the culture area and optimisation of cost-effective, high conversion practical feeds are highlighted. The areas which need focused research for successful hatchery production of lobster seeds, entry of private entrepreneurs in lobster breeding and seed production and need for ecolabelling of lobster fisheries for judicious harvest are highlighted at the end.

Gopalakrishnan A (✉) · E. V. Radhakrishnan
ICAR-Central Marine Fisheries Research Institute, Cochin, Kerala, India
e-mail: agopalkochi@gmail.com

B. F. Phillips
School of Molecular and Life Sciences, Curtin University, Perth, WA, Australia
e-mail: B.Phillips@curtin.edu.au

© Springer Nature Singapore Pte Ltd. 2019
E. V. Radhakrishnan et al. (eds.), *Lobsters: Biology, Fisheries and Aquaculture*,
https://doi.org/10.1007/978-981-32-9094-5_15

Keywords

Stock · Management · Genetics · Breeding · Aquaculture

15.1 Introduction

Lobsters are one of the widely researched group among the decapods probably due to higher commercial value of some species. They constitute economically important fisheries in many countries, and obviously due to higher demand from consumers of nations with higher economies, they have been the target of exploitation, subsequent overexploitation, resource depletion and even collapse of fisheries in some regions. Among the group, clawed and spiny lobsters have received more focus of research due to economic importance of certain species compared to the smaller and relatively rare scyllarid lobsters (Spanier and Lavalli 2007). Recently, scyllarids have also gained importance due to increased consumer acceptance and the potential for aquaculture of some of the tropical species. Research on nephropids and palinurids covered almost all aspects of biological research including taxonomy, evolution, morphology and anatomy, ecology, physiology, growth, reproduction, neuroendocrine system and molecular phylogeny. The spectrum of research on scyllarids were limited and covered mostly the taxonomy, distribution, larval ecology and biology of species which contribute to fisheries.

15.2 Fishery Database, Biology and Stock Assessment

Marine lobsters are highly valued and sustain some of the most profitable fisheries of the world with worldwide landings of 294,000 metric tonnes (t) with a value of around US 2800 million in 2012, with an additional 2000 t produced through aquaculture (FAO 2014). Several lobster stocks are near or above sustainable limits of exploitation and many lobster fisheries remain overexploited or near collapse (FAO 2011, Briones-Fourza'n and Lozano-A'lvarez 2015). Fisheries in some countries are well managed and annual production from wild fisheries are consistent. Canada, USA, Sweden, Australia, New Zealand and South Africa manage their fisheries with well-planned input and output controls. In many developing countries including India, lobster being a small volume fishery, the fishing regulations are not strictly enforced and catches are either stagnating or are on a declining trend. Due to the commercial nature of clawed and spiny lobster fisheries, reliable information on landings and biological data on major species constituting the fisheries is available. Among the 92 species of scyllarids known to date, 30 species are of some interest to fisheries (Holthuis 1991). Annual production of scyllarids in the world is just an average 4500 t. Fishery and biological data on scyllarid lobster fisheries is limited as the fishery in most of the countries are of minor importance.

In India, statistical data on lobster fisheries from landing centres across the Indian coast is collected by the Central Marine Fisheries Research Institute (CMFRI), Cochin, using FAO-recognized stratified random sampling technique since 1950. Assessment of stocks of major species forming fishery on the northwest coast of India (Kagwade 1987, 1993, 1994; Kagwade et al. 1991; Kabli and Kagwade 1996, Deshmukh 2001) and on the southwest coast of India (George 1967a, b, 1973; Kathirvel et al. 1989; Radhakrishnan et al. 2013; Thangaraja et al. 2015) has been carried out. However, at present information on stocks of *Panulirus polyphagus* and *T. unimaculatus* along the northwest coast is not available. Egg-bearing lobsters form a good percentage of the catch in Mumbai. Fishery for *T. unimaculatus* has already been collapsed in Mumbai, and catches are on a declining trend along the coast of Gujarat (see Chap. 7). Analysis of biological data on *T. unimaculatus* clearly shows that the species is overexploited. Intensive fishing of egg-bearing lobsters by mechanised trawlers during peak breeding season and destruction of the settling 'nisto' would have probably caused the collapse of the fishery. Recently, there is an emerging fishery for lobsters along West Bengal, on the northeast coast and no information on population dynamics of species landed is available. Current catch of the deep sea lobster *Puerulus sewelli* along the southwest coast of India is only a fraction of the potential estimate of the species. CMFRI continues to collect statistical data on catch and catch per unit effort (CPUE) of lobsters from major landing centres. Collection of biological data of major species needs to be continued as management decisions can be taken only on the basis of scientific data on stock position. Estimation of stock status, maximum sustainable yield and optimum fishing effort for *P. polyphagus* may be carried out using multi-species stock assessment model, a multivariate version of Schafer's model.

15.3 Management and Fishing Regulations

Lobster fisheries in India are characterised by the absence of or inadequate regulations in fishing or non-enforcement of existing regulations. The Central Government has notified minimum legal size (MLS) for major spiny and slipper lobster species for export (Radhakrishnan et al. 2005). However, fishing within the territorial limit (20 nautical miles from shore) is under the jurisdiction of the state governments. Lobster fishing by artisanal gears is within this region. Recently, Kerala state has notified MLS for fishing for *P. homarus homarus* (200 g) on the recommendation of CMFRI, which is in tune with the MLS notification by the Central Ministry of Commerce and Industry and this is a progressive step towards fisheries management. Thangaraja & Radhakrishnan (2017) have suggested that MLS for *P. homarus homarus* may be fixed at 65.0 mm CL (equivalent weight 192 g), as the size is above first sexual maturity of the species. Multiday and single-day trawlers land 67% and gillnets 28% of the total annual production. *P. polyphagus* and *T. unimaculatus* are a trawl-based fishery and these lobsters appear as a bycatch.

Management of the spiny and slipper lobster fishery in India is complicated by failure to enforce regulations due to the practical difficulties of implementing fishing restrictions, as the resource is a bycatch in trawlers. Mesh regulation and closed season for lobster fishing during the peak breeding months are not practical management measures, as the states have already been enforcing a fishing holiday. Though *P. polyphagus* is intensively fished, it may be interesting to study the resilience of the resource even though the current catches are much lower than that landed in 1980s. In the artisanal sector, fishery management regulations such as ban on using trammel nets, minimum legal size carapace lengths for all commercially important species and ban on retention of egg-bearing lobsters may be strictly enforced by the respective state governments. Radhakrishnan et al. (2005) suggested establishment of fishery management councils with participation of fishermen, local panchayats, state government officials, NGOs and scientists for managing the fishery in each region.

15.4 Climate Change, Population Genetics and Aquaculture Genetics

The greatest genetic threats in the marine ecosystem are the extinction of genetically unique sub-populations and loss of genetic diversity primarily through overfishing and climate change (Ayyappan et al. 2014). Climate change influences genetic structure of populations by reducing or eliminating local populations, by disrupting patterns of migration and by shifting geographical distributions. Maintaining the maximum level of genetic variations in lobster stocks is vital for the preservation of their genetic resources. Application of molecular marker techniques can provide information on genetic stock structure and can aid in genetic stock identification (GSI) that can be of direct management relevance. Scaling up genome coverage for weakly structured organisms like lobsters will be helpful in improved estimates of their population genetic parameters. The genetic data can be utilised in redefining management units as implemented in the case of Pacific salmon (Bernatchez et al. 2017). Intensive research with latest high-precision molecular tools like SNPs needs to be carried out for all commercially important lobster species available along the Indian coast, in a manner similar to the American lobster (Bernatchez 2017), to know their genetic structure and patterns of connectivity. These genetic data can be integrated with other types of biological and oceanographic information.

Molecular markers used in numerous population genetic and phylogeographical studies of marine species have detected population responses to climate change such as range expansions, adaptive shifts and declines or increases in abundance (Lo Brutto et al. 2011). Early life stages of invertebrates like lobsters are more vulnerable to climate changes and environmental stress. Climate change is found to have complex effects on populations of the American lobster (*Homarus americanus*) along the east coast of North America. Preliminary studies have found that warmer water temperatures and more acidic conditions seem to make lobster larvae grow more slowly. Use of genetic tools may help to monitor the changes in genes of lobster populations which may provide an early-warning system so that changes to populations may be anticipated. Only a few studies have concerned with lobsters on the effect of stressors like low pH, oxidative stress (hypoxia), genetic variation in ocean acidification tolerance etc. on embryonic physiology, larvae, juveniles and adults (Styf 2014). Studies regarding genetics of local adaptation (for example, how genes vary with sea temperature) and studies on how lobster may respond to climate change and ocean acidification need to be initiated in India.

Documentation of genetic variation and diversity are extremely significant to evolve conservation strategies with long-term impact. Establishing Marine Protected Areas (MPAs) or Marine Reserves where fishing is illegal or very restricted is an important ecosystem-based management approach to help improve the sustainability of the fishery. Because there are opportunity costs to conservation, there is a need for science-based assessment of MPAs (Moland et al. 2013) and the vital information concerning levels of population connectivity among MPAs can be achieved through genetic studies. A few markers may be derived from key genes to derive recommendations for conservation and management for Indian lobsters as implemented in *Panulirus argus* from Central America (Truelove et al. 2015).

Geographic origins of lobsters may be quickly identified using genetic databases developed for each region by sequencing genomes of many lobsters from different areas. Traceability (fishery 'branding', eco-labelling) of marine shellfish products can be applied either at species or population level. Marine Stewardship Council has established a programme of genetic testing of species/stocks in relation to food authenticity and compliance with eco-certification. Assignments based on a moderate number of non-neutral SNP markers identified using genomics, have been successful at differentiating populations of lobsters like *Jasus edwardsii* (Villacorta-Rath et al. 2016). Attempts to develop traceability markers (small SNP panels ~100 nos.), which can detect species and populations of Indian lobsters, may be tried for their sustainable utilisation and conservation.

Fisheries-induced evolution can cause changes in growth rate, age at reproduction and can affect the productivity of fisheries. Through scans of candidate genes or larger panels of markers, DNA-level evidence of fisheries-induced evolution in wild fish stocks has been detected (Bernatchez et al. 2017). Knowledge regarding genomic changes through fisheries-induced selection is critical for enabling accurate modelling of future stock abundances under size-selective harvesting. It is hightime for cross-border collaborations, discussions, sharing of data and funds for carrying out future genetic research in lobsters which has global distribution pattern.

The routine population genetic studies using neutral/type II molecular markers, though useful in devising conservation strategies, reveal little about the adaptive side of the evolutionary coin. There is a growing interest in demonstrating adaptive population divergence at the molecular level, as well as in identifying the genetic architecture of local adaptive traits conferring fitness advantages to resident individuals of species in aquaculture. The construction of microarray DNA chips, containing thousands of ESTs derived from the whole animal or from tissues with particular biological function allows more specific assessment of characteristics (Ayyappan et al. 2014). Microarray chips may be developed to identify adaptive traits like hypo-saline tolerance in the mud spiny lobster, *P. polyphagus*, which is considered as a candidate species for cage culture in North West coast of India.

Aquaculture and population enhancement remain the holy grails of lobster management (Briones-Fourza'n and Lozano-A'lvarez 2015). An important bottleneck in the development of aquaculture of spiny lobsters is larval feeding and nutrition as the composition of diet of larvae remains uncharacterized for all species (Jeffs

2007). The recent advances in diet analysis of phyllosoma through NGS techniques is sure to contribute to a better understanding of larval diet which can aid to overcome the hurdle of providing suitable feed in lobster larviculture. The genetic interventions can supplement future research on perfecting hatchery technology of lobsters. Nutrigenomic studies will have applications for improving larval survival of scyllarid lobsters like *Thenus unimaculatus* and *Petrarctus rugosus*, seed production technology of which was developed by CMFRI.

The transition from phyllosoma to puerulus, which may take months to years after hatching, is another major bottleneck for the development of commercially sustainable aquaculture of spiny lobsters (Ventura et al. 2015). One of the challenges is the lack of knowledge concerning the biochemical triggers and molecular basis that initiates this transition (McWilliam and Phillips 1997, 2007). Elucidating the molecular mechanisms underlying this transition in spiny lobsters has the potential to highlight key pathways and base a technology for synchronising it. Studies in this aspect have been undertaken in *Sagmariasus verreauxi* (Ventura et al. 2015). Use of SNP array and whole-genome resequencing to identify genome regions that have large effects on life-history traits was elucidated in Atlantic salmon (Aykanat et al. 2016; Barson et al. 2017). Similar studies in the above aspects may prove useful in warm water species of lobsters available along the Indian coast.

High cost of hatchery production due to the extended and complex larval cycle and poor survival during the many moult stages, especially at metamorphosis; poses another major impediment to spiny lobster aquaculture. Knowledge regarding quantitative genetic characters for emerging aquaculture species like *P. homarus* need to be developed for key traits like improvement of larval survival through selection as attempted for *S. verreauxi* in Australia (Nguyen et al. 2018).

The wide distribution of lobster stocks also holds particular challenges for the treatment of emerging diseases like Milky haemolymph syndrome. DNA-based assays developed from genomic technologies can simultaneously provide rapid diagnosis of multiple pathogens. Assessment of infectious agents using quantitative PCR platform may be developed using high-throughput genomic technologies.

Integration of genetic data into production practices, fisheries governance, management and policy decisions is essential to maintain a sustainable future for the seafood industry, but in several countries including India, the concept is still at infancy. Relevant legal bodies and authorities need to bring together all stakeholders (management practitioners, policy-makers, fisheries and genomic scientists) to facilitate interaction and promote synergistic activities aimed at improved stock monitoring and management of lobsters to ensure sustainability.

15.5 Lobster Breeding, Hatchery Production and Aquaculture

Prospects for profitable aquaculture of tropical species of spiny and slipper lobsters are far more promising than the temperate species due to greater availability of wild seed lobsters and faster growth rates. Farming in both indoor or raceway/

pond-based system and in floating or fixed cages in lagoons or bays is technically feasible, with the former technology more capital intensive. Though volume of information is available on farming lobsters, further research on seed collection and transportation, efficient cage/tank design, stocking density, shelter requirements, food and feeding regimes and health management is necessary. Tank or cage shape and size, fabrication material and cage density (number of cages) play a major role in successful cage farming. A database on carrying capacity of the culture area, water currents in the bay, climatic and hydrographic conditions in the culture area is to be prepared.

A few countries in the world (e.g. Vietnam and Indonesia) practised capture-based aquaculture (CBA) of lobsters on a commercial scale. Small-scale farming is also being carried out in some countries in Asia and Pacific region. The industry peaked with an annual production of 1900 t drastically declining to 720 t in 2008 due to disease outbreak but recovered to 1387 t in 2015 (FAO 2016). Indonesia has also ventured into lobster aquaculture with a peak production of 914 t in 2013, declining to 161 t in 2015. Wild-caught pueruli, post-pueruli and juveniles of the tropical spiny lobsters, *P. ornatus* and *P. homarus* are cultured in floating and fixed cages until they attain marketable size.

Feed (fresh) constitutes 60–70% of the production cost, and depending on trash fish or other molluscan meat for feeding lobsters is not ecologically and environmentally sustainable, and therefore cost-effective, high conversion practical feeds have to be developed. There are many commercial feeds available in the market, but their performance is less than optimum. Therefore, advanced formulation of nutritionally balanced artificial grow-out diets will have to be developed for each species. Optimum feeding regimes need to be standardised to minimise wastage of food and environmental contamination and improvement in the conversion rate. Co-culture of lobsters and mussel may maintain the environmental quality of the culture site by reducing the organic matter in the water and sediments. Accumulation of faecal and farm waste organic matter at the cage bottom may attract pathogenic microorganisms and parasites resulting in water quality deterioration, and instances of disease outbreaks and high mortality of cultured lobsters have been reported in Vietnam.

Commercial aquaculture with hatchery produced seeds is still in the developing stage though remarkable advances have been accomplished in controlled breeding and seed production of lobsters. In recent years, open life cycle production systems have given way to closed life cycle production systems, where there is no necessity to depend upon wild source of breeders. Many aquatic species have been completely domesticated, and quality breeders could be produced in biosecure systems, and therefore unsustainable dependency upon wild supply of breeders could be avoided.

Complete larval development of several tropical and temperate species of palinurid lobsters have been achieved on a laboratory scale by the Japanese researchers followed by Australia, New Zealand and USA. However, the prospects of developing mass larval culture systems were not very promising due to difficulties in maintaining the fragile phyllosoma larvae with unique feeding habits for a prolonged period under strict hygiene conditions and also due to fundamental lack of information on several aspects of larval biology. Among the 10 species of palinurids in which post-larval production was successfully completed, only three tropical, *P. ornatus, P. argus* and *P. homarus*, and one temperate species, *P. elephas*, have the potential to be considered for hatchery production, as the larval phase is relatively shorter than other species. Entry of a private entrepreneur, Lobster Harvest, a company established by M.G. Kailis Group, Australia in 2007, with massive investments in lobster breeding and seed production programme is a promising step towards commercialising lobster aquaculture technology. Hatchery production of the spiny lobster *Panulirus ornatus* with a survival of 4.0–4.4% has been achieved in production-scale trials. Recent production of F_2 generation progeny is a significant step towards achieving this goal. The company is planning to scale up the larval production tanks to mass production systems. Scaling up production from small tank systems to mass scale production systems will be challenging as entirely new information on hydrodynamics of larval culture tanks, including tank shape, water flow rate, light intensity, water quality maintenance, and health management protocols will be required.

In India, lobster culture experiments were initiated in 1976 by CMFRI. Later scientific institutions such as NIOT and the TANUVAS have also been associated with lobster research. Spiny lobsters were reared to sexual maturity and bred under captive conditions and partial success achieved in larval culture. Grow-out of juveniles procured from wild were carried out in indoor culture systems and floating cages, and the economics worked out. In 2004, the slipper lobster larvae were successfully reared to settlement on a laboratory scale. However, mass production of lobster seeds was not successful. The two potential species of spiny lobsters for aquaculture in India are *P. ornatus* and *P. homarus homarus*. Although seed production of these two species were achieved by Australia and Japan, respectively, the technology may not be available to other countries. Therefore, the breeding programme may be continued with the two spiny lobster species (*P. ornatus* and *P. homarus homarus*) and the slipper lobster *T. unimaculatus*. However, it may be a time bound project, as the investors may not be interested to invest in lobster breeding research programme for a prolonged period. Two areas which need focused research are development of larval feeds and health management of the larvae to succeed in hatchery production of lobster seeds.

15.6 Conclusions

Remarkable progress has been achieved with respect to the biological knowledge of nephropid, palinurid and scyllarid lobsters. Commercial aquaculture of the European clawed lobster, *H. gammarus* is being pursued in Norway. Though hatchery technology for the American lobster *H. americanus* was achieved a few decades back, profitable commercial culture was not pursued due to slow growth and cannibalism. Larvae of several species of spiny lobsters were reared to settlement and the major hurdle is prolonged larval phase and low survival rate. The scyllarid *Thenus* spp.

were successfully bred and seeds produced to commercial level. Commercial seed production technology for the tropical spiny lobster *P. ornatus* is in the pipeline and shows high promise. Therefore, lobster aquaculture with hatchery produced seeds is going to be a reality.

Lobster research is being pursued by several countries across the world and research data on biology, physiology, taxonomy, phylogeny, larval, juvenile and adult ecology of lobsters will be continuously generated. Climate change and anthropogenic interventions will be the major threat for sustainability of lobster fisheries. More and more fisheries are to be ecolabelled so that the resource may be expected to be harvested judiciously.

References

Aykanat, T., Lindqvist, M., Pritchard, V. L., & Primmer, C. R. (2016). From population genomics to conservation and management: A workflow for targeted analysis of markers identified using genome-wide approaches in Atlantic salmon *Salmo salar*. *Journal of Fish Biology, 89*(6), 2658–2679.

Ayyappan, S., Jena, J. K., & Gopalakrishnan, A. (2014). Molecular tools for sustainable management of aquatic germplasm resources of India. *Agricultural Research, 3*(1), 1–21.

Barson, N. J., Aykanat, T., Hindar, K., Baranski, M., Bolstad, G. H., Fiske, P., & Primmer, C. R. (2017). The genetic basis of sea-age in Atlantic salmon: a large-effect gene and some additional surprises. Conference session 'Genomics for improved fisheries management and conservation: have the promises been fulfilled?' at the 7th World Fisheries Congress in Korea in 2016.

Bernatchez, L. (2017). Population genomics for conservation and management of aquatic resources: have the promises been fulfilled? Conference session 'Genomics for improved fisheries management and conservation: have the promises been fulfilled?' at the 7th World Fisheries Congress in Korea in 2016.

Bernatchez, L., Wellenreuther, M., Araneda, C., Ashton, D. T., Barth, J. M., Beacham, T. D., Maes, G. E., Martinsohn, J. T., Miller, K. M., Naish, K. A., & Ovenden, J. R. (2017). Harnessing the power of genomics to secure the future of seafood. *Trends in Ecology & Evolution, 32*(9), 665–680.

Briones-Fourzán, P., & Lozano-Álvarez, E. (2015). Lobsters: Ocean icons in changing times. *ICES Journal of Marine Science, 72*(suppl_1), i1–i6.

Deshmukh, V. D. (2001). Collapse of sand lobster fishery in Bombay waters. *Indian Journal of Fisheries, 48*(1), 71–76.

FAO. (2011). *Review of the State of World Marine Fisheries Resources*. Marine and Inland Fisheries Service Fisheries and Aquaculture Resources Use and Conservation Division, FAO Fisheries and Aquaculture Department. FAO Fisheries and Aquaculture Technical paper 569.

FAO. (2014). *FAO yearbook. Fishery and aquaculture statistics 2012. Statistics and information branch of the fisheries and aquaculture department* (p. 76). Rome: FAO.

FAO. (2016). *The state of world fisheries and aquaculture 2016. Contributing to food security and nutrition for all* (p. 200). Rome: Food & Agriculture Org.

George, M. J. (1967a). The Indian Spiny Lobster. In: *Souvenir 20th Anniversary Central Marine Fisheries Research Institute*, 3 February 1967, Mandapam.

George, M. J. (1967b). Observations on the biology and fishery of the spiny lobster *Panulirus homarus (Linn.)*. In: *Proceedings of the Symposium on Crustacea*, Part 4, MBAI, 12–16 January 1965, Ernakulam.

George, M. J. (1973). The lobster fishery resources of India. In: *Proceedings of the symposium on living resources of the seas around India*, 1968, Mandapam Camp.

Holthuis, L. B. (1991). FAO species catalogue. Vol. 13. Marine lobsters of the world. An annotated and illustrated catalogue of species of interest to fisheries known to date. *FAO Fisheries Synopsis, 125*(13), 1–292.
Jeffs, A. (2007). Revealing the natural diet of the Phyllosoma larvae of spiny lobster. *Bulletin of Fisheries Research Agency Japan, 20*, 9–13.
Kabli, L. M., & Kagwade, P. V. (1996). Morphometry and conversion factors in the sand lobster *Thenus orientalis* (Lund) from Bombay waters. *Indian Journal of Fisheries, 43*(3), 249–254.
Kagwade, P. V. (1987). Age and growth of spiny lobster *Panulirus polyphagus* (Herbst) of Bombay waters. *Indian Journal of Fisheries, 34*(4), 389–398.
Kagwade, P. V. (1993). tock assessment of the spiny lobster *Panulirus polyphagus* (Herbst) off north-west coast of India. *Indian Journal of Fisheries, 40*(1&2), 63–73.
Kagwade, P. V. (1994). Estimates of the stocks of the spiny lobster *Panulirus polyphagus* (Herbst) in the trawling grounds off Bombay. *Journal of the Marine Biological Association of India, 36*(2), 161–167.
Kagwade, P. V., Manickaraja, M., Deshmukh, V. D., Rajamani, M., Radhakrishnan, E. V., Suresh, V., Kathirvel, M., Rao, & Sudhakara, G. (1991). Magnitude of lobster resources of India. *Journal of the Marine Biological Association of India, 33*(1&2), 150–158.
Kathirvel, M., Suseelan, C., Rao, & Vedavyasa, P. (1989). Biology, Population and Exploitation of the Indian Deep-Sea Spiny Lobster, *Puerulus sewelli* Ramadan. *Fishing Chimes, 8*(11), 16–25.
Lo Brutto, S., Arculeo, M., & Stewart Grant, W. (2011). Climate change and population genetic structure of marine species. *Chemical Ecology, 27*(2), 107–119.
McWilliam, P. S., & Phillips, B. F. (1997). Metamorphosis of the final phyllosoma and secondary lecithotrophy in the puerulus of *Panulirus cygnus* George: A review. *Marine and Freshwater Research, 48*, 783–790.
McWilliam, P. S., & Phillips, B. F. (2007). Spiny lobster development: Mechanisms inducing metamorphosis to the puerulus: A review. *Reviews in Fish Biology and Fisheries, 17*, 615–632.
Moland, E., Olsen, E. M., Knutsen, H., Garrigou, P., Espeland, S. H., Kleiven, A. R., André, C., & Knutsen, J. A. (2013). Lobster and cod benefit from small-scale northern marine protected areas: Inference from an empirical before–after control-impact study. *Proceedings of the Royal Society, 280*(1754), 20122679.
Nguyen, N. H., Fitzgibbon, Q. P., Quinn, J., Smith, G., Battaglene, S., & Knibb, W. (2018). Can metamorphosis survival during larval development in spiny lobster *Sagmariasus verreauxi* be improved through quantitative genetic inheritance? *BMC Genetics, 19*(1), 27.
Radhakrishnan, E. V., Deshmukh, V. D., Manisseri, M. K., Rajamani, M., Kizhakudan, J. K., & Thangaraja, R. (2005). Status of the major lobster fisheries in India. *New Zealand Journal of Marine and Freshwater Research, 39*, 723–732.
Radhakrishnan, E. V., Chakraborty, R. D., Baby, P. K., & Radhakrishnan, M. (2013). Fishery and population dynamics of the sand lobster *Thenus unimaculatus* (Burton & Davie, 2007) landed by trawlers at Sakthikulangara fishing harbour on the southwest coast of India. *Indian Journal of Fisheries, 60*(2), 7–12.
Spanier, E., & Lavalli, K. L. (2007). Slipper lobster fisheries—Present status and future perspectives. In *The biology and fisheries of the slipper lobster* (pp. 377–391). Boca Raton: CRC Press.
Styf, H. (2014). Climate Change and the Norway lobster - Effects of Multiple Stressors on Early Development. Doctoral thesis submitted at Faculty of Science, University of Gothenburg, Sweden.
Thangaraja, R., & Radhakrishnan, E. V. (2017). Reproductive biology and size at onset of sexual maturity of the spiny lobster *Panulirus homarus homarus* (Linnaeus, 1758) in Kadiyapattanam, southwest coast of India. *Journal of the Marine Biological Association of India, 59*(2), 19–28.
Thangaraja, R., Radhakrishnan, E. V., & Chakraborty, R. D. (2015). Stock and population characteristics of the Indian rock lobster *Panulirus homarus homarus* (Linnaeus, 1758) from Kanyakumari, Tamil Nadu, on the southern coast of India. *Indian Journal of Fisheries, 62*(3), 21–27.

Truelove, N. K., Griffiths, S., Ley-Cooper, K., Azueta, J., Majil, I., Box, S. J., Behringer, D. C., Butler, M. J., & Preziosi, R. F. (2015). Genetic evidence from the spiny lobster fishery supports international cooperation among central American marine protected areas. *Conservation Genetics, 16*(2), 347–358.

Ventura, T., Fitzgibbon, Q. P., Battaglene, S. C., & Elizur, A. (2015). Redefining metamorphosis in spiny lobsters: Molecular analysis of the phyllosoma to puerulus transition in *Sagmariasus verreauxi*. *Scientific Reports, 5*, 135–137.

Villacorta-Rath, C., Ilyushkina, I., Strugnell, J. M., Green, B. S., Murphy, N. P., Doyle, S. R., Hall, N. E., Robinson, A. J., & Bell, J. J. (2016). Outlier SNPs enable food traceability of the southern rock lobster, *Jasus edwardsii*. *Marine Biology, 163*, 223.

Bibliography of Lobster Fauna of India

Abraham, T. J., Rahman, K. M., & Mary Leema, T. J. (1996). Bacterial disease in cultured spiny lobster, *Panulirus homarus* (Linnaeus). *Journal of Aquaculture in the Tropics, 11,* 187–192.

Ajmal Khan, S. (2006). *Management of spiny lobster fishery resources.* Chennai: National Biodiversity Authority.

Alcock, A. (1901). *A descriptive catalogue of the Indian deep-sea Crustacea, Decapoda, Macrura and Anomala in the Indian Museum, being a revised account of the deep-sea species collected by the Royal Marine Ship 'Investigator', Calcutta* (pp. 286).

Alcock, A. (1906). *Catalogue of the Indian decapod Crustacea in the collections of the Indian Museum.* Part III. Macrura (pp. 1–55).

Alcock, A., & Anderson, A. R. S. (1894). Natural History notes from H.M. Indian Marine Survey Steamer "Investigator", Commander C.D. Oldham, R.N. Commanding, Series II, no 14. *An account of a recent collection of deep sea Crustacea from the Bay of Bengal and the Laccadive sea. Journal of the Asiatic Society of Bengal, 63,* pt 2(3), 141–185. pl.9.

Ali, D. M., Pandian, P. P., Somavanshi, V. S., John, M. E., & Reddy, K. S. N. (1994). Spear lobster, *Linuparus somniosus* Berry & George, 1972 (Fam. Palinuridae) in the Andaman Sea. *Occasional paper, Fishery Survey of India, Mumbai,* 6, 13pp.

Anbarasu M., Kirubagaran, R., & Vinithkumar, N. V. (2010). Diet and eyestalk ablation induced changes in lipid and fatty acid composition of *Panulirus homarus, International conference on recent advances in lobster biology, aquaculture and management,* 4–8 January, 2010, Chennai. *Abstract* (p. 111).

Anbarasu, M., Krubagaran, R., & Vinithkumar, N. V. (2012). Diet and eyestalk ablation induced changes in lipid and fattyacid composition of female spiny lobster *Panulirus homarus* (Linnaeus, 1758). *Indian Journal of Fisheries, 59*(4), 163–168.

Anil Kumar, P. K. (2002). *Biochemical studies and energetics of the spiny lobster Panulirus homarus (Linnaeus, 1758).* Versova, Mumbai: Central Institute of Fisheries Education.

Anon. (2009). *biosearch,* Marine Biodiversity database of India, National Institute of Oceanography, Goa.

Anrose, A., Selvaraj, P., Dhas, J. C., Prasad, G. V. A., & Babu, C. (2010). Distribution and abundance of deep sea spiny lobster *Puerulus sewelli* in the Indian exclusive economic zone. *Journal of the Marine Biological Association of India, 52*(2), 162–165.

Antony, M. A. (2003). *Carotenoid profiles in relation to maturation, moulting, food and habitat in the Indian spiny lobster Panulirus homarus (Linnaeus, 1758).* Mumbai: CIFE.

Anuraj, A., & Roy, S. D. (2016). New slipper lobster added to the Andaman and Nicobar Islands, India. *The Lobster News Letter, 29*(1), 23–24.

Anuraj, A., Kirubasankar, R., Kaliyamoorthy, M., & Roy, D. (2017). Genetic evidence and morphometry for Shovel-nosed lobster, *Thenus unimaculatus* from Andaman and Nicobar Islands, India. *Turkish Journal of Fisheries and Aquatic Sciences, 17,* 209–215.

Athithan, S., & Gopalakannan, A. (2015). Growth assessment of spiny lobster (*Panulirus homarus*) under open sea cage culture in Tharuvaikulam of Tamilnadu coast, South India. *The Bioscan, 10*(4), 1655–1658.

Ayyappan, M. (2002). *Studies on the moulting frequency, growth responses and biochemical changes in baby rock lobster Panulirus homarus (Linnaeus)*. Tirunelveli: Manonmaniam Sundaranar University.

Balaji, K., Thirumaran, G., Arumugam, R., Kumaraguruvasagam, K. R., & Anantharaman, P. (2007). Marine ornamental invertebrate resources of Parangipettai coastal waters (south east coast of India). *Journal of Fish and Aquatic Science Academy, 2*(5), 328–336. http://docsdrive.com/pdfs/academicjournals/jfas/2007/328-336.pdf.

Balasubramanyan, R. (1969). On the occurrence of palinurid spiny lobster in the Cochin backwaters. *Journal of the Marine Biological Association of India, 9*(2), 425–438.

Balasubramanyan, R., Satyanarayana, A. V. V., & Sadanandan, K. A. (1960). A preliminary account of the experimental rock lobster fishing conducted along the southwest coast of India, with bottom-set-gillnets. *Indian Journal of Fisheries, 7*, 405–422.

Balasubramanyan, R., Satyanarayana, A. V. V., & Sadanandan, K. A. (1961). A further account of the rock lobster fishing experiments with the bottom-set-gillnets. *Indian Journal of Fisheries, 8*, 269–290.

Banumathy, R., Padmaja, M., & Deecaraman, M. (2013). Biochemical changes during the reproductive activity in male sand lobster *Thenus orientalis* (Lund, 1793). *Indian Journal of Science and Technology, 6*(5), 4500–4504.

Borradaile, L. A. (1906). Marine crustaceans XIII. The Hippidae, Thalassinidea and Scyllaridea. In J. S. Gardiner (Ed.), *The fauna and geography of the Maldives and Laccadive Archipelago* (pp. 750–754). Cambridge: Cambridge University Press.

Chacko, S. (1967). The central nervous system of *Thenus orientalis* (Leach). *Marine Biology, 1*(2), 113–117.

Chacko, P. I., & Balakrishnan Nair, N. (1963). Size and sex compositions of catches of the lobster *Panulirus dasypus* (Latreille) along Kanyakumari district coast in 1960–1962. *Proceedings of the Indian Science Congress, 50*(3), 50.

Chacko, P. I., Srinivasan, R., & Anantanarayanan, R. (1969). On the lobster fishery trend and conservancy measure in Kanyakumari district, Madras State. *Madras Journal of Fisheries, 5*, 72–81.

Chakraborty, R. D., & Radhakrishnan, E. V. (2015). Taxonomy, biology and distribution of lobsters. In K. K. Joshi, M. S. Varsha, V. L. Sruthy, V. Susan, & P. V. Prathyusha (Eds.), *Summer school on recent advances in marine biodiversity conservation and management* (pp. 100–110). Cochin: CMFRI.

Chakraborty, K., Chakraborty, R. D., Radhakrishnan, E. V., & Vijayan, K. K. (2010). Fatty acid profiles of spiny lobster (*Panulirus homarus*) phyllosoma fed enriched *Artemia*. *Aquaculture Research, 41*(10), 393–403.

Chakraborty, R. D., Radhakrishnan, E. V., Sanil, N. K., Thangaraja, R., & Unnikrishnan, C. (2011). Mouthpart morphology of phyllosoma of the tropical spiny lobster *Panulirus homarus* (Linnaeus, 1758). *Indian Journal of Fisheries, 58*(1), 1–7.

Chakraborty, R. D., Maheswarudu, G., Radhakrishnan, E. V., Purushothaman, P., Kuberan, P., Jomon, G., Sebastian, & Thangaraja, R. (2014). Rare occurrence of blunthorn lobster *Palinustus waguensis* Kubo, 1963 from the southwest coast of India. *Marine Fisheries Information Service Technical and Extension Series*, (219), 25–26.

Chekunova, V. I. (1971). Distribution of commercial invertebrates on the shelf of India, the northeastern part of Bay of Bengal and the Andaman Sea. In S. A. Bogdanov (Ed.), *Soviet fisheries investigations in the Indian Ocean* (pp. 68–83). Jerusalem: Israel Programme for Scientific Translations.

Chhapgar, B. F., & Deshmukh, S. K. (1961). On the occurrence of the spiny lobster *Panulirus dasypus* (H. Milne Edwards) in Bombay waters with a note on the systematics of Bombay lobsters. *Journal of the Bombay Natural History Society, 58*(3), 632–638.

Chhapgar, B. F., & Deshmukh, S. K. (1964). Further records of lobsters from Bombay. *Journal of the Bombay Natural History Society, 61*, 203–207.

Chhapgar, B. F., & Deshmukh, S. K. (1971). Lobster fishery of Maharashtra. *Journal of the Indian Fisheries Association, 1*(1), 74–86.

Chopra, B. N. (1939). Some food prawns and crabs of India and their fisheries. *Journal of the Bombay Natural History Society, 41*, 221–234.

Clement, A. (1968). Observations on allometric growth and regeneration in palinurid lobsters. *Journal of the Marine Biological Association of India, 10*(2), 388–391.

CMFRI, Kochi. (1988). Deep-sea spiny lobster resource. *CMFRI Newsletter* No. 41, July–September 1988 (pp. 2–3).

CMFRI, Kochi. (2009). Open sea cage farming of spiny lobster by CMFRI–A success story. *CMFRI Newsletter* No. 121, April–June 2009 (pp. 3–5).

CMFRI, Kochi. (2010). Successful harvest of sea cage farmed spiny lobsters and ornamental fishes at Kanyakumari and Mandapam. *CMFRI Newsletter* No. 124, January–March 2010 (pp. 5–6).

Damodaran, D., Mohammed Koya, K., Mojjada, S. K., Devaji Lalaji, C., Dash, G., Vase, V. K., & Sreenath, K. R. (2018). Optimization of the stocking parameters for mud spiny lobster *Panulirus polyphagus* (Herbst, 1793) capture-based aquaculture in tropical open sea floating net cages. *Aquaculture Research, 49*(2), 1080–1086.

Deshmukh, S. (1966). The puerulus of the spiny lobster *Panulirus polyphagus* (Herbst) and its metamorphosis into post-puerulus. *Crustaceana, 10*, 137–150.

Deshmukh, S. (1968). On the first phyllosomae of the Bombay spiny lobsters (*Panulirus*) with a note on the unidentified first *Panulirus* phyllosoma from India (Palinuridae). *Crustaceana (Supplementary), 2*, 47–58.

Deshmukh, V. D. (2001). Collapse of sand lobster fishery in Bombay waters. *Indian Journal of Fisheries, 48*(1), 71–76.

Devaraj, M., & Pillai, N. G. K. (2001). An overview of the marine fisheries research in the Lakshadweep. *Geological Survey of India Special Publication, 56*, 83–94.

Dharani, G., Maitrayee, G. A., Karthikayulu, S., Kumar, T. S., Anbarasu, M., & Vijayakumaran, M. (2009). Identification of *Panulirus homarus* puerulus larvae by restriction fragment length polymorphism of mitochondrial cytochrome oxidase I gene. *Pakistan Journal of Biological Sciences, 12*(3), 281–285.

Dharani, G., Annapoorna, G., Yamini Lakshmi, M. K., Karthikayulu, S., Kumar, T.S., Anbarasu, M., Vijayakumaran, M., & Kirubagaran, R. (2010). Intra and interspecific restriction fragment length polymorphism in mitochondrial cytochrome oxidase i gene in six species of spiny lobster of the genus *Panuliru*s from Indian coast. In *International conference on recent advances in lobster biology, Aquaculture and Management*, 4–8 January, 2010, Chennai. *Abstract* (pp. 49).

Dineshbabu, A. P. (2008). Morphometric relationship and fishery of Indian Ocean lobsterette, *Nephropsis stewarti* Wood_Mason, 1873 along the southwest coast of India. *Journal of the Marine Biological Association of India, 50*(1), 113–116.

Dineshbabu, A. P., & Radhakrishnan, E. V. (2008). टिकाऊ चिंगट उत्पादन के लिए नदीमुख और तटीय आवास तंत्रों के परिरक्षण की परम आवश्यकता. तटीय मेखला प्रबंधन - विशिष प्रकाशन , *97*, 49–53.

Dineshbabu, A. P., Sreedhara, B., & Muniyappa, Y. (2001). New crustacean resources in the trawl fishery off Mangalore coast. *Marine Fisheries Information Service Technical and Extension Series, 170*, 3–5.

Dineshbabu, A. P., Durgekar, R. N., & Zacharia, P. U. (2011). Estuarine and marine decapods of Karnataka inventory. *Fishing Chimes, 30*, 10–11.

Dutt, S., & Ravindranath, K. (1975). Pueruli of *Panulirus polyphagus* (Herbst) (Crustacea, Decapoda, Palinuridae) from east coast of India with a key to known Indo-West Pacific pueruli of *Panulirus* White. *Proceedings of the Indian Academy of Sciences, 82B*(3), 100–107.

Dutta, S., Chakraborty, K., & Hazra, S. (2016). The status of the marine fisheries of West Bengal coast of the northern bay of Bengal and its management options: A review. *Proceedings of the Zoological Society, 69*(1), 1–8.

Fernandez, R. (2002). *Neuroendocrine control of vitellogenesis In the spiny lobster Panulirus homarus (Linnaeus, 1758)* (p. 189). Mumbai: CIFE.

Fernandez, R., & Radhakrishnan, E. V. (2010). Classification and mapping of neurosecretory cells in the optic, supraoesophageal and thoracic ganglia of the female spiny lobster *Panulirus homarus* (Linnaeus, 1758) and their secretory activity during vitellogenesis. *Journal of the Marine Biological Association of India, 52*(2), 237–248.

Fernandez, R., & Radhakrishnan, E. V. (2016). Effect of bilateral eyestalk ablation on ovarian development and moulting in early and late intermoult stages of female spiny lobster *Panulirus homarus* (Linnaeus, 1758). *Invertebrate Reproduction and Development, 60*(3), 238–242.

Fernando, A. S., & Fernando, O. J. (2002). *A field guide to the common invertebrates of the east coast of India* (pp. 1–258). Parangipettai: Centre of Advanced Study in Marine Biology.

Ganeshkumar, A., Baskar, B., Santhanakumar, J., Vinithkumar, N. V., Vijayakumaran, M., & Kirubagaran, R. (2010). Diversity and functional properties of intestinal microbial flora of the spiny lobster *Panulirus versicolor* (Latreille, 1804). *Journal of the Marine Biological Association of India, 52*(2), 282–285.

George, M. J. (1967a). Observations on the biology and fishery of the spiny lobster *Panulirus homarus* (Linnaeus). In *Proceeding of the symposium on Crustacea, Part IV* (pp. 1308–1316). Mandapam camp, India: Marine biological Association of India.

George, M. J. (1967b). The Indian spiny lobster. In *Souvenir 20th Anniversary Central Marine Fisheries Research Institute,* 3 February 1967, Mandapam, India.

George, M. J. (1967c). Two new records of scyllarid lobsters from the Arabian Sea. *Journal of the Marine Biological Association of India, 9*(2), 433–435.

George, M. J. (1973). The lobster fishery resources of India. In *Proceeding of the symposium living resources of the seas around India,* Cochin, December 1968, Special Publication, Central Marine Fisheries Research Institute (pp. 570–580).

George, M. J., & George, K. C. (1965). *Palinustus mossambicus* Barnard (Palinuridae: Decapoda), a rare spiny lobster from Indian waters. *Journal of the Marine Biological Association of India, 7*(2), 463–464.

George, M. J., & Rao, P. V. (1965a). On some decapod crustaceans from south west coast of India. In *Proceeding of the symposium on Crustacea. Part 1. Series 2* (pp. 327–336). Mandapam Camp, India: Marine Biological Association of India.

George, M. J., & Rao, P. V. (1965b). A new record of *Panulirus longipes* (Milne Edwards) from the southwest coast of India. *Journal of the Marine Biological Association of India, 7*(2), 461–462.

George, C. J., Reuben, N., & Muthe, P. T. (1955). The digestive system of *Panulirus polyphagus* (Herbst). *Journal of Animal Morphology and Physiology, 2,* 14–27.

George, K. C., Thomas, P. A., Appukuttan, K. K., & Gopakumar, G. (1986). Ancillary living marine resources of Lakshadweep. *Marine Fisheries Information Service Technical and Extension Series, 68,* 46–50.

Ghosh, S., Mohanraj, G., Asokan, P. K., Dhokia, H. K., Zala, M. S., & Bhint, H. M. (2008). Bumper catch of spiny lobsters by trawlers and gillnetters at Okha, Gujarat. *Marine Fisheries Information Service Technical and Extension Series, 198,* 16–17.

Girijavallabhan, K. G., & Devarajan, K. (1978). On the occurrence of puerulus of spiny lobster *Panulirus polyphagus (Herbst) a*long the Madras coast. *Indian Journal of Fisheries, 25*(1 & 2), 253–254.

Gopal, K., Sachithanandam, V., Dhivya, P., & Mohan, P. M. (2013). Current status of exported fishery resources in Andaman Islands. *Asian Journal of Marine Science, 1*(1), 12–16.

Goswami, B. C. B. (1992). Marine fauna of Digha coast. *Journal of the Marine Biological Association of India, 34*(1 & 2), 115–137.

Gulshad, M., & Ghosh, S. (2016). गुजरात के सूत्रपादा और दक्षि के घोघला में समुद्री पजिरों में शूली महाचगिट पानुलरिस पोलफिगस का पालन. समुद्री संवर्धन (120). pp. 95–101. ISSN 0972-2351

Gulshad, M., Rao, G. S., & Ghosh, S. (2010a). Aquaculture of spiny lobsters in sea cages in Gujarat, India. *Journal of the Marine Biological Association of India, 52*(2), 316–319.

Gulshad, M., Rao, G. S., & Ghosh, S. (2010b). Culture of spiny lobsters in small cages along Saurashtra coast, Gujarat. In: UNSPECIFIED (Ed.) *Coastal fishery resources of India – Conservation and sustainable utilization.* Society of Fisheries Technologists (pp. 115–119).

Holthuis, L. B. (1966). On spiny lobsters of the genera *Palinurellus, Linuparus* and *Puerulus* (Crustacea, Decapoda, palinuridae). In *Proceeding of the symposium crust*. Marine Biological Association of India, Part 1 (pp. 260–278).

Hossain, M. A. (1975). On the squat lobster, *Thenus orientalis* off Visakhapatnam (Bay of Bengal). *Current Science, 44*(5), 161–162.

Hossain, M. A. (1978). Appearance and development of sexual characters of sand lobster *Thenus orientalis* (Lund) (Decapoda: Scyllaridae) from the Bay of Bengal. *Bangladesh Journal of Zoology, 6*, 31–42.

Hossain, M. A., Shyamasundari, K., & Rao, K. H. (1975). On the landings of sand lobster *Thenus orientalis* (Lund). *Seafood Export Journal, 7*(4), 1–6.

Immanuel, G., Iyappa Raj, P., Esakki Raj, P., & Palavesam, A. (2006). Intestinal bacterial diversity in live rock lobster *Panulirus homarus* (Linnaeus) (Decapoda, Pleocyemata, Palinuridae) during transportation process. *Pan-American Journal of Aquatic Sciences, 1*(2), 69–73.

Jadhav, D. G., & Rao, R. B. (1998). On the unusual landings of lobster, *Panulirus polyphagus* at Borli Mandla, Raigad District, Maharashtra. *Marine Fisheries Information Service Technical and Extension Series, 154*, 1–4.

James, P. S. B. R., & Pillai, V. N. (1990). Fishable concentrations of fishes and crustaceans in the offshore and deep sea areas of the Indian exclusive economic zone based on observations made onboard FORV Sagar Sampada. In *Proceeding of the first workshop scientific result of FORV Sagar Sampada*, 5–7 June, 1989 (pp. 201–213).

Jawahar, A., Kaleemur, R., & Leema, J. (1996). Bacterial disease in cultured spiny lobster, *Panulirus homarus* (Linnaeus). *Journal of Aquaculture in the Tropics, 11*, 187–192.

Jayagopal, P. (2014). *Studies on stress during live transport of spiny lobsters-Panulirus spp.* Chepauk, Chennai: University of Madras.

Jayagopal, P., & Vijayakumaran, M. (2010). Studies on stress during live transport of spiny lobsters *Panulirus* sp. In *International conference on recent advances in lobster biology, aquaculture and management*, 5–8 January 2010, Chennai, India. Abstract (pp. 103–104).

Jayaprakash, A. A., Kurup, B. M., Venu, S., Thankappan, D., Pachu, A. V., Manjebrayakath, H., Thampy, P., & Sudhakar, S. (2006). Distribution, diversity, length-weight relationship and recruitment pattern of deep sea finfishes and shellfishes in the shelf-break area off southwest coast of Indian EEZ. *Journal of the Marine Biological Association of India, 48*(1), 56–61.

Jeena, N. S. (2013, May). *Genetic divergence in lobsters (Crustacea: Palinuridae and Scyllaridae) from the Indian EEZ* (pp. 153). Kochi: Cochin University of Science and Technology.

Jeena, N. S., Gopalakrishnan, A., Radhakrishnan, E. V., Kizhakudan, J. K., Basheer, V. S., Asokan, P. K., & Jena, J. K. (2015a). Molecular phylogeny of commercially important lobster species from Indian coast inferred from mitochondrial and nuclear DNA sequences. *Mitochondrial DNA, 27*, 2700–2709. https://doi.org/10.3109/19401736.2015.1046160.

Jeena, N. S., Gopalakrishnan, A., Kizhakudan, J. K., Radhakrishnan, E. V., Kumar, R., & Asokan, P. K. (2015b). Population genetic structure of the shovel-nosed lobster *Thenus unimaculatus* (Decapoda, Scyllaridae) in Indian waters based on RAPD and mitochondrial gene sequences. *Hydrobiologia, 766*, 1–12. https://doi.org/10.1007/s10750-015-2458-z.

Jha, D. K., Kumar, T. S., Nazar, A. K. A., Venkatesan, R., & Saravanan, N. (2007). Spiny lobster resources of North Andaman Sea: Preliminary observations. *Fishing Chimes, 27*(1), 138–141.

Jha, D. K., Vijaykumaran, M., Murugan, T. S., Santhanakumar, J., Kumar, T. S., Vinithkumar, N. V., & Kirubagaran, R. (2010). Survival and growth of early phyllosoma stages of *Panulirus homarus* under different salinity regimes. *Journal of the Marine Biological Association of India, 52*, 215–218.

John, C. C., & Kurian, C. V. (1959). A preliminary note on the occurrence of deep water prawn and spiny lobster off the Kerala coast. *Bulletin of Central Research Institute Trivandrum Series C, 7*(1), 155–162.

Jones, S. (1965). The crustacean fishery resources of India. In *Proceeding of the symposium on Crustacea*, Part IV, Series 2 (pp. 1328–1341).

Joseph, K. M. (1972). A profile of deep sea fishery resources off the southwest coast of India. *Seafood Export Journal, 4*(1), 97–104.

Joseph, K. M. (1974). Demersal fisheries resources off the North-west coast of India. *Bulletin of the Exploratory Fisheries Project, 1*, 1–45.

Joseph, K. M. (1986). Some observations on potential fishery resources from the Indian EEZ. *Fishery Survey of India Bulletin*, No. 14, 1–20.

Joseph, K. M., & John, M. E. (1987). Potential marine fishery resources. Seminar on Potential Marine Fishery Resources. *CMFRI Special Publication*, No. 30, 18–43.

Joseph, M. T., & Nair, A. K. K. (1979). On certain allometric relations of spiny lobster *Panulirus polyphagus* (Herbst). *Fishery Technology, 16*, 83–85.

Kabli, L. M., & Kagwade, P. V. (1996). Morphometry and conversion factors in the sand lobster *Thenus orientalis* (Lund) from Bombay waters. *Indian Journal of Fisheries, 43*(3), 259–254.

Kagwade, P. V. (1987a). Morphological relationships and conversion factors in spiny lobster *Panulirus polyphagus* (Herbst). *Indian Journal of Fisheries, 34*(3), 348–352.

Kagwade, P. V. (1987b). Age and growth of the spiny lobster *Panulirus polyphagus* (Herbst). *Indian Journal of Fisheries, 34*(4), 389–398.

Kagwade, P. V. (1988a). Reproduction in the spiny lobster *Panulirus polyphagus* (Herbst). *Journal of the Marine Biological Association of India, 30*(1 & 2), 37–46.

Kagwade, P. V. (1988b). Fecundity in the spiny lobster *Panulirus polyphagus* (Herbst). *Journal of the Marine Biological Association of India, 30*(1 & 2), 114–120.

Kagwade, P. V. (1993). Stock assessment of the spiny lobster *Panulirus polyphagus* (Herbst) off north-west coast of India. *Indian Journal of Fisheries, 40*(1 & 2), 63–73.

Kagwade, P. V. (1994). Estimates of the stocks of the spiny lobster *Panulirus polyphagus* (Herbst) in the trawling grounds of Bombay. *Journal of the Marine Biological Association of India, 36*(1 & 2), 161–167.

Kagwade, P. V., & Kabli, L. M. (1991). Embryonic development of larvae on the pleopods of the spiny lobster *Panulirus polyphagus* (Herbst) and the sand lobster *Thenus orientalis* (Lund) from Bombay waters. *Indian Journal of Fisheries, 38*(2), 73–82.

Kagwade, P. V., & Kabli, L. M. (1996a). Reproductive biology of the sand lobster *Thenus orientalis* (Lund) from Bombay waters. *Indian Journal of Fisheries, 43*(1), 13–25.

Kagwade, P. V., & Kabli, L. M. (1996b). Age and growth of the sand lobster *Thenus orientalis* (Lund) from Bombay waters. *Indian Journal of Fisheries, 43*(3), 241–247.

Kagwade, P. V., Manickaraja, M., Deshmukh, V. D., Rajamani, M., Radhakrishnan, E. V., Suresh, V., Kathirvel, M., & Rao, G. S. (1991). Magnitude of lobster resources of India. *Journal of the Marine Biological Association of India, 33*(1 & 2), 150–158.

Kalayanaraman, V., & Senthil Kumar, P. (2009). Effect of lead in the expression of nutritional content in edible lobster *Thenus orientalis* (Lund, 1793). *International Journal of Science and Technology, 2*(10), 17–22.

Kalidas, C., & Salim, S. S. (2005). Backyard lobster fattening unit Economic feasibility analysis. *Fishing Chimes, 24*(11), 22–26.

Kalidas, C., Ranjith, L., Kavitha, M., Jagadis, I., & Manojkumar, P. P. (2017). *Success stories: Lobster farming in model sea-cage farm at Sippikulam fisher village Thoothukudi District*. Cochin: Central Marine Fisheries Research Institute.

Kasim, H. M. (1986). Effect of salinity, temperature and oxygen partial pressure on the respiratory metabolism of *Panulirus polyphagus* (Herbst). *Indian Journal of Fisheries, 33*, 66–75.

Kathirvel, M. (1973). The growth and regeneration of an aquarium-held Spiny lobster, *Panulirus polyphagus* (Herbst) (Crustacea: Decapoda: Palinuridae). *Indian Journal of Fisheries, 20*(1), 219–221.

Kathirvel, M. (1975). On the occurrence of the puerulus larvae of the Indian spiny lobster, *Panulirus homarus* (Linn.), in Cochin backwater. *Indian Journal of Fisheries, 22*(1 & 2), 287–290.

Kathirvel, M. (1998). A pictorial guide for identification of Indian spiny lobsters. Part 1. *Fish and Fisheries, 18*, 4.

Kathirvel, M., & James, D. B. (1990). The phyllosoma larvae from Andaman and Nicobar Islands. In *Proceeding of the first workshop on scientific results of FORV Sagar Sampada*, DOD, CMFRI, CIFT, Cochin (pp. 147–150).

Kathirvel, M., & Nair, K. R. (2002). A new record of scyllarid lobster from southwest coast of India. *Fish and Fisheries, 32,* 4.

Kathirvel, M., Suseelan, C., & Rao, P. V. (1989, February). Biology, population and exploitation of the Indian deep sea lobster, *Puerlus sewelli. Fishing Chimes, 8,* 16–25.

Kathirvel, M., Thirumilu, P., & Gokul, A. (2007). Biodiversity and economical values of Indian lobsters. In *Zoological survey of India national symposium on conservation and valuation of marine biodiversity* (pp. 177–200).

Kirubagaran, R., Peter, S. M., Dharani, G., Vinithkumar, N. V., Sreeraj, G., & Ravindran, R. (2005). Changes in vertebrate type steroids and 5-hydroxytryptamine during ovarian recrudescence in the Indian spiny lobster *Panulirus homarus. New Zealand Journal of Marine and Freshwater Research, 39,* 527–537.

Kizhakudan, J. K. (2002). First report of the spiny lobster, *Panulirus versicolor* (Latreille, 1804) from trawl landings at Veraval. *Marine Fisheries Information Service Technical and Extension Series,* No. 172, 6–7.

Kizhakudan, J. K. (2005). Culture potential of the sand lobster *Thenus orientalis* (Lund). International Symposium on improved sustainability of fish production by appropriate technologies for utilization. *Sustain fish 2005* (pp. 16–18).

Kizhakudan, J. K. (2006). Culture potential of the sand lobster *Thenus orientalis* (Lund). In B. M. Kurup & K. Ravindran (Eds.), *Sustain fish.* Cochin: School of Industrial Fisheries, Cochin University of Science & Technology.

Kizhakudan, J. K. (2007). *Reproductive biology, ecophysiology and growth in the Mudspiny lobster, Panulirus polyphagus (Herbst, 1793) and the sand lobster, Thenus orientalis (Lund, 1793)* (pp. 169). Ph.D. Thesis. Bhavnagar University, Bhavnagar, Gujarat.

Kizhakudan, J. K. (2010). Comparative assessment of growth in the mud spiny lobster *Panulirus polyphagus* (Herbst, 1793) in the wild and in captivity. In *International conference on recent advances in lobster biology, aquaculture and management,* 4–8 January, 2010, Chennai. *Abstract* (pp. 23).

Kizhakudan, J. K. (2013). Effect of eyestalk ablation on moulting and growth in the mud spiny lobster *Panulirus polyphagus* (Herbst, 1793) held in captivity. *Indian Journal of Fisheries, 60*(1), 77–81.

Kizhakudan, J. K. (2014). Reproductive biology of the female shovel-nosed lobster *Thenus unimaculatus* (Burton & Davie, 2007) from north-west coast of India. *Indian Journal of Geo-Marine Sciences, 43*(6), 927–935.

Kizhakudan, J. K. (2016, January 5–25). Hatchery technology and seed production of lobsters. In: I. Joseph, & B. Ignatius (Eds.), *Course manual, winter school on technological advances in mariculture for production enhancement and sustainability* (pp. 106–113). Kochi: CMFRI.

Kizhakudan, J. K., & Krishnamoorthi, S. (2014). Complete larval development of Thenus *unimaculatus* Burton & Davie, 2007 (Decapoda, Scyllaridae). *Crustaceana, 87*(5), 570–584.

Kizhakudan, J. K., & Patel, S. K. (2009, February 9–12). Traditional knowledge and methods in lobster fishing in Gujarat. In: E. Vivekanandan, T. M. Najmudeen, T. S. Naomi, A. Gopalakrishnan, K. V. Jayachandran, & M. Harikrishnan (Eds.) *Marine ecosystems: Challenges and opportunities. Book of Abstracts* (pp. 92–93). Cochin: Marine Biological Association of India.

Kizhakudan, J. K., & Patel, S. K. (2010). Size at maturity in the mud spiny lobster *Panulirus polyphagus* (Herbst, 1793). *Journal of the Marine Biological Association of India, 52*(2), 170–179.

Kizhakudan, J. K., & Patel, S. K. (2011). Effect of diet on growth of the mud spiny lobster *Panulirus* polyphagus (Herbst, 1793) and the sand lobster *Thenus orientalis* (Lund, 1793) held in captivity. *Journal of the Marine Biological Association of India, 53*(2), 167–171.

Kizhakudan, J. K., & Rathinam, A. M. M. (2010). Juvenile rearing and fattening techniques in sand lobster aquaculture. *International Conference on Recent Advances in Lobster Biology, Aquaculture and Management,* 4–8 January, 2010, Chennai. *Abstract* (pp. 85).

Kizhakudan, J. K., & Thirumilu, P. (2006). A note on the bluntthorn lobsters from Chennai. *Journal of the Marine Biological Association of India, 48*(2), 260–262.

Kizhakudan, J. K., & Thumber, B. P. (2000). Report on a large sized spiny lobster *Panulirus polyphagus* (Herbst, 1793) landed at Veraval. *Marine Fisheries Information Service Technical and Extension Series, 164*, 25.

Kizhakudan, J. K., & Thumber, B. P. (2003). Fishery of marine crustaceans in Gujarat. In M. R. Boopendranath et al. (Eds.), *Sustainable fisheries development: Focus on Gujarat* (p. 207). Cochin: Society of Fisheries Technologists (India).

Kizhakudan, J. K., Radhakrishnan, E. V., George, R. M., Thirumilu, P., Rajapackiam, S., Manibal, C., & Xavier, J. (2004). Phyllosoma larvae of *Thenus orientalis* and *Scyllarus rugosus* reared to settlement. *The Lobster Newsletter, 17*(1).

Kizhakudan, J. K., Thirumilu, P., & Manibal, C. (2004a). Fishery of the sand lobster *Thenus orientalis* (Lund) by bottom-set gillnets along Tamilnadu coast. *Marine Fisheries Information Service Technical and Extension Series, 181*, 6–7.

Kizhakudan, J. K., Thirumilu, P., Rajapackiam, S., & Manibal, C. (2004b). Captive breeding and seed production of scyllarid lobsters- opening new vistas in crustacean aquaculture. *Marine Fisheries Information Service Technical and Extension Series,* (181), 1–4.

Kizhakudan, J. K., Margaret, A. M. R., & Kandasami, D. (2007). High density grow out techniques in tropical spiny lobsters. *8th International Conference and Workshop on Lobster Biology and management, Charlottetown, Canada* (pp. 39).

Kizhakudan, J. K., Vidya, J., & Krishnamoorthy, S. (2010a). Spermatophore structure in Indian spiny and scyllarid lobsters. *International conference on recent advances in lobster biology, aquaculture and management,* 4–8 January, 2010, Chennai. *Abstract* (pp. 97).

Kizhakudan, J. K., Jasper, B., Vivekanandan, E., & Kasim, M. H. (2010b). Effect of some environmental factors on larval progression and survival during hatchery rearing of spiny and sand lobsters. *International conference on recent advances in lobster biology, aquaculture and management,* 4–8 January, 2010, Chennai. *Abstract* (pp. 121).

Kizhakudan, J. K., Kripa, V., Rathinam, A. M. M., Manibal, C., Dhanpathi, V., Ghandhi, A. T., Leslie, V. A., & Ganesan, S. (2011). झींगों के साथ रेतीले झींगों के पालन की संभावना. सी आइ बी ए वशिष पुरकाशन 56 जलकृषि में आधुनकि वकिास (56), 31–32.

Kizhakudan, J. K., Krishnamoorthi, S., & Thiyagu, R. (2012a). Unusual landing of deep sea lobster *Palinustus waguensis* at Cuddalore. *Marine Fisheries Information Service Technical and Extension Series*, No. 211, 19.

Kizhakudan, J. K., Krishnamoorthi, S., & Thiyagu, R. (2012b). First record of the scyllarid lobster *Scyllarides tridacnophaga* from Chennai coast. *Marine Fisheries Information Service Technical and Extension Series*, No. 211, 13.

Kizhakudan, J. K., Kizhakudan, S. J., & Patel, S. K. (2013). Growth and moulting in the mud spiny lobster, *Panulirus polyphagus* (Herbst, 1793). *Indian Journal of Fisheries, 60*(2), 79–85.

Kumar, T. S., Vijayakumaran, M., Senthil Murugan, T., Jha, D. K., Sreeraj, G., & Muthukumar, S. (2009). Captive breeding and larval development of the scyllarinae lobster *Petrarctus rugosus*. *New Zealand Journal of Marine and Freshwater Research, 43*, 101–112.

Kumar, T. S., Jha, D. K., Syed Jahan, S., Dharani, G., Abdul Nazar, A. K., Sakthivel, M., Alagarraaja, K., Vijayakumaran, M., & Kirubagaran, R. (2010). Fishery resources of spiny lobsters in the Andaman Island, India. *Journal of the Marine Biological Association of India, 52*(2), 166–169.

Kurian, C. V. (1963a). Further observations on the deep water lobster *Puerulus sewelli* Ramadan off the Kerala coast. *Bulletin Department of Marine Biology and Oceanography of University of Kerala, 1*, 122–127.

Kurian, C. V. (1963b). Occurrence of lobster in the offshore waters of Kerala. *Seafood Trade Journal, 3*(6), 22–26.

Kurian, C. V. (1965). Deep water prawns and lobsters off the Kerala coast. *Fishery Technology, 2*(1), 51–53.

Kurian, C. V. (1967). Further observations on deep water lobsters in the collections of *R.V. Conch*. *Bulletin of Department of Marine Biology and Oceanography of University of Kerala, 3*, 131–135.

Kuthalingam, M. D. K., Luther, G., & Lazarus, S. (1980). Rearing of early juveniles of spiny lobster *Panulirus versicolor* (Latreille) with notes on the lobster fishery of Vizhinjam area. *Indian Journal of Fisheries, 27*(1 & 2), 17–23.

Lalithadevi, S. (1981). The occurrence of different stages of spiny lobsters in Kakinada region. *Indian Journal of Fisheries, 28*(1 & 2), 298–300.

Lamek Jayakumar, V. N., Ramanathan, M. J., Jeyaseelan, P., & Athithan, S. (2011). Growth performance of spiny lobster *Panulirus homarus* (Linnaeus) fed with natural animal food. *Indian Journal of Fisheries, 58*(3), 149–152.

Leslie, V. A., Rathinam, A. M. M., & Anusha, B. (2012). Distribution profile of *Vibrio harveyi* in *Panulirus homarus*. *International Research Journal of Biological Sciences, 1*(4), 61–64.

Lipton, A. P., Rao, G. S., Kingsly, H. J., Imelda, J., Mojjada, S. K., Rao, H. G., & Rajendran, P. (2010). Open sea floating cage farming of lobsters. Successful demonstration by CMFRI off Kanyakumari Coast. *Fishing Chimes, 30*(2), 11–13.

Maharajan, A., & Vijayakumaran, M. (2004). Copper toxicity in the spiny lobster, *Panulirus homarus* (Linnaeus). In *Proceedings of MBR 2004 national seminar on New Frontiers in Marine Bioscience Research* (pp. 205–212).

Maharajan, A., Rajalakshmi, S., & Vijayakumaran, M. (2012). Effect of copper in protein, carbohydrate and lipd contents of the juvenile lobster *Panulirus homarus homarus* (Linnaeus, 1758*). Sri Lanka Journal of Aquatic Sciences, 17,* 19–34.

Manickaraja, M. (2004). Lobster fishery by a modified bottom-set-gillnet at Kayalapattanam. *Marine Fisheries Information Service Technical and Extension Series,* No. 181, 7–8.

Manisseri, M. K. (1998). Crab and lobster culture. In K. Ravindran, S. Krishna, K. K. Kunjipalu, & V. Sasikumar (Eds.), *Kadalekum Kanivukal (Bounties of the Sea)* (pp. 71–75). Cochin: CIFT.

Mary Leema, J. T., Vijayakumaran, M., Kirubagaran, R., & Jayraj, K. (2010a). Effects of Artemia enrichment with microalgae on the survival and growth of *Panulirus homarus* phyllosoma larvae. *Journal of the Marine Biological Association of India, 52*(2), 208–214.

Mary Leema, J. T., Kirubagaran, R., Vijayakumaran, M., Remany, M. C., Kumar, T. S., & Babu, T. D. (2010b). Effects of intrinsic and extrinsic factors on the haemocyte profile of the spiny lobster, *Panulirus homarus* (Linnaeus, 1758) under controlled conditions. *Journal of the Marine Biological Association of India, 52*(2), 219–228.

Mathew, S., Ammu, K., Viswanathan Nair, P. G., & Devadasan, K. (1999). Cholesterol content of Indian fish and shellfish. *Food Chemistry, 66,* 455–461.

Meenakumari, B., & Mohan Rajan, K. V. (1985). Studies on materials for traps for spiny lobsters. *Fisheries Research, 3,* 309–321.

Meenakumari, B., Mohan Rajan, K. V., & Nair, K. (1986). Length weight and tail length relationship in *Panulirus homarus* (Linnaeus) of the southeast coast of India. *Fishery Technology, 23,* 84–87.

Meiyappan, M. M., & Kathirvel, M. (1978). On some new records of crabs and lobsters from Minicoy, Lakshadweep (Laccadives). *Journal of the Marine Biological Association of India, 20*(1 & 2), 116–119.

Miyamoto, H., & Shariff, A. T. (1961). Lobster fishery off the southwest coast of India-anchor hook and trap fisheries. *Indian Journal of Fisheries, 8,* 252–268.

Mohammed, K. H., & George, M. J. (1968). Results of the tagging experiments on the Indian spiny lobster, *Panulirus homarus* (Linnaeus)- movement and growth. *Indian Journal of Fisheries, 15,* 15–26.

Mohammed, K. H., Vedavyasa Rao, P., & Suseelan, C. (1971). The first phyllosoma stage of the Indian deep sea spiny lobster *Puerulus sewelli* Ramadan. *Proceedings of the Indian Academy of Sciences B, 74,* 208–215.

Mohammed, G., Rao, G. S., & Ghosh, S. (2010). Aquaculture of spiny lobsters in sea cages in Gujarat, India. *Journal of the Marine Biological Association of India, 52*(2), 316–319.

Mohan Rajan, K. V., & Meenakumari, B. (1982). Development of lobster traps:preliminary experiment with three new designs of Rectangular, Australian pot and Ink-well traps. *Fishery Technology, 19,* 83–87.

Mohanrajan, K. V. (1991). *Studies on spiny lobster fishery of southwest coast of India*. Cochin: Cochin University of Science and Technology.

Mohanrajan, K. V., Meenakumari, B., & Balasubramanyan, R. (1981). Spiny lobsters and their fishing techniques. *Fishery Technology, 18*, 1–11.

Mohanrajan, K. V., Meenakumari, B., Kandoran, M. K., & Balasubramanyan, R. (1984). Lobster fishing with modern traps- A viability report. *Fish Technology News letter, 3*(12).

Mohanrajan, K. V., Meenakumari, B., & Kesavan Nair, A. K. (1988). Development of an efficient trap for lobster fishing. *Fishery Technology, 25*, 1–4.

Mojjada, S. K., Imelda, J., Koya, M., Sreenath, K. R., Dash, G., Dash, S. S., Fofandi, M., Anbarasu, M., Bhint, H. M., Pradeep, S., Shiju, P., & Rao, G. S. (2012). Capture based aquaculture of mud spiny lobster, *Panulirus polyphagus* (Herbst, 1793) in open sea floating net cages off Veraval, north-west coast of India. *Indian Journal of Fisheries, 59*(4), 29–33.

Mullainadhan, P. (1990). Immunobiology of the Indian sand lobster *Thenus orientalis*. *The Lobster News Letter, 3*(2), 4–5.

Murugan, A., & Durgekar, R. (2008). *Beyond the Tsunami. Status of Fisheries in Tamilnadu, India: A snapshot of present and long-term trends*. UNDP/UNTRS, Chennai and ATREE, Bangalore, India, 75 pp.

Mushtaq, S. S., Sudhakaran, R., Balasubramanian, G., & Shahul Hameed, S. (2006). Experimental transmission and tissue tropism of white spot syndrome virus (WSSV) in two species of lobsters, *Panulirus homarus* and *Panulirus ornatus*. *Journal of Invertebrate Pathology, 93*, 75–80.

Mustafa, A. M. (1990). *Linuparus andamanensis*, a new spear lobster from Andaman. *Andaman Science Association, 6*(2), 177–180.

Muthukumar, S., Anbarasu, M., Vinithkumar, N. V., Vijayakumaran, M., & Kirubagaran, R. (2010). Systematics and phylogenetics of lobsters: A bioinformatics approach. In *International conference on recent advances in lobster biology, aquaculture and management*, 4–8 January, 2010, Chennai. Abstract (pp. 127).

Nair, R. V., Soundararajan, R., & Dorairaj, K. (1973). On the occurrence of *Panulirus longipes longipes, Panulirus penicillatus* and *Panulirus polyphagus* in the Gulf of Mannar with notes on the lobster fishery around Mandapam. *Indian Journal of Fisheries, 20*(2), 333–350.

Nair, R. V., Soundarajan, R., & Nandakumar, G. (1981). Observations on growth and moulting of spiny lobsters *Panulirus homarus* (Linnaeus), *P. ornatus* (Fabricius) and *P. penicillatus* (Oliver) in captivity. *Indian Journal of Fisheries, 28*, 25–35.

Nayak, S., & Umadevi, K. (2012). In silico comparative molecular phylogeny of mitochondrial 16S rRNA and COI genes of the spiny lobster genus *Panulirus* (Decapoda: Palinuridae). *Journal of Advanced Bioinformatics Applications and Research, 3*(3), 364–373.

Ninan, T. V., Basu, P. P., & Verghese, P. K. (1984). Observations on the demersal fishery resources along the Andhra Pradesh coast. *Fishery Survey of India Bulletin, 13*, 13–22.

Oommen, P. V. (1985). Deep sea resources of the southwest coast of India. *Bulletin IFP, 11*, 1–85.

Oommen, P. V., & Philip, K. P. (1976). Observations on the fishery and biology of the deep sea spiny lobster *Puerulus sewelli* Ramadan. *Indian Journal of Fisheries, 21*(2), 369–385.

Padmaja, M. (2004). *A study on the reproductive aspects of sand lobster "Thenus orientalis" of Royapuram coast, Chennai, Tamilnadu*. Tirunelveli, Tamilnadu: Manonmaniam Sundaranar University. 153 pp.

Padmaja, M., Deecaraman, M., Shettu, N., Jagannath Bose, M. T., & Sarojini, N. (2009). Biochemical changes during the reproductive activity in a female sand lobster *Thenus orientalis*. *Biomedical and Pharmacology Journal, 2*(2), 269–276.

Padmaja, M., Deecaraman, M., & Jagannath Bose, M. T. (2010). Study of neurosecretory cells in sand lobster *Thenus orientalis* of Royapuram coast. *World Journal of Fish and Marine Sciences, 2*(2), 82–85.

Parulekar, A. H. (1981). Marine fauna of Malawan, Central West Coast of India. *Mahasagar, 14*, 33–34.

Patil, A. M. (1953). Study of the marine fauna of Karwar coast and the neighbouring Islands. Part IV. Echinodermata and other groups. *Journal of the Bombay Natural History Society, 51*, 429–434.

Pearl, T. M., Sitarami Reddy, P., & Thangavelu, R. (1987). On the endopharyngeal skeletons of spiny lobsters *Panulirus homarus* and *P. ornatus*. *Indian Journal of Fisheries, 34*(2), 218–222.

Peter, M. D., Kirubagaran, R., Inbakandan D., Daniel, B., Mary Leema, J., Thilkam, D. G., Vinithkumar, N. V., Subramoniam, T., Vijayakumaran, M., & Atmanand. (2010). Changes in the biochemical composition during the molting cycle in Panulirus homarus (Linnaeus, 1758). In *International conference on recent advances in lobster biology, aquaculture and management*, 4–8 January, 2010, Chennai. Abstract (pp. 105).

Philip, K. P., Premchand, B., Avhad, G. K., & Joseph, P. S. (1984). A note on the deep sea demersal resources off Karnataka, North Kerala coast. *Fishery Survey of India Bulletin*, No. 13, 23–29.

Philipose, K. K. (1994). Lobster culture along the Bhavanagar coast. *Marine Fisheries Information Service Technical and Extension Series*, No. 130, 8–12.

Pillai, N. K. (1962). *Choniomyzon* gen. nov. (Copepoda: Choniostomatidae) associated with *Panulirus*. *Journal of the Marine Biological Association of India, 4*, 95–99.

Pillai, S. L. (2007). *Reproductive biology of the male lobster Panulirus homarus* (p. 119). Kerala, India: University of Calicut.

Pillai, V. N., & Ramachandran, V. S. (1972). Some trends observed in the deep sea lobster catches of the vessel Blue Fin during the period January 1969 to December 1971. *Seafood Export Journal, 4*(6).

Pillai, S. L., & Thirumilu, P. (2006). Unusual landing of *Palinustus waguensis* at Chennai fishing harbor by indigenous gear. *Marine Fisheries Information Service Technical and Extension Series*, No. 190, 25–26.

Pillai, S. L., & Thirumilu, P. (2007). Extension in the distributional range of long-legged spiny lobster *Panulirus longipes longipes* (A. Mile-Edwards, 1868) along the southeast coast of India. *Journal of the Marine Biological Association of India, 49*(1), 95–96.

Pillai, C. S. G., Mohan, M., & Kunhikoya, K. K. (1985). Observations on the lobsters of Minicoy Atoll. *Indian Journal of Fisheries, 32*(1), 112–122.

Pillai, S. L., Nasser, M., & Sanil, N. K. (2014). Histology and ultrastructure of male reproductive system of the Indian spiny lobster *Panulirus homarus* (Decapoda: Palinuridae). *International Journal of Tropical Biology and Conservation/Revista de Biologia Tropical, 62*(2), 533–541.

Prasad, R. R. (1983). Distribution and growth: Studies on the phyllosoma larvae from the Indian Ocean: I. *Journal of the Marine Biological Association of India, 20*, 143–156.

Prasad, R. R. (1986). Distribution, habits and habitats of palinurid lobsters and their larvae. *Marine Fisheries Information Service Technical and Extension Series, 70*, 8–15.

Prasad, R. R., & Nair, P. V. R. (1973). India and the Indian Ocean fisheries. *Journal of the Marine Biological Association of India, 15*(1), 1–19.

Prasad, R. R., & Tampi, P. R. S. (1957). Phyllosoma of Mandapam. *Proceedings of the National Institute of Sciences of India Section B, 23*, 48–67.

Prasad, R. R., & Tampi, P. R. S. (1959). On a collection of palinurid phyllosomas from the Laccadive seas. *Journal of the Marine Biological Association of India, 1*(2), 143–164.

Prasad, R. R., & Tampi, P. R. S. (1960). Phyllosoma of scyllarid lobsters from the Arabian sea. *Journal of the Marine Biological Association of India, 2*(2), 241–249.

Prasad, R. R., & Tampi, P. R. S. (1961). On the newly hatched phyllosoma of *Scyllarus sordidus* (Stimpson). *Journal of the Marine Biological Association of India, 2*, 250–252.

Prasad, R. R., & Tampi, P. R. S. (1968). On the distribution of palinurid and scyllarid lobsters in the Indian Ocean. *Journal of the Marine Biological Association of India, 10*(1), 78–87.

Prasad, R. R., Tampi, P. R. S., & George, M. J. (1980). Phyllosoma larvae from the Indian Ocean collected by dana expedition 1928–1930. *Journal of the Marine Biological Association of India, 2*, 56–107.

Prasath, E. B., Rahman, M. K., & Subramoniam, T. (1994). Egg incubation and brooding behaviour of the female sand lobster *Thenus orientalis* (Lund) (Crustacea: Decapoda). In *Proceeding of the third Asian fisheries forum 26–30 October 1992*, Singapore (pp. 480–483).

Princy, A. S., & Shrivastava, S. (2010). Casein and gelatinolytic activity of hepatopancreatic extracts from rock lobster, *Panulirus homarus*, *International conference on recent advances in lobster biology, aquaculture and management, 4–8 January, 2010*, Chennai. *Abstract* (pp. 101).

Purushottama, G. B., & Saravanan, R. (2014). A rare deep water Japanese Blunt-horn lobster *Palinustus waguensis* Kubo, 1963 (Decapoda: Palinuridae) found off Mangalore coast, central west coast of India. *Indian Journal of Geo-Marine Sciences, 43*(8), 1550–1553.

Radha, T., & Subramoniam, T. (1985). Origin and nature of spermatophoric mass of the spiny lobster *Panulirus homarus*. *Marine Biology, 86*, 13–19.

Radhakrishna, Y., & Ganapati, P. N. (1969). Fauna of Kakinada Bay. *Bulletin of the National Institute of Sciences of India, 38*, 689–699.

Radhakrishnan, E. V. (1977). Breeding of laboratory reared spiny lobster *Panulirus homarus* (Linnaeus) under controlled conditions. *Indian Journal of Fisheries, 24*(1 & 2), 269–270.

Radhakrishnan, E. V. (1989). *Physiological and biochemical studies on the spiny lobster Panulirus homarus*. Chennai: University of Madras.

Radhakrishnan, E. V. (1993). Sea ranching of spiny lobsters. *Marine Fisheries Information Service Technical and Extension Series, 124*, 5–8.

Radhakrishnan, E. V. (1994). Commercial prospects for farming spiny lobsters. In K. P. P. Nambiar, & T. Singh (Eds.) Aquaculture towards the 21st century. In *Proceedings of the INFOFISH-AQUATECH '94 Conference, Colombo, Sri Lanka*, 29–31 August 1994 (pp. 96–102).

Radhakrishnan, E. V. (1995). Lobster fisheries of India. *The Lobster Newsletter, 8*(2), 1.

Radhakrishnan, E. V. (1996). Lobster farming in India. *Bulletin CMFRI, Artificial Reefs and Seafarming Technologies, 48*, 96–98.

Radhakrishnan, E. V. (2001). Removal of nitrogenous wastes by seaweeds in closed lobster culture system. *Journal of the Marine Biological Association of India, 43*(1 & 2), 181–185.

Radhakrishnan, E. V. (2004). Prospects for grow-out of the spiny lobster, *Panulirus homarus* in indoor farming system. In *Program and Abstracts, 7th international conference and workshop on lobster biology and management*, February, Hobart, Tasmania (pp. 8–13).

Radhakrishnan, E. V. (2005a, October–December). Co-management: An alternative approach to lobster fisheries management. *CMFRI Newsletter*, (108), 2–3.

Radhakrishnan, E. V. (2005b). Breeding and hatchery technology development of spiny lobsters and crabs- A review. In *Proceedings International conference and exposition on marine living resources of India for food & medicine*, 27–29 February, 2004, Aquaculture Foundation of India (pp. 265–272).

Radhakrishnan, E. V. (2008, January). Status of crustacean fishery resources. In E. Vivekanandan & J. Jayasankar (Eds.), *Winter School on Impact of climate change on Indian marine fisheries* (pp. 22–31). Kochi: CMFRI.

Radhakrishnan, E. V. (2009). Overview of lobster farming. In K. Madhu & R. Madhu (Eds.), *Winter school on recent advances in breeding and larviculture of marine finfish and shellfish..* 30.12.2008 to 19.1.2009 (pp. 129–138). Kochi: CMFRI.

Radhakrishnan, E. V. (2010a). Lobster fishery and management. *Souvenir: Lobster research in India*. RALBAM (pp. 19–23).

Radhakrishnan, E.V. (2010b). Lobster fisheries in India- present status and future perspectives. In *International conference on recent advances in lobster biology, aquaculture and management*, 4–8 January, 2010, Chennai. *Abstract* (pp. 17).

Radhakrishnan, E. V. (2012). Review of prospects for lobster farming. In K. K. Philipose, J. Loka, K. Sharma, & D. Damodaran (Eds.), *Handbook on open sea cage culture*. CMFRI, Karwar Research Centre (pp. 96–111).

Radhakrishnan, E. V. (2013). Lobsters. In J. Jose & S. L. Pillai (Eds.), *Manual on Taxonomy and identification of commercially important crustaceans of India* (pp. 137–156). Cochin: Crustacean Fisheries Division, Central Marine Fisheries Research institute.

Radhakrishnan, E. V. (2015). Review of prospects for lobster farming. In S. Perumal, A. R. Thirunavukkarasu, & P. Perumal (Eds.), *Advances in marine and brackishwater aquaculture* (pp. 173–186). India: Springer.

Radhakrishnan, E. V., & Devarajan, K. (1986). Growth of the spiny lobster *Panulirus polyphagus* (Herbst) reared in the laboratory. In *Proceedings of the Symposium on Coastal Aquaculture, Part 4*; 12–18 January 1980, Cochin.

Radhakrishnan, E. V., & Jayasankar, P. (2014). First record of the reef lobster *Enoplometopus occidentalis* (Randall, 1840) from Indian waters. *Journal of the Marine Biological Association of India, 56*(2), 88–91.

Radhakrishnan, E. V., & Manisseri, M. K. (2001). Status and management of lobster fishery resources in India. *Marine Fisheries Information Service Technical and Extension Series, 169*, 1–3.

Radhakrishnan, E. V., & Manisseri, M. K. (2003). Lobsters. In M. J. Mohan Joseph & A. A. Jayaprakash (Eds.), *Status of exploited marine fishery resources of India* (pp. 195–202). Kochi: Central Marine Fisheries Research Institute.

Radhakrishnan, E. V., & Thangaraja, R. (2008). Sustainable exploitation and conservation of lobster resources in India-a participatory approach. *Glimpses of Aquatic Biodiversity-Rajiv Gandhi Chair Spl. Pub., 7* (pp. 184–192).

Radhakrishnan, E. V., & Vijayakumaran, M. (1982). Unprecedented growth induced in spiny lobsters. *Marine Fisheries Information Service Technical and Extension Series*, 43.

Radhakrishnan, E. V., & Vijayakumaran, M. (1984a). Effect of eyestalk ablation in the spiny lobster *Panulirus homarus* (Linnaeus) 1. On moulting and growth. *Indian Journal of Fisheries, 31*, 130–147.

Radhakrishnan, E. V., & Vijayakumaran, M. (1984b). Effect of eyestalk ablation in the spiny lobster *Panulirus homarus* (Linnaeus) 3. On gonadal maturity. *Indian Journal of Fisheries, 31*, 209–216.

Radhakrishnan, E. V., & Vijayakumaran, M. (1986). Observations on the feeding and moulting of laboratory reared phyllosoma larvae of the spiny lobster *Panulirus homarus* (Linnaeus) under different light regimes. In *Proceedings of the Symposium on Coastal Aquaculture, Part 4*, 12–18 January 1980, Cochin.

Radhakrishnan, E. V., & Vijayakumaran, M. (1990). An assessment of the potential of spiny lobster culture in India. In *CMFRI Bulletin-National symposium on research and development in marine fisheries* (Vol. 44 (2), pp. 416–426). Kochi: CMFRI.

Radhakrishnan, E. V., & Vijayakumaran, M. (1992). Eyestalk ablation and growth, molting and gonadal development in spiny lobsters. *The Lobster News Letter, 5*(2), 1,6.

Radhakrishnan, E. V., & Vijayakumaran, M. (1995). Early larval development of spiny lobster, *Panulirus homarus* (Linnaeus, 1758) reared in the laboratory. *Crustaceana, 68*(2), 151–159.

Radhakrishnan, E. V., & Vijayakumaran, M. (1998a). Observations on the moulting behaviour of the spiny lobster *Panulirus homarus* (Linnaeus). *Indian Journal of Fisheries, 45*(3), 331–338.

Radhakrishnan, E. V., & Vijayakumaran, M. (1998b). Bilateral eyestalk ablation induces morphological and behavioural changes in spiny lobsters. *The Lobster News Letter, 11*(1), 9–11.

Radhakrishnan, E. V., & Vijayakumaran, M. (2000). Problems and prospects for lobster farming in India. In V. N. Pillai & N. G. Menon (Eds.), *Marine Fisheries Research and Management* (pp. 753–764). Cochin, India: Central Marine Fisheries Research Institute.

Radhakrishnan, E. V., & Vijayakumaran, M. (2003). The status of lobster fishery in India and options for sustainable management. In V. S. Somavanshi (Ed.), *Large marine ecosystem: Exploration and exploitation for sustainable development and conservation of fish stocks* (pp. 294–311). Mumbai: Fishery Survey of India.

Radhakrishnan, E. V., & Vivekanandan, E. (2004). Prey preference and feeding strategies of the spiny lobster *Panulirus homarus* (Linnaeus) Predating on the Green Mussel *Perna viridis* (Linnaeus). In *Program and Abstracts of the 7th international conference and workshop on lobster biology and management*, 40 pp.

Radhakrishnan, E. V., Vijayakumaran, M., & Shahul Hameed, K. (1990a). On a record of the spiny lobster *Panulirus penicillatus* (Olivier) from Madras. *Indian Journal of Fisheries, 37*(1), 73–75.

Radhakrishnan, E. V., Vijayakumaran, M., & Shahul Hameed, K. (1990b). Incidence of pseudohermaphroditism in the spiny lobster *Panulirus homarus* (Linnaeus). *Indian Journal of Fisheries, 37*(2), 169–170.

Radhakrishnan, E. V., Kasinathan, C., & Ramamoorthy, N. (1995). Two new records of scyllarids from the Indian coast. *The Lobster Newsletter, 8*(1), 9.

Radhakrishnan, E. V., Koumudi Menon, K., & Lakshmi, S. (2000). Small-scale traditional spiny lobster fishery at Tikkoti. *Marine Fisheries Information Service Technical and Extension Series, 164,* 5–8.

Radhakrishnan, E. V., Pillai, S. L., Saleela, K. N., & Kizhakudan, J. K. (2004). शूली महा चगिटों के पुरजनन और सृफूटनशाला पुरौद्योगिकी के विकास में नई पुरगतियाँ. समुद्र कृषि की नई पुरागतियाँ - विशेष पुरकाशन, 80: 1–6.

Radhakrishnan, E. V., Deshmukh, V. D., Manisseri, M. K., Rajamani, M., Kizhakudan, J. K., & Thangaraja, R. (2005). Status of the major lobster fisheries in India. *New Zealand Journal of Marine and Freshwater Research, 39,* 723–732.

Radhakrishnan, E. V., Nandakumar, G., & Manisseri, M. K. (2006a). Sustainable production and management of crustacean fisheries of India. In M. J. Modayil & P. J. Sheela (Eds.), *Livelihood issues in fisheries and aquaculture, Special Publication* (Vol. 90, pp. 29–33). Cochin: Central Marine Fisheries Research Institute.

Radhakrishnan, E. V., Nandakumar, G., & Manisseri, M. K. (2006b). Environment and crustacean fisheries. In M. J. Modayil & P. J. Sheela (Eds.), *Matsyagandha CMFRI Special Publication* (Vol. 91, pp. 7–9).

Radhakrishnan, E. V., Manisseri, M. K., & Nandakumar, G. (2007a). Status of research on crustacean fishery resources. In M. J. Modayil & N. G. K. Pillai (Eds.), *Status and perspectives in marine fisheries research in India* (pp. 135–172). Kochi, India: CMFRI.

Radhakrishnan, E. V., Chakraborty, R. D., Thangaraja, R., & Unnikrishnan, C. (2007b). Development and morphological changes in larval stages of spiny lobster *Panulirus homarus* Linnaeus reared under laboratory conditions. The 8th Asian Fisheries Forum, 20–23 November 2007, Abstract. Kochi.

Radhakrishnan, E. V., Chakraborty, R. D., Thangaraja, R., & Unnikrishnan, C. (2009). Effect of *Nannochloropsis salina* on the survival and growth of phyllosoma of the tropical spiny lobster, *Panulirus homarus* L. under laboratory conditions. *Journal of the Marine Biological Association of India, 51*(1), 52–60.

Radhakrishnan, E. V., Meenakumari, B., Puthran, P., Thangaraja, R., Ramachandran, C., & Joseph, A. M. (2010). Conservation of lobster resources in India – A participatory approach. In *International conference on recent advances in lobster biology, aquaculture and management,* 4–8 January, 2010, Chennai. *Abstract* (pp. 31).

Radhakrishnan, E. V., Lakshmi Pillai, S., Shanis, R., & Radhakrishnan, M. (2011). First record of the reef lobster *Enoplometopus macrodontus* Chan & Ng, 2008 from Indian waters. *Journal of the Marine Biological Association of India, 53*(2), 264–267.

Radhakrishnan, E. V., Chakraborty, R. D., Baby, P. K., & Radhakrishnan, M. (2013). Fishery and population dynamics of the sand lobster *Thenus unimaculatus* Burton & Davie, 2007 landed by trawlers at Sakthikulangara fishing harbour on the southwest coast of India. *Indian Journal of Fisheries, 60*(2), 7–12.

Radhakrishnan, E. V., Thangaraja, R., & Vijayakumaran, M. (2015). Ontogenetic changes in morphometry of the spiny lobster, *Panulirus homarus homarus* (Linnaeus, 1758) from southern Indian coast. *Journal of the Marine Biological Association of India, 57*(1), 5–13.

Rahman, M. K. (1989). *Reproductive biology of female sand lobster Thenus orientalis (Lund) (Decapoda: Scyllaridae)*. Chennai: University of Madras.

Rahman, M. K., & Srikrishnadhas, B. (1992). Possibility of lobster culture in India. *The Lobster News Letter, 5*(2), 5–6.

Rahman, M. K., & Srikrishnadhas, B. (1994a). The potential for spiny lobster culture in India. *InfoFish International, 1,* 51–53.

Rahman, K. Md., & Srikrishnadhas, B. (1994b). Packing of live lobsters-the Indian experience. *InfoFish International,* 6/94, pp. 47–49.

Rahman, M. K., & Subramoniam, T. (1989). Molting and its control in the female sand lobster *Thenus orientalis* (Lund). *Journal of Experimental Marine Biology and Ecology, 128*(2), 105–115.

Rahman, M. K., Prakash, E. B., & Subramoniam, T. (1987). Studies on the egg development stages of sand lobster *Thenus orientalis* (Lund). In P. Natarajan, H. Suryanarayana, & P. K. A. Aziz

(Eds.), *Advances in aquatic biology and fisheries. Prof N. Balakrishnan Nair felicitation volume* (pp. 327–335). Trivandrum: Department of Aquatic Biology and Fisheries, University of Kerala.

Rahman, M. K., Joseph, M. T. L., & Srikrishnadhas, B. (1997). Growth performance of spiny lobster *Panulirus homarus* (Linnaeus) under mass rearing. *Journal of Aquaculture in the Tropics, 12*(4), 243–253.

Rai, H. S. (1933). Shell fisheries of Bombay presidency. Part II. *Journal of Natural History Society, 36*(4), 884–897.

Rajamani, M. (2001). Creation of artificial habitat for spiny lobsters in the sea off Vellapatti, Gulf of Mannar. In N. G. Menon & P. P. Pillai (Eds.), *Perspectives in mariculture* (pp. 131–138). Cochin: The Marine Biological Association of India.

Rajamani, M., & Manickaraja, M. (1991). On the collection of spiny lobsters by skin divers in the Gulf of Mannar off Tuticorin. *Marine Fisheries Information Service Technical and Extension Series, 113*, 17–18.

Rajamani, M., & Manickaraja, M. (1994). A spiny lobster *Panulirus ornatus* with antennule-like outgrowth in the place of eyestalk. *Journal of the Marine Biological Association of India, 36*(1 & 2), 302–303.

Rajamani, M., & Manickaraja, M. (1995). Fishery of the painted crayfish *Panulirus versicolor* in the Gulf of Mannar. *Marine Fisheries Information Service Technical and Extension Series*, No. 140, 6–7.

Rajamani, M., & Manickaraja, M. (1997a). On the fishery of the spiny lobster off Tharuvaikulam, Gulf of Mannar. *Marine Fisheries Information Service Technical and Extension Series*, No. 146, 7–8.

Rajamani, M., & Manickaraja, M. (1997b). The spiny lobster resources in the trawling grounds off Tuticorin. *Marine Fisheries Information Service Technical and Extension Series, 148*, 7–9.

Rajendran, K. V., Vijayan, K. K., Santiago, T. C., & Krol, R. M. (1999). Experimental host range and histopathology of white spot syndrome virus (WSSV) infection on shrimp, prawns, crabs, and lobsters from India. *Journal of Fish Diseases, 22*, 183–191.

Ramanadhan, R., & Chacko, P. I. (1962). A preliminary report on the lobster fishery of Kanyakumari coast. *Fisheries station reports and yearbook, Department of Fisheries, Madras*. April '57–March '58 (pp. 86–93).

Ramya, B., Vijaya, R., & Kirubagaran, R. (2010). Possible impacts of organometal pollution on developmental stages of lobsters: A review. In *International conference on recent advances in lobster biology, aquaculture and management*, 4–8 January, 2010, Chennai. *Abstract* (pp. 117–118).

Rao, P. V., & George, M. J. (1973). Deep-sea spiny lobster, *Puerulus sewelli* Ramadan: The commercial potentialities. In *Proceeding of the symposium living resources of the seas around India, Spl. Publ., CMFRI* (pp. 634–640).

Rao, P. V., & Kathirvel, M. (1971). On the seasonal occurrence of *Penaeus semisulcatus* De Haan, *Panulirus polyphagus* (Herbst) and *Portunus (P.) pelagicus* (Linn.) in the Cochin backwater. *Indian Journal of Fisheries, 18*(1 & 2), 129–134.

Rao, G. S., Suseelan, C., & Kathirvel. M. (1989). Crustacean resources of the Lakshadweep Islands. In *Marine living resources of the Union Territory of Lakshadweep- An indicator survey with suggestions for development. CMFRI Bulletin* No. 43, 72–76.

Rao, G. S., George, R. M., Anil, M. K., Saleela, K. N., Jasmine, S., Jose Kingsly, H., & Rao, G. H. (2010). Cage culture of the spiny lobster *Panulirus homarus* (Linnaeus) at Vizhinjam, Trivandrum along the southwest coast of India. *Indian Journal of Fisheries, 57*(1), 23–29.

Rasheeda, M. K., Anbarasu, M., Vedaprakash, L., Vijayakumaran, M., Subramoniam, T., & Kirubagaran, R. (2010). Immunological evidence for the presence of putative cytochrome p450 aromatase in *Panulirus homarus*. In *International conference on recent advances in lobster biology, aquaculture and management*, 4–8 January, 2010, Chennai. *Abstract* (pp. 99).

Rathinam, A. M. M., Kandasami, D., Kizhakudan, J. K., Leslie, V. A., & Gandhi, A. D. (2009). Effect of dietary protein on the growth of spiny lobster *Panulirus homarus* (Linnaeus). *Journal of the Marine Biological Association of India, 51*(1), 114–117.

Rathinam, A. M. M., Vijayagopal, P., Kizhakudan, J. K., Vijayan K. K., & Chakraborty, K. (2010). Efficacy of ingredients included in pellet feeds to stimulate intake by spiny lobsters reared in captivity. In *International conference on recent advances in lobster biology, aquaculture and management*, 4–8 January, 2010, Chennai. Abstract (pp. 93).

Rathinam, A. M. M., Kizhakudan, J. K., Vijayagopal, P., Vidya, J., Leslie, V. A., & Sundar, R. (2014). Effect of dietary protein levels in the formulated diets on growth and survival of juvenile spiny lobster *Panulirus homarus* (Linnaeus). *Indian Journal of Fisheries, 61*(2), 67–72.

Ratna Raju, M., & Babu, D. E. (2015). Rock lobster fishery from east coast of India. *Asian Journal of Experimental Sciences, 29*(1), 1–6.

Rejinie Mon, T. S. (2011). *Population genetics of spiny lobster Panulirus homarus (Linnaeus) in peninsular India*. Tirunelveli: Manonmaniam Sundaranar University.

Remany, M. C., Santhanakumar, J., Senthilmurugan T,. Dharani, G., Vijayakumaran, M., Vinithkumar N. V., & Kirubagaran, R. (2010). Evaluation of a commercial pelletized feed (shrimp feed) and semi moist feed on growth performance of spiny lobster *Panulirus homarus* (Linnaeus, 1758). In *International conference on recent advances in lobster biology, aquaculture and management*, 4–8 January, 2010, Chennai. Abstract (pp. 113).

Roy, M. K. (2010). Diversity and distribution of Crustacea fauna in wetlands of West Bengal. *Journal of Environment and Sociobiology.*. Social Environmental and Biological Association, Kolkata, India, 7(2), 147–187.

Saha, S. N., Vijayanand, P., & Rajagopal, S. (2009). Length-weight relationship and relative condition factor in *Thenus orientalis* (Lund, 1793) along east coast of India. *Current Research Journal of Biological Sciences, 1*(2), 11–14.

Saleela, K. N. (2015). *Nutritional studies on the spiny lobster Panulirus homarus from the south west coast of India* (p. 199). Tirunelveli, Tamil Nadu: Manonmaniam Sundaranar University.

Saleela, K. N., Beena, S., & Palavesam, A. (2015). Effects of binders on stability and palatability of formulated dry compounded diets for spiny lobster *Panulirus homarus* (Linnaeus, 1758). *Indian Journal of Fisheries, 62*(1), 95–100.

Samraj, Y. C. T., Jayagopal, P., & Kamalraj, K. (2010). Farming of spiny lobsters in onshore facility. In *Abstracts. International conference on Recent Advances in Lobster Biology, Aquaculture and Management (RALBAM 2010)*, 5–8 January, 2010, Chennai, India (pp. 69).

Sarasu, T. N. (1987). *Larval biology of the spiny lobsters of the genus Panulirus* (p. 123). Cochin: Cochin University of Science and Technology.

Sarkar, R. J., & Talukdar, S. (2003). Marine invertebrates of Digha coast and some recommendations on their conservation. *Records of the Zoological Survey of India, 101*(Part 3–4), 1–23.

Sarvaiya, R. T. (1987). Successful spiny lobster culture In Gujarat. *Fishing Chimes, 7*(7), 18–23.

Satyanarayana, A. V. V. (1961). A record of *Panulirus penicillatus* (Olivier) from the inshore waters off Quilon, Kerala. *Journal of the Marine Biological Association of India, 3*(1), 269–270.

Satyanarayana, A. V. V. (1987). *Panulirus versicolor* (Latrielle) A new record from Central Andhra Coast (Bay of Bengal). *Fishery Technology, 24*(2), 134–135.

Sawant, A. D., Dias, J. R., Waghmare, K. B., & Sundaram, S. (2006). High catch of rock lobster *Panulirus polyphagus* landed at New Ferry Wharf, Mumbai. *Marine Fisheries Information Service Technical and Extension Series, 190*, 19–20.

Selvarani, J. B. (2000). Likely impact of sea urchin removal as by-catch on lobster resources in Gulf of Mannar. *Marine Fisheries Information Service Technical and Extension Series, 166*, 15–17.

Senthil Murugan, T., Vijayakumaran, M., Remany, M. C., Mary Leema, T., Jha, D., Kumar, T. S., Santhanakumar, J., Sreeraj, G., & Venkatesan, R. (2004). Early phyllosoma larval stages of the sand lobster, *Thenus orientalis* (Lund, 1793). In S. A. H. Abidi, M. Ravindran, R. Venkatesan, & M. Vijayakumaran (Eds.), *Proceeding of the seminar on recent advances marine bioscience research* (pp. 161–168). Chennai: NIOT.

Senthil Murugan, T., Remany, M. C., Mary Leema, T., Jha, D. K., Santhanakumar, J., Vijayakumaran, M., Venkatesan, R., & Ravindran, M. (2005). Growth, repetitive breeding, and aquaculture potential of the spiny lobster, Panulirus ornatus. N Z. *Journal of Marine and Freshwater Research, 39*, 311–316.

Sewel, R. B. S. (1913). Notes on the biological work of R.I.M.S.S. 'Investigator' during survey seasons 1910–1911 and 1911–1912. *Journal and Proceedings of the Asiatic Society of Bengal, 9*, 329–390.

Shanmugham, S., & Kathirvel, M. (1983). Lobster resources and culture potential. *CMFRI Bulletin, 34*, 61–65.

Shyamasundari, K., & Rao, H. K. (1978). Studies on the Indian sand lobster *Thenus orientalis* (Lund): Mucopolysaccharides of the tegumental glands. *Folia Histochem Cytochem (Krakow), 16*(3), 247–254.

Silas, E. G. (1969). Exploratory fishing by *R. V. Varuna*. *Bulletin of the Central Marrine Fisheries Research Institute*, No. 12, 1–86.

Silas, E. G. (1982). Major breakthrough in spiny lobster culture. *Marine Fisheries Information Service Technical and Extension Series, 43*, 1–5.

Silas, E. G., & Alagarswamy, K. (1983). General considerations of mariculture potential of Andaman and Nicobar Islands. *CMFRI Bulletin, 34*, 104–107.

Sivaraj, P., Dhas, J. C., Prasad, G. V. A., & Babu, C. (2010). Distribution and abundance of deep sea spiny lobster, *Puerulus sewelli* in the Indian EEZ. In *International conference on recent advances in lobster biology, aquaculture and management*, 4–8 January, 2010, Chennai. Abstract (pp. 17).

Solanki, Y., Jetani, K. L., Khan, S. I., Kotiya, A. S., Makawana, N. P., & Rather, M. A. (2012). Effect of stocking density on growth and survival rate of spiny lobster (*Panulirus polyphagus*) in cage culture system. *International Journal of Aquatic Science, 3*(1), 3–14.

Somavanshi, V. S., & Bhar, P. K. (1984). A note on the demersal fishery resources of Gulf of Mannar. *Fishery Survey of India Bulletin*, No. 13, 12–17.

Srikrishnadhas, B., Sunderraj, V., & Kuthalingam, M. D. K. (1983). Cage culture of spiny lobster *Panulirus homarus*. In *Proceeding of the national Seminar on Cage and Pen culture* (pp. 103–106). Tuticorin: Fisheries College and Research Institute.

Srikrishnadhas, B., Rahman, M. K., & Anandasekharan, A. S. M. (1991). A new species of the scyllarid lobster *Scyllarus tutiensis* (Scyllaridae: Decapoda) from the Tuticorin Bay in the Gulf of Mannar. *Journal of the Marine Biological Association of India, 33*(1 & 2), 418–421.

Srinivasa Gopal, T. K., & Leela Edwin, P. (2013). Development of fishing industry in India. *Journal of Aquatic Biology and Fisheries, 1*(1 & 2), 38–53.

Srivastava, S. (2014). *Chemical Interference with molt Inhibiting hormone: Design, Synthesis and Validation of a Novel Growth Enhancer* (p. 188). Thanjavur: SASTRA University.

Srivastava, S., & Princy, S. A. (2015). Effect of a novel compound as dietary supplement on growth of decapod crustaceans. *RSC Advances, 00*, 1–7. https://doi.org/10.1039/x0xx00000x. www.rsc.org/

Subramaniam, V. T. (2004). Fishery of sand lobster *Thenus orientalis* (Lund) along Chennai coast. *Indian Journal of Fisheries, 51*(1), 111–115.

Subramoniam, T. (1999). Egg production of economically important crustaceans. *Current Science, 76*, 350–360.

Subramoniam, T., & Kirubagaran, R. (2010). Endocrine regulation of vitellogenesis in lobsters. *Journal of the Marine Biological Association of India, 52*(2), 229–236.

Sulochanan, P., & John, M. E. (1988). Offshore, deep sea and oceanic fishery resources off Kerala coast. *Bulletin of Fisheries Survey India*, No. 16, 27–48.

Suseelan, C. (1996). Crustacean biodiversity, conservation and management. In N. G. Menon & C. S. G. Pillai (Eds.), *Marine biodiversity, conservation and management* (pp. 41–68). Cochin: CMFRI.

Suseelan, C., & Pillai, N. N. (1993). Crustacean fishery resources of India: A review. *Indian Journal of Fisheries, 40*(1 & 2), 104–111.

Suseelan, C., Nandakumar, G., & Rajan, K. N. (1990a). Results of bottom trawling by *FORV Sagar Sampada* with special reference to catch and abundance of edible crustaceans. In *Proceedings of the first workshop on scientific results of FORV Sagar Sampada*, 5–7 June, 1989 (pp. 337–346).

Suseelan, C., Muthu, M. S., Rajan, K. N., Nandakumar, G., Neelakanta Pillai, N., Surendranatha Kurup, N., & Chellappan, K. (1990b). Results of an exclusive survey for deep sea crustaceans off southwest coast of India. In *Proceedings of the first workshop on scientific results of FORV Sagar Sampada*, 5–7 June 1989 (pp. 347–359).

Suseelan, C., Neelakanta Pillai, N., Radhakrishnan, E. V., Rajan, K. N., Manmadhan Nair, K. R., Sampson Manikam, P. E., & Saleela, K. N. (1992). *Handbook on aquafarming -shrimps, lobsters and mudcrabs* (pp. 72). Marine Products Export Development Authority.

Tami, P. R. S., & George, M. J. (1975). Phyllosoma larvae in the IIOE (1960–1965) collections: Systematics. *Mahasagar, 8*(1–2), 15–44.

Thangaraja, R. (2011). *Ecology, reproductive biology and hormonal control of reproduction in the female spiny lobster Panulirus homarus (Linnaeus, 1758)* (p. 172). Mangalore: Mangalore Biosciences University.

Thangaraja, R., & Radhakrishnan, E. V. (2012). Fishery and ecology of the spiny lobster *Panulirus homarus* (Linnaeus, 1758) at Khadiyapatanam in the southwest coast of India. *Journal of the Marine Biological Association of India, 54*(2), 69–79.

Thangaraja, R., Radhakrishnan, E. V., & Chakraborthy, R. D. (2015). Stock and population characteristics of the Indian rock lobster *Panulirus homarus homarus* (Linnaeus, 1758) from Kanyakumari, Tamilnadu, on the southern coast of India. *Indian Journal of Fisheries, 62*(3), 21–27.

Thiagarajan, R., Krishna Pillai, S., Jasmine, S., & Lipton, A. P. (1998). On the capture of a live South African Cape locust lobster at Vizhinjam. *Marine Fisheries Information Service Technical and Extension Series, 158*, 18–19.

Thirumilu, P. (2011). Largest recorded ridge-back lobsterette, *Nephropsis carpenteri* Wood-Mason, 1885 from Chennai coast. *Marine Fisheries Information Service Technical and Extension Series, 209*, 13–14.

Tholasilingam, T., & Rengarajan, K. (1986). Prospects on spiny lobster *Panulirus* spp. culture in the east coast of India. In *Proceedings of the symposium on coastal aquaculture, Part 4*, 12–18 January 1980, Cochin.

Tholasilingam, T., Venkatraman, G., Krishna Kartha, K. N., & Karunakaran Nair, P. (1968). Exploratory fishing off the southwest coast of India by *M.F.V. Kalava*. *Indian Journal of Fisheries, 11*, 547–558.

Thomas, P. A. (1975). A note on the occurrence of the puerulus of the spiny lobster *Panulirus polyphagus* (Herbst) In Zuari estuary in Goa. *Indian Journal of Fisheries, 22*(1 & 2), 299–300.

Thomas, M. M. (1979). On a collection of deep sea decapod crustaceans from the Gulf of Mannar. *Journal of the Marine Biological Association of India, 21*(1 & 2), 41–44.

Thurston, E. (1887). Preliminary report on the marine fauna of Rameshwaram and neighbouring Islands, Madras (Madras Government Museum). *Science Series, 1*, 1–41.

Thurston, E. (1894). Rameswaram Island and fauna of the Gulf of Mannar, Madras (Madras Government Museum). *Science Series, 11*, 98–138.

Tikader, B. K., & Das, A. K. (1985). *Glimpses of animal life in Andaman and Nicobar Islands* (p. 170). Calcutta: Zoological Survey of India.

Tikader, B. K., Daniel, A., & Subba Rao, N. V. (1986). *Seashore animals of Andaman and Nicobar Islands* (p. 188). Calcutta: Zoological Survey of India.

Trivedi, D. J., Trivedi, J. N., Soni, G. M., Purohit, B. D., & Vachhrajani, K. D. (2015). Crustacean fauna of Gujarat state of India: A review. *Electronic Journal of Environmental Science, 8*, 23–31.

Vaishanvi, V., & Mullainadhan, P. (2017). Physico-chemical characterization of a naturally occurring hemagglutinin in the serum of sand lobster, *Thenus orientalis* with affinity for n-acetylated aminosugars. *International Journal of Current Microbiology and Applied Sciences, 6*(6), 2163–2173.

Vaitheeswaran, T. (2015). A new record of Scyllarid lobster *Scyllarus batei batei* (Holthuis, 1946) (Family: Scyllaridae, Latreille, 852) (Crustacea: Decapoda: Scyllaridae) off Thoothukudi coast of Gulf of Mannar, southeast coast of India 08o52.6′N 78o16′E and 08o53.8′N 78o32′E (310 m). *International Journal of Marine Science, 5*(54), 1–2.

Vaitheeswaran, T. (2017). On rare occurrence of pronghorn spiny lobster, *Panulirus penicillatus* (Olivier, 1791) off Tuticorin coast, India (08°35.912′N and 78° 25.327′E) (25M). *Journal of Agricultural Science and Botany, 1*(1), 1–2.

Vaitheeswaran, T. Venkataramani, V. K., & Srikrishnadhas, B. (2002a, October & December). Length-weight relationship of spiny lobster *Panulirus homarus* (Linnaeus). *Cheiron, 31*(5 & 6), 143–145

Vaitheeswaran, T., Venkataramani, V. K., & Srikrishnadhas, B. (2002b). Length weight relationship of lobster panulirus versicolor (latreille, 1804) (family: palinuridae) off thoothukudi waters, southeast coast of India.

Vaitheeswaran, T., Jayakumar, N., & Venkataramani, V. K. (2012). Length-weight relationship of lobster *Panulirus versicolor* (Latreille, 1804) (Family: Palinuridae) off Thoothukudi waters, southeast coast of India. *Journal of Tamilnadu Veterinary & Animal Sciences, 8*(91), 54–59.

Verghese, B. (2003). *Some immunobiological aspects of the spiny lobster Panulirus homarus (Linnaeus, 1758)* (p. 118). Mumbai: CIFE.

Verghese, B., Radhakrishnan, E. V., & Padhi, A. (2007). Effect of environmental parameters on immune response of the Indian spiny lobster, *Panulirus homarus* (Linnaeus, 1758). *Fish & Shellfish Immunology, 23*, 928–936.

Verghese, B., Radhakrishnan, E. V., & Padhi, A. (2008). Effect of moulting, eyestalk ablation, starvation and transportation on the immune response of the Indian spiny lobster, *Panulirus homarus*. *Aquaculture Research, 39*(9), 1009–1013.

Vidya, K., & Joseph, S. (2012). Effect of salinity on growth and survival of juvenile Indian spiny lobster, *Panulirus homarus* (Linnaeus). *Indian Journal of Fisheries, 59*(1), 113–118.

Vijayakumaran, M. (1990). *Energetics of a few marine crustaceans.* Kochi: Cochin University of Science and Technology.

Vijayakumaran, M., & Radhakrishnan, E. V. (1984). Effect of eyestalk ablation in the spiny lobster *Panulirus homarus* (Linnaeus) 2. On food intake and conversion. *Indian Journal of Fisheries, 31*, 148–155.

Vijayakumaran, M., & Radhakrishnan, E. V. (1986). Effects of food density on feeding and moulting of phyllosoma larvae of the spiny lobster *Panulirus homarus* (Linnaeus). In *Proceedings of the symposium on Coastal Aquaculture,* Part 4, MBAI, 12–18 January 1980, Cochin.

Vijayakumaran, M., & Radhakrishnan, E. V. (1988). Ecological and behavioural studies of the pueruli of the palinurid lobsters along the Madras coast. *Abstract. Symposium on Tropical Marine Living Resources*, Cochin.

Vijayakumaran, M., & Radhakrishnan, E. V. (1997). Live transport and marketing of spiny lobster in India. *Marine and Freshwater Research, 48*, 823–828.

Vijayakumaran, M., & Radhakrishnan, E. V. (1998). Lobster culture and live transport. In M. Sakthivel, E. Vivekanandan, M. Rajagopalan, M. M. Meiyappan, R. Paulraj, S. Ramamurthy, & K. Alagaraja (Eds.), *Proceedings of the workshop national aquaculture week* (pp. 97–103). Chennai: The Aquaculture Foundation of India.

Vijayakumaran, M., & Radhakrishnan, E. V. (1999). Influence of size and maturity on food conversion efficiency in the spiny lobster *Panulirus homarus* (Linnaeus). In *Proceeding of the fourth Indian fisheries forum,* 24–28, November, Kochi (pp. 197–198).

Vijayakumaran, M., & Radhakrishnan, E. V. (2002). Changes in biochemical and mineral composition during ovarian maturation in the spiny lobster, *Panulirus homarus* (Linnaeus). *Journal of the Marine Biological Association of India, 44*(1 & 2), 85–96.

Vijayakumaran, M., & Radhakrishnan, E. V. (2003). Control of epibionts with chemical disinfectants in the phyllosoma larvae of the spiny lobster, *Panulirus homarus* (Linnaeus). In I. S. Bright Singh, S. Pai, R. Philip, & A. Mohandas (Eds.), *Aquaculture medicine* (pp. 69–72). Kochi: Centre for Fish Disease Diagnosis and Management, CUSAT.

Vijayakumaran, M., & Radhakrishnan, E. V. (2011). Slipper lobsters. In R. K. Fotedar & B. F. Phillips (Eds.), *Recent advances and new species in aquaculture* (pp. 85–114). Oxford: Blackwell Publishing Ltd..

Vijayakumaran, M., Senthilmurugan, T., Remany, M. C., Mary Leema, T., Dilip Kumar, J., Santhanakumar, T., Venkatesan, R., & Ravindran, M. (2005). Captive breeding of the spiny lobster, *Panulirus homarus*. *New Zealand Journal of Marine and Freshwater Research, 39*, 325–334.

Vijayakumaran, M., Venkatesan, R., Murugan, T. S., Kumar, T. S., Jha, D. K., Remany, M. C., Mary Leema, T., Syed Jahan, S., Dharani, G., Selvan, K., & Kathiroli, S. (2009). Farming of spiny lobsters in sea cages in India. *New Zealand Journal of Marine and Freshwater Research, 43*, 623–634.

Vijayakumaran M., Anbarasu, M., & Kumar, T.S. (2010a). What helps tropical spiny lobsters to grow better in communal rearing – Physical or chemical interactions? In *International conference on recent advances in lobster biology, aquaculture and management*, 4–8 January, 2010, Chennai. Abstract (pp. 41).

Vijayakumaran, M., Anbarasu, M., & Kumar, T. S. (2010b). Moulting and growth in communal and individual rearing of the spiny lobster, Panulirus homarus. *Journal of the Marine Biological Association of India, 52*(2), 274–281.

Vijayakumaran, M., Maharajan, A., Rajalakshmi, S., Jayagopal, P., Subramanian, M. S., & Remani, M. C. (2012). Fecundity and viability of eggs in wild breeders of spiny lobsters, *Panulirus homarus* (Linnaeus, 1758), *Panulirus versicolor* (Latrielle, 1804) and *Panulirus ornatus* (Fabricius, 1798). *Journal of the Marine Biological Association of India, 54*(2), 18–22.

Vijayakumaran, M., Maharajan, A., Rajalakshmi, S., Jayagopal, P., & Remani, M. C. (2014). Early larval stages of the spiny lobsters, *Panulirus homarus, Panulirus versicolor* and *Panulirus ornatus* cultured under laboratory conditions. *International Journal of Development Research, 4*(2), 377–383.

Vijayan, K. K., Sharma, K., Kizhaskuda, J. K., Sanil, N. K., Saleela, K. N., Alvandi, S. V., Margaret, M. R., & Radhakrishnan, E. V. (2010a). Gaffkemia (Red tail disease)- an emerging disease problem in lobster holding facilities in southern India. In *Abstracts. International conference on Recent Advances in Lobster Biology, Aquaculture and Management (RALBAM 2010)*, 5–8 January, 2010, Chennai, India (pp. 61).

Vijayan, K. K., Sanil, N. K., & Sharma, K. (2010b). Health management concepts in lobster mariculture. *RALBAM,* 2010.

Vijayanand, P., Murugan, A., Saravanakumar, K., Khan, S. A., & Rajagopal, S. (2007). Assessment of lobster resources along Kanyakumari (Southeast coast of India). *Journal of Fisheries and Aquatic Science, 2*, 387–394.

Vinithkumar, N. V., Remany, M. C., Dharani, G., Magesh Peter, D., Mary Leema, T., Babu, T. D., Kirubagaran, R., Senthil Murugan, T., Dalmin, G., Nair, K. V. K., Sampath, V., Subramoniam, T., & Ravindran, M. (2002). Biochemical composition of different live feeds and their influence on moulting during fattening of juvenile lobster *Panulirus homarus* (Linnaeus, 1758). In *Proceeding of the XX national symposium on Reproductive biology and Comparative endocrinology* held during 7–9 Jan, 2002 organized by Department of Animal Sciences, Bharathidasan University,Tiruchirappalli-620 024, Tamilnadu, India.

Wood-Mason, J. (1872). On *Nephropsis stewarti*, a new genus and species of macrurous crustaceans, dredged in deep water off the eastern coast of the Andaman Islands. *Proceedings of the Asiatic Society of Bengal, 1872,* 151.

Yang, C.-H., Bijukumar, A., & Chan, T.-Y. (2017). A new slipper lobster of the genus *Petrarctus* Holthuis, 2002 (Crustacea: Decapoda:Scyllaridae) from the southwest coast of India. *Zootaxa, 4329*(5), 477–486.

Zacharia, P. U., Krishnakumar, P. K., Dineshbabu, A. P., Vijayakumaran, K., Rohit, P., Thomas, S., Geetha, S., Kaladharan, P., Durgekar, R. N., & Mohamed, K. S. (2008). Species assemblage in the coral reef ecosystem of Netrani Island off Karnataka along the southwest coast of India. *Journal of the Marine Biological Association of India, 50*(1), 87–97.

Index

A

Abdomen, 12, 16, 22, 23, 71, 74, 81, 83, 84, 92, 93, 96–98, 107–109, 114, 179, 182, 280, 366, 374, 378, 385, 386, 392, 394, 395, 447, 474, 520, 564, 572, 575, 577, 578, 585, 592, 608, 616
Ablation, 186, 190, 197–200, 202–204, 396–398, 466, 467, 594
Abundance, 84, 86, 130, 152, 153, 166, 167, 169, 172, 177, 193–195, 244, 258, 263, 271, 280, 291, 309, 325, 339, 433, 434, 451, 543, 546, 548, 549, 556–558, 563, 566, 567, 638, 639
Acanthacaris tenuimana, 37, 39, 111, 304
Acrosome, 368
Adaptation, 9, 385, 434, 533, 563, 638
Aequorea coerulescens, 522
Aequorea macrodactyla, 522
Allometrically, 192, 269, 288, 366, 381, 385–387
Allopatric lineages, 129
Allozymes, 132, 133
Anchor hooks, 229–231, 272, 274, 279, 280
Andaman and Nicobar Islands, 69, 74, 78, 80, 81, 83, 84, 86, 97, 100, 103, 114, 115, 196, 235, 302, 309, 311
Annual landing, 112, 223, 226, 235, 237, 239, 246, 247, 253, 255, 258, 262, 264, 265, 267, 274, 275, 283, 286, 290, 291, 298, 299, 301, 311, 319–322, 324, 327, 333, 337, 338, 454
Anorexia, 579
Antenna, 12, 16, 17, 19, 24, 66, 70–72, 78, 80, 84–87, 91, 98, 100, 109, 111, 114, 127, 193, 235, 274, 391, 474, 574, 609
Antennules, 16, 17, 24, 71, 74, 86, 195, 393
Appendages, 9, 11, 15–17, 23, 24, 183, 375, 386, 420, 425, 435, 464, 473, 526, 589, 607

Aquaculture, vii–ix, 1–31, 139, 177, 206, 357–359, 398, 410–414, 419, 424, 431, 441, 443, 450, 451, 453, 462–465, 469, 475, 483, 484, 492–496, 541–601, 604, 605, 630, 636, 638–643
Arabian Sea, 67, 69, 75, 93, 98, 102, 103, 105, 110, 112, 114, 115, 136, 237, 243, 252, 262, 271, 302, 324, 326
Artemia nauplii, 419, 420, 423, 429–433, 435, 437, 439, 440, 449, 450, 575, 592
Arthrodial membrane, 22, 180, 182, 593
Arthropoda, 5, 70, 200
Artificial larval diets, 435
Artisanal fishing, 153, 169, 262, 324, 328
Association behaviour, 521, 523, 525, 526, 534
Asymptotic length, 260, 270, 285, 295
Atlantic, 11, 13, 14, 38, 42, 43, 52, 71, 89, 91, 103, 104, 107, 129, 135–138, 153–161, 164, 226, 336, 337, 354, 355, 543, 618, 640
Aurelia aurita, 432, 439, 440, 450, 523
Aurelia coerulea, 521, 522, 527
Auto-stocking, 468
Autotomy, 443

B

Bacteria, 196, 394, 415, 416, 418, 424, 428, 430, 432, 433, 435–437, 455, 456, 470, 475, 480, 485, 521, 564, 565, 571–575, 577–579, 586–588, 591
Bacteriophages, 437, 578
Bacterioplankton, 436
Bag net, 231, 234, 249, 312
Bahamas, 157, 159, 225–227, 353, 354, 359, 523, 524
Baja Red rock lobster, 337, 357
Bandhans, 234, 453, 466

Bathyarctus rubens, 93
Bay of Bengal, viii, 3, 42, 52, 67, 69, 85, 92, 98, 102, 103, 105–107, 112, 115, 237, 271, 297, 302, 388, 389
Benthic phase, 6, 520
Beroe cucumis, 522, 525
Berried, 261, 269–271, 286, 287, 311, 327, 378, 389, 397, 415, 416, 418, 442, 590
Bhavnagar, 76, 234, 235, 247–249, 312, 466, 473
Biarctus sordidus, 92, 274
Biochemical composition, 486
Bioeconomics, 358, 482
Biofilters, 444
Biogeographic barrier, 129
Biology, vii–ix, 2, 67, 128, 222, 364, 414, 521, 557, 606, 636
Biosecure, 641
Blackening, 578
Bolinopsis mikado, 522, 525
Bottlenecks, 138, 440, 441, 477, 483, 494, 639, 640
Bottom-set-gillnets, 296, 454, 607
Breeding, ix, 3, 4, 8, 9, 14, 24, 77, 79, 82, 138, 139, 167, 246, 249, 251, 252, 259–262, 268, 286, 287, 292, 305, 311–315, 338, 339, 364, 365, 376–378, 382, 384, 385, 388–390, 392, 393, 396–398, 409–540, 584, 588, 637, 640–642
Breeding larval rearing, 495
Breeding migrations, 311, 315, 389
Broodstock, 356, 377, 393, 397, 411, 415, 416, 436, 441–446, 577
Broodstock holding tanks (BHT), 416, 444
Bully nets, 229, 231, 247
Bycatch, 100, 169, 220, 234, 246, 251–253, 263, 265, 268, 275, 276, 289, 290, 294, 310, 312–315, 319, 326, 338, 357, 473, 475, 494, 548, 637

C
Cannibalism, 180, 414, 443, 457, 462–464, 472, 483, 530, 534, 607, 609, 642
Capture-based, 332, 412, 419, 482, 495, 545, 595, 641
Capture-based aquaculture (CBA), 595, 641
Capture fisheries, vii, 5, 223, 225, 226, 235, 315, 334, 335, 605
Carapace length (CL), 8, 12, 13, 15, 74, 84, 86, 89, 91, 93, 100, 113, 167, 180, 186, 188, 189, 203, 251, 269, 271, 280, 285, 288, 292, 296, 311, 314, 320, 322, 327, 366, 376, 378, 386, 387, 415–418, 458, 637

Carcinonemertean, 583–585
Caribbean, 6, 8, 12, 39, 40, 42, 50, 104, 129, 134, 135, 152, 157, 159, 223, 226, 353, 354, 370, 412, 452, 461, 557, 573, 584
Carina, 22, 93, 94, 96–98, 101–103, 106, 107, 112–114
Cast net, 233, 234, 263, 267
Catch per unit effort (CPUE), 246, 248, 254, 256, 266, 267, 275, 277, 283, 284, 290, 294, 295, 300, 303, 305, 310, 324, 553, 637
Catch rates, 247, 249–251, 254, 257, 260, 263, 265, 269, 270, 275, 290, 291, 295, 296, 309, 310, 314, 319, 454, 548, 551
Catostylus mosaicus, 522, 525
Cephalic shield, 14, 447
Cephalothorax, 12, 15, 17, 19, 197, 199, 592
Certification, 335–338, 354, 356–359, 617
Cervical groove, 19, 84–86, 89, 92, 94, 96, 98, 101–103, 106, 112, 113
Checklist, ix, 4, 35–64, 67, 69
Chemical sense, 525
Chemoattractants, 436, 477
Chemoprophylactic, 574
Chemotherapeutic, 437, 574
Chitinoclastic, 575
Chromosomes, 131
Chrysaora pacifica, 439, 450, 522, 525, 527
Ciliate protozoans, 437, 588, 591
Cladistics, 127
Clams, 193, 194, 416, 439–441, 443, 448–450, 456, 459, 460, 465, 468, 471, 476, 477, 479, 484, 485, 520, 521
Clawed lobsters, vii, 5, 6, 66, 130, 223, 336, 376, 381, 384, 410, 414, 486, 487, 490, 492, 543, 573, 585, 607–610, 612, 616, 619, 642
Climate change, 173, 221, 338, 355, 356, 358, 410, 483, 638, 643
Closed season, 220, 305, 312–314, 316, 320, 325, 329, 637
Clytia languida, 522
Cnidarians, 520, 522–527, 529–533
Co-management, 222, 312–314, 339
Conservation, 128, 129, 139, 141, 288, 304, 333, 398, 425, 639
Conspecifics, 169
Consumption rates, 434, 530, 534, 535
Copulation, 23, 367, 386, 391, 393, 394
Coxopodite, 261, 367
Crustaceans, vii, viii, 2, 66, 128, 169, 177, 221, 366, 411, 521, 547, 572, 605
Cryptic species, 129
Ctenophore, 424, 432, 440, 450, 520, 521, 525, 527, 531, 532

Cultures, vii–ix, 2, 11, 14, 139, 178, 234, 396, 411, 412, 414, 415, 419–430, 432–439, 441, 446, 447, 450, 451, 457, 458, 460–467, 469, 470, 472, 474, 477, 480–482, 484, 485, 492–496, 519–540, 545, 566, 575, 577–579, 585–587, 591, 594, 595, 639, 641, 642
Culture tanks, 422–424, 426–428, 430, 575, 585, 642
Cylindro-conical, 422, 425, 429, 430, 446
Cylindro-conical tanks, 422, 429, 430, 446
Cytogenetics, 131

D
Ddradseq, 138
Deep sea, viii, 3, 87, 88, 93, 127, 128, 139, 152, 164, 220, 221, 234, 240, 241, 262–265, 268, 269, 273, 276, 280, 290, 296, 300–303, 308–311, 313–315, 319, 325, 334, 338, 443, 607, 618, 620, 637
Deficiencies, 477, 490, 564, 572, 592, 593
Deformities, 577
Descendants, 126
Diagnosis, 70–72, 74, 75, 78, 80, 81, 83–87, 89, 90, 92–109, 111–114, 640
Diets, 3, 129, 130, 141, 178, 193, 194, 196, 412, 414, 429–432, 435, 436, 447, 450, 456, 459, 465, 477, 478, 480, 484, 495, 520, 531, 533, 534, 562, 565, 566, 595, 639–641
Dimorphism, 366, 387
Dinoflagellates, 580
Divergence, 127–129, 639
DNA barcode, 128
DNA barcoding, 14, 128, 524
Dol nets, 229, 231, 242, 243, 245, 247, 253, 254, 258

E
Ebi, 66
Ecdysis, 177, 179–185, 199–201, 418, 592
Echinoderms, 169, 170, 172, 178, 193, 196, 476
Eco-certification, 15
Ecolabelling, 222, 339, 615, 643
Ecology, vii, ix, 2–4, 6, 74, 77, 79, 81–84, 86, 87, 89, 91, 94, 96, 97, 100, 101, 112, 128, 130, 151–177, 195, 282, 328, 453, 484, 636, 643
Economics, vii, ix, 35, 67, 86, 91, 115, 173, 221, 222, 296, 310, 339, 352, 354–356, 414, 427, 438, 441, 463, 465, 469, 471, 480–482, 494, 495, 543–545, 563, 566, 567, 595, 636, 642

Ecosystems, vii, 3, 152, 153, 166, 170–173, 252, 262, 272, 297, 302, 323, 335, 354, 357, 358, 421, 427, 430, 493, 638, 639
Eduarctus martensii, 5, 96, 97, 129, 304, 521, 522
Effective population size, 132, 134, 138
Eggs, 6, 47, 79, 157, 178, 252, 364, 410, 520, 577, 631, 637
Ejaculatory, 366, 367, 381
Elliptical, 425, 426
Encapsulated, 436, 532, 593
Endemic, 115, 153, 164, 173, 228, 318, 565
Endopods, 12, 17, 23, 24, 107, 203, 261, 375, 383, 384, 387, 393, 394
Energy consumption, 526, 534
Energy storage, 432, 530
Enoplometopus macrodontus, 109, 110, 264, 268
Enoplometopus occidentalis, 108, 109, 131, 274, 293
Enterobacteria, 196, 575
Entrepreneur, 310, 495, 642
Environmental requirements, 6, 422, 461, 495
Epibionts, 586, 588, 590
Escape vents, 280, 339
Exopods, 12, 17, 23, 24, 74, 75, 81, 83, 92, 101, 104, 112, 114, 127, 366, 380, 387, 392, 395, 420, 526
Exoskeleton, 15, 119, 180, 182, 183, 186, 201, 416, 448, 475, 572, 574, 575, 579, 585, 589, 592, 593
Exploitation rates, 258, 269, 270, 285, 292, 305
Extant, vii, 5, 36, 50, 52, 66, 69, 86, 102, 107, 114, 115, 126, 127
Exuding, 579
Eyestalk-ablated, 180, 184–186, 195, 199–203, 205, 398, 416, 467
Eyestalk ablation, 190, 197–200, 202–204, 396–398, 466, 467, 594
Eyestalks, 105–107, 190, 197–205, 396–398, 466, 467, 594, 609

F
Faecal, 457, 477, 524, 577, 641
Family, 5, 24, 36, 38, 43, 44, 47, 50, 51, 66–70, 86, 88, 90, 91, 99, 101, 102, 108, 110, 127, 128, 153–156, 165, 166, 169, 198, 204, 261, 302, 483, 580, 582
Farming, 3, 4, 332, 333, 419, 427, 438, 465, 467–469, 471, 474, 480–482, 484, 485, 492–496, 542–547, 549, 552, 558, 559, 563–566, 595, 640, 641
Fattening, 412, 444, 466, 468, 477, 479, 480, 544, 545, 548

Feasibility, ix, 322, 410, 411, 426, 465, 467, 471, 481, 482
Fecundity, 9, 79, 81, 100, 261, 270, 292, 311, 314, 364, 365, 376, 377, 396, 398, 418, 442, 483
Feed conversion ratio (FCR), 465, 484
Fertilization, 8, 14, 178, 364, 369, 390, 392, 394, 416, 418
Filamentous bacteria, 416, 437, 586–590
Fisheries, vii–x, 2, 77, 126, 152, 220, 352, 410, 542, 572, 604, 636
Fisheries governance, 352, 640
Fishery collapse, vii, 139, 220, 240, 253, 255, 257, 262, 312, 313, 324, 358, 636, 637
Fishing effort, 172, 222, 273, 315, 324, 330, 352, 637
Fishing gears, 222, 228, 263, 272, 279, 305, 324, 326, 330, 352
Flagellum, 12, 24, 66, 70, 81, 83, 85, 92, 127
Flavobacteriaceae, 575
Flow-through, 416, 422, 427, 429, 455, 457, 458, 463–465, 481, 495, 630
Flow-through system, 422, 455, 457, 463
Fluidized bed filter, 441, 443
Food and Agriculture Organization (FAO), 2, 3, 5, 10, 16–23, 25, 66, 126, 128, 222–228, 272, 315, 317–335, 353, 354, 358, 410, 412, 413, 487, 490, 492, 604–606, 631, 636, 641
Foregut filling time, 478
Formulated diets, 432, 435, 436, 456, 477, 478, 495, 534
Fouling, 392, 416, 437, 468, 471, 551, 561, 586, 592
Fresh, Chilled Whole Lobster, 622
Frontal horns, 17, 70–72, 85–87
Frozen Lobster Head, 617
Frozen Lobster Meat, 617, 624, 625, 631
Frozen Rock and Slipper Lobster Tails, 623
Frozen whole lobster, 616, 617, 620, 622, 624
Functional maturity, 8, 288, 365, 366, 377, 378, 380, 381, 387

G
Gaffkemia, 455, 462, 470, 573, 574
Galearctus kitanoviriosus, 521, 523
Gelatinous, 8, 367–370, 391, 394, 424–426, 428, 433, 435, 436, 520–527, 529–535
Gelatinous zooplankton, 425, 426, 428, 433, 520–527, 529–535
Genera, vii, 5, 6, 12, 17, 19, 36, 39, 47, 52, 66–71, 86, 88, 91, 92, 102, 110, 115, 126–128, 130, 153–156, 164–166, 169, 173, 222, 227, 334, 352, 447, 520, 575, 591
Genetic stock identification, xiv, 638
Genome, 130, 638–640
Genomics, 130, 134, 138, 639, 640
Global and Domestic Market, 619
Glycine, 487, 525, 531
Glypheidea, 43, 67, 115, 126
Gonopores, 367, 381–383, 391
Gorgonians, 153, 169, 171
Growout, 412, 419, 438, 440, 443, 444, 450–465, 467–472, 475–477, 480–485, 492–496, 544, 545, 547, 558, 559, 561–564, 566, 574, 593
Growth, ix, 4, 77, 138, 167, 177, 220, 366, 423, 528, 559, 572, 604, 636

H
Habitats, 3, 6, 8, 15, 67, 74, 77, 79, 81–87, 89, 91–94, 96–98, 100–108, 110–114, 152–154, 164, 166–169, 173, 178, 189, 193–195, 221, 235, 243, 247, 272, 293, 328, 341, 377, 398, 410, 414, 450, 452, 466, 483, 484, 548–551, 558, 588
Haemocytes, 462, 572, 573, 578, 590, 593, 594
Haemolymph, 179, 186, 187, 198, 200–202, 204, 416, 475, 564, 573, 574, 579, 580, 582, 593, 609–613, 640
Handpicking, 229, 235, 248, 262, 303, 624
Harveyi, V., 436, 437, 475, 575, 577, 578
Hatchery production, viii, 3, 4, 138, 139, 409–517, 640, 642
Hatchery technology, 141, 411, 480, 481, 493, 543, 546, 640, 642
Hatching, 7–9, 11, 12, 14, 15, 188, 190, 202, 376, 388, 389, 392, 393, 395–397, 416–420, 426, 430, 442, 447–449, 527, 529, 534, 592, 640
Hepatopancreas, 12, 414, 432, 486, 487, 490, 492, 495, 564, 573, 575, 593, 610
Hermaphrodites, 6
High hydrostatic pressure processing (HPP), 612, 614
High-throughput genomic technologies, 640
Holding system, 443, 444, 455, 462–464
Homarus americanus, viii, 2, 17, 37, 131–133, 138, 153, 164, 167, 168, 181, 189, 223, 225, 226, 336, 338, 375, 410, 426, 479, 488–491, 543, 573, 605, 638
Homarus gammarus, 18, 131, 135–138, 153, 168, 488–491, 543
Hormiphora, 521, 522
Hypoxia, 610, 638

I

Ibacus ciliatus, 14, 48, 131, 439, 521, 522, 526–528
Ibacus novemdentatus, 40, 432, 439, 522, 525, 527, 528, 530, 535
Immunocompetence, 462
Impregnation, 370, 385, 386, 391, 392, 394
Incubation period, 7, 383, 392, 395, 396, 416, 417, 442, 585
Indian Ocean rim countries, ix, 4, 24, 219–350
Indonesia, vii, ix, 4, 37, 74, 157, 222, 353, 376, 413, 543, 641
Indoor culture systems, 412, 642
Indo-West Pacific, 74, 78, 80, 81, 83, 87, 89, 91, 93, 96, 100, 103, 107, 108, 111–114, 129, 136, 154–161, 164, 173, 483
Infestation, 418, 429, 437, 583, 584, 591
Inflammation, 578, 593
Input controls, 316, 339, 352, 356
Instars, 11, 12, 14, 420, 421, 427, 435, 438–440, 448, 449
Intermoult period, 178, 180, 186, 188, 200, 201, 203, 388, 422, 434, 435, 437, 449, 461
Invasions, 394, 575, 576
IQF products, 616, 620, 626
Island fisheries, 302, 304

J

Jasus caveorum, 160
Jasus edwardsii, 9, 38, 130, 134, 136, 138, 161, 168, 172, 195, 333, 334, 364, 370, 393, 419, 421, 480, 610, 639
Jasus frontalis, 136, 168, 337
Jasus lalandii, 135, 168, 225, 228, 317–319, 364, 368–370, 383, 392, 411, 419–421
Jasus paulensis, 44, 154, 156, 161, 168, 225, 227, 228
Jellyfish, 4, 424, 425, 432, 439, 440, 447, 450, 519–540
Justitia longimanus, 38, 154, 156, 159, 169
Juveniles, 6, 71, 138, 167, 178, 230, 365, 412, 520, 543, 574, 607, 638

K

Keystone, 5, 153, 170, 172, 178

L

Lagenidium, 437, 586
Lakshadweep, viii, 3, 69, 71, 74, 80, 82, 88, 91, 95, 97, 111, 113–115, 302–304
Langoustes, 66
Larval cycle, 422, 438, 448, 640
Larval diseases, 436
Larval rearing system, 425, 428, 429, 445
Larval rearing tank, 430, 445, 446
Larviculture, 130, 141, 480, 521, 527, 534, 640
Lethargy, 573, 579, 586
Life history, ix, x, 6, 9, 15, 152, 177, 178, 520, 640
Light intensity, 190, 416, 423, 424, 642
Linuparus meridionalis, 44, 86
Linuparus somniosus, 86, 158, 168, 303, 311, 319, 331, 332, 382, 387
Linuparus sordidus, 44, 158
Linuparus trigonus, 38, 44, 158, 368, 369
Lipid composition, 490, 532
Lipids, 414, 433, 436, 477–479, 490, 492, 494, 530, 532
Liriope tetraphylla, 522, 523
Live lobster, 5, 83, 274, 303, 313, 314, 480, 485, 493, 595, 605–610, 614, 618–621, 625, 627, 630, 631
Live transport, 221, 494, 565, 593, 594, 606, 607, 609–612, 620
Lobster Pizza, 618
Lobster Pricing, 618
Lobster traps, 229, 230, 262, 272, 305, 306, 322, 336, 337
Luminescent, 575, 578

M

Macroalgae, 167, 169, 196, 476
Macronutrients, 478
Macrura Reptantia, vii, 5, 36, 52, 66–68, 70, 115, 126
Management, vii–ix, 2–5, 24, 126, 129, 132, 134, 138, 139, 141, 152, 166, 219–339, 351–359, 365, 378, 398, 415, 416, 427, 437, 461, 462, 475, 495, 542, 550, 554, 556, 559, 564, 565, 571–595, 637–642
Mandibles, 16, 24, 194, 484
Mariculture, 409–517
Marine protected areas (MPAs), 134, 138, 172, 639
Marine Stewardship Council (MSC), 222, 335–338, 354, 356–358, 617, 639
Marketing, ix, 5, 37, 39, 66, 83, 139, 232, 267, 269, 272, 275, 279, 283, 298, 299, 304, 305, 312, 314, 315, 330, 333, 335, 336, 338, 354, 357, 359, 414, 419, 438, 441, 457, 467, 475, 477, 479–481, 483, 484, 493, 495, 496, 543–546, 558–560, 565–567, 580, 592, 595, 603–631, 641

Mating, 7, 132, 138, 178, 203, 204, 288, 366, 368, 376, 377, 381, 382, 385, 386, 389–393, 396, 397, 416, 418, 442, 443, 469, 579

Maturation, ix, 178, 204, 205, 364, 365, 367, 371, 373, 374, 377, 380, 386, 388, 396–398, 416, 441, 442, 445, 479, 492

Maxillae, 16, 24, 435

Maxilliped, 12, 14, 16, 24, 74, 75, 81, 83, 92, 97, 100, 102, 104, 105, 127, 194, 195, 379, 435

Maximum sustainable yield (MSY), 132, 220, 259, 356, 410, 637

Mechanized, 247, 265, 276, 277

Medusa, 432, 527, 529

Metabolism, 464, 609, 610, 612

Metamorphose, 11, 12, 420, 422, 431, 449, 450, 520, 529, 530

Metamorphosis, 9, 15, 427, 431, 432, 441, 448, 450, 527, 529, 530, 532, 534, 640

Metanephrops andamanicus, 114, 319, 323, 331, 332

Micro-algal, 419, 423, 428, 430

Microarray, 639

Micro-encapsulated diet, 436

Micronutrients, 479

Microsatellites, 130, 132, 134, 136–138, 141

Microsporidians, 580

Minerals, 180, 414, 479, 490–492, 532, 593, 604

Minimum legal size (MLS), viii, 220–222, 273, 304, 305, 312–314, 316, 318, 322–325, 329, 331, 333, 459, 467, 494, 554, 563, 566, 637

Mitochondrial, 11, 115, 128–130, 132, 136, 137, 139, 329

Mitogenome, 130–132

Modal length, 269, 270, 292

Modified Atmospheric Packaging, 615

Molecular markers, 66, 127, 129, 135, 141, 638, 639

Molluscs, 24, 169, 170, 172, 196, 443, 465, 476–478, 484, 560, 562, 565, 579, 605, 641

Moribund, 575, 620

Morphometry, 100, 253, 288, 301, 386

Morphotypes, 72, 74

Mortality, 130, 259, 270, 285, 389, 414, 424–427, 429, 437, 440, 447, 453, 455–457, 462, 463, 465, 466, 470, 475, 477, 482, 485, 495, 530, 532, 574–580, 583, 585, 587, 591, 590, 593–595, 607–610, 619, 641

Moult, 7, 12, 14, 177–183, 185–190, 192–194, 197–203, 364, 390–393, 396–398, 414, 418, 420, 422, 423, 431, 434, 436, 441, 443, 448–450, 458, 459, 461, 462, 486, 520, 577, 592, 640

Mouthpart, 24, 435, 520, 525, 581

Mucopolysaccharide, 368

Mud spiny lobster, 157, 234, 240, 244, 249, 255, 257, 419, 462, 472, 589, 616, 639

Muggiaea, 523

Multiday fishing, 235, 252, 263, 265, 268, 294, 298, 300

Multiple Paternity, 138

Multivariate, 637

Mussels, 153, 169, 193–196, 229–231, 266, 280, 416, 420, 427, 431–433, 436, 439, 441, 449, 456, 459, 465, 469–471, 476–480, 482, 484, 485, 494, 520, 521, 641

N

Nannochloropsis, 427–430

Naupliosoma, 9, 14, 420, 447

Near-infrared spectroscopy (NIRS), 414

Nematocysts, 524, 532, 533

Nephropidae, 5, 6, 37, 38, 68, 110, 126, 153–155, 165, 302

Nephropid lobsters, vii, ix, 3, 6, 9, 16, 273, 302

Nephropsis carpenteri, 111, 264, 294, 296

Nephropsis ensirostris, 37, 42, 113, 304

Nephropsis stewarti, 42, 111, 112, 264, 268, 273, 294, 296, 301, 302, 311, 319

Nephropsis suhmi, 42, 114

Nephropsis sulcata, 43, 113, 304

Nephrops norvegicus, 6, 7, 18, 42, 131–133, 135–138, 166, 168, 223, 225, 226, 228, 605, 610–612, 615

Neuroendocrine, 177–206, 636

Neuroendocrine hormones, 177–206

Next-generation sequencing (NGS), 126, 134, 138, 141

Nisto, 6, 15, 95, 438–441, 446–450, 483, 495, 520, 529, 530, 534, 535, 637

Northeast, 74, 76, 77, 86, 91, 114, 115, 135, 157, 159, 220, 237, 243, 252, 296–299, 319, 322, 324, 330, 365, 415, 421, 637

Northwest, viii, ix, 11, 40, 42, 52, 74, 76, 77, 81, 85, 113–115, 129, 135, 139, 189, 193, 220, 237, 240, 243, 271, 296, 299, 306, 319, 338, 467, 472, 477, 482, 494, 636, 637

Nupalirus chani, 44, 160
Nupalirus japonicus, 38, 44
Nupalirus vericeli, 44, 159
Nursery systems, 445, 559
Nutrient quality, 492
Nutrient resource, 534
Nutrigenomics, 640
Nutritional requirements, 178, 432, 433, 465, 495

O

Ocean acidification, 638
On-growing, 190, 193, 328, 412, 414, 423, 456, 460–462, 464, 465, 467, 470, 471, 477, 478, 485, 495, 543, 546, 549, 552, 558, 559, 563, 565, 567
Ontogenetic changes, 435
Ontogenetic shift, 167, 169, 178, 196
Oomycetes, 437, 585
Outboard Motor gillnet, 232, 242, 251, 254, 274, 275, 277, 290
Output controls, 222, 316, 352, 356, 636
Overexploitation, vii, 84, 132, 222, 312, 324, 335, 338, 636
Ovigerous, 7, 83, 96, 200, 269, 285, 288, 292, 312, 369, 375, 378, 381, 383, 384, 389, 392–396
Ovigerous setae, 269, 288, 375, 378, 381, 383, 384, 392–396
Oviposit, 202–204, 269, 375, 388, 392, 415–418
Oviposition, 202, 204, 269, 375, 388, 392, 416–418
Ozonation, 424, 585

P

Packing, 374, 455, 463, 474, 480, 551, 605, 607–612, 615–618, 621, 624–626, 630
Palibythus magnificus, 39, 42, 159
Palinurellus gundlachi, 45, 159
Palinurellus wienecki, 39, 45, 67, 131, 534
Palinuridae, 6, 38, 66, 126, 152, 195, 220, 369, 520, 542
Palinurid lobsters, vii, ix, 3, 6, 8, 10, 12–14, 19, 66, 67, 154, 162, 168, 178, 179, 199, 222, 253, 276, 308, 311, 315, 365, 396, 397, 411, 412, 419, 421, 433–437, 484, 607, 610, 641
Palinurus argus, 45
Palinurus barbarae, 45, 129, 159
Palinurus charlestoni, 45, 158
Palinurus delagoae, 11, 39, 45, 159, 164, 168, 228, 317–319
Palinurus elephas, 11, 13, 45, 132, 135, 136, 156, 161, 168, 195, 225, 226, 228, 366, 411, 419, 421, 422, 426, 431, 434, 492, 642
Palinurus gilchristi, 45, 135, 161, 164, 168, 225, 228, 317, 318, 364, 368, 369
Palinurus mauritanicus, 13, 45, 132, 135, 159, 228
Palinustus holthuisi, 45, 48, 50, 87, 160
Palinustus mossambicus, 45, 67, 88, 159
Palinustus truncatus, 45, 159
Palinustus unicornutus, 45, 159
Palinustus waguensis, 39, 45, 67, 87, 88, 139, 159, 169, 173, 264, 268, 273, 294, 296, 332, 470
Panmictic, 138
Panuirus brunneiflagellum, 45, 160
Panulirus, 6, 39, 68, 127, 152, 179, 220, 352, 364, 411, 542, 572, 605, 637
  P. cygnus, 8, 9, 13, 45, 130, 131, 136, 160, 164, 167, 189, 190, 192, 194–196, 202, 225, 228, 316, 333, 334, 336, 337, 353, 355, 356, 358, 359, 365, 379, 392, 397, 416, 417, 453, 456, 459, 463, 464, 470, 478, 482, 542, 580, 581, 584, 586, 593
  P. echinatus, 11, 45, 129, 134, 137, 157, 164, 168, 169, 193–195, 364, 371, 372
  P. femoristriga, 45, 83, 157, 332
  P. gracilis, 10, 45, 51, 106, 157, 225, 228, 364, 415
  P. guttatus, 11–13, 45, 134, 157, 379, 381
  P. homarus homarus, ix, 8, 45, 128, 164, 179, 228, 365, 411, 579, 637
  P. homarus rubellus, 8, 45, 72, 73, 75, 128, 129, 136, 160, 164, 168, 180, 189, 193–195, 287–289, 317, 319, 320, 368, 376–378, 380, 388, 390, 417, 458
  P. interruptus, 46, 134–137, 160, 164, 172, 195, 228, 316, 337, 353, 357–359, 432, 584
  P. japonicus, 11–13, 46, 48, 130–132, 135, 160, 364, 367, 371, 383, 388, 393, 419–421, 423, 425–427, 431, 435, 492, 585
  P. laevicauda, 46, 157, 168, 368, 371, 372
  P. longipes bispinosus, 46, 82, 83, 419, 421, 492
  P. longipes longipes, 39, 46, 82, 83, 128, 157, 164, 168, 240, 263, 273, 282, 293, 301–303, 320, 322, 332, 364, 573

P. marginatus, 46, 49, 132, 160, 164, 181, 192, 228, 358, 365, 379
P. meripurpuratus, 46, 71, 129, 157, 223, 225, 226, 228, 353
P. ornatus, viii, 8, 46, 72, 131, 157, 181, 225, 352, 370, 411, 543, 573, 607, 641
P. pascuensis, 46, 157
P. penicillatus, 8, 12, 13, 44, 46, 71, 81, 129, 135–137, 157, 164, 169, 173, 189, 190, 192, 228, 240, 263, 264, 268, 273, 281, 282, 293, 301–304, 320, 322, 323, 326, 329, 331, 332, 364, 367, 372, 379, 415, 419, 421, 422, 492, 546, 573, 592
P. polyphagus, 5, 46, 72, 157, 180, 225, 365, 412, 545, 574, 609, 637
P. regius, 46, 132, 135, 157, 226, 228, 364
P. stimpsoni, 46, 131, 160, 546
P. versicolor, 11, 46, 72, 131, 157, 196, 230, 369, 415, 546, 573, 609
Parasites, 418, 437, 571, 580, 581, 586, 589, 594, 641
Parentage analysis, 126, 138
Parribacus antarcticus, 14, 48, 90, 91, 273, 304, 320, 329, 331, 332, 370, 582
Participatory, 222, 304, 306–308, 314
Pathogens, 3, 5, 411, 415, 416, 432, 433, 436, 437, 462, 475, 483, 495, 521, 527, 564, 571–575, 578, 586, 589, 594, 595, 640, 641
Pelagia panopyra, 522
Pentacheles gibbus, 51, 105
Pentacheles laevis, 41, 51, 106
Pentacheles obscurus, 51, 105
Pereiopods, 12, 17, 19, 24, 83, 85, 100, 194, 195, 367, 381–383, 385–387, 391, 392, 394, 395, 420, 433–435, 447, 520, 525, 526, 533, 585–587
Peritrophic membrane, 533
Petrarctus demani, 50, 440, 523, 527
Petrarctus jeppiari, 95, 115, 164
Petrarctus rugosus, 41, 50, 94, 95, 274, 304, 371, 389, 394, 395, 438, 440, 443, 446, 449, 450, 584, 640
Phagocytosis, 572, 593
Photoperiod, 178, 364, 388, 416, 423, 431, 440, 441, 443
Photoperiodicity, 443
Phyllosoma, viii, 6, 67, 127, 179, 252, 388, 411, 520, 557, 575, 640
Phyllosoma larva, viii, 9, 67, 129, 179, 388, 411, 520, 557, 575, 641
Phylogenetic, 2–4, 36, 38, 43, 66, 67, 69, 115, 126–130

Phylogeny, ix, 11, 24, 36, 126, 128, 130, 139, 636, 643
Phylogeographical, 638
Phylogeography, 128, 130
Physiology, vii, ix, x, 2, 3, 199, 366, 423, 636, 638, 643
Plankton kreisel, 424, 426, 428, 431, 526, 528–530
Pleon, 15–17, 22, 23
Pleopods, 7, 9, 17, 23, 86, 180–183, 200, 203, 261, 286–288, 311, 366, 374–376, 378, 381, 383, 384, 386–388, 392–395, 442, 583, 590
Pleuron, 22, 101, 103–107, 110
Polycheles typhlops typhlops, 104
Polychelidae, 41, 44, 50, 51, 66, 68, 101, 154, 156, 166
Polymorphic, 134
Population genetics, 125–141, 156, 329, 638, 639
Postsettlement, 138
Potential estimate, 286, 310, 637
Predator, vii, 5, 24, 130, 153, 167, 172, 178, 195, 377, 418, 433, 524, 534, 583, 584
Premium Lobster Salad, 618
Prenaupliosoma, 9, 420
Processed Lobster Products, 612
Production, vii, 3, 138, 177, 222, 357, 367, 410, 520, 543, 580, 604, 636
Projasus bahamondei, 39, 46, 161
Projasus parkeri, 46, 161, 169, 319
Protein, 194, 199, 204, 411, 412, 414, 428, 434, 436, 476–479, 486–488, 492, 494, 531–533, 565, 573, 593, 604, 614
Protozoans, 429, 436, 437, 572, 588, 592, 594
Pueruli, 12, 13, 167, 300, 416, 417, 419, 421, 422, 426, 431, 450–453, 456, 457, 464, 465, 468–470, 480–482, 493, 495, 496, 544–546, 548–551, 556–558, 560, 563, 566, 585, 641
Puerulus, ix, 6, 39, 69, 127, 153, 179, 240, 355, 368, 410, 520, 544, 620, 637
  P. angulatus, 46, 84–86, 158, 303, 319, 322, 368, 369
  P. carinatus, 46, 158, 319, 323, 384
  P. collector, 410, 451
  P. gibbosus, 39, 46, 158
  P. mesodontus, 46, 158, 331, 332
  P. quadridenticus, 158
  P. richeri, 46, 158
  P. sericus, 46, 158
  P. sewelli, 46, 84–86, 158, 168, 240–242, 263–266, 268, 273, 276, 278, 290, 293,

294, 296, 300, 301, 303, 308–311,
  313–315, 323, 325, 338, 382, 620, 637
P. velutinus, 46, 158
Pyrosequencing, 130

Q
Quorum, 437, 578
Quorum sensing disruption, 437, 578

R
Raceways, 441, 445, 446, 457, 465, 485, 529, 640
RAD sequencing, 138
Range distribution, 74–76, 78, 80, 81, 83–87, 89, 91–94, 96, 98, 100, 102–108, 110–114
Recirculation, 416, 422, 425, 427, 429, 455, 470, 481, 493
Recreational fishing, 221, 318, 320, 339, 354
Recruitment, 6, 77, 100, 167, 220, 258, 262, 283, 295, 305, 311, 313, 315, 330, 339, 410, 453, 556, 558
Reptantia, vii, 5, 36, 52, 66–68, 70, 115, 126
Resilience, 172, 339, 356, 637
Resource, vii, 2, 84, 132, 152, 195, 221, 352, 398, 414, 521, 544, 636
Rock lobster, vii, 5, 131, 157, 222, 352, 365, 416, 542, 572, 606

S
Sagmariasus verreauxi, 5, 8, 46, 131, 161, 225, 227, 228, 419, 421, 422, 426, 433, 435–437, 447, 580, 640
Salpa fusiformis, 521, 522
Saurashtra, 76, 100, 243, 244, 469, 472
Scammarctus batei batei, 98
Schafer's Model, 637
Scoop nets, 229, 231, 272, 274, 279, 280
Scyllarides elisabethae, 89, 264, 319
Scyllarides tridacnophaga, 47, 89, 90, 264, 268, 273, 282, 293, 370, 384, 395
Scyllarids, vii, 4, 39, 66, 126, 127, 154, 203, 223, 365, 410, 420, 520, 521, 581, 636
Scyllarus chacei, 41, 50, 523, 524
Sea cages, ix, 412, 451, 457, 464, 465, 467, 469–471, 474, 477, 482, 493–495, 543–545, 566
Sea urchins, 153, 172, 195, 196
Seed fishing, 548, 551–556, 558, 563, 566, 567

Seed production, ix, 3, 141, 411, 414, 438, 450, 480, 483, 520, 534, 640–643
Seeds, ix, 3, 139, 396, 411, 520, 543, 640
Setation, 435
Sexual maturity, 8, 186, 286, 288, 311, 365, 374, 377, 380–387, 415, 416, 458, 493, 637, 642
Shelter, 7, 8, 74, 81, 152, 166–169, 180, 194, 305, 322, 328, 339, 443, 452, 456, 463–465, 472, 495
Siphonophora, 433
Site associated, 138
Skin diving, 229, 292
Slipper lobsters, viii, 2, 66, 126, 152, 222, 364, 411, 520, 582, 605, 637
SNP Marker, 138, 639
Somite, 15, 17, 22–24, 75, 76, 78, 80, 83, 85, 88–90, 92–96, 98, 108, 110–113
Southeast coast, 45, 74, 75, 79, 81–84, 87–89, 91, 94, 108, 139, 193, 195, 235, 271, 281, 288, 293, 296, 309, 310, 388, 450, 451, 457, 464, 466–468, 482, 494, 574, 575, 577, 581, 586, 587, 607
Southwest coast, viii, 67, 77, 88, 89, 95, 96, 98, 105, 110, 111, 152, 164, 169–171, 188, 189, 195, 229, 230, 271, 272, 276, 277, 283, 287, 296, 306, 307, 309, 310, 314, 339, 453, 467, 470, 471, 477, 482, 636, 637
Spawning peaks, 259, 271, 296, 311
Spawnings, 7, 8, 14, 74, 204, 235, 259, 261, 262, 271, 272, 295, 296, 311, 314, 365, 370, 372–374, 376, 377, 386, 388–393, 396–398, 416–418, 430, 469, 577
Spearing, 70, 86, 158, 229, 235, 248, 303, 322, 324, 327
Speciation, 129
Specific growth rate (SGR), 459–461, 464, 469, 474, 482
Spermatophore, 6, 8, 9, 14, 269, 311, 365, 367–370, 378, 385, 391–394, 396
Spermatophoric, 364, 367–370, 388, 390–392, 415, 416
Spermatozoa, 368, 370, 381, 394
Spiny lobsters, 2, 66, 126, 153, 178, 223, 353, 364, 411, 520, 543, 572, 605, 636
Sponges, 153, 169–172, 608
Stages, 6, 67, 129, 167, 179, 252, 368, 411, 520, 548, 573, 638
Stakeholders, 222, 307, 308, 313, 315, 323, 335, 339, 352, 357, 358, 640
Stake net, 229, 234, 245, 249, 312, 467, 473
Starvation, 434, 435, 576, 594

Stereomastis nana, 52, 102
Stereomastis phosphorus, 52, 102, 304
Stereomastis sculpta, 52, 103
Sternum, 8, 14, 22, 92, 93, 97, 311, 370, 385, 391, 393
Stock assessment, 139, 177, 188, 220, 255, 259, 269, 339, 355, 357, 358, 636, 637
Stocking density, 426, 428, 441, 443–445, 455, 456, 463–465, 468, 469, 472, 475, 482, 484, 493, 495, 641
Stratified random sampling technique, 636
Stress evaluation, 610
Stridulating organ, 17, 70, 71, 84, 86, 126, 127
Substrate, 5, 7, 8, 19, 77, 79, 82, 84, 89, 97, 98, 100, 152, 164, 166–169, 234, 396, 397, 443–445, 484, 585
Sub-tropical, 71, 152, 154, 156, 159, 161, 164–166, 173, 355, 388, 492
Supply Chain, 496, 630
Survival, 12, 180, 397, 419, 528, 552, 578, 640
Susceptibility, 411, 462, 495, 564, 565, 594
Sustainability, 2, 139, 152, 222, 259, 286, 313, 314, 335, 336, 339, 354, 355, 453, 493, 545, 547, 552, 556, 565, 567, 617, 639, 640, 643
Swarming, 428, 577
Swimming propulsion, 526
Symbiotic, 531, 572, 580
Sympatric species, 12
Symptoms, 436, 564, 565, 578, 579, 584, 592, 608
Synamorphic, 127
Syndrome, 564, 573–577, 579, 592, 593, 640
Synergistic, 640
Synonyms, 35, 36, 39, 44, 47, 51, 52, 67, 72, 74, 83, 92, 102, 127, 156, 179
Systematics, 66, 125–141, 263, 302, 307

T

Tactile sense, 524, 525
Tag-recovery, 191, 279, 283, 284
Taurine, 487, 531
Telson, 19, 23, 84, 100, 101, 103, 105–109, 113, 386, 585, 586
Temperate, 6, 9, 14, 127, 152–154, 156, 161, 164–166, 173, 178, 190, 355, 356, 370, 388, 418–420, 430, 434, 450, 459, 481, 492, 640–642
Tergite, 85, 95, 102, 103, 105–108, 110, 394, 593
Testis, 366–368

Thenus, 6, 41, 69, 128, 156, 189, 220, 365, 432, 523, 576, 627, 640
Thenus australiensis, 14, 40, 50, 432, 438, 440, 449, 450, 483, 523, 527–530, 584
Thenus indicus, 14, 50, 99–101, 115, 167, 170, 300, 331, 332, 438, 440, 483, 492, 523
Thenus unimaculatus, ix, 14, 50, 99, 133, 150, 234, 365, 412, 527, 577, 618, 637
Therapy, 578
Torres strait, 8, 50, 192, 334, 352, 353, 359, 389
Total length (TL), 5, 11–14, 74, 77, 79, 83, 84, 89, 91, 92, 100, 111–114, 188, 189, 191, 192, 252, 258–261, 269–271, 273, 284–286, 288, 290–293, 295, 297, 310, 311, 325, 328, 366, 376, 377, 379
Traceability, 138, 615, 639
Trammel nets, 89, 221, 229, 233, 272, 274, 280, 284, 288, 293, 294, 296, 297, 305, 330, 453–455, 467, 494, 637
Trash fish, 457, 465, 466, 471, 474, 476, 479, 480, 494, 579, 641
Trawl ban, 243, 253, 265, 276
Trawls, vii, viii, 6, 88, 89, 93, 95, 97, 100, 139, 189, 220, 221, 229, 234, 235, 242, 243, 245, 246, 248, 249, 251–256, 263, 265, 269, 275–278, 288–296, 298, 303, 305, 312–315, 319, 325, 328, 389, 454, 473, 483, 496, 607, 637
Trophic cascades, 172, 173
Tropical, ix, 3, 6, 8, 9, 11, 14, 71, 91, 138, 152–154, 157, 161, 164–166, 173, 191, 222, 228, 261, 315, 317, 323, 331, 333, 334, 352, 353, 364, 366, 369, 371, 375, 388, 396, 397, 411, 415, 418, 419, 422, 428, 434, 450, 459, 470, 481–483, 486, 492, 493, 543, 544, 546, 547, 557, 558, 564, 608, 609, 636, 640–643
Tubercles, 19, 22, 84, 86, 87, 90, 92–94, 96–98, 105, 106, 108, 111

U

Umbrella nets, 229, 234, 245, 248
Uropods, 7, 23, 101, 104, 107, 112, 114, 394, 585, 586

V

Vaccines, 574
Value-added products, 617, 631
Vibrio, 424, 430, 436, 437, 440, 475, 477, 564, 575–578

Vibrios, 424, 430, 436, 437, 440, 475, 477, 564, 575–578
Vibriosis, 436, 575, 578
Vietnam, ix, 4, 5, 13, 76, 78, 94, 96, 136, 157, 158, 160, 228, 412, 413, 431, 451, 457, 464, 465, 467–470, 475, 477, 480, 493, 494, 496, 541–570, 578–580, 620, 641
Virus, 456, 571–573, 595
Von Bertalanffy, 188–191, 295
Vulnerable, 132, 167, 173, 179, 180, 305, 312, 313, 335, 338, 339, 447, 483, 638

W

Water absorption, 119, 181
Western Australia, 13, 45, 136, 157, 158, 160, 164, 167, 189, 192, 228, 316, 353, 355, 356, 359, 431, 581, 584
Whole-genome, 640
Whole Lobster Cooking, 616
Willemoesia leptodactyla, 41, 107
Windows, 287, 288, 369, 370, 380, 381, 385, 386, 391, 396, 626
Worms, 8, 178, 420, 476, 581, 582, 584, 585, 589, 591

Y

Year class, 188, 269, 284

Z

Zoea, 432, 520
Zooplankton, 7, 130, 425, 426, 428, 432, 433, 443, 520–527, 529–535